Adaptive System Identification and Signal Processing Algorithms

Prentice Hall International Series in Acoustics, Speech and Signal Processing

Managing Editor: Professor M. J. Grimble, University of Strathclyde, UK

Series Editors:
Albert Benveniste, INRIA, France
Vito Capellini, University of Florence, Italy
A. G. Constantinides, Imperial College London, UK
Patrick Dewilde, University of Delft, The Netherlands
Tariq Durrani, University of Strathclyde, UK
Odile Macchi, Ecole Superieure d'Electricite, France

Other titles in the Series

Signals and Systems: An Introduction
L. Balmer

Signal Processing, Image Processing and Pattern Recognition
S. Banks

Randomized Signal Processing
I. Bilinskis, A. Mikelsons

Perturbation Signals for System Identification
K. Godfrey

Digital Image Processing Algorithms
I. Pitas

Digital Image Processing
J. Teuber

Programming Real-time Multicomputers for Signal Processing
U. Thoeni

Restoration of Lost Samples in Digital Signals
R. Veldhuis

Adaptive System Identification and Signal Processing Algorithms

edited by

N. Kalouptsidis
University of Athens

S. Theodoridis
University of Patras

Prentice Hall
New York London Toronto Sydney Tokyo Singapore

First published 1993 by
Prentice Hall International (UK) Limited
Campus 400, Maylands Avenue
Hemel Hempstead
Hertfordshire, HP2 7EZ
A division of
Simon & Schuster International Group

Typeset in 10/12pt Times
by PPS Limited, 174 London Road, Amesbury, Wilts.

Printed and bound in Great Britain
at the University Press, Cambridge

Library of Congress Cataloging-in-Publication Data

Adaptative System identification and signal processing algorithms / edited by N. Kalouptsidis, S. Theodoridis.
p. cm. -- (Prentice Hall international series in acoustics, speech, and signal processing)
Includes bibliographical references and index.
ISBN 0-13-006545-5
1. Adaptive signal processing. 2. Signal processing—Digital techniques. 3. Algorithms. I. Kalouptsidis, N. II. Theodoridis, S. III. Series.
TK5102.5.A29615 1993
621.382'2--dc20 92—40993
CIP

British Library Cataloguing in Publication Data

A catalogue record for this book is available from the British Library

ISBN 0-13-006545-5 (hbk)

---

1 2 3 4 5 97 96 95 94 93

Contents

List of Contributors

C.F.N. Cowan	Department of Electronic and Electrical Engineering, University of Technology, Loughborough
Simon Haykin	Communications Research Laboratory, McMaster University
Ronald A. Iltis	Department of Electrical and Computer Engineering, University of California, Santa Barbara
Thomas Kailath	Information Systems Laboratory, Department of Electrical Engineering, Stanford University
N. Kalouptsidis	Department of Informatics, University of Athens
Fuyun Ling	Motorola Inc., Corporate Research
Lennart Ljung	Department of Electrical Engineering, Linköping University
J.G. McWhirter	Royal Signals and Radar Establishment, Ministry of Defence
I.K. Proudler	Royal Signals and Radar Establishment, Ministry of Defence
William A. Sethares	Department of Electrical Engineering, University of Wisconsin at Madison
Dirk T.M. Slock	Eurecom, Sophia Antipolis
S. Theodoridis	Department of Computer Engineering, University of Patras
Andrew Ukrainec	Communications Research Laboratory, McMaster University

Preface

This book is addressed to senior undergraduate and postgraduate students, engineers and scientists interested in the fields of *signal processing* and *system identification.* It is assumed that the reader is aware of the basic concepts of the above disciplines, for which a number of very good introductory books exist.

The book focuses on the algorithmic issues of the above areas. This topic is becoming increasingly popular, following the advances in the development of commercially available, low-cost digital signal processors, as well as the possibilities offered by VLSI technology for the design of special-purpose processors. Hence, the development and implementation of efficient algorithms which meet the timing and performance requirements of the various applications has become a major issue.

The book provides a state-of-the-art account of the most important performance features of a widely used class of algorithmic families for adaptive system identification and signal processing. Important issues such as convergence, tracking, numerical accuracy and stability, computational complexity and parallelism are addressed and discussed in depth. The goal of each chapter is to discuss the basic related concepts and mathematical proofs in such detail so as to remain rigorous, on the one hand and, on the other, to make the text readable not only to the theoretician but to the user as well.

Although the book is a multiauthor text, a background theme runs through it, which binds it together as a coherent whole. Topics have been carefully selected and a detailed introductory chapter has been written by the editors to ensure unification and cohesiveness. All the authors have conformed to common notational rules. Furthermore, each chapter has been written by world experts in the particular topic. Specifically, Chapter 3 is written by Lennart Ljung and deals with an in-depth study of adaptation issues and tracking ability of time varying systems. William Sethares provides a comprehensive overview of the LMS family in Chapter 4. Chapter 5 is written by Dirk Slock and Thomas Kailath and presents a state of the art account of the RLS and its numerically stabilized fast versions. Lattice realizations are presented in a unified formulation in Chapter 6 by Fuyun Ling. The closely related *QR* family is treated systematically in Chapter 7 by John McWhirter and Ian Proudler. Chapter 8 provides various efficient computational schemes related to spectral analysis and has been worked out by the editors. The following three chapters are devoted to typical application areas. Colin Cowan, in Chapter 9, gives a broad overview of the major approaches followed in channel equalization. In Chapter 10 Fuyun Ling discusses echo cancellation and provides a detailed account of various design issues. An in-depth treatment of adaptive schemes used in the design of spread-spectrum receivers is the subject of Chapter 11 and is presented by Ronald Iltis. The last chapter provides a state-of-the-art description of typical structures of neural networks encountered in related signal processing applications and is written by Simon Haykin and Andrew Ukrainec.

We hope that the book will make the topic more widely known and understood in the applications community and at the same time motivate readers for further research in this challenging field.

Many colleagues and students from the Universities of Athens and Patras have made the editing of this book easier. Special acknowledgements should be made to Dr G. Glentis, Dr K. Berberidis, Professor G. Moustakides, Mr A. Liavas, Mr L. Kofidis, Mr V. Tsoulkas, Mr N. Koutsoulis and Mr Linardatos.

Finally, it should be said that the idea for this book was conceived in a Glasgow pub during the 1989 ICASSP.

1

Introduction

N. Kalouptsidis and S. Theodoridis

A discrete time *signal* represented by a sequence of values which a physical quantity takes as time (or the corresponding free variable) evolves in discrete manner. A discrete time *system* transforms a discrete time signal (input) to another one (output). The basic questions addressed in this book will be formulated in terms of system concepts. Questions regarding signals can be expressed in systems terminology once it is realized that a signal can be generated at the output of a system driven by some excitation.

Identification of an unknown system has been a central issue in the field of control for many years. To be able to force a signal to follow a predetermined trajectory presupposes knowledge of the related characteristics of the system to be driven. If the generating physical mechanisms of the system are not known, a model of the system must be employed. *Identification* is the procedure of specifying the unknown model in terms of experimental evidence, that is, a set of measurements of the input–output signals. *Recursive* or *adaptive identification* refers to a particular procedure where we learn more about the model as each new pair of measurements is received and we update our knowledge to incorporate the newly received information. This type of identification is particularly suitable when the system, and/or the environment in which it operates, is time varying; thus past measurements should somehow be forgotten in favour of the most recent evidence.

The processing of a signal refers to the sequence of transformations imposed on the signal with aims (a) to extract the useful information and discard unwanted signal components, known as noise or interference; and (b) to accentuate certain signal characteristics relevant to the useful information. Filtering, smoothing and prediction, formulated either in the time domain or in the frequency domain, are typical *signal processing* tasks. In model-based signal processing the signal is described as the output of a system excited by a known signal and the system is in turn modelled so that its output resembles the original signal in an optimal way. This type of approach is becoming increasingly popular and is well suited for the treatment of stochastic signals (Wiener and Kalman filtering belong to this category). Prediction, noise cancellation and spectral analysis, to name but a few, are areas where this methodology has found wide application. The underlying affinity of the model-based approach with system identification hardly needs any justification.

The most common approach to identification departs from a suitable model structure. The model structure is a set of candidate models parametrized by a finite number of unknown parameters. Each model takes the data sequence as input and produces an estimate of the unknown system output. A widely used class consists of *linear realizable filters*, that is, systems with proper rational transfer functions. In the non-linear case, *neural models* appear to be a promising paradigm. The basic unit of an artificial neural network tries to mimic the behaviour of a biological neuron. It is formed by the cascade of a linear combiner and a threshold non-linearity.

Once a model structure is decided, the particular model that best interprets the data is sought. The corresponding parameter of the optimum model is determined by optimization of a carefully selected cost function. The algebraic steps followed during this task constitute the *algorithm* which provides the solution. The choice of the (non-uniquely defined) algorithm is always the result of a trade-off between a number of factors. This choice may become critical in the case of real-time processing, where performance, implementation complexity and cost often trade off under rigid constraints. Figure 1.1 is an example illustrating the procedure described above. The model structure adopted for the unknown system is that of an IIR (Infinite Impulse Response) filter. That is, the output is a linear combination of the current and past input samples as well as past output values. A special case of such a model is an FIR (Finite Impulse Response) filter, where all the bs are zero. The difference (error) between the system's output and a desired response signal is sent to the 'algorithmic box' that performs the optimization. This (i.e. the optimization) consists of the estimation of the unknown system parameters (a and b) so that a cost function involving the error is minimized.

The performance of an algorithm can be measured by a number of factors quantifying (a) how close the resulting solution is to the theoretically expected setup and (b) in how many iterations, with respect to the number of measurement samples, this solution is obtained. The latter is particularly important when recursive identification is involved. In a stationary environment the algorithm is expected to converge (under a certain criterion) to the fixed optimal values starting from arbitrary initial conditions in a general set. The number of measurement samples which the algorithm needs to converge, that is, to learn the optimal parameters, is related to the *convergence rate*, which constitutes one of the basic performance measures of an algorithm. A closely related, yet different, performance measure is that of *tracking*. In a non-stationary environment the algorithm and the underlying model are expected to accommodate the time variation of the pertinent statistics. Tracking is related to the ability of the algorithm to track this time evolution of the physical system. The convergence and tracking requirements of a particular problem dictate not only the specific algorithm to be used but also the engineering aspects of choosing the correct

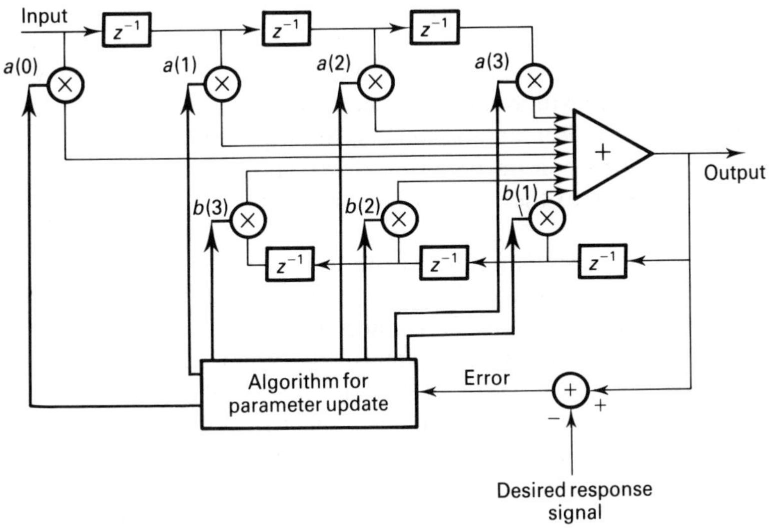

Figure 1.1 Block diagram for an adaptive IIR filter

internal parameters which control the behaviour of the adopted algorithm. This choice is also the result of some type of optimization or trade-off. For example, if an algorithm is trimmed to track fast time variations, then because of its necessarily low 'internal inertia' it will also attempt to track unwanted variations due to noise. This will result in an increase of the so-called 'self-noise' of the algorithm and consequently in an overall increase of the estimation error variance.

Another issue which determines the choice of an algorithm for a particular application is *computational complexity*, that is, the number of arithmetic operations required by the algorithm. For real-time applications this is a highly critical issue since the computations required by the algorithm to update the parameters must be completed before the next set of measurement samples is received. This may force the engineer to make concessions with respect to the algorithm best suited to his or her physical problem. The alternatives offered are: (a) use another type of algorithm that fits the sampling time constraints, at the possible expense of worse performance; and (b) derive 'performance equivalent' fast computational schemes by exploiting the specific nature of the algorithm and the structure of the data sequence upon which the algorithm operates. The latter has been a major research line over the last 15 years and a number of fast algorithmic schemes have flourished, achieving savings of an order of magnitude. The importance of this alternative is that computational reduction is achieved with no sacrifice in performance. This activity has been reinforced by the development in the 1980s of the digital signal processor (DSP). The use of a DSP has a twofold advantage. First it speeds up the execution of the algorithmic steps. This is due to the distinct architectural traits of a DSP with respect to a general-purpose computer, that is, the integration of a hardware multiplier/accumulator into the data path and the use of more than one memory bank in combination with heavily exploited pipelining. Today's DSPs can achieve a performance of the order of 40 Mflops. The second advantage is the possibility offered for fixed point arithmetic. For comparable technology this may offer a 50% speed advantage over floating point processors. However, the use of short wordlengths (16 or 32 bit DSPs) brings to the fore another performance measure, that of *numerical accuracy*.

Roundoff error accumulation drives the estimates of the algorithm away from its theoretically expected value. Different algorithms exhibit different degrees of robustness with respect to roundoff error accumulation. It may turn out that in some cases it is preferable to use a more robust algorithm with higher computational complexity since this can allow the use of shorter wordlengths and thus an overall saving in computational time. Another aspect, *numerical stability*, which is of primary importance, also arises from the roundoff error accumulation/propagation and manifests itself by the numerical explosion of certain algorithmic parameters which become unbounded. The use of shorter wordlengths accelerates the occurrence of this explosion, which is otherwise intrinsic to the algorithm's structure. The user should be alert to this phenomenon and either monitor and prevent the occurrence of this type of instability or use alternative, numerically stable algorithmic structures at the possible cost of higher complexity. The numerical stability of the algorithms has occupied the research community for the last decade and some interesting and useful results have appeared.

The concepts of *parallelism* and *pipelining* have offered a fertile direction in reducing the computational time and a variety of massively parallel, general-purpose machines have been designed, i.e. MIMD, SIMD. However, real-time signal processing has benefited little from these developments, basically because of excessive system overheads. The advent of systolic/wavefront arrays had a major impact in digital signal processing in general and the design of algorithms in particular. The notion of systolization, together with the advances in VLSI technology, have made it possible to design special-purpose array processors which efficiently exploit the major attributes possessed by an algorithm, thus overcoming the aforementioned inefficient overheads. This in turn has pushed the scientific community to design algorithms with structural features which make them amenable to VLSI systolic implementation. Such features include modularity, pipelineability and communications locality. A regular architecture uses processors of more or less the same type. Local communication requires that the movement of data in the network of the processors is confined to neighbouring processors only. Parallel algorithms offering the above features are termed *locally recursive algorithms*. Thus the heart-type rhythmical data pumping nature of systolic arrays can be realized. Wavefront arrays avoid global synchronization and can also be employed for the implementation of locally recursive algorithms. The design of highly parallel/pipelined algorithms has also benefited from the development of programmable processors specially designed for parallel implementations, such as transputers. Although these have not yet found the wide applicability originally anticipated, it is envisaged that they will have a dynamic impact in the future.

System identification and model-based signal processing have been widely used in many application areas, and the related driving algorithms are part of many existing systems. Adaptive control, image and speech processing, and digital communications are typical examples.

Control is concerned with the modification of a system by means of an appropriate feedback mechanism so that the output behaves in a desirable manner. Typical applications are encountered in the control of mechanical systems (robotic arms, space vehicles, helicopters, etc.), process control and manufacturing, and other areas. Adaptive control is initiated when the controlled system is partially or entirely unknown, or when the system parameters change with time. In such cases the controller structure is continuously updated by suitable adaptation algorithms which prudently use online input–output data and which inherently embody an adaptive identification scheme accounting for system variations and uncertainties. In this respect, identification is a prerequisite for adaptive control. The main algorithmic families discussed in this book can be used for adaptive control purposes. Adaptive control strategies based on neural network architectures have also been studied in order to meet the non-linear requirements of many practical systems.

Digital image processing is another application area where adaptive signal processing and system identification algorithms have been widely used. This is not unexpected as the primary one-dimensional signal processing issues, namely enhancement of certain signal characteristics and removal of noise, are manifest with equal propensity in image processing. In image filtering and restoration, we seek to minimize the noise and the effects caused by the limitations of the imaging system and its environment. Wiener and Kalman filters can be used to deal with these problems. Image compression is another area of major practical importance in image archiving as well as image transmission, due to the excessive number of pixels involved. The signal processing algorithms and techniques discussed in this book can be applied to exploit redundancy among pixels and to reduce the required number of bits stored or transmitted, with predictive coding techniques being typical examples. Computationally efficient schemes with enhanced parallelism are highly desirable for coping with the demands imposed by the large number of pixels involved.

Channel equalization and *echo cancellation* in digital communications are two closely related problems. The goal in both of them is to remove the unwanted contamination which a communication channel imposes on the transmitted information sequence. Channel equalization deals with the effects caused by the finite bandwidth and/or the existing multiple path of physical channels. The consequence of this is the spread in time of the originally transmitted information, causing the effect known as *intersymbol interference* when sampling is performed on the receiver side. Echo cancellation is a different type of channel contamination caused by delayed replicas of the transmitted information, which the sender receives back due to feedback sources within the channel. Telephone line networks, data modems and hands-free mobile radio are typical examples where the presence of echo disturbs communications or even makes them prohibitivily erroneous as in the data communications case. Model-based signal processing techniques have been used extensively in both of the above problems to remove channel impairments. The demand for reliable communications at heavily exploited channel bandwidths, for high data rates and in noisy, time-varying channels does not make the choice of a particular algorithm a straightforward task.

Spread-spectrum communications rely on using more bandwidth than would be necessary to transmit the basic information. This can be achieved in either of two possible ways. According to the first approach each information bit is encoded using a predetermined pseudorandom sequence, which is then transmitted using one of the standard modulation schemes (i.e. PSK). The alternative is to modulate each information bit by a pseudorandom sequence of frequencies, each one transmitted at a fraction of the information bit duration. Model-based adaptive signal processing techniques can be used as an integral part of such systems to reject interference and to synthesize the matched filter for demodulation. Spread-spectrum systems, originally used in military applications because of their low probability of interception, are becoming increasingly popular for commercial applications as an alternative to time division multiple access and frequency division multiple access systems. Spread-spectrum code division multiple access systems have been proposed for the next generation of mobile radio and personal communications networks in North America.

Besides the application disciplines summarized above, one can mention many others such as seismic signal processing, adaptive array beamforming, noise cancellation, signal detection, etc., where the system identification and model-based signal processing algorithms examined in this book have been applied. However, the limited size of the book has constrained us to focus on three application tasks, namely channel equalization, echo cancellation and spread-spectrum interference rejection. This choice was dictated by two things. First, these application areas are typical examples where the designer of an algorithm is challenged to trade off all the algorithmic issues raised above, due to the ever increasing demands of low cost, high performance and real-time operation, usually in a 'hostile environment'. The second thing is the timing of the book. Communications are about to take a qualitative leap within the technological revolution which is now emerging. Wireless personal communication networks and services (PCNs and PCSs) will make communications truly personal and allow anyone to exchange information with anyone else irrespective of time, distance and location, and it will have a dramatic impact on our society.

2

Basic Concepts and Algorithmic Schemes

N. Kalouptsidis and S. Theodoridis

2.1 Introduction

The intention of this chapter is to provide a brief review of the basic concepts the reader will encounter throughout the book. It also gives some motivation and background for the fundamental algorithmic families which form the core of the book. Finally, it standardizes notation. An extensive presentation of the topics covered in this chapter can be found in the books by Ljung and Soderstrom [1], Ljung [2], Soderstrom and Stoica [3], Haykin [4] and Goodwin and Sin [5]. Our primary aim in this book is the efficient implementation of system identification and signal processing algorithms, and their application in channel equalization, spectrum analysis, spread-spectrum and neural network synthesis. To provide sufficient insight into these algorithms we start the presentation with the system identification problem.

2.2 System identification

Identification is concerned with the determination of a system that best describes a given collection of data $z(t)$, $M \leqslant t \leqslant N$. We assume that the sequence $z(t)$ is split into two components:

$$z(t) = [y(t), u(t)]^{\mathrm{T}} \tag{2.1}$$

The signal $y(t)$ constitutes the output of an unknown system S resulting from the input signal $u(t)$ and a disturbance, or noise signal, $\eta(t)$. We write such a cause–effect relationship in the following descriptive format:

$$\mathbf{y} = S(\mathbf{u}, \boldsymbol{\eta}) \tag{2.2}$$

where vector quantities have been used to indicate that a number of successive samples are generally involved in (2.2). The identification task is to design an estimator

$$\hat{\mathbf{y}} = \hat{S}(\mathbf{z}) \tag{2.3}$$

which removes the noise and produces the output $\mathbf{y}$ when fed with the data sequence. $\hat{S}$ is chosen from a class of allowable models so that it interprets the data in an optimum fashion and hence matches S. Goodness of fit is formulated via optimization of some measure of the error signal

$$e(t) = y(t) - \hat{y}(t) \tag{2.4}$$

We observe that such a model $\hat{S}$ tries to reconstruct the data and simultaneously remove the noisy contribution.

Several possibilities exist for the selection of $\hat{S}$. A natural choice is the minimum variance estimator obtained via minimization of the mean squared error

$$E[(y(t) - g(Z_{t-1}))^2]$$

where

$$Z_{t-1} = [z(0), z(1), \ldots, z(t-1)]^{\mathrm{T}} \tag{2.5}$$

is the vector of past samples, and minimization is taken over all non-linear functions g; $E[\,]$ denotes expectation. As is well known [3], the solution to this problem is the conditional expectation

$$\hat{y}(t) = E(y(t) | Z_{t-1}) \tag{2.6}$$

We shall refer to the above estimator as the **minimum variance predictor**.

Among the various, particularly important estimation methods are the so-called **sequential** or **online** methods. These estimation techniques combine data acquisition and system estimation. This means that between successive samples an estimate of the system is obtained. In contrast, **batch** or **offline** or **block** estimation techniques view data collection and storage separately from the estimation process.

Despite its elegant form, the minimum variance predictor is not easily computed from raw data by means of statistical techniques based on the probability density function estimation, particularly in the case of online methods. To make the concept of real-time performance meaningful we further restrict our identification setup as follows. We assume that the predictor is factored in the following way:

$$\hat{S}(\mathbf{z}) = \hat{F}(\mathbf{z}, \boldsymbol{\theta}) \tag{2.7}$$

where $\hat{F}$ is a known transformation and $\boldsymbol{\theta}$ is a finite-dimensional parameter taking values in the parameter set

$$\boldsymbol{\theta} \in \Theta \subset R^p$$

As we shall see in subsequent sections, (2.7) holds if an analogous finite-dimensional parameterization of the unknown system (2.2) is assumed:

$$\mathbf{y} = S(\mathbf{u}, \boldsymbol{\eta}) = F(\mathbf{u}, \boldsymbol{\eta}, \boldsymbol{\theta})$$

Then the estimator $\hat{S}$, and hence the entire identification problem, reduces to that of determining $\boldsymbol{\theta}$. For emphatic purposes we refer to the latter case as *parametric system identification.*

To determine $\boldsymbol{\theta}$ we seek to minimize some measure of the error signal

$$e(t \mid \boldsymbol{\theta}) = y(t) - \hat{y}(t \mid \boldsymbol{\theta}) \tag{2.8}$$

with respect to $\boldsymbol{\theta}$. A popular performance criterion is the quadratic function

$$V_N(\boldsymbol{\theta}) = \sum_{t=M+p-1}^{N} e^2(t \mid \boldsymbol{\theta}) \tag{2.9}$$

The lower limit in the summation ensures that no data outside the given interval $[M, N]$ are involved. We often assume that input and output signals are causal, i.e. $u(t) = 0$, $y(t) = 0$, for $t < 0$. We refer to this case as the *prewindowed* case and the minimizing cost becomes

$$V_N(\boldsymbol{\theta}) = \sum_{t=0}^{N} e^2(t \mid \boldsymbol{\theta}) \tag{2.10}$$

It is often useful to attach different weights to the errors committed at various instants. We model this as

$$V_N(\boldsymbol{\theta}) = \sum_{t=M+p-1}^{N} w_N(t) e^2(t \mid \boldsymbol{\theta}) \tag{2.11}$$

The window function $w_N(t)$ imposed on the error signal reflects the time-varying significance of past and recent information. Consequently, it is of particular interest in the identification of time-varying dynamics. The algorithms that handle this situation are capable of adapting and tracking the dynamical changes of the system or signal. For this reason they are called *adaptive algorithms.* Time-varying dynamics and adaptive algorithms are treated in Chapter 3. A popular choice of window function is the exponential window

$$w_N(t) = \lambda^{N-t} \qquad 0 < \lambda < 1 \tag{2.12}$$

The exponential window allocates greater value to recent information and tends to suppress old data. An alternative is the sliding window

$$w_N(t) = \begin{cases} 1 & \text{if } N - L \leqslant t \leqslant N \\ 0 & \text{otherwise} \end{cases} \tag{2.13}$$

The sliding window utilizes only the most recent information displaced in the frame $N - L \leqslant t \leqslant N$. A more detailed discussion on these windows is presented in Chapter 5.

In what follows we shall adhere to the performance criterion (2.10) and assume that the optimization problem has a unique solution which we denote by $\boldsymbol{\theta}(N)$, i.e.

$$\boldsymbol{\theta}(N) = \text{argmin}_{\boldsymbol{\theta}} V_N(\boldsymbol{\theta}) \tag{2.14}$$

As the notation indicates, $\boldsymbol{\theta}(N)$ depends on the data length N. It represents our best guess or estimate of the unknown parameter up to time N. Now suppose the operation is taking place in adaptive mode. At time instant N the most recent data piece $z(N)$ is retrieved. Before the next data $z(N+1)$ becomes available, the estimation of $\boldsymbol{\theta}(N)$ must be completed. Then $z(N+1)$ is used together with the past estimate to determine $\boldsymbol{\theta}(N+1)$ and the process is repeated.

The above formulation can be expanded in several useful directions. Thus the performance criterion may employ a filtered version of the error

$$\varepsilon_F(t) = Q(e(t))$$

If the filter Q is a linear, time-invariant, single-input, single-output system and the predictor is also linear and time invariant, the above process amounts to prefiltering the data sequences. The purpose of such generalization is to concentrate on a certain frequency band of interest or to remove insignificant information. A further extension is to replace the square of the error in (2.10) with a general differentiable function and to minimize the cost

$$\sum_{t=0}^{N} l(\varepsilon_F(t, \boldsymbol{\theta}), t)$$

This enables us to incorporate other approaches, like maximum likelihood (see [1–3] for a detailed discussion).

The system identification formulation discussed above is closely related to several topics in digital signal processing. A brief discussion is given next.

2.3 Signal processing

We shall focus on two topics of generic importance for signal processing: filtering and signal modelling. We demonstrate that both of them fall within the system identification framework of the previous section.

Filtering aims to shape an input signal $\mathbf{u}$ so that the corresponding output $\hat{\mathbf{y}}$ tracks a desired response signal $\mathbf{y}$. As in the previous cases our knowledge base consists of data records of the input and desired response signals. The filter we seek to determine is essentially an estimate of the system that represents the cause–effect relationship between $\mathbf{u}$ and $\mathbf{y}$. Therefore, filtering is a form of system identification. To achieve finite-dimensional reduction, we constrain our search in filter structures of the form (2.7), F again being a known transformation, and we seek to determine $\boldsymbol{\theta}$ so that some measure of the error between the filter output and the desired response is optimized. Apparently, we are confronted with the same formalism as in system identification.

Signal modelling is conceived as the process of reconstructing a signal $\mathbf{y}$ from signal measurements in the presence of noise. We view the signal as being produced at the output of a system S subject to disturbance $\boldsymbol{\eta}$

$$\mathbf{y} = S(\boldsymbol{\eta}) \tag{2.15}$$

In effect, the system represents the source emitting the signal $\mathbf{y}$, and the modelling problem is the identification of the source by means of signal data $y(t)$. It is easy to translate the problem in the system identification setup. Note in this case that $z(t) = y(t)$. If the source is also excited by a measurable signal $\mathbf{u}$, the general system identification case applies.

2.4 Efficiency issues in adaptive algorithms

In the previous section we concluded that adaptive identification and the related issues of filtering and signal modelling are reduced to the efficient implementation of the discrete time system

$$z(t) \rightarrow \mathscr{A}(z(t)) = \boldsymbol{\theta}(t) = \text{argmin}_{\boldsymbol{\theta}} V_t(\boldsymbol{\theta}) \tag{2.16}$$

The system $\mathscr{A}$ reads the data sequence $z(t)$ at the input and produces the estimate $\boldsymbol{\theta}(t)$ at the output. $\mathscr{A}$ is expected to be capable of reproducing the actual system if we assume such a system exists. More generally, it is natural for $\mathscr{A}$ to share the following qualitative features:

- The estimate $\boldsymbol{\theta}(t) = \mathscr{A}(z(t))$ is bounded in amplitude. In fact in a stationary (or quasistationary) environment $\boldsymbol{\theta}(t)$ has a limit

$$\lim_{t \rightarrow \infty} \boldsymbol{\theta}(t) = \boldsymbol{\theta}^*$$

- Small perturbations of the data sequence do not cause large deviations in the resulting estimates.
- If there is a unique parameter $\boldsymbol{\theta}^0$ called the true parameter which represents the data sequence, that is,

$$\mathbf{y} = F(\mathbf{u}, \boldsymbol{\eta}, \boldsymbol{\theta}^0) \tag{2.17}$$

then $\boldsymbol{\theta}(t)$ asymptotically recovers $\boldsymbol{\theta}^0$, i.e.

$$\boldsymbol{\theta}^0 = \boldsymbol{\theta}^*$$

Development of conditions that ensure the validity of the above properties and thus offer provision of a well-posed structure is the main theme of the convergence theory of identification. A thorough exposition can be found in [1–4] and the references cited therein.

Let us next concentrate on algorithmic issues associated with the computation of $\mathscr{A}$. We observe that estimation of $\boldsymbol{\theta}(t)$ at time t requires all past data information $z(\tau)$:$\tau \leqslant t$. Therefore, a direct evaluation would need storage requirements and hence an operations count that would grow at least linearly with time. This is clearly prohibitive in real-time applications.

This difficulty is generally overcome by working with state-space realizations of (2.16). A finite state space realization of the system $\mathscr{A}$ is specified by the pair of equations

$$\begin{aligned} \mathbf{x}(t+1) &= f(\mathbf{x}(t), z(t)) \quad \mathbf{x}_0 \in R^p \\ \boldsymbol{\theta}(t) &= g(\mathbf{x}(t), z(t)) \end{aligned} \tag{2.18}$$

in the sense that the unique input–output map specified by the above equations and the initial state $\mathbf{x}_0$ coincides with $\mathscr{A}$.

Any such realization specifies a computational procedure or an algorithm that computes $\mathscr{A}$. Hence an algorithm is determined by the state map f, the output map g and the initial seed $\mathbf{x}_0$. The dimension of the state space determines the length of the register and hence the memory requirements. Apparently, if a system $\mathscr{A}$ is realizable by such a state-space model, memory is fixed in time and is proportional to p.

The system $\mathscr{A}$ is not in general realizable by a finite-dimensional, non-linear state-space model. An important exception which we discuss in the next section is the linear regression case where the predictor depends linearly on the unknown parameter.

Since in general $\mathscr{A}$ is not realizable, we look for approximate realizations, namely for representations of the form (2.18) whose input–output map is close to $\mathscr{A}$ in some appropriate sense, for instance

$$\hat{\boldsymbol{\theta}}(t, \mathbf{x}_0) - \mathscr{A}(z(t)) \to 0 \quad \text{as } t \to \infty$$

where $\hat{\boldsymbol{\theta}}(t, \mathbf{x}_0)$ represents the estimate produced by the given algorithm when the initial seed is $\mathbf{x}_0$ and the input is $z(t)$.

In section 2.6 we derive two such approximate realizations: the prediction error algorithm and the stochastic gradient algorithm.

STORAGE AND COMPUTATIONAL COMPLEXITY

Among the various realizations of $\mathscr{A}$ we wish to choose those that call for low storage and computational requirements.

Memory minimization is in essence a non-linear minimal realization problem. Thus we seek to determine, among all representations of the form (2.18) of $\mathscr{A}$, the one having the minimum dimension p.

Minimum complexity algorithms require a small number of basic operations. The majority of the algorithms in this book are rational, that is, they execute one of the ordinary real operations. The operations count then determines the complexity of the algorithm.

STABILITY

Given a particular algorithmic realization (2.18) of $\mathscr{A}$, the fact that the estimates it produces are close to the nominal estimates for a given initial seed is not enough. Indeed, the algorithm is intended to operate with finite arithmetic. As a result, deviations due to finite wordlength effects are bound to occur. To guarantee control of such numerical errors, we must necessarily confine ourselves to algorithms which are insensitive to initial state perturbations. Stated differently, the algorithm must be asymptotically stable in the vicinity of the nominal trajectory $\boldsymbol{\theta}(t)$.

CONVERGENCE AND TRACKING

In recursive system identification the algorithm kicks off from arbitrary initial values. Its convergence performance characterizes the speed with which it converges to its steady-state behaviour. In time-varying systems the true parameter vector is continuously varying and the adaptive algorithm must track this variation. The tracking perfor-mance of an adaptive algorithm is assessed by examining the behaviour of the tracking error

$$\boldsymbol{\theta}(t) - \boldsymbol{\theta}^0(t)$$

$\boldsymbol{\theta}(t)$ represents the estimate produced by the algorithm and $\boldsymbol{\theta}^0(t)$ the true value of the parameter at time t. A useful performance index is the error covariance matrix

$$\Pi(t) = E\,[\boldsymbol{\theta}(t) - \boldsymbol{\theta}^0(t)][\boldsymbol{\theta}(t) - \boldsymbol{\theta}^0(t)]^{\mathrm{T}}$$

Bounds on $\Pi(t)$ express the tracking ability of the algorithm and are derived in Chapters 3 and 5. It must be emphasized that convergence and tracking refer to two different, though related, properties of an algorithm. The former characterizes transient behaviour, in contrast to the latter, which describes steady-state performance. As we shall see in Chapters 3 and 5 the convergence and tracking performances of an algorithm are closely related to the algorithm's 'self'-noise, which manifests itself as a fluctuation of the resulting steady-state parameters. It is shown that convergence/tracking speed and this type of noise conflict and the algorithm design usually has to seek a reasonable compromise.

PARALLELISM

A wide range of applications require high-speed, real-time digital signal processing. Minimization of processing time is efficiently achieved in a parallel processing environment. Parallel processing is carried out either by decomposing the computation process into subprocesses, each executed independently and concurrently with the others by a respective processor, or by decomposing the process into many subprocesses which are pipelined through a number of processors in cascade and processing each subprocess one after the other (pipelining). In applications with intensive real-time requirements we must look for algorithms with maximum parallel potential.

LOCALLY RECURSIVE REALIZATIONS AND SYSTOLIC–WAVEFRONT ARCHITECTURES

Recent advances in technology and the availability of low-cost, high-density, high-speed VLSI devices, together with the emerging CAD tools, have made parallel processing an extremely powerful alternative in applications requiring intensive computations and real-time processing. There are three basic approaches in this direction, i.e. SIMD computers, MIMD machines and special-purpose VLSI array processors such as systolic and wavefront arrays. The third class is a very promising tool in real-time applications as it avoids the severe system overheads encountered in the domain of general-purpose machines to which the other two classes belong. Systolic–wavefront array processors share the very attractive features of modularity, regularity, pipeline ability and interprocessor communications locality. These make them attractive from a VLSI implementation point of view and the design of locally recursive algorithmic schemes, which can be mapped on such arrays, has become an important issue. Local recursivity indicates that once these algorithms are mapped to an array structure, communication is localized to neighbouring processors only.

2.5 Linear regression

Let us consider the minimization problem (2.10). To determine the minimum, we set the derivative of the cost with respect to $\boldsymbol{\theta}$ equal to zero. The resulting equations are non-linear and hence iterative techniques must be employed. We return to this point in the next section. Here we confine ourselves to a particular case where the optimization problem does lead to linear equations. This case appears when the predictor depends linearly on the unknown parameters, that is,

$$\hat{y}(t \mid \boldsymbol{\theta}) = \boldsymbol{\phi}^{\mathrm{T}}(t)\boldsymbol{\theta} \tag{2.19}$$

$\boldsymbol{\phi}(t)$ is a known function of the data up to time $t-1$ and is called the **regressor vector**. If (2.19) holds, setting the derivative of (2.10) equal to zero gives

$$0 = \sum_{t=0}^{N} \boldsymbol{\phi}(t)(y(t) - \boldsymbol{\phi}^{\mathrm{T}}(t)\boldsymbol{\theta}(N)) \tag{2.20}$$

or

$$R(N)\boldsymbol{\theta}(N) = \mathbf{d}(N) \tag{2.21}$$

where

$$R(N) = \sum_{t=0}^{N} \phi(t)\phi^{\mathrm{T}}(t) \qquad \mathbf{d}(N) = \sum_{t=0}^{N} \phi(t)y(t) \tag{2.22}$$

Our design problem is then reduced to finding efficient realizations of the input–output map

$$z(N) \rightarrow \mathscr{A}(z(N)) = \boldsymbol{\theta}(N) = R^{-1}(N)\mathbf{d}(N) \tag{2.23}$$

The resulting $\boldsymbol{\theta}(N)$ is also known as the *least-squares* (*LS*) *estimate*. Before we derive state-space representations of (2.23) it is important to examine how typical the predictor model linear in $\boldsymbol{\theta}$ is. The assumption of linearity in $\boldsymbol{\theta}$ is met when the unknown system is described by one of the following models.

FIR MODELS

Suppose the input–output representation of the unknown system has the form

$$y(t) + \sum_{k=1}^{p} c_i u(t-k) = \eta(t) \tag{2.24}$$

and the unknown parameter is

$$\boldsymbol{\theta} = -\mathbf{c} \qquad \mathbf{c} = [c_1, c_2, \ldots, c_p]^{\mathrm{T}}$$

The disturbance $\eta(t)$ has zero mean with uncorrelated values at different time instants, i.e.

$$E[\eta(t)] = 0 \qquad E[\eta(t)\eta(\tau)] = 0 \qquad t \neq \tau$$

Equation (2.24) defines a linear time-invariant system. The impulse response has finite support. Its first values are $c_1, c_2, \ldots, c_p$, while all others are zero. The FIR model provides an approximation to general BIBO (Bounded Input Bounded Output) stable convolutional models

$$y(t) = -\sum_{k=-\infty}^{t} c_k u(t-k) + \eta(t) \tag{2.25}$$

In this case the impulse response is absolutely summable, i.e.

$$\sum_{k=0}^{\infty} |c_k| < \infty$$

and consequently it satisfies $|c_k| \rightarrow 0$. Hence we may assume that, eventually, $c_k \approx 0$.

If we take conditional expectations on (2.24) we find the following expression for the predictor:

$$\hat{y}(t \mid \boldsymbol{\theta}) = -\sum_{k=1}^{p} c_k u(t-k) = \boldsymbol{\phi}^{\mathrm{T}}(t)\boldsymbol{\theta} \tag{2.26}$$

where

$$\boldsymbol{\phi}(t) = [u(t-1), u(t-2), \ldots, u(t-p)]^{\mathrm{T}} \tag{2.27}$$

AR MODELS

An autoregressive (AR) model has the form

$$y(t) + \sum_{k=1}^{p} a_k y(t - k) = \eta(t) \tag{2.28}$$

No measurable input is present in this case. We take the unknown parameter to be

$$\boldsymbol{\theta} = -\mathbf{a} \qquad \mathbf{a} = [a_1, a_2, \ldots, a_p]^{\mathrm{T}}$$

$\eta(t)$ represents a white-noise signal. Applying the conditional expectation on both sides of (2.28) we find

$$\hat{y}(t\,|\,\boldsymbol{\theta}) = -\sum_{k=1}^{p} a_k\, y(t-k) = \boldsymbol{\phi}^{\mathrm{T}}(t)\boldsymbol{\theta} \tag{2.29}$$

where

$$\boldsymbol{\phi}(t) = [\,y(t-1), y(t-2), \ldots, y(t-p)]^{\mathrm{T}} \tag{2.30}$$

The above cases can be merged into the following more general class.

ARX MODELS

The input–output relationship of an ARX model is

$$y(t) + \sum_{i=1}^{p} a_i\, y(t-i) + \sum_{k=1}^{l} b_k u(t-k) = \eta(t) \tag{2.31}$$

$\eta(t)$ is white noise and the unknown parameter is

$$-\boldsymbol{\theta} \equiv [a_1, a_2, \ldots, a_p, b_1, b_2, \ldots, b_l]^{\mathrm{T}}$$

The minimum variance predictor is given by

$$\hat{y}(t\,|\,\boldsymbol{\theta}) = \boldsymbol{\phi}^{\mathrm{T}}(t)\boldsymbol{\theta}$$

and the regressor vector is

$$\boldsymbol{\phi}(t) = [\,y(t-1), \ldots, y(t-p), u(t-1), \ldots, u(t-l)]^{\mathrm{T}} \tag{2.32}$$

Remark The ARX family is the basic class of models leading to predictor models linear in $\boldsymbol{\theta}$. It is by no means the only such case. Non-linear systems such as polynomial Volterra filters also lead to linear regressions.

2.5.1 Steady-state performance of the LR estimate and the Wiener filter

In the previous section we described various qualitative properties that convergence analysis attempts to study. To provide a glimpse of such results we highlight the linear regression (LR) case. For a thorough discussion the reader is referred to [1–5].

In what follows we shall assume that the following conditions for (2.22) are satisfied:

$$\frac{1}{N} R(N) \to R > 0 \qquad N \to +\infty \tag{2.33}$$

$$\frac{1}{N} \mathbf{d}(N) \to \mathbf{d} \qquad N \to +\infty \tag{2.34}$$

The positive definiteness of R from condition (2.33) guarantees that all the system modes have been excited and the experiment is sufficiently informative [2]. In the case of FIR modelling where the regressor vector is given by (2.27), Equation (2.33) holds if the input is sufficiently exciting, such as a sum of sufficiently many sinusoidal signals, or a white-noise signal. In the case of AR modelling, (2.33) is always valid. Finally, in the case of ARX identification, where the regressor vector is given by (2.32), Equation (2.33) is satisfied if the input signal is sufficiently persistently exciting and no overparameterization occurs. In all the above cases the input $\mathbf{u}$ and the noise signal $\boldsymbol{\eta}$ are uncorrelated.

Under suitable ergodic assumptions we have

$$R = E\,[\boldsymbol{\phi}(t)\boldsymbol{\phi}^{\mathrm{T}}(t)] \tag{2.35}$$

$$\mathbf{d} = E\,[\boldsymbol{\phi}(t)y(t)] \tag{2.36}$$

If we divide both sides of Equation (2.21) by N and take limits we obtain

$$R\boldsymbol{\theta} = \mathbf{d} \tag{2.37}$$

The limiting parameter $\boldsymbol{\theta}$ can be shown to be the minimizing value of

$$V(\boldsymbol{\theta}) = \lim_{N\to\infty} \frac{1}{N} \sum_{t=0}^{N} e^2(t\,|\,\boldsymbol{\theta}) \tag{2.38}$$

Under assumptions of ergodicity the above limit is equal to

$$E\,[e^2(t\,|\,\boldsymbol{\theta})]$$

We shall refer to $\boldsymbol{\theta}$ as the **Wiener** solution. If there is a true parameter $\boldsymbol{\theta}^0$, such that

$$y(t) = \boldsymbol{\phi}^{\mathrm{T}}(t)\boldsymbol{\theta}^0 + \eta(t) \tag{2.39}$$

and the disturbance is white noise, then

$$\boldsymbol{\theta} = \boldsymbol{\theta}^0$$

Indeed, we substitute (2.39) into (2.37) to obtain

$$R\boldsymbol{\theta} = E\,[\boldsymbol{\phi}(t)\,y(t)] = E\,[\boldsymbol{\phi}(t)[\boldsymbol{\phi}^{\mathrm{T}}(t)\boldsymbol{\theta}^0 + \eta(t)]] = R\boldsymbol{\theta}^0 + E\,[\boldsymbol{\phi}(t)\eta(t)]$$

Since $\eta(t)$ is uncorrelated to the regressor vector the last term in the above equation is zero and the claim follows.

2.5.2 State-space forms of the LR estimate and the RLS algorithm

Next we derive state-space representations of the linear regression estimate. A direct state space form is suggested by Equations (2.21) and (2.22) if they are written in the following format:

$$R(N) = R(N-1) + \boldsymbol{\phi}(N)\boldsymbol{\phi}^{\mathrm{T}}(N) \tag{2.40}$$

$$\mathbf{d}(N) = \mathbf{d}(N-1) + \boldsymbol{\phi}(N)y(N) \tag{2.41}$$

$$\boldsymbol{\theta}(N) = \mathrm{LINSOLV}(R(N), \mathbf{d}(N)) \tag{2.42}$$

LINSOLV represents an exact linear system solver, such as Cholesky's algorithm. The memory requirements of (2.40) are proportional to p^2, the dimension of $\boldsymbol{\theta}$. The computational complexity of the state equations (2.40) and (2.41) is of about the same size. The output equation (2.42) is, however, more demanding since a conventional linear solver requires $O(\dim\boldsymbol{\theta})^3$ multiplications and additions. As we shall see in Section 2.7, significant savings can be achieved if the special structure of matrix R is properly manipulated.

Under assumptions (2.33), $R(N) \to \infty$ and thus (2.40) is unstable. To avoid this unsatisfactory behaviour we define new state variables using scaling:

$$\tilde{R}(N) = \frac{1}{N} R(N) \tag{2.43}$$

$$\tilde{\mathbf{d}}(N) = \frac{1}{N} \mathbf{d}(N) \tag{2.44}$$

Then we readily obtain the following state-space representation of the linear regression (LR) estimate:

$$\tilde{R}(N) = \left(\frac{N-1}{N}\right) \tilde{R}(N-1) + \frac{1}{N} \boldsymbol{\phi}(N)\boldsymbol{\phi}^{\mathrm{T}}(N) \tag{2.45}$$

$$\tilde{\mathbf{d}}(N) = \left(\frac{N-1}{N}\right) \tilde{\mathbf{d}}(N-1) + \frac{1}{N} \boldsymbol{\phi}(N)y(N) \tag{2.46}$$

$$\boldsymbol{\theta}(N) = \mathrm{LINSOLV}(\tilde{R}(N), \tilde{\mathbf{d}}(N)) \tag{2.47}$$

An alternative way to counteract the instabilities of (2.40) is to use the inverse of $R(N)$

$$P^{-1}(N) = R(N) \tag{2.48}$$

as a state variable. The update of $P(N)$ is easily inferred from (2.40) and the matrix inversion lemma (see, for instance, Appendix A of Chapter 8). The resulting state-space form is

$$P(N) = P(N-1) - P(N-1)\boldsymbol{\phi}(N)\alpha(N)\boldsymbol{\phi}^{\mathrm{T}}(N)P(N-1) \tag{2.49}$$

$$\alpha(N) = \frac{1}{1 + \boldsymbol{\phi}^{\mathrm{T}}(N)P(N-1)\boldsymbol{\phi}(N)} \tag{2.50}$$

$$\tilde{\mathbf{d}}(N) = \frac{N-1}{N}\tilde{\mathbf{d}}(N-1) + \frac{1}{N}\boldsymbol{\phi}(N)y(N) \tag{2.51}$$

$$\boldsymbol{\theta}(N) = NP(N)\tilde{\mathbf{d}}(N) \tag{2.52}$$

Notice that the output equation (2.52) executes matrix and vector multiplication rather than a linear system solution. All the above representations involve approximately the same memory requirements. The computational performance of the form (2.49)–(2.52) is superior to those preceding it.

We conclude our discussion of exact realizations for the linear regression estimate with the most popular of them, the recursive least-squares (RLS) algorithm. It relies on the observation that the updating equations of $R(N)$ and $\mathbf{d}(N)$, (2.40) and (2.41), should somehow be inherited in $\boldsymbol{\theta}(N)$. Standard calculations lead to

$$\boldsymbol{\theta}(N) = \boldsymbol{\theta}(N-1) + P(N-1)\boldsymbol{\phi}(N)\alpha(N)e(N) \tag{2.53}$$

$$e(N) = y(N) - \boldsymbol{\phi}^{\mathrm{T}}(N)\boldsymbol{\theta}(N-1) \tag{2.54}$$

$$\alpha(N) = \frac{1}{1 + \boldsymbol{\phi}^{\mathrm{T}}(N)P(N-1)\boldsymbol{\phi}(N)} \tag{2.55}$$

$$P(N) = P(N-1) - P(N-1)\boldsymbol{\phi}(N)\alpha(N)\boldsymbol{\phi}^{\mathrm{T}}(N)P(N-1) \tag{2.56}$$

The RLS algorithm provides an exact state-space realization of the LR estimator if it is properly initialized. Indeed, let N_0 be an integer such that the matrix

$$R(N_0) = \sum_{t=0}^{N_0} \boldsymbol{\phi}(t)\boldsymbol{\phi}^{\mathrm{T}}(t)$$

is invertible. Clearly, $N_0 \geqslant p$. We set

$$P(N_0) = R^{-1}(N_0) \qquad \mathbf{d}(N_0) = \sum_{t=0}^{N_0} \boldsymbol{\phi}(t)y(t) \qquad \boldsymbol{\theta}(N_0) = P(N_0)\mathbf{d}(N_0)$$

and we run the RLS algorithm for $N \geqslant N_0$. Then the resulting sequence $\boldsymbol{\theta}(N)$ coincides with the LR estimate (2.21). The performance features of the RLS algorithm are summarized next.

The memory requirements of the RLS algorithm are characterized by Equations (2.53) and (2.56). Equation (2.56) is the most demanding. Since $P(N)$ is symmetric, $p(p+1)/2$ memory stages are required for (2.56) and p stages for (2.53). Equation (2.53) calls for $O(p)$ multiplications and additions, while (2.56) requires $O(p^2)$ operations. Reductions by an order of magnitude in both memory and computational complexity are ensured by the fast RLS (FRLS) implementations treated in Chapter 5.

Parallel implementations can be obtained via a Kalman filter interpretation of the RLS algorithm and subsequent use of locally recursive implementations of their square root variants [6].

We investigate the behaviour of the algorithm when different initial seeds P_0, $\boldsymbol{\theta}_0$ for (2.56) and (2.53), respectively, are used. Although we allow for arbitrary $\boldsymbol{\theta}_0$, we shall restrict P_0 to the class of positive definite matrices, since this is clearly a necessary condition. It is easy to see that the resulting sequence $P(N)$ consists of positive definite matrices, too. Moreover, the following closed form expression is easily derived for the solution $\hat{\boldsymbol{\theta}}(N)$ emanating from P_0, and $\boldsymbol{\theta}_0$:

$$\hat{\boldsymbol{\theta}}(N) = (P_0^{-1} + R(N))^{-1}(P_0^{-1}\boldsymbol{\theta}_0 + \mathbf{d}(N)) \tag{2.57}$$

where $R(N)$ and $\mathbf{d}(N)$ are given by the normal equations (2.22). Under assumptions (2.33) and (2.34), ensuring convergence of the LR estimate, we obtain

$$\hat{\boldsymbol{\theta}}(N) - \boldsymbol{\theta}(N) \to 0 \qquad N \to \infty$$

which means that the effect of initial conditions becomes negligible after sufficiently large N. In practice, P_0 is chosen to have a large value.

From (2.33) and (2.48) it is seen that $P(N) \to 0$ as $N \to \infty$. Equation (2.49) implies

$$P(N) = -\sum_{t=0}^{N} \mathbf{w}(t)\alpha(t)\mathbf{w}^{\mathrm{T}}(t)$$

where

$$\mathbf{w}(t) = -P(t-1)\boldsymbol{\phi}(t) \tag{2.58}$$

defines the so-called Kalman gain. It thus follows that $\mathbf{w}(N) \to 0$ and $\boldsymbol{\theta}(N)$ converges. This is in accordance with the implicit stationarity assumption made in (2.33). When the dynamics are time varying the Kalman gain should remain alert to track time variations. The prevention of $\mathbf{w}(N)$ going to zero is accomplished by incorporating the exponential window in (2.11). Following similar steps as for (2.53)–(2.56), we obtain the well-known form of the RLS algorithm given in Table 2.1. Note that the inverse of $\alpha(N)$, denoted $\alpha^*(N)$, has been used. We shall meet both of these variables in later chapters dealing with fast algorithms. $\alpha(N)$, being less than one, is usually referred to as the angle variable. A closely related variable is the so-called likelihood variable, because of its presence in the maximum likelihood identification problem, defined as

$$\gamma(N) = \boldsymbol{\phi}^{\mathrm{T}}(N)P(N)\boldsymbol{\phi}(N)$$

Table 2.1 The RLS algorithm

INITIALIZATION

$\boldsymbol{\theta}(-1) = \mathbf{0}$

$P(-1) = \delta I,\ \delta \gg 1$

FOR $N = 0, 1, 2, \ldots$

$e(N) = y(N) - \boldsymbol{\phi}^{\mathrm{T}}(N)\boldsymbol{\theta}(N-1)$

$\mathbf{w}(N) = -P(N-1)\boldsymbol{\phi}(N)$

$\alpha^*(N) = \lambda - \mathbf{w}^{\mathrm{T}}(N)\boldsymbol{\phi}(N)$

$\boldsymbol{\theta}(N) = \boldsymbol{\theta}(N-1) - \mathbf{w}(N)e(N)/\alpha^*(N)$

$P(N) = (1/\lambda)[P(N-1) - \mathbf{w}(N)\alpha^{*-1}(N)\mathbf{w}^{\mathrm{T}}(N)]$

END

These variables play an important part in monitoring numerical instability in the adaptive LS family. Appendix A of Chapter 8 provides further insight on these variables. Another notational convention results if we combine $\mathbf{w}(N)$ and $\alpha^*(N)$ in a single variable:

$$\tilde{\mathbf{w}}(N) \equiv \frac{\mathbf{w}(N)}{\alpha^*(N)}$$

Using the respective definitions and (2.56) it can be shown that

$$\tilde{\mathbf{w}}(N) = -P(N)\phi(N) \tag{2.59}$$

$\tilde{\mathbf{w}}(N)$ is usually known as the posterior Kalman gain and is sometimes used in certain forms of fast versions of the RLS algorithm.

The presence of $\lambda < 1$ in the algorithm of Table 2.1 prevents $P(N)$ and $\mathbf{w}(N)$ from going to zero. In practice the correct choice of λ is the result of a compromise between tracking speed and noise sensitivity. Indeed, for $\lambda \neq 1$, the algorithm introduces an error that forces the average parameter vector $\boldsymbol{\theta}$ away from the optimal Wiener solution. This is called misadjustment error and it increases with decreasing λ. On the other hand, the tracking ability of the algorithm to time variations is measured in terms of another error component known as lag error, which decreases with decreasing λ. The choice of λ is suggested by these confronting terms. These issues are discussed in Chapters 3 and 5.

Adaptive least-squares algorithms implemented in a finite precision environment suffer from two types of instability, owing to the accumulation of numerical errors, leading either to explosive divergence or to the so-called stalling or lock-up phenomenon. The former manifests itself by (a) a sudden increase of the norm of the parameter error vector (with respect to the finite precision value) and (b) the negative value of the angle variable. Stalling is caused when $P(N)$ becomes small compared with the accuracy determined by the register length. For the filtering case (i.e. Equation (2.26)) it is not difficult to show that for λ close to one [4]

$$E[P(N)] \simeq \frac{1-\lambda}{\sigma_u^2} R^{-1}$$

where σ_u^2 is the variance of the input $u(t)$ and R is the autocorrelation matrix, assuming stationarity. From the above equation it becomes apparent that for λ close to unity or large σ_u^2, $P(N)$ can become sufficiently small to be represented as a zero in the register. This obviously freezes adaptation. The explosive divergence phenomenon is more complicated and, as we shall see later on, is associated with the loss of the positive definiteness of the matrix $P(N)$. The treatment of this phenomenon for fast versions of RLS is even more involved and is discussed in Chapter 5, where stabilization schemes are also discussed.

2.5.3 Approximate realizations of the LR estimate and the LMS algorithm

As we pointed out above, the RLS algorithm converges under assumptions (2.33) and, for $\lambda = 1$, to the Wiener solution. Consequently, it can be viewed as an iterative scheme that in the limit solves the least mean square (LMS) criterion (2.38). The main RLS recursion (2.53) belongs to a general class of optimization algorithms aiming to minimize a cost functional $V(\mathbf{x})$. This class has the form

$$\mathbf{x}(N+1) = \mathbf{x}(N) + s(N)\mathbf{p}(N) \tag{2.60}$$

The sequence $\mathbf{x}(N)$ so constructed aims to capture a minimum of $V(\mathbf{x})$. At each step N, $\mathbf{p}(N)$ provides the search direction on which the new update $\mathbf{x}(N+1)$ is located. $s(N)$ determines the step size, namely how far along the direction of $\mathbf{p}(N)$ the new estimate will be. A typical family of algorithms (2.60) is based on the Picard scheme for the iterative solution of equations. Indeed, a local minimum of a smooth function $V(\mathbf{x})$ is necessarily a stationary point, i.e. it satisfies

$$\frac{\mathrm{d}V}{\mathrm{d}\mathbf{x}} = 0 \tag{2.61}$$

The Picard algorithm for the solution of (2.61) employs a recursion of the form

$$\mathbf{x}(N+1) = F(\mathbf{x}(N)) \tag{2.62}$$

where

$$F(\mathbf{x}) = \mathbf{x} - A(\mathbf{x})\frac{\mathrm{d}V}{\mathrm{d}\mathbf{x}} \tag{2.63}$$

$A(\mathbf{x})$ is a design parameter such that for each $\mathbf{x}$, $A(\mathbf{x})$ is an invertible matrix. The algorithm (2.62)–(2.63) converges near $\mathbf{x}^*$ to a local minimum of $V(\mathbf{x})$ if $A(\mathbf{x})$ is chosen so that the eigenvalues of the linear part of $F(\mathbf{x})$ at $\mathbf{x}^*$

$$\left.\frac{\partial F}{\partial \mathbf{x}}\right|_{\mathbf{x}^*} = I - A(\mathbf{x}^*)R(\mathbf{x}^*) \tag{2.64}$$

where $R(\mathbf{x})$ is the Hessian matrix

$$R(\mathbf{x}) = \left(\frac{\partial^2 V}{\partial^2 \mathbf{x}}\right) \tag{2.65}$$

are inside the unit circle. One choice that ensures local convergence is

$$A(\mathbf{x}) = \left(\frac{\partial^2 V}{\partial^2 \mathbf{x}}\right)^{-1} \tag{2.66}$$

This gives rise to the Gauss–Newton or Newton–Raphson search direction. This is precisely the direction employed by the RLS algorithm. Indeed, it can be shown ([1], page 47) that for the linear regression case the covariance matrix coincides with the second derivative of the quadratic cost function. The above choice offers the fastest possible rate of convergence in the sense that the eigenvalues of the linear part of the algorithm are placed at the origin. This superior convergence performance is counterbalanced by heavy computational and memory requirements.

Inspection of Equations (2.63)–(2.66) shows that the computation of the search direction, and in particular the estimation of the inverse of the Hessian matrix, is the most demanding portion of the algorithm. To come up with variants of the form (2.63) offering an improved trade-off between operations count and storage requirements on one hand, and convergence speed on the other, alternative choices of $A(\mathbf{x})$ must be tested. To ensure a reduction of complexity, we try $A(\mathbf{x}) = \mu I$ and we check if the resulting algorithm is stable. The linear part (2.64) of F at $\mathbf{x}^*$ is $I - \mu R$, $R = R(\mathbf{x}^*)$, and its eigenvalues are inside the unit circle provided

$$\mu < \frac{2}{\lambda_{\max}} \tag{2.67}$$

$\lambda_{\max}$ denotes the maximum eigenvalue of R. One such μ is

$$\mu = \frac{1}{\operatorname{tr}(R)} \tag{2.68}$$

where $\operatorname{tr}(\cdot)$ is the trace of R. Thus the above choice leads to a convergent algorithm with a reduced computational complexity, albeit slower convergence speed. The main recursion (2.63) takes the form

$$\mathbf{x}(N + 1) = \mathbf{x}(N) - \mu \nabla V(\mathbf{x}(N)) \tag{2.69}$$

Let us apply the above scheme in the context of linear regression and for the LMS criterion (2.38):

$$V(\boldsymbol{\theta}) = E[e^2(t \mid \boldsymbol{\theta})]$$

The gradient of $V(\boldsymbol{\theta})$ is

$$\nabla V(\boldsymbol{\theta}) = -2E[e(t \mid \boldsymbol{\theta})\boldsymbol{\phi}(t)] \tag{2.70}$$

and the gradient algorithm (2.69) becomes

$$\boldsymbol{\theta}(N) = \boldsymbol{\theta}(N-1) + \mu E[e(N \mid \boldsymbol{\theta}(N-1))\boldsymbol{\phi}(N)] \tag{2.71}$$

To alleviate computational complexity further, we employ the approximation

$$E[e(N, \boldsymbol{\theta}(N-1))\boldsymbol{\phi}(N)] = e(N, \boldsymbol{\theta}(N-1))\boldsymbol{\phi}(N)$$

That is, we replace the mean of the relevant random variable with the random variable itself. This can be justified by an averaging argument [1–5]. With the above approximation we arrive at the so-called *LMS algorithm* of Widrow and Hoff:

$$\boldsymbol{\theta}(N) = \boldsymbol{\theta}(N-1) + \mu e(N \mid \boldsymbol{\theta}(N-1))\boldsymbol{\phi}(N) \tag{2.72}$$

Frequently, use of a time-varying step size $\mu(N)$, rather than a constant step, gives superior performance, particularly when the algorithm operates in a time-varying environment. A reasonable choice of $\mu(N)$, supported by (2.68), is

$$\mu(N) = \frac{1}{r(N)} \qquad r(N) = r(N-1) + \boldsymbol{\phi}^{\mathrm{T}}(N)\boldsymbol{\phi}(N)$$

Note that $r(N)$ realizes the trace of $R(N)$, which under ergodic assumptions tends to the trace of R.

A variant of the LMS algorithm which is widely used in practice is the *normalized LMS* (NLMS) algorithm. In this scheme the gradient is normalized with the norm of the regressor vector so as to minimize the effect of signal amplitude in the correction of the parameter vector. The resulting algorithm is

$$e(N) = y(N) - \boldsymbol{\phi}^{\mathrm{T}}(N)\boldsymbol{\theta}(N-1) \tag{2.73}$$

$$\boldsymbol{\theta}(N) = \boldsymbol{\theta}(N-1) + \frac{\mu}{a + \|\boldsymbol{\phi}(N)\|^2} e(N)\boldsymbol{\phi}(N) \tag{2.74}$$

The initial value of $\boldsymbol{\theta}$ is commonly set to zero. a is a small positive constant to safeguard against possible division with very small signal value. It is interesting to note that for $a = 0$ the above recursions result from minimization of the cost $\|\boldsymbol{\theta}(N) - \boldsymbol{\theta}(N-1)\|^2$, subject to the constraint $y(N) = \boldsymbol{\theta}^{\mathrm{T}}(N-1)\boldsymbol{\phi}(N)$. In other words the choice of the time-varying step size ensures that successive estimates are optimally close. Important variants of the LMS algorithm such as the sign LMS algorithm are discussed in Chapter 4.

Let us next summarize the performance features of the family of stochastic gradients.

We focus attention on the operations count of the LMS algorithm. The number of multiplications is $O(p)$ for the computation of the inner product appearing in the error equation, plus $O(p)$ for the multiplication of the scalar by the vector $\boldsymbol{\phi}(N)$. Analogous to this is the number of additions. Hence the total number of operations is proportional to $2p$ multiplications and additions (MADs). Since there is only one recursive relation involved, the storage requirements are proportional to p. We conclude that the gradient algorithms offer significant computational and memory savings compared with the RLS method.

A task of major importance for practical applications is the comparison of performance of the LMS algorithm (and its variants) and the RLS family with respect to convergence and tracking for time-varying dynamics. This is still an ongoing area of research [7,8]. It is highlighted in Chapters 3 to 6.

2.5.4 The instrumental variable method

A serious limitation of the linear regression estimate is that consistency is achieved if $\eta(t)$ is uncorrelated with past data (i.e. see p. 17). This requirement is essentially satisfied when $\eta(t)$ is white noise. This is so because the regressor vector carries past output values $y(t-i)$, which in turn depend on past noise values $\eta(t-i)$. Here we refer to the ARX case. If $\eta(t)$ is not white noise, the predictor fails to extract all the relevant information incorporated into $\eta(t)$. The obvious approach then is to modify the predictor structure so that this extra information inherited into $\eta(t)$ is accommodated. We follow this line in section 2.6.1. As we shall see, it turns out that if the data are generated by a rational linear system, the predictor is described by a linear rational system as well, with the data sequences as inputs. Nevertheless, it does exhibit a non-linear dependence on the unknown parameter $\boldsymbol{\theta}$. As a consequence, the least-squares optimization problem leads to non-linear equations. An alternative approach which maintains the linear system of equations for the determination of the unknown parameters is the instrumental variable method and its variants. The instrument is a multichannel sequence $\zeta(t)$ which is much 'like' the regressor sequence $\boldsymbol{\phi}(t)$, and at the same time it is uncorrelated with the noise sequence $\eta(t)$. Leaving aside for a while the choice of the instrument, we determine $\boldsymbol{\theta}$ so that the error in (2.8) is uncorrelated with past information, which is now represented by the instrument, i.e.

$$\sum_{t=0}^{N} \zeta(t)e(t) = \mathbf{0}$$

This leads to the linear system

$$R_{iv}(N)\boldsymbol{\theta}_{iv}(N) = \mathbf{r}_{iv}(N)$$

where

$$R_{iv}(N) = \sum_{t=0}^{N} \zeta(t)\boldsymbol{\phi}^{\mathrm{T}}(t) \qquad \mathbf{r}_{iv}(N) = \sum_{t=0}^{N} \zeta(t)y(t)$$

If we carry out the same analysis as in the linear regression case we find that $\boldsymbol{\theta}_{iv}(N)$ yields the true parameter asymptotically if the following conditions hold:

$$\det \lim_{N\to\infty} \frac{1}{N} R_{iv}(N) \neq 0$$

$$\lim_{N\to\infty} \frac{1}{N} \sum_{t=0}^{N} \zeta(t)\eta(t) = \mathbf{0}$$

The second condition states that the instrument is uncorrelated with the noise sequence $\eta(t)$, and the first condition requires that the instrument is sufficiently correlated to the data.

Let us suppose that data are generated by an ARX model of the form (2.31). To guarantee the validity of the above requirements we generate $\zeta(t)$ by filtering the data sequence. A general and common prescription is

$$\zeta(t) = (x(t-1) \ldots x(t-n_x) \qquad u(t-1) \ldots u(t-n_u))^{\mathrm{T}}$$

where

$$N(q)x(t) = M(q)u(t)$$

where $q^{-l}x(t) \equiv x(t-l)$ and $N(q)$, $M(q)$ polynomials of q^{-1}.

Assuming that the input $u(t)$ is uncorrelated with $\eta(t)$ (this precludes the use of feedback), we infer that $x(t)$ is also uncorrelated with $\eta(t)$. Furthermore, it is obtained by linear filtering of the data sequence, so we expect that the first of the above conditions should hold generically. The filter coefficients N and M can be chosen in many ways. A common choice is $N(q) = 1$, $M(q) = q^{-l}$. Another possibility is first to apply the least-squares method on the given data sequences and then use the resulting estimates for $N(q)$ and $M(q)$.

In an adaptive context the instrument is made to depend on the parameter. Thus it can be generated with $N(q)$ and $M(q)$ being formed from the previous estimate $\boldsymbol{\theta}(t-1)$ [2,3].

2.6 General prediction models

2.6.1 The general LS optimization problem and the prediction error method

In the previous section we focused our interest on the special case of linear regression where the linear dependence in $\boldsymbol{\theta}$ of the predictor leads to linear equations. Next we consider the general optimization problem (2.10). To determine the minimizing solution $\boldsymbol{\theta}(N)$ we set the derivative of $V_N(\theta)$ equal to zero. This leads to

$$\sum_{t=0}^{N} e(t \mid \boldsymbol{\theta})\Psi(t \mid \boldsymbol{\theta}) = 0$$

where

$$\Psi^{\mathrm{T}}(t \mid \boldsymbol{\theta}) \equiv \frac{\partial e(t \mid \boldsymbol{\theta})}{\partial \boldsymbol{\theta}} = \frac{\partial \hat{y}(t \mid \boldsymbol{\theta})}{\partial \boldsymbol{\theta}}$$

Unless $\Psi(t \mid \boldsymbol{\theta})$ depends linearly on $\boldsymbol{\theta}$, the above equations are non-linear with respect to $\boldsymbol{\theta}$. Hence we must resort to iterative techniques for the determination of $\boldsymbol{\theta}(N)$. This means that at each step N an iterative algorithm

$$\boldsymbol{\theta}_{k+1}(N) = f(\boldsymbol{\theta}_k(N))$$

must be employed which, as $k \to \infty$, will provide $\boldsymbol{\theta}(N)$. Apparently, we are confronted with a costly, time-consuming operation which is hard to use in an adaptive real-time situation. To accomplish satisfactory computational performance we prefer to work suboptimally at a local level. Thus at each time N we determine an estimate $\hat{\boldsymbol{\theta}}(N)$ which is suboptimal with respect to $V_N(\boldsymbol{\theta})$ yet is close to the actual minimizing parameter $\boldsymbol{\theta}(N)$. We do, however, expect to obtain asymptotic equivalence, that is,

$$\boldsymbol{\theta}(N) - \hat{\boldsymbol{\theta}}(N) \to 0 \qquad N \to \infty$$

Two popular approaches for the construction of such estimates are the prediction error method and the stochastic gradient method. We outline the basic ideas below.

THE PREDICTION ERROR ALGORITHM

It is convenient to divide the computational procedure into three parts: the basic underlying RLS structural module, the computation of the predictor, and the computation of the gradient.

THE UNDERLYING RLS STRUCTURE

Suppose we have an estimate $\boldsymbol{\theta}(N)$. We determine $\boldsymbol{\theta}(N+1)$ as follows. We expand $V_{N+1}(\boldsymbol{\theta})$ in a second-order Taylor series around $\boldsymbol{\theta}(N)$ and assign $\boldsymbol{\theta}(N+1)$ as the minimum of the approximating quadratic form. The resulting estimate will satisfy a system of linear equations of the form (2.21) and can thus be determined by the RLS algorithm. It is convenient to develop the second-order Taylor approximation of $V_{N+1}(\boldsymbol{\theta})$ through a first-order Taylor expansion of the error $e(t \mid \boldsymbol{\theta})$ with respect to $\boldsymbol{\theta}$ (recall that $V_{N+1}(\boldsymbol{\theta})$ is the total squared error (2.10)). Thus we have

$$\begin{aligned} e(t \mid \boldsymbol{\theta}) &= y(t) - \hat{y}(t \mid \boldsymbol{\theta}) \\ &= y(t) - \hat{y}(t \mid \boldsymbol{\theta}(N)) - \boldsymbol{\psi}^{\mathrm{T}}(t \mid \boldsymbol{\theta}(N))(\boldsymbol{\theta} - \boldsymbol{\theta}(N)) + \text{higher-order terms} \end{aligned}$$

where $\Psi(t \mid \boldsymbol{\theta}(N))$ is the gradient vector evaluated at $\boldsymbol{\theta}(N)$:

$$\boldsymbol{\psi}(t \mid \boldsymbol{\theta}(N)) = \left. \frac{\partial \hat{y}(t \mid \boldsymbol{\theta})}{\partial \boldsymbol{\theta}} \right|_{\boldsymbol{\theta} = \boldsymbol{\theta}(N)}$$

If we employ the approximation

$$e(t \mid \boldsymbol{\theta}) \approx \tilde{e}(t \mid \boldsymbol{\theta}) = y(t) - \hat{y}(t \mid \boldsymbol{\theta}(N)) - \boldsymbol{\psi}^{\mathrm{T}}(t \mid \boldsymbol{\theta}(N))(\boldsymbol{\theta} - \boldsymbol{\theta}(N))$$

we obtain a linear dependence in $\boldsymbol{\theta}$ and the resulting approximating cost functional

$$\sum \tilde{e}^2(t \mid \boldsymbol{\theta})$$

becomes quadratic in $\boldsymbol{\theta}$. Therefore, the minimizing value $\boldsymbol{\theta}(N+1)$ satisfies a linear system of equations of the form (2.21) where

$$R(N+1) = \sum_{t=0}^{N+1} \boldsymbol{\psi}(t \mid \boldsymbol{\theta}(N)) \boldsymbol{\psi}^{\mathrm{T}}(t \mid \boldsymbol{\theta}(N)) \tag{2.75}$$

$$\mathbf{d}(N+1) = \sum_{t=0}^{N+1} \boldsymbol{\psi}(t \mid \boldsymbol{\theta}(N))[\, y(t) - \hat{y}(t \mid \boldsymbol{\theta}(N)) + \boldsymbol{\psi}^{\mathrm{T}}(t \mid \boldsymbol{\theta}(N)) \boldsymbol{\theta}(N)] \tag{2.76}$$

Now the main RLS recursion (2.53) applies. Moreover, the matrix $R(N)$ is the second derivative of the above approximate cost, and since we assumed the cost is quadratic, $R(N)$ is approximately constant with respect to $\boldsymbol{\theta}$. Thus its inverse is updated as in (2.56). Let us next proceed with the determination of the predictor $\hat{y}(t \mid \boldsymbol{\theta}(N))$.

COMPUTATION OF THE PREDICTOR

Predictor models and their properties are discussed thoroughly in [1–3]. Predictor models are typically chosen as stable, linear and rational transformations of the input–output data smoothly dependent on the unknown $\boldsymbol{\theta}$. We thus write

$$\hat{y}(t \mid \boldsymbol{\theta}) = W_u(q \mid \boldsymbol{\theta})u(t) + W_y(q \mid \boldsymbol{\theta})y(t) \tag{2.77}$$

where q stands for the shift operator and $W_u(q \mid \boldsymbol{\theta})$, $W_y(q \mid \boldsymbol{\theta})$ are ratios of polynomials in q, with the corresponding polynomial coefficients smoothly depending on $\boldsymbol{\theta}$.

Predictor representations (2.77) result when the unknown system is a realizable, linear, time-invariant system driven by white noise. In this case the input–output data sequences satisfy the **Box–Jenkins model**

$$y(t) = G(q \mid \boldsymbol{\theta})u(t) + H(q \mid \boldsymbol{\theta})\eta(t) \tag{2.78}$$

where η is a sequence of identically distributed zero-mean uncorrelated random variables and $G(q \mid \boldsymbol{\theta})$, $H(q \mid \boldsymbol{\theta})$ are rational functions:

$$G(q \mid \boldsymbol{\theta}) = \frac{B_u(q \mid \boldsymbol{\theta})}{A_u(q \mid \boldsymbol{\theta})} \qquad H(q \mid \boldsymbol{\theta}) = \frac{B_\eta(q \mid \boldsymbol{\theta})}{A_\eta(q \mid \boldsymbol{\theta})} \tag{2.79}$$

If we multiply both sides of (2.78) with the inverse of $H(q \mid \boldsymbol{\theta})$

$$H^{-1}(q \mid \boldsymbol{\theta}) = \frac{A_\eta(q \mid \boldsymbol{\theta})}{B_\eta(q \mid \boldsymbol{\theta})}$$

we obtain

$$H^{-1}(q \mid \boldsymbol{\theta})y(t) = H^{-1}(q \mid \boldsymbol{\theta})G(q \mid \boldsymbol{\theta})u(t) + \eta(t)$$

or

$$y(t) = [1 - H^{-1}(q \mid \boldsymbol{\theta})]y(t) + H^{-1}(q \mid \boldsymbol{\theta})G(q \mid \boldsymbol{\theta})u(t) + \eta(t) \tag{2.80}$$

We assume that the leading coefficient in $H^{-1}(q \mid \boldsymbol{\theta})$ is one. If the above equation is evaluated at time t, the first term on the right-hand side depends only on past output values $y(t-1)$, $y(t-2)$, and so forth. Therefore the minimum variance predictor is

$$\hat{y}(t \mid \boldsymbol{\theta}) = [1 - H^{-1}(q \mid \boldsymbol{\theta})]y(t) + H^{-1}(q \mid \boldsymbol{\theta})G(q \mid \boldsymbol{\theta})u(t) \tag{2.81}$$

If we substitute (2.79) into (2.81) we establish that the predictor is a rational linear system (2.77) given by

$$W_u(q \mid \boldsymbol{\theta}) = \frac{A_\eta(q \mid \boldsymbol{\theta})B_u(q \mid \boldsymbol{\theta})}{B_\eta(q \mid \boldsymbol{\theta})A_u(q \mid \boldsymbol{\theta})} \qquad W_y(q \mid \boldsymbol{\theta}) = \frac{B_\eta(q \mid \boldsymbol{\theta}) - A_\eta(q \mid \boldsymbol{\theta})}{B_\eta(q \mid \boldsymbol{\theta})} \tag{2.82}$$

We can also write (2.77) as

$$D(q \mid \boldsymbol{\theta})\hat{y}(t \mid \boldsymbol{\theta}) = N_u(q \mid \boldsymbol{\theta})u(t) + N_y(q \mid \boldsymbol{\theta})y(t) \tag{2.83}$$

where D, N_u and N_y are polynomials in q with coefficients smoothly depending on $\boldsymbol{\theta}$.

We often find it more convenient to work with the error

$$e(t \mid \boldsymbol{\theta}) = y(t) - \hat{y}(t \mid \boldsymbol{\theta}) \tag{2.84}$$

rather than the predictor $\hat{y}(t \mid \boldsymbol{\theta})$. In fact it is the error which is propagated by the prediction error algorithm – equation (2.76). We readily infer from (2.84) and (2.81) that the error obeys the following rational linear model:

$$e(t \mid \boldsymbol{\theta}) = H^{-1}(q \mid \boldsymbol{\theta})y(t) - H^{-1}(q \mid \boldsymbol{\theta})G(q \mid \boldsymbol{\theta})u(t)$$

$$= \frac{A_\eta(q \mid \boldsymbol{\theta})}{B_\eta(q \mid \boldsymbol{\theta})} y(t) - \frac{A_\eta(q \mid \boldsymbol{\theta})}{B_\eta(q \mid \boldsymbol{\theta})} \frac{B_u(q \mid \boldsymbol{\theta})}{A_u(q \mid \boldsymbol{\theta})} u(t) \tag{2.85}$$

EXAMPLE: PREDICTOR DYNAMICS FOR ARMAX MODELS

An ARMAX model is a generalization of the ARX model introduced in section 2.5 and involves past noise samples as well, that is,

$$A(q)y(t) + B(q)u(t) = C(q)\eta(t) \tag{2.86}$$

or

$$y(t) + \sum_{i=1}^{n_a} a_i y(t-i) = -\sum_{i=1}^{n_b} b_i u(t-i) + \sum_{i=0}^{n_c} c_i \eta(t-i)$$

with c_0 being equal to one. We define the unknown parameter $\boldsymbol{\theta}$ to be

$$\boldsymbol{\theta} = (a_1, a_2, \ldots, a_{n_a}, b_1, b_2, \ldots, b_{n_b}, c_1, c_2, \ldots, c_{n_c})^{\mathrm{T}} \tag{2.87}$$

If all the b_i are identically zero the resulting model is known as ARMA.

To determine the predictor, we divide both sides of (2.86) by A or directly use (2.81). In this way we find

$$C(q)\hat{y}(t \mid \boldsymbol{\theta}) = (C(q) - A(q))y(t) - B(q)u(t) \tag{2.88}$$

The parameter set Θ consists of vectors (2.87) for which the zeros of the polynomial $C(q)$ are outside the unit circle.

In a similar fashion the following model governs the behaviour of the error signal:

$$C(q)e(t \mid \boldsymbol{\theta}) = A(q)y(t) + B(q)u(t) \tag{2.89}$$

In the light of the analysis of the linear regression case it is instructive to seek expressions of the form (2.19). The error equation (2.89) is very convenient for this purpose. Indeed, we have

$$e(t) + \sum_{i=1}^{n_c} c_i e(t-i) = y(t) + \sum_{i=1}^{n_a} a_i y(t-i) + \sum_{i=1}^{n_b} b_i u(t-i)$$

or

$$e(t) = y(t) - \boldsymbol{\theta}^{\mathrm{T}} \boldsymbol{\phi}(t \mid \boldsymbol{\theta}) \tag{2.90}$$

where

$$\boldsymbol{\phi}(t \mid \boldsymbol{\theta}) = (-y(t-1), \ldots, -y(t-n_a), -u(t-1), \ldots, -u(t-n_b), e(t-1), \ldots, e(t-n_c))^{\mathrm{T}} \tag{2.91}$$

Finally, we deduce from (2.90) and (2.84)

$$\hat{y}(t \mid \boldsymbol{\theta}) = \boldsymbol{\theta}^{\mathrm{T}} \boldsymbol{\phi}(t \mid \boldsymbol{\theta}) \tag{2.92a}$$

This equation determines the so-called pseudolinear form of the predictor for the ARMAX system. Apparently, it is non-linear in $\boldsymbol{\theta}$ because $\boldsymbol{\phi}(t \mid \boldsymbol{\theta})$ depends on $\boldsymbol{\theta}$. A remarkable fact is that the pseudolinear expression (2.92a) remains true for general Box–Jenkins models.

Equation (2.83) demonstrates the computation of the predictor model required in the application of the prediction error algorithm. We observe that for each time N, $\hat{y}(t \mid \boldsymbol{\theta}(N))$ must be determined for all values of $0 \leqslant t \leqslant N$. To determine a true finite-dimensional state-space representation of the prediction error algorithm we must further impose the approximation

$$D^i(\boldsymbol{\theta}(N)) = D^i(\boldsymbol{\theta}(N-i)) \quad N_u^i(\boldsymbol{\theta}(N)) = N_u^i(\boldsymbol{\theta}N-i)) \quad N_y^i(\boldsymbol{\theta}(N)) = N_y^i(\boldsymbol{\theta}(N-i)) \tag{2.92b}$$

$D^i(\boldsymbol{\theta}(N))$ denotes the ith coefficient of the polynomial D, and similarly for the others. This is a reasonable assumption if $\boldsymbol{\theta}(N)$ and $\boldsymbol{\theta}(N-i)$ are close and allows the use of the previously computed $\hat{y}(t-i \mid \boldsymbol{\theta}(N-i))$ in place of $\hat{y}(t-i \mid \boldsymbol{\theta}(N))$.

COMPUTATION OF THE GRADIENT

Let us next consider the gradient vector

$$\boldsymbol{\psi}(t \mid \boldsymbol{\theta}) = \frac{\partial \hat{y}(t \mid \boldsymbol{\theta})}{\partial \boldsymbol{\theta}}$$

If the predictor is represented by the rational model (2.83), a similar representation is valid for the gradient. Indeed, we differentiate (2.83) with respect to $\boldsymbol{\theta}$ to obtain (we use the notation $\dot{\mathbf{x}} = \partial \mathbf{x}/\partial \boldsymbol{\theta}$ and suppress the arguments $\boldsymbol{\theta}$ and q)

$$\dot{D}\hat{y} + D\boldsymbol{\psi} = \dot{N}_u u + \dot{N}_y y \tag{2.93}$$

This is a vector equation. To eliminate $\hat{y}$ we multiply both sides by D and make use of (2.83):

$$D^2\boldsymbol{\psi} + \dot{D}D\hat{y} = D\dot{N}_u u + D\dot{N}_y y$$

or

$$D^2\boldsymbol{\psi} + \dot{D}(N_u u + N_y y) = D\dot{N}_u u + D\dot{N}_y y$$

or

$$D^2\boldsymbol{\psi} = (D\dot{N}_u - \dot{D}N_u)u + (D\dot{N}_y - \dot{D}N_y)y \tag{2.94}$$

This is the desired rational model. It is always stable since its dynamics are governed by the roots of D^2 and, by assumption, the predictor (2.83) is always stable. We illustrate next the gradient predictor model (2.94) for ARMAX systems.

EXAMPLE: GRADIENT DYNAMICS FOR ARMAX SYSTEMS

We consider the ARMAX predictor model (2.86) where $\boldsymbol{\theta}$ is given by the polynomial coefficients (2.87). We introduce the notation

$$\mathbf{q}_n = [1, q, q^2, \ldots, q^n]^{\mathrm{T}} \tag{2.95}$$

Differentiation with respect to $\boldsymbol{\theta}$ or direct evaluation of (2.94) leads to the following forms:

$$C^2\boldsymbol{\psi} \equiv C^2\begin{pmatrix} \partial\hat{y}/\partial A \\ \partial\hat{y}/\partial B \\ \partial\hat{y}/\partial C \end{pmatrix} = \left(C\begin{pmatrix} 0 \\ -\mathbf{q}_{\mathbf{nb}} \\ 0 \end{pmatrix} + \begin{pmatrix} 0 \\ 0 \\ \mathbf{q}_{\mathbf{nc}} \end{pmatrix}B\right)u$$

$$+\left[C\begin{pmatrix} -\mathbf{q}_{\mathbf{na}} \\ 0 \\ \mathbf{q}_{\mathbf{nc}} \end{pmatrix} - (C-A)\begin{pmatrix} 0 \\ 0 \\ \mathbf{q}_{\mathbf{nc}} \end{pmatrix}\right]y$$

More explicitly, we have

$$C\,\frac{\partial\hat{y}}{\partial A} = -\mathbf{q}_{\mathbf{na}}\,y$$

$$C\,\frac{\partial\hat{y}}{\partial B} = -\mathbf{q}_{\mathbf{nb}}\,u$$

$$C^2\,\frac{\partial\hat{y}}{\partial C} = -\mathbf{q}_{\mathbf{nc}}(-Bu - Ay) = \mathbf{q}_{\mathbf{nc}}\,Ce$$

Therefore, using (2.91) we obtain

$$\boldsymbol{\psi}(t \mid \boldsymbol{\theta}) = \frac{1}{C}\,\boldsymbol{\phi}(t \mid \boldsymbol{\theta})$$

or

$$\boldsymbol{\psi}(t \mid \boldsymbol{\theta}) + \sum_{i=1}^{n_c} c_i\boldsymbol{\psi}(t-i \mid \boldsymbol{\theta}) = \boldsymbol{\phi}(t \mid \boldsymbol{\theta}) \tag{2.96}$$

As we pointed out in the computation of the predictor, the rational model (2.94) does not yield a finite-dimensional state-space representation of the prediction error algorithm. We employ the same approximation technique (see (2.92b)) as we did with the predictor, this time for the polynomial coefficients of the gradient dynamics. Thus it is not difficult to see that (2.96) leads to

$$\begin{aligned}\boldsymbol{\psi}(t \mid \boldsymbol{\theta}) = (&-y^{\mathrm{f}}(t-1), \ldots, -y^{\mathrm{f}}(t-n_a), -u^{\mathrm{f}}(t-1), \ldots, \\ &-u^{\mathrm{f}}(t-n_b), e^{\mathrm{f}}(t-1), \ldots, e^{\mathrm{f}}(t-n_c))^{\mathrm{T}}\end{aligned}$$

where

$$\begin{aligned} y^{\mathrm{f}}(t) &= y(t) - c_1\,y^{\mathrm{f}}(t-1) - \ldots - c_{n_c}\,y^{\mathrm{f}}(t-n_c) \\ u^{\mathrm{f}}(t) &= u(t) - c_1\,u^{\mathrm{f}}(t-1) - \ldots - c_{n_c}\,u^{\mathrm{f}}(t-n_c) \\ e^{\mathrm{f}}(t) &= e(t) - c_1\,e^{\mathrm{f}}(t-1) - \ldots - c_{n_c}\,e^{\mathrm{f}}(t-n_c) \end{aligned}$$

Combining the above and the RLS algorithm for solving (2.21) for the case of (2.75) and (2.76), results in the algorithm of Table 2.2.

Table 2.2 Recursive prediction error algorithm for the ARMAX model

$\boldsymbol{\theta}(-1) = \mathbf{0}$

$P(-1) = \delta I, \ \delta \gg 1$

FOR $N = 0, 1, 2, 3, \ldots$

$\Psi(N) = (-y^{\mathrm{f}}(N-1), \ldots, y^{\mathrm{f}}(N-n_a), -u^{\mathrm{f}}(N-1), \ldots, -u^{\mathrm{f}}(N-n_b), e^{\mathrm{f}}(N-1), \ldots, e^{\mathrm{f}}(N-n_c))$

$e(N) = y(N) - \boldsymbol{\theta}^{\mathrm{T}}(N-1)\boldsymbol{\phi}(N)$

$\mathbf{w}(N) = -P(N-1)\Psi(N)$

$\alpha^*(N) = \lambda - \mathbf{w}(N)^{\mathrm{T}}\Psi(N)$

$\boldsymbol{\theta}(N) = \boldsymbol{\theta}(N-1) - \mathbf{w}(N)\alpha^{*-1}(N)e(N)$

$P(N) = (1/\lambda)[P(N-1) - \mathbf{w}(N)\alpha^{*-1}(N)\mathbf{w}^{\mathrm{T}}(N)]$

$y^{\mathrm{f}}(N) = y(N) - \Sigma_{i=1}^{n_c} c_i(N)y^{\mathrm{f}}(N-i)$

$u^{\mathrm{f}}(N) = u(N) - \Sigma_{i=1}^{n_c} c_i(N)u^{\mathrm{f}}(N-i)$

$e^{\mathrm{f}}(N) = e(N) - \Sigma_{i=1}^{n_c} c_i(N)e^{\mathrm{f}}(N-i)$

END

$\boldsymbol{\theta} \equiv (a_1, a_2, \ldots, a_{n_a}, b_1, b_2, \ldots, b_{n_b}, c_1, c_2, \ldots, c_{n_c})^{\mathrm{T}}$

$\boldsymbol{\phi}(N) \equiv (-y(N-1), \ldots, -y(N-n_a), -u(N-1), \ldots, -u(N-n_b), e(N-1), \ldots, e(N-n_c))^{\mathrm{T}}$

The analysis leading to the prediction error algorithm is now complete. The task of synthesis to establish algorithm convergence is hard and a subject of ongoing research. The interested reader may consult [1]–[5] and the references therein.

PSEUDOLINEAR REGRESSION ALGORITHM

We pointed out that if the predictor is given by (2.81), and $\boldsymbol{\theta}$ consists of the coefficients of the rational functions, the predictor is alternatively expressed by the pseudolinear form (2.92a). The pseudolinear regression algorithm is a simplified variant of the prediction error algorithm and alleviates computational complexity by means of the approximation

$$\boldsymbol{\psi}(t \mid \boldsymbol{\theta}) = \boldsymbol{\phi}(t \mid \boldsymbol{\theta}) \tag{2.97}$$

This follows if we differentiate (2.92a) and assume that the derivative of $\boldsymbol{\phi}(t \mid \boldsymbol{\theta})$ with respect to $\boldsymbol{\theta}$ is small.

Thus in the pseudolinear regression algorithm the predictor is determined from (2.92a) and the computation of the gradient is suppressed because of (2.97).

STOCHASTIC GRADIENT ALGORITHMS

To relieve the computational effort associated with the prediction error algorithm, a gradient approach similar to the linear regression case can be invoked. To obtain such a gradient algorithm for the general problem we shall follow the argument we described in the linear regression case. Thus we replace the Gauss–Newton search direction (Kalman gain) with the approximate gradient. This amounts to setting

$$P(N) = \mu(N)I \tag{2.98}$$

The resulting recursion has the form

$$\begin{aligned} \boldsymbol{\theta}(N+1) &= \boldsymbol{\theta}(N) + \boldsymbol{\psi}(N \mid \boldsymbol{\theta}(N))\mu(N)e(N+1) \\ e(N+1) &= y(N+1) - \hat{y}(N+1 \mid \boldsymbol{\theta}(N)) \end{aligned} \tag{2.99}$$

The prediction error $e(N+1) = e(N+1 \mid \boldsymbol{\theta}(N))$ and its gradient $\boldsymbol{\psi}(N \mid \boldsymbol{\theta}(N))$ are computed by (2.85), (2.94).

We close this section by noting that the derivation of the preceding algorithms extends to multiple input–output systems without difficulty.

2.6.2 Predictors in state-space form and the Kalman filter

In the preceding discussion we expressed the predictor as a rational filter acting on the data sequence with coefficients depending on the unknown parameter. In some applications an equivalent state-space representation of the predictor filter is more natural. Such a state representation arises in a natural way when the unknown system is expressed in state-space form. Indeed, suppose the data sequence is generated by the model

$$\begin{aligned} \mathbf{x}(t+1) &= A(\boldsymbol{\theta})\mathbf{x}(t) + \mathbf{B}(\boldsymbol{\theta})u(t) + \boldsymbol{\eta}(t) \\ y(t) &= \mathbf{c}^{\mathrm{T}}(\boldsymbol{\theta})\mathbf{x}(t) + v(t) \end{aligned}$$

Let us assume that $\boldsymbol{\eta}(t)$, $v(t)$ are Gaussian white-noise processes with covariances

$$\begin{aligned} E[\boldsymbol{\eta}(t)\boldsymbol{\eta}^{\mathrm{T}}(t)] &= R_1(\boldsymbol{\theta}) \\ E[v(t)v(t)] &= R_2(\boldsymbol{\theta}) \\ E[\boldsymbol{\eta}(t)v(t)] &= \mathbf{R}_{12}(\boldsymbol{\theta}) \end{aligned}$$

Both noise signals are uncorrelated to the initial state $\mathbf{x}_0$, which is a Gaussian random variable with mean and covariance

$$E[\mathbf{x}_0] = \mathbf{m}_0(\boldsymbol{\theta}) \qquad E[\mathbf{x}_0 - \mathbf{m}_0(\boldsymbol{\theta})][\mathbf{x}_0 - \mathbf{m}_0(\boldsymbol{\theta})]^{\mathrm{T}} = P_0(\boldsymbol{\theta})$$

If we take conditional expectations on both sides of the output equation we find

$$\hat{y}(t, \boldsymbol{\theta}) = \mathbf{c}^{\mathrm{T}}(\boldsymbol{\theta})\hat{\mathbf{x}}(t, \boldsymbol{\theta})$$

where

$$\hat{\mathbf{x}}(t, \boldsymbol{\theta}) = E[\mathbf{x}(t) \mid Z_{t-1}]$$

The above state estimator as a function of past data is represented by the celebrated Kalman filter, i.e. the linear time-varying state-space representation

$$\hat{\mathbf{x}}(t+1, \boldsymbol{\theta}) = A(\boldsymbol{\theta})\hat{\mathbf{x}}(t, \boldsymbol{\theta}) + \mathbf{B}(\boldsymbol{\theta})u(t) + \mathbf{w}(t, \boldsymbol{\theta})[y(t) - \mathbf{c}^{\mathrm{T}}(\boldsymbol{\theta})\hat{\mathbf{x}}(t, \boldsymbol{\theta})]$$

or

$$\hat{\mathbf{x}}(t+1, \boldsymbol{\theta}) = [A(\boldsymbol{\theta}) - \mathbf{w}(t, \boldsymbol{\theta})\mathbf{c}^{\mathrm{T}}(\boldsymbol{\theta})]\hat{\mathbf{x}}(t, \boldsymbol{\theta}) + \mathbf{B}(\boldsymbol{\theta})u(t) + \mathbf{w}(t, \boldsymbol{\theta})\, y(t)$$

The Kalman filter is driven by the data sequence $u(t)$, $y(t)$. It is time varying even if the original system is time invariant because of the time-varying parameter $\mathbf{w}(t, \boldsymbol{\theta})$, called the Kalman gain. The Kalman gain is computed by the following linear system (for simplicity we ignore the dependence on $\boldsymbol{\theta}$):

$$\mathbf{w}(t) = [AP(t)\mathbf{c} + \mathbf{R}_{12}][\mathbf{c}^{\mathrm{T}}P(t)\mathbf{c} + R_2]^{-1}$$

The matrix $P(t)$, which is responsible for the time variation of the Kalman gain, coincides with the covariance of the error

$$P(t) = E[\hat{\mathbf{x}}(t, \boldsymbol{\theta}) - \mathbf{x}(t)][\hat{\mathbf{x}}(t, \boldsymbol{\theta}) - \mathbf{x}(t)]^{\mathrm{T}}$$

and is estimated by the Riccati difference equation

$$P(t+1) = AP(t)A^{\mathrm{T}} + R_1 - \mathbf{w}(t)[AP(t)\mathbf{c} - \mathbf{R}_{12}]^{\mathrm{T}}$$

initialized at $P_0(\boldsymbol{\theta})$. Under weak conditions it can be shown that the Riccati equation converges. The limit is an equilibrium point and thus it satisfies the algebraic Riccati equation

$$P = APA^{\mathrm{T}} + R_1 - \mathbf{w}[AP\mathbf{c} + \mathbf{R}_{12}]^{\mathrm{T}}$$

where

$$\mathbf{w} = [AP\mathbf{c} + \mathbf{R}_{12}][\mathbf{c}^{\mathrm{T}}P\mathbf{c} + R_2]^{-1}$$

Insertion of the constant gain $\mathbf{w}$ in the Kalman filter equation yields a time-invariant predictor.

2.7 Fast algorithms for Wiener filtering

In the previous sections we presented the basic methods for parameter estimation in system identification and the related topics of signal processing and modelling. This and the next sections focus on the efficiency issues of computational complexity and parallelism. The current section concentrates on the steady-state limiting form of the linear regressor given in section 2.5.1. Our aim is to present the philosophy behind the basic algorithmic families associated with the computationally efficient solution of (2.37). Thus we shall restrain ourselves to the simplest case of single-channel FIR filtering. Multichannel extensions of these algorithmic schemes will be commented upon later in this chapter. It must be emphasized that ARX models can also be seen as two-channel FIR filters with a different number of tap delays per channel. Hence, provided (2.33) and (2.34) are satisfied, multichannel extensions of the schemes to be presented here cover ARX models, too.

We established in section 2.5 that in the case of FIR filtering the parameter vector $\boldsymbol{\theta}$ consists of the filter taps $-c_1, -c_2, \ldots, -c_p$, the regressor vector $\boldsymbol{\phi}(t)$ consists of the input samples $u(t)$, $u(t-1), \ldots, u(t-p+1)$ and the Wiener equations are

$$R_p\mathbf{c}_p = -\mathbf{d}_p \tag{2.100}$$

where R_p is now the input's autocorrelation matrix of order p, $\mathbf{d}_p$ is the input–output cross-correlation vector and $\mathbf{c}_p^{\mathrm{T}} = [c_1, c_2, \ldots, c_p]$. The matrix dimension p is explicitly indicated in the notation since it will be of major importance from now on. The above limiting form of the linear regressor is called the known statistics case since the computation of the estimates assumes that the autocorrelation and the input–desired output cross-correlation are available.

2.7.1 The Levinson algorithm

The essence of Levinson's algorithm [9] is the exploitation of the special properties of the autocorrelation matrix of any order R_m, $m = 1, 2, \ldots, p$. In particular, R_m is (a) positive definite, (b) symmetric, i.e. $R_m^{\mathrm{T}} = R_m$, (c) Toeplitz, i.e. all the elements across each diagonal are equal, namely $\rho_{ij} = \rho_{i+k,j+k}$. As result the following is true:

$$JR_mJ = R_m^{\mathrm{T}}$$

where $J \equiv J_m$ is the $m \times m$ exchange matrix with ones in the antidiagonal and zeros elsewhere

$$J = \begin{bmatrix} O & & 1 \\ & 1 & \\ 1 & & O \end{bmatrix}$$

with O denoting zero elements. It is mainly this last property whose exploitation leads to the derivation of fast computational schemes.

Levinson's algorithm solves the linear system of equations (2.100) in $O(p^2)$ multiplications and additions (MADs) as opposed to $O(p^3)$ MADs required by conventional schemes. Although Levinson's algorithm is well documented we shall highlight the basic steps of its derivation. This will help us to set up the basic ideas behind all subsequent fast algorithmic schemes.

Let us consider the autocorrelation matrix of some order m of a stationary scalar real input:

$$R_m = \begin{bmatrix} \rho_0 & \rho_1 & \cdots & \rho_{m-1} \\ \rho_1 & \rho_0 & \cdots & \rho_{m-2} \\ \vdots & \vdots & \ddots & \vdots \\ \rho_{m-1} & \rho_{m-2} & \cdots & \rho_0 \end{bmatrix} \tag{2.101}$$

It is readily apparent how R_{m-1}, of order $m - 1$, is nested within R_m, both in its upper and lower partitionings. Indeed, for $m = 1, 2, \ldots, p$

$$R_m = \begin{bmatrix} \rho_0 & \mathbf{r}_{m-1}^{\mathrm{T}} \\ \mathbf{r}_{m-1} & R_{m-1} \end{bmatrix} \tag{2.102}$$

$$R_m = \begin{bmatrix} R_{m-1} & J\mathbf{r}_{m-1} \\ \mathbf{r}_{m-1}^{\mathrm{T}} J & \rho_0 \end{bmatrix} \tag{2.103}$$

where

$$\mathbf{r}_{m-1}^{\mathrm{T}} \equiv [\rho_1, \rho_2, \ldots, \rho_{m-1}] \tag{2.104}$$

Obviously $J\mathbf{r}_{m-1}$ is the reverse of $\mathbf{r}_{m-1}$, i.e.

$$J\mathbf{r}_{m-1} = \begin{bmatrix} \rho_{m-1} \\ \rho_{m-2} \\ \vdots \\ \rho_1 \end{bmatrix} \tag{2.105}$$

Having established the above nested form of the autocorrelation matrices of two successive orders, we then encounter two questions:

1. Can the solution $\mathbf{c}_m = -R_m^{-1}\mathbf{d}_m$ be expressed in terms of $\mathbf{c}_{m-1} = -R_{m-1}^{-1}\mathbf{d}_{m-1}$?
2. Does such a scheme lead to a computationally more attractive algorithm than solving (2.100) in a conventional manner?

The answer to both questions is in the affirmative. Recalling the matrix inversion lemma for partitioned matrices and applying it to (2.102) we obtain

$$R_m^{-1} = \begin{bmatrix} O & \mathbf{0}^{\mathrm{T}} \\ \mathbf{0} & R_{m-1}^{-1} \end{bmatrix} + \begin{bmatrix} 1 \\ -R_{m-1}^{-1}\mathbf{r}_{m-1} \end{bmatrix} \frac{1}{\alpha_{m-1}^{\mathrm{f}}} [1 - \mathbf{r}_{m-1}^{\mathrm{T}} R_{m-1}^{-1}] \tag{2.106}$$

$$\alpha_{m-1}^{\mathrm{f}} = \rho_0 - \mathbf{r}_{m-1}^{\mathrm{T}} R_{m-1}^{-1} \mathbf{r}_{m-1} \tag{2.106a}$$

In this equation the vector

$$\mathbf{a}_{m-1} = -R_{m-1}^{-1}\mathbf{r}_{m-1} \tag{2.107}$$

appears with elements $a_{m-1}^1, \ldots, a_{m-1}^{m-1}$ and has a special and important meaning. It is the Wiener optimal forward predictor of order $m-1$, as one can easily deduce by comparing (2.107) and the Wiener equation. At this point it must be emphasized that an optimal prediction problem will always be present, in one way or another, in all fast schemes treated in this book. This is a consequence of the nesting property of the input signal correlation sequence. α_m^{f} also has a special meaning: it is the prediction error power, $E[(u(t) + \mathbf{a}_{m-1}^{\mathrm{T}}\mathbf{u}_{m-1}(t-1))^2]$, corresponding to the optimal predictor $\mathbf{a}_{m-1}$. Sometimes the predictor $\mathbf{a}_{m-1}$ and its associated prediction error power are combined together in a single equation, as can be seen from (2.106) and (2.107):

$$R_m \begin{bmatrix} 1 \\ \mathbf{a}_{m-1} \end{bmatrix} \equiv R_m \mathbf{A}_m = \begin{bmatrix} \alpha_{m-1}^{\mathrm{f}} \\ \mathbf{0} \end{bmatrix} \tag{2.108}$$

Application of the matrix inversion lemma to (2.103) leads to

$$R_m^{-1} = \begin{bmatrix} R_{m-1}^{-1} & \mathbf{0} \\ \mathbf{0}^{\mathrm{T}} & O \end{bmatrix} + \begin{bmatrix} -R_{m-1}^{-1}J\mathbf{r}_{m-1} \\ 1 \end{bmatrix} \frac{1}{\alpha_{m-1}^{\mathrm{b}}} [-R_{m-1}^{-1}J\mathbf{r}_{m-1} \;\; 1] \tag{2.109}$$

$$\alpha_{m-1}^{\mathrm{b}} = (\rho_0 - \mathbf{r}_{m-1}^{\mathrm{T}} J R_{m-1}^{-1} J\mathbf{r}_{m-1}) \tag{2.109a}$$

The vector

$$\mathbf{b}_{m-1} = -R_{m-1}^{-1} J\mathbf{r}_{m-1} \tag{2.110}$$

can easily be identified as the Wiener optimal backward predictor of order $m-1$. Furthermore, due to the Toeplitz as well as the symmetry properties of the autocorrelation matrix, the following is true:

$$\mathbf{b}_{m-1} = J\mathbf{a}_{m-1} \tag{2.111}$$

and for the backward predictor error power $\alpha_{m-1}^{\mathrm{b}}$

$$\alpha_{m-1}^{\mathrm{b}} = \alpha_{m-1}^{\mathrm{f}} \tag{2.112}$$

Thus the optimum backward predictor is the reverse of the forward predictor and the associated error powers are equal. A formulation similar to that in (2.108) is also easily established, i.e.

$$R_m \begin{bmatrix} \mathbf{b}_{m-1} \\ 1 \end{bmatrix} \equiv R_m \mathbf{B}_m = \begin{bmatrix} \mathbf{0} \\ \alpha_{m-1}^{\mathrm{b}} \end{bmatrix} \tag{2.113}$$

Returning now to the original Wiener equation we observe that the upper partitioning of the right-hand-side vector also results in a cross-correlation vector related to the lower-order Wiener solution

$$\mathbf{d}_m = \begin{bmatrix} \mathbf{d}_{m-1} \\ d_m \end{bmatrix} \tag{2.114}$$

where $d_m = E[u(t-m+1)\,y(t)]$. Combining (2.100), (2.109) and (2.114) results in

$$\mathbf{c}_m = \begin{bmatrix} \mathbf{c}_{m-1} \\ 0 \end{bmatrix} + \begin{bmatrix} J\mathbf{a}_{m-1} \\ 1 \end{bmatrix} k_{m-1}^{c} \tag{2.115}$$

$$k_{m-1}^{c} = \frac{-1}{\alpha_{m-1}^{b}} (d_m + \mathbf{r}_{m-1}^{\mathrm{T}} J\mathbf{c}_{m-1}) \equiv -\frac{\beta_{m-2}^{c}}{\alpha_{m-1}^{b}} \tag{2.116}$$

Coefficient k_{m-1}^{c}, also denoted as k_{m-1}^{y} in some of the chapters to emphasize its relationship with the desired response sequence, is known as the reflection coefficient and we shall return to it soon. Equation (2.115) is known as the two-term order recursion (two successive orders m, $m-1$ are involved) providing the optimum solution of order m in terms of its $m-1$ predecessor. For the algorithm to be complete we need a similar set of recursions for the predictor $\mathbf{a}_m$ filter. From (2.107) and for order m the right-hand-side vector is also partitioned as

$$\mathbf{r}_m = \begin{bmatrix} \mathbf{r}_{m-1} \\ \rho_m \end{bmatrix} \tag{2.117}$$

Following similar arguments one obtains

$$\mathbf{a}_m = \begin{bmatrix} \mathbf{a}_{m-1} \\ 0 \end{bmatrix} + \begin{bmatrix} J\mathbf{a}_{m-1} \\ 1 \end{bmatrix} k_{m-1} \tag{2.118}$$

$$k_{m-1} = \frac{-\beta_{m-1}}{\alpha^{\mathrm{f}}_{m-1}} \equiv -\frac{\rho_m + \mathbf{r}^{\mathrm{T}}_{m-1}\,\mathbf{J}\mathbf{a}_{m-1}}{\alpha^{\mathrm{f}}_{m-1}} \tag{2.119}$$

Starting from its definition, (2.106), (2.117) and (2.118), it is easily shown that the prediction error power is updated as

$$\alpha^{\mathrm{f}}_m = \alpha^{\mathrm{f}}_{m-1}(1 - k^2_{m-1}) = \alpha^{\mathrm{b}}_m \tag{2.120}$$

Since both α^{f}_m and α^{b}_m are powers, and hence non-negative, $-1 \leqslant k_m \leqslant 1$. The above recursions (2.115)–(2.120) are true for all orders $m = 1, 2, \ldots, p$. The sequence of (2.119), (2.118), (2.120), (2.116) and (2.115) for $m = 1, 2, \ldots, p$ constitutes Levinson's algorithm. From its definition α^{f}_0 is initialized to ρ_0 and all inner products of zero order are identically zero. The computational complexity is $1m$ MADs for each of (2.115), (2.118) and also $1m$ MADs for each of (2.116), (2.119). Thus the total complexity of the algorithm is $2p^2$ MADs. If only the predictors are of interest, (2.115) and (2.116) are not required and complexity is reduced to p^2 MADs.

The above treatment was for a real scalar input time series. Generalizations to complex and/or vector (matrix) inputs are also possible and the interested reader is referred to [10,11]. A notable difference between single and multichannel signals is that the optimum forward and backward predictors are no longer the reverse of each other. Thus two sets of recursions like (2.118) are required, one for the forward and one for the backward predictor.

2.7.2 The split Levinson algorithm

Recently it has been pointed out that in the case of single-channel signals, Levinson's algorithm is redundant in complexity in its prediction part [12,13]. The so-called split Levinson algorithm reduces the required amount of multiplications in the prediction part of the algorithm by half. This is achieved by imposing symmetry on to the problem. To this end the following symmetric vector is defined:

$$\mathbf{p}^s_{m+1} = \begin{bmatrix} \mathbf{A}_m \\ 0 \end{bmatrix} + \begin{bmatrix} 0 \\ J\mathbf{A}_m \end{bmatrix} \tag{2.121}$$

where superscript s stands for symmetry. From the above definition, (2.108) and (2.118) one can show that

$$\mathbf{A}_{m+1} + J\mathbf{A}_{m+1} = (1 + k_{m-1})\mathbf{p}^s_{m+1} \tag{2.122}$$

Combining (2.121) and (2.122) for the right order results in

$$\begin{bmatrix} \mathbf{A}_m \\ 0 \end{bmatrix} - \begin{bmatrix} 0 \\ \mathbf{A}_m \end{bmatrix} = \begin{bmatrix} 1 \\ \mathbf{a}_{m-1} \\ 0 \end{bmatrix} - \begin{bmatrix} 0 \\ 1 \\ \mathbf{a}_{m-1} \end{bmatrix}$$

$$= \mathbf{p}^s_{m+1} - \begin{bmatrix} 0 \\ \mathbf{p}^s_m \end{bmatrix} (1 + k_{m-2}) \tag{2.123}$$

Equation (2.123) is the first interesting point of the split Levinson algorithm. Once the right-hand side, involving symmetric vectors, has been computed, predictor $\mathbf{a}_{m-1}$ can be obtained with no multiplications, due to the special form of the left-hand side. Also note that the multiplicative factor $1 + k_{m-2}$ may be computed by summing both sides of (2.123). It turns out that $1 + k_{m-2}$ is given as the ratio of the sum of the coefficients of $\mathbf{p}^s_{m+1}$ over that of $\mathbf{p}^s_m$.

The second interesting point is the order update of $\mathbf{p}^s_m$. Combining (2.121), (2.122) and (2.118) gives [12]

$$\mathbf{p}^s_{m+2} = \begin{bmatrix} \mathbf{p}^s_{m+1} \\ 0 \end{bmatrix} + \begin{bmatrix} 0 \\ \mathbf{p}^s_{m+1} \end{bmatrix} + \begin{bmatrix} 0 \\ \mathbf{p}^s_m \\ 0 \end{bmatrix} k^s_m \tag{2.124}$$

$$k^s_m = (k_{m-2} + 1)(k_{m-1} - 1) \tag{2.125}$$

Equation (2.124) is a three-term recursion involving vectors of three successive orders. Furthermore, only symmetric vectors are involved; thus only half of the operations need to be performed, hence the computational saving. k^s_m can also be obtained in terms of symmetric vectors. Indeed, from (2.125), (2.106), (2.119), (2.120) and (2.121) we can show that

$$k^s_m = -\frac{\alpha^{\mathrm{f}}_{m-1} + \beta_{m-1}}{\alpha^{\mathrm{f}}_{m-2} + \beta_{m-2}} = -\frac{[\mathbf{r}^{\mathrm{T}}_m, \rho_0]\mathbf{p}^s_{m+1}}{[\mathbf{r}^{\mathrm{T}}_{m-1}, \rho_0]\mathbf{p}^s_m} \tag{2.126}$$

Observe that each of the above inner products can be computed in terms of symmetric vectors by noting that

$$\tau_m \equiv [\mathbf{r}^{\mathrm{T}}_m, \rho_0]\mathbf{p}^s_{m+1} = \tfrac{1}{2}\{[\mathbf{r}^{\mathrm{T}}_m, \rho_0] + [\rho_0, J\mathbf{r}^{\mathrm{T}}_m]\}\mathbf{p}^s_{m+1}$$

Thus, overall, the order propagation of $\mathbf{p}^s_m$ needs half the multiplications required for the propagation of $\mathbf{a}_m$ in the Levinson algorithm. The number of additions, however, remains the same. In contrast the derivation of such three-term recursive algorithms for the case of multichannel signals, where the optimal forward and backward predictors are no longer the reverse of each other, has not led so far to computational savings [14,15]. Table 2.3 summarizes the split Levinson algorithm. The predictor of order p can then be computed from (2.123) via $O(p)$ operations. The last equation of Table 2.3 is a rearrangement of (2.125).

Table 2.3 The split Levinson algorithm

	Adds	Mults
INITIALIZATION		
$k_0 = -\rho_1/\rho_0,\ p_3^{s3} = p_3^{s1} = 1$		
$p_3^{s2} = 2k_0,\ p_2^{s1} = p_2^{s2} = 1$		
$\tau_1 = \rho_0 + \rho_1$		
FOR $m = 2, 3, \ldots, p$		
$\mathbf{r}_{m+1}^s = \begin{bmatrix} \mathbf{r}_m \\ \rho_0 \end{bmatrix} + \begin{bmatrix} \rho_0 \\ J\mathbf{r}_m \end{bmatrix}$	$0.5m$	
$\tau_m = \mathbf{r}_{m+1}^{sT}\mathbf{p}_{m+1}^s$	$0.5m$	$0.5m$
$k_m^s = -\tau_m/\tau_{m-1}$		
$\mathbf{p}_{m+2}^s = \begin{bmatrix} \mathbf{p}_{m+1}^s \\ 0 \end{bmatrix} + \begin{bmatrix} 0 \\ \mathbf{p}_{m+1}^s \end{bmatrix} + \begin{bmatrix} 0 \\ \mathbf{p}_m^s \\ 0 \end{bmatrix} k_m^s$	m	$0.5m$
$k_{m-1} = 1 + k_m^s/(1 + k_{m-2})$		
END		

2.7.3 The lattice family

So far we have been concerned with the transversal implementation of a system. That is, the output of the system is expressed directly as a linear combination of delayed input samples. In what follows we shall look at the problem in a different way, leading to an alternative parameterization of the unknown system. Let us denote by $e_m^{\mathrm{f}}(t)$ and $e_m^{\mathrm{b}}(t)$ the output error signals at time t corresponding to the optimum forward and backward predictors of order m. Then

$$e_m^{\mathrm{f}}(t) = u(t) + \mathbf{a}_m^{\mathrm{T}}\mathbf{u}_m(t-1) \tag{2.127}$$

$$e_m^{\mathrm{b}}(t) = u(t-m) + \mathbf{a}_m^{\mathrm{T}} J\mathbf{u}_m(t) \tag{2.128}$$

Using the nesting property of the input vector, i.e.

$$\begin{aligned} \mathbf{u}_m(t) &\equiv [u(t), u(t-1), \ldots, u(t-m+1)] \\ &= [\mathbf{u}_{m-1}(t), u(t-m+1)] \\ &= [u(t), \mathbf{u}_{m-1}(t-1)] \end{aligned}$$

and the Levinson recursion (2.118), we obtain the following:

$$e_m^{\mathrm{f}}(t) = e_{m-1}^{\mathrm{f}}(t) + k_{m-1}e_{m-1}^{\mathrm{b}}(t-1) \tag{2.129}$$

$$e_m^{\mathrm{b}}(t) = e_{m-1}^{\mathrm{b}}(t-1) + k_{m-1}e_{m-1}^{\mathrm{f}}(t) \tag{2.130}$$

The above recursions are known as lattice equations. Figure 2.1 shows the evolution of the errors in order and time within the lattice structure. It is obvious that such a

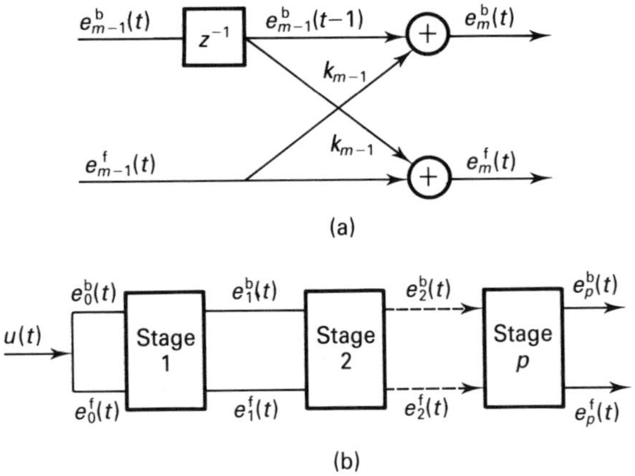

Figure 2.1 The lattice structure

structure is completely parameterized in terms of the reflection coefficients instead of the tap weights of the forward predictor $\mathbf{a}_m$.

An alternative interpretation of the lattice predictor is to look at it as a Gram–Schmidt orthogonalizer. Indeed, looking at the samples $u(t), \ldots, u(t-m)$ as vectors in a signal subspace it can easily be shown that $e^b_0(t), \ldots, e^b_m(t)$ constitute an orthogonal vector basis in the same subspace. Two zero-mean random variables are orthogonal if they are uncorrelated. Thus

$$E[e^b_i(t)e^b_k(t)] = \begin{cases} \alpha^b_i & i = k \\ 0 & i \neq k \end{cases} \tag{2.131}$$

This is an implication of the geometric properties of the least mean square estimator. Indeed, from the definition (2.128) $e^b_m(t)$ is the error in approximating $u(t-m)$ in terms of $u(t), \ldots, u(t-m+1)$ in the least mean square optimal sense. Thus $e^b_m(t) \perp u(t), u(t-1), \ldots, u(t-m+1)$, and hence to any of their linear combinations. Thus $e^b_m(t) \perp e^b_{m-1}(t), \ldots, e^b_0(t)$. In the same way e^b_{m-1} can be shown to be orthogonal to all lower-order errors, etc. Approximating (projecting) the unknown desired response vector $y(t)$, in the signal subspace, in terms of this orthogonal basis we get

$$\hat{y}(t) = -\sum_{m=0}^{p-1} \hat{k}_m e^b_m(t) \tag{2.132}$$

Using standard arguments and the orthogonality property (2.131) we can show that

$$\hat{k}_m = k^c_m \qquad m = 1, 2, \ldots, p-1$$

In other words the reflection coefficients k^c_m, $m = 0, 1, \ldots, p-1$, appearing in the Levinson algorithm and the tap coefficients $\mathbf{c}_p$ constitute different sets of parameters

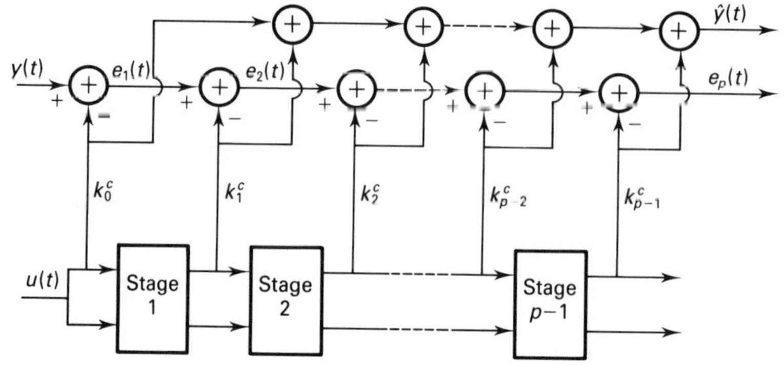

Figure 2.2 The lattice–ladder structure

expressing the orthogonal projection of signal vector $y(t)$ on to a subspace in terms of the two basis sets $\{e_0^b(t), e_1^b(t), \ldots, e_{p-1}^b(t)\}$ and $\{u(t), u(t-1), \ldots, u(t-p+1)\}$. The Levinson algorithm provides a fast scheme to obtain one set of parameters from the other. From a system identification point of view this implies that the unknown system can be modelled either as a transversal structure, parameterized in terms of $\mathbf{c}_p$, or as a lattice structure parameterized in terms of the equivalent reflection coefficients. Figure 2.2 shows the combined lattice–ladder structure, known as the joint process estimator, since it performs two estimations jointly. It is straightforward to show, either from (2.132) or (2.115), that

$$\begin{aligned} y(t) - \hat{y}(t) &\equiv e_m(t) \\ &= e_{m-1}(t) + k_{m-1}^c\, e_{m-1}^b(t) \end{aligned} \tag{2.133}$$

Having discussed the geometric implications of the prediction part we shall now point out some of its algebraic consequences. From the definition of the backward prediction errors we can write

$$\begin{bmatrix} e_0^b(t) \\ e_1^b(t) \\ \vdots \\ e_{m-1}^b(t) \end{bmatrix} = \begin{bmatrix} 1 & 0 & \ldots & 0 \\ a_1^1 & 1 & \ldots & 0 \\ \vdots & \vdots & \ddots & \vdots \\ a_{m-1}^{m-1} & a_{m-1}^{m-2} & \ldots & 1 \end{bmatrix} \begin{bmatrix} u(t) \\ u(t-1) \\ \vdots \\ u(t-m+1) \end{bmatrix} \tag{2.134}$$

or

$$\mathbf{e}_m^b(t) = U^{\mathrm{T}}\mathbf{u}_m(t)$$

From (2.134) we obtain

$$E[\mathbf{e}_m^b(t)\mathbf{e}_m^{b\mathrm{T}}(t)] \equiv D = U^{\mathrm{T}} R_m U \tag{2.135}$$

where

$$D = \mathrm{diag}[\alpha_0^b, \alpha_1^b, \ldots, \alpha_{m-1}^b] \tag{2.136}$$

and finally

$$R_m^{-1} = UD^{-1}U^{\mathrm{T}} \equiv (UD^{-1/2})(UD^{-1/2})^{\mathrm{T}} \tag{2.137}$$

In other words, the prediction error powers and prediction transversal parameters provide the UDU^{T} and, consequently, Cholesky factorization of the positive definite matrix R_m^{-1}; U is obviously an upper-triangular matrix. Thus the Levinson algorithm is a fast scheme for Cholesky factorization of the inverse autocorrelation matrix. It has recently been pointed out that the split Levinson algorithm induces a so-called bowtie factorization of R_m^{-1} (R_m) [16].

2.7.4 The Schur algorithm

The Levinson algorithm consists of two types of operation, namely (2.118) and the inner product in (2.119). In a parallel processor environment the first operation may be completed in the time taken for 1 MAD. In contrast, the best one can achieve with the inner product operation is to reduce the time to $O(\log_2 m)$ time units. Thus this operation poses as the bottleneck in the flow of the algorithm. An alternative scheme that overcomes the bottleneck and enhances the underlying parallelism is to look at each of the inner products as the output of a linear transversal filter with appropriately defined input.

Indeed, consider the system described by the transversal parameter set $(\mathbf{c}_m, \mathbf{a}_m)$ or, equivalently, the lattice–ladder set $(k_m^c, k_m, m = 0, 2, \ldots, p-1$, assumed to be Wiener optimum with respect to the signal pair $(u(t), y(t))$. If we now feed the system with the autocorrelation sequence as input and consider the cross-correlation sequence as the desired response the respective errors are

$$\tilde{e}_m(n) = \mathbf{r}_m^{\mathrm{T}}(n)\mathbf{c}_m + d_{n+1} \tag{2.138}$$

$$\tilde{e}_m^{\mathrm{b}}(n) = \mathbf{r}_m^{\mathrm{T}}(n)J\mathbf{a}_m + \rho_{n-m} \tag{2.139}$$

$$\tilde{e}_m^{\mathrm{f}}(n) = \mathbf{r}_m^{\mathrm{T}}(n-1)\mathbf{a}_m + \rho_n \tag{2.140}$$

with

$$\mathbf{r}_m^{\mathrm{T}}(n) \equiv [\rho_n, \rho_{n-1}, \cdots, \rho_{n-m+1}] \tag{2.141}$$

From the discussion in the previous paragraph it is straightforward to see that these errors can be obtained at the corresponding stages of the equivalent lattice–ladder structure fed with

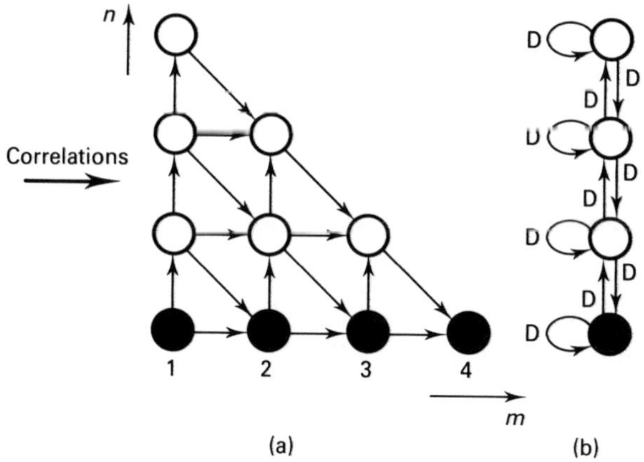

Figure 2.3 (a) DG and (b) SFG for the Schur algorithm

$$\tilde{e}_0^{\mathrm{f}}(n) = \tilde{e}_0^{\mathrm{b}}(n) = \rho_n \tag{2.142}$$

$$\tilde{e}_0(n) = d_{n+1} \tag{2.143}$$

In other words, (2.129), (2.130) and (2.133) hold for $\tilde{e}_m^{\mathrm{f}}(n)$, $\tilde{e}_m^{\mathrm{b}}(n)$ and $\tilde{e}_m(n)$. It is a matter of simple algebra to show that [17]

$$\alpha_m^{\mathrm{b}} = \tilde{e}_m^{\mathrm{b}}(m) \tag{2.144}$$

$$k_m = \tilde{e}_m^{\mathrm{f}}(m+1)/\alpha_m^{\mathrm{b}} \tag{2.145}$$

$$k_m^c = \tilde{e}_m(m)/\alpha_m^{\mathrm{b}} \tag{2.146}$$

We observe that the required inner products result from these error signals for a particular combination of time instant n and order m. Furthermore, each of these errors can be obtained with a complexity of 1 MAD, i.e. independent of m. On the other hand, for each order m more than one, errors have to be computed corresponding to the different time instants n, which will be used at later stages. In a uniprocessor environment this algorithm results in $5m$ MAD per recursion, that is, $1m$ more compared with Levinson's. However, in a multiprocessor environment the error computations for the various time instants n at a particular order m can take place concurrently [17,18]. Thus a saving of an order of magnitude results. The origin of this algorithm, summarized in Table 2.4, can be traced back to the pioneering work of Schur in 1917 [19]. An interesting aspect is that the involved error sequences (the $\tilde{e}$) are bounded variables [20,21], an attractive feature when the algorithm is implemented in fixed point arithmetic.

Table 2.4 The Schur algorithm

INITIALIZATION
  $\tilde{e}_0^b(n) = \tilde{e}_0^f(n) = \rho_n$
  $\tilde{e}_0(n) = d_{n+1}$
FOR $m = 0, 1, \ldots, p-1$
  IN PARALLEL DO τ_1
  BEGIN
    $k_m^c = -\tilde{e}_m(m)/\tilde{e}_m^b(m)$
    $k_m = -\tilde{e}_m^f(m+1)/\tilde{e}_m^b(m)$
  END
  IN PARALLEL DO $\tau_2 + \tau_3$
    FOR $n = m+1$ TO p IN PARALLEL DO
    BEGIN
      $\tilde{e}_{m+1}^b(n) = \tilde{e}_m^b(n-1) + \tilde{e}_m^f(n)k_m$
      $\tilde{e}_{m+1}^f(n) = \tilde{e}_m^f(n) + \tilde{e}_m^b(n-1)k_m$
      $\tilde{e}_{m+1}(n-1) = \tilde{e}_m(n-1) + \tilde{e}_m^b(n-1)k_m^c$
    END
    FOR $i = 1$ TO m IN PARALLEL DO
    BEGIN
      $a_{m+1}^i = a_m^i + a_m^{m-i+1}k_m$
      $c_{m+1}^i = c_m^i + a_m^{m-i+1}k_m^c$
    END
    $a_{m+1}^{m+1} = k_m$
    $c_{m+1}^{m+1} = k_m^c$
  END
END

A closer look at Table 2.4 reveals that for each order m, $p - m$ processors are required to compute the three errors in parallel. Thus the number of processors involved decreases as order increases. Figure 2.3(a) shows the dependence graph (DG) of the algorithm for order $p = 4$. Each node corresponds to a processing element (PE) performing a lattice-type operation (Figure 2.1). In addition the lower (dark) nodes perform the divisions leading to the reflection coefficients. Note that the updates for $\tilde{e}_{m+1}^f(m+1)$, $\tilde{e}_{m+1}(m)$, in the lower PE, and for $\tilde{e}_{m+1}^b(p)$ in the top PE, need not be performed, since they will not be used later on. Once the reflection coefficients have been computed, they are propagated to the rest of the processors to perform the lattice operations. When these are completed each processor passes the $\tilde{e}^f$s and $\tilde{e}$s to its lower neighbour (diagonal arrows). Thus at the beginning of each recursion $\tilde{e}_m^f(m+1)$ and $\tilde{e}_m(m)$ reside at the lowest processor. Figure 2.3(b) shows the resulting one-dimensional signal flow graph (SFG) after projection of the DG across the default schedule [6]. In the case of high-order systems, where the propagation of the reflection coefficients from the lowest to the rest of the nodes (processors) may not be desirable, due to delays, systolization techniques can be applied on the resulting SFG for temporal localization [18].

The computations related to the transversal parameters follow a similar mode of operation. This is apparent from Table 2.3 if a third (redundant) equation is added for the computation of the reverse forward predictor. The only difference between the error and transversal parameter updates is that, in contrast to the former, the number of processors required for the latter increases with order m. Thus in a combined mode of operation the total number of processors utilized per recursion remains constant and equal to p. After $O(p)$ time units the computed transversal filter parameters reside at the processor elements ([21], [22] and Chapter 8).

The Schur algorithm has an interesting algebraic implication. It can be shown that the errors $\tilde{e}_m^{\mathrm{b}}(n)$ are the matrix elements in the LDL^{T} factorization of R_m, where L is a lower-triangular matrix. Indeed, it can be seen from the prediction error formulation, Equation (2.113), that

$$\begin{bmatrix} \rho_0 & \rho_1 & \cdots & \rho_i \\ \rho_1 & \rho_0 & \cdots & \rho_{i-1} \\ \vdots & \vdots & \ddots & \vdots \\ \rho_{i-1} & \rho_{i-2} & \cdots & \rho_1 \end{bmatrix} \begin{bmatrix} a_i^i \\ a_i^{i-1} \\ \vdots \\ 1 \end{bmatrix} = \begin{bmatrix} 0 \\ 0 \\ \vdots \\ 0 \end{bmatrix} \tag{2.147}$$

Combining (2.147) with (2.139) results in

$$\begin{bmatrix} \tilde{e}_0^{\mathrm{b}}(0) & 0 & 0 & \cdots & 0 \\ \tilde{e}_0^{\mathrm{b}}(1) & \tilde{e}_1^{\mathrm{b}}(1) & 0 & \cdots & 0 \\ \vdots & \vdots & \vdots & \ddots & \vdots \\ \tilde{e}_0^{\mathrm{b}}(m-1) & \tilde{e}_1^{\mathrm{b}}(m-1) & \tilde{e}_2^{\mathrm{b}}(m-1) & \cdots & \tilde{e}_{m-1}^{\mathrm{b}}(m-1) \end{bmatrix}$$

$$= \begin{bmatrix} \rho_0 & \rho_1 & \rho_2 & \cdots & \rho_{m-1} \\ \rho_1 & \rho_0 & \rho_1 & \cdots & \rho_{m-2} \\ \vdots & \vdots & \vdots & \ddots & \vdots \\ \rho_{m-1} & \rho_{m-2} & \rho_{m-3} & \cdots & \rho_0 \end{bmatrix} \begin{bmatrix} 1 & a_1^1 & a_2^2 & \cdots & a_{m-1}^{m-1} \\ 0 & 1 & a_2^1 & \cdots & a_{m-1}^{m-2} \\ \vdots & \vdots & \vdots & \ddots & \vdots \\ 0 & 0 & 0 & \cdots & 1 \end{bmatrix} \tag{2.148}$$

or

$$L = R_m U$$

We already know that

$$\begin{aligned} R_m^{-1} &= UD^{-1}U^{\mathrm{T}} \Rightarrow \\ I &= R_m UD^{-1}U^{\mathrm{T}} = LD^{-1}U^{\mathrm{T}} = R_m R_m^{-1} \Rightarrow \\ R_m &= LD^{-1}U^{\mathrm{T}}R_m = LD^{-1}L^{\mathrm{T}} \end{aligned} \tag{2.149}$$

From (2.149) and (2.135) we deduce the following interesting relation:

$$DL^{-1} = U^{\mathrm{T}}$$

2.8 Fast algorithms for linear regression

Let us now come back to the linear regressor defined in section 2.5. We shall again constrain ourselves to single-channel FIR filtering and try to highlight computationally efficient schemes for the solution of (2.21) and (2.22). The comments concerning the ARX model made in the previous section are obviously valid for the (least-squares) linear regressor of (2.21), which for our special case of interest becomes

$$\left(\sum_{t=0}^{N} \mathbf{u}_p(t)\mathbf{u}_p^{\mathrm{T}}(t)\right)\mathbf{c}_p(N) = -\sum_{t=0}^{N} \mathbf{u}_p(t)y(t) \tag{2.150}$$

or

$$R_p(N)\mathbf{c}_p(N) = -\mathbf{d}_p(N) \tag{2.151}$$

From the nesting properties of the regressor vector $\mathbf{u}_m(t)$ it is not difficult to see that the positive definite symmetric matrix $R_m(N)$, for any $m = 1, 2, \ldots, p$, is partitioned as [23]

$$R_m(N) = \begin{bmatrix} * & * \\ * & R_{m-1}(N-1) \end{bmatrix} \tag{2.152}$$

$$= \begin{bmatrix} R_{m-1}(N) & * \\ * & * \end{bmatrix} \tag{2.153}$$

$$\mathbf{d}_m(N) = \begin{bmatrix} \mathbf{d}_{m-1}(N) \\ * \end{bmatrix} \tag{2.154}$$

where * denotes elements of no interest at present. The nesting property which was exploited in the Levinson algorithm is also present here, although in a slightly disguised form. The resulting matrix $R_{m-1}(N-1)$ is different from $R_{m-1}(N)$ yet closely related. It is not difficult to see that they are time-shifted versions of each other. Indeed,

$$R_{m-1}(N-1) = R_{m-1}(N) - \mathbf{u}_{m-1}(N)\mathbf{u}_{m-1}^{\mathrm{T}}(N) \tag{2.155}$$

Thus $R_{m-1}(N)$ is nested within $R_m(N)$ in both upper and lower partitionings via its time-shifted version, from which it is obtained via a correcting matrix of rank at most rank 1. The presence of the correcting matrices will complicate the procedure but an $O(p^2)$ algorithm for the solution of (2.151) is still possible [23]. Its form, although more complicated, is similar to that of Levinson and the solution is obtained in an order recursive manner via the LS forward and backward predictors. However, in this case the predictors are no longer the reverse of each other and order update equations for both backward and forward predictors are needed. They have the form

$$\mathbf{b}_{m+1}(N) = \begin{bmatrix} 0 \\ \mathbf{b}_m(N-1) \end{bmatrix} + \begin{bmatrix} 1 \\ \mathbf{a}_m(N) \end{bmatrix} k_m^{\mathrm{b}}(N) \tag{2.156}$$

$$\mathbf{a}_{m+1}(N) = \begin{bmatrix} \mathbf{a}_m(N) \\ 0 \end{bmatrix} + \begin{bmatrix} \mathbf{b}_m(N-1) \\ 1 \end{bmatrix} k_m^{\mathrm{f}}(N) \tag{2.157}$$

Reflection coefficients $k_m^f(N)$ and $k_m^b(N)$ are obtained via inner products and, as with the Toeplitz case, they may provide an equivalent lattice parameterization of the predictor for the same set of data. The above algorithm can be employed to effect the LINSOLV procedure in the state-space realization (2.42). The resulting algorithm has memory and complexity requirements proportional to the RLS algorithm.

It is worth pointing out that a counterpart of the split Levinson algorithm, with reduced complexity, is still not available for the LR case.

Schur-type algorithms with enhanced parallelism have also been derived for the linear regressor case [17,21]. In a parallel environment with $O(p)$ processors, the required computational time is reduced to $O(p)$ time units. As was the case for Toeplitz matrices, inner product computations are overcome by feeding appropriately defined error signals into appropriate lattice-type recursions. These algorithms are treated in more detail in Chapter 8. The resulting Schur algorithm can effect the LINSOLV in (2.42) and in a parallel processing environment a significant reduction in computational time is achieved.

2.8.1. Fast RLS algorithms

The RLS algorithm has been presented in section 2.5.2 and is summarized in Table 2.1. In this section we shall illustrate how the basic nesting properties of the covariance matrix, given by (2.153) and (2.154), can be combined to reduce the complexity of the RLS from $O(p^2)$ to $O(p)$ per time iteration. The simple FIR case is again assumed. It is already known that the gain $\mathbf{w}_p(N)$ of the RLS algorithm is given by

$$\mathbf{w}_p(N) = -R_p^{-1}(N-1)\mathbf{u}_p(N)$$

Combining (2.153) and (2.154) with the two possible partitions of $\mathbf{u}_p(t)$, one can derive the following two steps known as

STEP UP

$$\mathbf{w}_{p+1}(N+1) = \begin{bmatrix} 0 \\ \mathbf{w}_p(N) \end{bmatrix} + \begin{bmatrix} 1 \\ \mathbf{a}_p(N) \end{bmatrix} k_1^* \tag{2.158}$$

STEP DOWN

$$\mathbf{w}_{p+1}(N+1) = \begin{bmatrix} \mathbf{w}_p(N+1) \\ 0 \end{bmatrix} + \begin{bmatrix} \mathbf{b}_p(N) \\ 1 \end{bmatrix} k_2^* \tag{2.159}$$

where k_1^*, k_2^* are quantities that do not need to be specified at present. The above two steps provide $\mathbf{w}_p(N+1)$ from $\mathbf{w}_p(N)$ in $O(p)$ MADs. It also turns out that the rest of the variables involved can be updated with $O(p)$ complexity. Such an $O(p)$ scheme was first presented by Ljung *et al.* in their pioneering paper [24] and later improved in [25]. It is interesting to note that the same algorithm can be derived by exploiting the geometric properties of the least-squares estimation as first pointed out in [26]. The basic problem associated with fast RLS schemes is their numerical stability properties [27], which have limited their practical applications. A solution to this problem was first given in [28]. In a clever way redundancy was artificially introduced into the algorithm by computing the same quantity (backward error) in two different ways and forcing their difference to be minimum. Improved stable versions were later introduced in [29,30]. Chapter 5 deals with the fast RLS scheme and its stabilized form and also provides a theoretical insight into its numerical behaviour.

2.8.2 Square root QR and lattice algorithms

The problem of numerical stability is not only of concern in the fast versions of RLS. What is dramatic with these versions is the speed of occurrence of their catastrophic effects, usually after a few hundred iterations. As already discussed, RLS in its classical form (Table 2.1) also suffers from the effects of numerical error propagation, although these become noticeable usually after thousands of iterations [31]. In [32,33] a fixed point error analysis involving second-order terms reveals that error accumulation leads to loss of the positive definite property of $P_m(N)$ ($R_m(N)$) and this results in explosive divergence. Square root algorithms provide a means of ensuring the positive definiteness of the covariance matrix, hence improving the numerical behaviour at no extra computational cost [34]. It must be emphasized that although square root algorithms improve the numerical performance, they do not alleviate the problem completely, due to error accumulation in the filter update equation (2.53) (the standard form of LMS also exhibits similar behaviour [35]). The essence of the square root algorithms is that instead of time-updating the matrix $P_m(N)$ directly, one chooses to propagate the upper-triangular factor U and diagonal factor D of the UDU^{T} factorization of $P_m(N)$. This has the advantage of exploiting better the available dynamic range of the registers, since square roots of quantities require half the dynamic range with respect to their squares [31]. In section 2.7.1 we established the $UD^{-1}U^{\mathrm{T}}$ factorization of the inverse autocorrelation matrix in terms of the backward predictors' parameters. This was the consequence of the backward errors' orthogonality. Following similar arguments one can show that under the least-squares optimization, backward errors remain orthogonal, where orthogonality is now meant in the total least-squares sense, that is,

$$\sum_{t=0}^{N} \lambda^{N-t} e_i^{\mathrm{b}}(t) e_j^{\mathrm{b}}(t) = \delta_{ij} \alpha_i^{\mathrm{b}}(N) \tag{2.160}$$

where δ_{ij} is Kronecker's delta. Following similar arguments as for (2.137) the LS counterpart results

$$P_m(N) \equiv R_m^{-1}(N) = U_m(N) D_m^{-1}(N) U_m^{\mathrm{T}}(N) \qquad m = 1, 2, \ldots, p \tag{2.161}$$

where

$$U_m(N) = \begin{bmatrix} 1 & b_1^1(N) & b_2^1(N) & \cdots & b_{m-1}^1(N) \\ 0 & 1 & b_2^2(N) & \cdots & b_{m-1}^2(N) \\ 0 & 0 & 1 & \cdots & b_{m-1}^3(N) \\ \vdots & \vdots & \vdots & \ddots & \vdots \\ 0 & 0 & 0 & \cdots & 1 \end{bmatrix}$$

$$= \begin{bmatrix} U_{m-1}(N) & \mathbf{b}_{m-1}(N) \\ & 1 \end{bmatrix} \tag{2.162}$$

and

$$D_m(N) = \text{diag}[\alpha_0^{\text{b}}(N), \alpha_1^{\text{b}}(N), \ldots, \alpha_{m-1}^{\text{b}}(N)] \tag{2.163}$$

$\alpha_i^{\text{b}}(N)$ is the backward error power of order i defined in (2.160). Hence propagation of the U and D matrices becomes equivalent to updating in time the backward predictors $\mathbf{b}_i(N)$, $i = 1, 2, \ldots, m - 1$, and their powers. From the above established relations it is apparent that

$$U_m^{\text{T}}(N-1)\boldsymbol{\phi}_m(N) \equiv [e_0^{\text{b}}(N), e_1^{\text{b}}(N), \ldots, e_{m-1}^{\text{b}}(N)]^{\text{T}} \equiv \mathbf{e}_m^{\text{b}}(N)$$

$$= [\mathbf{e}_{m-1}^{\text{bT}}(N), e_{m-1}^{\text{b}}(N)]^{\text{T}} \tag{2.164}$$

Define

$$D_m^{-1}(N-1)\mathbf{e}_m^{\text{b}}(N) \equiv \mathbf{g}_m(N) \equiv [g_1(N), g_2(N), \ldots, g_m(N)]^{\text{T}}$$

$$= [\mathbf{g}_{m-1}^{\text{T}}(N), g_m(N)]^{\text{T}} \tag{2.165}$$

which is a set of normalized backward errors. To adapt the elements $[u_{ij}]$ of the matrix $U_m(N)$, that is, the backward predictors' elements, we adopt the RLS algorithm of Table 2.1. Using (2.161), (2.164) and (2.165) it is easy to show that

$$u_{ij}(N) = u_{ij}(N-1) - w_{j-1}^{i}\frac{e_j^{\text{b}}(N)}{\alpha_{j-1}^{*}} \qquad i = 1, 2, \ldots, j-1 \tag{2.166}$$

$$j = 1, 2, \ldots, p$$

Note that the order of the $\alpha^*(N)$ factor, in the RLS algorithm, has been given explicitly above. This is because all lower-order problems up to p are present. This can be exploited to compute the α^* factor in an order recursive manner resulting in a computational saving. Following the respective definitions it is not difficult to see that

$$\alpha_{j-1}^{*}(N) = \lambda + \sum_{k=1}^{j-1} e_k^{\text{b}}(N)g_k(N)$$

$$= \alpha_{j-2}^{*}(N) + e_{j-1}^{\text{b}}(N)g_{j-1}(N) \tag{2.167}$$

and w_{j-1}^{i} is the ith element of the Kalman gain $\mathbf{w}_{j-1}(N)$ given by

$$\mathbf{w}_{j-1}(N) = -U_{j-1}(N-1)\mathbf{g}_{j-1}(N) \tag{2.168}$$

The derived algorithm for updating the gain $\mathbf{w}_m(N)$ via its U and D^{-1} constituents is summarized in Table 2.5 and is known as Bierman's algorithm [1]. The algorithm

Table 2.5 Bierman's algorithm for the Kalman gain time update

COMPUTE

1. $\mathbf{e}_p^{\mathrm{b}}(N) = U_p^{\mathrm{T}}(N-1)\boldsymbol{\phi}_p(N)$
2. $\mathbf{g}_p(N) = -D_p^{-1}(N-1)\mathbf{e}_p^{\mathrm{b}}(N)$
3. $a_0^*(N) = \lambda$

FOR $j = 1, 2, \ldots, p$

FOR $i = 1, 2, \ldots, j,\ j \neq 1$

4. $u_{ij}(N) = u_{ij}(N-1) - w_{j-1}^{i}(N)\, e_j^{\mathrm{b}}(N)/\alpha_{j-1}^*(N)$
5. $w_j^i(N) = w_{j-1}^i(N) + u_{ij}(N-1)g_j(N)$

END

6. $w_j^j(N) = g_j(N)$
7. $\alpha_j^*(N) = \alpha_{j-1}^*(N) + e_j^{\mathrm{b}}(N)g_j(N)$
8. $[D^{-1}(N)]_{jj} = \alpha_{j-1}^*[D^{-1}(N-1)]_{jj}/\lambda\alpha_j^*$

END

9. $\mathbf{w}_p(N) = (w_p^1(N), \cdots, w_p^p(N))^{\mathrm{T}}$
10. $\alpha^*(N) = \alpha_p^*(N)$

guarantees the positiveness of the elements of the diagonal matrix D; hence it guarantees that $P(N)$ remains positive definite [34]. Recursions 5 and 6 of Table 2.5 are the direct result of (2.168), (2.165) and the special structure of $U_m(N-1)$ in (2.162). From (2.167) and the definitions in (2.164) and (2.165) it is easy to see that

$$\alpha_j^*(N) = \alpha_{j-1}^*(N) + \frac{(e_j^{\mathrm{b}}(N))^2}{\alpha_j^{\mathrm{b}}(N)} \tag{2.169}$$

It can be shown (i.e. [11]) that

$$\alpha_j^{\mathrm{b}}(N) - \lambda\,\frac{(\alpha_j^{\mathrm{b}}(N-1))^2}{\alpha_j^*(N)} \tag{2.170}$$

Combining the above leads to step 8 of Table 2.5. The algorithm provides at its output the gain vector of order p, $\mathbf{w}_p(N)$ and $\alpha_p^*(N)$, required for the update of $\boldsymbol{\theta}$ in the RLS algorithm. The complexity of Bierman's algorithm is $O(3p^2/2)$, that is, the same as that of RLS. Thus a better numerical performance is achieved at no computational cost.

The family of algorithms based on the propagation of the UDU^{T} factorizations of $R_m^{-1}(N)$ is closely related to another family of algorithms whose starting point is the orthogonal factorization of the input data matrix $\mathcal{U}_m(N)$ defined as

$$\mathscr{U}_m(N) = \begin{bmatrix} \mathbf{u}_m^{\mathrm{T}}(N) \\ \mathbf{u}_m^{\mathrm{T}}(N-1) \\ \vdots \\ \mathbf{u}_m^{\mathrm{T}}(0) \end{bmatrix}$$

Indeed, if

$$\mathscr{U}_m(N) = Q\tilde{R} \tag{2.171}$$

and since the covariance matrix is the Grammian of the input matrix $\mathscr{U}_m(N)$, we have

$$(\mathscr{U}_m^{\mathrm{T}}(N)\mathscr{U}_m(N)) = \tilde{R}^{\mathrm{T}}\tilde{R} = (LD^{-1/2})(LD^{1/2})^{\mathrm{T}} = LD^{-1}L^{\mathrm{T}} \tag{2.172}$$

Thus the orthogonal factorization of the data matrix is directly related to the Cholesky factorization of its Grammian. This property has also led to fast order recursive schemes for the orthogonal factorization of the data matrix, i.e. [16,36]. Take, for example, the special case where the Grammian is Toeplitz. It is apparent from (2.172) that the $\tilde{R}$ factor in the QR decomposition is directly related to the errors involved in the Schur algorithm (Equations (2.148) and (2.149)). On the other hand, it can be shown that the orthogonal Q matrix is related to the backward errors, the result of a Gram–Schmidt (lattice) orthogonalization procedure.

Adaptive LS lattice schemes belong to this QR algorithmic family. It has been pointed out above that an alternative parameterization of the LR predictors is provided by the set of reflection coefficients, leading to lattice–ladder structures, with error sequences being propagated from one order section to another. In the LS case the corresponding reflection coefficients, as well as the error signals, are updated at each time instant directly from the input samples, as they become available. By exploiting the nesting property of the input data schemes with complexity of $O(p)$ MADs per iteration can be derived. The LS lattice part provides the backward errors of all intermediate orders up to p. These errors are orthogonal, as pointed out above. LS lattice algorithms are treated in Chapter 6.

It is well known from linear algebra that an alternative to the Gram–Schmidt procedure for the orthogonal factorization of a matrix is possible via a series of Givens rotations. Givens rotations are orthogonal, thus inheriting the good numerical properties associated with such operations. Time iterative $O(p^2)$ schemes to obtain LR solutions based on the QR factorization of the input data matrix via Givens rotations were introduced in [37] and later in [38]. It is only recently that the nested property of the input vector (shift invariance), wherever applicable, has been exploited to derive fast versions with complexity $O(p)$ [39,40]. This algorithmic family, known as QR, is treated in Chapter 7, where the interesting close relationship between the respective fast versions and the previously discussed LS lattice structures is also established.

2.8.3 The multichannel case

In this section we briefly outline the multichannel extensions of the previous setup. This is important for two reasons. First, it enables us to deal with multichannel signals and multiple input–output systems. Second, it can be utilized to convert a single-channel problem into a multichannel formulation and thereby reduce computational complexity, as we shall explain below. We shall first discuss the FIR filtering problem as it forms the basis for more general cases.

Suppose we have k input channels $u_1(n), u_2(n), \ldots, u_k(n)$ and q desired response signals $y_1(n), y_2(n), \ldots, y_q(n)$. The class of models in which the search for the optimum filter is conducted presumes that each output value is determined by a fixed finite number, p_i, of past values of each input signal, and this number, p_i, named *the order* of the filter with respect to input i, is in general different for each input signal channel. Thus, we write

$$\begin{aligned}\hat{y}_l(n) = &- c_{1l}(1)u_1(n) - c_{1l}(2)u_1(n-1) - \ldots - c_{1l}(p_1)u_1(n-p_1+1)\\ &- c_{2l}(1)u_2(n) - c_{2l}(2)u_2(n-1) - \ldots - c_{2l}(p_2)u_2(n-p_2+1)\\ &\ldots\\ &- c_{kl}(1)u_k(n) - c_{kl}(2)u_k(n-1) - \ldots - c_{kl}(p_k)u_k(n-p_k+1) \qquad l = 1, 2, \ldots, q\end{aligned} \tag{2.173}$$

and we note that the filter orders $p_1, p_2, \ldots, p_k$ are generally unequal, i.e. $p_i \neq p_j, i \neq j$.

For each input i and output j, let us define the coefficients vector $\mathbf{c}_{ij}(p_i) = [c_{ij}(1), c_{ij}(2), \ldots, c_{ij}(p_i)]^{\mathrm{T}}$ of dimension $p_i \times 1$. For each input channel i, consider the regressor vector $\mathbf{u}^i_{p_i}(n) = [u_i(n)u_i(n-1)\ldots u_i(n-p_i+1)]^{\mathrm{T}}$ of dimension $p_i \times 1$. We insert these vectors into the block vectors

$$\mathbf{c}_{\mathbf{p}_k} = [c_{ij}(p_i)]_{\substack{i=1,2,\ldots,k\\ j=1,2,\ldots,q}} \tag{2.174}$$

$$\mathbf{u}_{\mathbf{p}_k}(n) = [\mathbf{u}^{1\mathrm{T}}_{p1p2}(n), \mathbf{u}^{2\mathrm{T}}_{p1p2}(n), \ldots, \mathbf{u}^{k\mathrm{T}}_{p_k}(n)]^{\mathrm{T}} \tag{2.175}$$

where the multi-index $\mathbf{p}_k$ consists of the individual filter orders

$$\mathbf{p}_k = [p_1, p_2, \ldots, p_k]$$

Note that $\mathbf{u}_{\mathbf{p}_k}(n)$ is a block vector of block order k, with entries vectors of dimensions $p_i \times 1$. Note that $\mathbf{c}_{\mathbf{p}_k}$ is also a block vector of block order k, with entries matrices of dimensions $p_i \times q$. Clearly, $\mathbf{u}_{\mathbf{p}_k}(n)$ has dimension $P \times 1$, where P is the sum of entries dimensions, $P = \Sigma_{i=1}^{k} p_i$, while $\mathbf{c}_{\mathbf{p}_k}$ has dimensions $P \times q$.

Next we write (2.173) compactly as

$$\hat{y}_q(n) = -\mathbf{c}^{\mathrm{T}}_{\mathbf{p}_k}\mathbf{u}_{\mathbf{p}_k}(n) \tag{2.176}$$

where $\hat{y}_q(n) = [\hat{y}_1(n), \hat{y}_2(n), \ldots, \hat{y}_q(n)]^{\mathrm{T}}$. [Note that in this and subsequent chapters vectors corresponding to multichannel signals will be denoted with plain letters. We have reserved bold characters to denote vectors of the system order dimension.] We assume all relevant signals are causal, i.e. $u_i(n) = 0, i = 1, 2 \ldots k$, and $y_i(n) = 0, i = 1,$

$2 \ldots q$, $\forall n < 0$, and that the data records of the input signal $\{u_1(n), u_2(n) \ldots u_k(n)\}$, and the desired response signal $\{y_1(n), y_2(n) \ldots y_q(n)\}$, are given over the interval $0 \leqslant n \leqslant N$.

We form the error signal $e_{\mathbf{p}_k}(n) = y_q(n) - \hat{y}_q(n) = y_q(n) + \mathbf{c}_{\mathbf{p}_k}^{\mathrm{T}} \mathbf{u}_{\mathbf{p}_k}(n)$, where $y_q(n) = [y_1(n), y_2(n), \ldots, y_q(n)]^{\mathrm{T}}$ is the desired output at time n; $\mathbf{e}_{\mathbf{p}_k}(n)$ is a $q \times 1$ vector, and the multi-index $\mathbf{p}_k$ associated with it indicates that it is the modelling error with respect to a filter of order $\mathbf{p}_k$. The total squared error of order $\mathbf{p}_k$ is $E_{\mathbf{p}_k}(N) = \Sigma_{n=0}^{N} e_{\mathbf{p}_k}^{\mathrm{T}}(n) e_{\mathbf{p}_k}(n)$ and the optimal filter which minimizes the above cost satisfies the linear system of equations

$$R_{\mathbf{p}_k}(N)\mathbf{c}_{\mathbf{p}_k}(N) = -\mathbf{d}_{\mathbf{p}_k}(N) \tag{2.177}$$

where

$$R_{\mathbf{p}_k}(N) = \sum_{n=0}^{N} \mathbf{u}_{\mathbf{p}_k}(n)\mathbf{u}_{\mathbf{p}_k}^{\mathrm{T}}(n) \tag{2.178}$$

$$\mathbf{d}_{\mathbf{p}_k}(N) = \sum_{n=0}^{N} \mathbf{u}_{\mathbf{p}_k}(n) y_q(n) \tag{2.179}$$

Equation (2.177) determines the optimal FIR filter in the LS sense. The pertinent filter is obtained as the solution of the linear system of equations (2.177). The structure of matrix $R_{\mathbf{p}_k}(N)$ enables the development of efficient algorithms for the computation of the optimal filter $\mathbf{c}_{\mathbf{p}_k}(N)$. The derivation of such fast algorithms is based on the nesting properties of matrix $R_{\mathbf{p}_k}(N)$, resulting from the possible partitionings of the block input vector. Take, for example, the case of the two-channel input vector

$$\mathbf{u}_{2(m+1)}(t) \equiv [u_1(t), \ldots, u_1(t-m), u_2(t), \ldots, u_2(t-m)] \tag{2.180}$$

From the above definition the following results:

$$\begin{aligned} \mathbf{u}_{2(m+1)} &= T\begin{bmatrix} u_1(t) \\ u_2(t) \\ \mathbf{u}_{2m}(t-1) \end{bmatrix} \\ &= S\begin{bmatrix} \mathbf{u}_{2m}(t) \\ u_1(t-m) \\ u_2(t-m) \end{bmatrix} \end{aligned} \tag{2.181}$$

where T, S are appropriately defined permutation matrices [41]. In the above partitioning the order was increased by two (the number of channels) in one step. Another possibility could be to increase the order by one at a time in a two-step procedure (k steps in the more general k-channel case) [42]. This results in the so-called staircase algorithmic family and only scalar quantities are engaged [43–46]. A closely related family is that of circular lattices [47]. Multichannel algorithms are discussed in Chapters 5–7.

The family of multichannel algorithms can be employed for the efficient computation of single-channel problems by suitably converting them into multichannel format. Consider as an example ARX identification and the linear system of equations (2.21) and (2.22) with the regressor vector given by (2.32). We readily see that (2.22) takes the form (2.177)–(2.179) by embedding ARX into a two-channel filtering format via the assignment

$$u_1(n) = u(n-1) \qquad u_2(n) = y(n-1)$$

The recursive prediction error algorithm of Table 2.2 can be embedded in a three-channel formulation, $u_1(t) = y^{\mathrm{f}}(t)$, $u_2(t) = u^{\mathrm{f}}(t)$, $u_3(t) = e^{\mathrm{f}}(t)$, and the fast transversal schemes of Chapter 5 can be employed. More generally, the prediction error algorithm for the Box–Jenkins model can be implemented by a fast adaptive realization, once we observe that the involved Kalman gain is determined by a linear system solver whose associated matrix derives from a regressor vector $\boldsymbol{\psi}(t, \boldsymbol{\theta})$ that can be embedded in a four-channel signal.

Another example of a single-channel problem which can be treated efficiently as a multichannel one is non-linear Volterra filtering. Given the input $u(t)$ the output of the system is given by

$$\begin{aligned}\hat{y}(t) = c_0 &+ \sum_{i=0}^{p_{i,1}} c_i u(t-i)\\ &+ \sum_{i=0}^{p_{i,2}} \sum_{j=0}^{p_{j,2}} c_{ij} u(t-i) u(t-j)\\ &+ \sum_{i=0}^{p_{i,3}} \sum_{j=0}^{p_{j,3}} \sum_{k=0}^{p_{k,3}} c_{ijk} u(t-i) u(t-j) u(t-k)\\ &+ \dots\end{aligned}$$

The task, again, is to estimate the unknown coefficients c_i, c_{ij}, c_{ijk} so that $\hat{y}(t)$ tracks a desired response sequence $y(t)$ in an optimal way. The respective multichannel formulation results by treating the input sample products as different input channels. Take, for example, $u(t-1)u(t-2)$, $u(t-2)u(t-3)$, $\dots$. These can be thought of as delayed versions of the signal channel $u(t)u(t-1)$. An overview of efficient computational schemes for Volterra-type modelling can be found in [48] and the references therein.

References

[1] L. Ljung and T. Soderstrom, *Theory and Practice of Recursive Identification*, MIT Press: Cambridge, MA, 1982.

[2] L. Ljung, *System Identification – Theory for the User*, Prentice Hall: Englewood Cliffs, NJ, 1987.

[3] T. Soderstrom and P. Stoica, *System Identification*, Prentice Hall: UK, 1989.

[4] S. Haykin, *Adaptive Filter Theory* (2nd edn), Prentice Hall: Englewood Cliffs, NJ, 1991.

[5] G. Goodwin and K. Sin, *Adaptive Filtering, Prediction and Control*, Prentice Hall: Englewod Cliffs, NJ, 1984.

[6] S.Y. Kung, *VLSI Array Processors*, Prentice Hall: Englewood Cliffs, NJ, 1989.

[7] O. Macchi and N. Bershad, 'Adaptive recovery of a chirped sinusoid in noise. Part I: Performance of the RLS algorithm', *IEEE Trans. Acoust., Speech, Signal Process.*, vol. ASSP-39, pp. 583–95, 1991.

[8] N. Bershad and O. Macchi, 'Adaptive recovery of a chirped sinusoid in noise. Part II: Performance of the LMS algorithm', *IEEE Trans. Acoust., Speech, Signal Process.*, vol. ASSP-39, pp. 595–603, 1991.

[9] N. Levinson, 'The Wiener error criterion in filter design and prediction', *J. Math. Phys.*, vol. 25, pp. 261–78, 1947.

[10] R.A. Wiggins and E.A. Robinson, 'Recursive solution to the multichannel filtering problem', *J. Geophys. Res.*, vol. 70, pp. 1885–91, 1965.

[11] N. Kalouptsidis, D. Manolakis and G. Carayannis, 'Computationally efficient algorithms for multichannel signal processing', *Signal Process.*, pp. 5–19, 1983.

[12] P. Delsarte and Y. Genin, 'The split Levinson algorithm', *IEEE Trans. Acoust., Speech, Signal Process.*, vol. ASSP-34, pp. 470–8, 1986.

[13] P. Delsarte and Y. Genin, 'On the splitting of classical algorithms in linear prediction', *IEEE Trans. Acoust., Speech, Signal Process.*, vol. ASSP-35, pp. 645–53, 1987.

[14] P. Delsarte and Y. Genin, 'Multichannel singular predictor polynomials', *IEEE Trans. Circuits Syst.*, vol. CAS-35, pp. 190–200, 1988.

[15] A. Yagle, 'A new multichannel split Levinson algorithm for block Hermitian–Toeplitz matrices', *IEEE Trans. Circuits Syst.*, vol. CAS-36, pp. 928–31, 1989.

[16] C. Deumeure, 'Bowtie factors of Toeplitz matrices by means of split Algorithms', *IEEE Trans. Acoust., Speech, Signal Process.*, vol. ASSP-35, pp. 1601–3, 1987.

[17] N. Kalouptsidis and S. Theodoridis, 'Parallel implementation of efficient LS algorithms for filtering and prediction', *IEEE Trans. Acoust., Speech, Signal Process.*, vol. ASSP-35, pp. 1565–9, 1987.

[18] S.Y. Kung and Y.H. Hu, 'A highly concurrent algorithm and pipelined architecture for solving Toeplitz systems', *IEEE Trans. Acoust., Speech, Signal Process.*, vol. ASSP-31, pp. 66–76, 1983.

[19] J. Schur, 'Uber Potenzreihen die im Innern des Einheitskreises beschrankt sind', *J. Reine Angew. Math.*, vol. 147, pp. 205–32, 1917.

[20] J. Le Roux and C.J. Geugeun, 'A fixed point computation of partial correlation coefficients', *IEEE Trans. Acoust., Speech, Signal Process.*, vol. ASSP-25, pp. 257–9, 1977.

[21] S. Theodoridis, 'Pipelined architecture for block adaptive filtering and prediction', *IEEE Trans. Acoust., Speech, Signal Process.*, vol. ASSP-38, pp. 81–91, 1990.

[22] S. Theodoridis, N. Kalouptsidis, D. Bakirtzis, 'Pipelined algorithms for LS FIR filters with symmetric impulse response', *IEEE Trans. Acoust., Speech, Signal Process.*, ASSP-38, pp. 260–71, Nov. 1990.

[23] N. Kalouptsidis, G. Carayannis, D. Manolakis and E. Koukoutsis, 'Efficient recursive in order LS FIR filtering and prediction', *IEEE Trans. Acoust., Speech, Signal Process.*, vol. ASSP-33, pp. 1175–89, 1985.

[24] L. Ljung, M. Morf and D. Falconer, 'Fast calculation of gain matrices for recursive estimation schemes', *Int. J. Control.*, vol. 27, pp. 1–19, 1978.

[25] G. Carayannis, D. Manolakis and N. Kalouptsidis, 'A fast sequential algorithm for the LS filtering and prediction', *IEEE Trans. Acoust., Speech, Signal Process.*, vol. ASSP-31, pp. 1394–402, 1983.

[26] J. Cioffi and T. Kailath, 'Fast recursive LS transversal filters for adaptive processing', *IEEE Trans. Acoust., Speech, Signal Process.*, vol. ASSP-32, pp. 304–37, 1984.

[27] L. Ljung and L. Ljung, 'Error propagation properties of recursive LS adaptation algorithms', *Automatica*, vol. 21, pp. 157–67, 1985.

[28] J. Botto and G. Moustakides, 'Stabilizing the fast Kalman algorithm', *IEEE Trans. Acoust., Speech, Signal Process.*, vol. ASSP-37, pp. 1342–8, 1989.

[29] D.T.M. Slock and T. Kailath, 'Numerically stable fast RLS transversal filters,' *Proc. ICASSP-88 Conf.*, pp. 1365–8. New York, 1988.

[30] A. Benallal and A. Gilloire, 'A new method to stabilize fast RLS algorithms in transversal adaptive filters', *Proc. ICASSP-88 Conf.* pp. 1373–6, Apr. 1988.

[31] J.M. Cioffi, 'Limited precision effects in adaptive filtering', *IEEE Trans. Circuits Syst.*, vol. CAS-34, pp. 821–33, 1987.

[32] G. Bottomley and S.T. Alexander, 'A theoretical basis for the divergence of conventional recursive LS filters', *Proc. ICASSP – 1989, Glasgow*, pp. 908–11.

[33] G. Bottomley and S.T. Alexander, 'A novel approach for stabilizing RLS filters', *IEEE Trans. Acoust., Speech, Signal Process.*, vol. ASSP-39, pp. 1170–80, 1991.

[34] F.M. Hsu, 'Square root Kalman filtering for high speed data received over fading dispersive channels', *IEEE Trans. Inf. Theory*, vol. IT-28, pp. 753–63, 1982.

[35] S.H. Ardalan, 'Floating point roundoff error analysis of the RLS and LMS adaptive algorithms', *IEEE Trans. Circuit Syst.*, vol. CAS-33, pp. 1192–208, 1986.

[36] C.P. Rialan and L.L. Scharf, 'Fast algorithms for computing QR and Cholesky factors of Toeplitz operators', *IEEE Trans. Acoust., Speech, Signal Process.*, vol. ASSP-36, pp. 1740–9, 1988.

[37] W.M. Gentleman and H.T. Kung, 'Matrix triangularization by systolic arrays', *Proc. SPIE (Real-Time Signal Processing)*, vol. 298, pp. 19–26, 1981.

[38] J.G. McWhirter, 'Recursive LS minimization using systolic array', *Proc. SPIE (Real-Time Signal Processing)*, vol. 431, pp. 105–12, 1983.

[39] J.M. Cioffi, 'The fast adaptive rotor's RLS algorithm', *Proc. ICASSP-88 Conf.*, pp. 1584–7, 1988.

[40] I.K. Proudler, J.G. McWhirter and T.J. Shepherd, 'QRD based lattice filter algorithms', *Proc. SPIE*, 1152, *Advanced Algorithms and Architectures for Signal Processing*, 1989.

[41] D.D. Falconer and L. Ljung, 'Application of fast Kalman estimation to adaptive equalization', *IEEE Trans. Commun.*, vol. COM-26, pp. 1439–46, 1978.

[42] E. Karlsson and M.H. Hayes, 'Least squares ARMA modeling of linear time-varying systems: Lattice filter structures and fast RLS algorithms', *IEEE Trans. Acoust., Speech, Signal Processing*., vol. ASSP-35, pp. 994–1014, 1987.

[43] S. Caraboyas, N. Kalouptsidis and C. Caroubalos, 'Efficient ARX identification algorithms with full parallelism', *IEEE Trans. Acoust., Speech, Signal Process.*, ASSP-38, pp. 1902–13, Nov. 1990.

[44] G.O. Glentis and N. Kalouptsidis, 'Efficient order recursive algorithms for multichannel LS filtering', IEEE, SP-40, no. 6, pp. 1354–75, 1992.

[45] G. Glentis and N. Kalouptsidis, 'Efficient adaptive algorithms for multichannel least squares filtering', *IEEE ISCAS-90. New Orleans.*

[46] A. Liavas and S. Theodoridis, 'Efficient Levinson and Schur type algorithms for block e-Toeplitz systems of equations', *Signal Processing*, vol. 35, no. 3, 1994.

[47] H. Sakai, 'Circular lattice filtering using Pagano's method', *IEEE Trans. Acoust., Speech, Signal Process.*, vol. ASSP-30, pp. 279–87, 1982.

[48] V.J. Mathews, 'Adaptive polynomial filters', *IEEE Signal Process. Mag.*, vol. 8, pp. 10–28, 1991.

3

General Structure of Adaptive Algorithms: Adaptation and Tracking

Lennart Ljung

3.1 Introduction

Adaptation and adaptability are desired features in most systems' behaviour (including human relations!). In technical systems dealing with signal processing – in a broad sense – adaptive properties are manifested in such concepts as 'adaptive control', 'adaptive filtering', 'adaptive prediction', and so on.

The main feature in any adaptation mechanism is a *tracking facility*, which, explicitly or implicitly, tracks the time-varying properties of the signal or system, to which we want to adapt. Tracking a system's properties is always a question of critically evaluating the observations obtained from the process in question: do they contain information about changes in the process or are they just dominated by random fluctuations? Thus even in a non-mathematical setting, *adaptation and tracking are always characterized by a trade-off between tracking ability* (dare to believe signs of process changes in the measurements!) and *noise sensitivity* (don't get confused by random fluctuations!). We shall see this fundamental trade-off show up in various formalized ways in the course of this chapter.

One focus of our discussion will be how to translate certain assumptions about the system's behaviour and criteria for good tracking to optimal algorithms. We shall then also see that many common *ad hoc* algorithms can be interpreted as corresponding to certain assumptions about the system's behaviour.

Another focus of our discussion is to outline basic procedures for the evaluation of analytic performance of the various algorithms.

We shall mostly confine ourselves to the case where the underlying system or signal model can be formulated as a linear regression. See also the survey in [1].

The organization of the chapter is as follows:

1. Structure of adaptation algorithms: We describe the basic setup of how to derive adaptation algorithms under varying assumptions. This is covered in sections 3.2–3.5.
2. Asymptotic behaviour of algorithms under decreasing gain: The situation where the true system to be identified is constant leads to algorithms with gains that tend to zero. Some general results about the asymptotic properties of the estimates thus obtained are quoted in section 3.6.
3. Analyzing the tracking ability of the algorithms under non-decreasing gain: The real use of adaptation algorithms is to track time-varying properties. In sections 3.7–3.9 we outline the basic results and tools for the analysis of the algorithms' tracking properties.

3.2 Optimal algorithms for tracking drifting parameters

3.2.1 A basic signal model

We shall use the following linear regression signal model:

$$y(t) = \boldsymbol{\varphi}^{\mathrm{T}}(t)\boldsymbol{\theta} + \eta(t) \tag{3.1}$$

where $\{y(t)\}$ and $\{\boldsymbol{\varphi}(t)\}$ are observed signals. The vector $\boldsymbol{\theta}$ contains the unknown parameters which are to be estimated by the tracker.

The most common application of (3.1) in control and signal processing is when the regression vector $\boldsymbol{\varphi}(t)$ consists of lagged outputs and inputs

$$\boldsymbol{\varphi}^{\mathrm{T}}(t) = (-y(t-1), \ldots, u(t-m)) \tag{3.2}$$

In this case (3.1) and (3.2) correspond to a linear difference relationship between the input and the output. In case there is no 'input' signal $\{u(t)\}$ we have the well-known AR model for the signal $\{y(t)\}$.

We now assume that there is a true – and time-varying – value $\theta_0(t)$ for the parameters and that these develop over time as a random walk. This means that the 'true' description of the signals $\{y(t)\}$ and $\{\varphi(t)\}$ becomes

$$\theta_0(t) = \theta_0(t-1) + w(t) \tag{3.3}$$

$$y(t) = \theta_0^{\mathrm{T}}(t)\varphi(t) + \eta(t) \tag{3.4}$$

We here assume $\{\eta(t)\}$ to be white Gaussian noise with variance $R_2(t)$, while $\{w(t)\}$ is white Gaussian noise with covariance matrix $R_1(t)$ independent of $\{\eta(t)\}$. It is then well known (see e.g. section 2.3 in [2]) that the estimate $\hat{\theta}(t)$ that minimizes the conditional expectation, given past observations

$$\Pi(t) = E(\hat{\theta}(t) - \theta_0(t))(\hat{\theta}(t) - \theta_0(t))^{\mathrm{T}} \tag{3.5}$$

(even in a matrix sense), is given by the Kalman filter

$$\hat{\theta}(t) = \hat{\theta}(t-1) + w(t)e(t) \tag{3.6}$$

$$e(t) = y(t) - \hat{\theta}^{\mathrm{T}}(t)\varphi(t-1) \tag{3.7}$$

where the gain vector $\mathbf{w}(t)$ is given by

$$\mathbf{w}(t) = \frac{P(t-1)\varphi(t)}{\hat{R}_2(t) + \varphi^{\mathrm{T}}(t)P(t-1)\varphi(t)} \tag{3.8}$$

and the matrix $P(t)$ is updated according to

$$\begin{aligned} P(t) &= P(t-1) - \frac{P(t-1)\varphi(t)\varphi^{\mathrm{T}}(t)P(t-1)}{\hat{R}_2(t) + \varphi^{\mathrm{T}}(t)P(t-1)\varphi(t)} + \hat{R}_1(t) \\ P(0) &= P_0 \end{aligned} \tag{3.9}$$

We have used here the notations $\hat{R}_1(t)$ and $\hat{R}_2(t)$ to indicate that the values used in the algorithm may very well differ from the true values $R_1(t)$ and $R_2(t)$. In the case $\hat{R}_1(t) \equiv R_1(t)$ and $\hat{R}_2(t) \equiv R_2(t)$, however, $\hat{\theta}(t)$ is the conditional expectation of $\theta_0(t)$, given the observations $\{u(k), y(k)\}$, $k \leqslant t$, and $P(t)$ is the conditional covariance matrix of the parameter estimation error.

Note also that if $R_1(t)$ is known then (3.6)–(3.9) is the optimal algorithm also for abrupt changes in θ_0. (Take $R_1(t) = 0$ except when a jump occurs, say, for $t \in T_1$; then take $R_1(t) = R_1$.) However, this requires the time instants for the jumps to be known – not too realistic an assumption.

Remark In fact, the problem of recursive parameter estimation can be seen as a special case of *non-linear filtering*. The parameters are then interpreted as states. There are consequently several important links to the wide literature on non-linear filtering. The reader may consult [2], section 2.3 of [3], and [4] for some aspects of this. In the current context, though, the dynamics in (3.3) are linear, and, under Gaussian noise sources, the non-linear filtering problem specializes to a linear one.

In the algorithm (3.9) it follows that, after a transient, the size of $P(t)$ will be like the square root of $\hat{R}_1$. (This will be shown formally in (3.74) below.) For slowly changing systems, P will thus be small. To show this explicitly it is useful to scale P so as to rewrite (3.6) as

$$\hat{\boldsymbol{\theta}}(t) = \hat{\boldsymbol{\theta}}(t-1) + \mu(t)P_t L(\boldsymbol{\varphi}(t))(y(t) - \boldsymbol{\varphi}^{\mathrm{T}}(t)\hat{\boldsymbol{\theta}}(t-1)) \tag{3.10}$$

We have allowed here a possible non-linear transformation L (such as normalization) of $\boldsymbol{\varphi}(t)$. We shall regard (3.10) as the archetypical algorithm for adaptive parameter estimation. The link to (3.6)–(3.9) can be made explicit by associating

$$L(\boldsymbol{\varphi}(t)) = \boldsymbol{\varphi}(t)$$

$$P_t = \frac{1}{\mu(t)\hat{R}_2(t)}(P(t) - \hat{R}_1(t))$$

$$\mu^2(t) \approx \| \hat{R}_1(t) \|$$

3.2.2 A signal model with global and local trends

In some cases we may know that the parameter changes typically show trends, so that they continue for a while in a certain direction. To capture this we may model them as

$$\boldsymbol{\theta}_0(t) = \boldsymbol{\theta}_0(t-1) + \mathbf{v}(t) + \boldsymbol{\delta}(t) \tag{3.11}$$

where $\{\mathbf{v}(t)\}$ is a correlated stochastic process and $\boldsymbol{\delta}(t)$ is a deterministic or slowly varying stochastic vector. The term $\boldsymbol{\delta}(t)$ models the global trends while $\{\mathbf{v}(t)\}$ describes the local trends, with the amount of correlation in $\{\mathbf{v}(t)\}$ determining the duration of the local trends.

When $\boldsymbol{\delta}(t)$ can be described as a random walk (possibly with zero increments) and $\mathbf{v}(t)$ can be modelled as a filtered white-noise equation then we can write

$$\mathbf{x}_1(t+1) = \mathbf{x}_1(t) + \mathbf{r}_1(t)$$

$$\mathbf{x}_2(t+1) = \alpha_2\mathbf{x}_2(t) + \mathbf{r}_2(t)$$

$$\boldsymbol{\delta}(t) + \mathbf{v}(t) = \mathbf{x}_1(t) + D_2\mathbf{x}_2(t) = [I\ D_2]\begin{bmatrix}\mathbf{x}_1(t)\\ \mathbf{x}_2(t)\end{bmatrix}$$

for some matrix $\tilde{A}_2$. Combining this with (3.11) we obtain

$$\mathbf{X}(t) = A(t)\mathbf{X}(t-1) + \mathbf{r}(t) \tag{3.12}$$

$$\boldsymbol{\theta}_0(t) = C\mathbf{X}(t) \tag{3.13}$$

$$E[\mathbf{r}(t)\mathbf{r}^T(s)] = \begin{cases} R_1(t) & t = s \\ 0 & t \neq s \end{cases} \tag{3.14}$$

where

$$\mathbf{X}(t) = \begin{pmatrix} \boldsymbol{\theta}_0(t) \\ \mathbf{x}_1(t) \\ \mathbf{x}_2(t) \end{pmatrix} \tag{3.15}$$

Furthermore,

$$A(t) = \begin{pmatrix} I & D(t) \\ 0 & \begin{pmatrix} I & 0 \\ 0 & \tilde{A}_2 \end{pmatrix} \end{pmatrix} \qquad R_1(t) = \begin{pmatrix} 0 & 0 \\ 0 & \bar{R}_1(t) \end{pmatrix} \qquad C = [I \; 0] \tag{3.16}$$

where the matrix elements $\bar{R}_1(t)$ come from the description of $\bar{w}$. Clearly, (3.3) is a special case of (3.11)–(3.13). Combining this description with (3.4) gives

$$\mathbf{X}(t) = A(t)\mathbf{X}(t-1) + \mathbf{r}(t) \tag{3.17}$$

$$y(t) = [\varphi^T(t) \; 0]\mathbf{X}(t) + \eta(t) \tag{3.18}$$

This is still an estimation problem for which the Kalman filter gives the optimal solution (provided $\mathbf{r}$ and η are Gaussian with known covariances). One can immediately write down the filter and read the $\hat{\theta}(t)$ update formula from the upper part of the $\hat{\mathbf{X}}(t)$ expression. This approach has been termed multistep algorithms by [5], and [6] and [7]. See also [8].

3.3 *Some* ad hoc *algorithms for tracking drifting parameters*

The basic formulation (3.3) and (3.4) with the optimal algorithm (3.6)–(3.9) is quite powerful. It can deal with both slowly drifting parameters and sudden changes, by assigning proper values to the covariance matrix $\hat{R}_1(t)$ and the variance $\hat{R}_2(t)$. The main shortcoming is then that these values will rarely be known to the user. One approach to deal with this problem is to choose some *ad hoc* values for $\hat{R}_1(t)$. We will discuss two such *ad hoc* choices below.

3.3.1 The RLS algorithm

A popular approach for dealing with time-varying linear regressions is to minimize a weighted criterion

$$V_t(\boldsymbol{\theta}) = \sum_{k=1}^{t} \beta(t, k)(y(k) - \boldsymbol{\theta}^T\varphi(k))^2 \tag{3.19}$$

where

$$\beta(t, k) = \prod_{j=k+1}^{t} \lambda(j) \tag{3.20}$$

and where $|\lambda(j)| \leqslant 1$ denotes the forgetting factor.

From [2] we have that this is accomplished by the recursive least-squares (RLS) algorithm, which is given by (3.6)–(3.8) with $\mathbf{w}(t)$ chosen as

$$\mathbf{w}(t) = \frac{P(t-1)\boldsymbol{\varphi}(t)}{\lambda(t) + \boldsymbol{\varphi}^{\mathrm{T}}(t)P(t-1)\boldsymbol{\varphi}(t)} \tag{3.21}$$

and

$$P(t) = \frac{1}{\lambda(t)}\left[P(t-1) - \frac{P(t-1)\boldsymbol{\varphi}(t)\boldsymbol{\varphi}^{\mathrm{T}}(t)P(t-1)}{\lambda(t) + \boldsymbol{\varphi}^{\mathrm{T}}(t)P(t-1)\boldsymbol{\varphi}(t)}\right] \tag{3.22}$$

We note that this is a special case of (3.6)–(3.9), corresponding to the choices

$$\begin{aligned} \hat{R}_1(t) &= \left(\frac{1}{\lambda(t)} - 1\right) \\ &\quad \times \left(P(t-1) - \frac{P(t-1)\boldsymbol{\varphi}(t)\boldsymbol{\varphi}^{\mathrm{T}}(t)P(t-1)}{\lambda(t) + \boldsymbol{\varphi}^{\mathrm{T}}(t)P(t-1)\boldsymbol{\varphi}(t)}\right) \approx \left(\frac{1}{\lambda(t)} - 1\right)P(t-1) \\ \hat{R}_2(t) &= \lambda(t) \end{aligned} \tag{3.23}$$

(The approximation follows in the typical case where $\| P \| \ll 1$.)

For future use we also note that

$$P(t) = \left(\sum_{k=1}^{t} \beta(t, k)\boldsymbol{\varphi}(k)\boldsymbol{\varphi}^{\mathrm{T}}(k)\right)^{-1} \tag{3.24}$$

The connection to the archetypical algorithm (3.10) is given by

$$\mu(t) = \left(\sum_{k=1}^{t} \beta(t, k)\right)^{-1} \tag{3.25}$$

which gives

$$\begin{aligned} P_t &= \frac{1}{\mu(t)\hat{R}_2(t)}(P(t) - \hat{R}_1(t)) \\ &\approx \frac{1}{\lambda(t)\mu(t)}\left(2 - \frac{1}{\lambda(t)}\right)P(t) \approx \frac{1}{\mu(t)} P(t) \end{aligned}$$

In the first approximation we set $P(t) \approx P(t-1)$ and in the second one we used $2\lambda - 1 \approx \lambda^2$, which holds for λ close to 1. This gives

$$P_t = \left[\left(\sum_{k=1}^{t} \beta(t, k)\right)^{-1} \cdot \sum_{k=1}^{t} \beta(t, k)\varphi(k)\varphi^{\mathrm{T}}(k)\right]^{-1}$$

which shows that

$$P_t \approx \{E[\varphi(t)\varphi^{\mathrm{T}}(t)]\}^{-1}$$

by a weighted sample sum approximation. Consequently, the normalization with μ makes P_t of a size that does not depend on λ. We shall later also use the expression

$$R(t) = \mu(t)P_t^{-1} \approx E[\varphi(t)\varphi^{\mathrm{T}}(t)] \tag{3.26}$$

3.3.2 The LMS algorithm

Widrow's least mean squares algorithm (see e.g. [9]) is a commonly used tool for adaptation. It is given by

$$\mathbf{w}(t) = \mu\varphi(t) \tag{3.27}$$

The LMS algorithm can also be formulated in a normalized variant

$$\mathbf{w}(t) = \frac{\mu\varphi(t)}{1 + \mu|\varphi(t)|^2} \tag{3.28}$$

Again, we may verify that (3.6) and (3.28) are a special case of the basic algorithm (3.6)–(3.9) corresponding to

$$\hat{R}_1(t) = \mu^2 \frac{\varphi(t)\varphi^{\mathrm{T}}(t)}{1 + \mu|\varphi(t)|^2} \tag{3.29}$$

$$\hat{R}_2(t) = 1 \tag{3.30}$$

$$P(0) = \mu \cdot I \tag{3.31}$$

from which it follows that $P(t) = \mu I$.

See the excellent overview by Sethares in this book (Chapter 4) for more information about LMS.

3.3.3 Estimating the unknown covariances

A more systematic approach to deal with the problem of unknown $R_1(t)$ and $R_2(t)$ values is of course to estimate them. We shall now discuss a few possibilities of this kind.

Let us consider the case where the parameters are slowly drifting and the values of $R_1(t) \equiv R_1$ and $R_2(t) \equiv R_2$ are nearly constant over extended periods of time. It is then feasible to devise efficient methods for estimating R_1 and R_2. Techniques for this go back to the literature on adaptive filtering. See, for example, [10–13]. Isaksson [14,15] has developed this approach further and also tested the feasibility of such methods. The idea can be described as a least-squares method applied to a linear regression model for the covariances. A variant is given in [16]. Isaksson [15] also contains a survey of other approaches to estimate R_1 and R_2. See also [17–19].

Another common approach is to use the RLS algorithm (3.21) and (3.22) and adjust the size of the forgetting factor $\lambda(t)$. Several ways to do this can be conceived. Fortesque *et al.* have devised one method that is based on monitoring the residual variance $e^2(t)$, ($e(t)$ defined in (3.7)). When this increases, $\lambda(t)$ is decreased. From (3.23) we see that methods to adjust $\lambda(t)$ can be seen as ways to estimate the 'size' of $R_1(t)$, while direction information is neglected.

A third family of approaches that can be seen as adjustments or selections of $R_1(t)$ can be summarized under the name 'directional forgetting'. The prime idea is then to select $\hat{R}_1(t)$ in (3.9) not based on estimates of $R_1(t)$ but as a means to keep $P(t)$ well-conditioned. One interpretation is that we forget information only in the 'direction' where the new one is obtained. Examples of such strategies are given in [21–24].

3.4 Algorithms for tracking abruptly changing parameters

3.4.1 Formulation

A typical situation may be that the dynamics remain constant for a while and suddenly go through a change at a random time instant. To capture this we may describe $\boldsymbol{\theta}_0(t)$ as

$$\boldsymbol{\theta}_0(t) = \boldsymbol{\theta}_0(t - 1 + \mathbf{r}(t) \tag{3.32}$$

$$\mathbf{r}(t) = \begin{cases} \mathbf{0} & \text{with probability } 1 - \gamma^2 \\ \mathbf{v} & \text{with probability } \gamma^2 \end{cases} \tag{3.33}$$

where $\mathbf{v}$ is a random variable with some distribution. Furthermore, $\mathbf{r}(t)$ and $\mathbf{r}(s)$ are assumed to be independent for $t \neq s$. If $\mathbf{v}$ has zero mean with covariance matrix R_1, $\mathbf{r}(t)$ will have the covariance matrix $\gamma^2 R_1$. This type of behaviour occurs, for example, in signal segmentation problems.

3.4.2 Detection algorithms

One possibility to deal with systems subject to abrupt changes is to use the formulation (3.6)–(3.9). The fundamental problem then is that we do not know the time instants T_1 when the jumps occur. Estimating $R_1(t)$ thus becomes a problem of estimating T_1, which really is a detection problem. Detecting the time instants when the system parameters jump has been discussed extensively [25]. Hägglund [26] has used carefully designed change detection algorithms to supply (3.9) with as correct $\hat{R}_1(t)$ matrices as possible, and Holst and Pouksen [27] discuss how to estimate R_1 at the jumps.

3.4.3 ML-type algorithms

Let us now turn to another way of dealing with abrupt system changes, that is not based on the direct estimation of $R_1(t)$ (or T_1) in (3.9). Consider the formulation (3.32) for sudden changes in the parameters. If $\mathbf{v}$ is described as a Gaussian random variable with zero mean and covariance R_1, we can describe $\mathbf{v}(t)$ as a sequence of Gaussian random variables with covariances $R_1(t)$, where $R_1(t)$ is either 0 or R_1, but we do not know when. We do know, however, that for N data points, the true sequence $R_1(t)$ is one of 2^N possible combinations of 0 and R_1. In principle, we could run all the 2^N possible versions of (3.6)–(3.9), and we would know that the optimal $\hat{\boldsymbol{\theta}}(t)$ would be one of the obtained 2^N variants. How would we know which one? It is reasonable to assume that it would be the one that produced the smallest sum of squared prediction errors, $e(t) = y(t) - \boldsymbol{\varphi}^{\mathrm{T}}(t)\hat{\boldsymbol{\theta}}(t-1)$, $t - 1, \ldots, N$. That would at least be the maximum likelihood estimate among this finite collection of possibilities.

Let us introduce a slight reformulation of the problem (3.32) to the case where

$$\boldsymbol{\theta}_0(t) = \begin{cases} \boldsymbol{\theta}_0(t-1) & w.p. \quad 1 - \gamma^2 \\ \mathbf{v} & w.p. \quad \gamma^2 \end{cases} \tag{3.34}$$

This way of describing the abrupt change will be quite an acceptable alternative to (3.32) in most cases. Gustafsson [28] has shown that under (3.34) the ML estimate of the jump instants can be computed by examining only N (rather than 2^N) of the possible values. Further reductions to a constant number of branches can be obtained at the price of a certain risk of missing the global ML estimate. However, it is always possible to perform a test to ensure that the estimate obtained is indeed the global ML one.

3.5 Algorithms for general non-linear regressions

Most models for dynamical systems can be cast into the form

$$y(t) = \hat{y}(t \mid \boldsymbol{\theta}) + \eta(t) \tag{3.35}$$

where $\hat{y}(t \mid \boldsymbol{\theta})$ is a general function of input–output data and of the parameter vector $\boldsymbol{\theta}$. The notation $\hat{y}$ emphasizes the interpretation of this quality as a predictor. See also section 2.2 of the previous chapter. Ljung [29] contains many examples on how different model descriptions fit into the format (3.35). We note in passing that also multilayered perceptrons (neural networks) are special cases of (3.35). (To see this, we realize that the multilayered perceptron is just a parameterized – $\boldsymbol{\theta}$ being the weighting coefficients in the interconnections – non-linear map from the input layer containing the data $\boldsymbol{\varphi}(t)$ and the output layer $\hat{y}(t \mid \boldsymbol{\theta}) = g(\boldsymbol{\theta}, \boldsymbol{\varphi}(t)$.) See, among many references, [30] for some connections to standard estimation results.

Based on the general model (3.35) we can form a weighted prediction error criterion

$$
\begin{aligned}
V_t(\theta) &= \sum_{k=1}^{t} \beta(t, k) l(e(k, \boldsymbol{\theta}), k) \\
e(t, \boldsymbol{\theta}) &= y(t) - \hat{y}(t \mid \boldsymbol{\theta})
\end{aligned}
\tag{3.36}
$$

(See also Equation (2.11) in Chapter 2.) Here $l(\cdot)$ is a differentiable scalar-valued function that – in some sense – measures the 'size' of the prediction error e.

In the offline case (3.36) is typically minimized by an iterative search, e.g. of the Gauss–Newton type. A basic approach to adaptation is to perform one iteration for the minimization of (3.36) at the same time as one more observation (t increased one unit) is obtained. This approach is detailed in Chapter 11 of [29].

If

$$\beta(t, k) = \prod_{j=k+1}^{t} \lambda(j) \tag{3.37}$$

the resulting algorithm is of the form

$$\hat{\theta}(t) = \hat{\theta}(t-1) + R^{-1}(t)\boldsymbol{\psi}(t) l'_e(e(t), t) \tag{3.38}$$

$$R(t) = \lambda(t)R(t-1) + \boldsymbol{\psi}(t) l''_{ee}(e(t), t)\boldsymbol{\psi}^{\mathrm{T}}(t) \tag{3.39}$$

(See (11.52) of [29].) Here $\boldsymbol{\psi}(t)$ is an approximation of the gradient

$$\boldsymbol{\psi}(t, \hat{\boldsymbol{\theta}}(t-1)) = \frac{\mathrm{d}}{\mathrm{d}\boldsymbol{\theta}} \hat{y}(t \mid \boldsymbol{\theta}) \Big|_{\theta = \hat{\theta}(t-1)} \tag{3.40}$$

and $e(t)$ is an approximation of

$$e(t, \hat{\boldsymbol{\theta}}(t-1)) = y(t) - \hat{y}(t \mid \hat{\theta}(t-1)) \tag{3.41}$$

Moreover, l'_e and l''_{ee} are the derivatives of l with respect to e. In the special case, where $\hat{y}(t \mid \boldsymbol{\theta})$ is a linear regression

$$\hat{y}(t \mid \boldsymbol{\theta}) = \boldsymbol{\varphi}^{\mathrm{T}}(t)\boldsymbol{\theta}$$

and the norm l is quadratic

$$l(e) = e^2$$

we recognize in (3.38)–(3.39) the RLS algorithm.

To put the general model (3.35) more in line with the linear regression case, treated in sections 3.2–3.4, we can make an approximate derivation of a general algorithm as follows.

Consider the general structure (3.35) together with a random walk model for the variation of the 'true parameter vector'

$$
\begin{aligned}
\boldsymbol{\theta}_0(t) &= \boldsymbol{\theta}_0(t-1) + \mathbf{v}(t) \\
y(t) &= \hat{y}(t \mid \boldsymbol{\theta}_0(t)) + e(t)
\end{aligned}
\tag{3.42}
$$

Suppose that we have an approximation $\boldsymbol{\theta}_*(t)$ of $\boldsymbol{\theta}_0(t)$ available. We can then write, using the mean value theorem,

$$\hat{y}(t \mid \boldsymbol{\theta}_0(t)) = \hat{y}(t \mid \boldsymbol{\theta}_*(t)) + (\boldsymbol{\theta}_0(t) - \boldsymbol{\theta}_*(t))^{\mathrm{T}} \boldsymbol{\psi}(t, \xi(t)) \tag{3.43}$$

where $\xi(t)$ is a value 'between' $\boldsymbol{\theta}_*(t)$ and $\boldsymbol{\theta}_0(t)$. Here $\boldsymbol{\psi}(t, \boldsymbol{\theta})$ is the gradient of $\hat{y}(t \mid \boldsymbol{\theta})$, as defined in (3.40). Normally, $\boldsymbol{\psi}(t, \xi(t))$ would not be known, but we may assume that an approximation

$$\boldsymbol{\psi}(t) \approx \boldsymbol{\psi}(t, \xi(t)) \tag{3.44}$$

is available. Introduce the known variable

$$z(t) = y(t) - \hat{y}(t \mid \boldsymbol{\theta}_*(t)) + \boldsymbol{\theta}_*^{\mathrm{T}}(t)\boldsymbol{\psi}(t) \tag{3.45}$$

Subject to the approximation (3.44) we can then rewrite (3.42) as

$$\begin{aligned} \boldsymbol{\theta}_0(t) &= \boldsymbol{\theta}_0(t-1) + \mathbf{v}(t) \\ z(t) &= \boldsymbol{\theta}_0^{\mathrm{T}}(t)\boldsymbol{\psi}(t) + \eta(t) \end{aligned} \tag{3.46}$$

and we are back to the situation of section 3.2.1. A natural choice $\boldsymbol{\theta}_*(t)$ of a good approximation of $\boldsymbol{\theta}_*(t)$ would be the previous estimate $\boldsymbol{\theta}_*(t) = \hat{\boldsymbol{\theta}}(t-1)$. We then obtain algorithms of the recursive prediction error type since

$$z(t) - \hat{\boldsymbol{\theta}}^{\mathrm{T}}(t-1)\boldsymbol{\psi}(t) = y(t) - \hat{y}(t \mid \hat{\boldsymbol{\theta}}(t-1)) \tag{3.47}$$

As $\hat{\boldsymbol{\theta}}(t-1)$ comes closer to $\boldsymbol{\theta}_*(t)$, the approximation involved in going from (3.42) to (3.46) will become arbitrarily good. This shows that an asymptotic theory of tracking parameters in arbitrary model structures can be developed from the linear regression case.

It should also be noted that in the non-linear regression case (3.35), it may be beneficial to let the gain matrix P in (3.9) be affected also by cross-terms that reflect the uncertainty of the estimates of internal 'states'. If the prediction/parameter estimation problem inherent in (3.42) is described by an extended state vector (containing both the system's states and the vector $\boldsymbol{\theta}$) we obtain a description like

$$\begin{pmatrix} \mathbf{x}(t+1) \\ \boldsymbol{\theta}(t+1) \end{pmatrix} = \begin{pmatrix} A(\boldsymbol{\theta}(t))\mathbf{x}(t) \\ \boldsymbol{\theta}(t) \end{pmatrix} + \begin{pmatrix} B(\boldsymbol{\theta}(t))u(t) \\ 0 \end{pmatrix} \begin{pmatrix} \mathbf{v}(t) \\ \mathbf{r}(t) \end{pmatrix} \tag{3.48}$$

$$y(t) = C(\boldsymbol{\theta}(t))\mathbf{x}(t) + e(t) \tag{3.49}$$

The estimation of the extended state

$$\mathbf{X}(t) = \begin{pmatrix} \mathbf{x}(t) \\ \boldsymbol{\theta}(t) \end{pmatrix} \tag{3.50}$$

can now be approached by non-linear filtering techniques, such as the extended Kalman filter. A careful analysis shows that the resulting algorithm for updating $\hat{\boldsymbol{\theta}}(t)$ is of the recursive prediction error family (3.38)–(3.39) (provided the dependence of the 'Kalman gain' on $\boldsymbol{\theta}$ is properly accounted for). See [2] for such a discussion.

However, the filtering approach gives a more complicated expression for $R^{-1}(t) = P(t)$ in (3.39) in that the cross-covariance matrix for $\hat{\mathbf{x}}$ and $\hat{\boldsymbol{\theta}}$ also enters. While these terms have no asymptotic effect as the gain tends to zero, they may very well have positive transient effects.This still has to be carefully analyzed.

3.6 Asymptotic properties of the decreasing gain case

The actual use of the adaptive algorithms is to track the time-varying properties of a system or a signal. Still, a natural first question is to ask how well the algorithms are capable of handling a time-invariant system. This corresponds to the special case $R_1(t) = \hat{R}_1(t) = 0$ in (3.9) and (3.4) or $\lambda(j) \equiv 1$ in (3.20) or (3.37).

A substantial part of [2] is devoted to such analysis, and here we shall only quote the bottom lines:

1. A recursive prediction error algorithm (3.38) will, as t tends to infinity and the gain tends to zero, converge to a local minimum of the expected loss function

$$\bar{V}(\boldsymbol{\theta}) = E[l(e(t, \boldsymbol{\theta}), t)] \tag{3.51}$$

   i.e.

$$\hat{\boldsymbol{\theta}}(t) \rightarrow \arg\min \bar{V}(\boldsymbol{\theta}) \quad \text{w.p. 1 as } t \rightarrow \infty \tag{3.52}$$

2. If, in addition, the Gauss–Newton search direction (3.39) is used, and asymptotically equal weighting is used ($\lambda(j) \equiv 1$), then the asymptotic accuracy

$$\bar{P} = \lim_{t \rightarrow \infty} tE[(\hat{\boldsymbol{\theta}}(t) - \boldsymbol{\theta}_0)(\hat{\boldsymbol{\theta}}(t) - \boldsymbol{\theta}_0)^{\mathrm{T}}]$$

   will be the same as for the corresponding offline estimation method.

These asymptotic properties are thus the best one could ask for. It remains,though, to study how the algorithms actually can cope with time-varying systems. This is the question we turn to next.

3.7 Tracking ability of the algorithms

In the analysis of the tracking ability we will only study algorithms for linear regressions. We first develop an exact expression for the parameter error.

Let us consider the description (3.3)–(3.4) for the behaviour of the true system together with the generic parameter estimation algorithm (3.6) and (3.8)

$$\theta_0(t+1) = \theta_0(t) + \gamma\mathbf{r}(t) \tag{3.53}$$

$$y(t) = \varphi^{\mathrm{T}}(t)\boldsymbol{\theta}_0(\mathrm{t}) + \eta(\mathrm{t}) \tag{3.54}$$

$$\hat{\theta}(t) = \hat{\theta}(t-1) + \mathbf{w}(t)e(t) \tag{3.55}$$

$$e(t) = y(t) - \boldsymbol{\varphi}^{\mathrm{T}}(t)\hat{\theta}(t-1) \tag{3.56}$$

Introduce the parameter error

$$\tilde{\theta}(t) = \hat{\theta}(t) - \theta_0(t+1) \tag{3.57}$$

Remark The variable γ is used to treat scaling of the parameter changes easily. The time indexing here may seem somewhat peculiar, but it will simplify the expressions to follow. From an expression for the covariance of $\tilde{\theta}(t)$ we can exactly derive, for example, the covariance of $\hat{\theta}(t) - \theta_0(t)$. □

Then

$$\tilde{\theta}(t) = [I - \mathbf{w}(t)\boldsymbol{\varphi}^{\mathrm{T}}(t)]\tilde{\theta}(t-1) + \mathbf{w}(t)\eta(t) - \gamma\mathbf{r}(t) \tag{3.58}$$

The parameter error thus obeys a linear, time-varying difference equation. Notice that the $\mathbf{w}(t)$ is always of the form

$$\mathbf{w}(t) = P(t)\boldsymbol{\varphi}(t) \tag{3.59}$$

for some matrix $P(t)$. Solving (3.58) gives

$$\tilde{\theta}(t) = \Phi(t, 0)\tilde{\theta}(0) + \sum_{k=1}^{t} \Phi(t, k)[P(k)\boldsymbol{\varphi}(k)\eta(k) - \gamma\mathbf{r}(k)] \tag{3.60}$$

where

$$\Phi(t, k) = \prod_{j=k}^{t} [I - P(j)\boldsymbol{\varphi}(j)\boldsymbol{\varphi}^{\mathrm{T}}(j)] \tag{3.61}$$

Expressions (3.58) and (3.60) form the basis for all analysis of the performance of the algorithm, and they hold for any sequences $\{\boldsymbol{\varphi}(t)\}$, $\{\eta(t)\}$ and $\{\mathbf{r}(t)\}$. The difficulty in the analysis lies in the complicated expression for $\Phi(t, k)$. Its properties depend entirely on the sequence $\{\boldsymbol{\varphi}(t)\}$, but they are inherited in a fairly complicated way. We shall be interested in the properties of $\tilde{\theta}(t)$ as the gain $\mathbf{w}(t)$ becomes small. We therefore write

$$\mathbf{w}(t) = \mu P_t \boldsymbol{\varphi}(t) \tag{3.62}$$

where μ is a positive scaling parameter (see (3.10)), and obtain

$$\tilde{\theta}(t) = [I - \mu P_t \boldsymbol{\varphi}(t)\boldsymbol{\varphi}^{\mathrm{T}}(t)]\tilde{\theta}(t-1) + \mu P_t \boldsymbol{\varphi}(t)\eta(t) - \gamma\mathbf{r}(t) \tag{3.63}$$

The quantity that we are interested in is the size of the error $\tilde{\theta}(t)$ as measured by the covariance matrix

$$\Pi(t) = E[\tilde{\theta}(t)\tilde{\theta}^{\mathrm{T}}(t)] \tag{3.64}$$

Here expectation E is over $\{\eta(\mathrm{t})\}$, $\{\mathbf{r}(t)\}$ as well as over any random components of $\{\boldsymbol{\varphi}(t)\}$. The exact expression for $\Pi(t)$ follows a somewhat complex equation. Our goal is to show that $\Pi(t)$ is well approximated by $\hat{\Pi}(t)$, defined by

$$\hat{\Pi}(t) = [I - \mu \bar{P}_t Q(t)]\hat{\Pi}(t-1)[I - \mu \bar{P}_t Q(t)] + \mu^2 \bar{P}_t Q(t) \bar{P}_t \cdot R_2(t) + \gamma^2 R_1(t) \tag{3.65}$$

$$\hat{\Pi}(t_0) = \Pi(t_0) \tag{3.66}$$

Here

$$\bar{P}_t = [EP_t] \tag{3.67}$$

$$Q(t) = E[\boldsymbol{\varphi}(t)\boldsymbol{\varphi}^{\mathrm{T}}(t)] \tag{3.68}$$

$$R_1(t) = E[\mathbf{r}(t)\mathbf{r}^{\mathrm{T}}(t)] \tag{3.69}$$

$$R_2(t) = E[\eta^2(t)] \tag{3.70}$$

In essence, (3.65) is obtained from (3.63) by squaring it and applying expectation, neglecting certain dependencies between random variables.

There are several possibilities to establish that Π and $\hat{\Pi}$ are close, and in the next four sections we shall show one fairly straightforward way to do so.

Before that, however, let us briefly discuss the implications of the expression (3.65). There is a substantial number of papers that discuss such implications, e.g. [31–34]. We shall only comment on the case of RLS with forgetting factor $\lambda = 1 - \mu$. This gives

$$\bar{P}_t = \bar{P} = Q^{-1}$$

$$Q(t) = Q$$

$$\hat{\Pi}(t) = \hat{\Pi}(t-1) - 2\mu\hat{\Pi}(t-1) + \mu^2\hat{\Pi}(t-1) + \mu^2 Q^{-1} \cdot R_2 + \gamma^2 R_1 \tag{3.71}$$

As $t \to \infty$ we find that

$$\hat{\Pi}(t) \to \hat{\Pi}$$

where

$$\hat{\Pi} = \frac{1}{2}\left(\mu Q^{-1} R_2 + \frac{\gamma^2}{\mu} R_1\right) \tag{3.72}$$

(neglecting the term $\mu^2\hat{\Pi}$ which for small μ is of an order of magnitude less than the other terms).

This expression shows clearly the trade-off in the choice of step size (adaptation gain) μ (or forgetting factor $\lambda = 1 - \mu$). A small μ gives a small influence from the noise $\{e(t)\}$ in the term $\mu Q^{-1} R_2$ and a large tracking error from the term $(\gamma^2/\mu)R_1$, and vice versa for a large μ.

Other specific algorithms, such as LMS, show similar trade-offs. We may note, in the general case, as t tends to infinity,that $\hat{\Pi}(t)$ will converge to the solution $\hat{\Pi}$ of

$$\bar{P}Q\hat{\Pi} + \Pi Q\bar{P} = \mu\bar{P}Q\bar{P}R_2^0 + \frac{\gamma^2}{\mu} R_1^0 \tag{3.73}$$

(where we assume $\bar{P}$, Q, R_1 and R_2 to be time invariant). If $P(t)$ obeys (3.9) and $\hat{R}_1(t) = \mu^2 \hat{R}_1$ is small and constant (or averages around such a value), similar arguments will show that $P(t) \approx \bar{\bar{P}}$ for small μ where

$$\bar{\bar{P}} = \bar{\bar{P}} - \frac{\bar{\bar{P}} Q \bar{\bar{P}}}{\hat{R}_2 + \mu \bar{\bar{P}} Q} + \mu^2 \hat{R}_1$$

If we scale $P(t)$ as in (3.59), (3.62),

$$P_t = \mu P(t)$$

we find that $P_t \approx \bar{P}$, which, neglecting small terms (tr $\bar{\bar{P}}Q$ is neglected compared with R_2), is given by

$$\bar{P} Q \bar{P} = \hat{R}_1 \hat{R}_2 \tag{3.74}$$

We refer to the references mentioned above for further discussion. In section 3.11 we will develop expressions like (3.72) for the error in the estimated frequency functions of linear systems. These are more transparent in the general case.

3.8 A useful lemma

We first give a technical lemma that is useful for the analysis of the tracking capabilities of the algorithms we have studied.

LEMMA 3.8.1
Let $\Pi(t)$ be defined by

$$\Pi(t) = A_\mu \Pi(t-1) A_\mu^{\mathrm{T}} + \mu x + \rho(t) \qquad \Pi(0) = \Pi_0 \tag{3.75}$$

where A_μ is a stable matrix

$$|A_\mu^t| \leqslant C_A (1 - \alpha\mu)^{t/2} \qquad \alpha > 0 \tag{3.76}$$

and

$$|\rho(t)| \leqslant \mu \cdot \sigma(\mu)(|x| + C_\Pi \max_{t-\tau \leqslant k \leqslant t} |\Pi(k)|) \tag{3.77}$$

for some decreasing non-negative function $\sigma(\mu)$. Let $\hat{\Pi}(t)$ be defined as in (3.75) but without the term $\rho(t)$. Let $\mu_0 > 0$ be defined by $\sigma(\mu_0) \leqslant \frac{1}{2} C_A^2 C_\Pi / \alpha$. Then for $\mu \leqslant \mu_0$

$$|\Pi(t) - \hat{\Pi}(t)| \leqslant C^* \sigma(\mu) |x| + C_* \sigma(\mu)[\sigma(\mu) + \mu t (1 - \alpha\mu)^{t-\tau}] |\Pi_0| \tag{3.78}$$

where

$$C^* = \frac{C_A^2}{\alpha}\left(1 + 4 C_\Pi \frac{C_A^2}{\alpha}\right) \tag{3.79}$$

$$C_* = C_A^4 C_\Pi [1 + 2(C_A/\alpha)^2 C_\Pi] \tag{3.80}$$

Remarks Note that the 'size' of $\hat{\Pi}(t)$ is

$$|\hat{\Pi}(t)| \sim |x| + (1 - \alpha\mu)^t |\Pi_0| \tag{3.81}$$

so (3.78) tells us that the relative approximation of $\Pi(t)$ by $\hat{\Pi}(t)$ improves with the factor $\sigma(\mu)$ as μ decreases.

Proof Let

$$\tilde{\Pi}(t) = \Pi(t) - \hat{\Pi}(t) \tag{3.82}$$

$$\Pi_s(t) = \Pi(t) - A_\mu^t \Pi_0 (A_\mu^{\mathrm{T}})^t \tag{3.83}$$

$$\hat{\Pi}_s(t) = \hat{\Pi}(t) - A_\mu^t \Pi_0 (A_\mu^{\mathrm{T}})^t \tag{3.84}$$

Also define

$$m(t) = \max_{t-\tau \leqslant k \leqslant t} |\Pi(k)| \tag{3.85}$$

$$\tilde{m}(t) = \max_{k \leqslant t} |\Pi_s(\mathrm{k})| \tag{3.86}$$

Then

$$m(t) \leqslant \tilde{m}(t) + C_A^2(1 - \mu\alpha)^{t-\tau} |\Pi_0| \tag{3.87}$$

$$\tilde{\Pi}(t) = \Pi_s(t) - \hat{\Pi}_s(t) \tag{3.88}$$

Thus

$$\tilde{m}(t) \leqslant \max_{k \leqslant t} |\hat{\Pi}_s(\mathrm{k})| + \max_{k \leqslant t} |\tilde{\Pi}(k)| \tag{3.89}$$

Now,

$$\begin{aligned} |\tilde{\Pi}(t)| &= \left| \sum_{k=1}^{t} A_\mu^{t-k} \rho(k) (A_\mu^{\mathrm{T}})^{t-k} \right| \\ &\leqslant \sum_{k=1}^{t} C_A^2 (1 - \alpha\mu)^{t-k} \sigma(\mu) (\mu \cdot |x| + C_\Pi \cdot \mu \cdot \tilde{m}(k) \\ &\quad + C_\Pi \mu \cdot C_A^2 (1 - \alpha\mu)^{k-\tau} |\Pi_0|) \end{aligned} \tag{3.90}$$

using (3.77) and (3.87). Thus

$$|\tilde{\Pi}(t)| \leqslant \frac{C_A^2}{\alpha} \sigma(\mu)(|x| + C_\Pi \cdot \tilde{m}(t)) + \sigma(\mu)\mu C_\Pi C_A^4 t (1 - \alpha\mu)^{t-\tau} |\Pi_0| \tag{3.91}$$

Moreover,

$$|\hat{\Pi}_s(\mathrm{t})| = \left| \sum_{k=1}^{t} A_\mu^{t-k} \mu x (A_\mu^{\mathrm{T}})^{t-k} \right| \leqslant \frac{C_A^2}{\alpha} |x| \tag{3.92}$$

We also have that

$$\max_{k \leqslant t} k(1-\alpha\mu)^k \leqslant \frac{1}{\alpha\mu}(1-\alpha\mu)^{\frac{1}{\alpha\mu}} \leqslant \frac{1}{\alpha\mu}$$

(assuming $\tau < 1/\alpha\mu$). Inserting (3.91) and (3.92) into (3.89) gives

$$\tilde{m}(t) \leqslant \frac{C_A^2}{\alpha}|x|(1+\sigma(\mu)) + \frac{C_A^2}{\alpha}\cdot C_\Pi \sigma(\mu)\cdot \tilde{m}(t) + \frac{1}{\alpha}\sigma(\mu)C_\Pi C_A^4|\Pi_0| \tag{3.93}$$

Let μ_0 be defined by

$$\sigma(\mu_0) = \frac{1}{2}\frac{C_A^2}{\alpha}C_\Pi \tag{3.94}$$

Then, for $\mu \leqslant \mu_0$,

$$\tilde{m}(t) \leqslant 2\frac{C_A^2}{\alpha}|x|(1+\sigma(\mu)) + 2\sigma(\mu)C_\Pi C_A^4|\Pi_0|/\alpha \tag{3.95}$$

Inserting this into (3.91) now gives the desired result. □

3.9 The tracking error for M-dependent regressor sequences

To outline the tools for performance analysis we shall study the archetypical algorithm (3.10)

$$\hat{\theta}(t) = \hat{\theta}(t-1) + \mu P_t L(\varphi(t))(y(t) - \varphi^{\mathrm{T}}(t)\hat{\theta}(t-1)) \tag{3.96}$$

The true system is assumed to satisfy (3.3)–(3.4):

$$y(t) = \boldsymbol{\varphi}^{\mathrm{T}}(t)\boldsymbol{\theta}_0(t) + \eta(t) \tag{3.97}$$

$$\boldsymbol{\theta}_0(t+1) = \boldsymbol{\theta}_0(t) + \gamma\mathbf{r}(t) \tag{3.98}$$

We also assume the following:

1. $\{\eta(t)\}$ and $\{\mathbf{r}(t)\}$ are independent sequences of independent random vectors and zero mean:

$$E[\eta^2(t)] = R_2(t) \qquad E[\mathbf{r}(t)\mathbf{r}^{\mathrm{T}}(t)] = R_1(t) \qquad |R_1(t)| + |R_2(t)| \leqslant C_R \tag{3.99}$$

2.

$$P_t \text{ is a bounded, deterministic sequence of matrices} \tag{3.100}$$

3.

$$|P_t L(\boldsymbol{\varphi}(t))\boldsymbol{\varphi}^{\mathrm{T}}(t)| \leqslant C_\varphi \tag{3.101}$$

For this section we also introduce the following assumption.

4.

$$\begin{array}{c}\varphi(t) \text{ and } \varphi(s) \text{ are independent for } |t-s| > M. \\ \text{They are also independent of } \{\eta(t)\} \text{ and } \{\mathbf{r}(t)\}\end{array} \tag{3.102}$$

Let us now consider the expression for the tracking error $\tilde{\boldsymbol{\theta}}(t) = \hat{\boldsymbol{\theta}}(t) - \boldsymbol{\theta}_0(t)$:

$$\tilde{\boldsymbol{\theta}}(t) = [I - \mu P_t L(\boldsymbol{\varphi}(t))\boldsymbol{\varphi}^{\mathrm{T}}(t)]\tilde{\boldsymbol{\theta}}(t-1) + \mu P_t L(\boldsymbol{\varphi}(t))\eta(t) - \gamma\mathbf{r}(t) \tag{3.103}$$

Squaring both sides and taking expectations gives

$$\begin{aligned}\Pi(t) = \Pi(t-1) &- \mu E[P_t L(\boldsymbol{\varphi}(t))\boldsymbol{\varphi}^{\mathrm{T}}(t)\tilde{\boldsymbol{\theta}}(t-1)\tilde{\boldsymbol{\theta}}^{\mathrm{T}}(t-1)] \\ &- \mu E[\tilde{\boldsymbol{\theta}}(t-1)\tilde{\boldsymbol{\theta}}^{\mathrm{T}}(t-1)\boldsymbol{\varphi}(t)L(\boldsymbol{\varphi}(t))^{\mathrm{T}}P_t] \\ &+ \mu^2 E[P_t L(\boldsymbol{\varphi}(t))\boldsymbol{\varphi}^{\mathrm{T}}(t)\tilde{\boldsymbol{\theta}}(t-1)\tilde{\boldsymbol{\theta}}^{\mathrm{T}}(t-1)\boldsymbol{\varphi}(t)L(\boldsymbol{\varphi}(t))^{\mathrm{T}}P_t] \\ &+ E[\mu^2 P_t L(\boldsymbol{\varphi}(t))L(\boldsymbol{\varphi}(t))^{\mathrm{T}}P_t R_2(t)] + \gamma^2 R_1(t)\end{aligned} \tag{3.104}$$

Using

$$|P_t L(\boldsymbol{\varphi}(t))\boldsymbol{\varphi}^{\mathrm{T}}(t)| < C_\varphi$$

gives immediately

$$|\Pi(t) - \Pi(t-1)| \leqslant (2\mu C_\varphi + \mu^2 C_\varphi^2)|\Pi(t-1)| + \mu^2 C_\varphi C_R + \gamma^2 C_R \tag{3.105}$$

Introduce

$$Q = E[P_t L(\boldsymbol{\varphi}(t))\boldsymbol{\varphi}^{\mathrm{T}}(t)] \tag{3.106}$$

$$\tilde{Q} = E[P_t L(\boldsymbol{\varphi}(t))L(\boldsymbol{\varphi}(t))^{\mathrm{T}}P_t] \tag{3.107}$$

Then we can write (3.104) as

$$\begin{aligned}\Pi(t) = (I - \mu Q)\Pi(t-1)(I-\mu Q)^{\mathrm{T}} + \mu^2\tilde{Q}\cdot R_2 + \gamma^2 R_1 \\ - \mu\alpha(t) - \mu\alpha^{\mathrm{T}}(t) + \mu^2\beta(t)\end{aligned} \tag{3.108}$$

where

$$\alpha(t) = E[P_t L(\boldsymbol{\varphi}(t))\boldsymbol{\varphi}^{\mathrm{T}}(t)\tilde{\boldsymbol{\theta}}(t-1)\tilde{\boldsymbol{\theta}}^{\mathrm{T}}(t-1) - Q\Pi(t-1)] \tag{3.109}$$

$$\begin{aligned}\beta(t) = E[&P_t L(\boldsymbol{\varphi}(t))\boldsymbol{\varphi}^{\mathrm{T}}(t)\tilde{\boldsymbol{\theta}}(t-1)\tilde{\boldsymbol{\theta}}^{\mathrm{T}}(t-1)\boldsymbol{\varphi}(t) \\ &\times L(\boldsymbol{\varphi}(t))^{\mathrm{T}}P_t - Q\Pi(t-1)Q]\end{aligned} \tag{3.110}$$

Clearly,

$$|\beta(t)| \leqslant C_\varphi^2 |\Pi(t-1)| \tag{3.111}$$

Now consider $\alpha(t)$. We write

$$\begin{aligned}\tilde{\boldsymbol{\theta}}(t-1) = \psi(t-1, t-M)\tilde{\boldsymbol{\theta}}(t-M) \\ + \sum_{k=t-M}^{t-1} \psi(t-1, k)(\mu P_k L(\boldsymbol{\varphi}(k))\eta(k) - \gamma\mathbf{r}(k))\end{aligned} \tag{3.112}$$

where

$$\psi(t, k) = \prod_{j=k}^{t} [I - \mu P_j L(\varphi(j)]\varphi^T(j)) \tag{3.113}$$

We note that for $t - M \leqslant k \leqslant t - 1$

$$|\psi(t-1, k)| \leqslant (1 + \mu C_\varphi)^{t-k} \leqslant (1 + \mu C_\varphi)^M \triangleq C_\psi(M) \tag{3.114}$$

We now insert (3.112) into (3.109) for both expressions of $\tilde{\theta}(t-1)$. When taking expectations all cross-terms arising from (3.112) disappear since $\eta(k)$ and $\mathbf{r}(k)$ are independent of all the other variables involved there, including $\tilde{\theta}(t-M)$. We thus have

$$\begin{aligned} \alpha(t) = \tilde{\alpha}(t) &+ E[P_t L(\varphi(t))\varphi^T(t)\tilde{\theta}(t-M)\tilde{\theta}^T(t-M)] \\ &- Q\Pi(t-1) + E[P_t L(\varphi(t))\varphi^T(t)(\psi(t-1, t-M) - I) \\ &\times \tilde{\theta}(t-M)\tilde{\theta}^T(t-M)\psi^T(t-1, t-M)] \end{aligned} \tag{3.115}$$

with

$$\begin{aligned} |\tilde{\alpha}(t)| = \Bigg| \sum_{k=t-M}^{t-1} & E[\psi(t-1, k)(\mu^2 P_k L(\varphi(k))\eta^2(k) \\ &\times L(\varphi(k))^T P_k] + \gamma^2 \mathbf{r}^2(k))\psi^T(t-1, k) \Bigg| \leqslant C_\psi^2(M) \cdot C_\alpha(\mu^2 + \gamma^2) \cdot M \end{aligned} \tag{3.116}$$

where $C_\alpha = C_\varphi \cdot C_R$. Note that the second term on the right-hand side of (3.115) equals $Q\Pi(t-M)$, by the independence between $\varphi(t)$ and $\tilde{\theta}(t-M)$. Let us now consider $\psi(t-1, t-M) - I$. By expanding the product in (3.113), subtracting the identity matrix and then reassembling the product it follows that

$$|\psi(t-1, t-M) - I| \leqslant (1 + \mu C_\varphi)^M - 1 \leqslant e^{\mu M C_\varphi} - 1 \leqslant 2C_\varphi \cdot \mu M \tag{3.117}$$

where the last inequality follows for

$$\mu \leqslant 1/C_\varphi M \tag{3.118}$$

For the last term of (3.115) we have that it is bounded by

$$\begin{aligned} & C_\varphi |\psi(t-1, t-M) - I| \cdot |\psi(t-1, t-M)| \cdot |\Pi(t-M)| \\ & \leqslant 2C_\varphi^2 \mu \cdot M \cdot C_\psi(M) \cdot |\Pi(t-M)| \end{aligned} \tag{3.119}$$

using (3.117) and (3.114). Collecting all this gives for (3.115)

$$\begin{aligned} |\alpha(t)| &\leqslant Q|\Pi(t-M) - \Pi(t-1)| \\ &+ \mu \cdot (2M \cdot C_\varphi^2 \cdot C_\psi(\mu))|\Pi(t-M)| \\ &+ (\mu^2 + \gamma^2) C_\psi^2(M) \cdot C_\alpha M \\ &\leqslant \mu \cdot C_1(M)|\Pi(t-M)| + (\mu^2 + \gamma^2) \cdot C_2(M) \end{aligned} \tag{3.120}$$

The last step follows using (3.105) M times. We have also introduced the constants C_1 and C_2, which depend on M as follows:

$$C_1(M) = C_1^* \cdot M \cdot (1 + \mu C_\varphi)^M \tag{3.121}$$

$$C_2(M) = C_2^* \cdot M \cdot (1 + \mu C_\varphi)^{2M} \tag{3.122}$$

Returning to (3.108) we see that all the last three terms are bounded by

$$\mu^2\left(C_1(M) \sup_{0<j\leqslant M} |\Pi(t-j)| + C_2(M)\,\frac{\mu^2+\gamma^2}{\mu}\right) \tag{3.123}$$

We can thus apply Lemma 3.8.1 with $\sigma(\mu) = \mu$ and $x = \tilde{Q}R_2 + (\gamma^2/\mu)R_1$ to conclude the following theorem.

THEOREM 3.9.1
Let $\hat{\Pi}(t)$ be defined by

$$\hat{\Pi}(t) = (I - \mu Q)\hat{\Pi}(t-1)(I - \mu Q)^{\mathrm{T}} + \mu^2\tilde{Q}R_2 + \gamma^2 R_1 \tag{3.124}$$

$$\hat{\Pi}(0) = \Pi_0$$

with Q and $\tilde{Q}$ defined by (3.107). Let $\Pi(t) = E[\tilde{\boldsymbol{\theta}}(t)\tilde{\boldsymbol{\theta}}^{\mathrm{T}}(t)]$ with expectation over $\{\boldsymbol{\varphi}(t)\}$, $\{\eta(t)\}$, $\{\mathbf{r}(t)\}$ and $\boldsymbol{\theta}_0(0)$. Here $\tilde{\boldsymbol{\theta}}(t)$ is the tracking error (3.103). Assume that $Q > \alpha I$ and that (3.99)–(3.102) hold. Then there is a $\mu_0 > 0$ such that for $\mu < \mu_0$

$$|\Pi(t) - \hat{\Pi}(t)| \leqslant C_3\sigma(\mu)\cdot\left(\mu + \frac{\gamma^2}{\mu}\right) + C_4\sigma(\mu)[\sigma(\mu) + \mu(1-\alpha\mu)^{t-M}t]\,|\Pi_0| \tag{3.125}$$

where C_3 and C_4 depend on M (in assumption (3.102)) in the same way as (3.121) and $\sigma(\mu) = \mu$. The constants μ_0, C_3 and C_4 can be explicitly calculated from the bounds in the assumptions. □

Note that

$$|\hat{\Pi}(t)| \approx C\left(\mu + \frac{\gamma^2}{\mu}\right) + (1-\alpha\mu)^t|\Pi_0| \tag{3.126}$$

so the relative degree in the approximation of $\Pi(t)$ by $\hat{\Pi}(t)$ improves like μ.

The equation (3.124) is easy to analyze, as we saw in section 3.7, so the trade-off between noise sensitivity and tracking ability can be easily analyzed in terms of this equation. Many studies of this character have been published. See, among many references [31–35].

3.10 The tracking error for mixing regressor sequences

Suppose now that we relax assumption (3.102) in the following way:

4.

$$\begin{aligned}&\text{The sequence of vectors } \{\boldsymbol{\varphi}(t)\} \text{ is } \phi\text{-mixing}\\&\text{with a decaying dependence } \phi(M).\\&\{\boldsymbol{\varphi}(t)\} \text{ is also independent of } \{\eta(t)\} \text{ and } \{\mathbf{r}(t)\}\end{aligned} \tag{3.127}$$

Roughly speaking, ϕ-mixing means that $\boldsymbol{\varphi}(t)$ and $\boldsymbol{\varphi}(t+M)$ become more and more independent as M increases. The 'amount of dependence' decreases like the function $\phi(M)$. See e.g. [36] for a formal definition.

Let us go through the calculations in the previous section under this relaxed asssumption. The only change is in (3.120), where we obtain a remainder term

$$|\,E[P_t L(\boldsymbol{\varphi}(t))\boldsymbol{\varphi}(t)\tilde{\boldsymbol{\theta}}(t-M)\tilde{\boldsymbol{\theta}}^{\mathrm{T}}(t-M)] - Q\Pi(t-M)\,| \leqslant C_{\varphi}\phi(M)$$

using the fact that $\tilde{\boldsymbol{\theta}}(t-M)$ depends only on $\boldsymbol{\varphi}(k)$, $k \leqslant t-M$.

Equation (3.123) will thus continue to hold if we take

$$C_t(M) = C_1^*[1 + \mu(1 + \mu C_{\varphi})^M](\mu \cdot M + \phi(M))/\mu \tag{3.128}$$

Now let

$$\sigma^*(\mu) = \min_M \, (\mu \cdot M + \phi(M)) \tag{3.129}$$

We can thus still apply Lemma 3.8.1, with $\sigma(\mu) = \sigma^*(\mu)$, and conclude that (3.125) still holds, now with $\sigma(\mu) = \sigma^*(\mu)$ as defined above.

If the dependence between the regressors decreases exponentially, i.e. $\phi(m) = \lambda^m$, $\lambda < 1$, we find that $\sigma^*(\mu)$ decays with μ like

$$\sigma^*(\mu) \sim \mu/\log \mu$$

which is almost as good as in the M-dependent case. (See [37] for more details of this result.)

3.11 Evaluation of the error in the frequency domain

The expressions for the mean square error that we derived in the previous section are somewhat implicit. In [38] and [39], explicit expressions for the mean square error of a corresponding transfer function estimate were derived. The results can be summarized as follows. Consider an FIR model where $\boldsymbol{\varphi}(t)$ contains only lagged inputs

$$y(t) = \boldsymbol{\varphi}^{\mathrm{T}}(t)\boldsymbol{\theta} = \sum_{k=1}^{d} g_k u(t-k) \tag{3.130}$$

The corresponding transfer function then is

$$G(\mathrm{e}^{\mathrm{j}\omega}) = \sum_{k=1}^{d} g_k \,\mathrm{e}^{\mathrm{j}k\omega} = \mathbf{W}_d^*(\omega)\boldsymbol{\theta} \tag{3.131}$$

where

$$\mathbf{W}_d(\omega) = [e^{j\omega} \ldots e^{dj\omega}]^T \tag{3.132}$$

and where '*' denotes transpose and complex conjugate.

The mean square error of the transfer function estimate at frequency ω is then

$$\pi_d(\omega) = \mathbf{W}_d^*(\omega)\hat{\Pi}\mathbf{W}_d(\omega) \tag{3.133}$$

where $\hat{\Pi}$ is the mean square error matrix for the parameters, as derived in sections 3.7–3.10. The key properties to be used are as follows.

Let A and B be $d \times d$ Toeplitz-like matrices that satisfy some regularity conditions, see [39]. We can then define the scalar functions $a(\omega)$ and $b(\omega)$ by

$$\frac{1}{d}\mathbf{W}_d^*(\omega)A\mathbf{W}_d(\omega) \to a(\omega) \text{ as } d \to \infty \tag{3.134}$$

and

$$\frac{1}{d}\mathbf{W}_d^*(\omega)B\mathbf{W}_d(\omega) \to b(\omega) \text{ as } d \to \infty \tag{3.135}$$

Furthermore, it can be shown that

$$\frac{1}{d}\mathbf{W}_d^*(\omega)AB\mathbf{W}_d(\omega) \to a(\omega)b(\omega) \text{ as } d \to \infty \tag{3.136}$$

and

$$\frac{1}{d}\mathbf{W}_d^*(\omega)A^{-1}\mathbf{W}_d(\omega) \to \frac{1}{a(\omega)} \text{ as } d \to \infty \tag{3.137}$$

When applying this operation to the covariance matrix

$$Q = E[\boldsymbol{\varphi}(t)\boldsymbol{\varphi}^T(t)]$$

with

$$\boldsymbol{\varphi}^T(t) = (u(t-1), \ldots, u(t-d))$$

(cf. (3.68)) we get

$$\frac{1}{d}\mathbf{W}_d^*(\omega)Q\mathbf{W}_d(\omega) \to \Phi_u(\omega) \text{ as } d \to \infty \tag{3.138}$$

where $\Phi_u(\omega)$ is the spectrum of the input $\{u(t)\}$.

We are now going to apply these results to the general expression (3.73) by evaluating

$$\bar{\pi}(\omega) = \lim_{d\to\infty} \frac{1}{d}\pi_d(\omega) = \lim_{d\to\infty} \frac{1}{d}\mathbf{W}_d^*(\omega)\hat{\Pi}\mathbf{W}_d(\omega) \tag{3.139}$$

as the order, d, of the FIR model (3.130) tends to infinity. For large-order models we will thus have that the mean square error of the transfer function estimate at frequency ω is given by

$$\pi_d(\omega) \approx \mathrm{d} \cdot \bar{\pi}(\omega) \tag{3.140}$$

Introduce

$$\begin{aligned} p(\omega) &= \lim_{d \to \infty} \frac{1}{d} \mathbf{W}_d^*(\omega) P \mathbf{W}_d(\omega) \\ \hat{r}_1(\omega) &= \lim_{d \to \infty} \frac{1}{d} \mathbf{W}_d^*(\omega) \hat{R}_1 \mathbf{W}_d(\omega) \\ r_1^0(\omega) &= \lim_{d \to \infty} \frac{1}{d} \mathbf{W}_d^*(\omega) R_1^0 \mathbf{W}_d(\omega) \end{aligned} \tag{3.141}$$

(Recall that the normalization is such that the actual parameter change covariance matrix is $\gamma^2 R_1^0$ and that the corresponding assumed covariance in (3.9) is $\mu^2 \hat{R}_1$.)

From (3.74) we then find, by applying the limiting procedure to both members

$$p^2(\omega)\Phi_u(\omega) = \hat{r}_1(\omega) \tag{3.142}$$

or

$$p(\omega) = \sqrt{\left(\frac{\hat{r}_1(\omega)}{\Phi_u(\omega)}\right)} \tag{3.143}$$

Similarly, (3.73) gives

$$2p(\omega)\Phi_u(\omega)\bar{\pi}(\omega) = \mu R_2^0 \hat{r}_1(\omega) + \frac{\gamma^2}{\mu} r_1^0(\omega) \tag{3.144}$$

or

$$\bar{\pi}(\omega) = \frac{1}{2}\sqrt{\left(\frac{\hat{r}_1(\omega)}{\Phi_u(\omega)}\right)}\left(\mu \cdot R_2^0 + \frac{\gamma^2}{\mu}\frac{r_1^0(\omega)}{\hat{r}_1(\omega)}\right) \tag{3.145}$$

Expressions (3.140) and (3.145) give an explicit and useful description of how the accuracy of the estimate varies with frequency and with the design variables $\hat{r}_1(\omega)$ and μ.

It is easy to minimize (3.145) explicitly with respect to these variables, and this gives, as it should (if $R_2^0 = 1$),

$$\hat{r}(\omega) = r_1^0(\omega) \qquad \mu = \gamma \tag{3.146}$$

We also obtain for the LMS algorithm from (3.145) with $p(\omega) \equiv 1$

$$\bar{\pi}(\omega) = \frac{1}{2}\left(\mu \cdot R_2^0 + \frac{\gamma^2}{\mu}\frac{r_1^0(\omega)}{\Phi_u(\omega)}\right) \tag{3.147}$$

and for the RLS algorithm

$$\bar{\pi}(\omega) = \frac{1}{2}\left(\mu \frac{R_2^0}{\Phi_u(\omega)} + \frac{\gamma^2}{\mu}\cdot r_1^0(\omega)\right). \tag{3.148}$$

The results (3.145)–(3.148) thus describe how the basic recursive identification algorithm performs under small gain and under steady parameter drift. A further discussion of these aspects in contained in [39].

3.12 Conclusions

We have outlined how to approach the problem of deriving or constructing adaptation algorithms for tracking time-varying systems. We have, among other things, stressed how the Kalman filter provides a natural starting point for the derivations. We have also stressed how common *ad hoc* approaches can be interpreted as special cases corresponding to specific assumptions about the behaviour of the true parameters.

The analysis of the tracking ability of adaptation algorithms is of great interest. We have shown the archetypical result where the true covariance matrix for the parameter error can be approximated by an expression that is simpler to study. This study brings out the basic trade-off between tracking ability and noise sensitivity. We have shown how this trade-off becomes especially explicit when evaluated in the frequency domain for linear systems and models.

References

[1] L. Ljung and S. Gunnarsson, 'Adaptive tracking in system identification – A survey', *Automatica*, vol. 26, pp. 7–22, 1990.

[2] L. Ljung and T. Söderström, *Theory and Practice of Recursive Identification*, MIT Press: Cambridge, MA, 1983.

[3] A. Jazwinski, *Stochastic Process and Filtering Theory*, vol. 64 of *Mathematics in Science and Engineering*, Academic Press: New York, 1970.

[4] B.D.O. Anderson and J.B. Moore, *Optimal Filtering*, Prentice Hall: Englewood Cliffs, NJ, 1979.

[5] A.P. Korostelev, 'Multistep procedures of stochastic optimization', *Avtomatika i Telemekhanika*, (5), pp. 82–90, 1981.

[6] S.V. Shilman and A.I. Yastrebov, 'Convergence of a class of multistep stochastic adaptation algorithms', *Avtomatikha i Telemekhanika*, vol. 8, pp. 111–18, 1976.

[7] S.V. Shilman and A.I. Yastrebov, 'Properties of a class of multistep gradient and pseudogradient algorithms of adaptation and learning', *Avtomatikha i Telemekhanika*, vol. 4, pp. 95–104, 1978.

[8] A. Benveniste, 'Design of adaptive algorithms for the tracking of time-varying systems', *Int. J. Adaptive Control Signal Process.*, vol. 1, pp. 3–29, 1987.

[9] B. Widrow and S. Stearns, *Adaptive Signal Processing*, Prentice Hall: Englewood Cliffs, NJ, 1985.

[10] R.K. Mehra, 'On the identification of variances and adaptive Kalman filtering', *IEEE Trans. Autam. Control*, vol. AC-15, pp. 175–84, 1970.

[11] J.C. Shellenbarger, 'Estimation of covariance parameter for an adaptive Kalman filter', *Proc. National Electronics Conf., 1966*, pp. 698–702.

[12] J.C. Shellenbarger, 'A multivariance learning technique for improved dynamics system performance', *Proc. National Electronics Conf., 1967*, pp. 146–51.

[13] P.R. Belanger, 'Estimation of noise covariance matrices for a linear time-varying stochastic process', *Automatica*, vol. 10, pp. 267–75, 1974.

[14] A. Isaksson, 'Identification of time-varying systems through adaptive Kalman filtering', *Preprints 10th IFAC World Congress, Munich, 1987*, pp. 306–11.

[15] A. Isaksson, 'On system identification in one and two dimensions with signal processing applications', PhD Thesis, Department of Electrical Engineering, Linköping University, 1988.

[16] J.G. Wang and Z.L. Deng, 'Simulation of a newly designed adaptive controller', *IFAC Symp. on Simulation of Control Systems, Vienna, 1986*, pp. 93–106.

[17] I.M. Weiss, 'A survey of discrete Kalman–Bucy filtering with unknown noise covariances', *AIAA Guidance, Control and Flight Mechanics Conf., 1970*.

[18] A.P. Sage and G.W. Husa, 'Adaptive filtering with unknown prior statistics', *Proc. 1969 Joint Automatic Control Conf.*, pp. 760–9, 1969.

[19] A.P. Sage and G.W. Husa, 'Algorithms for sequential adaptive estimation of prior statistics', *Proc. 8th IEEE Symp. on Adaptive Processes, Pennsylvania State University, University Park 1969.*

[20] T.R. Fortesque, L.S. Kershenbaum and B.F. Ydstie, 'Implementation of self-tuning regulators with variable forgetting factors', *Automatica*, vol. 17, pp. 831–5, 1981.

[21] R. Kulhavy and M. Karny, 'Tracking of slowly varying parameters by directional forgetting', *Proc. 9th IFAC World Congress, Budapest, 1984*, vol. X, pp. 78–83.

[22] T. Hägglund, 'New estimation techniques for adaptive control', PhD Thesis, Department of Automatic Control, Lund University, 1983.

[23] R. Kulhavy, 'Restricted exponential forgetting in real-time identification', *Automatica*, vol. 23, pp. 589–600, 1987.

[24] G.C. Goodwin, M.E. Salgado and R.H. Middleton, 'Modified least squares algorithm incorporation resetting and forgetting', *Int. J. Control*, vol. 47, pt 2, pp. 477–91, 1988.

[25] M. Basseville and A. Benveniste, *Detection of Abrupt Changes in Signals and Dynamical Systems*, Lecture Notes in Control and Information Sciences, Springer: Berlin, 1986.

[26] T. Hägglund, 'Adaptive control of systems subject to large parameter changes', *Proc. 9th IFAC World Congress, Budapest, Hungary, 1984*, pp. 993–8.

[27] J. Holst and N.K. Poulsen, 'Self tuning control of plants with abrupt changes', *Proc. 9th IFAC World Congress, Budapest, 1984*, vol. VII, pp. 144–9.

[28] F. Gustafsson, 'Optimal segmentation of linear regression parameters', Tekn. lic. Thesis, Department of Electrical Engineering, Linköping University, 1990.

[29] L. Ljung, *System Identification – Theory for the User*, Prentice Hall, Englewood Cliffs, NJ, 1987.

[30] J. Sjöberg and L. Ljung, 'Overtraining, regularization, and searching for minimum in neural networks', *Proc. Symp. on Adaptive Systems in Control and Signal Processing, Grenoble, 1992.*

[31] B. Widrow, J.M. McCool, M.G. Larimore and C.R. Johnson Jr, 'Stationary and nonstationary learning characteristics of the lms adaptive filter', *Proc. IEEE*, vol. 64, pp. 1151–62, 1976.

[32] D.C. Farden, 'Tracking properties of adaptive signal processing algorithms', *IEEE Trans. Acoust, Speech, Signal Process.* vol. ASSP-29, pp. 439–46, 1981.

[33] O. Macchi and E. Eweda, 'Second-order convergence analysis of stochastic adaptive linear filtering', *IEEE Trans. Autom. Control*, vol. AC-28, pp. 76–85, 1983.

[34] W. Gardner, 'Nonstationary learning, characteristics of the LMS algorithms: A general study, analysis and critique', *IEEE Trans. Circuits Syst.*, vol. CAS-34, pp. 1199–207, 1987.

[35] W. Gardner, 'Learning characteristics of stochastic-gradient-descent algorithms', *Signal Process.*, (6), pp. 113–33, 1984.

[36] P. Hall and C.C. Heyde, *Martingale Limit Theory and Its Applications*, Academic Press: New York, 1980.

[37] L. Ljung and P. Priouret, 'A result of the mean square error obtained using general tracking algorithms', *Int. J. Adaptive Control*, vol. 4, pp. 231–50, 1991.

[38] S. Gunnarsson, 'Frequency domain aspects of modeling and control in adaptive systems', PhD Thesis, Department of Electrical Engineering, Linköping University, 1988.

[39] S. Gunnarsson and L. Ljung, 'Frequency domain tracking characteristics of adaptive algorithms', *IEEE Trans. Acoust., Speech, Signal Process.*, vol. ASSP-37, pp. 1072–84, 1989.

4

The Least Mean Square Family

William A. Sethares

In the beginning, Gauss created Least Squares, and he saw that it was good. So he said unto his algorithm, 'be fruitful, and multiply'. And it was so. Least Squares begat the Least Mean Squares (LMS), whose years were plenty. And Least Mean Squares begat Normalized LMS, whose fame was projected throughout the land. And Normalized LMS begat a Leaky LMS, whose offspring were the Signed LMS and the Quantized LMS. And these children of the Sign proliferated.

This is the story of LMS and its children.

4.1 Introduction

In Chapter 2, the least mean square (LMS) algorithm was introduced as a way to adjust recursively the parameters $\theta(k)$ of a linear filter with the goal of minimizing the error between a given desired signal and the output of the linear filter. LMS is one of many related algorithms which are appropriate for this task and a whole family of algorithms has been developed which can address a variety of problem settings, computational restrictions, and minimization criteria. This chapter begins by deriving the LMS as an instantaneous approximation to the steepest descent minimization of a cost function $V(\theta(k))$, which results in a simple recursive scheme of the form

$$\left\{\begin{matrix}\text{new}\\ \text{parameter}\\ \text{estimate}\end{matrix}\right\} = \left\{\begin{matrix}\text{old}\\ \text{parameter}\\ \text{estimate}\end{matrix}\right\} + \{\text{step size}\}\left\{\begin{matrix}\text{new}\\ \text{information}\end{matrix}\right\} \tag{4.1}$$

where the new information is a function of the past and present inputs (often concatenated into a vector called the *regressor* vector), as well as the error between the output of the linear filter and the desired signal. Some general observations are made regarding the positive (and negative) aspects of the performance of LMS, and a variety of 'children of LMS' are introduced as attempts to alleviate certain problems or to fine-tune some aspect of the algorithm's performance. These algorithms have the general recursive form

$$\boldsymbol{\theta}(k) = \boldsymbol{\theta}(k-1) + \mu(k-1)F(\boldsymbol{\phi}(k-1))g(e(k)) \tag{4.2}$$

where $\boldsymbol{\theta}(k)$ represents the new parameter estimate at time k, $F(\cdot)$: $\Re^m \rightarrow \Re$ and $g(\cdot)$: $\Re \rightarrow \Re$ are functions of the regressor vector $\boldsymbol{\phi}(k)$ and the error signal $e(k)$, respectively, and the step size $\mu(k-1)$ is a parameter chosen by the user that may vary with time.

All of the algorithms of the 'LMS family' are special cases of (4.2). This usage is somewhat unfortunate, however, because many of these variants of the LMS algorithm do not actually minimize the least mean square error. The signed-error variant, for instance, tends to minimize the absolute value of the error. With a leakage parameter, the algorithm tends to minimize a linear combination of the least mean square error and the squared error away from some nominal $\boldsymbol{\theta}^0$. Other variants do not admit a minimization interpretation at all. A large body of literature has been devoted to the analysis of the behaviour of the various members of the LMS family. Sections 4.3–4.5 discuss three major branches of this investigation.

The first of these methods, an attempt to develop a 'statistical theory of adaptation' was initiated by Widrow and his co-authors in [1,2]. By examining the expected behaviour of the algorithm under a variety of assumptions on the input and desired signals, they develop useful guidelines for implementing the algorithms (including optimal choice of step size), and derive expressions describing the statistical performance of the algorithms.

A second analytical technique, the 'deterministic approach', treats the parameter error update equation as a non-linear dynamical system, and uses the tools of stability theory to examine the convergence and stability characteristics of the algorithms. Observe that (4.2) achieves an 'averaged equilibrium' whenever $\text{avg}\{F(\boldsymbol{\phi})g(e)\} = 0$. Assuming that there is some ideal system $\boldsymbol{\theta}^*$ that can exactly match the dynamics of the desired signal, suitable conditions (known as 'persistence of excitation' conditions) on the character of the input sequences can be derived which imply that the parameter estimates $\theta(k)$ converge to this ideal θ^*, and that this convergence occurs in a neighbourhood of the desired equilibrium. Interestingly, the PE conditions are different for the various members of the LMS family. For instance, there are inputs for which the signed-error algorithm converges yet the signed-regressor variant diverges. The ideas of total stability extend these convergence/divergence results to the more realistic 'non-ideal' scenario, when disturbances are present; that is, when the desired input–output mapping cannot be represented exactly as a linear system parameterized by θ^*.

The third analytical technique is the 'stochastic approximation' approach, pioneered by L. Ljung in [3], in which the parameter error update equation is examined indirectly by studying a related ordinary differential equation (ODE). In particular, local stability of the ODE implies weak convergence of the algorithm. Though the analysis in [3] requires a vanishing step size (where $\mu \to 0$ as time $\to \infty$), this restriction may be removed as shown in [4] and [5]. Relating the motion of the parameter error to an associated *forced* differential equation allows the convergent (stationary) distributions to be expressed in a concrete manner. Think of it this way: the generic behaviour of LMS and its variants is that the parameter estimates converge to a region about their final value, and then 'rattle around' this value as a result of unavoidable noises and disturbances. The beauty of the stochastic approximation approach is that this rattling behaviour can often be described in terms of specific probability distributions.

While other ways of understanding the various members of the LMS family exist, these three are (so far) the most widely known and the most powerful. The analytical techniques are complementary, and each offers unique insights into the behaviour and performance of the algorithms. Section 4.6 shows how to apply the various analytical techniques to the children of LMS, and section 4.7 wraps up the discussion by posing a number of open questions.

4.2 LMS and its children

The least mean square (LMS) algorithm, popularized by Widrow [1,2], has become one of the standard techniques of adaptive filtering. The LMS algorithm is a form of steepest (or gradient) descent that attempts to minimize a cost function $V(\theta(k))$ at each time step k by a suitable choice of the parameter vector $\theta(k)$. The strategy is to update the parameter estimate proportional to the instantaneous gradient value $\mathrm{d}V(\theta(k-1))/\mathrm{d}\theta(k-1)$, that is,

$$\theta(k) = \theta(k-1) - \mu \frac{\mathrm{d}V(\theta(k-1))}{\mathrm{d}\theta(k-1)} \tag{4.3}$$

where μ is a small positive step size, and the minus sign ensures that the parameter estimates descend (rather than climb) the error surface. (Throughout this chapter, the time index k is used for discrete time processes while the variable t is reserved for continuous time processes.)

If the adaptive filter has a linear structure, then its output can be expressed as

$$\hat{y}(k) = \boldsymbol{\phi}^{\mathrm{T}}(k-1)\boldsymbol{\theta}(k-1) \tag{4.4}$$

where $\boldsymbol{\phi}(k-1)$ is a vector of past and present inputs (and possibly past outputs). For instance, given an input sequence $u(k-1)$, the input vector is $\boldsymbol{\phi}(k-1) = (u(k-1), u(k-2), \ldots, u(k-m))^{\mathrm{T}}$. Choosing the cost function to be one-half the square of the error between the output of the adaptive filter and the desired signal

$$V(\boldsymbol{\theta}(k-1)) = \tfrac{1}{2}(y(k) - \hat{y}(k))^2 = \tfrac{1}{2}(y(k) - \boldsymbol{\phi}^{\mathrm{T}}(k-1)\boldsymbol{\theta}(k-1))^2 \tag{4.5}$$

the gradient is

$$\frac{\mathrm{d}V(\boldsymbol{\theta}(k-1))}{\mathrm{d}\boldsymbol{\theta}(k-1)} = -(y(k) - \hat{y}(k))\boldsymbol{\phi}(k) \tag{4.6}$$

The LMS algorithm is then

$$\boldsymbol{\theta}(k) = \boldsymbol{\theta}(k-1) + \mu\boldsymbol{\phi}(k-1)(y(k) - \hat{y}(k)) \tag{4.7}$$

If there is a fixed vector $\boldsymbol{\theta}^*$ such that the desired signal $y(k)$ is generated from a linear system with parameterization $\boldsymbol{\theta}^*$

$$y(k) = \boldsymbol{\phi}^{\mathrm{T}}(k-1)\boldsymbol{\theta}^* + \xi(k) \tag{4.8}$$

where $\xi(k)$ represents the noise component that cannot be modelled in the desired signal, then (4.7) can be rewritten

$$\tilde{\boldsymbol{\theta}}(k) = \tilde{\boldsymbol{\theta}}(k-1) - \mu\boldsymbol{\phi}(k-1)\boldsymbol{\phi}^{\mathrm{T}}(k-1)\tilde{\boldsymbol{\theta}}(k-1) + \mu\boldsymbol{\phi}(k-1)\xi(k) \tag{4.9}$$

$$= [I - \mu\boldsymbol{\phi}(k-1)\boldsymbol{\phi}^{\mathrm{T}}(k-1)]\tilde{\boldsymbol{\theta}}(k-1) + \mu\boldsymbol{\phi}(k-1)\xi(k) \tag{4.10}$$

where $\tilde{\boldsymbol{\theta}}(k) = \boldsymbol{\theta}^* - \boldsymbol{\theta}(k)$ is the parameter estimate error. This is called the *error system* and is primarily useful for analysis, since the behaviour of the parameter error $\tilde{\boldsymbol{\theta}}(k)$ about the origin describes exactly the behaviour of the parameter estimates $\boldsymbol{\theta}(k)$ about the true parameterization $\boldsymbol{\theta}^*$.

The LMS algorithm (4.7) has been successfully used in numerous applications throughout the years [6–8], and it has been analyzed extensively [9,10]. Some notable aspects of its performance are the following:

- LMS tends to reject noisy data due to the smoothing action of the small step-size parameter μ.
- LMS can track slowly time-varying systems, and is often useful in non-stationary environments.
- The LMS error function has a unique global minimum, and hence the algorithm does not tend to get stuck at undesirable local minima.

- LMS is computationally simple (m multiplies and m adds per iteration) and memory efficient (only one m-vector must be stored).
- The convergence of LMS is often slow (it may take hundreds or thousands of iterations to converge from an arbitrary initialization).
- LMS is susceptible to problems with noise during periods when the input fails to excite all the modes of the system.

Successful algorithms tend to breed closely related variants which attempt to alleviate problems or to fine-tune some aspect of performance. LMS is no exception. These variants range from the normalized LMS (designed to speed convergence) to leakage (which combats potential numerical problems during periods of high noise or low excitation) to the signed algorithms (which further simplify the numerical requirements) to the dual-sign algorithm (and other quantized versions which attempt to simplify the numerics without sacrificing convergence speed) to 'high-order' algorithms (which minimize l^p norms for p greater than 2) to the median LMS and other order statistic algorithms (which attempt to optimize LMS for use in impulsive environments).

4.2.1 Normalized LMS

The desire to have fast convergence from an arbitrary initial state requires a large step size. This conflicts with the desire to have significant smoothing of the noise signal in steady state, which requires a small step size. An obvious algorithm modification uses a large step size initially and then switches to a small step size when in the region of the correct solution. The normalized LMS (NLMS) algorithm provides one way to automate this choice of varying step size.

It is easy to see from (4.10) that the step size μ must be less than $2/\boldsymbol{\phi}^{\mathrm{T}}(k-1)\boldsymbol{\phi}(k-1)$ at each time step k, or instability may result since the term $[I-\mu\boldsymbol{\phi}(k-1)\boldsymbol{\phi}^{\mathrm{T}}(k-1)]$ will be an expansion rather than a contraction, and the solution of the difference equation (4.10) will tend to diverge. At each time step, the algorithm moves a distance $\mu(y(k)-\hat{y}(k))$ in the $\boldsymbol{\phi}(k-1)$ direction. An 'optimal' distance to move would be to set $\mu(k-1)=1/\boldsymbol{\phi}^{\mathrm{T}}(k-1)\boldsymbol{\phi}(k-1)$, since then the term

$$[I-\mu\boldsymbol{\phi}(k-1)\boldsymbol{\phi}^{\mathrm{T}}(k-1)]=\left(I-\frac{\boldsymbol{\phi}(k-1)\boldsymbol{\phi}^{\mathrm{T}}(k-1)}{\boldsymbol{\phi}^{\mathrm{T}}(k-1)\boldsymbol{\phi}(k-1)}\right)$$

is maximally contractive (with an eigenvalue exactly equal to zero) in the direction of the eigenvector $\boldsymbol{\phi}(k-1)$. As a practical matter, the step size is often set to $\mu(k-1)=\mu/1+\mu\boldsymbol{\phi}^{\mathrm{T}}(k-1)\boldsymbol{\phi}(k-1)$, where μ is chosen small enough to encourage smoothing in the steady state and the 1 avoids division by zero in the event that $\boldsymbol{\phi}^{\mathrm{T}}(k-1)\boldsymbol{\phi}(k-1)=0$. This leads to the update

$$\boldsymbol{\theta}(k)=\boldsymbol{\theta}(k-1)+\frac{\mu\boldsymbol{\phi}(k-1)(y(k)-\hat{y}(k))}{1+\mu\boldsymbol{\phi}^{\mathrm{T}}(k-1)\boldsymbol{\phi}(k-1)} \tag{4.11}$$

Thus, rather than taking small steps as in (4.7), the parameter estimates of (4.11) are projected on to the subspace complementary to $\boldsymbol{\phi}(k-1)$. Consequently, the normalized LMS is also called the 'projection algorithm'.

4.2.2 Leakage

The possibility of sensitivity to roundoff errors and other parasitic disturbances exists because the LMS update equation (4.7) is essentially an integrator. The introduction of a small *leakage* parameter $\lambda \in (0, 1)$

$$\boldsymbol{\theta}(k) = (1-\lambda)\boldsymbol{\theta}(k-1) + \mu\boldsymbol{\phi}(k-1)(y(k) - \hat{y}(k)) \tag{4.12}$$

can guard against such numerical problems. The effect of λ on the behaviour of the algorithm is seen most clearly by transforming (4.12) into its error system,

$$\tilde{\boldsymbol{\theta}}(k) = [(I-\lambda)I - \mu\boldsymbol{\phi}(k-1)\boldsymbol{\phi}^{\mathrm{T}}(k-1)]\tilde{\boldsymbol{\theta}}(k-1) + \mu\boldsymbol{\phi}(k-1)\xi(k) + \lambda\boldsymbol{\theta}^* \tag{4.13}$$

which should be compared with (4.10). For any bounded regressor $\boldsymbol{\phi}(k-1)$ and disturbance $\xi(k)$, the step size μ can be chosen small enough so that the bracketed term in (4.13) is exponentially contractive and the error system is bounded-input bounded-output stable. Thus leakage provides an exponential 'safety net' from which the parameter estimates cannot escape. The price of this extra degree of stability is that the estimates will be biased away from their true values, that is, $\tilde{\boldsymbol{\theta}}(k) = 0$ is no longer a solution to (4.13), even in the absence of disturbance. This bias will be proportional to λ and to the unknown $\boldsymbol{\theta}^*$.

An alternative way to look at (4.13) is to suppose that the true parameter $\boldsymbol{\theta}^*$ is known to lie near some nominal value $\boldsymbol{\theta}^0$. Such a priori knowledge can be incorporated into the algorithm by considering the cost function

$$V(\boldsymbol{\theta}(k-1)) = \tfrac{1}{2}(y(k) - \hat{y}(k))^2 + \frac{\lambda}{2\mu}(\boldsymbol{\theta}(k-1) - \boldsymbol{\theta}^0)^{\mathrm{T}}(\boldsymbol{\theta}(k-1) - \boldsymbol{\theta}^0) \tag{4.14}$$

The gradient of $V(\boldsymbol{\theta}(k-1))$ is

$$\frac{\mathrm{d}V(\boldsymbol{\theta}(k-1))}{\mathrm{d}\boldsymbol{\theta}(k-1)} = -(y(k) - \hat{y}(k))\boldsymbol{\phi}(k-1) + \frac{\lambda}{\mu}\boldsymbol{\theta}(k-1) - \frac{\lambda}{\mu}\boldsymbol{\theta}^0 \tag{4.15}$$

and the gradient algorithm that tends to minimize this cost is

$$\boldsymbol{\theta}(k) = (1-\lambda)\boldsymbol{\theta}(k-1) + \mu\boldsymbol{\phi}(k-1)(y(k) - \hat{y}(k)) + \lambda\boldsymbol{\theta}^0 \tag{4.16}$$

Thus the leakage algorithm (4.12) introduced above in an *ad hoc* manner as a method to combat potential numerical problems is identical to the algorithm (4.16) in the special case when the nominal value $\boldsymbol{\theta}^0$ is assumed to be the origin.

4.2.3 Dead zone

Small errors may reflect disturbances or noises, or may result from numerical problems. Large errors, on the other hand, are likely to be caused by poor parameter estimates. A member of the LMS family designed to combat numerical problems from small error signals forbids updates when the error signal is below some user-defined threshold. The dead zone non-linearity

$$g(x) = \begin{cases} x - d & x > d > 0 \\ 0 & -d < x < d \\ x + d & x < -d \end{cases} \tag{4.17}$$

when applied to the error signal, converts the LMS update (4.7) to

$$\boldsymbol{\theta}(k) = \boldsymbol{\theta}(k-1) + \mu\boldsymbol{\phi}(k-1)g(y(k) - \hat{y}(k)) \tag{4.18}$$

As with leakage, this variant can be viewed as a modification to the cost function. Let

$$V(\boldsymbol{\theta}(k-1)) = \begin{cases} \frac{1}{2}(y(k) - \hat{y}(k))^2 - d(y(k) - \hat{y}(k)) & \text{if } y(k) - \hat{y}(k) > d \\ 0 & \text{if } -d \leqslant y(k) - \hat{y}(k) \leqslant d \\ \frac{1}{2}(y(k) - \hat{y}(k))^2 + d(y(k) - \hat{y}(k)) & \text{if } y(k) - \hat{y}(k) < -d \end{cases} \tag{4.19}$$

Then $\mathrm{d}V/d\boldsymbol{\theta} = \boldsymbol{\phi}g(y - \hat{y})$, and (4.18) is the gradient algorithm that minimizes this cost function V.

4.2.4 Signed-error LMS

Although LMS is computationally quite simple, there are always applications for which even m multiplies are too many. It is reasonable to suppose that as long as the correct gradient direction is maintained, the exact length of the step size is unimportant. This suggests the use of $\mathrm{sgn}(y - \hat{y})$ in place of the $(y - \hat{y})$ term in (4.7), and gives the signed-error (SE) algorithm

$$\boldsymbol{\theta}(k) = \boldsymbol{\theta}(k-1) + \mu\boldsymbol{\phi}(k-1)\mathrm{sgn}(y(k) - \hat{y}(k)) \tag{4.20}$$

where

$$\mathrm{sgn}(x) = \begin{cases} 1 & x > 0 \\ 0 & x = 0 \\ -1 & x < 0 \end{cases} \tag{4.21}$$

If μ is chosen to be a multiple of two, then the term $\mu\boldsymbol{\phi}(k-1)\mathrm{sgn}(y(k) - \hat{y}(k))$ can be computed directly with bit shifts, and *no* multiplications are necessary, significantly reducing the computational burden of adaptation. What is sacrificed in terms of performance for this simplification?

In one sense, it is obvious that the SE algorithm (4.20) will tend to converge slower than the LMS algorithm (at least in the initial phase when the parameter estimates are poor) because the motion of the parameter updates will be smaller for a given step size μ. On the other hand, for small errors, the SE algorithm will tend to react faster. The simplest way to compare the two is to compare their cost functions. Consider

$$V(\boldsymbol{\theta}(k-1)) = |\, y(k) - \hat{y}(k)\,| \tag{4.22}$$

Then

$$\frac{\mathrm{d}V(\boldsymbol{\theta}(k-1))}{\boldsymbol{\theta}(k-1)} = -\boldsymbol{\phi}(k-1)\mathrm{sgn}(y(k) - \hat{y}(k)) \tag{4.23}$$

modulo some ambiguity at $y(k) = \hat{y}(k)$. Accordingly, the SE can be viewed as an approximate gradient descent method that attempts to minimize the least absolute value of the error. In no sense, then, is the SE a 'degraded version of LMS', as is sometimes stated. Rather, it is a valid minimization scheme operating optimally on its own cost function. For a problem with disturbances that consist of large outliers, the SE might well tend to return a parameter estimate with smaller variance than LMS, which exaggerates the importance of outliers.

4.2.5 Signed-regressor LMS

An alternative way to reduce the numerical complexity is to apply the signum function element by element to the regressor vector $\boldsymbol{\phi}$, leading to the signed-regressor (SR) variant of LMS:

$$\boldsymbol{\theta}(k) = \boldsymbol{\theta}(k-1) + \mu\ \mathrm{sgn}(\boldsymbol{\phi}(k-1))(y(k) - \hat{y}(k)) \tag{4.24}$$

As in the previous section, if μ is chosen to be a power of two, the update term can be calculated by replacing multiplications with bit shifts. Moreover, the SR algorithm has the capability to react quickly to large errors, unlike the SE algorithm. The SR algorithm was first proposed by Moschner [11] and has been implemented in several successful applications [12]. Claasen and Mecklenbrauker noted in [13] that the direction of the update can be significantly different from the true gradient direction (e.g. the vector (100, 0.01, -0.01) points in a radically different direction from its signed version (1, 1, -1)), and hypothesized that this might cause the algorithm to climb, rather than descend, the gradient. This was debated in [14], where it was shown that, on average, updates tend to proceed in a reasonable approximation to the downhill direction, at least when the inputs are Gaussian. Then, in [15], it was shown that certain classes of inputs can actually cause divergence of the parameter estimates to infinity (at least theoretically). Though such inputs are somewhat unlikely in a typical application, situations exist where they may cause catastrophic failure of the adaptive element. Consequently, more information about the environment in which the algorithm will operate is necessary before applying the SR algorithm than in using some of the other variants. These divergence examples also explain why no gradient minimization interpretation is possible for the SR algorithm – there is no sensible cost function $V(\boldsymbol{\theta})$ which allows infinite $\boldsymbol{\theta}$ as its minimum solution.

4.2.6 Sign–sign LMS

Another way of reducing the numerical complexity of LMS is to incorporate the signum function on both the error and the regressor

$$\boldsymbol{\theta}(k) = \boldsymbol{\theta}(k-1) + \mu \operatorname{sgn}(\boldsymbol{\phi}(k-1))\operatorname{sgn}(y(k) - \hat{y}(k)) \tag{4.25}$$

which is often called the sign–sign (SS) variant of LMS. Again, no multiplications are necessary, and even the n additions can be simplified by judicious choice of μ. This is the oldest of the variants of LMS, first used by Lucky in 1966 [16]. A recent adaptive differential pulse code modulation (ADPCM) standard [17] utilizes signum functions on both the regressor and the error signals (the standard incorporates several other interesting features as well). Despite its status as the first born of the signed children of LMS, it has remained the least understood, probably due to the extra complexity of its twin non-linearities. The enigma of the SS algorithm has recently begun to clear. An elegant example in [18] shows that the algorithm can diverge if its input is suitably pathological. This sparked a flurry of activity attempting to define precisely the class of signals which could cause the algorithm to misbehave [18,19]. This has finally been resolved in [20] and [18], and will be discussed further in the examples.

4.2.7 Quantized state LMS

The signed variants of LMS succeeded in reducing the (already low) numerical complexity of LMS in exchange for an even slower convergence rate. Is there a way to maintain the numerical simplicity of the signed algorithms without this sacrifice?

One of the simpler proposals in this direction is the 'dual-sign' algorithm of [21] which utilizes two step sizes: a large step size μ_L in the initial phase when the error is large, and a small step size μ_S in the converged phase to encourage sufficient smoothing. If both μ_L and μ_S are powers of two, then there is no significant increase in the numerical complexity over the signed algorithms. One reasonable generalization of this idea is [19]

$$\boldsymbol{\theta}(k) = \boldsymbol{\theta}(k-1) + \mu Q_1(\boldsymbol{\phi}(k-1))Q_2(y(k) - \hat{y}(k)) \tag{4.26}$$

where Q_1 and Q_2 are quantization functions applied to the regressor and error signals respectively. This generalization transforms the choice of μ_L and μ_S (and the error values at which they switch) to a choice of the appropriate quantization functions. As will be shown, the introduction of such quantization functions does not add significant complexity to the analysis of the algorithm (over the SS algorithm) except that the issue of how to choose the optimal Q must be addressed.

4.2.8 Least mean fourth

Closely related to LMS are algorithms intended to minimize higher powers of the error

$$V(\boldsymbol{\theta}(k-1)) = (y(k) - \hat{y}(k))^q \tag{4.27}$$

for integer powers of q. The case $q = 4$ has been explicitely used in [22], and is often called the least mean fourth (LMF) algorithm. For even q, it is easy to derive the algorithm

$$\boldsymbol{\theta}(k) = \boldsymbol{\theta}(k-1) + \mu\boldsymbol{\phi}(k-1)(y(k) - \hat{y}(k))^{q-1} \tag{4.28}$$

which will tend to minimize the least mean qth estimates of $\boldsymbol{\theta}$. For large q, one should expect numerical problems when dealing with large errors, though this effect can be ameliorated by choosing a judicious normalization of the step size.

4.2.9 Median LMS

The performance of LMS and its children often degrades badly when subjected to input signals that are corrupted by impulsive noise. One modification designed to combat this problem is the median LMS [23]

$$\boldsymbol{\theta}(k) = \boldsymbol{\theta}(k-1) + \mu med_3\{\boldsymbol{\phi}(k-1)e(k), \boldsymbol{\phi}(k-2)e(k-1), \boldsymbol{\phi}(k-3)e(k-2)\} \tag{4.29}$$

where the 'median' function med_3 is applied element by element to the update vectors $\boldsymbol{\phi}e$. For example, if the numbers a, b and c are ordered from smallest to largest, then $med_3\{a, b, c\} = b$. Of course, median functions of all different lengths, or other order statistic functions, may be used as well. The median function is interesting because it tends to reject single occurrences of large spikes of noise, and these spikes are not passed into the parameter estimates.

4.3 Expected behaviour approach

This section briefly describes how we expect the children of LMS to behave, and highlights some of the benefits and pitfalls of the 'expected value' approach.

In an offline technique, when the parameter estimates can be calculated in closed form, no questions of stability or convergence occur. For instance, with the least-squares approach, the parameter estimate is calculated

$$\boldsymbol{\theta} = (\Phi^{\mathrm{T}}\Phi)^{-1}\Phi^{\mathrm{T}}Y \tag{4.30}$$

where $\Phi = [\boldsymbol{\phi}(k-1), \boldsymbol{\phi}(k-2), \ldots, \boldsymbol{\phi}(k-m)]$, $Y = [y(k), y(k-1), \ldots, y(k-m+1)]^{\mathrm{T}}$, and the solution $\boldsymbol{\theta}$ is the vector that minimizes the summed least-squares error. In adaptive online schemes such as LMS, however, the parameter estimates are made via a recursion like (4.7), (4.20), (4.24), or (4.28), and it becomes crucial to determine the behaviour of the recursion as it evolves in time. Typically, the parameter estimates begin at some setting, move slowly to a region about some final value, and then 'bounce around' this final value. But what can be said concretely about this behaviour?

One approach is to assume that the parameter estimates are already in the converged region (that they are stationary), and that the input $\phi(k)$ is stationary. Taking the expected value of both sides of (4.7) gives

$$E[\theta(k)] = E[\theta(k-1)] + \mu E[\phi(k-1)(y(k) - \hat{y}(k))] \tag{4.31}$$

From the stationarity assumptions, $E[\theta(k)] = E[\theta(k-1)]$, and (4.31) can be rewritten

$$E[\phi(k-1)y(k)] = E[\phi(k-1)\phi^{\mathrm{T}}(k-1)\theta(k-1)] \tag{4.32}$$

If $\phi(k-1)$ is statistically independent of $\theta[k-1]$, then $E[\phi\phi^{\mathrm{T}}\theta] = E[\phi\phi^{\mathrm{T}}]E[\theta]$ and (4.32) can be solved for $E[\theta]$ assuming that $E[\phi\phi^{\mathrm{T}}]$ is invertible

$$E[\theta(k-1)] = [E[\phi\phi^{\mathrm{T}}]]^{-1}E[\phi(k-1)y(k)] \tag{4.33}$$

This is formally analogous to (4.30) and provides evidence that the LMS algorithm (4.7) is likely to return the same answer, on average, as the least-squares method.

Unfortunately, the independence assumption used to derive (4.33) from (4.32) is virtually always false. To see this, recall that $\phi(k-1) = [u(k-1), u(k-2), \ldots, u(k-m)]^{\mathrm{T}}$ is typically a regressor vector of past inputs $u(k-i)$. Hence $\phi(k-2) = [u(k-2), u(k-3), \ldots, u(k-m-1)]^{\mathrm{T}}$. Since $\theta(k-1)$ is explicitly a function of $\phi(k-2)$ (from (4.7)), $\theta(k-1)$ and $\phi(k-1)$ cannot be independent, except perhaps in the scalar one-parameter case. Nevertheless, this is a very common assumption, since it often leads to useful guidelines for implementation issues. An important attempt to justify this assumption formally, based on a small step-size assumption μ, can be found in [24].

An alternative approach [2,7] is to retain the dynamics of the adapted system by defining a parameter error vector $\tilde{\theta}(k)$ as in (4.10) but without assuming that steady state has been achieved. Taking the expected value of both sides of (4.10), and assuming that

- the input is stationary,
- the disturbance term $\xi(k)$ is independent of the input $\phi(k-1)$, and $E[\phi(k-1)\xi(k)] = 0$,
- $\phi(k-1)$ is independent of $\tilde{\theta}(k-1)$,

the error system (4.10) can be rewritten

$$E[\tilde{\theta}(k)] = \{I - \mu E[\phi(k-1)\phi^{\mathrm{T}}(k-1)]\}E[\tilde{\theta}(k-1)] \tag{4.34}$$

Since the autocorrelation matrix $E[\phi\phi^{\mathrm{T}}]$ is symmetric and non-negative definite due to its structure as an outer product of ϕ with itself, it can be diagonalized $E[\phi\phi^{\mathrm{T}}] = QDQ^{\mathrm{T}}$, where $QQ^{\mathrm{T}} = I$ and $D = \mathrm{diag}(d_1, d_2, \ldots, d_n)$ is a diagonal matrix with all entries real and non-negative. Thus

$$E[\tilde{\theta}(k)] = (QQ^{\mathrm{T}} - \mu QDQ^{\mathrm{T}})E[\tilde{\theta}(k-1)] \tag{4.35}$$

$$E[\tilde{\theta}(k)] = Q(I - \mu D)Q^{\mathrm{T}}E[\tilde{\theta}(k-1) \tag{4.36}$$

Defining a transformed (expected) parameter error vector $\theta_Q(k) = Q^{\mathrm{T}}E[\tilde{\theta}(k)]$, and multiplying both sides of (4.36) by Q^{T}, yields

$$\theta_Q(k) = (I - \mu D)\theta_Q(k-1) \tag{4.37}$$

Since D is diagonal, this is simply m copies of the scalar equation

$$\phi_i(k) = (1 - \mu d_i)\phi_i(k-1) \tag{4.38}$$

which decreases exponentially to zero as long as $\mu d_i < 2$. Consequently, if μ is chosen small enough so that $\mu < 2/\lambda_{\max}(R)$ (where $\lambda_{\max}(R)$ indicates the largest eigenvalue of R), then all modes of (4.37) are stable and $\theta_Q(k) \to 0$ as $k \to \infty$. This implies that $E[\tilde{\theta}(k)] \to 0$, which in turn implies that $E[\theta(k)] \to \theta^*$ at a rate proportional to the size of the eigenvalues of the correlation matrix.

Many useful results are possible from this style of analysis:

- Expressions for 'misadjustment' (ratio of excess mean squared error to the minimum mean square error) can be derived as in [7].
- Time constants of convergence rates can be shown to be proportional to the magnitude of the eigenvalues of $E[\phi\phi^{\mathrm{T}}]$, and to the eigenvalue 'spread' (the ratio of the largest to the smallest eigenvalue). See [2].
- In some cases (especially when the inputs are Gaussian [9]), expressions for second- and higher-order moments are feasible.
- The analysis applies (via an extension of Price's theorem [25]) to the signed algorithms [26], and to other algorithms which incorporate quantization functions in their error updates [27].
- The method can be extended to examine the non-stationary case (when θ^* itself is time varying) or when the statistics of ϕ are changing with time; see [28] and [29].
- Optimum step sizes can often be calculated in terms of (possibly) available quantities as in [28] and [30].

Despite the fact that this style of analysis is non-rigorous in a formal mathematical sense (for instance, the independence assumption on $\theta(k-1)$ and $\phi(k-1)$), there is quite close agreement between the conclusions of the analysis, simulations of the algorithms, and the behaviour of the algorithm in applications. Indeed, it has taken more careful analysts over a decade to verify rigorously but a small part of the conclusions of this rougher analysis. While most of the results of the careful analysis were anticipated, there are situations in which the conclusions of this non-rigorous analysis can be misleading. Possibly the most dramatic of these differences involves the actual stability of certain of the variants of LMS, especially those which manipulate the regressor vector so that the update can point away from the 'downhill' gradient direction.

Consider the signed regressor algorithm (4.24). Following the technique and assumptions of (4.31)–(4.33) yields

$$E[\theta(k-1)] = [E[\mathrm{sgn}(\phi)\phi^{\mathrm{T}}]]^{-1}E[\mathrm{sgn}(\phi)y(k)] \tag{4.39}$$

Consequently, one expects that, as long as the matrix $E[\text{sgn}(\phi)\phi^T]$ is invertible, $E[\theta(k-1)]$ takes on a single, well-defined value.

On the other hand, following the techniques and assumptions of (4.34)–(4.37) yields

$$E[\tilde{\theta}(k)] = \{I - \mu E[\text{sgn}(\phi)\phi^T]\}E[\tilde{\theta}(k-1)] \tag{4.40}$$

The matrix $E[\text{sgn}(\phi)\phi^T]$ is not symmetric, and may have real or complex eigenvalues, with positive or negative real parts. Suppose that the stochastic process $\phi(k)$ is chosen so as to cause $E[\text{sgn}(\phi)\phi^T]$ to have an eigenvalue with a negative real part. Then there is a direction (the eigenvector associated with this negative eigenvalue) in which the scalar analog (4.38) is exponentially unstable, since $d_i < 0$. This implies that the parameter estimates will be driven away from the 'correct' solution (4.39).

Clearly, the conclusion of (4.39) is diametrically opposed to the conclusion of (4.40). Hopefully, the careful reader will spot the reason for this discrepancy – the assumption of stationarity in the θ process in (4.33) and (4.39) is tantamount to an assumption of stability of (4.34) and (4.40). The intent here is to highlight the need for a more careful analysis, in which the ramifications of the various assumptions are pursued vigorously. The next two sections present two approaches to a more rigorous analysis: the 'deterministic approach' and the 'stochastic approximation' approach.

4.4 The deterministic approach

The deterministic approach uses the tools of non-linear system theory to examine LMS and its children. The generic adaptive update form (4.2) can be interpreted as the state equation of a non-linear and time-varying system. This system can be linearized and averaged to derive conditions under which the various algorithms can be expected to succeed in their identification task.

The conditions are stated in terms of a persistence of excitation which, in the ideal case (with no disturbances), must be satisfied in order to guarantee exponential convergence of the parameter estimates to their true values. When bounded disturbances are present, the conditions guarantee convergence to a small region about the true value. The excitation conditions involve the non-linear functions of the data and the error signal, F and g of (4.2), but the non-linearities enter in different ways. Sign-preserving *error* non-linearities are essentially benign in terms of stability of the adaptive system, while even modest data non-linearities can cause stability problems.

These results have a simple geometrical interpretation in terms of descending an error surface. Recall that LMS is an approximate gradient descent method utilizing the squared error as a cost function. At each update instant, the vector of input data points in the 'downhill' direction, while the error signal scales the motion in that direction. The effect of a non-linearity on the data vector (such as in the signed-regressor (4.24), the sign–sign (4.25), or the quantized state (4.26) algorithms) is to cause motion in a direction that is not necessarily 'downhill'. Is it surprising that for certain data sequences, this misalignment from the actual gradient direction can cause the algorithm to climb, rather than descend, the error surface? The effect of an error non-linearity (such as in the dead-zone (4.18), signed-error (4.20), or the least mean pth (4.28) algorithms) is subtler. It changes the cost function that will be minimized. Each of the latter three algorithms has a simple interpretation as an approximate gradient method on some cost surface. Thus the presence of sign-preserving error non-linearities is transparent in terms of system stability, though the various non-linearities behave somewhat differently in terms of convergence rate and minimization properties.

4.4.1 Analytical background

The key ideas of the deterministic approach are linearization, the slow time variation lemma [31], averaging [32], and total stability [33]. Linearization is used to examine the stability of the algorithm (4.2) operating in a region about its equilibrium. This linearization is time varying (due to the data signal), and a slow time-variation result can be used to relate the stability of the time-varying system to the stability of the related frozen systems. The slowness is a consequence of the small value which the step size μ is assumed to have. Averaging is used to derive conditions under which the frozen systems are locally exponentially stable. The total stability theorem then translates the exponential stability result into robustness of the adaptive system to small disturbances, including small measurement noises, small non-linearities, and slow parameter variation.

LINEARIZATION

Consider the discrete time system

$$\mathbf{z}(k) = \mathcal{F}(k-1, \mathbf{z}(k-1)) \tag{4.41}$$

where $\mathbf{z}(k)$ is a state vector in $\Re^m$, and $\mathcal{F}$ is a vector function $\Re^m \rightarrow \Re^m$ defining the evolution of the state. The states $\mathbf{z}^*$ for which $\mathcal{F}(k, \mathbf{z}^*) = \mathbf{z}^*$ for all k are the equilibria of (4.41), which we may assume without loss of generality to be located at the origin. $\mathcal{F}$ is linearized at the equilibrium $\mathbf{z}^* = 0$ via the Jacobian $A(k) = D\mathcal{F}\,|_{\mathbf{z}^*=\mathbf{0}}$. The linearization theorem (Lyapunov's indirect method [34]) asserts that the behaviour of (4.41) near $\mathbf{z}^*$ is dictated by the behaviour of the related linear system

$$\mathbf{y}(k) = A(k-1)\mathbf{y}(k-1) \tag{4.42}$$

that is, if the linearized state equation (4.42) is exponentially asymptotically stable (e.a.s.), then (4.41) is also e.a.s. The theorem holds assuming that $A(k)$ is bounded, and assuming that the norm of the difference $\mathcal{F}(k, \mathbf{z}) - A(k)\mathbf{z}$ is uniformly bounded in time. Formally, this requires that

$$\lim_{\| z \| \to 0} \max_k \frac{\| \mathscr{F}(k, \mathbf{z}) - A(k)\mathbf{z} \|}{\| \mathbf{z} \|} = 0 \tag{4.43}$$

which essentially guarantees that time variation in the non-linear terms of the Taylor series do not become arbitrarily large as time progresses.

SLOW TIME VARIATION AND AVERAGING

The task of showing stability for the adaptive system is therefore translated to the simpler problem of finding conditions under which the linear time-varying system (4.42) is e.a.s. One approach is to use the 'slow time variation lemma' of [31] which asserts that if the change in A is slow enough (that is, $\| A(k) - A(k-1) \|$ is small), then exponential stability of each $A(j)$ (uniformly in j) is enough to imply that the time-varying system (4.42) is e.a.s.

Unfortunately, the $A(k)$ matrices from the adaptive systems of interest are virtually never exponentially stable due to the structure of the problem. This implies that the desired systems $A(j)$ fail to be e.a.s. An alternative approach [32] is to take the time average of (4.42) and to define the *sliding average*

$$\bar{A}(k, m) = \frac{1}{m} \sum_{i=1}^{m} A(k+i) \tag{4.44}$$

If the eigenvalues of $\bar{A}(k, m)$ are (uniformly in k) less than one in magnitude for some m, and if the $\bar{A}(k, m)$ vary slowly enough, then it can be shown that the averaged system

$$\bar{\mathbf{y}}(k) = \bar{A}(k-1)\bar{\mathbf{y}}(k-1) \tag{4.45}$$

and the related (4.42) are both e.a.s. Fortunately, the sliding averages can be exponentially stable even when the $A(k)$ are not.

TOTAL STABILITY

The final step in the argument is to relax the assumption that there are no disturbances. The total stability theorem of [32] relates the behaviour of the unforced system (4.41) to the behaviour of

$$\bar{\mathbf{z}}(k) = \mathscr{F}(k-1, \bar{\mathbf{z}}(k-1)) + \mathscr{G}(k-1, \bar{\mathbf{z}}(k-1)) \tag{4.46}$$

where $\mathscr{G}$ is some small disturbance term that may depend on the state. Assuming that $\mathscr{F}$ is Lipschitz continuous, the difference between the state $\mathbf{z}$ of (4.41) and the state $\bar{\mathbf{z}}$ of (4.46) can be bounded when $\mathscr{F}$ is known to be e.a.s. by requiring that $\mathscr{G}$ be suitably small and that the initial difference is small. Formally, for every ε, there is a δ_1 and δ_2 such that $\| \bar{\mathbf{z}}(0) - \mathbf{z}(0) \| < \delta_1$ and $\| \mathscr{G}(k, \bar{\mathbf{z}}(k)) \| < \delta_2$ for every k imply that $\| \mathbf{z}(k) - \bar{\mathbf{z}}(k) \| < \varepsilon$ for every k. Thus, the system no longer converges to its equilibrium; rather, it converges to a ball about the equilibrium and then 'rattles around'. This disturbance term can be used formally to consider measurement disturbances, small non-linearities, slow time variation of the parameters, and other small 'non-idealities' that may arise.

4.4.2 Persistence of excitation

The above ideas can be used to examine the stability of the generic adaptive algorithm

$$\tilde{\boldsymbol{\theta}}(k) = \tilde{\boldsymbol{\theta}}(k-1) - \mu F(\boldsymbol{\phi}(k-1))g(\tilde{\boldsymbol{\theta}}^{\mathrm{T}}(k-1)\boldsymbol{\phi}(k-1)) \tag{4.47}$$

which is derived from (4.2) by the introduction of the parameter estimate error $\tilde{\boldsymbol{\theta}}(k) = \boldsymbol{\theta}^* - \boldsymbol{\theta}(k)$. The following assumptions are made about the non-linear functions F and g:

1. F and g are sign preserving.
2. F and g are memoryless.
3. $g(\cdot)$ is differentiable at the origin.

Assumption (1) is fundamental in the sense that if F or g were not sign preserving, this is equivalent to designing an algorithm to climb rather than to descend the error surface. This is also equivalent to reversing the sign of the step size μ. Assumption (2) is implicit in the formulation of F and g as functions of their specified arguments, but it is worth while noting because there is, perhaps, some interest in considering functions with memory. The linear case with memory is dealt with in [31] via similar techniques to those used here, and others have attacked this situation in other ways; see [35] and [36]. Assumption (3) assures that the linearization step is possible. Note that no differentiability (or continuity) is required on F, nor on g anywhere but at the origin. Most of the non-linear variants of LMS fulfil these requirements, though the 'signed-error' algorithm (4.20) fails condition (3).

LINEARIZATION

Define the vectors $\tilde{\boldsymbol{\theta}}(k) = [\theta_1(k), \theta_2(k), \ldots, \theta_m(k)]^{\mathrm{T}}$, $\boldsymbol{\phi}(k) = [x_1(k), x_2(k), \ldots, x_m(k)]^{\mathrm{T}}$, and the vector function $F(\boldsymbol{\phi}(k)) = [f_1(\boldsymbol{\phi}(k)), f_2(\boldsymbol{\phi}(k)), \ldots, f_m\boldsymbol{\phi}(k))]^{\mathrm{T}}$. Typically, $\boldsymbol{\phi}(k)$ consists of a 'regressor' vector of time-shifted versions of a scalar sequence $x(k)$, that is, $x_i(k) = x_{i-1}(k-1)$ for $i = 2, \ldots, m$, but this is not necessary. Identify the function F of (4.41) with the right-hand side of (4.47), and let

$$H(k) = F(\boldsymbol{\phi}(k))g(\tilde{\boldsymbol{\theta}}^{\mathrm{T}}(k)\boldsymbol{\phi}(k)) = \begin{Bmatrix} f_1(\boldsymbol{\phi}(k))g(\theta_1(k)x_1(k) + \theta_2(k)x_2(k) + \theta_m(k)x_m(k)) \\ f_2(\boldsymbol{\phi}(k))g(\theta_1(k)x_1(k) + \theta_2(k)x_2(k) + \theta_m(k)x_m(k)) \\ \vdots \\ f_m(\boldsymbol{\phi}(k))g(\theta_1(k)x_1(k) + \theta_2(k)x_2(k) + \theta_m(k)x_m(k)) \end{Bmatrix} \tag{4.48}$$

Then the Jacobian $\mathrm{d}H(k)/\mathrm{d}\tilde{\theta}(k)$ can be calculated as

$$\begin{pmatrix} f_1(\boldsymbol{\phi}(k))\theta_1(k)g'(\tilde{\boldsymbol{\theta}}^{\mathrm{T}}(k)\boldsymbol{\phi}(k)) & f_1(\boldsymbol{\phi}(k))\theta_2(k)g'(\tilde{\boldsymbol{\theta}}^{\mathrm{T}}(k)\boldsymbol{\phi}(k)) & \cdots & f_1(\boldsymbol{\phi}(k))\theta_m(k)g'(\tilde{\boldsymbol{\theta}}^{\mathrm{T}}(k)\boldsymbol{\phi}(k)) \\ f_2(\boldsymbol{\phi}(k))\theta_1(k)g'(\tilde{\boldsymbol{\theta}}^{\mathrm{T}}(k)\boldsymbol{\phi}(k)) & f_2(\boldsymbol{\phi}(k))\theta_2(k)g'(\tilde{\boldsymbol{\theta}}^{\mathrm{T}}(k)\boldsymbol{\phi}(k)) & \cdots & f_2(\boldsymbol{\phi}(k))\theta_m(k)g'(\tilde{\boldsymbol{\theta}}^{\mathrm{T}}(k)\boldsymbol{\phi}(k)) \\ \vdots & \vdots & \vdots & \vdots \\ f_m(\boldsymbol{\phi}(k))\theta_1(k)g'(\tilde{\boldsymbol{\theta}}^{\mathrm{T}}(k)\boldsymbol{\phi}(k)) & f_m(\boldsymbol{\phi}(k))\theta_2(k)g'(\tilde{\boldsymbol{\theta}}^{\mathrm{T}}(k)\boldsymbol{\phi}(k)) & \cdots & f_m(\boldsymbol{\phi}(k))\theta_m(k)g'(\tilde{\boldsymbol{\theta}}^{\mathrm{T}}(k)\boldsymbol{\phi}(k)) \end{pmatrix} \tag{4.49}$$

When evaluated at the equilibrium $\boldsymbol{\theta}^* = 0$, this simplifies to

$$B(k) = \left.\frac{\mathrm{d}H(k)}{\mathrm{d}\tilde{\boldsymbol{\theta}}(k)}\right|_{\theta^* = 0} = g'(0)F(\boldsymbol{\phi}(k))\boldsymbol{\phi}^{\mathrm{T}}(k)) \tag{4.50}$$

and the linearized system is

$$\mathbf{y}(k) = [I - \mu B(k-1)]\mathbf{y}(k-1) \tag{4.51}$$

The linearization results shows that if (4.51) is exponentially stable, then the original non-linear system (4.47) is also exponentially stable.

SLOW TIME VARIATION AND AVERAGING

Note that by choosing the step-size parameter μ small, the time variation of the transition matrix $[I - \mu B(k-1)]$ is slowed. In fact, as $\mu \to 0$, $\| [I - \mu B(k)] - [I - \mu B(k-1)] \| \to 0$. Consequently, the exponential stability of the time-varying linearized system can be translated via the slow time variation lemma to the exponential stability of the frozen (or time-invariant) systems $[I - \mu B(j)]$, for each j.

Unfortunately, due to the structure of $B(k)$ as a scaled product of two vectors, each $B(k)$ has rank 1, at most, and so has $m - 1$ zero eigenvalues. This implies that $[I - \mu B(k)]$ has $m - 1$ unity eigenvalues, and hence is not exponentially stable. To overcome this, define the sliding average $\bar{B}(k, l)$ over the time window l as in (4.44). Then the averaging theorem demonstrates that exponential stability of

$$\bar{\mathbf{y}}(k) = [I - \mu\bar{B}(k-1)]\bar{\mathbf{y}}(k-1) \tag{4.52}$$

implies exponential stability of (4.51), and hence (4.47). Define the *excitation matrix*

$$M_s = \sum_{k=1}^{s} F(\boldsymbol{\phi}(k))\boldsymbol{\phi}^{\mathrm{T}}(k) \tag{4.53}$$

which, for s-periodic inputs, is equal to the sliding average, as is done in [19]. Then the magnitude of all eigenvalues of $(I - \mu M_s)$ can be guaranteed to be less than one as long as M_s has all eigenvalues with positive real part, and as long as μ is chosen small enough. Gathering the above results together shows the following.

THEOREM 4.1

(Persistence of excitation theorem) Consider the algorithm (4.47) with s-periodic input data $\boldsymbol{\phi}(k)$ and non-linear elements F and g, under assumptions 2 and 3. If there are $\alpha > 0$ and $\beta > 0$ such that

$$\beta > g'(0)\mathrm{Re}\lambda_i(\mathrm{M}_s) > \alpha \quad \text{for every } i \tag{4.54}$$

then there is a μ^* such that for every μ in $(0, \mu^*)$, the algorithm (4.47) is locally exponentially stable about its equilibrium $\boldsymbol{\theta}^* = 0$. Conversely, if $g'(0)\mathrm{Re}\lambda_i(\mathrm{M}_s)$ is negative for some i, then the algorithm (4.47) is locally unstable about its equilibrium at $\boldsymbol{\theta}^* = 0$.

(The notation $\text{Re}\lambda_i(M)$ means the real part of the ith eigenvalue of the matrix M.) An input which fulfils (4.54) for a particular algorithm is said to be *persistently exciting* for the algorithm. Equivalently, the algorithm is *persistently excited* by the input.

Some remarks follow:

1. Local exponential stability of the algorithm implies that the parameter estimate error $\tilde{\boldsymbol{\theta}}(k)$ converges to zero if it is initialized in some region about zero. Convergence of the parameter estimate error to zero is equivalent to the convergence of the parameter estimates $\boldsymbol{\theta}(k)$ to their true values $\boldsymbol{\theta}^*$. Local instability implies that there are arbitrarily small perturbations that can drive the parameter estimates away from $\boldsymbol{\theta}^*$. This does not necessarily imply divergence to infinity of the parameter estimates.
2. The condition (4.54) is called the *persistence of excitation* (PE) condition for the LMS algorithm with non-linearities F and g. Note that the condition involves the input data sequence $\boldsymbol{\phi}(k)$ as well as the data non-linearity F and the derivative of the error non-linearity g at the origin.
3. The importance of the sign preservation property of F is apparent from the persistence of excitation condition, since if F reverses the sign of the data, then the right-hand inequality of (4.54) fails. Similarly, $g'(0)$ must be positive.
4. If $g'(0) = \infty$ then assumption 3 and the left-hand inequality of (4.54) fail. In particular, this averaging approach is inapplicable to the signed-error algorithm with $g(e) = \text{sgn}(e)$. An extended Lyapunov approach for this algorithm can be found in [19].
5. The convergence rate of the averaged system (4.52) (and hence the convergence rate of the algorithm (4.47)) is proportional to the size of the real part of the smallest eigenvalue of (4.53). Thus, given an input sequence $\boldsymbol{\phi}(k)$, if α is chosen as large as possible, the convergence rate is dictated by α. Since $g'(0)$ is directly proportional to α, increasing the slope of g near the origin will tend to increase the convergence rate, if other parameters are held fixed, provided that the left-hand inequality in (4.54) is not violated.
6. The periodicity assumption is not necessary, and can be relaxed to 'almost periodic' inputs as in [37] at the expense of a large amount of technical detail.
7. The fact that (4.54) depends on the function g only at the origin emphasizes the local nature of the results; initial conditions must be chosen so that g remains in this ball about the origin.

Suppose that $g'(0) = 0$, as occurs in the dead-zone algorithm (4.18), in the least mean qth algorithms (4.28) and in the quantized state algorithms (4.26) for certain quantization functions Q_2. Then the right-hand side of the persistence of excitation condition (4.54) fails, and the algorithm is not exponentially stable about $\boldsymbol{\theta}^* = 0$. If, however, g is non-decreasing, continuous and differentiable at the end points of some region R, then there is hope that the parameter estimate errors will converge to the region R rather than to $\boldsymbol{\theta}^*$ itself. To make this notion more precise, consider the following definition.

DEFINITION 4.4.1

The system $\mathbf{x}(k) = f(k-1, \mathbf{x}(k-1))$ is said to be (uniformly) locally exponentially stable to the compact region R contained in B if there exists a $\gamma \in (0, 1)$ and an $N > 0$ such that $\forall \mathbf{x}(0) \in B$, $d(\mathbf{x}(k), R) < N \| \mathbf{x}(0) \| \gamma^k \; \forall k$, where the distance from the point $\mathbf{x}(k)$ to the set R is defined as $d(\mathbf{x}(k), R) = \min_{r \in R} \| \mathbf{x}(k) - \mathbf{r} \|$.

Note that this minimum exists when R is compact, and that the definition reduces to the standard definition of (local, uniform) exponential stability when R consists of an isolated equilibrium. The following corollary simply extends the theorem to include the case of convergence to a region, rather than a point.

COROLLARY 4.4.1

Consider the algorithm (4.47) with s-periodic input data $\boldsymbol{\phi}(k)$ and non-linear elements F and g under assumptions 2 and 3. Suppose, further, that g is non-decreasing and continuous in a region $R = [-r, r]$, that $g'(0) = 0$, that $g'(r)$ and $g'(-r)$ exist and are positive, and that there are $\alpha > 0$ and $\beta > 0$ such that $\beta > \text{Re}\lambda_i(M_s) > \alpha \; \forall i$. Then there is a μ^* such that for every $\mu \in (0, \mu^*)$, the algorithm is locally exponentially stable to the region R.

TOTAL STABILITY

The final step is to remove the 'ideal' assumption, and to suppose that some small non-idealities are present. The $\mathscr{F}$ and $\mathscr{G}$ of (4.46) may be related to the various versions of LMS by identifying the state $\bar{\mathbf{z}}(k)$ with the parameter estimate errors $\tilde{\boldsymbol{\theta}}(k)$, and $\mathscr{G}$ with the disturbance term. Assuming that the input data fulfil the PE condition (4.54), then the homogeneous system (4.47) (and (4.41) with $\mathscr{F}$ identified as the right-hand side of (4.47), is exponentially stable. Consequently, the total stability theorem asserts that for small disturbances $\mathscr{G}$, the perturbed system will remain within an ε ball about the origin. This has several implications:

1. Robustness to small measurement noises. Suppose that a bonded measurement disturbance $\xi(k)$ corrupts the prediction error $e(k) = y(k) - \hat{y}(k)$. Then $\mathscr{G}(k-1, \tilde{\boldsymbol{\theta}}(k-1)) = \mu F(\boldsymbol{\phi}(k-1))[g(e(k) + \xi(k)) - g(e(k))]$, and the norm of $\mathscr{G}$ can be bounded in terms of μ, $\| F \|$, $\| \boldsymbol{\phi}(k-1) \|$, and the smoothness of g. Hence, if $|\xi(k)|$ is small enough so that $\| \mathscr{G} \| < \delta_2$, the total stability theorem shows that an algorithm that is exponentially stable cannot be destabilized by arbitrarily small measurement biases or inaccuracies.
2. Robustness to undermodelling. Suppose that the n-dimensional $\boldsymbol{\theta}^*$ is only an approximation to the 'true' plant, which is $n + m$ dimensional. If this undermodelling is not too severe (if there is an n-dimensional $\boldsymbol{\theta}^*$ that is a good approximation to the true plant), then the algorithm retains stability. In this case, $\xi(k)$ represents the difference between the output of the true $(n + m)$-dimensional plant and the output due to $\boldsymbol{\theta}^*$. As in (4.1), if this $\xi(k)$ is small, then the perturbed system is stable.
3. Robustness to small non-linearities. Suppose that the linear $\boldsymbol{\theta}^*$ is only an approximation to the 'real' plant which contains small non-linearities. If $\xi(k)$

represents the output due to these non-linearities, and if this is kept small, then the algorithm retains stability.

4. Robustness to slow time variations. The 'real' plant may actually vary with time. If these time variations are slow enough, then the exponentially stable algorithm will track the motion and remain stable. Let $\boldsymbol{\theta}^*(k)$ represent the time-varying plant, and suppose that $\| \boldsymbol{\theta}^*(k) - \boldsymbol{\theta}^*(k-1) \|$ is small. The error system becomes

$$\tilde{\boldsymbol{\theta}}(k) = \tilde{\boldsymbol{\theta}}(k-1) - \mu F(\boldsymbol{\phi}(k-1))g(\tilde{\boldsymbol{\theta}}^{\mathrm{T}}(k-1)\boldsymbol{\phi}(k-1)) + \boldsymbol{\theta}^*(k) - \boldsymbol{\theta}^*(k-1) \quad (4.55)$$

Letting $\mathscr{G} = \boldsymbol{\theta}^*(k) - \boldsymbol{\theta}^*(k-1)$, and bounding the rate of variation by $\| \mathscr{G} \| < \delta_2$, shows that the algorithm retains stability.

INTERPRETATION OF THE EXCITATION CONDITIONS

This section compares the persistence of excitation (PE) condition for LMS with non-linearities F and g to the PE condition for LMS by showing that it is strictly more difficult to fulfil the PE condition for the non-linear variants of LMS than the PE condition for (linear) LMS. The standard PE condition for LMS [38] (without non-linearities), when excited by s-periodic inputs $\boldsymbol{\phi}(k)$, is that there exist $\alpha > 0$ and $\beta > 0$ such that

$$\beta I > \sum_{k=1}^{s} \boldsymbol{\phi}(k)\boldsymbol{\phi}^{\mathrm{T}}(k) > \alpha I \quad (4.56)$$

As above, this implies local exponential stability of the error system. Since the matrix in (4.56) is symmetric, all eigenvalues are real, and the notation '>' means positive definite. How does (4.56) compare to (4.54)?

LEMMA 4.4.1

Suppose that (4.54) holds for a given F, g and input sequence $\boldsymbol{\phi}(k)$, and supose that $F(\boldsymbol{\phi})$ does not vanish as $\boldsymbol{\phi} \to \infty$. Then (4.56) also holds.

Proof By contradiction. There are two possibilities:

1. If (4.56) fails the upper bound, then $\boldsymbol{\phi}(k)$ must be diverging. By assumption, this implies that $F(\boldsymbol{\phi}(k))\boldsymbol{\phi}^{\mathrm{T}}(k)$ must diverge. Hence the left-hand inequality of (4.54) is violated.
2. If (4.56) fails the lower bound for every positive α, then there must be a zero eigenvalue of $\Sigma_{k=1}^{s} \boldsymbol{\phi}(k)\boldsymbol{\phi}^{\mathrm{T}}(k)$. Consequently, there must be a non-zero eigenvector $\mathbf{v}$ such that $\mathbf{v}^{\mathrm{T}}\boldsymbol{\phi}(k) = 0$ for every k. This implies that $(\Sigma_{k=1}^{s} F(\boldsymbol{\phi}(k))\boldsymbol{\phi}^{\mathrm{T}}(\mathrm{k}))\mathbf{v} = 0$, and so there is a zero eigenvalue of the matrix in (4.54).

This says that if the non-linear LMS algorithm is persistently excited (4.54), then the standard LMS algorithm is also persistently excited (4.56). For a large class of F, the reverse implication is false, since one can easily construct examples for which (4.54) has eigenvalues with negative real parts. Such examples can be found in [15] and a generic counterexample is constructed in [30] to demonstrate that whenever

F is non-linear, there are input sequences that will fail the PE condition and destabilize the algorithm.

4.5 The stochastic approximation approach

The stochastic approximation approach to the analysis of adaptive algorithms was pioneered by L. Ljung [3], and has been extended over the years by several researchers, most notably Kushner and Schwartz [40] and Benveniste *et al.* [5]. The approach relates the motion of the parameter estimate errors of the algorithms to the behaviour of an unforced deterministic ordinary differential equation (ODE) by showing that local stability of the ODE implies weak convergence of the algorithm. More recent is the observation that the ODE can be unstable, which implies non-convergence of the parameter estimates [20]. These results are stated in terms of the eigenvalues of a correlation-like matrix which may be thought of as the stochastic analog of the PE condition of the previous section. When the recursion is stable, the asymptotic distribution of the parameter trajectories is a Gauss–Markov process under very general assumptions on the statistics of the inputs and disturbances. It is not necessary to assume independence of $\tilde{\theta}(k-1)$ and $\phi(k-1)$.

The ability of the algorithms to track moving parameterizations (when θ^* is a function of time) can be analyzed in a similar manner, by relating the time-varying system to a *forced* ODE. The asymptotic distribution about the forced ODE is again a Gauss–Markov process, whose properties can be described in a straightforward manner. This allows a comparison of the various adaptive algorithms in terms of their convergence and tracking ability.

The analysis is carried out by examining the general recursive form

$$\tilde{\theta}(k) = \tilde{\theta}(k-1) - \mu H(\tilde{\theta}(k-1), \phi(k-1), \xi(k)) \tag{4.57}$$

which is a slight rewriting of (4.2) and (4.47) that captures the children of LMS by suitable choice of $H(\cdot)$. As before, $\tilde{\theta}(k)$ represents the parameter estimate errors, $\phi(k)$ is (usually) a vector of inputs, $\xi(k)$ is a disturbance process that represents all non-idealities such as measurement and modelling errors, and μ is a small, positive, constant step size. As in the previous section, convergence of the process $\tilde{\theta}(k)$ to a stationary distribution about zero is equivalent to convergence of the adaptive filter parameter estimates to a region about their optimal values. Two important questions concerning the behaviour of $\tilde{\theta}(k)$ arise:

- Under what conditions is the process stable?
- When do stationary distributions for $\tilde{\theta}(k)$ exist, and how can these stationary distributions be characterized?

One way to answer these questions is to relate the behaviour of the adaptive algorithm (4.57) for small μ to the behaviour of the associated deterministic ODE:

$$\tilde{\theta}(t) = \tilde{\theta}(0) - \int_0^t \hat{H}(\tilde{\theta}(s))\mathrm{d}s \tag{4.58}$$

where $\hat{H}(\cdot)$ is a smoothed version of $H(\cdot, \cdot, \cdot)$. The fundamental issue is to determine the relationship between $\tilde{\theta}(k)$ and $\tilde{\theta}(t)$. Recall that processes using the time index k are discrete, while processes with time index t are continuous. The two questions about (4.57) translate into analogous questions concerning (4.58):

- Under what conditions is the ODE stable or unstable?
- How closely does the algorithm (4.57) track the behaviour of the ODE (4.58)?

If the ODE is stable, then the algorithm (4.57) is stable (indicating probable success of the adaptive scheme), while if (4.58) is unstable, then (4.57) is also unstable, and the adaptive algorithm fails. For instance, if the correlation matrix of the input process $E[\boldsymbol{\phi}\boldsymbol{\phi}^{\mathrm{T}}]$ is positive definite, then (for small enough μ) the parameter estimate errors of the LMS algorithm converge in distribution to a region about the origin [2,10]. The same matrix $E[\boldsymbol{\phi}\boldsymbol{\phi}^{\mathrm{T}}]$ appears here as the linearization of $\hat{H}(\tilde{\boldsymbol{\theta}})$, and is called the *stochastic excitation matrix*. Positive definiteness of this matrix implies local stability of the ODE, while a negative eigenvalue would imply local instability. Of course, due to its structure as a correlation matrix, $E[\boldsymbol{\phi}\boldsymbol{\phi}^{\mathrm{T}}]$ is always at least non-negative definite, and the instability cannot occur.

Certain of the children of LMS are not so fortunate. The analogous stochastic excitation condition for the signed regressor algorithm, for instance, requires that all eigenvalues of $E[\mathrm{sgn}(\boldsymbol{\phi})\boldsymbol{\phi}^{\mathrm{T}}]$ have positive real parts [15]. As before, this same matrix appears in the present analysis as the linearization of $\hat{H}(\tilde{\boldsymbol{\theta}})$, and positivity of the real parts of the eigenvalues of $E[\mathrm{sgn}(\boldsymbol{\phi})\boldsymbol{\phi}^{\mathrm{T}}]$ implies stability, while an eigenvalue with negative real part implies instability. In this case, there are non-trivial input distributions which cause instability of the associated ODE, and hence of the signed-regressor algorithm.

The relation between the adaptive algorithm (4.57) and the ODE (4.58) may be thought of as a type of 'law of large numbers'. To investigate how close the behaviour of the algorithm is to the deterministic trajectory of the ODE, one desires a corresponding 'central limit theorem'. Consider the time-scaled process $\tilde{\boldsymbol{\theta}}_{[t/\mu]}(t)$ where $[z]$ represents the integer part of z. The martingale central limit theorem can be exploited to show that the error process

$$V_\mu = \frac{1}{\sqrt{\mu}}(\tilde{\boldsymbol{\theta}}_{[t/\mu]}(t) - \tilde{\boldsymbol{\theta}}(t)) \tag{4.59}$$

converges to a forced ODE that is driven by a sum of independent, mean-zero Brownian motions. Significantly, under mild assumptions on the input and disturbance processes, the limiting distribution is a Gauss–Markov process, with known mean and variance.

In practical terms, this convergence has two implications. First, for a given algorithm, it is easy to calculate the parameters of the convergent distribution in terms of the properties of the inputs and disturbances, and hence to give a measure of the performance of the algorithm. Second, this allows a fair comparison between competing adaptive schemes by calculating the mean and variance of the convergent distributions for the various algorithms. Equivalently, this allows a fair comparison of the convergence speed of the various algorithms.

Finally, the ability of the adaptive algorithms to track a slowly moving parameterization $\boldsymbol{\theta}^*$ can be examined by following essentially the same programme as above. The asymptotic distributions of the appropriate error process can be related to a forced ODE, where the forcing term is directly related to the motion of the underlying parameterization, and the asymptotics once again prove to be a Gauss–Markov process. In contrast to the convergence speed, there is little difference between the various algorithms in terms of their ability to track slowly moving targets.

4.5.1 Theoretical development

This section presents the limit theorems which relate the behaviour of the adaptive algorithm (4.57) to the ODE (4.58), basically following the presentation in [20]. The adaptive update term $H(\cdot)$ in (4.57) has three arguments:

- $\tilde{\boldsymbol{\theta}}(k-1)$ is the parameter estimate error.
- $\boldsymbol{\phi}(k-1)$ is the input to the adaptive filter.
- $\xi(k)$ is the disturbance term.

The process $\{\tilde{\boldsymbol{\theta}}(k-1), \boldsymbol{\phi}(k-1, \xi(k)\}$ takes on values in $\Re^m \times E_1 \times E_2$, where m is the number of adaptive parameters, and E_1 and E_2 are the appropriate state spaces on which $\boldsymbol{\phi}(k)$ and $\xi(k)$ evolve. $\{\tilde{\boldsymbol{\theta}}(k-1), \boldsymbol{\phi}(k-1), \xi(k)\}$ is adapted to a filtration $\{\mathscr{F}_k\}$ (usually one takes $\mathscr{F}_k$ to be the σ-algebra generated by the random variables $(\tilde{\boldsymbol{\theta}}(l-1), \boldsymbol{\phi}(l-1), \xi(l))_{l=-\infty}^{k}$). We assume that there exists a transition function $\eta(\tilde{\boldsymbol{\theta}}, \boldsymbol{\phi}, C)$ such that $P(\xi(k) \in C \mid \mathscr{F}_k) = \eta(\tilde{\boldsymbol{\theta}}(k), \boldsymbol{\phi}(k), C)$ and that H is integrable with respect to $\eta(\tilde{\boldsymbol{\theta}}, \boldsymbol{\phi}, \cdot)$ for each $(\tilde{\boldsymbol{\theta}}, \boldsymbol{\phi}) \in \Re^m \times E_1$. Define

$$\bar{H}(\tilde{\boldsymbol{\theta}}, \boldsymbol{\phi}) = \int_{E_2} H(\tilde{\boldsymbol{\theta}}, \boldsymbol{\phi}, \xi)\eta(\tilde{\boldsymbol{\theta}}, \boldsymbol{\phi}, \mathrm{d}\xi) \tag{4.60}$$

The probability distribution η of the disturbance term is used in (4.60) to smooth out, or average, H through the action of the integral. Of most significance for the present purpose is that $\bar{H}$ can be continuous even when H is not.

We assume the following:

C1. $\{\boldsymbol{\phi}(k)\}$ is stationary and ergodic, there is a sequence of i.i.d. E_3-valued random variables $\{\psi(k)\}$, independent of $\{\boldsymbol{\phi}(k)\}$, and a measurable function q: $\Re^d \times E_1 \times E_3 \to E_2$ such that $\xi(k) = q(\tilde{\boldsymbol{\theta}}(k-1), \boldsymbol{\phi}(k-1), \psi(k))$, and $\tilde{\boldsymbol{\theta}}(0)$ is independent of $\{(\boldsymbol{\phi}(k-1), \psi(k))\}$. And $v_\phi \in \mathscr{P}(E_1)$ will denote the distribution of $\boldsymbol{\phi}(k-1)$.

C2. $\bar{H}$ is continuous, and for compact $K \subset \Re^d$

$$\begin{aligned} &E[\sup\nolimits_{\tilde{\boldsymbol{\theta}} \in K} \mid H(\tilde{\boldsymbol{\theta}}, \boldsymbol{\phi}(k-1), q(\tilde{\boldsymbol{\theta}}, \boldsymbol{\phi}(k-1), \psi(k))) \mid] < \infty \\ &E[\sup\nolimits_{\tilde{\boldsymbol{\theta}} \in K} \mid \bar{H}(\tilde{\boldsymbol{\theta}}, \boldsymbol{\phi}(k-1)) \mid] < \infty \end{aligned} \tag{4.61}$$

Note that there no assumptions on the autocorrelations of the inputs or disturbances. H is allowed to be discontinuous, provided that the expectation over η is smooth enough to make $\bar{H}$ continuous. Just as $\bar{H}$ averages H, the distribution of $\boldsymbol{\phi}(k)$ is used to average $\bar{H}$ over the inputs $\boldsymbol{\phi}(k)$, and the doubly averaged quantity

$$\hat{H}(\tilde{\boldsymbol{\theta}}) = \int \bar{H}(\bar{\boldsymbol{\theta}}, \boldsymbol{\phi}) v_{\phi}(\mathrm{d}\boldsymbol{\phi}) \tag{4.62}$$

is the key ingredient in the ODE and to the questions of stability.

THEOREM 4.5.1
(Stochastic excitation theorem) Let $\tilde{\boldsymbol{\theta}}_{\mu}(t) = \tilde{\boldsymbol{\theta}}([t/\mu])$, and for compact $K \subset \Re^m$, define $\tau_{\mu}^{K} = \inf\{t : \tilde{\boldsymbol{\theta}}_{\mu}(t) \notin K\}$. Denote the minimum of a and b as $a \wedge b$ and let $\tilde{\boldsymbol{\theta}}_{\mu}^{\tau_{\mu}^{K}}(\cdot) = \tilde{\boldsymbol{\theta}}_{\mu}(\cdot \wedge \tau_{\mu_K})$ define the 'stopped' process. Assume C1 and C2, and that $\tilde{\boldsymbol{\theta}}_{\mu}(0) \to \tilde{\boldsymbol{\theta}}_0$ as $\mu \to 0$. Then for each K, $\{\tilde{\boldsymbol{\theta}}_{\mu}^{\tau_{\mu}^{K}}, \mu > 0\}$ is relatively compact, and every limit point a(as $\mu \to 0$) satisfies

$$\tilde{\boldsymbol{\theta}}(t) = \tilde{\boldsymbol{\theta}}_0 - \int_0^t \hat{H}(\tilde{\boldsymbol{\theta}}(s))\mathrm{d}s \tag{4.63}$$

for $t < \tau^K = \inf\{t : \tilde{\boldsymbol{\theta}}(t) \notin K\}$.

COROLLARY 4.5.1
Define $\check{H}(\tilde{\boldsymbol{\theta}}, \boldsymbol{\phi}, \mathbf{z}) = H(\tilde{\boldsymbol{\theta}}, \boldsymbol{\phi}, q(\tilde{\boldsymbol{\theta}}, \boldsymbol{\phi}, \mathbf{z}))$. Let $C = \{(\tilde{\boldsymbol{\theta}}, \boldsymbol{\phi}, \mathbf{z}) : \check{H}$ is continuous at $(\tilde{\boldsymbol{\theta}}, \boldsymbol{\phi}, \mathbf{z})\}$, and let I_C be the indicator function of the set C. If $\iint I_C(\tilde{\boldsymbol{\theta}}, \boldsymbol{\phi}, \mathbf{z}) v_{\phi}(\mathrm{d}\boldsymbol{\phi}) v_{\psi}(d\mathbf{z}) = 1$, for every $\tilde{\boldsymbol{\theta}}$, the assumption of the continuity of $\bar{H}$ can then be dropped, and if in addition the solution of (4.63) is unique, the convergence of $\tilde{\boldsymbol{\theta}}_{\mu}$ to $\tilde{\boldsymbol{\theta}}$ is almost sure.

The theorem and corollary are proven in [20].

Consider the various elements of this theorem. A new process $\tilde{\boldsymbol{\theta}}_{\mu}(t)$ is defined as a time-scaled version of the original $\tilde{\boldsymbol{\theta}}(k)$ for each step-size μ. The time scaling compresses the original process variation into a smaller time frame. In effect, the gross motion of the parameter estimate errors of $\tilde{\boldsymbol{\theta}}_{\mu}$ remains unchanged for various μ. Large μ imply larger steps of $\tilde{\boldsymbol{\theta}}_{\mu}$, but fewer steps are taken. Smaller μ imply smaller updates of $\tilde{\boldsymbol{\theta}}_{\mu}$, but more steps are taken. Thus the result is applicable to reasonable step sizes due to the time rescaling, even though it is exact only asymptotically in μ.

The stopping time τ_{μ}^{K} measures how long it takes the time-scaled process $\tilde{\boldsymbol{\theta}}_{\mu}(t)$ to reach the edge of some closed and bounded set K. The stopped process $\{\tilde{\boldsymbol{\theta}}_{\mu}^{\tau_{\mu}^{K}}(t)\}$ is defined to be equal to $\tilde{\boldsymbol{\theta}}_{\mu}(t)$ from time 0 to the stopping time τ_{μ}^{K} and is then held constant for all $t > \tau_{\mu}^{K}$. The theorem asserts that for any given compact set K, every possible sequence (as $\mu \to 0$) of the stopped process $\{\tilde{\boldsymbol{\theta}}_{\mu}^{\tau_{\mu}^{K}}(t)\}$ contains a convergent subsequence, and that every limit of these subsequences is a process that satisfies the ODE (4.63), at least up until the stopping time. If the solution to the differential equation is unique, then the sequence actually converges (not just as a convergent subsequence).

The stability of the ODE can be determined by linearizing (4.63) about $\tilde{\theta} = 0$, that is, by calculating $M = d\hat{H}(\tilde{\theta})/d\tilde{\theta}\,|_{\tilde{\theta}=0}$. If all eigenvalues of the resulting matrix M have positive real parts, then the ODE is exponentially stable (note the minus sign before the integral in (4.63)), while if some eigenvalue has a negative real part, then the ODE is unstable. These stability and instability results translate directly into convergence and divergence results for the algorithms.

The ramifications of the stochastic excitation theorem for particular adaptive algorithms will be examined in the next section. Note that the theorem is a form of 'law of large numbers' where the time-scaled process $\tilde{\theta}_\mu(t)$ plays the role of 'observations' and the convergent process $\tilde{\theta}(t)$ plays the role of the 'expected value' to which the $\tilde{\theta}_\mu(t)$ converge as the number of observations t/μ increases. To investigate how this convergence occurs, the corresponding 'central limit theorem' describes the weak convergence of the error process

$$V_\mu(t) = \frac{1}{\sqrt{\mu}}(\tilde{\theta}_\mu(t) - \tilde{\theta}(t)) \tag{4.64}$$

where the scaling factor $1/\sqrt{\mu}$ expands V_μ to compensate for the time compression of $\tilde{\theta}_\mu(t)$. The next theorem shows that the error process V_μ converges to a forced ODE that is driven by the sum of two independent, mean-zero Brownian motions. One driving term accounts for the error introduced by the smoothing with the disturbance $(H - \bar{H})$, while the other $(\hat{H} - \bar{H})$ accounts for the error when averaging over the inputs.

To understand this, imagine that the solution trajectory of the ODE is a smooth curve in $\Re^m$. The stochastic excitation theorem asserts that trajectories of the algorithm tend to follow this curve, though any particular trajectory will make occasional excursions which wiggle about the curve. The central limit theorem below describes how this wiggling occurs by showing that it can be described in terms of a stationary Gauss–Markov process. This is what is meant by the statement that the algorithm converges to a stationary distribution about the solution of the ODE.

Assume that H is square integrable with respect to $\eta(\tilde{\theta}, \phi, \cdot)$ for each pair $(\tilde{\theta}, \phi) \in \Re^d \times E_1$. Let $G(\tilde{\theta}, \phi, \xi) = (H(\theta, \phi, \xi) - \bar{H}(\tilde{\theta}, \phi))(H(\tilde{\theta}, \phi, \xi) - \bar{H}(\tilde{\theta}, \phi))^{\mathrm{T}}$ be the matrix that represents the deviation of H from its smoothed version $\bar{H}$, and define a smoothed version of G as

$$\bar{G}(\tilde{\theta}, \phi) = \int_{E_2} G(\tilde{\theta}, \phi, \xi)\eta(\tilde{\theta}, \phi, d\xi) \tag{4.65}$$

Averaging over all inputs yields

$$\hat{G}(\tilde{\theta}) = \int \bar{G}(\tilde{\theta}, \phi)v_\phi(d\phi) \tag{4.66}$$

The various Gs play a similar role in the central limit theorem that the Hs play in the previous theorem. In addition to C1 and C2, we make the further assumptions:

C3. $\bar{H}$ is differentiable as a function of $\tilde{\theta}$, $\bar{G}$ and $\partial_{\tilde{\theta}}\bar{H}$ are continuous, and for compact $K \subset \Re^d$

$$E[\sup_{\tilde{\theta} \in K} |H(\tilde{\theta}, \phi(k), q(\tilde{\theta}, \phi(k), \psi(k)))|^2] < \infty$$
$$E[\sup_{\tilde{\theta} \in K} |\bar{G}(\tilde{\theta}, \phi(k))|] < \infty$$
$$E[\sup_{\tilde{\theta} \in K} |\partial_{\tilde{\theta}}\bar{H}(\tilde{\theta}, \phi(k))|] < \infty$$

Note that C3 implies $\hat{H}$ is locally Lipschitz (in fact continuously differentiable), so the solution of (4.63) is unique and hence $V_\mu(t)$ is well defined. For simplicity (so we do not have to stop our process outside of a compact set), we assume that the solution exists for all $t \geqslant 0$. Define

$$\tilde{\mathbf{M}}_\mu(t) = \sum_{k=0}^{[t/\mu]-1} (H(\tilde{\theta}(k-1), \phi(k-1), \xi(k)) - \bar{H}(\tilde{\theta}(k-1), \phi(k-1)))\sqrt{\mu} \quad (4.67)$$

and

$$\mathbf{L}_\mu(t) = \sum_{k=0}^{[t/\mu]-1} (\bar{H}(\tilde{\theta}(k\mu), \phi(k)) - \hat{H}(\tilde{\theta}(k\mu)))\sqrt{\mu} \quad (4.68)$$

There are a variety of different conditions (for example, mixing conditions on $\{\phi(k)\}$) that imply $\{L_\mu\}$ converges in distribution to a (time inhomogeneous) Brownian motion. We simply assume this convergence.

C4. $L_\mu \Rightarrow L$.

THEOREM 4.5.2
(Central limit theorem) Assume C1–C4, that $\tilde{\theta}_\mu(0) \rightarrow \tilde{\theta}_0$, that the solution of (4.63) exists for all $t \geqslant 0$, and that $\mathbf{V}_\mu(0) \rightarrow v_0$. Then $\tilde{\mathbf{M}}_\mu \Rightarrow \tilde{\mathbf{M}}$ where $\tilde{\mathbf{M}}$ is a mean-zero Brownian motion independent of L with

$$E[\tilde{\mathbf{M}}(t)\tilde{\mathbf{M}}(t)^{\mathrm{T}}] = \int_0^t \hat{G}(\tilde{\theta}(s))\mathrm{d}s$$

and $\mathbf{V}_\mu \Rightarrow \mathbf{V}$ satisfying

$$\mathbf{V}(t) = \mathbf{v}_0 + \tilde{\mathbf{M}}(t) + \mathbf{L}(t) - \int_0^t \partial_{\tilde{\theta}}\hat{H}(\tilde{\theta}(s))\mathbf{V}(s)\mathrm{d}s \quad (4.69)$$

This theorem is taken from [20] where a proof can be found.

These results can be extended in a variety of directions with little or no change in the hypotheses. For example, consider the asymptotics of the 'tracking problem' for adaptive filters. Let $\tilde{\theta}^*(k)$ denote the time-varying 'correct' filter coefficients that the adaptive filter is attempting to track, and let $\theta(k)$ be the parameter estimates. The parameter estimate error is then $\tilde{\theta}(k) = \tilde{\theta}^*(k) - \theta(k)$, which evolves according to

$$\tilde{\theta}(k) = \tilde{\theta}(k-1) - \mu H(\tilde{\theta}(k-1), \phi(k-1), \xi(k)) + (\tilde{\theta}^*(k) - \tilde{\theta}^*(k-1)) \quad (4.70)$$

Clearly, some restrictions must be placed on the possible motion of the filter $\tilde{\theta}^*$. One possibility is to assume the following:

C5. $\tilde{\boldsymbol{\Psi}}^*(k) = \Psi(k\mu)$ where $\boldsymbol{\Psi}(k\mu)$ where $\boldsymbol{\Psi}$ is a differentiable function with derivative denoted by ψ. It is then easy to show that Equation (4.63) can be replaced by

$$\mathrm{d}\tilde{\theta}(t) = -\psi \mathrm{d}t - \hat{H}(\tilde{\theta}(t))\mathrm{d}t$$

or

$$\tilde{\theta}(t) = \tilde{\theta}_0 - \int_0^t \psi(s)\mathrm{d}s - \int_0^t \hat{H}(\tilde{\theta}(t))\mathrm{d}t \tag{4.71}$$

The implications of (4.71), in terms of the tracking capabilities of the various adaptive algorithms, are explored in the next section. Note that in books such as [41], the martingale convergence theorem is viewed as an alternative to the ODE approach. Here (and in [20] and [5]) the two approaches are married.

4.6 Examples, comparisons and discussion

Several of the children of LMS are examined concretely in light of the deterministic and stochastic approaches of the previous sections. The intent is to compare the algorithms with each other and to compare the types of conclusions possible from the analytical methods. Progress in the analysis of adaptive algorithms has often alternated between the deterministic and stochastic realms. The deterministic approach typically assumes that the disturbances are identically zero, proves an exponential stability result, and then uses some form of total stability to guarantee robustness to disturbances. Speaking loosely, the deterministic 'persistence of excitation condition' tends to function analogously to the stochastic excitation conditions derived via linearization of $\hat{H}$. For example, the LMS algorithm is exponentially stable when $\Sigma\phi\phi^{\mathrm{T}}$ is positive definite, which clearly parallels the stochastic excitation condition which requires that $E[\phi\phi^{\mathrm{T}}]$ be positive definite. One need only replace the Greek letter Σ with the Latin letter E!

What is the relation between these conditions? The deterministic condition (even when relaxed for general non-periodic inputs) is strictly stronger than the stochastic condition, since it requires uniform positive definiteness of $\Sigma_s\phi\phi^{\mathrm{T}}$ over *every* window of some length s. Most stochastic processes will fail this for some windows, albeit with small probability for large s. Thus the hypotheses required to demonstrate exponential stability via the deterministic approach are stronger. So, too, are the conclusions. Even when stable, there exist sample paths that cause the parameter errors to attain arbitrarily large values under the stochastic assumptions, though these events are of vanishing probability [42]. In contrast, the total stability conclusions of the deterministic approach are absolute; the parameter errors *never* leave the appropriate δ ball. The disadvantage is that it is hard to determine, a priori, a reasonable bound for δ, and it is impossible to say anything at all about the behaviour of the parameter estimates inside the δ ball. In contrast, the stochastic method gives a stationary distribution that describes how the parameter errors 'rattle around' their converged values. This distribution can often be calculated, although it is only exact asymptotically (as $\mu \to 0$). To apply the deterministic method:

- Determine the appropriate F and g (4.47).
- Derive the relevant PE condition (4.53).
- Determine stability/instability and convergence rates based on the PE condition and the type of inputs expected in the given application.

To apply the stochastic approximation approach:

- Define appropriate $\boldsymbol{\phi}(k)$ (input) and H (update term).
- Find the unforced ODE (4.57) by calculating the smoothed versions $\bar{H}$ and $\hat{H}$.
- Check local stability of the ODE by linearizing $\hat{H}$ about the equilibrium $\tilde{\boldsymbol{\theta}} = 0$ (recall that $\tilde{\boldsymbol{\theta}} = 0$ precisely when the algorithm has achieved its optimum performance).
- Examine the forced ODE (4.69) to determine the convergent distribution of the algorithm.

4.6.1 LMS

Analysis of LMS does not require all the machinery of the last two sections, but it is an important special case. In the absence of disturbances, the deterministic approach shows that the LMS parameter estimates converge exponentially to $\boldsymbol{\theta}^*$ if the step size is chosen 'small enough', and if the excitation matrix $M = \Sigma\boldsymbol{\phi}\boldsymbol{\phi}^{\mathrm{T}}$ (4.53) is positive definite. Convergence occurs at a rate proportional to the smallest eigenvalue α of M. When disturbances are present, the convergence is to a δ ball about $\boldsymbol{\theta}^*$, where the size of the ball is proportional to the disturbance $\xi(k)$ and inversely proportional to α and μ. Unfortunately, there is no easy way to determine the constants of proportionality or to state explicitly how large 'small' can be.

The condition that $E[\boldsymbol{\phi}\boldsymbol{\phi}^{\mathrm{T}}]$ is positive definite implies convergence in distribution is well known [2], though it appeared that the limiting distribution was strongly dependent on the input distribution [10]. Our theorem demonstrates that the limiting distribution is closely approximated by a Gauss–Markov process irrespective of the input, assuming sufficiently smooth disturbances, mixing and sufficiently small step size. This result was foreshadowed in [9] (under the condition that the inputs are Gaussian), and the result is implicit in [5] and [40].

From (4.7) and (4.62), the smoothed version of H can be shown to be

$$\hat{H}(\tilde{\boldsymbol{\theta}}) = E[\boldsymbol{\phi}\boldsymbol{\phi}^{\mathrm{T}}]\tilde{\boldsymbol{\theta}} \tag{4.72}$$

assuming that the disturbances have zero mean. Since (4.72) is already linear, its 'linearization' is

$$\frac{\partial}{\partial \tilde{\boldsymbol{\theta}}} \hat{H}(\tilde{\boldsymbol{\theta}}) = E[\boldsymbol{\phi}\boldsymbol{\phi}^{\mathrm{T}}] \tag{4.73}$$

For the 'central limit theory' we may easily verify $\bar{G} = \sigma_u^2 \boldsymbol{\phi}\boldsymbol{\phi}^{\mathrm{T}}$, $G(\bar{\boldsymbol{\theta}}) = \sigma_{\phi^2} I \sigma_u^2$, and $L(t) = 0$. Define $\alpha = \sigma_{\phi^2}$ and $\sigma^2 = \sigma_{\phi^2}\sigma_u^2$. Then the stationary distribution of $\tilde{\boldsymbol{\theta}}_{[t/\mu]}$ is (approximately) $N(0, \mu\sigma^2/2\alpha) = N(0, \mu\sigma_u^2/2)$.

4.6.2 Normalized LMS

The simplest way to analyze the NLMS algorithm is to rewrite (4.11) as

$$\tilde{\boldsymbol{\theta}}(k) = \tilde{\boldsymbol{\theta}}(k-1) - \mu\boldsymbol{\phi}(k-1)\boldsymbol{\phi}^{\mathrm{T}}(k-1)\tilde{\boldsymbol{\theta}}(k) \tag{4.74}$$

as is done in [43]. Note that (4.74) has an 'a posteriori' error $\boldsymbol{\phi}^{\mathrm{T}}(k-1)\tilde{\boldsymbol{\theta}}(k)$ in the rightmost term in contrast to the more common 'a priori' error $\boldsymbol{\phi}^{\mathrm{T}}(k-1)\tilde{\boldsymbol{\theta}}(k-1)$ of (4.10). None the less, all the analysis of the previous sections applies without change to (4.74) because the a priori and a posteriori parameter errors are virtually the same for small μ. As expected, the conditions for stability are that $\Sigma\boldsymbol{\phi}\boldsymbol{\phi}^{\mathrm{T}}(E[\boldsymbol{\phi}\boldsymbol{\phi}^{\mathrm{T}}])$ be positive definite for the deterministic (stochastic) analysis.

4.6.3 Leakage

Although leakage is one of the most used variants of LMS it is surprisingly difficult to analyze its behaviour precisely. At a global level, it is easy to see that (4.12) is bounded-input bounded-output stable for small μ, since all eigenvalues of $(1-\lambda)I$ are strictly within the unit circle. The problem arises because the equilibrium of the system is not independent of the inputs. Consider a scalar version of (4.13) with disturbance $\xi(k) = 0$ and with constant input $\boldsymbol{\phi}(k) = \boldsymbol{\phi}$

$$\tilde{\boldsymbol{\theta}}(k) = [(1-\lambda) - \mu\boldsymbol{\phi}^2]\tilde{\boldsymbol{\theta}}(k-1) + \lambda\tilde{\boldsymbol{\theta}}^* \tag{4.75}$$

which has an equilibrium at $\tilde{\boldsymbol{\theta}} = \lambda\boldsymbol{\theta}^*/(\lambda + \mu\boldsymbol{\phi}^2)$. Clearly, this equilibrium changes for different values of $\boldsymbol{\phi}$. Moreover, $\tilde{\boldsymbol{\theta}}$ is biased away from zero, and hence $\boldsymbol{\theta}(k)$ is biased away from the desired parameterization $\boldsymbol{\theta}^*$. Because of this dependence of the equilibrium point on the input, it is difficult to carry out the linearization in either the deterministic or stochastic approaches. It remains an open issue how to deal with this situation.

4.6.4 Dead zone

The dead-zone algorithm (4.18) cannot have an asymptotically stable equilibrium at the origin because the dead-zone non-linearity $g(x)$ of (4.17) is insensitive to small perturbations of x about zero, that is, (4.18) implies that the algorithm behaves like

$\boldsymbol{\theta}(k) = \boldsymbol{\theta}(k-1)$ near zero. Rather, the algorithm is locally exponentially stable to the region $R = [-d, d]$ as long as $\Sigma\ \boldsymbol{\phi}\boldsymbol{\phi}^{\mathrm{T}}$ is positive definite. As usual, the stochastic analog requires that $E[\boldsymbol{\phi}\boldsymbol{\phi}^{\mathrm{T}}]$ be positive definite.

4.6.5 Signed error

The signed-error algorithm fails assumption 3 of the deterministic approach (which requires differentiability of g at the origin), and the linearization and averaging approach fails. In fact, the equilibrium at $\tilde{\boldsymbol{\theta}} = 0$ is unstable in the sense of Lyapunov since the signum function has 'infinite gain'. However, a different line of deterministic reasoning, an extended Lyapunov approach, can be used as in [10] to demonstrate that the algorithm is totally stable (convergent to a small ball about the origin) when $\Sigma\ \boldsymbol{\phi}\boldsymbol{\phi}^{\mathrm{T}}$ is positive definite.

Using an expected value approach (and assuming the independence of $\tilde{\boldsymbol{\theta}}(k-1)$ from $\boldsymbol{\phi}(k-1)$), Gersho [29] shows that the signed-error algorithm converges in distribution to the optimal solution plus a term dependent on the step size when the inputs are jointly Gaussian. The stochastic approximation approach does not require this independence assumption, nor does it require differentiability of H because the disturbance ξ smoothes H enough so that $\hat{H}$ can be differentiated. Suppose that the probability distribution function η is absolutely continuous with density f_ξ. Then conditions C1 and C2 (and hence the theorem) hold, and the corresponding linearization is

$$\frac{\partial}{\partial \tilde{\boldsymbol{\theta}}} \hat{H}(0) = 2f_\xi(0)E[\boldsymbol{\phi}\boldsymbol{\phi}^{\mathrm{T}}] \tag{4.76}$$

The signum function has essentially been 'smoothed away' by the averaging effect of the disturbance. The central limit results follow easily, with $\bar{G}(\tilde{\boldsymbol{\theta}}, \boldsymbol{\phi}) = \boldsymbol{\phi}\boldsymbol{\phi}^{\mathrm{T}}\{1 - [1 - 2\eta(-\boldsymbol{\phi}^{\mathrm{T}}\tilde{\boldsymbol{\theta}})]^2\}$, $\hat{G}(\tilde{\boldsymbol{\theta}}) = \sigma_\varphi^2 I$. Since $\bar{H}(0, \boldsymbol{\phi}(k)) = \hat{H}(0, \boldsymbol{\phi}(k)) = 0$, $\mathbf{L}(t) = 0$, and the limiting stochastic differential equation (4.69) is

$$\mathbf{V}(t) = \mathbf{v}_0 + \tilde{\mathbf{M}}(t) - 2f_\xi(0)E[\boldsymbol{\phi}\boldsymbol{\phi}^{\mathrm{T}}] \int_0^t \mathbf{V}(s)\mathrm{d}s \tag{4.77}$$

Again, under the assumptions of an independent input sequence and symmetric noise, $E[\boldsymbol{\phi}\boldsymbol{\phi}^{\mathrm{T}}] = \sigma_\phi^2 I$, $\eta(0) = 1/2$. Now define $\alpha = 2f_\xi(0)\sigma_\phi^2$ and $\sigma^2 = \sigma_\phi^2$. Hence, $\tilde{\boldsymbol{\theta}}_\mu(t) = \tilde{\boldsymbol{\theta}}_{[t/\mu]}$ has (approximately) a $N(0, \mu\sigma^2/2\alpha) = N(0, \mu/4f_\xi(0))$ density.

4.6.6 Signed regressor

The original analysis of the signed-regressor algorithm [11] assumed that the entries of the input are Gaussian, and also assumed the independence of the input $\boldsymbol{\phi}(k-1)$ and the parameter estimates $\boldsymbol{\theta}(k-1)$. Combined with the small step-size assumptions, this is enough to demonstrate mean convergence of the parameter estimates. The present approach removes these assumptions and sharpens the results.

The deterministic approach shows that if all eigenvalues of $\Sigma \text{sgn}(\boldsymbol{\phi})\boldsymbol{\phi}^{\text{T}}$ have positive real parts, then the algorithm is exponentially stable. Though 'most' sequences fulfil this condition, it is fairly easy to construct short periodic sequences that violate this condition, as is done in [15]. Such inputs cause the parameter estimates to diverge away from the optimal $\boldsymbol{\theta}^*$, no matter how small the step size.

The stochastic approach defines

$$\hat{H}(\tilde{\boldsymbol{\theta}}) = \iint \text{sgn}(\boldsymbol{\phi})(\boldsymbol{\phi}^{\text{T}}\tilde{\boldsymbol{\theta}} + \xi)\text{d}\eta(\xi)\text{d}F(\boldsymbol{\phi})$$

Assuming that the disturbance is symmetric with zero mean, this can be rewritten

$$\int \text{sgn}(\boldsymbol{\phi})(\boldsymbol{\phi}^{\text{T}}\tilde{\boldsymbol{\theta}})\text{d}F(\boldsymbol{\phi})$$

which can be linearized as

$$\frac{\partial}{\partial \tilde{\boldsymbol{\theta}}} \hat{H}(\tilde{\boldsymbol{\theta}}) = E[\text{sgn}(\boldsymbol{\phi})\boldsymbol{\phi}^{\text{T}}] \tag{4.78}$$

The signed-regressor algorithm was proved stable in [15] if all eigenvalues of $E[\text{sgn}(\boldsymbol{\phi})\boldsymbol{\phi}^{\text{T}}]$ have positive real parts, and instability was conjectured if an eigenvalue has negative real parts. The stochastic approximation approach shows that this instability conjecture is true, at least locally. Examples of non-trivial stochastic processes for which $E[\text{sgn}(\boldsymbol{\phi})\boldsymbol{\phi}^{\text{T}}]$ has negative eigenvalues were calculated in [15]. Such inputs destabilize the signed-regressor algorithm.

When the inputs cause the algorithm to be stable, the stochastic approximation theorem describes the limiting distributions. The 'central limit' results define $\bar{G}(\tilde{\boldsymbol{\theta}}, \boldsymbol{\phi}) = \text{sgn}(\boldsymbol{\phi})\text{sgn}(\boldsymbol{\phi}^{\text{T}})\sigma_{\xi^2}$, $\hat{G}(\tilde{\boldsymbol{\theta}}) = \mathbf{I}\sigma_{\xi^2}$, and hence $\mathbf{L}(t) = 0$. Then Equation (4.69) becomes

$$\mathbf{V}(t) = \mathbf{v}_0 + \tilde{\mathbf{M}}(t) - E[\text{sgn}(\boldsymbol{\phi})\boldsymbol{\phi}^{\text{T}}] \int_0^t \mathbf{V}(s)\text{d}s \tag{4.79}$$

Let $\alpha = E[\text{sgn}(\phi_1)\phi_1]$ and $\sigma^2 = \sigma_{\xi^2}$. Then $\tilde{\boldsymbol{\theta}}_{[t/\mu]}$ has (approximately) a $N(0, \mu\sigma^2/2\alpha) = N(0, \mu\sigma_u^2/2E[X_1\text{sgn}(X_1)])$ density.

4.6.7 Sign–sign

The first example of instability in an adaptive FIR filter demonstrated that the three periodic input sequence $\{3, -1, -1, 3, -1, -1, 3, -1, -1, \ldots\}$ can drive the parameter estimates of the sign–sign algorithm to infinity. This was shown by a simple inductive argument in [18] and did not give any method to distinguish between the class of signals that stabilize the algorithm and the class of signals that are destabilizing.

The stochastic approach allows such classification of signals in terms of the stochastic excitation matrix. Suppose that $\xi(k)$ is a real-valued i.i.d. disturbance with probability distribution function $\eta(\cdot)$. Define $\bar{\boldsymbol{\phi}} = (\boldsymbol{\phi}, \text{sgn}(\boldsymbol{\phi}))$. Then

$$\bar{H}(\tilde{\boldsymbol{\theta}}, \bar{\boldsymbol{\phi}}) = \text{sgn}(\boldsymbol{\phi}) \int \text{sgn}(\boldsymbol{\phi}^{\text{T}}\tilde{\boldsymbol{\theta}} + \xi)\text{d}\eta(\xi) = \text{sgn}(\boldsymbol{\phi})[1 - 2\eta(-\boldsymbol{\phi}^{\text{T}}\tilde{\boldsymbol{\theta}})] \tag{4.80}$$

is continuous in $(\tilde{\boldsymbol{\theta}}, \bar{\boldsymbol{\phi}})$ if, for example, $\eta(\cdot)$ is absolutely continuous with density $f_\xi(\cdot)$. Consequently, conditions C1 and C2 (and hence the theorem) hold. If $F(\cdot)$ denotes the marginal distribution function of $\{X_k\}$, then

$$\hat{H}(\tilde{\boldsymbol{\theta}}) = \int \text{sgn}(\boldsymbol{\phi})[1 - 2\eta(-\boldsymbol{\phi}^{\text{T}}\tilde{\boldsymbol{\theta}})]\text{d}F(\boldsymbol{\phi}) \tag{4.81}$$

which can be linearized about the equilibrium $\tilde{\boldsymbol{\theta}} = 0$ as

$$\frac{\partial}{\partial\tilde{\boldsymbol{\theta}}}\hat{H}(0) = 2f_\xi(0)E[\text{sgn}(\boldsymbol{\phi})\boldsymbol{\phi}^{\text{T}}] \tag{4.82}$$

For the central limit results, note that

$$\bar{G}(\tilde{\boldsymbol{\theta}}, \bar{\boldsymbol{\phi}}) = \text{sgn}(\boldsymbol{\phi})\text{sgn}(\boldsymbol{\phi}^{\text{T}})\{1 - [1 - 2\eta(-\tilde{\boldsymbol{\theta}}^{\text{T}}\boldsymbol{\phi})]^2\} \tag{4.83}$$

For non-trivial, symmetric i.i.d. inputs, $E[\text{sgn}(\boldsymbol{\phi}\boldsymbol{\phi}^{\text{T}})] = I$, and

$$\hat{G}(\tilde{\boldsymbol{\theta}}) = I - E_\phi[\text{sgn}(\boldsymbol{\phi}\boldsymbol{\phi}^{\text{T}})[1 - 2\eta(-\boldsymbol{\phi}^{\text{T}}\tilde{\boldsymbol{\theta}})]^2] \tag{4.84}$$

or $\hat{G}(0) = \{1 - [1 - 2\eta(0)]^2\}I = 4[\eta(0) - \eta(0)^2]I = I$ for the case of symmetric noise. Define $\alpha = 2f_\xi(0)E[\phi_1\text{sgn}(\phi_1)]$ and $\sigma^2 = 1$. Practically speaking, this implies that for small μ, the approximation $\mathbf{V}_\mu(t) = (1/\sqrt{\mu})(\tilde{\boldsymbol{\theta}}_\mu(t) - \tilde{\boldsymbol{\theta}}(t)) \approx \mathbf{V}(t)$, where $\mathbf{V}(t)$ has a $N(0, \sigma^2/2\alpha)$ density, and $\tilde{\boldsymbol{\theta}}(t) \approx 0$. Hence $\tilde{\boldsymbol{\theta}}_\mu(t) = \tilde{\boldsymbol{\theta}}_{[t/\mu]}$ has (approximately) a $N(0, \mu\sigma^2/2\alpha) = N(0, \mu/4f_\xi(0)E[\phi_1\text{sgn}(\phi_1)])$ density.

The form of the stochastic excitation matrix $E[\text{sgn}(\boldsymbol{\phi})\boldsymbol{\phi}^{\text{T}}]$ ties the stability properties of the sign–sign algorithm to the stability properties of the signed-regressor algorithm, and it is reasonable to anticipate that a condition on the positivity of the (real parts of the) eigenvalues of $\Sigma\ \text{sgn}(\boldsymbol{\phi})\boldsymbol{\phi}^{\text{T}}$ will be the correct deterministic criterion for stability of the sign–sign algorithm. This is, however, false. Consider the 12 periodic input sequence $\{3, -1, -1, 3, -1, -1, 3, -1, -1, 3, -1, -7\}$. This can be shown (via an inductive argument) to destabilize the three-dimensional sign–sign algorithm just as the example in [18], but all eigenvalues of $\Sigma\ \text{sgn}(\boldsymbol{\phi})\boldsymbol{\phi}^{\text{T}}$ have positive real parts. Hence this input stabilizes the signed–regressor algorithm, but destabilizes the sign–sign algorithm. Thus $\Sigma\ \text{sgn}(\boldsymbol{\phi})\boldsymbol{\phi}^{\text{T}}$ is *not* the correct stability criterion for the deterministic sign–sign algorithm. Yet we have shown that both signed regressor and sign–sign algorithms are locally stable exactly when the real parts of the eigenvalues of $E[\text{sgn}(\boldsymbol{\phi})\boldsymbol{\phi}^{\text{T}}]$ are positive. The explanation of this apparent contradiction is simple, though somewhat surprising. Throughout the stochastic approximation approach, we have assumed that the disturbance term is 'smooth' enough to 'average out' the discontinuities. An identically zero disturbance does not give enough smoothing!

Thus the presence of disturbances is crucial to being able to state a concise condition for the stability of the algorithm. As evidence that this is the correct interpretation, one can resimulate the sign–sign algorithm with the same 12 periodic sequence above, but adding a small disturbance. The algorithm stabilizes, converging to a small ball about the optimal parameterization. This is discussed further in [20].

4.6.8 Quantized state

The quantization functions Q_1 and Q_2 of the QS algorithm (4.26) are typically 'staircase' functions which may be zero in some region about the origin, or they may be discontinuous at the origin. If Q_2 is differentiable at the origin, then the deterministic approach shows that $\Sigma\, Q_1(\phi)\phi^{\mathrm{T}}$ is the appropriate persistence of excitation condition. If Q_2 is discontinuous at the origin, then an extended Lyapunov approach can be used as in [19] to show total stability (though not exponential stability). As in the sign–sign algorithm, the stochastic approximation approach does not require continuity at the origin due to the smoothing of the disturbance. As expected, the stochastic analog of the PE condition is that all eigenvalues of $E[Q_1(\phi)\phi^{\mathrm{T}}]$ have positive real parts.

4.6.9 Least mean fourth

Deriving excitation conditions for the least mean qth algorithm (4.28) is straightforward in both the deterministic and stochastic settings. Since $F(\phi) = \phi$ and $g(e) = e^{q-1}$, $g'(0) = 0$ and the deterministic PE condition requires that $\Sigma\ \phi\phi^{\mathrm{T}}$ be positive definite. By the corollary, convergence is exponential to the region $[-r, r]$, and $r > 0$ can be chosen arbitrarily small. As expected, the equivalent stochastic condition is that $E[\phi\phi^{\mathrm{T}}]$ be positive definite.

4.6.10 Median LMS

The median LMS was included in this chapter as an example of an LMS variant that cannot be fully analyzed via any of the known methods. The approach in [23] is very much in the spirit of the 'expected value' analyses, and is restricted to showing mean convergence when the input and parameter vectors are assumed independent.

The median LMS cannot be analyzed in the deterministic framework because it does not have the 'nice' form of (4.2) which distinguishes between the function F on the regressor and the function g on the error. Moreover, the median function is not memoryless. It operates explicitly on 'old' values of both the regressor and the error. The stochastic approximation approach also cannot be applied directly to the median LMS due to its memory. Of course, there is nothing inherent in the approaches that forbids the analysis of such non-linearities with memory, but a significant effort may be required to extend the approaches to handle updates containing medians and other order statistic non-linearities.

4.6.11 Convergence and tracking of LMS and variants

One implication of the central limit theorem is that the signed variants of LMS converge to a Gaussian distribution with known mean and variance. A fair comparison of the convergence speed of the algorithms can be made by adjusting the step size so that the final distributions of all four algorithms are identical, and then to explore the convergence rates of the algorithms. Suppose the disturbance has mean zero and is symmetric with distribution $\eta(\cdot)$ and density $f_\xi(\cdot)$, and that the input has zero mean i.i.d. (Note that we do not assume that the regressor vector $\phi(k)$ is i.i.d.) Then the discussion of the previous sections shows that the convergent distribution of the algorithms is $N(0, \sigma^2\mu/2\alpha)$ where

- sign–sign: $\sigma^2 = 4(\eta(0) - \eta^2(0))$ and $\alpha = 2f_u(0)E[\text{sgn}(\phi)\phi^{\text{T}}]_{ii}$
- signed error: $\sigma^2 = 4\sigma_{\phi^2}(\eta(0) - \eta^2(0))$ and $\alpha = 2f_u(0)\sigma_{\phi^2}$
- signed regressor: $\sigma^2 = \sigma_u^2$ and $\alpha = E[\text{sgn}(\phi)\phi^{\text{T}}]_{ii}$
- LMS: $\sigma^2 = \sigma_{\phi^2}\sigma_u^2$ and $\alpha = \sigma_{\phi^2}$

where $E[\text{sgn}(\phi)\phi^{\text{T}}]_{ii}$ represents a diagonal term of the matrix $E[\text{sgn}(\phi)\phi^{\text{T}}]$. Suppose the input is i.i.d. uniform $[-0.5, 0.5]$ (which fulfils both stability criteria $E[\phi\phi^{\text{T}}]$ and $E[\text{sgn}(\phi)\phi^{\text{T}}]$), the distribution of the disturbance is $(1/10)N(0, 1)$, and the desired variance is 0.0025. This can be achieved by choosing

- $\mu = 0.01$ for sign–sign
- $\mu = 0.04$ for signed error
- $\mu = \sqrt{2\pi}/20$ for signed regressor
- $\mu = \sqrt{2\pi}/5$ for LMS

Using these four values, all four signed algorithms have the same convergent distribution. This is verified experimentally in Figure 4.1 (all figures are taken from [20]) which shows the four simulated densities for the LMS, signed-regressor, signed-error, and sign–sign algorithms. The data were gathered over 1 million iterations, and correspond remarkably well with the predicted Gaussian density. The theory asserts that these simulated densities must converge as μ vanishes; the diagram is striking because the step sizes are not 'small' compared with the sizes typically used in applications.

It is now possible to compare fairly the convergence speed and tracking abilities of the algorithms. Figure 4.2, for instance, shows the trajectories of the four algorithms with the same input and step sizes as above. All four are initialized at the same 'wrong' answer and converge towards zero (the desired answer). Typically, the algorithms which can respond to large errors by taking a larger step (LMS and signed regressor) converge faster than the algorithms which must react through the signum

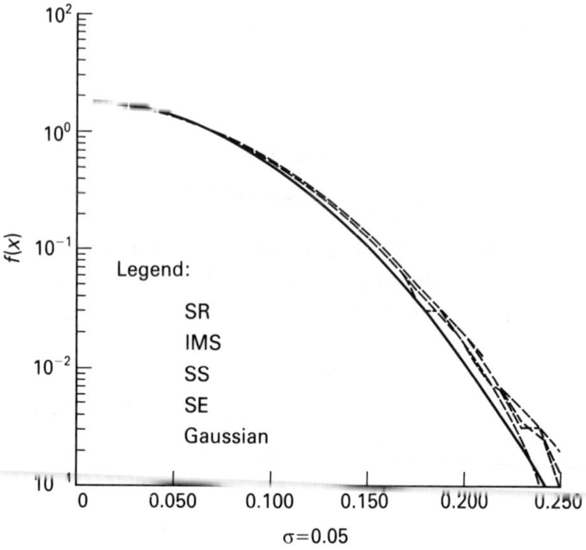

Figure 4.1 Predicted and actual error densities (From [19])

function of the error. This may not always be the case, however, since the relative performance of the algorithms may differ depending on the distributions of the input and disturbance processes. The importance of the central limit theorem in this regard is that it shows how to conduct such a study fairly, allowing a more knowledgeable choice of algorithm and step size for a given application setting.

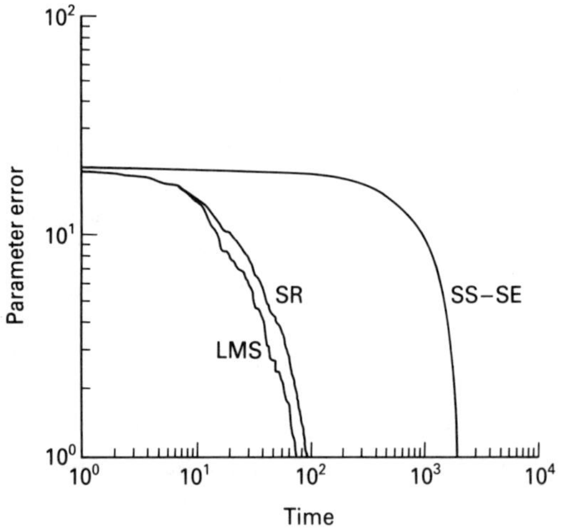

Figure 4.2 'Fair' convergence behaviour simulation (From [19])

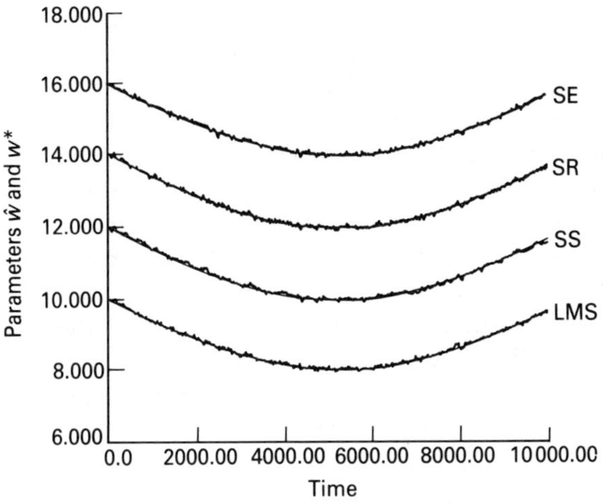

Figure 4.3 Tracking behaviour simulation (From [19])

A second important area in terms of performance is the algorithm's ability to track a moving parameterization. Reconsider (4.71). This ODE is forced by the term $\int \psi$, which represents the motion of the parameters that the algorithm is trying to identify. The term $\int \hat{H}(\tilde{\theta})$ represents the exponentially stable transient part (assuming all eigenvalues of linearization have positive real parts) that dies away as the algorithm converges to a region about the current $\tilde{\theta}^*$. Since (4.71) is essentially the same in all four cases (except for the details of the linearization), this implies that all four algorithms have roughly the same performance in terms of tracking ability, presuming the motion of $\tilde{\theta}^*$ is slow enough. Figure 4.3 shows simulations of the four algorithms tracking a slowly moving parameter (the four are offset from each other so that they can be distinguished in the diagram). The step-sizes are chosen as above so that the final convergent distributions match. As implied by the ODE analysis, it does not matter which algorithm is used when tracking a slowly moving parameterization. Differences in tracking performance would undoubtedly arise when the motion of $\tilde{\theta}^*$ became rapid. In this case, algorithms which converge faster will likely have an advantage over those (such as sign–sign) which have a bounded rate of change.

4.7 Conclusion

The three branches of analysis are complementary in the sense that they provide different insights into the behaviours of the various algorithms. The expected value approach gives usable answers and guidelines for implementation that are often quite straightforward, especially when the inputs are Gaussian. The deterministic results provide convergence and divergence proofs in terms of the eigenvalues of the excitation matrix, and total stability in the case of bounded disturbances. The stochastic convergence theorems are weaker, but they may allow calculation of the final distribution of the parameter estimates about their optimal value in terms of the distributions of the inputs and disturbances. On the other hand, the expected value approach suffers from problems with rigour, the deterministic approach is riddled with εs and δs that are hard to quantify, and the stochastic approximation approach can become tangled with an unmanageable number of expressions for densities and distributions, all defined in terms of one another.

There are, of course, numerous other approaches to the analysis of LMS and its variants. Lyapunov theory can often be applied to show convergence (or at least stability) in the ideal noise-free case. When possible, this is a powerful method because it can give global results, rather than local like the present averaging approach. Large deviations theory can be applied to answer questions about the expected length of time until the parameter estimates reach a certain bound, and may give insights into the expected time until failure of adaptive applications which contain a feedback of the error path back into the input of the adaptive element [44,45].

Recently, the Poisson clumping heuristic has been applied to describe the probability of reaching large bounds b as being distributed in a Poisson manner with parameter λ_b [42]. Together with the stochastic approximation results, this characterizes the behaviour of the parameter estimates quite fully. Near the equilibrium, the process behaves in an essentially Gaussian manner, while far from the equilibrium, it is Poisson.

Even after all these years and all these papers, there are still new questions to be asked about the LMS family. Some issues are the following:

- How can the approaches be generalized to apply to the median and other order statistic algorithms?
- What happens to the deterministic and stochastic excitation conditions when the adaptive element is enclosed in a feedback loop?
- The extension of the approaches to the IIR (Infinite Impulse Response) adaptive filters is sometimes straightforward, though often such extensions require new insights.
- How can the εs and δs of the deterministic approach be quantified?
- How can the bias inherent in algorithms such as the LMS with leakage be precisely analyzed?
- What is the actual excitation condition for the deterministic sign–sign algorithm?

References

[1] B. Widrow *et al.* 'Adaptive noise cancelling: Principles and applications', *Proc. IEEE*, vol. 63, pp. 1692–716, 1975.

[2] B. Widrow, J.M. McCool, M.G. Larimore and C.R. Johnson Jr 'Stationary and nonstationary learning characteristics of the LMS adaptive filter', *Proc. IEEE*, vol. 64, pp. 1151–62, 1976.

[3] L. Ljung, 'Analysis of recursive stochastic algorithms', *IEEE Trans. Autom. Control*, vol. 22, pp. 551–75, Aug. 1977.

[4] H.J. Kushner and H. Huang, 'Asymptotic properties of stochastic approximations with constant coefficients', *SIAM J. Control Optimization*, vol. 19, Jan, 1981.

[5] A. Benveniste, M. Goursat and G. Ruget, 'Analysis of stochastic approximation schemes with discontinuous and dependent forcing terms with applications to data communication algorithms', *IEEE Trans. Autom. Control*, vol. AC-25, pp. 1042–58, 1980.

[6] B. Widrow and S.D. Stearns, *Adaptive Signal Processing*, Prentice Hall: Englewood Cliffs, NJ, 1985.

[7] S. Haykin, *Adaptive Filter Theory*, Prentice Hall: Englewood Cliffs, NJ, 1986.

[8] Jayant and Noll, *Digital Coding of Waveforms*, Prentice Hall: Englewood Cliffs, NJ, 1984.

[9] N.J. Bershad and L.Z. Qu, 'On the probability density function of the complex scalar LMS adaptive weights', *IEEE Trans. Acoust., Speech, Signal Process.*, vol. ASSP-37, Jan., 1989.

[10] R.R. Bitmead, 'Convergence in distribution of LMS-type adaptive parameter estimates', *IEEE Trans. Autom. Control*, vol. AC-28, Jan., 1983.

[11] J.L. Moschner, 'Adaptive equalization via fast quantized state methods', Tech. Dept. 6796-1, Information Systems Laboratory, Stanford University, 1970.

[12] C.F.N. Cowan and P.M. Grant, *Adaptive Filters*, Prentice Hall: Englewood Cliffs, NJ, 1985.

[13] T.A.C.M. Classen and W.F.G. Mecklenbrauker, 'Comparison of the convergence of two algorithms for adaptive FIR digital filters', *IEEE Trans. Acoust., Speech, Signal Process.*, vol. ASSP-29, June 1981.

[14] N.J. Bershad, 'Comments on "Comparison of the convergence of two algorithms for adaptive FIR digital filters"', *IEEE Trans. Acoust., Speech, Signal Process.*, vol. ASSP-33, Dec., 1985.

[15] W.A. Sethares, I.M.Y. Mareels, B.D.O. Anderson and C.R. Johnson Jr 'Excitation conditions for sign-regressor LMS', *IEEE Trans. Circuits Syst.*, vol. 35, pp. 613–25, June, 1988.

[16] R.W. Lucky, 'Techniques for adaptive equalization of digital communication systems', *Bell Syst. Tech. J.*, vol. 45, Feb., 1966.

[17] *CCIT Red Book*, Recommendation G721, vol. III-3, Oct., 1984.

[18] S. Dasgupta and C.R. Johnson Jr 'Some comments on the behaviour of sign–sign adaptive identifiers', *Syst. Control Lett.*, vol. 7, pp. 75–82, 1986.

[19] W.A. Sethares and C.R. Johnson Jr. 'A comparison of two quantized state adaptive algorithms', *IEEE Trans. Acoust., Speech, Signal Process.*, vol. 37, pp. 138–43, Jan., 1989.

[20] J.A. Bucklew, T. Kurtyz and W.A. Sethares, 'Results on local stability of fixed step size recursive algorithms', *Proc. 1992 IEEE Conf. on Acoustics, Speech and Signal Processing, San Francisco.* Also submitted as 'Local stability and tracking properties of adaptive algorithms', to the *IEEE Trans. Inf. Theory.*

[21] C.P. Kwong, 'Dual sign algorithm for adaptive filtering', *IEEE Trans. Commun.*, vol. COM-34, Dec. 1986.

[22] D.L. Duttweiler, 'Adaptive filter performance with nonlinearities', *IEEE Trans. Acoust., Speech, Signal Process.*, vol. 30, Aug., 1982.

[23] G.A. Williamson, P.M. Clarkson and W.A. Sethares, 'Performance characteristics of the

median adaptive filter', Submitted to *IEEE Trans. Acoust., Speech, Signal Process.*

[24] J.E. Mazo, 'On the Independence theory of equalizer convergence', *Bell Syst. Tech. J.*, vol. 58, pp. 963–93, May, 1979.

[25] R. Price, 'A useful theorem for nonlinear devices having Gaussian inputs', *IRE Trans. Inf. Theory*, vol. IT-4, pp. 69–72, 1958.

[26] V.J. Matthews and C.H. Cho, 'Improved convergence analysis of stochastic gradient adaptive filters using the sign algorithm', *IEEE Trans. Acoust., Speech, Signal Process.*, vol. ASSP-35, pp. 450–4, Apr., 1987.

[27] J.B. Evans, P. Xue and B.D. Liu, 'Analysis and implementation of variable step-size adaptive algorithms', To appear in *IEEE Trans. Acoust., Speech, Signal Process.*

[28] E. Eweda, 'Optimum step size of sign algorithm for nonstationary adaptive filtering', *IEEE Trans. Acoust., Speech, Signal Process.*, vol. ASSP-38, pp. 1897–1902, Nov., 1990.

[29] A. Gersho, 'Adaptive filtering with binary reinforcement', *IEEE Trans. Inf. Theory*, vol. IT-30, pp. 191–8, 1984.

[30] R.Y. Chen and C.L. Wang, 'On the optimal step size for the adaptive sign and LMS algorithms', *IEEE Trans. Circuits Syst.*, June, 1990.

[31] W.A. Sethares, B.D.O. Anderson and C.R. Johnson Jr 'Adaptive algorithms with filtered regressor and filtered error', *Math. Control, Signals, Syst.*, vol. 2, pp. 381–403, 1989.

[32] R.R. Bitmead and C.R. Johnson Jr 'Discrete averaging principles and robust adaptive identification', *Control and Dynamic Systems: Advances in Theory and Applications*, vol. 24, ed. C.T. Leondes, Academic Press: London, 1986.

[33] W. Hahn, *Stability of Motion*, Springer, 1967.

[34] M. Vidyasagar, *Nonlinear Systems Analysis*, Prentice Hall, Englewood Cliffs, NJ, 1978.

[35] I.D. Landau, *Adaptive Control: The Model Reference Approach*, Marcel Dekker: New York, 1979.

[36] D.A. Lawrence and C.R. Johnson Jr 'Recursive parameter identification algorithm stability analysis via p-sharing', *IEEE Trans. Autom. Control*, vol. AC-25, Jan. 1986.

[37] B.D.O. Anderson, I.M.Y. Mareels, W.A. Sethares and C.R. Johnson, 'Averaging theory for sign–sign LMS', *Proc. 26th Ann. Allerton Conf. on Communication, Control, and Computing, September, 1988.*

[38] R.R. Bitmead, 'Persistence of excitation conditions and the convergence of adaptive systems', *IEEE Trans. Inf. Theory*, vol. IT-30, 183–91, 1984.

[39] W.A. Sethares, 'Adaptive algorithms with nonlinear data and error functions', Submitted to *IEEE Trans. Acoust., Speech, Signal Process.*

[40] H.JU. Kushner and A. Schwartz, 'An invariant measure approach to the convergence of stochastic approximations with state dependent noise', *SIAM J. Control Optimization*, vol. 22, Jan., 1984.

[41] G.C. Goodwin and K.S. Sin, *Adaptive Filtering, Prediction, and Control*, Prentice Hall: Englewood Cliffs, NJ, 1984.

[42] W.A. Sethares and J.A. Bucklew, 'Excursions of adaptive algorithms via the poisson clumping heuristic', Submitted to *IEEE Trans. Acoust., Speech, Signal Process.*

[43] C.R. Johnson Jr, *Lectures on Adaptive Parameter Estimation*, Prentice Hall: Englewood Cliffs, NJ, 1988.

[44] T. Brennan, 'Large deviation theory and the asymptotics of convergent LMS algorithms', *1990 Int. Symp. on Information Theory, San Diego, CA.*

[45] R.R. Bitmead and P.E. Caines, 'Escape time formulation of robust stochastic adaptive control', *Proc. 27th Conf. on Decision and Control, Austin TX, 1988.*

5

Fast Transversal RLS Algorithms*

Dirk T. M. Slock and Thomas Kailath

5.1 *Introduction*

Adaptive filters are a necessary ingredient in many signal processing applications (see [1,2]). They can tailor a filter's response to the specific needs of an unknown environment, or they can adapt to slow changes in non-stationary environments. There are two broad classes of adaptation algorithms. One is the least mean square (LMS) algorithm, which is based on a stochastic gradient approach (see the previous

* This work was supported in part by the US Army Research Office under Contract DAAL03-90-K-0109 and the ARO-SDIO under Contract DAAL03-90-G-0108.

This manuscript is submitted for publication with the understanding that the US Government is authorized to reproduce and distribute reprints for Government purposes notwithstanding any copyright notation thereon.

The views and conclusions contained in this document are those of the authors and should not be interpreted as necessarily representing the official policies or endorsements, either express or implied, of the Air Force Office of Scientific Research or the US Government.

chapter). The other is the recursive least-squares (RLS) algorithm, which minimizes a deterministic sum of squared errors. For comparable steady-state filter accuracy in a stationary environment, the RLS algorithm usually converges faster and reacts faster to environmental changes than the LMS algorithm (see [3–5] and chapter 3).

The computational complexity of the LMS algorithm, though, is $O(p)$ (p denotes the FIR filter order), whereas the conventional RLS algorithm shows a complexity of $O(p^2)$. However, for FIR filtering in time series applications, consecutive regression vectors show a shift-invariant structure. This allows for the derivation of fast RLS algorithms with a complexity of $O(p)$. These algorithms come in two groups, corresponding to two filter structures: the transversal filter and the lattice filter. Though both groups have a complexity of $O(p)$, the fast transversal RLS (FTRLS) algorithms have challenged the popularity of their lattice counterparts (FLA–FQR algorithms), based on a lower coefficient of p in the computations count. Furthermore, the recent numerical stabilization of the FTRLS algorithms [6–9] has significantly increased their potential applicability.

Adaptive FIR filters can be found in a wide range of applications [10] such as high-resolution spectrum estimation, noise cancellation, speech and biomedical signal processing, etc. Apart from these single-channel applications, there are many problems which involve several FIR filters and lead to a multichannel filtering formulation. Examples are the identification of systems described by difference equations with multiple polynomials (ARMA, ARX and ARMAX systems [11]), adaptive minimum variance control [12, section 6.3], fractionally spaced and decision feedback equalizers [13], frequency domain adaptive filtering [14], multirate signal processing [15, page 271] image restoration and enhancement [16], and adaptive beamforming with antenna arrays [17].

NUMERICAL CONSIDERATIONS

A recent survey of some numerical issues in adaptive filtering algorithms is given in [18]. The LMS and the conventional RLS and FLA–FQR algorithms have been found to be exponentially numerically stable (for recursive algorithms, by numerical stability we mean the stability of the error propagation system). Some pertinent early references are [19,20] for LMS, [21–23] for RLS, [21,24] for FLA, and [25] for FQR. Even though FTRLS algorithms were proposed in [26], numerical problems were reported only several years later [27, 4]. In order to arrive at a working algorithm, certain remedies were proposed in the form of rescuing devices [28,4]: a quantity indicative of the level of the numerical errors is observed until a threshold is reached, at which point the algorithm is restarted. Even though the restart can include the current solution as prior information by means of a soft-constrained initialization, such a restart leads to suboptimal performance because of a significant discontinuity in the sample covariance matrix in the normal equations that determine the solution.

The exponential divergence of numerical errors in a FTRLS algorithm was demonstrated in [21] for one particular input signal, considering the original so-called *fast Kalman* version [26] of the FTRLS algorithms. In [21], the following statement appears: 'There is nothing in terms of numerical tricks and improved organization of calculations that can be done to remove the bad error propagation properties. That can only be done by changing the algorithm itself.' This means that if one increases the precision of the computations by, for instance, doubling the wordlength, then numerical errors will still diverge. It will only take longer before the errors become noticeable. Stabilization of the algorithm really requires an actual change in the algorithm so that the dynamics of the error propagation are modified.

Earlier on, in [4], the following heuristic explanation for the numerical instability had been given. In the FTF algorithm, the backward prediction error filter $\mathbf{B}_p(N)$ and the Kalman gain $\mathbf{w}_p(N)$ are jointly updated via a so-called hyperbolic rotation.

$$\begin{bmatrix} \mathbf{B}_p(N) \\ [\mathbf{w}_p(N)\ 0] \end{bmatrix} = \mathscr{H}_p(N) \begin{bmatrix} \mathbf{B}_p(N-1) \\ \mathbf{w}_{p+1}(N) \end{bmatrix} \tag{5.1}$$

where $\mathscr{H}_p(N)$ is a hyperbolic rotation, or, more precisely,[1]

$$\begin{bmatrix} -\alpha_p^{\mathrm{b}}(N) & 0 \\ 0 & \alpha_p(N) \end{bmatrix} = \mathscr{H}_p(N) \begin{bmatrix} -\lambda\alpha_p^{\mathrm{b}}(N-1) & 0 \\ 0 & \alpha_{p+1}(N) \end{bmatrix} \mathscr{H}_p^{\mathrm{H}}(N) \tag{5.2}$$

where λ, α^{b} and α are positive quantities. Intuitively, since the hyperbolic rotation has an eigenvalue greater than one in magnitude, we can expect that certain error components on the filters will be amplified by the above transformation. However, this explanation is incomplete since in the algorithms the hyperbolic rotation itself is a function of the vector that is going to be rotated. So, one does not obtain the new filter estimates from the old ones via linear transformations, but through some more involved non-linear transformation. Hence, the errors on the filter estimates are not simply multiplied by hyperbolic transformations. It is precisely some degree of freedom in the structure of the non-linear transformation that will allow us to stabilize the error propagation. Another reason why the hyperbolic rotation picture is misleading is that it only focuses on part of the algorithm. Nevertheless it is true that the presence of hyperbolic rotations is strongly indicative of possible numerical difficulties.

More recently, efforts to attempt to stabilize the FTRLS algorithms have intensified. In the *leaky* LMS algorithm [18], certain leakage factors are introduced in order to stabilize the LMS algorithm in case the signal covariance matrix is singular. Cioffi [29] has studied leaky FTF algorithms by considering a so-called multiexperiment version [30] and introducing an artificial experiment to control the magnitude of the filter estimates: however, he found that too much performance had to be sacrificed to achieve stability. In contrast, the stabilized FTRLS algorithms discussed here do not introduce leakage factors and still solve the RLS problem exactly.

REDUNDANCY AND ERROR FEEDBACK

It has been noticed that by introducing computational *redundancy* in a certain sense (namely, computing certain quantities in two different ways), one can make specific measurements of the numerical errors available. These measurements can be *fed back* appropriately to modify the dynamics of the propagation of errors. In [31,32,6,33] and [7] the availability of redundancy in the FTF algorithm was recognized and utilized; however, all the algorithms therein turn out to be only partial solutions to the stabilization problem (in that they use only one redundancy instead of two and hence cannot completely stabilize the error propagation system). We should emphasize that the possibility of redundancy is a key feature of the FTRLS algorithms considered here. If one attempted to introduce redundancy in the LMS or conventional RLS algorithms, by scheduling the order of multiplications and additions in different ways and the like, one would not accomplish any change in the dynamics of the propagation of numerical errors (due to lack of observability). The redundancy in the FTRLS algorithm is of a more fundamental nature. We shall pair the availability of redundancy in the FTRLS algorithm with the idea of error feedback to open the way to a complete solution. Apart from a solution to the problem of numerical instability in FTRLS algorithms, we shall also provide an analysis of the solution, which demonstrates, among other things, the instability of the original $7p$ algorithm and its variants. By exploiting redundancy properly, we arrive at a numerically stable FTRLS algorithm, of which the computational complexity of $8p$ shows a 14% increase over the basic $7p$ (see also [8,9]).

SEQUENTIAL PROCESSING AND MODULAR (CIRCULAR) ALGORITHMS

Multichannel, fast RLS algorithms require vectors of forward and backward prediction errors. These algorithms can be obtained as straightforward extensions of the single-channel version by processing the error vectors *en bloc*. Another type of block processing appearing in least-squares parameter estimation is multiexperiment filtering [30]. In this case we obtain data from different simultaneous experiments with the same model. In each experiment, different data sets are feeding the same unknown system. Such a situation arises, for instance, in autoregressive modelling for failure detection, where several measurements are available, all containing contributions from the eigenmodes of the physical system under consideration. Again, the classical approach [30] is to process the measurements *en bloc* by defining the measurements to be vectors, leading to an algorithm that is notationally a straightforward extension of the single-experiment version.

Block processing occurs naturally in the Kalman filter when the measurements are vectors. If we specialize the Kalman filter to the RLS parameter estimation problem, then this block processing coincides with the multiexperiment filtering mentioned above. An alternative to block or simultaneous processing, well known in Kalman filter theory [34], is *sequential processing* of the components of the measurement vector. In the context of Kalman filtering, sequential processing has sometimes been justified on the following basis [34]: if the processor runs out of processing time (e.g. an interrupt occurs), then at least part of the data has been appropriately used to update the estimates. Another aspect that has perhaps not received much attention in this context is that sequential processing corresponds to propagating the factors in so-called lower–diagonal–upper (LDU) or upper–diagonal–lower (UDL) triangular factorizations of the covariance matrix of the innovations. This latter aspect has received significant attention, though, in factorized estimation or square root filtering [35,36,34], which in some sense corresponds to sequential processing of the state vector. In factorized estimation, the factors in an LDU or UDL factorization of the state error covariance matrix are propagated. In [36,37], plenty of simulation evidence has been given of numerical benefits that result from the factorized approach, while a theoretical motivation has been provided only recently [38]. These numerical considerations were also one of the driving factors in [39] for similar developments for the special case of RLS parameter estimation. Furthermore, the factorized estimation approach turned out to lead to algorithms that are more amenable to parallel implementation on processor arrays; see [39,40] for RLS and [41] for Kalman filtering.

In [42], a general principle for the modular decomposition of multichannel recursions was outlined and then applied to multichannel lattice algorithms; modular multichannel lattice algorithms were also derived in [43–45] (see Chapter 7 also). The modular decomposition principle was restated in [46] and then applied to the derivation of modular multichannel FTRLS algorithms. Modular multiexperiment algorithms were also described to complete a picture of dual factorization issues in the two time-updating strategies of FTRLS algorithms. Here we present a fully modular FTRLS algorithm for the combined multichannel, multiexperiment filtering problem with possibly indefinite weighting of the multiexperiment error vector. The modular algorithm will turn out to correspond to a doubly intertwined set of single-channel, single-experiment FTRLS algorithms (henceforth often called scalar algorithms). One of the consequences of this correspondence will be that stabilization techniques for scalar FTRLS algorithms can be straightforwardly applied to modular block processing algorithms.

COVARIANCE ALGORITHMS

The modular FTRLS algorithm will be derived with the prewindowing assumption (in which all data before time 0 are assumed to be zero), and an exponential weighting will be applied to the error signal to make the effective window length finite. The covariance (or unwindowed) method [47,48] is an alternative approach in which no assumptions are made on unavailable data. So the covariance method only starts considering regression vectors from the moment they can be completely filled with available data. Covariance methods can perform noticeably better than prewindowed algorithms for short window lengths. They come in two varieties, the growing-window and the sliding-window covariance algorithms (often considered) as alternatives for prewindowed algorithms without or with exponential weighting respectively. The derivation of fast covariance algorithms has always been something of an art, in that their derivation requires additional insight which goes a step beyond that required for prewindowed algorithms. In [49], a unified geometric theory is presented which tries to cover the derivation of all fast RLS algorithms. However, only the prewindowed case falls out nicely from this theory. In another attempt at unification, the two covariance cases were embedded into prewindowed problems in [50,30]. However, the algorithms resulting from the prewindowed embedding did not coincide with the existing covariance algorithms, and were in fact computationally more complex. In

[51], we also consider the prewindowed embedding framework of [50,30]. However, applying the modular prewindowed algorithm to these embedded problems enables us to recover the existing covariance algorithms, modulo the numerical stabilization part, of course. Apart from providing a unified prewindowing framework for the derivation of covariance algorithms, this approach also enables us straightforwardly to extend the numerical stabilization from prewindowed to covariance algorithms. Then taking the modularity one level of granularity further, we can easily obtain modular multichannel covariance algorithms. Exponential weighting will be superimposed on all windowing schemes considered, and some numerical consequences of this will be discussed.

5.2 Adaptive filtering algorithms

Adaptive algorithms are intended for long-term, possibly continuous, use. This mode of operation is desired in situations where one expects the signal characteristics to change in time. This implies that the optimal Wiener filter is time varying. The task of the adaptive algorithm will be, based on the given data, to adapt to the changes and track the time-varying optimal solution as well as possible. Another quality closely related to this tracking issue is that of initial convergence speed as the algorithm starts processing the data. The prototype adaptive filtering system is depicted in Figure 5.1 where we have explicitly indicated that the error signal is continuously steering the adapted filter. The particular error signal indicated there is the so-called a posteriori error $\varepsilon(t) = y(t) + \mathbf{c}^H(t)\mathbf{u}(t)$, which is computed with the latest update $\mathbf{c}(t)$ of the filter estimate.

To analyze the tracking characteristics of adaptive filtering algorithms, we shall concentrate here on a particular model for time-varying parameters that has become very popular over the last decade [52,53,3]. See Figure 5.2 for a schematic rendering of this model, which is governed by the following string of equations:

$$e(t) = y(t) + \mathbf{c}^H(t-1)\mathbf{u}(t) \tag{5.3}$$

$$y(t) = \eta(t) - \mathbf{c}^{oH}(t-1)\mathbf{u}(t) \qquad E[\eta(t)\eta^H(t)] = V^0 \tag{5.4}$$

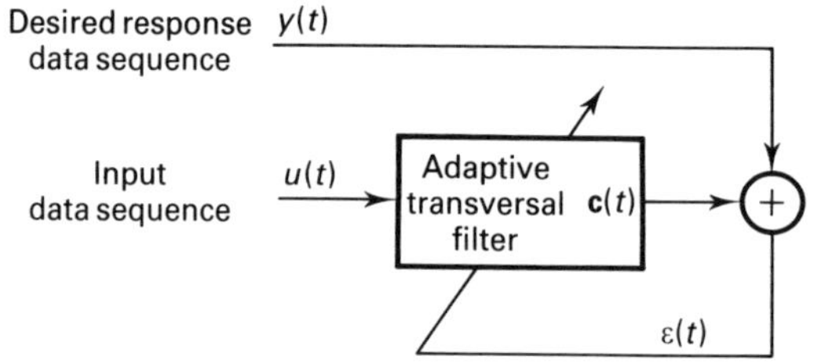

Figure 5.1 The adaptive transversal filtering system

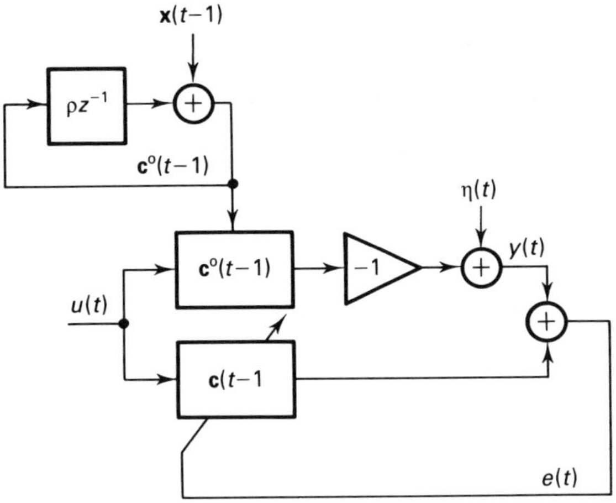

Figure 5.2 Setup for the analysis of adaptive algorithms regarding the tracking of time-varying parameters

$$\mathbf{c}^o(t) = \rho\mathbf{c}^o(t-1) + \mathbf{x}(t) \qquad E[\mathbf{x}(t)\mathbf{x}^H(t)] = Q \tag{5.5}$$

$$\tilde{\mathbf{c}}(t) \triangleq \mathbf{c}^o(t) - \mathbf{c}(t) \tag{5.6}$$

$$e(t) = \eta(t) - \tilde{\mathbf{c}}^H(t-1)\mathbf{u}(t) \tag{5.7}$$

where $\eta(\cdot)$, $u(\cdot)$ and $\mathbf{x}(\cdot)$ are independent stationary processes, and $\eta(\cdot)$ and $\mathbf{x}(\cdot)$ are white. Note that in the tracking analysis it is desirable to work with the so-called a priori error signal $e(t) = y(t) + \mathbf{c}^H(t-1)\mathbf{u}(t)$, which is computed with the previous update $\mathbf{c}(t-1)$ of the filter estimate. The desired response $y(\cdot)$ consists of the output of an optimal filter $\mathbf{c}^o(\cdot)$ plus a white measurement noise $\eta(\cdot)$, and hence $\mathbf{c}^o(\cdot)$ is the Wiener solution, yielding the MMSE. The components of $\mathbf{c}^o(\cdot)$ are time varying. In particular, each component is modelled as a first-order AR process. For stability reasons, we need $|\rho| < 1$. However, we shall consider a ρ which is so close to 1 that for all further considerations it can be considered equal to 1 (time constant of the parameter variation $(1/1-\rho)$ is much larger than the time constant(s) of the adaptive algorithm, see [52]). $\tilde{\mathbf{c}}(t)$ is the deviation of the filter estimate $\mathbf{c}(t)$ from the Wiener solution $\mathbf{c}^o(t)$. Let $R = E[\mathbf{u}(t)\mathbf{u}^H(t)]$ and $COV(t) \triangleq E[\tilde{\mathbf{c}}(t)\tilde{\mathbf{c}}^H(t)]$. If we take the variance of both sides of Equation (5.7) and neglect dependencies between $\tilde{\mathbf{c}}(t-1)$ and $\mathbf{u}(t)$ (the 'independence assumption', see [5]), then we get (with an obvious notation)

$$V(t) = E\| e(t) \|^2 = V^o + V^e(t) \qquad V^e(t) = \mathrm{tr}\{R\, COV(t-1)\} \tag{5.8}$$

where tr denotes trace, $V(t)$ is the average or mean squared error, V^o is the minimum MSE (the filter (adapted or not) can never do better than this!) and $V^e(t)$ is the excess MSE. Note that the above manipulation, invoking the independence assumption, reveals why it is desirable to consider the MSE as the variance of the a priori error signal. The a posteriori error signal is not meaningful here since it can be made identically zero in certain algorithms without implying that the filter estimates are perfect or that V^o is zero. The variance of the a posteriori error is often lower than V^o, whereas the variance of the a priori error is always higher: $V(t) > V^o$, as can be verified from (5.8). A quantity often used in this context is the normalized excess MSE, usually called the misadjustment

$$\mathcal{M} = V^e / V^o \tag{5.9}$$

The misadjustment should be small w.r.t. one for the adaptive algorithm to work well.

5.2.1 The LMS algorithm

The least mean square (LMS) algorithm attempts to adapt to the Wiener solution on the basis of the given data by replacing the stochastic LS criterion by an instantaneous deterministic LS criterion. So we have the following criterion modification

$$E \,\|\, \varepsilon(t) \,\|^2 \rightarrow \|\, \varepsilon(t) \,\|^2 \tag{5.10}$$

where $\| \cdot \|^2$ now denotes the Euclidean squared norm $\varepsilon^{\mathrm{H}}(t)\varepsilon(t)$. The gradient of this criterion is

$$\nabla \,\|\, \varepsilon(t) \,\|^2 = 2\mathbf{u}(t)\varepsilon^{\mathrm{H}}(t) \tag{5.11}$$

When we apply a steepest descent method to the instantaneous LS criterion, we get a so-called stochastic gradient method (stochastic because the resulting filter estimates are random variables). The LMS algorithm is a stochastic gradient method applied to the linear filtering problem. So, taking small steps in the direction of the negative gradient (evaluated at the previous filter estimate), we get

$$\mathbf{c}(t) = \mathbf{c}(t-1) - \mu\mathbf{u}(t)e^{\mathrm{H}}(t) \tag{5.12}$$

The idea is to take small steps along the instantaneous gradient direction in the hope that these steps average out to a step in the true gradient direction (of the stochastic LS criterion). We find for the parameter estimation error

$$\begin{aligned} \tilde{\mathbf{c}}(t) &= \tilde{\mathbf{c}}(t-1) + \mu\mathbf{u}(t)e^{\mathrm{H}}(t) + \mathbf{x}(t) \\ &= (I - \mu\mathbf{u}(t)\mathbf{u}^{\mathrm{H}}(t)]\tilde{\mathbf{c}}(t-1) + \mu\mathbf{u}(t)\eta^{\mathrm{H}}(t) + \mathbf{x}(t) \end{aligned} \tag{5.13}$$

For small step size μ, the system matrix $[I - \mu\mathbf{u}(t)\mathbf{u}^{\mathrm{H}}(t)]$ is close to an identity matrix. A result from averaging analysis [54] states that in this case, the solution $\tilde{\mathbf{c}}(t)$ of the stochastic difference equation (5.13) is 'close' to the solution of another stochastic difference equation (but we shall keep the same notation here) obtained from (5.13) by replacing the system matrix by its average $(I - \mu R)$, that is,

$$\tilde{\mathbf{c}}(t) = (I - \mu R)\tilde{\mathbf{c}}(t-1) + \mu\mathbf{u}(t)\eta^{\mathrm{H}}(t) + \mathbf{x}(t) \tag{5.14}$$

Even though, strictly speaking, this approximation only holds for small μ, conclusions based on this approximation seem to hold in practice over a fairly wide range for μ [5]. In particular, from (5.14) it is clear that the LMS algorithm converges (if $\eta(\cdot) \equiv 0$, $\mathbf{x}(\cdot) \equiv 0$) or the estimates retain a bounded variance if and only if the system eigenvalues are in the range $(-1, 1)$ or

$$0 < \mu < \frac{2}{\lambda_{\max}(R)} \tag{5.15}$$

where the trace may be taken as a conservative estimate for the maximum eigenvalue: $\lambda_{\max}(R) \leqslant \operatorname{tr}(R) = P\sigma_u^2$. Of course, estimates such as (5.15) for the stable range of μ, based on the averaging technique, can only be rough approximations since the averaging technique only works for small μ. A more refined analysis of the stable range for μ has been discussed in the previous chapter.

It is clear that the system (5.14) has widely different modes if the eigenvalues of R vary over a wide range. Taking the covariance of both sides of (5.14) yields

$$COV(t) = (I - \mu R)COV(t-1)(I - \mu R) + \mu^2 V^{o} R + Q \tag{5.16}$$

of which the steady-state solution COV satisfies approximately (the approximation, involving dropping the term $\mu^2 R\, COV R$, is justified for the calculation of $\mathscr{M}_{\text{LMS}}^{\text{opt}}$ below, when $\mathscr{M}_{\text{LMS}}^{\text{opt}} \ll 1$, or hence when the parameters vary very slowly)

$$R\, COV + COV\, R = \mu V^{o} R + \frac{1}{\mu} Q \tag{5.17}$$

Using (5.8) and (5.9), we find for the misadjustment

$$\mathscr{M}_{\text{LMS}} = \frac{\mu}{2} \operatorname{tr}(R) + \frac{1}{2\mu V^{o}} \operatorname{tr}(Q) \tag{5.18}$$

The misadjustment is seen to consist of two terms whose minimization leads to opposite requirements on the step size μ. The first term, 'estimation noise', is due to the fact that, effectively, only a finite number of data are used in the estimation of the parameters (the algorithm has finite time constants, obviously, since it should converge (exponentially)). The second term, 'lag noise', is due to the fact that the parameter estimates represent an average value of the true parameters over the data window, and thus are lagging behind the current true values. Clearly, an optimal value for the step size is obtained by minimizing the misdadjustment, which leads to equality of the two excess terms, that is,

$$\begin{aligned} \mu^{\text{opt}} &= \sqrt{\frac{\operatorname{tr}(Q)}{V^{o}\operatorname{tr}(R)}} \\ \mathscr{M}_{\text{LMS}}^{\text{opt}} &= \sqrt{\frac{\operatorname{tr}(R)\operatorname{tr}(Q)}{V^{o}}} \end{aligned} \tag{5.19}$$

The averaging arguments we have used above lead to reasonably accurate results when a small step size is used. This is the case for the steady-state situation of tracking slowly varying parameters (see [55] for the most recent results on this aspect). However, the larger step-size values that one may wish to use for the initial convergence period require a more accurate analysis (see the previous chapter or [56,57] and references therein).

5.2.2 The RLS algorithm

All RLS algorithms (conventional or fast) solve the same problem and hence have identical tracking characteristics. So we shall analyze these characteristics here in terms of the conventional $O(p^2)$ RLS algorithm. RLS algorithms can be made adaptive by the introduction of a finite effective window length. One way to achieve this is to introduce an exponential weighting in the LS criterion. So consider the following weighted criterion

$$\min_{\mathbf{c}(N)} \left\{ \sum_{t=0}^{N} \lambda^{N-t} \| e(t \mid \mathbf{c}(N)) \|^2 + \lambda^{N+1} r_0 \| \mathbf{c}(N) - \mathbf{c}_0 \|^2 \right\} \tag{5.20}$$

where the weighting factor λ is chosen in the range (0, 1), $\mathbf{c}_0$ is some prior guess for $\mathbf{c}(N)$ (possibly obtained from previous estimation efforts), and r_0 is a weighting factor that reflects the level of confidence in the prior guess. The inclusion of the (formal) prior information in fact facilitates (regularizes) the initialization of the algorithm and ensures that the LS problem is overdetermined from the first sample onwards. As an alternative to the exponential window, one could use a rectangular window of (constant) length L which leads to the sliding window covariance (SWC) algorithm. However, as is shown in [58], both windowing methods lead to an equivalent filtering performance for small $1 - \lambda$ and large L. In particular, most of the (exponential window) tracking performance formulae obtained below also hold for the SWC algorithm if one substitutes λ in terms of L according to

$$L = \frac{1 + \lambda}{1 - \lambda} \approx \frac{2}{1 - \lambda}$$

for small $1 - \lambda$ or large L.

Interestingly, one may remark that the LMS algorithm may be regarded as a special case of the RLS algorithm for the extreme case in which the window length shrinks down to one sample ($L = 1$ or $\lambda = 0$). In this case, the estimate obtained at time $N - 1$ can obviously be taken as the prior information for the estimation problem at time N. So consider the LS problem

$$\min_{\mathbf{c}(N)} \left\{ \| y(N) + \mathbf{c}(N)^{\mathrm{H}} \mathbf{u}(N) \|^2 + \left(\frac{1}{\mu} - \| \mathbf{u}(N) \|^2 \right) \| \mathbf{c}(N) - \mathbf{c}(N-1) \|^2 \right\} \tag{5.21}$$

Its solution leads exactly to the LMS recursion (5.12).

The RLS algorithm for criterion (5.20) updates the filter estimate according to (see Table 2.1 (for $\boldsymbol{\theta} = -\mathbf{c}$) and equation (2.59) in Chapter 2)

$$\mathbf{c}(t) = \mathbf{c}(t-1) - R^{-1}(t)\mathbf{u}(t)e^{\mathrm{H}}(t) \tag{5.22}$$

and the inverse of the sample covariance matrix

$$R(N) = \sum_{t=0}^{N} \lambda^{N-t}\mathbf{u}(t)\,\mathbf{u}^{\mathrm{H}}(t) + \lambda^{N+1}r_0 I$$

is recursively updated using the *Riccati* equation

$$R^{-1}(t) = \lambda^{-1}R^{-1}(t-1) - \mathbf{w}(t)\alpha(t)\mathbf{w}^{\mathrm{H}}(t) \tag{5.23}$$

where[2]

$$\begin{aligned} \mathbf{w}(t) &= -\lambda^{-1}R^{-1}(t-1)\mathbf{u}(t) \\ \alpha^*(t) &\equiv 1 - \mathbf{w}^{\mathrm{H}}(t)\mathbf{u}(t) = \alpha^{-2}(t) \end{aligned} \tag{5.24}$$

The initial conditions are obviously $\mathbf{c}(-1) = \mathbf{c}_0$, $R^{-1}(-1) = (1/r_0)I$. We find for the parameter estimation error

$$\begin{aligned} \tilde{\mathbf{c}}(t) &= \tilde{\mathbf{c}}(t-1) + R^{-1}(t)\mathbf{u}(t)e^{\mathrm{H}}(t) + \mathbf{x}(t) \\ &= [I - R^{-1}(t)\mathbf{u}(t)\mathbf{u}^{\mathrm{H}}(t)]\tilde{\mathbf{c}}(t-1) + R^{-1}(t)\mathbf{u}(t)\eta^{\mathrm{H}}(t) + \mathbf{x}(t) \end{aligned} \tag{5.25}$$

and after averaging (assuming small $1 - \lambda$), we get (note: $I - R^{-1}(t)\mathbf{u}(t)\mathbf{u}^{\mathrm{H}}(t) = \lambda R^{-1}(t)R(t-1)$)

$$\tilde{\mathbf{c}}(t) = \lambda\tilde{\mathbf{c}}(t-1) + (1-\lambda)R^{-1}\mathbf{u}(t)\eta^{\mathrm{H}}(t) + \mathbf{x}(t). \tag{5.26}$$

So we see that there is no problem of convergence here and all system modes are equal to λ. Taking the covariance of both sides of (5.26) yields

$$COV(t) = \lambda^2 COV(t-1) + (1-\lambda)^2 V^{\circ}R^{-1} + Q \tag{5.27}$$

of which the steady-state solution COV satisfies approximately ($1 + \lambda \approx 2$)

$$COV = \frac{1-\lambda}{2}V^{\circ}R^{-1} + \frac{1}{2(1-\lambda)}Q \tag{5.28}$$

Using (5.8) and (5.9), we find for the misadjustment

$$\mathscr{M}_{\mathrm{RLS}} = \frac{1-\lambda}{2}p + \frac{1}{2(1-\lambda)\,V^{\circ}}\operatorname{tr}(R\,Q) \tag{5.29}$$

The interpretation for the two terms contributing to the misadjustment is parallel to the LMS case. We can consider optimizing $\mathscr{M}_{\mathrm{RLS}}$ with respect to λ by equating the two terms and we find

$$\lambda^{\mathrm{opt}} = 1 - \sqrt{\frac{\operatorname{tr}(R\,Q)}{p\,V^{\circ}}}$$

$$\mathscr{M}_{\mathrm{RLS}}^{\mathrm{opt}} = \sqrt{\frac{p\ \mathrm{tr}(R\,Q)}{V^{\mathrm{o}}}}\ . \tag{5.30}$$

5.2.3 Comparison of the tracking performance of LMS and RLS

Comparing the misadjustment for LMS and RLS with optimized μ and λ respectively, we find

$$\frac{\mathscr{M}_{\mathrm{LMS}}^{\mathrm{opt}}}{\mathscr{M}_{\mathrm{RLS}}^{\mathrm{opt}}} = \sqrt{\frac{\mathrm{tr}(R)\ \mathrm{tr}(Q)}{p\ \mathrm{tr}(R\,Q)}} \tag{5.31}$$

Depending on the relation between R and Q, either one of the LMS and RLS algorithms may be the better one. Here are two examples:

$$\begin{array}{ll} Q = R & \text{LMS is better} \\ Q = R^{-1} & \text{RLS is better} \end{array} \tag{5.32}$$

We may note that for a given model for the time-varying parameters as shown in (5.5), the Kalman filter is the optimal adaptive algorithm (in the Gaussian case), and the LMS and RLS algorithms are just two different approximations of it (see also [60,61]). More recently, the relative tracking performance of the LMS and RLS algorithms has been compared for a case of deterministic parameter variations, such as the variation of the frequency in a chirp signal [62,63]. Also in this case, one or the other algorithm can be the better one, with the chirp rate being the determining factor.

If we take $Q = qI$, then we find $\mathscr{M}_{\mathrm{LMS}}^{\mathrm{opt}} = \mathscr{M}_{\mathrm{RLS}}^{\mathrm{opt}}$. This has sometimes led people to state [52] that LMS and RLS can have identical performance for the case under study, namely tracking drifting parameters with $Q = qI$. However, the above results only hold asymptotically for small q, i.e. very slowly varying parameters. For strongly varying parameters, the above analysis does not hold and, in particular, the value of μ that minimizes $\mathscr{M}_{\mathrm{LMS}}$ may exceed its stability bound. In practice, the stability requirement and the uncertainty about the statistics of the signal $u(\cdot)$ often force the algorithm designer to assign quite a conservative value to μ, limiting the tracking capabilities of the LMS algorithm.

A special instance of strongly varying parameters appears at the initial start of the adaptation, when the optimal parameters jump from their prior guess to their true value. The problem of analyzing the convergence speed during this transient period cannot be handled with the same tools as the problem of slowly drifting parameters addressed above. Indeed, during the initial convergence (and at times of severe non-stationarity), the averaging analysis cannot be applied. However, one can

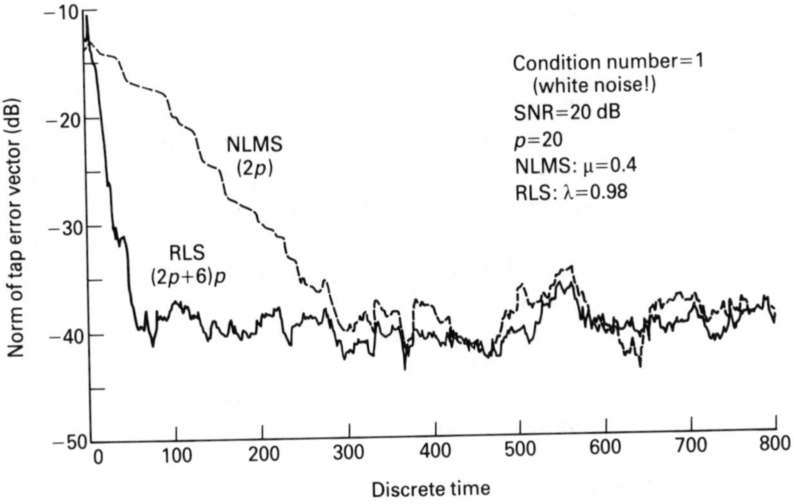

Figure 5.3 Comparison of the convergence behaviour of the LMS and RLS algorithms

make the following observations. If we let the SNR tend to infinity ($V^{\circ} \to 0$), then for the resulting noise-free problem, the deterministic LS approach may coincide with the stochastic LS problem as soon as the number of observations has reached the number p. Since the deterministic LS problem is quadratic and RLS solves quadratic problems exactly, the RLS algorithm may converge in p steps. This indicates that at high and medium SNR, the RLS algorithm may converge substantially faster than the LMS algorithm, but this difference may fade away if the SNR is low. In Figure 5.3, a simulation example is shown, comparing the initial convergence behaviour of the RLS algorithm and the so-called normalized LMS (NLMS) algorithm (see e.g. [64], and Chapter 2). The input signal is taken to be white noise, which is normally assumed to be the most favourable case for the (N)LMS algorithm.

For completeness, though, we should state that the LMS algorithm is generally believed to be more *robust* than the RLS algorithm, and this robustness attribute may comprise quite a variety of qualities of sorts. One such robustness feature may be revealed if we analyze what happens if the input signal $u(\cdot)$ lacks *persistency of excitation*, i.e. its covariance matrix R is singular. In this case, the modes for the tracking errors $\tilde{\mathbf{c}}(t)$ in the unexcited subspace (null space of R) are 1 (marginally unstable) for both the LMS and the RLS algorithms, but the corresponding modes for $R^{-1}(t)$ in the Riccati equation (5.23) are $1/\lambda$, which is exponentially unstable [8]. Hence, whereas it has been found that keeping the input signal persistently exciting is a crucial ingredient of practical implementations of the LMS algorithm [18], this same excitation issue is of paramount importance in implementations of any RLS algorithm (especially when exponential weighting is used). This issue re-emerges when the influence of numerical errors is considered.

5.2.4 Comparison of various RLS algorithms

So the RLS algorithm is preferable over the LMS algorithm when tracking performance and especially speed of convergence are critical. However, the disadvantage of the conventional RLS algorithm is that it is an order of magnitude more complex than the LMS algorithm. Fast RLS algorithms such as the FTRLS, FLA and fast RLS QR (FQR) algorithms were devised to alleviate this problem. These algorithms take $O(p)$ computations versus $O(p^2)$ for the conventional RLS algorithm. Among these fast RLS algorithms, the fixed-order (FTRLS) algorithms have the lowest complexity in the sense that they have a lower coefficient of p in their computations count than the order recursive solutions (FLA, FQR). These order recursive algorithms not only need $O(p)$ multiplications (and additions) but also $O(p)$ divisions (and $O(p)$ square roots if they are normalized). The fixed-order solutions only require a fixed small number of divisions (and square roots, if they are normalized). All other aspects being equal, the fixed-order algorithms are hence clearly the preferable adaptive algorithms, since they give the exact RLS performance at the lowest computational cost (it should be said, though, that with $O(p)$ processors available, the throughput of the order recursive algorithms can be increased by a factor of $O(p)$ (at the expense of a pipelining delay of $O(p)$), whereas this is not possible for the fixed-order algorithms). However, if we finally consider the aspect of numerical stability, then the early fixed-order solutions (fast Kalman [26], FAEST [65], FTF [4]) lose their favourable position. Indeed, those algorithms are exponentially unstable in the sense that an isolated roundoff error produces an exponentially growing numerical error as the recursions are proceeding. However, we have been able to pin down this instability mechanism and devise a solution for this problem that requires only p additional multiplications [8,9]. The thus *stabilized FTRLS* algorithm emerges as a desirable adaptive algorithm on all accounts of convergence speed, computational cost and numerical stability. Some of these issues are sketched in Table 5.1.

Table 5.1 Comparisonof the LMS and various RLS algorithms

	Problem solved	Convergence speed	Computational complexity (×)	Numerical stability
LMS	LMS	Slow	$2p$	Exponential
RLS	RLS	Fast	$2p^2 + 6p$	Exponential
FLA	RLS	Fast	$14p(\times) + 2p(\div)$ $\approx 30p(\times)$	Exponential
FTRLS	RLS	Fast	$7p$	Unstable
SFTRLS	RLS	Fast	$8p$	Exponential

5.3 Windowing issues in fast LS algorithms

To emphasize the special character of the linear regression problem considered here, we have redrawn the adaptive filtering problem in Figure 5.4. Assume at first $y(\cdot) \equiv 0$ in Figure 5.4. The output signal of the filter is given in terms of the input signal $u(\cdot)$ as

$$\sum_{i=0}^{p-1} c_p^{iH}(t)u(t-i) = \mathbf{c}_p^H(t)\mathbf{u}_p(t) \tag{5.33}$$

where $\mathbf{c}_p(t) = [c_p^{0H}(t), \ldots, c_p^{p-1}(t)]^H$ and $\mathbf{u}_p(t) = [u^H(t), \ldots, u^H(t-p+1)]^H$. This particular filter structure is called a *transversal* filter and it is perhaps the most natural one among equivalent filter structures realizing the same input–output relation since it is parameterized in terms of the samples of the impulse response. We focus here on a filter with finite impulse response (FIR) (of length p). The purpose of the filter is for (the negative of) its output to approximate a desired response signal $y(\cdot)$ as well as possible. So we want to minimize in some sense the error signal

$$\varepsilon_p(t) = y(t) + \mathbf{c}_p^H(t)\mathbf{u}_p(t) \tag{5.34}$$

For a general desired response, this is called the *joint-process* filtering problem. The special case $y(t) = u(t+1)$ is called the (*forward*) *prediction* problem, while the case $y(t) = u(t-p)$ is called the *backward prediction* problem, as has been pointed out in Chapter 2. All processes considered here are assumed to have zero mean. In statistics, this approximation problem is referred to as a regression of y on $\mathbf{u}_p$ and then $\mathbf{u}_p$ is called the regressor.

Consider the LS criterion (5.20). To simplify the present discussion, we shall discard the initialization issue: $r_0 = 0$. This issue will be taken up again later. Then the sum of squared errors $\Sigma_{t=N-L+1}^{N} \lambda^{N-t} \| \varepsilon_p(t) \|^2$ can be seen to be the squared norm of one large error vector

$$\boldsymbol{\varepsilon}_{p,L}(N) = \mathcal{Y}_L(N) + \mathcal{U}_{p,L}(N)\mathbf{c}_{p,L}(N) \tag{5.35}$$

where

$$\mathcal{Y}_L(N) = [y(N), \lambda^{1/2}y(N-1), \ldots, \lambda^{(L-1)/2}y(N-L+1)]^H \tag{5.36}$$

is the desired response data vector and

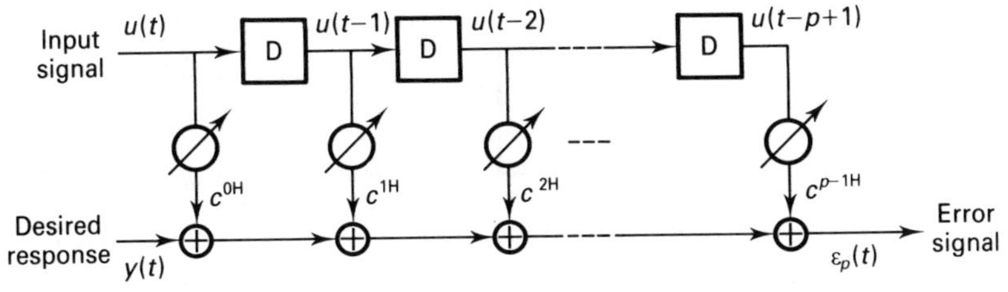

Figure 5.4 The linear transversal FIR filter

$$\mathcal{U}_{p,L}(N) = [\boldsymbol{\mathcal{U}}_L(N), \boldsymbol{\mathcal{U}}_L(N-1), \ldots, \boldsymbol{\mathcal{U}}_L(N-p+1)] \tag{5.37}$$

is called the (input) data matrix ($\boldsymbol{\mathcal{U}}_L(t)$ is defined similarly as $\boldsymbol{\mathcal{Y}}_L(t)$). The way we have written the data matrix leads to a *Hankel* matrix structure, i.e. elements along antidiagonals are equal. The normal equations now have the sample covariance matrix $R_{p,L}(N) = \mathcal{U}^{\mathrm{H}}_{p,L}(N)\mathcal{U}_{p,L}(N)$ as the matrix of coefficients.

Assume that the samples of the signal u are available in the time range $[0, N]$. The samples outside this range are by default assumed to be zero. Different *windowing* strategies are now possible:

- **Covariance:** $\mathcal{U}^{\mathrm{cov}}_p(N) = \mathcal{U}_{p,N-p+2}(N)$. In this case we want the data matrix to contain only actual data, and the window length should be as long as possible. This leads to the relation $L = N - \mathrm{p} + 2$. This implies that the window length shrinks as the order grows.
- **Postwindowed:** $\mathcal{U}^{\mathrm{po}}_p(N) = \lim_{t\to\infty} \mathcal{U}_{p,t-p+2}(t)$. The data matrix is filled up with zeros after time N.
- **Prewindowed:** $\mathcal{U}^{\mathrm{pre}}_p(N) = \lim_{L\to\infty} \mathcal{U}_{p,L}(N)$. The data matrix is filled up with zeros before time 0.
- **Pre- and postwindowed:** $\mathcal{U}^{\mathrm{prepo}}_p(N) = \lim_{t\to\infty}\lim_{L\to\infty}\mathcal{U}_{p,L}(t)$. The data matrix is filled up with zeros before time 0 and after time N and hence is a banded Hankel matrix. It is easy to show that $R^{\mathrm{prepo}}_p(N)$ is Toeplitz, i.e. its entries are constant along any diagonal.

These windowing issues are perhaps more easily understood from the following picture (in which u^{H} has been simplified to u and $\lambda = 1$ for simplicity):

$$\left[\begin{array}{cccc}
 & & & \vdots \\
 & & & 0 \\
 & \vdots & \iddots & u(N) \\
\vdots & 0 & \iddots & u(N-1) \\
0 & u(N) & \iddots & \vdots \\
\hline
u(N) & u(N-1) & & u(N-p+1) \\
\vdots & & & \vdots \\
u(p-1) & u(p-2) & \cdots & u(0) \\
\hline
 & \vdots & \iddots & 0 \\
\vdots & u(0) & \iddots & \vdots \\
u(0) & 0 & & \\
0 & \vdots & & \\
\vdots & & &
\end{array}\right]
\begin{array}{l}
\mathcal{U}^{\mathrm{cov}}_p(N) \text{ (middle block)} \\
\mathcal{U}^{\mathrm{po}}_p(N) \text{ (upper and middle blocks)} \\
\mathcal{U}^{\mathrm{pre}}_p(N) \text{ (middle and lower blocks)} \\
\mathcal{U}^{\mathrm{prepo}}_p(N) \text{ (all blocks)}
\end{array} \tag{5.38}$$

The pre- and postwindowed method has been very popular in the so-called *linear predictive coding* (LPC) method (the main ingredient of which is *autoregressive* (AR) modelling) in speech processing [66,67] since the Levinson or Schur algorithms can be used to solve the LS problem [68–70]. These are *fast* algorithms that apply to the Toeplitz case and that allow the solution of a set of p equations with p unknowns in $O(p^2)$ operations, instead of the usual $O(p^3)$ in the general case (see Chapter 2). The *lattice* filter structure appearing in these algorithms even has a direct link to a certain acoustic model for the vocal tract [66]. The covariance method was nevertheless also often mentioned in the speech processing community in the 1970s. The main reason for that was that the covariance method makes no assumptions on the data outside the given window $[0, N]$, whereas the windowing in the pre- and postwindowed method may have significant (disturbing) effects for short (w.r.t. p) data records. However, it was long generally assumed that no special structure could be exploited in the covariance method and hence that $O(p^3)$ operations were necessary to use it. This situation brought motivation to the development of the so-called *α-stationarity* theory. In a statistical problem description, the Toeplitz covariance matrix case corresponds to the case of a stationary process. It turns out that the computational savings for the stationary case can also be found for certain types of non-stationary processes. Consider the ranks of the following *displacements* of $R_p = \mathcal{U}_p^H \mathcal{U}_p$:

$$\alpha \triangleq \text{rank}(R_p - S_p R_p S_p^H)$$
$$\alpha_0 \triangleq \text{rank}(S_p S_p^H R_p S_p S_p^H - S_p R_p S_p^H) \tag{5.39}$$

where S_p is a shift matrix (ones on the first subdiagonal, zeros elsewhere). Usually, the number α is called the *displacement rank*, while the number α_0 is called the *Toeplitz distance*. Since α_0 is zero for stationary processes and positive for any non-stationary process, it has been used as a measure of how 'close to stationary' the process u is [71–74]. Processes for which α_0 remains finite (and small) as p tends to infinity are called *α-stationary* processes. It turns out that the Levinson and Schur algorithms can be extended to α-stationary processes and these extensions are called the *generalized Levinson algorithm* [73,75] and the *generalized Schur algorithm* [76] (early versions of these algorithms may be be found in [77]; see [78] for the potential of the Schur-type algorithms for parallel implementation). This class of algorithms is discussed in more detail in Chapter 8. Though α_0 may be an appropriate measure for non-stationarity, the parameter α and its associated displacement are more closely related to the structure of the generalized algorithms, whose computational complexity is $O(\alpha p^2)$. However, it is clear that the difference between α and α_0 is at most two. In particular, for the stationary case, $\alpha = 2$.

In the deterministic case, the notions of stationarity or α-stationarity no longer apply in a strict sense. However, deterministic LS problems may still lead to normal equations in which the matrix of coefficients has a displacement structure. For a displacement rank α, the generalized Levinson or generalized Schur algorithm can be used to solve the normal equations in $O(\alpha p^2)$ operations. In particular, we have for the different windowing methods considered above: $\alpha^{\text{cov}} = 4$, $\alpha^{\text{po}} = 3$, $\alpha^{\text{pre}} = 3$, $\alpha^{\text{prepo}} = 2$.

5.3.1 Recursive least squares (RLS)

When new data arrive at time $N + 1$, one may wish to update the LS solution to incorporate the new data. One way to do so would be to solve a new batch processing problem for the data up to time $N + 1$. Alternatively, one may be able to reduce the computational load by solving the problem recursively in time, incorporating the solution of time N. A reduction in computations can indeed be found for the prewindowed and covariance cases, where the corresponding update for the sample covariance matrix involves modifications of rank 1 or 2, that is,

$$\begin{aligned} R_{p,L}(N+1) &= R_{p,L-1}(N) + \mathbf{u}_p(N+1)\mathbf{u}_p^{\mathrm{H}}(N+1) \\ &= R_{p,L-1}(N+1) + \mathbf{u}_p(N-L+2)\mathbf{u}_p^{\mathrm{H}}(N-L+2) \end{aligned} \tag{5.40}$$

Exploiting no special structure in the problem, the recursive solution leads to the conventional RLS algorithms with $O(p^2)$ computations per time update, as compared with $O(p^3)$ computations if one were to solve a new batch processing problem. Furthermore, exploiting the displacement structure of the covariance matrix will lead to fast RLS algorithms with a complexity of $O(p)$ per time update.

ORDER RECURSIVE SOLUTIONS

One can combine the rank 1 update mechanism with the generalized Levinson or generalized Schur-based LS lattice algorithm and obtain a fast RLS lattice (FLA) algorithm; see [79,80]. The FLA algorithm can alternatively be derived using geometric arguments; see [81,47,49]. The FLA algorithm can be viewed as a fast algorithm for recursively computing the so-called QR factorization of the data matrix. Starting from this point of view, FLA algorithms with a different structure have been obtained, which are called FQR algorithms [82]. The fast adaptive rotors RLS (FRO) algorithm in [83] belongs to this class of algorithms. See [84] for a unified derivation of the FLA and FQR algorithms, and also Chapters 6–8.

FIXED-ORDER SOLUTIONS

Just as the convectional $O(p^2)$ RLS algorithm is a particular instance of the Kalman filter, specialized to the case of parameter estimation, there is a fast RLS algorithm corresponding to the Chandrasekhar equations, which are a fast form of the Kalman filter when a displacement structure is exploited. This fast RLS algorithm can be called the fast transversal RLS (FTRLS) algorithm. Whereas the Kalman filter works with linear time-varying state-space models, the application of the Chandrasekhar equations to state estimation requires a time-invariant, state-space model. The particular parameter estimation problem considered here (LS estimation of the impulse response of an FIR filter) may be formulated using an infinite-dimensional but time-invariant state-space model. This observation was first made in [85, first draft]. The same model was proposed independently in [86,87]. So, with a time-invariant, state-space model available, one can apply the Chandrasekhar equations to arrive at a FTRLS-type algorithm. In [85, first draft], however (and in fact also in [86]), the displacement rank considered was one too high (probably due to the covariance formulation, omitting the prewindowing simplification). This may have led to some confusion and the omission of any Chandrasekhar considerations in the final version of [85].

In [85,26], the FTRLS algorithm was introduced as the fast Kalman algorithm and was derived by considering the last step in the Levinson recursions, combined with the rank 1 time update (see [88] also). In [89–92], FTRLS algorithms have been derived from the Chandrasekhar point of view, with an emphasis on regularized filtering problems (leading to higher values of the displacement rank), while in [93–95] 2-D applications to image restoration are discussed. Though the algorithms obtained in these references are of similar complexity ($O(p)$ per data sample) as the fast Kalman algorithm, no detailed comparisons between these two have been made. In [96, chapter 7], it was shown how the fast Kalman algorithm (or its improved versions discussed below) follows from the Chandrasekhar point of view by taking the split Chandrasekhar equations (positive and negative inertia in the displacement structure are separated) which were first derived in [97].

In [65], some redundancies were eliminated in the fast Kalman algorithm (of complexity $10p$), leading to the fast, a posteriori estimation sequential technique (FAEST) algorithm with complexity $7p$. This same redundancy elimination was done in [4], leading to the fast transversal filter (FTF) algorithm, which was derived using the geometric approach that had been introduced before conveniently to derive the FLA algorithms. Below, we shall adhere to this last approach with some minor modifications.

ALGORITHMS THAT ARE INTERMEDIATE BETWEEN LMS AND RLS

More recently, new adaptive filtering algorithms have been developed that are close to the LMS algorithm in terms of computational complexity, but that are less sensitive to the eigenvalue spread of the input covariance matrix. As a result, these algorithms often converge almost as fast as RLS. Two approaches can be distinguished in this class of algorithms.

In the fast Newton transversal filter (FNTF) algorithm [98,99], the order of the prediction part of the FTF algorithm is decoupled from the length of the filter to be adapted. In the typical application of acoustic echo cancellation, the required filter length can be very long due to the duration of the acoustic impulse response which can be a large number (hundreds or thousands) of sampling periods. The input signal being speech, only a low (say ten) filter order is needed to whiten the input signal. Therefore, in the FNTF algorithm, the sample covariance matrix, of which the inverse appears in the Kalman gain, is replaced by another matrix. This matrix equals the sample covariance matrix in a narrow band (small number of sub- and superdiagonals), but the rest of the matrix is determined by a non-stationary version of the maximum entropy extension principle. The resulting matrix has a banded inverse. It turns out that the resulting gradient for the FNTF algorithm can still be computed using a FTF algorithm, but this time of low dimension, namely of the size of (half) the bandwidth m of the banded inverse. The FNTF algorithm has a complexity of $2p + 5m$.

In a different approach, the least-squares criterion with a rectangular window is considered, but this time with a window length L smaller than the filter order p. The motivation behind this is the following. Least squares with a window length $L = p$ produce unbiased (but noisy) filter estimates. A window of length p is the shortest window length that allows one to obtain the fast convergence properties of least squares. In order to reduce computational complexity, consider now requiring only $L < p$ a posteriori errors to be zeroed instead of p. This leads to the block underdetermined covariance FTF algorithm [100,101] with average complexity $2p + 5.5L$ per sample. This algorithm has convergence properties that are close to those of the RLS algorithm when the input covariance matrix has L or less dominating eigenvalues.

5.4 Prewindowed FTRLS algorithm derivation, definition and interpretation of algorithmic quantities

The FIR linear filtering problem considered here has been formulated in section 5.3. Some relevant notation and the concept of prewindowing have been introduced there also. It has been indicated that least-squares problems naturally lead to a vector space formulation. In the general case of r channels of input signals, q experiments (this terminology will be clarified later) and multiple (s) desired response signals, the signals $u(t)$ and $y(t)$ are actually an $r \times q$ and an $s \times q$ matrix, respectively, with complex entries. The filter $\mathbf{c}_p(N)$ then is an $s \times pr$ matrix, though we will continue referring to it as a row vector (which it is in the case of scalar signals ($q = r = s = 1$)). The derivation given below is valid in the general case (except for a detail in the numerical stabilization mechanism to be discussed later). We may explicitly recall that in the prewindowed LS problem with exponential weighting, the filter coefficients $\mathbf{c}_p(N)$ are obtained as the solution of the following weighted quadratic minimization problem:

$$V_p(N) = \min_{\mathbf{c}_p(N)} \left\{ \sum_{i=0}^{N} \lambda^{N-t} \| e(t \mid \mathbf{c}_p(N)) \|^2 \right\} \equiv \alpha_p^y(N) \tag{5.41}$$

and are given by

$$\mathbf{c}_p(N) = -\left(\sum_{t=0}^{N} \lambda^{N-t} \mathbf{u}_p(t) \mathbf{u}_p^{\mathrm{H}}(t) \right)^{-1} \left(\sum_{t=0}^{N} \lambda^{N-t} \mathbf{u}_p(t) y^{\mathrm{H}}(t) \right) \tag{5.42}$$

5.4.1 Vector space formulation and projection updating

Fast LS algorithms based on a shift-invariance structure can be conveniently derived in a geometric framework with the decomposition of orthogonal projection operators

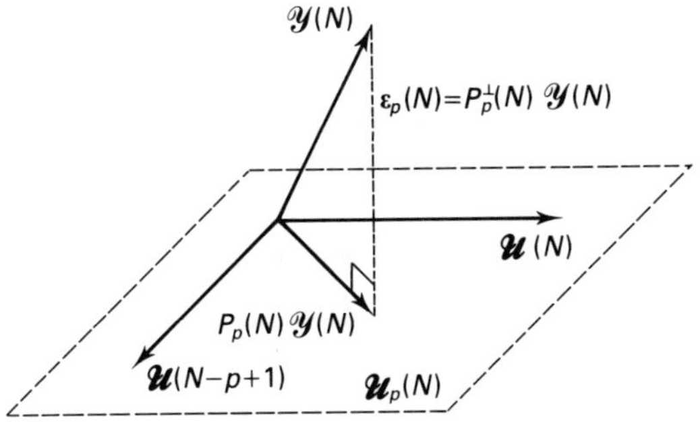

Figure 5.5 Vector space interpretation illustrated

as the basic tool. Consider a vector space containing the data vectors $\mathscr{U}^{\rm pre}(t)$, $\mathscr{Y}^{\rm pre}(t)$ defined in the discussion following Equation (5.35). We shall drop the superscript 'pre' in this discussion of the prewindowed case. The vector space is provided with the inner product

$$\langle \mathscr{U}, \mathscr{V} \rangle \triangleq \mathscr{U}^{\rm H}\mathscr{V} \tag{5.43}$$

The geometric interpretation of LS estimation leads to the orthogonality principle (see [102]), and therefore considers the projection operator on to the column space of the data matrix $\mathscr{U}$: $P_{\mathscr{U}} \triangleq \mathscr{U}(\mathscr{U}^{\rm H}\mathscr{U})^{-1}\mathscr{U}^{\rm H}$. This projection matrix satisfies the following properties:

$$\begin{aligned} P_{\mathscr{U}}^{\rm H} &= P_{\mathscr{U}} \quad \text{self-adjoint (Hermitian)} \\ P_{\mathscr{U}}P_{\mathscr{U}} &= P_{\mathscr{U}} \quad \text{idempotent} \end{aligned} \tag{5.44}$$

The projection matrix on to the orthogonal complement of the column space of $\mathscr{U}$ is $P_{\mathscr{U}}^{\perp} \triangleq I - P_{\mathscr{U}}$. One can readily verify the orthogonality property $\langle P_{\mathscr{U}}^{\perp}, P_{\mathscr{U}} \rangle = 0$. It will be convenient to introduce the shorthand notation $P_p(N) \triangleq P_{\mathscr{U}_p(N)}$, and similarly for related quantities. The least-squares criterion in (5.41) can now be seen to be the squared norm of the error vector $\boldsymbol{\varepsilon}_p(N)$:

$$\begin{aligned} \boldsymbol{\varepsilon}_p(N) &\triangleq \mathscr{Y}(N) + \mathscr{U}_p(N)\mathbf{c}_p(N) = P_p^{\perp}(N)\mathscr{Y}(N) \\ \alpha_p^y(N) &= \| \boldsymbol{\varepsilon}_p(N) \|^2 = \mathscr{Y}^{\rm H}(N)P_p^{\perp}(N)\mathscr{Y}(N) \end{aligned} \tag{5.45}$$

This geometric interpretation is illustrated in Figure 5.5.

We also introduce the sample covariance matrix $R_{\mathscr{U}} \triangleq \mathscr{U}^{\rm H}\mathscr{U}$, and the filter operator (left inverse of $\mathscr{U}$) $K_{\mathscr{U}} \triangleq (\mathscr{U}^{\rm H}\mathscr{U})^{-1}\mathscr{U}^{\rm H} = R_{\mathscr{U}}^{-1}\mathscr{U}^{\rm H}$. Note that $P_{\mathscr{U}}\mathscr{Y} = \mathscr{U}(K_{\mathscr{U}}\mathscr{Y})$. So $K_{\mathscr{U}}\mathscr{Y}$ is the filter that, when operating on $\mathscr{U}$, will produce the projection $\mathscr{U}(K_{\mathscr{U}}\mathscr{Y}) = P_{\mathscr{U}}\mathscr{Y}$ of $\mathscr{Y}$ on to $\mathscr{U}$. The orthogonality principle of LS states that the linear combination that will result in the error vector of smallest norm corresponds to the orthogonal projection of the vector to be approximated on the subspace spanned by the vectors involved in the linear combination. Hence the solution $\mathbf{c}_p(N)$ to the LS problem satisfies the orthogonality condition

$$\langle \mathcal{U}_p(N), \boldsymbol{\varepsilon}_p(N) \rangle = \mathbf{0} \tag{5.46}$$

and thus we find

$$\mathbf{c}_p(N) = -K_p(N)\boldsymbol{\mathcal{Y}}(N) \tag{5.47}$$

The updating of the LS solution naturally corresponds to the updating of a projection operator. So consider the basic projection updating formula

$$P^{\perp}_{[\mathcal{U}\mathcal{V}]} = P^{\perp}_{\mathcal{U}} - P^{\perp}_{\mathcal{U}\mathcal{V}} = P^{\perp}_{\mathcal{U}} - P^{\perp}_{\mathcal{U}}\mathcal{V}(\mathcal{V}^{\mathrm{H}}P^{\perp}_{\mathcal{U}}\mathcal{V})^{-1}\mathcal{V}^{\mathrm{H}}P^{\perp}_{\mathcal{U}} \tag{5.48}$$

This update is often called an order update, since it corresponds to an increase of the order of the filter producing the projection. A particular application of this formula ill lead to time-updating identities. Consider the pinning vector $\boldsymbol{\sigma} \triangleq [I\,0\,0\ldots]^{\mathrm{H}}$ [49,9] to pin down the most recent data sample: $y(t) = \langle \boldsymbol{\mathcal{Y}}(t), \boldsymbol{\sigma} \rangle$. The matrix S (ones on the first subdiagonal, zeros elsewhere) is the shift matrix corresponding to the pinning vector $\boldsymbol{\sigma}$ (namely, $P^{\perp}_{\boldsymbol{\sigma}} = SS^{\mathrm{H}}$), and has the shifting property $S^{\mathrm{H}}\boldsymbol{\mathcal{Y}}(t) = \lambda^{1/2}\boldsymbol{\mathcal{Y}}(t-1)$. Two applications of (5.48) now yield a time update formula:

$$\begin{aligned} P^{\perp}_{\mathcal{U}} - P^{\perp}_{\mathcal{U}}\boldsymbol{\sigma}(\boldsymbol{\sigma}^{\mathrm{H}}P^{\perp}_{\mathcal{U}}\sigma)^{-1}\boldsymbol{\sigma}^{\mathrm{H}}P^{\perp}_{\mathcal{U}} &= P^{\perp}_{[\mathcal{U}\sigma]} \\ = P^{\perp}_{[\sigma\mathcal{U}]} = P^{\perp}_{\boldsymbol{\sigma}} - P^{\perp}_{\boldsymbol{\sigma}}\mathcal{U}(\mathcal{U}^{\mathrm{H}}P^{\perp}_{\boldsymbol{\sigma}}\mathcal{U})^{-1}\mathcal{U}^{\mathrm{H}}P^{\perp}_{\boldsymbol{\sigma}} &= SP^{\perp}_{S^{\mathrm{H}}\mathcal{U}}S^{\mathrm{H}} \end{aligned} \tag{5.49}$$

The translation of this projection time update to the time update of filters (of the form $K_{\mathcal{U}}\boldsymbol{\mathcal{V}}$) involves the Kalman gain $\tilde{\mathbf{w}}_{\mathcal{U}}$, which has an interesting interpretation (in unnormalized form [4]) as the optimal LS estimation filter for the pinning vector, that is,

$$\min_{\tilde{\mathbf{w}}_{\mathcal{U}}} \|\boldsymbol{\sigma} + \mathcal{U}\tilde{\mathbf{w}}_{\mathcal{U}}\|^2 = \alpha_{\mathcal{U}} = \boldsymbol{\sigma}^{\mathrm{H}}P^{\perp}_{\mathcal{U}}\boldsymbol{\sigma} \tag{5.50}$$

resulting in $\tilde{\mathbf{w}}_{\mathcal{U}} = -K_{\mathcal{U}}\boldsymbol{\sigma}$. Then one can also translate (5.48) into an order update formula for the Kalman gain as in T5.3-(10,11) (Equations (10) and (11) in Table 5.3). A complete set of updating identities required for the derivation of FTRLS algorithms is given in Table 5.3 (the superscript # denotes an arbitrary generalized inverse). These identities are based on the more fundamental identities of Table 5.2. A derivation

Table 5.2 Geometric updating identities

Equation	Identity
	$P_{\mathcal{U}} \triangleq \mathcal{U}\{\mathcal{U}^{\mathrm{H}}\mathcal{U}\}^{\#}\mathcal{U}^{\mathrm{H}}, \quad P^{\perp}_{\mathcal{U}} \triangleq I - P_{\mathcal{U}}$
(1)	$P^{\perp}_{\mathcal{U},\mathcal{V}} = P^{\perp}_{\mathcal{U}} - P^{\perp}_{\mathcal{U}}\mathcal{V}\ \{\mathcal{V}^{\mathrm{H}}P^{\perp}_{\mathcal{U}}\mathcal{V}\}^{\#}\ \mathcal{V}^{\mathrm{H}}P^{\perp}_{\mathcal{U}} = P^{\perp}_{\mathcal{V},\mathcal{U}}$
(2)	$P^{\perp}_{\mathcal{U}} = SP^{\perp}_{S^{\mathrm{H}}\mathcal{U}}S^{\mathrm{H}} + P^{\perp}_{\mathcal{U}}\boldsymbol{\sigma}\ \{\boldsymbol{\sigma}^{\mathrm{H}}P^{\perp}_{\mathcal{U}}\boldsymbol{\sigma}\}^{\#}\ \boldsymbol{\sigma}^{\mathrm{H}}P^{\perp}_{\mathcal{U}}$
	$R_{\mathcal{U}} \triangleq \mathcal{U}^{\mathrm{H}}\mathcal{U}$
(3)	$R^{\#}_{\mathcal{U},\mathcal{V}} = \begin{bmatrix} R^{\#}_{\mathcal{U}} & 0 \\ 0 & 0 \end{bmatrix} + \begin{bmatrix} -K_{\mathcal{U}}\mathcal{V} \\ I \end{bmatrix} \{\mathcal{V}^{\mathrm{H}}P^{\perp}_{\mathcal{U}}\mathcal{V}\}^{\#} \begin{bmatrix} -K_{\mathcal{U}}\mathcal{V} \\ I \end{bmatrix}^{\mathrm{H}}$
(4)	$R^{\#}_{\mathcal{V},\mathcal{U}} = \begin{bmatrix} 0 & 0 \\ 0 & R^{\#}_{\mathcal{U}} \end{bmatrix} + \begin{bmatrix} I \\ -K_{\mathcal{U}}\mathcal{V} \end{bmatrix} \{\mathcal{V}^{\mathrm{H}}P^{\perp}_{\mathcal{U}}\mathcal{V}\}^{\#} \begin{bmatrix} I \\ -K_{\mathcal{U}}\mathcal{V} \end{bmatrix}^{\mathrm{H}}$
(5)	$R^{\#}_{\mathcal{U}} = R^{\#}_{S^{\mathrm{H}}\mathcal{U}} - K_{\mathcal{U}}\boldsymbol{\sigma}\ \{\boldsymbol{\sigma}^{\mathrm{H}}P^{\#}_{\mathcal{U}}\boldsymbol{\sigma}\}^{\#}\boldsymbol{\sigma}^{\mathrm{H}}K^{\mathrm{H}}_{\mathcal{U}}$
	$K_{\mathcal{U}} \triangleq \{\mathcal{U}^{\mathrm{H}}\mathcal{U}\}^{\#}\mathcal{U}^{\mathrm{H}}$
(6)	$K_{\mathcal{U},\mathcal{V}} = \begin{bmatrix} K_{\mathcal{U}} \\ 0 \end{bmatrix} + \begin{bmatrix} -K_{\mathcal{U}}\mathcal{V} \\ I \end{bmatrix} \{\mathcal{V}^{\mathrm{H}}P^{\perp}_{\mathcal{U}}\mathcal{V}\}^{\#}\mathcal{V}^{\mathrm{H}}P^{\perp}_{\mathcal{U}}$
(7)	$K_{\mathcal{V},\mathcal{U}} = \begin{bmatrix} 0 \\ K_{\mathcal{U}} \end{bmatrix} + \begin{bmatrix} I \\ -K_{\mathcal{U}}\mathcal{V} \end{bmatrix} \{\mathcal{V}^{\mathrm{H}}P^{\perp}_{\mathcal{U}}\mathcal{V}\}^{\#}\mathcal{V}^{\mathrm{H}}P^{\perp}_{\mathcal{U}}$
(8)	$K_{\mathcal{U}} = K_{S^{\mathrm{H}}\mathcal{U}}S^{\mathrm{H}} + K_{\mathcal{U}}\boldsymbol{\sigma}\ \{\boldsymbol{\sigma}^{\mathrm{H}}P^{\perp}_{\mathcal{U}}\boldsymbol{\sigma}\}^{\#}\ \boldsymbol{\sigma}^{\mathrm{H}}P^{\perp}_{\mathcal{U}}$

of these relations can be found in [4,96,103]. To give an indication of such a derivation, consider for simplicity the case when $R_{\mathcal{U},\mathcal{V}}$ is positive definite. Then the generalized inverse reduces to the usual inverse (see [47, Appendix B] for extensions to the singular case). T5.2-(1,2) have been derived in (5.48) and (5.49). For the remainder of Table 5.2, we start from the basic projection updating formula

$$P_{\mathcal{U},\mathcal{V}} = P_{\mathcal{U}} + P^{\perp}_{\mathcal{U}}\mathcal{V}(\mathcal{V}^{\mathrm{H}}P^{\perp}_{\mathcal{U}}\mathcal{V})^{-1}\mathcal{V}^{\mathrm{H}}P^{\perp}_{\mathcal{U}} \tag{5.51}$$

From the definition of $P_{\mathcal{U}}$, $K_{\mathcal{U}}$, the following identities readily follow:

$$P_{\mathcal{U},\mathcal{V}} = [\mathcal{U}, \mathcal{V}]R^{-1}_{\mathcal{U},\mathcal{V}}\begin{bmatrix} \mathcal{U}^{\mathrm{H}} \\ \mathcal{V}^{\mathrm{H}} \end{bmatrix} = [\mathcal{U}, \mathcal{V}]K_{\mathcal{U},\mathcal{V}} \tag{5.52}$$

$$P_{\mathcal{U}} = [\mathcal{U}, \mathcal{V}]\begin{bmatrix} R^{-1}_{\mathcal{U}} & 0 \\ 0 & 0 \end{bmatrix}\begin{bmatrix} \mathcal{U}^{\mathrm{H}} \\ \mathcal{V}^{\mathrm{H}} \end{bmatrix} = [\mathcal{U}, \mathcal{V}]\begin{bmatrix} K_{\mathcal{U}} \\ 0 \end{bmatrix} \tag{5.53}$$

$$P^{\perp}_{\mathcal{U}}\mathcal{V} = \mathcal{V} - \mathcal{U}K_{\mathcal{U}}\mathcal{V} = [\mathcal{U}, \mathcal{V}]\begin{bmatrix} -K_{\mathcal{U}}\mathcal{V} \\ I \end{bmatrix} \tag{5.54}$$

By substituting (5.52), (5.53) and (5.54) into (5.51), one derives T5.2-(3,6). T5.2-(4,7) follow from T5.2-(3,6) by a simple permutation of rows and columns. To get T5.2-(5,8), substitute $[\mathscr{U}\ \mathscr{V}] = [\mathscr{U}\ \boldsymbol{\sigma}]$ in T5.2-(3,6) and $[\mathscr{V}\ \mathscr{U}] = [\boldsymbol{\sigma}\ \mathscr{U}]$ in T5.2-(4,7). Equate the (1,1) blocks in the two expressions for $R^{-1}_{\mathscr{U},\boldsymbol{\sigma}}$ thus obtained, namely

$$R^{-1}_{\mathscr{U},\boldsymbol{\sigma}} = \begin{bmatrix} R^{-1}_{\mathscr{U}} & 0 \\ 0 & 0 \end{bmatrix} + \begin{bmatrix} -K_{\mathscr{U}}\boldsymbol{\sigma} \\ I \end{bmatrix} \{\boldsymbol{\sigma}^{\mathrm{H}} P^{\perp}_{\mathscr{U}} \boldsymbol{\sigma}\}^{-1} \begin{bmatrix} -K_{\mathscr{U}}\boldsymbol{\sigma} \\ I \end{bmatrix}^{\mathrm{H}}$$
$$= \begin{bmatrix} 0 & 0 \\ 0 & I \end{bmatrix} + \begin{bmatrix} I \\ -\boldsymbol{\sigma}^{\mathrm{H}}\mathscr{U} \end{bmatrix} R^{-1}_{S^{\mathrm{H}}\mathscr{U}} \begin{bmatrix} I \\ -\boldsymbol{\sigma}^{\mathrm{H}}\mathscr{U} \end{bmatrix}^{\mathrm{H}} \tag{5.55}$$

to get T5.2-(5) and equate the (1,1) blocks in the two expressions for $K_{\mathscr{U}}\boldsymbol{\sigma}$ to get T5.2-(8).

Table 5.3 Updating identities for prewindowed FTRLS algorithms

Equation	Identity
	Kalmain gain:
(1)	$\tilde{\mathbf{w}}_{\mathscr{U}} \triangleq -K_{\mathscr{U}}\boldsymbol{\sigma}$
(2)	$\mathbf{w}_{\mathscr{U}} \triangleq -R^{\#}_{S^{\mathrm{H}}\mathscr{U}} \mathscr{U}^{\mathrm{H}}\boldsymbol{\sigma}$
(3)	$\alpha_{\mathscr{U}} \triangleq \boldsymbol{\sigma}^{\mathrm{H}} P^{\perp}_{\mathscr{U}} \boldsymbol{\sigma} = I + \boldsymbol{\sigma}^{\mathrm{H}}\mathscr{U}\tilde{\mathbf{w}}_{\mathscr{U}} = \{I - \boldsymbol{\sigma}^{\mathrm{H}}\mathscr{U}\mathbf{w}_{\mathscr{U}}\}^{\#}$
(4)	$\tilde{\mathbf{w}}_{\mathscr{U}} = \mathbf{w}_{\mathscr{U}}\alpha_{\mathscr{U}}$
	(A posteriori) residuals and predicted (a priori) residuals:
(5)	$(\mathscr{V}^{\mathrm{H}} P^{\perp}_{\mathscr{U}}\boldsymbol{\sigma}) = [I,\ -\mathscr{V}^{\mathrm{H}} K^{\mathrm{H}}_{\mathscr{U}}][\mathscr{V},\ \mathscr{U}]^{\mathrm{H}}\ \boldsymbol{\sigma} = [-\mathscr{V}^{\mathrm{H}} K^{\mathrm{H}}_{\mathscr{U}},\ I]\ [\mathscr{U},\ \mathscr{V}]^{\mathrm{H}}\ \boldsymbol{\sigma}$
(6)	$(\mathscr{V}^{\mathrm{H}} P^{\perp}_{\mathscr{U}}\boldsymbol{\sigma})^{p} \triangleq [I,\ -\mathscr{V}^{\mathrm{H}} S K^{\mathrm{H}}_{S^{\mathrm{H}}\mathscr{U}}][\mathscr{V},\ \mathscr{U}]^{\mathrm{H}}\ \boldsymbol{\sigma} = [-\mathscr{V}^{\mathrm{H}} S K^{\mathrm{H}}_{S^{\mathrm{H}}\mathscr{U}},\ I][\mathscr{U},\ \mathscr{V}]^{\mathrm{H}}\ \boldsymbol{\sigma}$
(7)	$(\mathscr{V}^{\mathrm{H}} P^{\perp}_{\mathscr{U}}\boldsymbol{\sigma}) = (\mathscr{V}^{\mathrm{H}} P^{\perp}_{\mathscr{U}}\boldsymbol{\sigma})^{p}\ \alpha_{\mathscr{U}}$
	Residual energy order-and-time update:
(8)	$\{\alpha_{\mathscr{U},\mathscr{V}}\}^{\#} = \{\alpha_{\mathscr{U}}\}^{\#} + (\mathscr{V}^{\mathrm{H}} P^{\perp}_{\mathscr{U}}\boldsymbol{\sigma})^{p\mathrm{H}}\ \{\mathscr{V}^{\mathrm{H}} S P^{\perp}_{S^{\mathrm{H}}\mathscr{U}} S^{\mathrm{H}}\mathscr{V}\}^{\#}\ (\mathscr{V}^{\mathrm{H}} P^{\perp}_{\mathscr{U}}\boldsymbol{\sigma})^{p} = \{\alpha_{\mathscr{V},\mathscr{U}}\}^{\#}$
(9)	$\mathscr{V}^{\mathrm{H}} P^{\perp}_{\mathscr{U}}\mathscr{V} = \mathscr{V}^{\mathrm{H}} S P^{\perp}_{S^{\mathrm{H}}\mathscr{U}} S^{\mathrm{H}}\mathscr{V} + (\mathscr{V}^{\mathrm{H}} P^{\perp}_{\mathscr{U}}\boldsymbol{\sigma})\{\alpha_{\mathscr{U}}\}^{\#}\ (\mathscr{V}^{\mathrm{H}} P^{\perp}_{\mathscr{U}}\boldsymbol{\sigma})^{\mathrm{H}}$
	Kalman gain order update:
(10)	$\mathbf{w}_{\mathscr{U},\mathscr{V}} = \begin{bmatrix} \mathbf{w}_{\mathscr{U}} \\ 0 \end{bmatrix} - \begin{bmatrix} -K_{S^{\mathrm{H}}\mathscr{U}} S^{\mathrm{H}}\mathscr{V} \\ I \end{bmatrix} \{\mathscr{V}^{\mathrm{H}} S P^{\perp}_{S^{\mathrm{H}}\mathscr{U}} S^{\mathrm{H}}\mathscr{V}\}^{\#}\ (\mathscr{V}^{\mathrm{H}} P^{\perp}_{\mathscr{U}}\boldsymbol{\sigma})^{p}$
(11)	$\mathbf{w}_{\mathscr{V},\mathscr{U}} = \begin{bmatrix} 0 \\ \mathbf{w}_{\mathscr{U}} \end{bmatrix} - \begin{bmatrix} I \\ -K_{S^{\mathrm{H}}\mathscr{U}} S^{\mathrm{H}}\mathscr{V} \end{bmatrix} \{\mathscr{V}^{\mathrm{H}} S P^{\perp}_{S^{\mathrm{H}}\mathscr{U}} S^{\mathrm{H}}\mathscr{V}\}^{\#}\ (\mathscr{V}^{\mathrm{H}} P^{\perp}_{\mathscr{U}}\boldsymbol{\sigma})^{p}$
	Filter time update:
(12)	$[-K_{\mathscr{U}}\mathscr{V}] = [-K_{S^{\mathrm{H}}\mathscr{U}} S^{\mathrm{H}}\mathscr{V}] + [\mathbf{w}_{\mathscr{U}}]\ (\mathscr{V}^{\mathrm{H}} P^{\perp}_{\mathscr{U}}\boldsymbol{\sigma})^{\mathrm{H}}$

Consider now Table 5.3. With the definition of the Kalman gain in unnormalized ($\tilde{\mathbf{w}}_{\mathscr{U}}$) and overnormalized ($\mathbf{w}_{\mathscr{U}}$) form in T5.3-(1,2), the first identity in T5.3-(3) follows from (5.54), while the second one is obtained by equating the (2,2) blocks on both sides of (5.55). Note that the variable $\alpha_{\mathscr{U}}$ can at the same time be interpreted as the squared error in the estimation of the pinning vector (see (5.50)), or as the output of the Kalman gain $\tilde{\mathbf{w}}_{\mathscr{U}}$ (see T5.3-(3)). Equation T5.3-(4) follows from equating the (1,2) blocks in (5.55). An overnormalized filter [48] is a filter whose coefficients have been divided by the variance of the output signal of the filter. Hence, $\mathbf{w}_{\mathscr{U}}$ is the overnormalized Kalman gain. Equation T5.3-(5) follows from (5.54), T5.3-(9) from T5.2-(2), and T5.3-(12) from T5.13-(8). A posteriori residuals as in T5.3-(5) are obtained using the latest filter estimate $K_{\mathscr{U}}\mathscr{V}$. They will be denoted as ε. A priori residuals as in T5.3-(6) are obtained using the previous filter estimate $K_{S^{\mathrm{H}}\mathscr{U}}S^{\mathrm{H}}\mathscr{V}$. They will be denoted as $e = \varepsilon^p$. To obtain T5.3-(7), use T5.3-(12) in T5.3-(5), and also use T5.3-(3,6), that is,

$$(\mathscr{V}^{\mathrm{H}}P^{\perp}_{\mathscr{U}}\boldsymbol{\sigma}) = [I, \; -\mathscr{V}^{\mathrm{H}}SK^{\mathrm{H}}_{S^{\mathrm{H}}\mathscr{U}}][\mathscr{V}, \mathscr{U}]^{\mathrm{H}}\boldsymbol{\sigma} + (\mathscr{V}^{\mathrm{H}}P^{\perp}_{\mathscr{U}}\boldsymbol{\sigma})\mathbf{w}^{\mathrm{H}}_{\mathscr{U}}\mathscr{U}^{\mathrm{H}}\boldsymbol{\sigma}$$

$$= (\mathscr{V}^{\mathrm{H}}P^{\perp}_{\mathscr{U}}\boldsymbol{\sigma})^p + (\mathscr{V}^{\mathrm{H}}P^{\perp}_{\mathscr{U}}\boldsymbol{\sigma})\{I - \alpha^{-1}_{\mathscr{U}}\} \tag{5.56}$$

from which T5.3-(7) follows. To obtain T5.3-(10), substitute $[\mathscr{U}, \mathscr{V}]$ by $[S^{\mathrm{H}}\mathscr{U}, S^{\mathrm{H}}\mathscr{V}]$ in T5.2-(3), postmultiply with $-[\mathscr{U}, \mathscr{V}]^{\mathrm{H}}\boldsymbol{\sigma}$, and use T5.3-(6). T5.3-(11) follows from T5.3-(10) by a permutation of the subspaces $\mathscr{U}$ and $\mathscr{V}$. Finally, to obtain T5.3-(8), use the second identity in T5.3-(3), together with T5.3-(10) or T5.3-(11), and T5.3-(6).

5.4.2 Algorithm derivation (steady state)

In this section we show how careful exploitation of the updating identities in Table 5.3 will lead to a computationally efficient algorithm. The derivation of the allgorithm will generate a number of new quantities, which will be defined as they emerge. A summary of the definition and transversal filter computation of the different quantities can be found in Table II of [4]. The resulting algorithm (steady state) is given in Table 5.4. The computational cost indicated in Table 5.4 is the number of multiplications. The number of additions differs from this by a small amount (that is independent of p). Initialization issues will be discussed later. We shall show how different filters naturally come about, and how their time-updating recursions can be conveniently derived.

Table 5.4 Multichannel complex FTRLS algorithm for prewindowing with exponential weighting (computations count holds for $q = 1$)

Equation	Computation	Cost
	Prediction problem:	
(1)	$e_p^f(N) = \mathbf{A}_p^H(N-1)\mathbf{u}_{p+1}(N)$	pr^2
(2)	$w_{p+1}^0(N) = -\lambda^{-1}\alpha_p^{-f}(N-1)e_p^f(N)$	$1.5r^2 + 0.5r$
(3)	$\mathbf{w}_{p+1}(N) = \begin{bmatrix} 0 \\ \mathbf{w}_p(N-1) \end{bmatrix} + \mathbf{A}_p(N-1)w_{p+1}^0(N)$	pr^2
(4)	$\alpha_{p+1}^{-1}(N) = \alpha_p^{-1}(N-1) - w_{p+1}^{0H}(N)e_p^f(N)$	r
(5)	$e_p^b(N) = -\lambda\alpha_p^b(N-1)w_{p+1}^p(N)$	$1.5r^2 + 0.5r$
(6)	$\begin{bmatrix} \mathbf{w}_p(N) \\ 0 \end{bmatrix} = \mathbf{w}_{p+1}(N) - \mathbf{B}_p(N-1)w_{p+1}^p(N)$	pr^2
(7)	$\alpha_p^{-1}(\mathrm{N}) = \alpha_{p+1}^{-1}(\mathrm{N}) + \mathbf{w}_{p+1}^{pH}(N)e_p^b(N)$	r
(8)	$\varepsilon_p^f(N) = e_p^f(N)\alpha_p(N-1)$	r
(9)	$\mathbf{a}_p(N) = \mathbf{a}_p(N-1) + \mathbf{w}_p(N-1)\varepsilon_p^{fH}(N)$	pr^2
(10)	$\alpha_p^{-f}(N) = \lambda^{-1}\alpha_p^{-f}(N-1) - w_{p+1}^0(N)\alpha_{p+1}(N)w_{p+1}^{0H}(N)$	$r^2 + r$
(11)	$\varepsilon_p^b(N) = e_p^b(N)\alpha_p(N)$	r
(12)	$\mathbf{b}_p(N) = \mathbf{b}_p(N-1) + \mathbf{w}_p(\mathbf{N})\varepsilon_p^{bH}(N)$	pr^2
(13)	$\alpha_p^b(N) = \lambda\alpha_p^b(N-1) + \varepsilon_p^b(N)e_p^{bH}(N)$	r^2
	Joint-process extension:	
(14)	$e_p(N) = y(N) + \mathbf{c}_p^H(N-1)\mathbf{u}_p(N)$	prs
(15)	$\varepsilon_p(N) = e_p(N)\alpha_p(N)$	s
(16)	$\mathbf{c}_p(N) = \mathbf{c}_p(N-1) + \mathbf{w}_p(N)\varepsilon_p^H(N)$	prs
	Single real channel total (two divides)	$7p + 12$
	Multichannel total (two divides)	$(5r + 2s)rp + 5r^2 + 6r + s$

However, before going into the details, we shall outline the two main updating mechanisms present in the FTRLS algorithm (below, the same simplifications as in (5.38) are invoked):

- First time-updating mechanism: conventional RLS

$$\begin{bmatrix} y(N) \\ \hline y(N-1) \\ y(N-2) \\ \vdots \\ y(0) \\ 0 \\ \vdots \end{bmatrix} + \begin{bmatrix} u(N) & u(N-1) & \cdots & u(N-p+1) \\ \hline u(N-1) & u(N-2) & \cdots & u(N-p) \\ u(N-2) & \cdots & & \vdots \\ \vdots & & & u(0) \\ u(0) & & \cdot^{\cdot^{\cdot}} & 0 \\ 0 & & & \vdots \\ \vdots & & & \end{bmatrix} \mathbf{c}_p(N)$$

$$\mathbf{c}_p(N) = \mathbf{c}_p(N-1) + \mathbf{w}_p(N)\varepsilon_N^{\mathrm{H}}(N)$$

$$\mathbf{a}_p(N) = \mathbf{a}_p(N-1) + \mathbf{w}_p(N-1)\varepsilon_p^{\mathrm{fH}}(N)$$

$$\mathbf{b}_p(N) = \mathbf{b}_p(N-1) + \mathbf{w}_p(N)\varepsilon_p^{\mathrm{bH}}(N)$$

where $\mathbf{a}_p(N)$ and $\mathbf{b}_p(N)$ are the forward and backward prediction filters defined in Chapter 2.

- Second time-updating mechanism: shift invariance (Hankel data matrix)

$$[\boldsymbol{\mathscr{U}}(N), \mathscr{U}_p(N-1)] = \mathscr{U}_{p+1}(N) = [\mathscr{U}_p(N), \boldsymbol{\mathscr{U}}(N-p)]$$

where

$$\mathscr{U}_{p+1}(N) = \left[\begin{matrix} u(N) \\ u(N-1) \\ u(N-2) \\ \vdots \end{matrix} \; \begin{bmatrix} u(N-1 & \cdots & u(N-p+1) \\ u(N-2) & \cdots & u(N-p) \\ \vdots & & u(N-p-1) \\ & & \vdots \end{bmatrix} \; \begin{matrix} u(N-p) \\ u(N-p-1) \\ \vdots \\ \end{matrix}\right]$$

$$\mathbf{w}_{p+1}^{\mathrm{H}}(N) = [0, \mathbf{w}_p^{\mathrm{H}}(N-1)] + w_{p+1}^{0\mathrm{H}}(N)\mathbf{A}_p^{\mathrm{H}}(N-1)$$

$$[\mathbf{w}_p^{\mathrm{H}}(N), 0] = \mathbf{w}_{p+1}^{\mathrm{H}}(N) - w_{p+1}^{p\mathrm{H}}(N)\mathbf{B}_p^{\mathrm{H}}(N-1)$$

The first time-updating mechanism is that of the conventional RLS algorithm and hence consists of simply adding a new top entry to the shifted-down versions of all data vectors. To have a fast RLS algorithm, there has to be a second time-updating mechanism. This second mechanism will allow us to circumvent the Riccati equation (5.23) for the update of the Kalman gain. Such a mechanism is obtained by exploiting the Hankel structure of the data matrix. This structure allows us to obtain the subspace $\mathscr{U}_p(N)$ from the subspace $\mathscr{U}_p(N-1)$ by taking the direct sum with the vector $\boldsymbol{\mathscr{U}}(N)$ and the direct difference with the vector $\boldsymbol{\mathscr{U}}(N-p)$. This direct sum and difference are most easily effected in terms of the forward and backward residuals $\varepsilon_p^{\mathrm{f}}(N) = P_p^{\perp}(N-1)\boldsymbol{\mathscr{U}}(N)$ and $\varepsilon_p^{\mathrm{b}}(N) = P_p^{\perp}(N)\boldsymbol{\mathscr{U}}(N-p)$. These projections form the orthogonal complements of the subspaces $\mathscr{U}_p(N-1)$ and $\mathscr{U}_p(N)$, respectively, within the subspace $\mathscr{U}_{p+1}(N)$. So if $\{\mathscr{U}\}$ denotes the subspace spanned by the columns of $\mathscr{U}$, then

$$\{\mathscr{U}_{p+1}(N)\} = \{\mathscr{U}_p(N)\} \oplus \{\boldsymbol{\varepsilon}_p^{\mathrm{b}}(N)\} = \{\boldsymbol{\varepsilon}_p^{\mathrm{f}}(N)\} \oplus \{\mathscr{U}_p(N-1)\} \tag{5.57}$$

where $\oplus$ denotes the direct sum of vector spaces. The forward and backward residuals are obtained via the forward and backward prediction error filters $\mathbf{A}_p(N)$ and $\mathbf{B}_p(N)$:

$\varepsilon_p^f(N) = \mathscr{U}_{p+1}(N)\mathbf{A}_p(N)$ and $\varepsilon_p^b(N) = \mathscr{U}_{p+1}(N)\mathbf{B}_p(N)$. The time update of these filters then in turn involves the Kalman gain via the first mechanism, as for the joint-process filter $\mathbf{c}_p(N)$. This yields a complete set of time updates. Here are the details.

JOINT-PROCESS EXTENSION FILTER $\mathbf{c}_p(N)$
Adding a new top element to the shifted versions of all the data vectors is precisely the mechanism involved in T5.3-(12). Thus, using T5.3-(12) with $\mathscr{U} = \mathscr{U}_p(N)$ and $\mathscr{V} = \mathscr{Y}(N)$ yields T5.4-(16), that is,

$$\mathbf{c}_p(N) = \mathbf{c}_p(N-1) + \mathbf{w}_p(N)\varepsilon_p^{\mathrm{H}}(N) \tag{5.58}$$

T5.4-(14,15) can be obtained from T5.3-(6,7) with the same substitutions for $\mathscr{U}$ and $\mathscr{V}$. Using T5.3-(9), the value of the cost function can also be updated if so desired,

$$\alpha_p^y(N) = \lambda\alpha_p^y(N-1) + \varepsilon_p^{\mathrm{H}}(N)e_p(N) \tag{5.59}$$

UPDATE OF THE KALMAN GAIN $\mathbf{w}_p(N)$
The time update of $\mathbf{w}_p(N)$ (or $\tilde{\mathbf{w}}_p(N)$) can clearly not be done using T5.3-(12), because $S^{\mathrm{H}}\boldsymbol{\sigma} = 0$ ($\mathscr{V} = \boldsymbol{\sigma}$ in T5.3-(1)). The efficient update of the Kalman gain requires another strategy, involving the exploitation of the low displacement rank of the Gram matrix $R_{p+1}(N)$. While the conventional time-updating mechanism goes from $\mathscr{U}_p(N-1)$ to $\mathscr{U}_p(N)$ by taking $P_{\boldsymbol{\sigma}}^{\perp}\mathscr{U}_p(N) = SS^{\mathrm{H}}\mathscr{U}_p(N) = \lambda^{1/2}S\mathscr{U}_p(N-1)$ and adding a new most recent data vector $P_{\boldsymbol{\sigma}}\mathscr{U}_p(N) = \boldsymbol{\sigma}\langle\boldsymbol{\sigma}, \mathscr{U}_p(N)\rangle = \boldsymbol{\sigma}\mathbf{u}_p^{\mathrm{H}}(N)$ to the top row, the second mechanism combines the following order updating and order downdating steps:

$$[\boldsymbol{\mathscr{U}}(N)\ \mathscr{U}_p(N-1)] = \mathscr{U}_{p+1}(N) = [\mathscr{U}_p(N)\ \boldsymbol{\mathscr{U}}(N-p)] \tag{5.60}$$

It is exactly the existence of this second possibility which makes a fast update of the Kalman gain, and hence a fast algorithm, possible. Taking the above two partitions of the space spanned by the columns of $\mathscr{U}_{p+1}(N)$ and substituting in T5.3-(11,10) and T5.3-(8) yields

$$\begin{aligned}
\mathbf{w}_{p+1}(N) &= \begin{bmatrix} 0 \\ \mathbf{w}_p(N-1) \end{bmatrix} + \mathbf{A}_p(N-1)\lambda^{-1}\alpha_p^{-\mathrm{f}}(N-1)e_p^{\mathrm{f}}(N) \\
\mathbf{w}_{p+1}(N) &= \begin{bmatrix} \mathbf{w}_p(N) \\ 0 \end{bmatrix} - \mathbf{B}_p(N-1)\lambda^{-1}\alpha_p^{-\mathrm{b}}(N-1)e_p^{\mathrm{b}}(N) \\
\alpha_{p+1}^{-1}(N) &= \alpha_p^{-1}(N-1) + e_p^{\mathrm{fH}}(N)\lambda^{-1}\alpha_p^{-\mathrm{f}}(N-1)e_p^{\mathrm{f}}(N) \\
\alpha_{p+1}^{-1}(N) &= \alpha_p^{-1}(N) + e_p^{\mathrm{bH}}(N)\lambda^{-1}\alpha_p^{-\mathrm{b}}(N-1)e_p^{\mathrm{b}}(N)
\end{aligned} \tag{5.61}$$

where

$$\begin{aligned}
\mathbf{A}_p(N) &= \begin{bmatrix} I_r \\ \mathbf{a}_p(N) \end{bmatrix}, \qquad \mathbf{a}_p(N) = -K_p(N-1)\boldsymbol{\mathscr{U}}(N) \\
\mathbf{B}_p(N) &= \begin{bmatrix} \mathbf{b}_p(N) \\ I_r \end{bmatrix}, \qquad \mathbf{b}_p(N) = -K_p(N)\boldsymbol{\mathscr{U}}(N-p)
\end{aligned} \tag{5.62}$$

are forward and backward prediction error filters, respectively, and $e_p^{\mathrm{f}}(N)$, $e_p^{\mathrm{b}}(N)$ are their outputs at time N (see T5.3-(5,6)). The forward and backward parts appear slightly asymmetrical in the following sense. The a priori forward residual $e_p^{\mathrm{f}}(N)$ can be computed as the output at time N of the forward prediction filter $\mathbf{A}_p(N-1)$ at time $N-1$ (see T5.4-(1)). The a priori backward residual $e_p^{\mathrm{b}}(N)$ can be computed similarly. However, this inner product can be avoided using some redundancy in the order downdate of $\mathbf{w}_{p+1}(N) = [w_{p+1}^{0\mathrm{H}}(N) \ \ldots \ w_{p+1}^{p\mathrm{H}}(N)]^{\mathrm{H}}$. Indeed, equating the last components in (5.61) gives

$$w_{p+1}^{p}(N) = 0 - \lambda^{-1}\alpha_p^{-\mathrm{b}}(N-1)e_p^{\mathrm{b}}(N) \tag{5.63}$$

and $w_{p+1}^{p}(N)$ $(r \times q)$ was already computed in the order update. This yields T5.4-(5).

FORWARD AND BACKWARD PREDICTION PROBLEMS

The filters $\mathbf{A}_p(N)$ and $\mathbf{B}_p(N)$ are the solutions to the following forward and backward prediction problems

$$\min_{A_p^0(N)=I_r} \{\mathbf{A}_p^{\mathrm{H}}(N)R_{p+1}(N)\mathbf{A}_p(N)\} = \boldsymbol{\varepsilon}_p^{\mathrm{fH}}(N)\boldsymbol{\varepsilon}_p^{\mathrm{f}}(N) = \alpha_p^{\mathrm{f}}(N) = \mathcal{U}^{\mathrm{H}}(N)P_p^{\perp}(N-1)\mathcal{U}(N)$$

$$\min_{B_p^p(N)=I_r} \{\mathbf{B}_p^{\mathrm{H}}(N)R_{p+1}(N)\mathbf{B}_p(N)\} = \boldsymbol{\varepsilon}_p^{\mathrm{bH}}(N)\boldsymbol{\varepsilon}_p^{\mathrm{b}}(N) = \alpha_p^{\mathrm{b}}(N) = \mathcal{U}^{\mathrm{H}}(N-p)P_p^{\perp}(N)\mathcal{U}(N-p) \tag{5.64}$$

The time update for $\mathbf{a}_p(N)$ and $\mathbf{b}_p(N)$ progresses in the same way as for $\mathbf{c}_p(N)$. To obtain equations T5.4-(8–13), we use T5.3-(7,12,9) with the two partitionings for $\mathcal{U}_{p+1}(N)$ mentioned in (5.60). From T5.3-(9), we get in one case

$$\alpha_p^{\mathrm{f}}(N) = \lambda\alpha_p^{\mathrm{f}}(N-1) + e_p^{\mathrm{f}}(N)\alpha_p(N-1)e_p^{\mathrm{fH}}(N) \tag{5.65}$$

However, we need $\alpha_p^{-\mathrm{f}}(N)$. So it is better (at least in the multichannel case, in which case $\alpha_p^{\mathrm{f}}(N)$ is an $r \times r$ matrix, leading to a matrix inversion problem in T5.4-(2)) to propagate $\alpha_p^{-\mathrm{f}}(N)$. We can arrive at T5.4-(10) using the matrix inversion lemma [104, page 656, also Appendix A of Chapter 8] on (5.65) and exploiting T5.4-(4).

5.4.3 *Initialization and restarting*

Now we show how the fast algorithm should be initialized in order to achieve the exact solution of the RLS problem. Two cases are discussed: soft-constraint initialization and initialization with zero soft constraint. Although the latter is formally included in the former, a special procedure is needed in the case of a zero soft constraint. Furthermore, considering the zero constraint as a limiting case of the arbitrary soft constraint case would lead to ill-conditioning during the initialization period. Hence, it is better to treat the zero constraint case separately. However, this separate treatment necessitates a separate program to be executed during the first p

samples, while the initialization with non-zero soft constraint allows the steady-state algorithm to run from the first sample onwards, and it also allows for prior information to be incorporated and for the start-up period to be regularized. However, if the prior information is incorrect, its incorporation slows down the initial convergence.

ZERO SOFT-CONSTRAINT INITIALIZATION

For this case of no a priori information about the unknown optimal $\mathbf{c}^o$, fast, exact initialization will be discussed for the single-channel, single-experiment case[3]. Consider the normal equations $R_p(N)\mathbf{c}_p(N) = -\langle \mathcal{U}_p(N), \mathcal{Y}(N) \rangle$. During the initialization period, when $N + 1 < p$, the RLS problem is underdetermined. Only $N + 1$ coefficients can be determined, namely the $N + 1$ coefficients that multiply non-zero data, and the remaining coefficients can be chosen arbitrarily. Using the pseudoinverse corresponds to choosing the minimum-norm-filter solution and therefore to putting the remaining coefficients equal to zero. This leads to a simultaneous order-and-time update strategy (see Table 5.5), solving consecutive, exactly determined problems (see [4,96] for a derivation).

NON-ZERO SOFT-CONSTRAINT INITIALIZATION

Non-zero initial conditions can arise from additional side information, previous use of other (adaptive) algorithms, or when stopping and restarting the algorithm (e.g. to change the weighting factor λ). In this section we show how appending a soft constraint to the basic cost function (5.41) results in a specific initialization of the algorithm. The augmented cost function becomes (see (5.20))

$$\min_{\mathbf{c}_p(N)} \operatorname{tr}\{\boldsymbol{\varepsilon}_p^{\mathrm{H}}(N)\boldsymbol{\varepsilon}_p(N) + \lambda^{N+1}(\mathbf{c}_p(N) - \mathbf{c}_0)^{\mathrm{H}} r_0 (\mathbf{c}_p(N) - \mathbf{c}_0)\} \tag{5.66}$$

where $\mathbf{c}_0$ represents the initial condition and

$$r_0 = r_0^{1/2}\bar{\Lambda} r_0^{\mathrm{H}/2} \qquad r_0^{1/2} \triangleq \begin{bmatrix} \overline{r_0} & & 0 \\ & \ddots & \\ 0 & & \overline{r_0} \end{bmatrix} (pr \times pr), \qquad \overline{r_0}\ (r \times r) \tag{5.67}$$

and

$$\bar{\Lambda} \triangleq \operatorname{diag}\{\lambda^p, \lambda^{p-1}, \ldots, \lambda\} \otimes \Lambda, \qquad \Lambda \triangleq \operatorname{diag}\{1, \lambda^{p+1}, \ldots, \lambda^{(r-1)(p+1)}\} \tag{5.68}$$

The positive definite matrix r_0 is used to weight the effect of the initial condition upon the cost function.

The initialization can be effected concisely by augmenting the time series $u(\cdot)$ and $y(\cdot)$, and modifying the vector space appropriately. So one can reduce the augmented criterion in (5.66) to the form of the original criterion, $\operatorname{tr}\{\boldsymbol{\varepsilon}_p^{\mathrm{H}}(N)\boldsymbol{\varepsilon}_p(N)\}$, again with a Hankel data matrix, by introducing a non-causal part in the signals (an isolated pulse occurring before time 0, see [4,96] or the discussion of initialization issues in section 5.7.4 for more details). One can now easily find the following initial conditions at time -1 with which to start the steady-state algorithm (Table 5.4) at time 0:

Table 5.5 Fast zero soft constraint initialization of the single-channel FTRLS algorithm

Equation	Computation
$N=0$:	$\mathbf{A}_0(0)=1,\ \alpha_0^{\mathrm{f}}(0)=\lvert u(0)\rvert^2,\ \mathbf{w}_0(0),\ \mathbf{c}_0(-1),\ \mathbf{u}_0(0){:}\,0$ dimension
	$\alpha_0(0)=1,\ e_0(0)=y(0)$
	$1\leqslant N\leqslant p$: Simultaneous order-and-time update
	Forward prediction problem:
(1)	$e_{N-1}^{\mathrm{f}}(N)=\mathbf{A}_{N-1}^{\mathrm{H}}(N-1)\mathbf{u}_N(N)$
(2)	$\mathbf{A}_N(N)=\begin{bmatrix}\mathbf{A}_{N-1}(N-1)\\ -e_{N-1}^{\mathrm{fH}}(N)/u^{\mathrm{H}}(0)\end{bmatrix}$
(3)	$\alpha_N^{\mathrm{f}}(N)=\lambda\alpha_{N-1}^{\mathrm{f}}(N-1)$
(4)	$\mathbf{w}_N(N)=\begin{bmatrix}0\\ \mathbf{w}_{N-1}(N-1)\end{bmatrix}-\mathbf{A}_{N-1}(N-1)\alpha_N^{-\mathrm{f}}(N)e_{N-1}^{\mathrm{f}}(N)$
(5)	$\alpha_N^{-1}(N)=\alpha_{N-1}^{-1}(N-1)+e_{N-1}^{\mathrm{fH}}(N)\alpha_N^{-\mathrm{f}}(N)e_{N-1}^{\mathrm{f}}(N)$
	Joint-process extension:
(6)	$\mathbf{c}_N(N-1)=\begin{bmatrix}\mathbf{c}_{N-1}(N-2)\\ -e_{N-1}^{\mathrm{H}}(N-1)/u^{\mathrm{H}}(0)\end{bmatrix}$
(7)	$e_N(N)=y(N)+\mathbf{c}_N^{\mathrm{H}}(N-1)\mathbf{u}_N(N)$
(8)	$\varepsilon_N(N)=e_N(N)\alpha_N(N)$
	$N=p$: Interface with steady-state algorithm of Table 5.4
	Backward prediction problem:
(9)	$\mathbf{b}_p(p)=\mathbf{w}_p(p)\alpha_p(p)u^{\mathrm{H}}(0)$
(10)	$\alpha_p^{\mathrm{b}}(p)=\lvert u(0)\rvert^2\alpha_p(p)$
	Joint-process extension:
(11)	$\mathbf{c}_p(p)=\mathbf{c}_p(p-1)+\mathbf{w}_p(p)\varepsilon_p^{\mathrm{H}}(p)$

$$
\begin{aligned}
&\mathbf{a}_p(-1)=\mathbf{0} \quad \alpha_p^{\mathrm{f}}(-1)=\lambda^p\overline{r_0}\,\Lambda\,\overline{r_0}^{\mathrm{H}}\\
&\mathbf{b}_p(-1)=\mathbf{0} \quad \alpha_p^{\mathrm{b}}(-1)=\overline{r_0}\,\Lambda\,\overline{r_0}^{\mathrm{H}}\\
&\mathbf{w}_p(-1)=\mathbf{0} \quad \alpha_p(-1)=I_q\\
&\mathbf{c}_p(-1)=\mathbf{c}_0
\end{aligned}
\tag{5.69}
$$

Appropriate choices for $\overline{r_0}$ have been discussed in [4,103].

Note that the communication between the prediction and joint-process extension parts in the FTRLS algorithm is unidirectional: the joint-process part gets the Kalman gain and likelihood variable from the prediction part, but no information goes the other way. Now, a restart includes emptying the data vector $\mathbf{u}_p(N)$ (prewindowing method!). This introduces discontinuities in the performance of the (prewindowed) algorithm at the time of a restart. One would like to introduce modifications to smooth out the effects of the restart, but it is important to stick to the precise details of the prediction part in order not to introduce errors in the propagation of the Kalman gain. The joint-process part, however, can be modified. Simulations [103] show that it is beneficial to keep on computing the error signal $\varepsilon_p(N)$ for the joint-process part with the unmodified data vector $\mathbf{u}_p(N)$. This is intuitively acceptable. So the data vector $\mathbf{u}_p(N)$ should be emptied only for the prediction part. And even that does not have to be done explicitly since with a restart of the type (5.69), the length of the non-zero part of the filters $\mathbf{a}$, $\mathbf{b}$ and $\mathbf{w}$ is growing from 0 to p during the first p samples following the restart.

5.5 Computational redundancies and numerical stabilization

5.5.1 Hyperbolic rotations and numerical instability

In the derivation of the FTRLS algorithm above, we have considered the updates of all filters one by one. It turns out that there is another convenient way of deriving the algorithm which is based on 2×2 rotations (or $(q + r) \times (q + r)$ rotations for the q-experiment, r-channel case, see [96,105]). The prediction part of the FTRLS algorithm rearranged from this rotation point of view is depicted in Table 5.6. This table contains all key equations of the algorithm: updates for the filters and updates for the residual covariances. What is not explicitly shown in the table is the computation of the prediction errors.

In this organization, the FTRLS algorithm can be seen to consist of one orthogonal rotation (at least, when normalized [4] and $\lambda = 1$) and one hyperbolic rotation. An orthogonal rotation is numerically well behaved since its eigenvalues are on the unit circle. A hyperbolic rotation, on the other hand, always has one eigenvalue bigger than one, which indicates exponential instability. Though a discussion of the propagation of numerical errors on this level is far from accurate, it indicates that numerical difficulties can be expected, associated with the backward prediction quantities. A detailed analysis of this error propagation can be found in [8,96,9] and will be summarized below. A more recent alternative approach can be found in [105,38,106].

Table 5.6 Rotation structure of the FTRLS algorithm

Prediction problem

Order update — Orthogonal rotation

$$\left[\begin{array}{c|c|c} \mathbf{w}_{p+1}^{\mathrm{H}}(N) & \alpha_{p+1}^{-1}(N) & 0 \\ \mathbf{A}_p^{\mathrm{H}}(N) & 0 & \alpha_p^{\mathrm{f}}(N) \end{array}\right] =$$

$$\begin{bmatrix} I_q & -e_p^{\mathrm{fH}}(N)\lambda^{-1}\alpha_p^{-\mathrm{f}}(N-1) \\ \varepsilon_p^{\mathrm{f}}(N) & I_r \end{bmatrix} \left[\begin{array}{c|c|c} [0_{q\times r}\ \mathbf{w}_p^{\mathrm{H}}(N-1)] & \alpha_p^{-1}(N-1) & e_p^{\mathrm{fH}}(N) \\ \mathbf{A}_p^{\mathrm{H}}(N-1) & -e_p^{\mathrm{f}}(N) & \lambda\alpha_p^{\mathrm{f}}(N-1) \end{array}\right]$$

Order downdate — Hyperbolic rotation

$$\left[\begin{array}{c|c|c} [\mathbf{w}_p^{\mathrm{H}}(N)\ 0_{q\times r}] & \alpha_p^{-1}(N) & e_p^{\mathrm{bH}}(N) \\ \mathbf{B}_p^{\mathrm{H}}(N) & 0 & \alpha_p^{\mathrm{b}}(N) \end{array}\right] =$$

$$\begin{bmatrix} I_q & 0 \\ \varepsilon_p^{\mathrm{b}}(N) & I_r \end{bmatrix} \begin{bmatrix} I_q & e_p^{\mathrm{bH}}(N)\lambda^{-1}\alpha_p^{-\mathrm{b}}(N-1) \\ 0 & I_r \end{bmatrix} \left[\begin{array}{c|c|c} \mathbf{w}_{p+1}^{\mathrm{H}}(N) & \alpha_{p+1}^{-1}(N) & 0 \\ \mathbf{B}_p^{\mathrm{H}}(N-1) & -e_p^{\mathrm{b}}(N) & \lambda\alpha_p^{\mathrm{b}}(N-1) \end{array}\right]$$

5.5.2 An approach to the analysis of roundoff errors

Adaptive filtering algorithms can be viewed as *non-linear discrete time systems*, and a certain set of algorithmic quantities can be identified as constituting the *state* $\mathbf{\Theta}(N)$ of the algorithm. So, the adaptive filter can be written as

$$\begin{aligned} \mathbf{\Theta}(N) &= f(\mathbf{\Theta}(N-1), \mathbf{u}_p(N), y(N)) \\ e_p(N) &= g(\mathbf{\Theta}(N-1), \mathbf{u}_p(N), y(N)) \end{aligned} \tag{5.70}$$

The actual *implemented* algorithm performs these recursions up to some roundoff error and the corresponding state-space equation becomes

$$\hat{\mathbf{\Theta}}(N) = f(\hat{\mathbf{\Theta}}(N-1), \mathbf{u}_p(N), y(N)) + \mathbf{v}'(N) \tag{5.71}$$

where $\hat{\mathbf{\Theta}}(N)$ is the computed value of $\mathbf{\Theta}(N)$, as given by the implemented algorithm, and $\mathbf{v}'(N)$ represents the roundoff noise generated in the recursion at time N. The numerical error on $\mathbf{\Theta}(N)$ is $\Delta\mathbf{\Theta}(N) = \mathbf{\Theta}(N) - \hat{\mathbf{\Theta}}(N)$. Assuming the numerical errors to be small, we can linearize the system around the error-free trajectory, which leads to

$$\Delta\mathbf{\Theta}(N) = F(N)\Delta\mathbf{\Theta}(N-1) + \mathbf{v}(N) \tag{5.72}$$

where

$$F(N) = \nabla_{\mathbf{\Theta}} f\,|_{\mathbf{\Theta}=\mathbf{\Theta}(N-1)}$$

A complete study of the roundoff errors in the adaptive algorithm implementation requires the study of [21] the error *propagation* (the initial state response of the homogeneous $\Delta\mathbf{\Theta}(N)$ system), the error *generation* (properties of $\mathbf{v}(N)$) and the error *accumulation* (compounded error $\Delta\Theta(N)$). We shall restrict our attention to the error propagation issue. One approach here is to consider the signals $\{u(\cdot)\}$, $\{y(\cdot)\}$ as stochastic processes so that $F(N) = F(\mathcal{U}(N), \mathcal{Y}(N))$ becomes a stochastic matrix. A useful result then is that the process $\Delta\Theta(\cdot)$ in (5.72) converges weakly to the process $\Xi(\cdot)$ which satisfies [54]

$$\Xi(N) = E[F(N)]\Xi(N-1) + \mathbf{v}(N) \quad \text{for } I - F(N) = O(1-\lambda) \tag{5.73}$$

when $F(N)$ converges to an identity matrix as $\lambda \to 1$. So the error propagation dynamics can be described via the averaged linearized dynamics $E[F(N)]$. We shall illustrate the resulting approach via the LMS and conventional RLS algorithms:

- LMS algorithm: $\Theta(N) = \mathbf{c}(N)$

$$\Delta\mathbf{c}(N) = [I - \mu\mathbf{u}(N)\mathbf{u}^{\mathrm{H}}(N)]\Delta\mathbf{c}(N-1)$$

$$E[(I - \mu\mathbf{u}(N)\mathbf{u}^{\mathrm{H}}(N))] = I - \mu R$$

- RLS algorithm: $\Theta(N) = [\mathbf{c}^{\mathrm{H}}(N), \mathrm{vec}^{\mathrm{H}}(R^{-1}(N))]^{\mathrm{H}}$

$$\Delta\mathbf{c}(N) = F_1(N)\Delta\mathbf{c}(N-1) \qquad \text{where } F_1(N) = R^{-1}(N)\lambda R(N-1)$$

$$\Delta R^{-1}(N) = \tfrac{1}{\lambda}\, F_1(N)\, \Delta R^{-1}(N-1)F^{\mathrm{H}}(N) \qquad E[F_1(N)] = \lambda I$$

Since we restrict our attention to the error propagation dynamics, we have suppressed the driving terms in the state equations. So for the LMS algorithm, numerical stability goes hand in hand with convergence. In the RLS algorithm, errors decay exponentially with base λ (at least, when $\{u(\cdot)\}$ is persistently exciting, i.e. $R > 0$; see section 5.2.3 for some comments on the case of a singular R).

5.5.3 Introducing redundancies in the FTRLS algorithm

The state of the FTRLS algorithm is given by

$$\Theta(N) = [\mathbf{a}_p^{\mathrm{H}}(N), \alpha_p^{\mathrm{f}}(N), \mathbf{b}_p^{\mathrm{H}}(N), \alpha_p^{\mathrm{b}}(N), \mathbf{w}_P^{\mathrm{H}}(N), \alpha_P^{-1}(n) \,\|\, \mathbf{c}_P^{\mathrm{H}}(N)]^{\mathrm{H}} \tag{5.74}$$

The first six block components are from the prediction part, while the last component corresponds to the joint-process part. Since there is no feedback from the joint-process part to the prediction part, both parts can be considered separately. The joint-process part can then be seen to have the same dynamics as in the conventional RLS algorithm. We shall show below that the prediction part of the original FTRLS algorithm is unstable. A solution to this instability problem is illustrated in Figure 5.6 [8].

The implemented algorithm produces a series of computed quantities which are the sum of their infinite precision value and accumulated roundoff error. Normally, it is impossible to separate these two additive components. However, we introduce *redundancy* in the algorithm by computing a certain quantity in two different ways. The difference between the two computed values then consists of pure numerical error and hence constitutes a measurement (output) of the numerical error system.

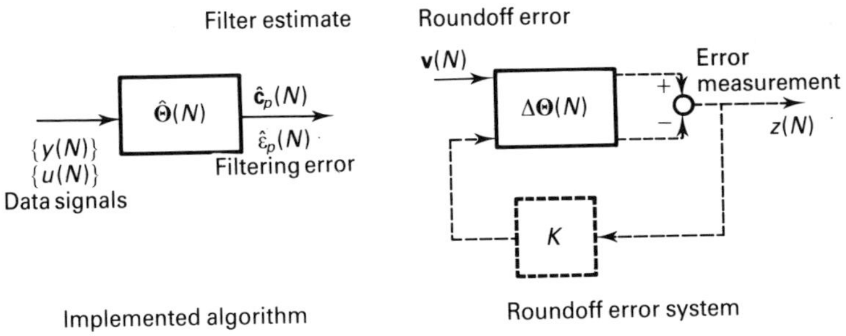

Figure 5.6 The implemented FTRLS algorithm and its associated error system

The particular quantity involved is the backward prediction error. One way of computing $e_p^b(N)$, which we shall denote by $e_p^{bF}(N)$ (superscript F denotes filtering), is cby exploiting its definition (see (5.64), T5.3-(6)) as the output of the backward prediction filter: $e_p^{bF}(N) = \mathbf{B}_p^H(N-1)\mathbf{u}_{p+1}(N)$. There is another possibility that follows from Equation (5.63). This yields an alternative way of computing $e_p^b(N)$ as in T5.4-(5). We shall denote this second computation by $e_p^{bS}(N)$ (superscript S denotes computation by manipulation of scalars). So we have

$$\left.\begin{aligned} e_p^{bS}(N) &= -\lambda\alpha_p^b(N-1)w_{p+1}^p(N) \\ e_p^{bF}(N) &= \mathbf{B}_p^H(N-1)\mathbf{u}_{p+1}(N) \end{aligned}\right\} z(N) = e_p^{bF}(N) - e_p^{bS}(N) = 0 + \text{numerical error} \tag{5.75}$$

We can feed the numerical error $z(N)$ back to try to stabilize the error system dynamics. The question then arises which input to use for the error system. Since the output $z(N)$ is zero in an infinite precision environment, we can add $z(N)$ to any quantity in the algorithm without modifying its true RLS character. The particular input chosen comes about as follows: take as the final value for $e_p^b(N)$ a convex combination of the two computed values, that is,

$$e_p^{b(i)}(N) = e_p^{bS}(N) + K_i z(N) = K_i e_p^{bF}(N) + (1-K_i)e_p^{bS}(N) \qquad i = 1, 2 \tag{5.76}$$

These convex combinations are then used further in the algorithm. Note that using different feedback (convex combination) parameter K_i values at different instances in the algorithm where $e_p^b(N)$ is needed leads to additional degrees of freedom that turn out to be useful for the stabilization. Again, when infinite precision is used, the two ways of computing $e_p^b(N)$ will yield identical answers. Therefore, the $7p$ form of the FTRLS algorithm [4,65] only uses $e_p^{bS}(N)$ and avoids the inner product of the filtering operation needed to compute $e_p^{bF}(N)$.

The quantity $\alpha_p^{-1}(N)$ plays a similar, though less dramatic, role. $\alpha_p(N)$ can be computed in three ways: once as the output $\alpha_p^F(N)$ of the filter $\mathbf{w}_p(N)$ (see T5.3-(3)), or as $\alpha_p^S(N)$ (see (5.50)) by using its interpretation as the inverse of the error covariance in estimating $\boldsymbol{\sigma}$ (see T5.4-(4,7)), or also alternatively as $\alpha_p^A(N)$, which comes about as follows. Using the update relations for α^f, α^b and α, one may show the following identities:

$$\det \alpha_{p+1}^{-1}(N) = \det \alpha_p^{-1}(N-1) \frac{\det \alpha_p^{\mathrm{f}}(N)}{\det \lambda\alpha_p^{\mathrm{f}}(N-1)} = \det \alpha_p^{-1}(N) \frac{\det \alpha_p^{\mathrm{b}}(N)}{\det \lambda\alpha_p^{\mathrm{b}}(N-1)} \tag{5.77}$$

which lead to

$$\det \alpha_p(N) \frac{\det \alpha_p^{\mathrm{f}}(N)}{\det \alpha_p^{\mathrm{b}}(N)} = \det \alpha_p(N-1) \frac{\det \alpha_p^{\mathrm{f}}(N-1)}{\det \alpha_p^{\mathrm{b}}(N-1)}$$

$$= \ldots = \det \alpha_p(-1) \frac{\det \alpha_p^{\mathrm{f}}(-1)}{\det \alpha_p^{\mathrm{b}}(-1)} = \lambda^{pr} \tag{5.78}$$

In the case of a single experiment ($q = 1$), $\alpha_p^{-1}(N)$ is a scalar quantity and the above equation may be used as an alternative way of computing it. In the multiexperiment case, the easiest way of incorporating the information provided by (5.78) is by using the modular multiexperiment form (see further).

5.5.4 Compact representation of the FTRLS algorithm

We can formalize the rotation representation of the FTRLS algorithm as given in Table 5.6, which will be especially useful when introducing (and deriving) more complex forms of the FTRLS algorithm (such as the covariance forms). So we consider now cutting the FTRLS algorithm of Table 5.4 into its three essential portions, while at the same time introducing the redundant computations for numerical stabilization. This leads to the order update operator f_{U} defined in Table 5.7, the order downdate operator f_{D} defined in Table 5.8, and the joint-process operator f_{J} defined in Table 5.9. Having introduced these three operators, the stabilized FTRLS algorithm for the single-experiment case can be represented compactly as in Table 5.10.

Table 5.7 Definition of the update transformation f_{U}

$$(\mathbf{a}_p(N),\ \alpha_p^{-\mathrm{f}}(N),\ \mathbf{w}_{p+1}(N),\ \alpha_{p+1}^{-1}(N))$$
$$= f_{\mathrm{U}}\left(\mathbf{a}(N-1),\ \alpha_p^{-f}(N-1),\ \begin{bmatrix} 0 \\ \mathbf{w}_p(N-1) \end{bmatrix},\ \alpha_p^{-1}(N-1),\ \mathbf{u}_{p+1}(N)\right)$$

Equation	Computation	Cost (×)
(1)	$e_p^{\mathrm{f}}(N) = \mathbf{A}_p^{\mathrm{H}}(N-1)\mathbf{u}_{p+1}(N)$	p
(2)	$w_{p+1}^{0}(N) = -\alpha_p^{-f}(N-1)e_p^{\mathrm{f}}(N)$	1
(3)	$\mathbf{w}_{p+1}(N) = \begin{bmatrix} 0 \\ \mathbf{w}_p(N-1) \end{bmatrix} + \mathbf{A}_p(N-1)w_{p+1}^{0}(N)$	p
(4)	$\alpha_{p+1}^{-1}(N) = \alpha_p^{-1}(N-1) - w_{p+1}^{0\mathrm{H}}(N)e_p^{\mathrm{f}}(N)$	1
(5)	$\varepsilon_p^{\mathrm{f}}(N) = e_p^{\mathrm{f}}(N)\alpha_p(N-1)$	1
(6)	$\mathbf{a}_p(N) = \mathbf{a}_p(N-1) + \mathbf{w}_p(N-1)\varepsilon_p^{\mathrm{fH}}(N)$	p
(7)	$\alpha_p^{-\mathrm{f}}(N) = \alpha_p^{-\mathrm{f}}(N-1) - w_{p+1}^{0}(N)\alpha_{p+1}(N)w_{p+1}^{0\mathrm{H}}(N)$	2
	Total cost (one division)	$3p + 5$

Table 5.8 Definition of the downdate transformation f_D

$$\left(\mathbf{b}_p(N), \alpha_p^{b}(N), \begin{bmatrix} \mathbf{w}_p(N) \\ \mathbf{0} \end{bmatrix}, \alpha_p^{-1}(N)\right) = f_D(\mathbf{b}_p(N-1), \alpha_p^{b}(N-1), \mathbf{w}_{p+1}(N), \alpha_{p+1}^{-1}(N), \mathbf{u}_{p+1}(N))$$

Equation	Computation	Cost (×)
(1)	$e_p^{bF}(N) = \mathbf{B}_p^{H}(N-1)\mathbf{u}_{p+1}(N)$	p
(2)	$e_p^{bS}(N) = -\alpha_p^{b}(N-1)w_{p+1}^{p}(N)$	1
(3)	$e_p^{b(k)}(N) = K_k e_p^{bF}(N) + (1-K_k)e_p^{bS}(N), \quad k = 1, 2$	2
(4)	$\begin{bmatrix} \mathbf{w}_p(N) \\ 0 \end{bmatrix} = \mathbf{w}_{p+1}(N) - \mathbf{B}_p(N-1)w_{p+1}^{p}(N)$	p
(5)	$\alpha_p^{-1}(N) = \alpha_{p+1}^{-1}(N) + w_{p+1}^{pH}(N)e_p^{bF}(N)$	1
(6)	$\varepsilon_p^{b(k)}(N) = e_p^{b(k)}(N)\alpha_p(N), \quad k = 1, 2$	2
(7)	$\mathbf{b}_p(N) = \mathbf{b}_p(N-1) + \mathbf{w}_p(N)\varepsilon_p^{b(1)H}(N)$	p
(8)	$\alpha_p^{b}(N) = \alpha_p^{b}(N-1) + \varepsilon_p^{b(2)}(N)e_p^{b(2)H}(N)$	1
	Total cost (one division)	$3p + 7$

Table 5.9 Definition of the joint-process extension f_J

$$(\mathbf{c}_p(N), \varepsilon_p(N)) = f_J(\mathbf{c}_p(N-1), \mathbf{w}_p(N), \alpha_p(N), y(N), \mathbf{u}_p(N))$$

Equation	Computation	Cost (×)
(1)	$e_p(N) = y(N) + \mathbf{c}_p^{H}(N-1)\mathbf{u}_p(N)$	p
(2)	$\varepsilon_p(N) = e_p(N)\alpha_p(N)$	1
(3)	$\mathbf{c}_p(N) = \mathbf{c}_p(N-1) + \mathbf{w}_p(N)\varepsilon_p^{H}(N)$	p
	Total cost	$2p + 1$

The numerical stabilization mechanism thus introduced captures the essential features of the solution proposed in [9]. Some possible variations for the scalar algorithm are considered in [9]. The $7p$ form of the FTRLS algorithm can be obtained straightforwardly from Table 5.10 by putting all (explicit and implicit) feedback coefficients K_k equal to zero in all relevant tables.

Table 5.10 Stabilized multichannel FTRLS algorithm (compact form)

Equation	Computation	Cost (×)
	Prediction problem:	
(1)	$(\mathbf{a}_p(N), \alpha_p^{-f}(N), \mathbf{w}_{p+1}(N), \bar{\alpha}_{p+1}^{S}(N))$	$(1\div)$
	$= f_U\left(\mathbf{a}_p(N-1), \lambda^{-1}\alpha_p^{-f}(N-1), \begin{bmatrix} 0_{r\times 1} \\ \mathbf{w}_p(N-1) \end{bmatrix},\right.$	
	$\left.\alpha_p^{-1}(N-1), \mathbf{u}_{p+1}(N)\right)$	$3r^2p + 2.5r^2 + 3.5r$
(2)	$\left(\mathbf{b}_p(N), \alpha_p^{b}(N), \begin{bmatrix} \mathbf{w}_p(N) \\ 0_{r\times 1} \end{bmatrix}, \bar{\alpha}_p^{S}(N)\right)$	$(1\div)$
	$= f_D(\mathbf{b}_p(N-1), \lambda\alpha_p^{b}(N-1), \mathbf{w}_{p+1}(N), \bar{\alpha}_{p+1}^{S}(N),$	
	$\mathbf{u}_{p+1}(N))$	$3r^2p + 2.5r^2 + 5.5r$
(3)	$\alpha_p(N) = \alpha_p^{A}(N) = \lambda^{pr} \det(\alpha_p^{b}(N)) \det(\alpha_p^{-f}(N))$	$(1\div)$ $r + 11$
	Joint-process extension:	
(4)	$(\mathbf{c}_p(N), \varepsilon_p(N)) = f_J(\mathbf{c}_p(N-1), \mathbf{w}_p(N), \alpha_p(N), y(N), \mathbf{u}_p(N))$	$2pr + 1$
	r-channel total cost (three divisions)	$(6r+2)pr + 5r^2 + 10r + 12$

5.5.5 Averaging analysis of the error propagation

The results presented here and in the next subsection assume $q = t = 1$. The feedback of $z(N) = \mathbf{H}(N)\Delta\mathbf{\Theta}(N-1)$ thus leads to a closed-loop system matrix $F^{c}(N) = F(N) + G(N)K\mathbf{H}(N)$ with averaged value

$$E[F^{c}(N)] = \begin{bmatrix} \lambda I & * & * & * & 0 & * \\ 0 & \lambda & * & * & 0 & * \\ 0 & 0 & [(1/\lambda) - K_1(1-\lambda)]I & * & 0 & * \\ 0 & 0 & * & (1/\lambda) - 2K_2(1-\lambda) & 0 & * \\ 0 & 0 & 0 & 0 & S_p^{H} - \mathbf{u}_p\mathbf{b}_p^{H} & 0 \\ * & * & * & * & 0 & 0 \end{bmatrix}^{H} \tag{5.79}$$

valid for $\lambda \in [1-\varepsilon, 1]$ (ε being some small positive real number and $\mathbf{u}_p = [0_{1\times(p-1)}, 1]^{H}$). Note that $S_p^{H} - \mathbf{u}_p\mathbf{b}_p^{H}$ is a bottom-companion matrix [104, page 659] with eigenvalues equal to the zeros of the backward prediction polynomial. So

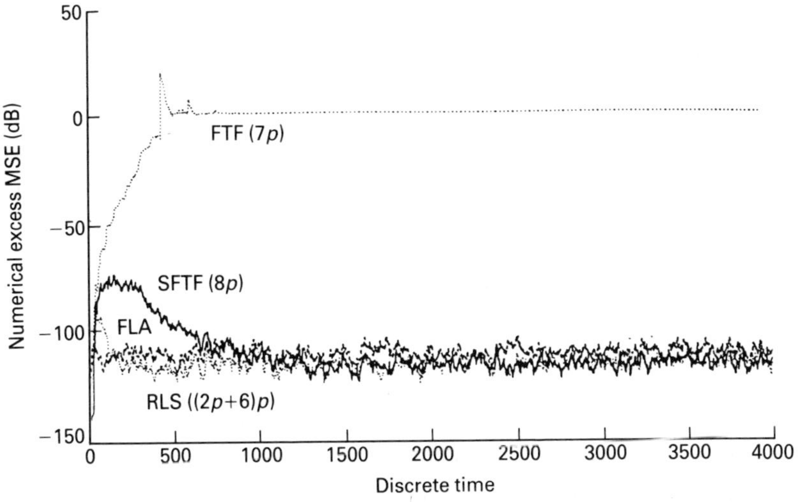

Figure 5.7 A simulation illustrating the numerical behaviour of several RLS algorithms. Condition number = 1, SNR = 20 dB, $p = 20$, $\lambda = 0.98$

we see that the original FTRLS algorithm (putting all $K_k = 0$) is exponentially unstable. The feedback parameter K_1 stabilizes the error propagation associated with $\mathbf{b}_p(N)$, while K_2 stabilizes the error propagation associated with $\alpha_p^{\mathrm{b}}(N)$. The choice $K_1 = 1.5$, $K_2 = 2.5$ turns out to stabilize the error propagation over a wide range of operating conditions. However, the feedback mechanism introduced above is only able to stabilize for a weighting factor $\lambda \in (1 - (1/2p), 1)$ which will be motivated below. This range covers some useful choices one may make in practice. We may also note that if $\alpha_p^A(N)$ were replaced by $\alpha_p^{\mathrm{S}}(N)$ in the algorithm, then the error propagation associated with $\alpha_p^{-1}(N)$ would become unstable (F_{66} would change from 0 to 1). With the use of $\alpha_p(N) = \alpha_p^A(N)$ in the stabilized algorithm, the sixth block component $\alpha_p^{-1}(N)$ can actually be removed from the algorithm state $\boldsymbol{\Theta}$.

The numerical behaviour of different RLS algorithms is illustrated in Figure 5.7. There the normalized difference between single- and double-precision computations of the filtering error $\varepsilon_p(N)$ is shown. After an initial transient, the stabilized FTRLS algorithm joins the low error level of the RLS algorithm, which is slightly below the error level of the fast lattice algorithm (with time updates for α^{f}, α^{b}).

5.5.6 The limited range for λ and a possible way out

Limitations of the averaging approach are as follows:

- It says something about the whole distribution of the errors (a strong statement).
- Its results only hold for a very small range for $\varepsilon = 1 - \lambda$.
- The FTRLS error system is really a two-timescale system:

$$F = \begin{bmatrix} I + \varepsilon Q & O(\sqrt{\varepsilon}) \\ O(\sqrt{\varepsilon}) & S - \mathbf{b}_p \mathbf{u}_p^{\mathrm{H}} \end{bmatrix}$$

The dynamics of the $I + \varepsilon Q$ part (the most critical part!) are slow indeed (allowing averaging), but the dynamics of $S - \mathbf{b}_p u_p^{\mathrm{H}}$ are arbitrary.

The *independence assumption*, widely used in analysis of the LMS algorithm, seems to hold for a wider range of $1 - \lambda$ than the averaging results. The independence assumption applied to the system (5.72) would mean that $F(N)$ and $\Delta\boldsymbol{\Theta}(N-1)$ are assumed to be independent. Using this assumption, we get the following results for the $\Delta\boldsymbol{\Theta}(N)$ system:

1. Stability of the mean: $|\lambda_i\{E[F(N)]\}| < 1$ (eigenvalues)
2. Stability of the variance: $\lambda_i\{E[F^{\mathrm{H}}(N)F(N)]\} < 1$ (singular values) (sufficient condition, stronger than condition 1). Note that, invoking a central limit theorem argument, satisfaction of condition 2 would imply stability of all moments.

Consider the application of condition 2 to the critical subsystem $\Delta\mathbf{b}_p(N)$ separately:

$$\begin{aligned} F_{33}^{\mathrm{H}}(N) &= \frac{\alpha_p^{\mathrm{b}}(N)}{\lambda\alpha_p^{\mathrm{b}}(N-1)} I + K_1 \mathbf{u}_p(N)\mathbf{w}_p^{\mathrm{H}}(N) \\ &\approx \frac{1}{\lambda} I - (1-\lambda)K_1\mathbf{u}_p(N)\mathbf{u}_p^{\mathrm{H}}(N)R_p^{-1} \\ &= R_p^{1/2}\left(\frac{1}{\lambda} I - (1-\lambda)K_1\mathbf{v}_p(N)\mathbf{v}_p^{\mathrm{H}}(N)\right)R_p^{-1/2}, \quad \mathbf{v}_p(N) \triangleq R_p^{-1/2}\mathbf{u}_p(N) \end{aligned} \tag{5.80}$$

Applying condition 2 to $F_{33}(N)$ only, we get

$$1 - \lambda \leqslant \frac{0.5}{p + \nu_v - 1} \quad \text{where } \nu_v = \frac{E[v^4(t)]}{\{E[v^2(t)]\}^2} \tag{5.81}$$

Hence this very approximate reasoning (on F_{33} considered separately [7]) provides an accurate prediction of the stable range for $1 - \lambda$, as observed in practice.

There are ways around this lower bound on λ. To understand this, one should realize that the essence of the FTRLS algorithm is a fast mechanism to compute the Kalman gain. The details of the computation of the Kalman gain in the prediction part are very critical. However, what one does with the Kalman gain thus obtained in the joint-process part does not necessarily have to be according to what the RLS algorithm dictates, in order to obtain a good adaptive algorithm (see, for example, also the improved restart procedures mentioned before). Hence, to modify the tracking capabilities of the FTRLS algorithm, we could very well replace T5.9-(3) by [9]:

$$\mathbf{c}_p(N) = \mathbf{c}_p(N-1) + \zeta(N)\mathbf{w}_p(N)\varepsilon_p^{\mathrm{H}}(N) \tag{5.82}$$

where $\zeta(N)$ is an arbitrary gain factor (a similar mechanism was proposed independently in [107]). Using averaging analysis, the tracking characteristics for the modified FTRLS algorithm are determined by the eigenmode of the following system matrix (assuming constant ζ):

$$E[I - \zeta \mathbf{u}_p(N)\mathbf{u}_p^{\mathrm{H}}(N)R_p^{-1}(N)] \approx [1 - \zeta(1 - \lambda)]I \approx \lambda^{\zeta} I \tag{5.83}$$

where the first approximation holds for $(1 - \lambda) \ll 1$ and the second approximation holds for $\zeta(1 - \lambda) \ll 1$. Hence the modified algorithm with (5.82) has approximately (up to first order in $1 - \lambda$) the same tracking characteristics as an RLS algorithm with exponential weighting factor equal to $1 - \zeta(1 - \lambda) \approx \lambda^{\zeta}$. So one can, for instance, choose $\lambda = 1 - (0.4/p)$ for good numerical behaviour, and adjust ζ to get the desired effective weighting factor (if this one would be smaller than $1 - (0.4/p)$). Note that one can also make ζ time varying at no extra cost. This is useful to, for example, shut off the adaptation ($\zeta = 0$) during double-talk in an echo cancellation application in a full duplex communication system.

5.5.7 Wordlength considerations

In general, the error level increases with the condition number of the data matrix. One can conclude that any condition number can be handled, given that the wordlength used is long enough. Conversely, for a given wordlength it is possible to create sufficiently 'non-persistently exciting' signals that will make the algorithm unstable. This is true for both RLS and FTRLS. Therefore, regularization of ill-conditioned problems (by, for example, adding white noise to the regressor signal u as described in [4]) is desirable.

In an attempt to quantify the size of the stable local region in which the results of the linearization approach would be valid, we can say that elementary analytical considerations and simulation experience indicate that non-linearities would come into effect when the numerical errors on the backward prediction error $e_p^{\mathrm{b}}(N)$ grow to become of the same order of magnitude as the true values of those prediction errors. Simulations also seem to indicate that those non-linearities, when they come into effect, are destabilizing.

5.5.8 Divisions

The computations counts given in Tables 5.7–5.9 are for the single-channel case, while the one given in Table 5.10 is for the multichannel case. The first two divisions that are mentioned in Table 5.10 are a '$1/x$' operation for $\alpha_p^{-\mathrm{S}}(N)$ (in f_{D}) and $\alpha_{p+1}^{-\mathrm{S}}(N)$ (in f_{U}). The algorithm also requires $\alpha_p^{-1}(N)$ (to be used in T5.7-(4)), which we can approximately compute from $\alpha_p(N)$ and $\alpha_p^{-\mathrm{S}}(N)$ as follows. Since $\alpha^{\mathrm{S}} - \alpha^{A}$ is presumably very small compared with α^{S}, one can use a first-order Taylor series expansion for computing $\alpha_p^{-1}(N)$, that is,

$$\alpha_p^{-1}(N) \approx \alpha_p^{-\mathrm{S}}(N)[2 - \alpha_p(N)\alpha_p^{-\mathrm{S}}(N)] \tag{5.84}$$

requiring two multiplications. The explanation for the computations count in T5.10-(3) can be found in [9].

5.6 Growing- and sliding-window covariance FTRLS algorithms

In this section we shall consider the covariance or 'unwindowed' formulation of least-squares filtering (for the case of a single experiment, $q = 1$). In the prewindowed case, we work with semi-infinite strings of data which have only one end point at which something happens as time proceeds. However, at time N we do not have the infinite past available in general, and so the way the prewindowing method gets around the infinity issue is by assuming that all data before time $t = 0$ are zero (only in the case when this assumption coincides with reality do we have the infinite past available). Using this assumption, the effective number of terms in the LS cost function (5.41) reduces from infinity to $N + 1$. In the covariance method, on the other hand, no extra assumptions on the data are made and the entries in all regression vectors $\mathbf{u}_p(t)$ used in the cost function are actual available data. Since, in practice, we can only have a finite window of data at our disposal, strings of used data are now marked by two end points.

For a window length equal to L, the LS cost function for the covariance method is (with exponential weighting superimposed) [48,108]

$$\alpha^{y}_{p,L}(N) = \min_{\mathbf{c}_{p,L}(N)} \left\{ \sum_{t=N-L+1}^{N} \lambda^{N-t} \| e(t \mid \mathbf{c}_{p,L}(N)) \|^2 \right\} \tag{5.85}$$

We can reformulate the LS problem in an L-dimensional vector space in which we shall consider vectors of the form

$$\mathscr{U}_L(N) \triangleq [u(N), \lambda^{1/2}u(N-1), \ldots, \lambda^{(L-1)/2}u(N-L+1)]^{\mathrm{H}} \tag{5.86}$$

with similar definitions for other vectors and data matrices. In this notation, we can write

$$\mathbf{c}_{p,L}(N) = -K_{p,L}(N)\mathscr{Y}_L(N), \quad \boldsymbol{\varepsilon}_{p,L}(N) = P^{\perp}_{p,L}(N)\mathscr{Y}_L(N), \quad \alpha^{y}_{p,L}(N) = \boldsymbol{\varepsilon}^{\mathrm{H}}_{p,L}(N)\boldsymbol{\varepsilon}_{p,L}(N) \tag{5.87}$$

We will invariably denote the pinning vector by $\boldsymbol{\sigma}$ and the shift matrix by S regardless of whether we are considering the infinite-dimensional vector space of the prewindowed problem or the L-dimensional vector space of the covariance problem.

In the covariance algorithms, we will need to pin down data at the other end point of the window as well, and hence we introduce a second $L \times 1$ pinning vector $\boldsymbol{\pi} \triangleq [0, \ldots, 0, \lambda^{(L-1)/2}]^{\mathrm{H}}$ with the property $\langle \mathscr{U}_L(N), \boldsymbol{\pi} \rangle = \lambda^{L-1}u(N-L+1)$. The shift matrix corresponding to $\boldsymbol{\pi}$ is S^{H}: $P^{\perp}_{\pi} = S^{\mathrm{H}}S$. Note that $R_{S^{\mathrm{H}}\mathscr{U}_{p,L}(N)} = \lambda R_{p,L-1}(N-1)$ and $R_{S_{\mathscr{U}p,L}(N)} = R_{p,L-1}(N)$. We can introduce a dual Kalman gain $\tilde{\mathbf{v}}_{\mathscr{U}}$ and associated variable $\delta_{\mathscr{U}}$, which can be defined (just as $\alpha_{\mathscr{U}}$) as both a prediction error and an error covariance. The dual version of Equations T5.3-(1–12) and (5.50) holds and one can find it by making the substitutions $(\boldsymbol{\sigma}, S, \tilde{\mathbf{w}}, \alpha) \leftrightarrow (\boldsymbol{\pi}, S^{\mathrm{H}}, \tilde{\mathbf{v}}, \delta)$. In particular, T5.3-(1–4) now become

$$\tilde{\mathbf{v}}_{\mathscr{U}} \triangleq -K_{\mathscr{U}}\boldsymbol{\pi}$$
$$\mathbf{v}_{\mathscr{U}} \triangleq -R^{\#}_{S\mathscr{U}}\mathscr{U}^{\mathrm{H}}\boldsymbol{\pi}(\boldsymbol{\pi}^{\mathrm{H}}\boldsymbol{\pi})^{-1}$$
$$\delta_{\mathscr{U}} \triangleq \boldsymbol{\pi}^{\mathrm{H}}P^{\perp}_{\mathscr{U}}\boldsymbol{\pi} = \boldsymbol{\pi}^{\mathrm{H}}\boldsymbol{\pi} + \boldsymbol{\pi}^{\mathrm{H}}\mathscr{U}\tilde{\mathbf{v}}_{\mathscr{U}} = \{I - \boldsymbol{\pi}^{\mathrm{H}}\mathscr{U}\mathbf{v}_{\mathscr{U}}\}^{\#}\boldsymbol{\pi}^{\mathrm{H}}\boldsymbol{\pi}$$
$$\tilde{\mathbf{v}}_{\mathscr{U}} = \mathbf{v}_{\mathscr{U}}\delta_{\mathscr{U}} \tag{5.88}$$

The predicted (a priori) residuals $(\cdot)^p$ in T5.3-(6) are now backwards predicted, instead of forwards, and we will therefore call them smoothed residuals and denote them by $(\cdot)^S$ (as in [48]).

The single-experiment covariance quantities are described in Table 5.11. Note that because of our compact algorithm description via f_{U}, f_{D} and f_{J}, it is, strictly speaking, not necessary to name all forward and backward prediction errors arising in the algorithm. In Table 5.11, we only give a detailed account of the error signals arising in the joint-process filtering. The description and interpretation of the prediction errors arising in the forward and backward prediction problems run totally parallel.

Table 5.11 Definition of covariance FTRLS variables

Variable	Definition	Transversal filter computation
Kalman gains:		
$\tilde{\mathbf{w}}_{p,L}(N)$	$-K_{p,L}(N)\boldsymbol{\sigma}$	
$\mathbf{w}_{p,L}(N)$	$-\lambda^{-1}R^{-1}_{p,L-1}(N-1)\mathbf{u}_p(N)$	$\tilde{\mathbf{w}}_{p,L}(N)\alpha^{-1}_{p,L}(N)$
$\tilde{\mathbf{v}}_{p,L}(N)$	$-K_{p,L}(N)\boldsymbol{\pi}$	
$\mathbf{v}_{p,L}(N)$	$-R^{-1}_{p,L-1}(N)\mathbf{u}_p(N-L+1)$	$\tilde{\mathbf{v}}_{p,L}(N)\delta^{-1}_{p,L}(N)$
Likelihood related variables:		
$\alpha_{p,L}(N)$	$\boldsymbol{\sigma}^{\mathrm{H}}P^{\perp}_{p,L}(N)\boldsymbol{\sigma}$	$1 + \tilde{\mathbf{w}}^{\mathrm{H}}_{p,L}(N)\mathbf{u}_p(N)$
$\alpha^{-1}_{p,L}N)$		$1 - \mathbf{w}^{\mathrm{H}}_{p,L}(N)\mathbf{u}_p(N)$
$\delta_{p,L}(N)$	$\boldsymbol{\pi}^{\mathrm{H}}P^{\perp}_{p,L}(N)\boldsymbol{\pi}$	$\lambda^{L-1}[1 + \tilde{\mathbf{w}}^{\mathrm{H}}_{p,L}(N)\mathbf{u}_p(N-L+1)]$
$\delta^{-1}_{p,L}(N)$		$\lambda^{1-L} - \mathbf{v}^{\mathrm{H}}_{p,L}(N)\mathbf{u}_p(N-L+1)$
$\eta_{p,L}(N)$	$\boldsymbol{\pi}^{\mathrm{H}}P^{\perp}_{p,L}(N)\boldsymbol{\sigma}$	$\lambda^{L-1}\mathbf{u}^{\mathrm{H}}_p(N-L+1)\tilde{\mathbf{w}}_{p,L}(N)$ $= \tilde{\mathbf{v}}^{\mathrm{H}}_{p,L}(N)\mathbf{u}_p(N)$
$\eta^{p}_{p,L-1}(N)$	Predicted	$\tilde{\mathbf{v}}^{\mathrm{H}}_{p,L-1}(N-1)\mathbf{u}_p(N)$

$\eta^{s}_{p,L-1}(N)$	smoothed	$\mathbf{u}^{\mathrm{H}}_{p}(N-L+1)\tilde{\mathbf{w}}_{p,L-1}(N)$
Filters:		
$\mathbf{A}_{p,L}(N)$	$\begin{bmatrix} I_r \\ \mathbf{a}_{p,L}(N) \end{bmatrix}$, $\mathbf{a}_{p,L}(N) = K_{p,L}(N-1)\boldsymbol{\mathcal{U}}_L(N)$	
$\mathbf{B}_{p,L}(N)$	$\begin{bmatrix} \mathbf{b}_{p,L}(N) \\ I_r \end{bmatrix}$, $\mathbf{b}_{p,L}(N) = K_{p,L}(N)\boldsymbol{\mathcal{U}}_L(N-p)$	
$\mathbf{c}_{p,L}(N)$	$-K_{p,L}(N)\boldsymbol{\mathcal{Y}}_L(N)$	
Error covariances:		
$\alpha^{\mathrm{f}}_{p,L}(N)$	$\boldsymbol{\mathcal{U}}^{\mathrm{H}}_{L}(N)P^{\perp}_{p,L}(N-1)\boldsymbol{\mathcal{U}}_L(N)$	$\boldsymbol{\mathcal{U}}^{\mathrm{H}}_{L}(N)\mathcal{U}_{p+1,L}(N)\mathbf{A}_{p,L}(N)$
$\alpha^{\mathrm{b}}_{p,L}(N)$	$\boldsymbol{\mathcal{U}}^{\mathrm{H}}_{L}(N-p)P^{\perp}_{p,L}(N)\boldsymbol{\mathcal{U}}_L(N-p)$	$\boldsymbol{\mathcal{U}}^{\mathrm{H}}_{L}(N-p)\mathcal{U}_{p+1,L}(N)\mathbf{B}_{p,L}(N)$
$\alpha^{y}_{p,L}(N)$	$\boldsymbol{\mathcal{Y}}^{\mathrm{H}}_{L}(N)P^{\perp}_{p,L}(N)\boldsymbol{\mathcal{Y}}_L(N)$	$\boldsymbol{\mathcal{Y}}^{\mathrm{H}}_{L}(N)\boldsymbol{\mathcal{Y}}_L(N) + \boldsymbol{\mathcal{Y}}^{\mathrm{H}}_{L}(N)\mathcal{U}_{p,L}(N)\mathbf{c}_{p,L}(N)$
Joint-process errors:		
$\varepsilon_{p,L}(N)$	$\boldsymbol{\mathcal{Y}}^{\mathrm{H}}_{L}(N)P^{\perp}_{p,L}(N)\sigma$	$y(N) + \mathbf{c}^{\mathrm{H}}_{p,L}(N)\mathbf{u}_p(N)$
$e_{p,L-1}(N)$	Predicted	$y(N) + \mathbf{c}^{\mathrm{H}}_{p,L-1}(N-1)\mathbf{u}_p(N)$
$v_{p,L}(N)$	$\boldsymbol{\mathcal{Y}}^{\mathrm{H}}_{L}(N)P^{\perp}_{p,L}(N)\pi$	$\lambda^{L-1}[y(N-L+1) + \mathbf{c}^{\mathrm{H}}_{p,L}(N)\mathbf{u}_p(N-L+1)]$
$v^{s}_{p,L-1}(N)$	Smoothed	$y(N-L+1) + \mathbf{c}^{\mathrm{H}}_{p,L-1}(N)\mathbf{u}_p(N-L+1)$

5.6.1 The growing-window covariance (GWC) FTRLS algorithm

The GWC method considers a window length that is growing with time, as in the prewindowing method. However, the window only starts at the first point in time when the regression vector $\mathbf{u}_p(t)$ can be completely filled up with available data. Assuming that at time N we can dispose of the data in $[0, N]$, the GWC method takes $L = N - p + 2$. The GWC FTRLS algorithm is given in Table 5.12. A derivation can be found in [48,108] or [96,51], where the GWC algorithm is derived by embedding the GWC problem into a prewindowed problem with an extra channel.

NON-ZERO SOFT-CONSTRAINT INITIALIZATION

As in the prewindowed case, we can augment the GWC cost function with a soft constraint to make the problem overdetermined from the first sample onwards and have a convenient way of initializing the algorithm. Here are the initial values at time $N = p - 1$ with which to start the algorithm of Table 5.12 at time $N = p$ (for the

Table 5.12 GWC FTRLS algorithm

Equation	Computation	Cost (×)
	Prediction problem: $L = N - p + 2$	
(1)	$(\mathbf{a}_{p,L-1}(N), \alpha_{p,L-1}^{-\mathrm{f}}(N), \mathbf{w}_{p+1,L-1}(N), \bar{\alpha}_{p+1,L-1}^{\mathrm{S}}(N))$	$(1 \div)$
	$= f_{\mathrm{U}}\left(\mathbf{a}_{p,L-2}(N-1), \frac{1}{\lambda}\alpha_{p,L-2}^{-\mathrm{f}}(N-1), \begin{bmatrix} 0_{r\times 1} \\ \mathbf{w}_{p,L-1}(N-1) \end{bmatrix},\right.$	
	$\left.\alpha_{p,L-1}^{-1}(N-1), \mathbf{u}_{p+1}(N)\right)$	$3r^2p + 2.5r^2 + 3.5r$
(2)	$\left(\mathbf{b}_{p,L-1}(N), \alpha_{p,L-1}^{\mathrm{b}}(N), \begin{bmatrix} \mathbf{w}_{p,L-1}(N) \\ 0_{r\times 1} \end{bmatrix}, \alpha_{p,L-1}^{-\mathrm{S}}(N)\right)$	$(1 \div)$
	$= f_{\mathrm{D}}(\mathbf{b}_{p,L-2}(N-1), \lambda\alpha_{p,L-2}^{\mathrm{b}}(N-1), \mathbf{w}_{p+1,L-1}(N),$	
	$\alpha_{p+1,L-1}^{-\mathrm{S}}(N), \mathbf{u}_{p+1}(N))$	$3r^2p + 2.5r^2 + 5.5r$
(3)	$\mathbf{w}_{2,2p+1}^{2p}(N) = -\mathbf{u}_p^{\mathrm{H}}(p-1)\mathbf{w}_{p,L-1}(N)$	pr
(4)	$\left(\tilde{\mathbf{v}}_{p,L}(N), \delta_{p,L}(N), \begin{bmatrix} \mathbf{w}_{p,L}(N) \\ 0 \end{bmatrix}, \alpha_{p,L}^{-\mathrm{S}}(N)\right)$	
	(see stability discussion also)	$(1 \div)$
	$= f_{\mathrm{D}}\left(\tilde{\mathbf{v}}_{p,L-1}(N-1), \lambda\delta_{p,L-1}(N-1), \begin{bmatrix} \mathbf{w}_{p,L-1}(N) \\ \mathbf{w}_{2,2p+1}^{2p}(N) \end{bmatrix},\right.$	
	$\left.\alpha_{p,L-1}^{-\mathrm{S}}(N), \begin{bmatrix} \mathbf{u}_p(N) \\ 0 \end{bmatrix}\right)$	$3rp + 8$
(5)	$\alpha_{p,L}(N) = \alpha_{p,L}^{A}(N) = \lambda^{2pr-N-1}\delta_{p,L}(N)\det(\alpha_{p,L-1}^{\mathrm{b}}(N))$	
	$\det(\alpha_{p,L-1}^{-\mathrm{f}}(N))$ $(1 \div)$	$r + 13$
	Joint-process extension:	
(6)	$(\mathbf{c}_{p,L}(N), \varepsilon_{p,L}(N)) = f_{\mathrm{J}}(\mathbf{c}_{p,L-1}(N-1), \mathbf{w}_{p,L}(N),$	
	$\alpha_{p,L}(N), y(N), \mathbf{u}_p(N))$	$2pr + 1$
	r-channel total cost (four divisions)	$(6r+6)pr + 5r^2 + 10r + 22$

single-channel case; see [96, chapter 2] for some insight into the proper modifications (especially of $R_{p,p-2}$) for the multichannel case):

$$\begin{aligned}
&\mathbf{a}_{p,0}(p-1) = \mathbf{0} && \alpha^{\mathrm{f}}_{p,0}(p-1) = \lambda^p r_0 \\
&\mathbf{b}_{p,0}(p-1) = 0 && \alpha^{\mathrm{b}}_{p,0}(p-1) = r_0 \\
&\mathbf{w}_{p,1}(p-1) = -\lambda^{-1} R^{-1}_{p,p-2}\mathbf{u}_p(p-1) && \alpha^{-1}_{p,1}(p-1) = 1 - \mathbf{w}^{\mathrm{H}}_{p,1}(p-1)\mathbf{u}_p(p-1) \qquad (5.89)\\
&\tilde{\mathbf{v}}_{p,1}(p-1) = \mathbf{w}_{p,1}(p-1)\alpha_{p,1}(p-1) && \delta_{p,1}(p-1) = \alpha_{p,1}(p-1)
\end{aligned}$$

$$(\mathbf{c}_{p,1}(p-1), \varepsilon_{p,1}(p-1)) = f_{\mathrm{J}}(\mathbf{c}_0, \mathbf{w}_{p,1}(p-1), \alpha_{p,1}(p-1), y(p-1), \mathbf{u}_p(p-1))$$

where

$$R_{p,p-2} = r_0 \operatorname{diag}\{\lambda^{p-1}, \lambda^{p-2}, \ldots, 1\} \tag{5.90}$$

ZERO SOFT-CONSTRAINT INITIALIZATION

In this case, we have to wait until $L = p$ (again, in the single-channel case) to have an exactly determined system of equations. Before $N = 2p - 2$, we have underdetermined systems of equations. One approach consists of working with these underdetermined systems and using the minimum-norm solution [109, 100]. Another approach to arrive with the proper initial conditions at time $N = 2p - 2$ is to let the filter order grow with time so that at each time we have an exactly determined system. This can be done with the version of the Levinson algorithm that applies to the non-Hermitian case (see Chapter 8 also). With the time axis being shifted so that $u(1-p)$ is the first available sample, consider for $m = 0, \ldots, p-1$:

$$\begin{aligned}
&\begin{bmatrix} \mathbf{A}^{\mathrm{H}}_{m+1,m+1}(m+1) & \varepsilon^{fs}_{m+1,m+1}(m+1,\ 0) & 0 \\ \mathbf{B}^{\mathrm{H}}_{m+1,m+1}(m) & 0 & \varepsilon^{bp}_{m+1,m+1}(m+1,\ m+1) \end{bmatrix} \\
&= \begin{bmatrix} 1 & -\varepsilon^{fp}_{m,m}(m+1,\ m+1)\varepsilon^{-bp}_{m,m}(m,\ m) \\ -\varepsilon^{bs}_{m,m}(m-1,\ -1)\varepsilon^{-fs}_{m,m}(m,\ 0) & 1 \end{bmatrix} \\
&\times \begin{bmatrix} [\mathbf{A}^{\mathrm{H}}_{m,m}(m)\ 0] & \sqrt{\lambda}\varepsilon^{fs}_{m,m}(m,\ 0) & \varepsilon^{fp}_{m,m}(m+1,\ m+1) \\ [0\ \mathbf{B}^{\mathrm{H}}_{m,m}(m-1)] & \sqrt{\lambda}\varepsilon^{bs}_{m,m}(m-1,\ -1) & \varepsilon^{bp}_{m,m}(m,\ m) \end{bmatrix}
\end{aligned} \tag{5.91}$$

with

$$\begin{aligned}
\varepsilon^{fp}_{m,m}(m+1,\ m+1) &= \mathbf{A}^{\mathrm{H}}_{m,m}(m)\mathbf{u}_{m+1}(m+1) \\
\varepsilon^{bs}_{m,m}(m-1,\ -1) &= \lambda^{m/2}\mathbf{B}^{\mathrm{H}}_{m,m}(m-1)\mathbf{u}_{m+1}(-1)
\end{aligned} \tag{5.92}$$

The computational cost for these recursions is $4m$ (plus constant) per time update. Some alternative computations are

$$\varepsilon^{fs}_{m,m}(m,\ 0) = \lambda^{m/2}\mathbf{A}^{\mathrm{H}}_{m,m}(m)\mathbf{u}_{m+1}(0) \qquad \varepsilon^{bp}_{m,m}(m,\ m) = \mathbf{B}^{\mathrm{H}}_{m,m}(m-1)\mathbf{u}_{m+1}(m) \tag{5.93}$$

These recursions have to be augmented with recursions for the Kalman gain $\mathbf{w}$ and the filter $\tilde{\mathbf{v}}$. And some further processing is needed at the end of the recursions to obtain the proper quantities with which to start the steady-state algorithm in Table 5.12.

NUMERICAL STABILITY CONSIDERATIONS

The numerical stabilization mechanism introduced in [9] and discussed before works best in a stationary environment. As mentioned before, the GWC problem can be embedded into an $(r + 1)$-channel prewindowed problem where the signal in the artificial channel $r + 1$ is an isolated pulse at time $t = -1$ [51,96]. However, this impulse in channel $r + 1$ represents anything but a stationary signal. Indeed, the whole purpose of this channel is to take care of some initial conditions, whose influence is very much of a transient nature. Especially in the presence of exponential weighting, the quantities $\delta_{p,L}(N)$ and $\tilde{\mathbf{v}}_{p,L}(N)$ decay exponentially as λ^N. Therefore, the influence of these quantities on the rest of the algorithm state becomes negligible after a few time constants $1/(1 - \lambda)$, at which point one could put these quantities equal to zero, which is equivalent to switching to the (r-channel) prewindowed algorithm.

More precisely, the quantity $\delta_{p,L}(N)$ behaves as $\delta_{p,L}(N) = \lambda^{L-1} + O(\lambda^{2L})$, where $L = N - p + 2$ in the context of the GWC algorithm. Now, the open-loop ($K_1 = 0$) eigenvalue associated with $\tilde{\mathbf{v}}_{p,L}(N)$, the backward prediction filter for channel $r + 1$, is given by $\delta_{p,L}(N)/\lambda\delta_{p,L-1}(N - 1)$. In a normal channel, this value would average out to $1/\lambda$, the familiar unstable mode. However, because of the exponential decay of $\delta_{p,L}(N)$, the open-loop eigenvalue averages out to one! Hence it is very easy for the feedback loop (involving K_1) to stabilize this marginal instability, and we can conclude that the non-stationarity in channel $r + 1$ is actually beneficial for its numerical behaviour.

However, T5.12-(5) reveals a problem that renders the algorithm as presented in Table 5.12 not quite amenable to a practical implementation. Indeed, the quantity λ^{-N}, which diverges to infinity, has to be multiplied with $\delta_{p,L}(N)$, which converges to zero. This becomes a numerically unstable operation as time grows (apart from overflow/underflow). The GWC FTRLS algorithm presented in [48] suffers from the same problem. It is clearly desirable to work instead with the scaled quantity

$$\breve{\delta}_{p,L}(N) \triangleq \lambda^{-(L-1)}\delta_{p,L}(N) = 1 + O(\lambda^L) \tag{5.94}$$

which is initialized as $\breve{\delta}_{p,1}(p - 1) = \delta_{p,1}(p - 1) < 1$ and converges to one as time grows. Introducing $\breve{\delta}_{p,L}(N)$ into the algorithm requires some changes in the handling of channel $r + 1$. The following rearrangement leads to a mere change of the feedback coefficients in f_D. Replace T5.12- (4,5) by

$$\text{T5.12-(4'):}\quad \left(\tilde{\mathbf{v}}_{p,L}(N), \breve{\delta}_{p,L}(N), \begin{bmatrix} \mathbf{w}_{p,L}(N) \\ 0 \end{bmatrix}, \alpha_{p,L}^{-\mathrm{S}}(N)\right)$$

$$= f_{D'}\left(\tilde{\mathbf{v}}_{p,L-1}(N-1),\ \breve{\delta}_{p,L-1}(N-1), \begin{bmatrix} \mathbf{w}_{p,L-1}(N) \\ \mathbf{w}_{2,p+1}^{2p}(N) \end{bmatrix}, \alpha_{p,L-1}^{-\mathrm{S}}(N), \begin{bmatrix} \mathbf{u}_p(N) \\ 0 \end{bmatrix}\right) \tag{5.95}$$

$$\text{T5.12-(5'):}\quad \alpha_{p,L}(N) = \alpha_{p,L}^{A}(N) = \lambda^{pr}\breve{\delta}_{p,L}(N)\det(\alpha_{p,L-1}^{\mathrm{b}}(N))\det(\alpha_{p,L-1}^{-\mathrm{f}}(N))$$

The downdate operator $f_{D'}$ is defined as f_D in Table 5.8 except for the following changes related to K_1, K_2:

$$\varepsilon_p^{b(1)}(N) = \varepsilon_p^{bF}(N) = e_p^{bF}(N)\alpha_p(N)$$
$$\alpha_p^{b}(N) = \alpha_p^{b}(N-1) + \varepsilon_p^{bF}(N)e_p^{bSH}(N)$$
$$\text{if } 1 - \breve{\delta}_{p,L}(N) < \varepsilon^{mp} \text{ then}$$
$$\tilde{\mathbf{v}}_{p,L}(N) = 0$$
$$\breve{\delta}_{p,L}(N) = 1$$
$$\text{end if} \tag{5.96}$$

where ε^{mp} is the machine precision. The above changes correspond to taking $K_1 = 1$, which is clearly sufficient here for stabilizing the roundoff error on $\tilde{\mathbf{v}}_{p,L}(N)$. The error propagation associated with $\breve{\delta}_{p,L}(N)$, on the other hand, is exponentially unstable. From (5.95) and (5.96) we get

$$\frac{\Delta\breve{\delta}_{p,L}(N)}{\Delta\breve{\delta}_{p,L-1}(N-1)} = \frac{\breve{\delta}_{p,L}(N)}{\breve{\delta}_{p,L-1}(N-1)} = 1 - \tilde{\mathbf{v}}_{p,L-1}^{H}(N-1)\mathbf{u}_p(N)\mathbf{w}_{2,2p+1}^{2p}(N)$$
$$= 1 + O(\lambda^N) > 1 \tag{5.97}$$

However, since $\breve{\delta}_{p,L}(N)$ does not have a constant average value during the transient period considered here, it is more relevant to consider the relative errors, and (5.97) leads immediately to $\Delta \log \breve{\delta}_{p,L}(N) = \Delta \log \breve{\delta}_{p,L-1}(N-1)$. So we have a random walk for the relative errors, a mild instability. The quantity $\breve{\delta}_{p,L}(N)$ increases to one during the transient period. Due to roundoff errors, the built-in switch to the prewindowed algorithm (see (5.96)) may occur a bit prematurely, but the effect of this will be negligible for reasonable values of ε^{mp}. Simulation experience with the algorithm indicates a stable behaviour.

In the case of $\lambda = 1$, the quantities $\delta_{p,L}(N)$ and $\tilde{\mathbf{v}}_{p,L}(N)$ decay as $1/N$. In this case, however, the numerical error propagation shows a random walk behaviour (digital integrator, see [9]), irrespective of the feedback coefficient values. So $\lambda < 1$ is desired for numerical stability.

5.6.2 The sliding-window covariance (SWC) FTRLS algorithm

The SWC method considers a fixed window length L. It offers an alternative to the prewindowed method with exponential weighting for achieving a finite effective window length. In [58], it was shown that under certain conditions both windowing methods have an identical performance for corresponding effective window length. However, this equivalence only holds asymptotically for large window lengths. For short windows, the SWC method might still be preferable for certain types of non-stationarities since its window really cuts off the past beyond time $N - L$ (see [110] for some advantages of the finite length window in decision feedback equalization).

EMBEDDING INTO A TWO-EXPERIMENT PREWINDOWED PROBLEM

It is easy to see that the cost function for the SWC problem is related to the cost functions of two related prewindowed problems, with a similar relationship for the sample covariance matrices, that is,

$$\sum_{t=N-L}^{N} \lambda^{N-t} \| e(t \mid \mathbf{c}) \|^2 = \sum_{t=0}^{N} \lambda^{N-t} \| e(t \mid \mathbf{c}) \|^2 - \lambda^L \sum_{t=0}^{N-L} \lambda^{N-L-t} \| e(t \mid \mathbf{c}) \|^2 \tag{5.98}$$

$$R_{p,L}(N) = R_p(N) - \lambda^L R_p(N-L)$$

When time progresses from $N-1$ to N, the sample covariance matrix $R_{p,L}(N-1)$ now undergoes a rank 2 modification:

$$R_{p,L}(N) = \lambda R_{p,L}(N-1) + \mathbf{u}_p(N)\mathbf{u}_p^{\mathrm{H}}(N) - \lambda^L \mathbf{u}_p(N-L)\mathbf{u}_p^{\mathrm{H}}(N-L) \tag{5.99}$$

Hence, an embedding of the SWC problem into a two-experiment prewindowed problem comes naturally. The SWC FTRLS algorithm is given in Table 5.13. A derivation can be found in [48,96,51]. It turns out to be convenient to introduce the following scaled quantities:

$$\hat{\tilde{\mathbf{v}}}_{p,L}(N) \triangleq \lambda^{1-L}\tilde{\mathbf{v}}_{p,L}(N) \quad \hat{\mathbf{v}}_{p,L}(N) \triangleq \lambda^{L-1}\mathbf{v}_{p,L}(N) \quad \hat{\delta}_{p,L}(N) \triangleq \lambda^{-2(L-1)}\delta_{p,L}(N) \tag{5.100}$$

which can also be found from (5.88) by replacing π by $\hat{\pi} \triangleq \lambda^{1-L}\pi$ in the definition of the unscaled quantities. We may remark that the number of multiplications mentioned in T5.13-(5,6) reduces to five in the single-channel case ($r = 1$) (a similar remark holds for the GWC algorithm in Table 5.12).

Table 5.13 SWC FTRLS algorithm

Equation	Computation	Cost (×)
	Prediction problem:	
(1)	$(\mathbf{a}_{p,L}(N), \alpha_{p,L}^{-f}(N), \mathbf{w}_{p+1,L}(N), \bar{\alpha}_{p+1,L}^{\mathrm{S}}(N))$	$(1 \div)$
	$= f_{\mathrm{U}}\left(\mathbf{a}_{p,L-1}(N-1), \lambda^{-1}\alpha_{p,L-1}^{-\mathrm{f}}(N-1), \begin{bmatrix} 0_{r\times 1} \\ \mathbf{w}_{p,L}(N-1) \end{bmatrix},\right.$	
	$\left.\alpha_{p,L}^{-1}(N-1), \mathbf{u}_{p+1}(N)\right)$	$3r^2p + 2.5r^2 + 3.5r$
(2)	$\left(\mathbf{b}_{p,L}(N), \alpha_{p,L}^{\mathrm{b}}(N), \begin{bmatrix} \mathbf{w}_{p,L}(N) \\ 0_{r\times 1} \end{bmatrix}, \bar{\alpha}_{p,L}^{\mathrm{S}}(N)\right)$	$(1 \div)$
	$= f_{\mathrm{D}}(\mathbf{b}_{p,L-1}(N-1), \lambda\alpha_{p,L-1}^{\mathrm{b}}(N-1), \mathbf{w}_{p+1,L}(N),$	
	$\bar{\alpha}_{p+1,L}^{\mathrm{S}}(N), \mathbf{u}_{p+1}(N))$	$3r^2p + 2.5r^2 + 5.5r$
(3)	$(\mathbf{a}_{p,L-1}(N), \alpha_{p,L-1}^{-\mathrm{f}}(N), \hat{\tilde{\mathbf{v}}}_{p+1,L}(N), -\hat{\delta}_{p+1,L}^{\mathrm{S}}(N))$	$(1 \div)$

	$= f_{\mathrm{U}}\left(\mathbf{a}_{p,L}(N), \alpha_{p,L}^{-\mathrm{f}}(N), \begin{bmatrix} 0_{r\times 1} \\ \hat{\tilde{\mathbf{v}}}_{p,L}(N-1) \end{bmatrix}, -\hat{\delta}_{p,L}(N-1), \mathbf{u}_{p+1}(N-L+1)\right)$	$3r^2p + 2r^2 + 3r$
(4)	$\left(\mathbf{b}_{p,L-1}(N), \alpha_{p,L-1}^{\mathrm{b}}(N), \begin{bmatrix} \hat{\tilde{\mathbf{v}}}_{p,L}(N) \\ 0_{r\times 1} \end{bmatrix}, -\hat{\delta}_{p,L}^{\mathrm{S}}(N)\right) = f_{\mathrm{D}}(\mathbf{b}_{p,L}(N), \alpha_{p,L}^{\mathrm{b}}(N), \hat{\tilde{\mathbf{v}}}_{p+1,L}(N), -\hat{\delta}_{p+1,L}^{\mathrm{S}}(N), \mathbf{u}_{p+1}(N-L+1))$	$(1\div)$ $3r^2p + 2r^2 + 5r$
(5)	N even: $\hat{\delta}_{p,L}(N) = \hat{\delta}_{p,L}^{\mathrm{S}}(N)$ $\alpha_{p,L}(N) = \alpha_{p,L}^{A}(N) = \lambda^{pr+L-1}\hat{\delta}_{p,L}^{\mathrm{S}}(N) \det(\alpha_{p,L-1}^{\mathrm{b}}(N)\alpha_{p,L-1}^{-\mathrm{f}}(N))$	$(1\div)\ r + 12$
(6)	N odd: $\alpha_{p,L}^{-1}(N) = \alpha_{p,L}^{-\mathrm{S}}(N)$ $\hat{\delta}_{p,L}^{-1}(N) = \hat{\delta}_{p,L}^{-A}(N) = \lambda^{pr+L-1}\bar{\alpha}_{p,L}^{\mathrm{S}}(N) \det(\alpha_{p,L-1}^{\mathrm{b}}(N)\alpha_{p,L-1}^{-\mathrm{f}}(N))$	$(1\div)\ r + 12$
	Joint-process extension:	
(7)	$(\mathbf{c}_{p,L}(N), \varepsilon_{p,L}(N)) = f_{\mathrm{J}}(\mathbf{c}_{p,L-1}(N-1), \mathbf{w}_{p,L}(N), \alpha_{p,L}(N), y(N), \mathbf{u}_p(N))$	$2pr + 1$
(8)	$(\mathbf{c}_{p,L-1}(N), -\lambda^{L-1}v_{p,L-1}^{s}(N)) = f_{\mathrm{J}}(\mathbf{c}_{p,L}(N), \hat{\tilde{\mathbf{v}}}_{p,L}(N), -\hat{\delta}_{p,L}^{-1}(N), y(N-L+1), \mathbf{u}_p(N-L+1))$	$2pr + 1$
	r-channel total cost (six divisions)	$(12r+4)pr + 9r^2 + 19r + 26$

INITIALIZATION

Two strategies are possible: a prewindowing initialization, which will have a prewindowing effect for a duration of $L + p - 1$ samples (in the SWC method, the emphasis is not usually on initialization effects, but on the rectangular window shape in steady-state operation), or a GWC initialization, enforcing the covariance character from the very start. In the first strategy, the non-zero, soft-constraint initialization yields

$$
\begin{aligned}
\mathbf{a}_{p,L-1}(-1) &= 0 & \alpha_{p,L-1}^{\mathrm{f}}(-1) &= \lambda^N r_0 \\
\mathbf{b}_{p,L-1}(-1) &= 0 & \alpha_{p,L-1}^{\mathrm{b}}(-1) &= r_0 \\
\mathbf{w}_{p,L}(-1) &= 0 & \alpha_{p,L}(-1) &= 1 \\
\hat{\tilde{\mathbf{v}}}_{p,L}(-1) &= 0 & \hat{\delta}_{p,L}^{-1}(-1) &= \lambda^{L-1} \\
\mathbf{c}_{p,L-1}(-1) &= \mathbf{c}_0 & &
\end{aligned}
\tag{5.101}
$$

This initialization allows one to start the SWC algorithm of Table 5.13 at time $N = 0$. In the second strategy, one starts to run the GWC algorithm of Table 5.12 and lets the window length grow until time $N = L + p - 2$, at which point the desired length L is reached, and then one switches to the SWC algorithm of Table 5.13, starting at time $N = L + p - 1$. At time $N = L + p - 2$, one will have to scale the quantities $\tilde{\mathbf{v}}$, δ to $\hat{\tilde{\mathbf{v}}}$, $\hat{\delta}$, but there is time for that since the GWC algorithm takes less computation per time step than the SWC algorithm.

NUMERICAL STABILITY CONSIDERATIONS

For the SWC algorithm, it turns out that constant feedback coefficients in the stabilization scheme do not work. This is related to the fact that the embedding of the SWC problem into a two-experiment prewindowed problem leads to a situation in which the contributions of the two experiments have an opposite sign (see (5.98)). In the downdate transformation f_D, one can take the feedback coefficients as

$$K_1(N) = K_2(N) - 1 = 0.75 \text{ sign } \{\alpha_p(N)\} \tag{5.102}$$

where $\alpha_p(N)$ is an argument of the function f_D. Note especially the alternating sign of K_1, $K_2 - 1$ as alternating experiments get processed.

Parelleling some of the analysis above for the prewindowed FTRLS algorithm, one can show that with the choice of feedback coefficients as in (5.102), the numerical error propagation in the SWC algorithm is exponentially stable for $\lambda = 1$. This is in sharp contrast with the prewindowed and GWC algorithms, where the absence of exponential weighting leads to random-walk-type numerical errors, irrespective of the feedback coefficients. We had originally expected to find the same phenomenon for the SWC algorithm and had therefore added the exponential weighting factor. Though the LS cost function would then be influenced by two parameters, L and λ, we had expected to determine the (effective) window length mainly via L and to adjust λ within a range very close to unity for optimal numerical performance. However, it can be shown that, for a given L, the optimal numerical stability is obtained for $\lambda = 1$.

The property of exponential stability for $\lambda = 1$ is actually a great asset of the SWC algorithm, which shows in ill-conditioned cases. Indeed, when we have non-persistent excitation, the rank 1 feedback term in the error propagation system matrix is inactive in the null space of the input covariance matrix. This means that, with $\lambda < 1$, there will be exponential error blow-up in this subspace (as in the conventional RLS algorithm [8]). The SWC FTRLS algorithm, which has a finite memory length even for $\lambda = 1$, will show the much milder random walk behaviour in this subspace (for $\lambda = 1$), just like the robust LMS algorithm. Even with $\lambda = 1$, though, the proposed feedback mechanism is crucial to obtain a stable numerical behaviour in the SWC algorithm (an improper choice of the feedback coefficient values can lead to exponential divergence, even when $\lambda = 1$). Summarizing, we may compare the SWC algorithm with the prewindowed algorithm with exponential weighting, and conclude that the SWC algorithm is a more numerically robust algorithm for comparable (if not more desirable) tracking characteristics, at twice the computational cost.

5.7 Modular (circular) multichannel and/or multiexperiment FTRLS algorithms

In this section, we consider RLS problems with multiple channels and/or experiments. *Sequential processing* of the different channels and experiments decomposes the multichannel, multiexperiment algorithm into a set of intertwined, single-channel, single-experiment algorithms, resulting in a *modular* algorithm structure. The sequential processing strategy corresponds to a *triangular factorization* of error covariance matrices, and numerical benefits accrue from this approach. Also, the numerical stabilization techniques considered above can straightforwardly be incorporated. This section is based on [111]. Related developments can be found in [112–119].

It turns out that, using an embedding into multichannel and multiexperiment prewindowed problems, the FTRLS algorithms for the GWC and SWC cases straightforwardly fall out of a modular prewindowed algorithm [96,51].

5.7.1 The multichannel RLS problem

The general multichannel adaptive transversal filtering system is shown in Figure 5.8. In this system, the desired response signal is approximated by the (negative of the) sum of the outputs of multiple (r) FIR filters. For the purpose of a conventional RLS algorithm, there would be no reason to consider the r FIR filters separately and one could easily lump the r corresponding regression vectors into one big regression vector. However, to exploit the shift-invariance structure in a fast RLS algorithm, one should carefully take the structural details of the regression vector into account.

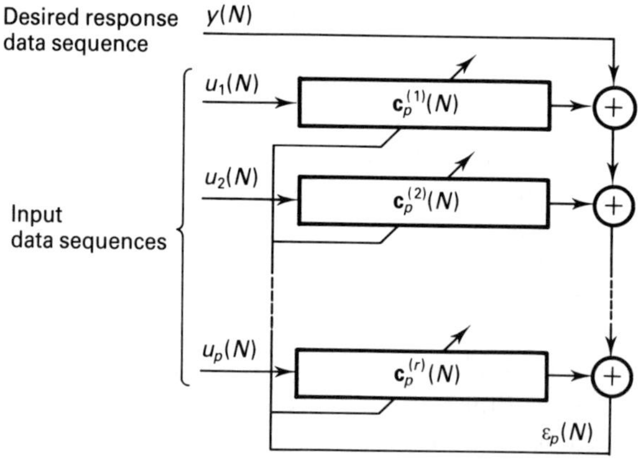

Figure 5.8 Multichannel adaptive transversal filter

5.7.2 The multiexperiment RLS problem

Another type of block processing appearing in least-squares parameter estimation is multiexperiment filtering [30]. This type of filtering arises when we have simultaneous measurements available from different replicas of the same model (different experiments). All signals available will yield valuable information about the same unknown system, and it will be to our advantage to combine these measurements in the estimation of the system parameters. An application might be the identification of the eigenmodes of a large vibrating structure with several sensors attached. Another application arises in multirate signal processing when the ratio of the sampling frequencies of the input and the desired response signals is a rational number. Finally, we may recall that the SWC problem can be viewed as a prewindowed problem with two experiments.

5.7.3 The multichannel multiexperiment RLS problem

The most general case consists of combining the multichannel and the multiexperiment aspects. We shall treat this case in more detail. The details for the special cases of multichannel or multiexperiment can easily be found from the general case. We may mention that one last further generalization would be to consider the case of multiple desired response signals ($s > 1$, see the discussion at the beginning of section 5.4).

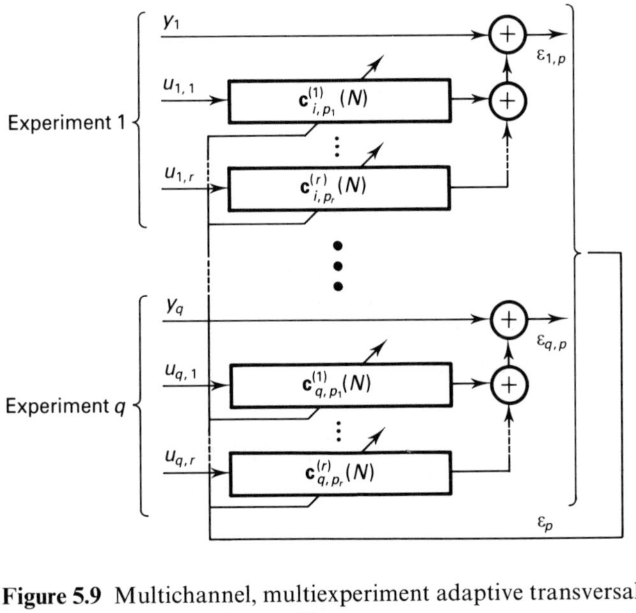

Figure 5.9 Multichannel, multiexperiment adaptive transversal filter

However, the way to solve the RLS problem with multiple desired response signals turns out to be to consider s separate RLS problems, each involving one of the s desired response signals (see [46]). So we shall take $s = 1$.

The multichannel, multiexperiment RLS problem is depicted in Figure 5.9. In order to describe in more detail the resulting modular FTRLS algorithm, it is necessary to introduce some notation. In experiment i, an error signal $\varepsilon_{i,p}(N)$ appears, resulting from the approximation of the current desired response sample $y_i(N)$ by the sum of the outputs of r transversal filters. Let $p = \Sigma_{j=1}^{r} p_j$ denote the sum of the orders of the r transversal filters. We may emphasize that the filter orders in the different channels may be taken to be different. So far, we had formulated the multichannel problem as one in which the input signal $u(N)$ is an r-dimensional vector (when $q = 1$). Such a formulation only works for multichannel problems with equal filter order p in all channels. The transversal filter with impulse response $\mathbf{c}_{i,p_j}^{(j)}(N)$, operating on channel j, forms a linear combination of the current and $p_j - 1$ previous samples of the input signal $u_{i,j}(N)$ (a complex scalar). $\mathbf{c}_{i,p}(N) = [\mathbf{c}_{i,p_1}^{(1)\mathrm{H}}(N) \ldots \mathbf{c}_{i,p_r}^{(r)\mathrm{H}}(N)]^{\mathrm{H}}$ contains a concatenation of the time-varying impulse responses of the r FIR filters. All q experiments are a copy of the first experiment and they all employ the same multichannel filter $\mathbf{c}_{i,p}(N)$. However, they operate on different desired response and input signals. In the RLS method considered here, the impulse response $\mathbf{c}_{i,p}(N)$ is obtained as the solution of a least-squares (LS) minimization problem, involving all data up to experiment i at time N.

The error signal in experiment i can be written as

$$e_{i,p}(N \mid \mathbf{c}) \triangleq y_i(N) + \mathbf{c}^{\mathrm{H}}\mathbf{u}_{i,p}(N) \tag{5.103}$$

where

$$\mathbf{u}_{i,p}(N) \triangleq [\mathbf{u}_{i,p_1}^{(1)\mathrm{H}}(N) \ldots \mathbf{u}_{i,p_r}^{(r)\mathrm{H}}(N)]^{\mathrm{H}} \tag{5.104}$$

with

$$\mathbf{u}_{i,p_j}^{(j)}(N) \quad \triangleq [\mathbf{u}_{i,j}^{\mathrm{H}}(N) \ldots u_{i,j}^{\mathrm{H}}(N - p_j + 1)]^{\mathrm{H}}$$

is the composite regressor vector for experiment i. Let $\mathbf{e}_p(t) \triangleq [e_{q,p}(t) \ldots e_{1,p}(t)]$, and let $\boldsymbol{\varepsilon}_p(t)$, $\mathbf{y}(t)$, $\mathbf{u}^{(j)}(t)$ and $\mathbf{u}_p(t)$ be similarly defined. The filter estimate $\mathbf{c}_p(N)$ $(=\mathbf{c}_{q,p}(N))$, based on all data up to time N, is obtained as the solution of the following quadratic minimization problem (prewindowing is assumed here!):

$$\alpha_p^y(N) = \min_{\mathbf{c}_p(N)} \sum_{t=0}^{N} \lambda^{N-t}\mathbf{e}_p(t \mid \mathbf{c}_p(N))\omega\mathbf{e}_p^{\mathrm{H}}(t \mid \mathbf{c}_p(N)) \tag{5.105}$$

and is given by

$$\mathbf{c}_p(N) = -\left(\sum_{t=0}^{N} \lambda^{N-t}\mathbf{u}_p(t)\omega\mathbf{u}_p^{\mathrm{H}}(t)\right)^{-1}\left(\sum_{t=0}^{N} \lambda^{N-t}\mathbf{u}_p(t)\omega\mathbf{y}(t)\right) \tag{5.106}$$

with $\omega \triangleq \text{diag}\{\omega_q, \ldots, \omega_1\}$ for some non-zero ω_i. Even though we have a possibly indefinite weighting matrix ω appearing in (5.105), we will assume that the quadratic cost function is nevertheless positive definite (meaning that the matrix being inverted in (5.106) above is positive definite).

In the geometric interpretation of the above prewindowed LS problem, we shall again consider infinite-dimensional vectors. In particular, let (see also [4,103])

$$\mathcal{Y}_i(N) \triangleq \begin{bmatrix} y_i^H(N) \\ \vdots \\ y_1^H N) \\ \mathbf{y}^H(N-1) \\ \vdots \\ \mathbf{y}^H(\mathbf{0}) \\ 0 \\ \vdots \end{bmatrix} \qquad \mathcal{U}_i^{(j)}(N) \triangleq \begin{bmatrix} u_{i,j}^H(N) \\ \vdots \\ u_{1,j}^H(N) \\ \mathbf{u}^{(j)H}(N-1) \\ \vdots \\ \mathbf{u}^{(j)H}(0) \\ 0 \\ \vdots \end{bmatrix} \tag{5.107}$$

and let $\boldsymbol{\mathcal{Y}}(N) \triangleq \boldsymbol{\mathcal{Y}}_q(N)$ $(\mathcal{Y}_0(N) = \boldsymbol{\mathcal{Y}}(N-1))$ and $\boldsymbol{\mathcal{U}}^{(j)}(N) \triangleq \boldsymbol{\mathcal{U}}_q^{(j)}(N)$ $(\boldsymbol{\mathcal{U}}_0^{(j)}(N) = \boldsymbol{\mathcal{U}}^{(j)}(N-1))$. We will consider the following data matrices:

$$\mathcal{U}_{i,p_j}^{(j)}(N) \triangleq [\boldsymbol{\mathcal{U}}_i^{(j)}(N), \ldots, \mathcal{U}_i^{(j)}(N-p_j+1)] \tag{5.108}$$

$$\mathcal{U}_{i,j,p}(N) \triangleq [\mathcal{U}_{i,p1}^{(1)}(N), \ldots, \mathcal{U}_{i,p_{j-1}}^{(j-1)}(N), \mathcal{U}_{i,p_j}^{(j)}(N), \mathcal{U}_{i,p_{j+1}}^{(j+1)}(N-1), \cdots, \mathcal{U}_{i,p_r}^{(r)}(N-1)] \tag{5.109}$$

$$\mathcal{U}_{i,j,p+1}(N) \triangleq [\mathcal{U}_{i,p_1}^{(1)}(N), \ldots, \mathcal{U}_{i,p_{j-1}}^{(j-1)}(N), \mathcal{U}_{i,p_{j+1}}^{(j)}(N), \mathcal{U}_{i,p_{j+1}}^{(j+1)}(N-1) \ldots \mathcal{U}_{i,p_r}^{(r)}(N-1)] \tag{5.110}$$

We also define the matrices $\mathcal{U}_{i,p}(N) \triangleq \mathcal{U}_{i,r,p}(N)$, $\mathcal{U}_p(N) \triangleq \mathcal{U}_{q,p}(N)$. These definitions imply that $\mathcal{U}_{i,0,p}(N) = \mathcal{U}_{i,p}(N-1)$, $\mathcal{U}_{0,p}(N) = \mathcal{U}_p(N-1)$. Let $\Sigma_j \triangleq \Sigma_{k=1}^j p_k$, so $p = \Sigma_r$. Finally,[4] we introduce a set of $(p+1 \times p+1)$ permutation matrices $\mathcal{P}_j \triangleq [I_{\Sigma_j} \oplus S_{p+1-\Sigma_j}] + \mathbf{u}_{\Sigma_j+1}\mathbf{u}_{p+1}^H$, where I_{Σ_j} and $S_{N+1-\Sigma_j}$ are unity and shift matrices of the indicated dimension and $\mathbf{u}_j$ is the jth unit vector (one in the jth position and zeros elsewhere). The permutation matrix $\mathcal{P}_j$ is such that the zero column in $[\mathcal{U}_{i,j+1,p}(N)\ 0]\ \mathcal{P}_j^H$ is permuted to the position of column $\boldsymbol{\mathcal{U}}_i^{(j+1)}(N)$ in $\mathcal{U}_{i,j+1,p+1}(N)$ or column $\boldsymbol{\mathcal{U}}_i^{(j)}(N-p_j)$ in $\mathcal{U}_{i,j,p+1}(N)$.

5.7.4 The modular multichannel multiexperiment FTRLS algorithm

We shall give here just a flavour of the idea behind the sequential processing strategy. The modular decomposition principle can be succinctly illustrated in the geometric picture. We have seen that in the geometric approach, the update of the projection operator is the basis for all recursive updating identities

$$P_{[\mathcal{U}\mathcal{V}]} = P_{\mathcal{U}} + P_{P_{\mathcal{U}}^{\perp}\mathcal{V}} \tag{5.111}$$

$$= P_{\mathcal{U}} + P_{\mathcal{U}}^{\perp}\mathcal{V}(\mathcal{V}^{\mathrm{H}}P_{\mathcal{U}}^{\perp}\mathcal{V})^{-1}\mathcal{V}^{\mathrm{H}}P_{\mathcal{U}}^{\perp}$$

In words, if one makes an orthogonal decomposition

$$\{\mathcal{U}, \mathcal{V}\} = \{\mathcal{U}\} \oplus \{P_{\mathcal{U}}^{\perp}\mathcal{V}\} \tag{5.112}$$

of the space spanned by the columns of the data matrices $\mathcal{U}$ and $\mathcal{V}$, then the projection on to this space is equal to the sum of the projections on to the two orthogonal subspaces. The fast transversal and fast lattice (see Chapter 6) RLS algorithms, and even the ordinary RLS algorithm, consist of an interconnection of building blocks, where each building block corresponds to an update $P_{\mathcal{U}} \to P_{[\mathcal{U}\mathcal{V}]}$ or downdate $P_{[\mathcal{U}\mathcal{V}]} \to P_{\mathcal{U}}$.

Now, suppose we decompose the column space of $\mathcal{V}$ into k subspaces $\mathcal{V}_i$, that is,

$$\mathcal{V} = [\boldsymbol{\mathcal{V}}_1, \boldsymbol{\mathcal{V}}_2, \ldots, \boldsymbol{\mathcal{V}}_k] \tag{5.113}$$

Then the orthogonal decomposition in (5.111) can be done recursively as

$$P_{[\mathcal{U}\mathcal{V}]} = P_{\mathcal{U}} + P_{P_{\mathcal{U}}^{\perp}\boldsymbol{\mathcal{V}}_1} + P_{P_{[\mathcal{U}\boldsymbol{\mathcal{V}}_1]}^{\perp}\boldsymbol{\mathcal{V}}_2} + \ldots + P_{P_{[\mathcal{U}\boldsymbol{\mathcal{V}}_1\cdots\boldsymbol{\mathcal{V}}_{k-1}]}^{\perp}\boldsymbol{\mathcal{V}}_k} \tag{5.114}$$

where the set of subspaces $\{P_{\mathcal{U}}^{\perp}\mathcal{V}\} = \{P_{\mathcal{U}}^{\perp}\mathcal{V}_i, i = 1, \ldots, k\}$ is replaced recursively by an orthogonalized set, that is,

$$\{P_{\mathcal{U}}^{\perp}\mathcal{V}\} = \{P_{\mathcal{U}}^{\perp}\boldsymbol{\mathcal{V}}_1\} \oplus \{P_{[\mathcal{U}\boldsymbol{\mathcal{V}}_1]}^{\perp}\boldsymbol{\mathcal{V}}_2\} \oplus \ldots \oplus \{P_{[\mathcal{U}\boldsymbol{\mathcal{V}}_1\ldots\boldsymbol{\mathcal{V}}_{k-1}]}^{\perp}\boldsymbol{\mathcal{V}}_k\} \tag{5.115}$$

By comparing (5.114) with (5.111), we find

$$P_{\mathcal{U}}^{\perp}\mathcal{V}(\mathcal{V}^{\mathrm{H}}P_{\mathcal{U}}^{\perp}\mathcal{V})^{-1}\mathcal{V}^{\mathrm{H}}P_{\mathcal{U}}^{\perp} = [P_{\mathcal{U}}^{\perp}\boldsymbol{\mathcal{V}}_1 \cdots P_{[\mathcal{U}\boldsymbol{\mathcal{V}}_1\cdots\boldsymbol{\mathcal{V}}_{k-1}]}^{\perp}\mathcal{V}_k]$$

$$\times \begin{bmatrix} (\boldsymbol{\mathcal{V}}_1^{\mathrm{H}}P_{\mathcal{U}}^{\perp}\boldsymbol{\mathcal{V}}_1)^{-1} & & \mathrm{O} \\ & \ddots & \\ \mathrm{O} & & (\boldsymbol{\mathcal{V}}_k^{\mathrm{H}}P_{[\mathcal{U}\boldsymbol{\mathcal{V}}_1\cdots\boldsymbol{\mathcal{V}}_{k-1}]}^{\perp}\boldsymbol{\mathcal{V}}_k)^{-1} \end{bmatrix} \begin{bmatrix} \boldsymbol{\mathcal{V}}_1^{\mathrm{H}}P_{\mathcal{U}}^{\perp} \\ \vdots \\ \boldsymbol{\mathcal{V}}_k^{\mathrm{H}}P_{[\mathcal{U}\boldsymbol{\mathcal{V}}_1\cdots\boldsymbol{\mathcal{V}}_{k-1}]}^{\perp} \end{bmatrix} \tag{5.116}$$

which shows that the recursive Gram–Schmidt orthogonalization procedure in (5.114) corresponds to an LDU factorization of the error covariance matrix $\mathcal{V}^{\mathrm{H}}P_{\mathcal{U}}^{\perp}\mathcal{V} = (P_{\mathcal{U}}^{\perp}\mathcal{V})^{\mathrm{H}}(P_{\mathcal{U}}^{\perp}\mathcal{V})$, that is,

$$\mathcal{V}^{\mathrm{H}}P_{\mathcal{U}}^{\perp}\mathcal{V} = L_k D_k L_{\mathrm{k}}^{\mathrm{H}} \tag{5.117}$$

$D_k = \mathrm{diag}\{\boldsymbol{\mathcal{V}}_1^{\mathrm{H}}P_{\mathcal{U}}^{\perp}\boldsymbol{\mathcal{V}}_1, \ldots, \boldsymbol{\mathcal{V}}_k^{\mathrm{H}}P_{[\mathcal{U}\boldsymbol{\mathcal{V}}_1\ldots\boldsymbol{\mathcal{V}}_{k-1}]}^{\perp}\boldsymbol{\mathcal{V}}_k\}$ is the diagonal matrix replacing $\mathcal{V}^{\mathrm{H}}P_{\mathcal{U}}^{\perp}\mathcal{V}$ in a sequential updating procedure, and the lower-triangular factor L_k is absorbed in the orthogonalized elements, that is,

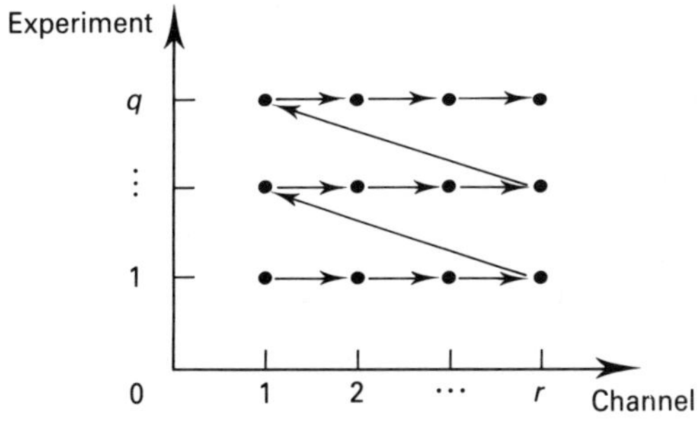

Figure 5.10 Sequential processing schedule of channels and experiments

$$[P_{\mathscr{U}}^{\perp}\mathscr{V}_1 \ldots P_{\mathscr{U}}^{\perp}\mathscr{V}_k]\, L_k^{-\mathrm{H}} = [P_{\mathscr{U}}^{\perp}\mathscr{V}_1 \ldots P_{[\mathscr{U}\mathscr{V}_1\ldots\mathscr{V}_{k-1}]}^{\perp}\mathscr{V}_k] \tag{5.118}$$

Note that

$$[P_{\mathscr{U}}^{\perp}\mathscr{V}_1 \ldots P_{[\mathscr{U}\mathscr{V}_1\ldots\mathscr{V}_{k-1}]}^{\perp}\mathscr{V}_k]\; L_k^{\mathrm{H}}$$

is the unnormalized QR factorization of $P_{\mathscr{U}}^{\perp}\mathscr{V}$, in which the columns of Q are orthogonal but not orthonormal and R is an upper-triangular matrix with a unit diagonal.

The modular FTRLS algorithm handles all experiment updates and the order up- and downdates in the r channels individually in a sequential way. One issue arising from this approach of sequential processing is the freedom of choice of the ordering of the operations. The particular strategy we propose is depicted in Figure 5.10. On the highest level, it involves processing the different experiments sequentially. Within each experiment, the strategy involves an intertwining of order up- and downdates, or, in other words, each channel is processed sequentially with one update followed by one downdate per channel. This has the advantage that the processing of any given channel then corresponds exactly to the processing in a single-channel FTRLS algorithm. Apart from regularity in the implementation, this means that all (numerical) expertise on single-channel algorithms can be carried over in a straightforward way to the modular algorithm. Also, the filter order (p) of the resulting r forward and backward prediction filters is constant for each channel (a sequential processing strategy involving processing all r-order downdates first would lead to filter orders ranging from p to $p - r + 1$, which would lead to an overall computations count that is, strictly speaking, a bit lower, but then the structure of intertwined single-channel FTRLS algorithms would be lost). The resulting modular FTRLS algorithm is shown in Table 5.14. See [96,111] for a derivation. In [51], one can also find how to obtain the GWC and SWC algorithms (even in modular multichannel form) from the prewindowing framework considered here. Note that we have

Table 5.14 Modular multichannel multiexperiment FTRLS algorithm

Equation	Computation	Cost (×)
	Stepping from $N-1$ to N:	
	$\mathbf{c}_{0,p}(N)=\mathbf{c}_{q,p}(N-1)$ and $\mathbf{A}_{0,j,p}(N)=\mathbf{A}_{q,j,p}(N-1)$,	
	$\mathbf{B}_{0,j,p}(N-1)=\mathbf{B}_{q,j,p}(N-1)$	
(0)	$\alpha^{-\mathrm{f}}_{0,j,p}(N)=\lambda^{-1}\alpha^{-\mathrm{f}}_{q,j,p}(N-1)$, $\alpha^{\mathrm{b}}_{0,j,p}(N)=\lambda\alpha^{\mathrm{b}}_{q,j,p}(N-1)$,	
	$j=1,\ldots,r$	$2r$
	Prediction problem:	
	FOR $i=1,\ldots,q$ DO	
	FOR $j=1,\ldots,r$ DO	
(1)	$(\mathbf{a}_{i,j,p}(N),\ \alpha^{-\mathrm{f}}_{i,j,p}(N),\ \mathbf{w}_{i,j,p+1}(N),\ \bar{\alpha}^{\mathrm{S}}_{i,j,p+1}(N))$	$(1\div)$
	$=f_{\mathrm{U}}\left(\mathbf{a}_{i-1,j,p}(N),\ \alpha^{-\mathrm{f}}_{i-1,j,p}(N),\ \mathscr{P}_{j-1}\begin{bmatrix}\mathbf{w}_{i,j-1,p}(N)\\ 0\end{bmatrix},\right.$	
	$\left.\alpha^{-\mathrm{S}}_{i,j-1,p}(N),\ \mathbf{u}_{i,j,p+1}(N)\right)$	$3p+5$
(2)	$\left(\mathbf{b}_{i,j,p}(N),\ \alpha^{\mathrm{b}}_{i,j,p}(N),\ \mathscr{P}_{j}\begin{bmatrix}\mathbf{w}_{i,j,p}(N)\\ 0\end{bmatrix},\ \bar{\alpha}^{\mathrm{S}}_{i,j,p}(N)\right)$	$(1\div)$
	$=f_{\mathrm{D}}(\mathbf{b}_{i-1,j,p}(N),\ \alpha^{\mathrm{b}}_{i-1,j,p}(N),\ \mathbf{w}_{i,j,p+1}(N),\ \alpha^{-\mathrm{S}}_{i,j,p+1}(N),\ \mathbf{u}_{i,j,p+1}(N))$	$3p+7$
	END DO $\mathbf{w}_{i,p}(N)=\mathbf{w}_{i,r,p}(N)=\mathbf{w}_{i,0,p}(N+1)$	
	END DO	
(3)	FOR $i=i_N\triangleq 1+N \bmod q$: $\alpha_{i_N,p}(N)=\alpha^{A}_{i_N,p}(N)=\alpha^{\mathrm{S}}_{i_N,0,p}(N+1)$	
	$=\lambda^{p}\left(\prod_{i=1}^{q}\omega_i\right)\left(\prod_{\substack{i=1\\ i\neq i_N}}^{q}\alpha^{\mathrm{S}}_{i,p}(N)\right)\left(\prod_{j=1}^{r}(\alpha^{\mathrm{b}}_{q,j,p}(N)\alpha^{-\mathrm{f}}_{q,j,p}(N))\right)$	$2r+q+1$
	FOR $i=1,\ldots,q\neq i_N$: $\alpha_{i,p}(N)=\alpha^{\mathrm{S}}_{i,r,p}(N)=\alpha^{\mathrm{S}}_{i,0,p}(N+1)$	
	Joint-process extension:	
	FOR $i=1,\ldots,q$ DO	

(4) $(\mathbf{c}_{i,p}(N), \varepsilon_{i,p}(N)) = f_J(\mathbf{c}_{i-1,p}(N), \mathbf{w}_{i,p}(N), \alpha_{i,p}(N), y_i(N),$
$\mathbf{u}_{i,p}(N))$ $2p+1$

END DO $\mathbf{c}_p(N) = \mathbf{c}_{q,p}(N)$

r-channel q-experiment total cost ($2rq$ divisions)
$(6r+2)qp + 12rq + 4r + 2q + 1$

$$\mathbf{A}_{i,j,p}(N) = \mathscr{P}_{j-1}\begin{bmatrix}\mathbf{a}_{i,j,p}(N)\\ 1\end{bmatrix} \quad \mathbf{B}_{i,j,p}(N) = \mathscr{P}_j\begin{bmatrix}\mathbf{b}_{i,j,p}(N)\\ 1\end{bmatrix} \tag{5.119}$$

So when f_U in Table 5.14 is shown as a function of the product $[\mathbf{w}^H_{i,j-1,p}(N), 0]\,\mathscr{P}^H_{j-1}$, it is in fact a function of $[\mathbf{w}^H_{i,j-1,p}(N), 0]$ and $\mathscr{P}_{j-1}$ separately (f_U needs to be able to identify the position of '1' in $\mathbf{A}_{i,j,p}(N)$). A similar remark holds for f_D. If all the ω_is are equal, then the usual feedback coefficient values $K_1 = 1.5$, $K_2 = 2.5$ in f_D work. If there is substantial variation in the ω_is (especially differing signs), then the feedback coefficients have to be varied with the experiment i (see [111] for details).

INITIALIZATION

Appending a soft constraint to the basic cost function (5.105), the augmented cost function becomes

$$\min_{\mathbf{c}_p(N)} \left\{ \sum_{t=0}^{N} \lambda^{N-t}\mathbf{e}_p(t\,|\,\mathbf{c}_p(N))\omega\mathbf{e}_p^H(t\,|\,\mathbf{c}_p(N)) + \lambda^{N+1}(\mathbf{c}_p(N)-\mathbf{c}_0)^H R_p(-1)(\mathbf{c}_p(N)-\mathbf{c}_0)\right\} \tag{5.120}$$

where $\mathbf{c}_0$ represents the a priori estimate of the filter $\mathbf{c}$ and $R_p(-1) = \mathscr{U}_p^H(-1)\Omega\mathscr{U}_p(-1)$ for some data matrix $\mathscr{U}_p(-1)$ which should have (full) rank pr (Ω is a diagonal matrix explicitly representing the combination of exponential and ω weighting in the LS problem: $\langle \mathscr{U}, \mathscr{V} \rangle = \mathscr{U}^H\Omega\mathscr{V}$). The matrix $R_p^{-1}(-1)$ can be interpreted as the covariance matrix of the a priori estimate $\mathbf{c}_0$. Even if a priori knowledge about $\mathbf{c}$ is not available (in which case one usually takes $\mathbf{c}_0 = 0$), the augmented cost function is positive definite and hence its solution is well defined starting at time $N = 0$. One can absorb the soft constraint into the time series $y_i(\cdot)$ and $u_{i,j}(\cdot)$ if $\mathscr{U}_p(-1)$ conforms to the structure indicated in (5.109), and if one takes $\mathscr{Y}(-1) = -\mathscr{U}_p(-1)\mathbf{c}_0$. In this way, the augmented cost function (5.120) reduces in form to the original cost function (5.105) and can still be minimized recursively using the FTRLS algorithm. The soft-constraint part just leads to a specific initialization for the algorithmic quantities $\{\mathbf{a}_{q,j,p}(-1), \mathbf{b}_{q,j,p}(-1), \alpha^f_{q,j,p}(-1), \alpha^b_{q,j,p}(-1), j = 1,\ldots, r\}$, $\{\mathbf{w}_{i,0,p}(0), \alpha_{i,0,p}(0), i = 1,\ldots, q\}$ and $\mathbf{c}_p(-1) = \mathbf{c}_0$. One can in principle choose any $\mathscr{U}_p(-1)$ and work out the corresponding initial conditions for the above quantities. The following is an easy choice, corresponding to a diagonal matrix $R_p(-1)$. Consider the non-causal part of the signals: $\mathbf{u}^{(j)}(t) = \mathbf{r}_{0j}\delta(t + j + \Sigma_j)$, $j = 1,\ldots, r$, $t < 0$, with $\delta(t)$ denoting the unit pulse at time

$t = 0$ and the $\mathbf{r}_{0j}$ are some $1 \times q$ vectors. The resulting initial conditions at time -1 with which to start the steady-state algorithm (Table 5.14) at time 0 are

$$\begin{aligned}
\mathbf{A}^{\mathrm{H}}_{q,j,p}(-1) &= [0\ldots0\,1]\mathscr{P}^{\mathrm{H}}_{j-1} & \alpha^{\mathrm{f}}_{q,j,p}(-1) &= \lambda^{j-1+\Sigma_j}\mathbf{r}_{0j}\omega\mathbf{r}^{\mathrm{H}}_{0j} & & \\
\mathbf{B}^{\mathrm{H}}_{q,j,p}(-1) &= [0\ldots0\,1]\mathscr{P}^{\mathrm{H}}_{j} & \alpha^{\mathrm{b}}_{q,j,p}(-1) &= \lambda^{j-1+\Sigma_{j-1}}\mathbf{r}_{0j}\omega\mathbf{r}^{\mathrm{H}}_{0j} & j &= 1,\ldots,r \\
\mathbf{w}_{i,0,p}(0) &= [0\ldots0]^{\mathrm{H}} & \alpha_{i,0,p}(0) &= \omega_i & i &= 1,\ldots,q \\
\mathbf{c}_p(-1) &= \mathbf{c}_0 & & & &
\end{aligned} \tag{5.121}$$

Appropriate choices for the $\mathbf{r}_{0j}$ have been discussed in [4,103]. The weights $\mathbf{r}_{0j}$ should in principle reflect the level of confidence in the initial estimate $\mathbf{c}_0$. However, they should not be chosen too small to avoid ill-conditioning in the initial convergence period.

5.7.5 Triangular factorization

As was indicated before, the sequential processing of the modular approach corresponds to triangular factorization of error covariance matrices. We will show that the modular FTRLS algorithm propagates the factors in triangular factorizations of $\alpha^{\mathrm{f}}_{i,p}(N)$, $\alpha^{\mathrm{b}}_{i,p}(N)$ and $\alpha_p(N)$.

In the conventional multichannel, multiexperiment FTRLS algorithm, $\alpha^{\mathrm{f}}_p(N)$ and $\alpha^{\mathrm{b}}_p(N)$ are $r \times r$ matrices and $\alpha_p(N)$ is a $q \times q$ matrix. These matrices obey recursions of the following form:

$$\alpha_p^{-\mathrm{f}}(N) = \lambda^{-1}\alpha_p^{-\mathrm{f}}(N-1) - w^0_{p+1}(N)\alpha_{p+1}(N)w^{0\mathrm{H}}_{p+1}(N) \tag{5.122}$$

$$\alpha_p^{-\mathrm{b}}(N) = \lambda^{-1}\alpha_p^{-\mathrm{b}}(N-1) - w^p_{p+1}(N)\alpha_{p+1}(N)w^{p\mathrm{H}}_{p+1}(N) \tag{5.123}$$

$$\alpha_p^{-1}(N) = \alpha^{-1}_{p+1}(N) - e^{b\mathrm{H}}_p(N)\lambda^{-1}\alpha_p^{-\mathrm{b}}(N-1)e^{\mathrm{b}}_p(N) \tag{5.124}$$

Equations (5.122) and (5.123) are identical in form to the Riccati equation in the conventional RLS algorithm. Here we are just considering two higher steps of granularity: the experiment (q) and channel (r) dimensions instead of the regressor dimension (p). As for the RLS algorithm, it is essential for the proper functioning of the FTRLS algorithm that the three matrices mentioned above (which are inverses of covariance matrices) are Hermitian and positive definite (see [22,120,4]). In a finite-wordlength environment, however, these matrices may not maintain these properties, especially since their updates are obtained via a subtraction between positive (semi-) definite matrices. Therefore, working with LDU or UDL decompositions has certain numerical advantages: the decomposition forces the matrices to be Hermitian, and it allows for easy monitoring of the positive definiteness (and possibly reinforcing it by a simple manipulation of the diagonal factors). Furthermore, working with the factors of the LDU decomposition is a better conditioned procedure (see [121]); if a covariance matrix is ill-conditioned, then the (unit diagonal) triangular factor usually is much better conditioned, while the ill-conditioning is mostly absorbed into the (more easily controlled) diagonal factor. The ratio of the maximum to minimum element of the diagonal factor gives a good indication about the conditioning of the covariance matrix (see [122]).

Let us now take a closer look at the displacement structure of the inverse of a sample covariance matrix. From T5.2-(3,4), one can easily show that

$$R_{i,j,p+1}^{-1}(N) = \mathscr{P}_{j-1}\begin{bmatrix} R_{i,j-1,p}^{-1}(N) & 0 \\ 0 & 0 \end{bmatrix}\mathscr{P}_{j-1}^{\mathrm{H}} + \mathbf{A}_{i,j,p}(N)\alpha_{i,j,p}^{-\mathrm{f}}(N)\mathbf{A}_{i,j,p}^{\mathrm{H}}(N)$$

$$= \mathscr{P}_{j}\begin{bmatrix} R_{i,j,p}^{-1}(N) & 0 \\ 0 & 0 \end{bmatrix}\mathscr{P}_{j}^{\mathrm{H}} + \mathbf{B}_{i,j,p}(N)\alpha_{i,j,p}^{-\mathrm{b}}(N)\mathbf{B}_{i,j,p}^{\mathrm{H}}(N) \qquad (5.125)$$

Let $\underline{\mathbf{A}}_{i,j,p}(N)$ and $\underline{\mathbf{B}}_{i,j,p}(N)$ be straightforward $(p + r) \times 1$ extensions of $\mathbf{A}_{i,j,p}(N)$ and $\mathbf{B}_{i,j,p}(N)$, obtained by inserting $r - 1$ zeros in appropriate places so that the inner product of $\underline{\mathbf{A}}_{i,j,p}(N)$ and $\underline{\mathbf{B}}_{i,j,p}(N)$ with $\mathscr{U}_{i,p+r}(N) = [\mathscr{U}_{i,p_1+1}^{(1)} \cdots \mathscr{U}_{i,p_r+1}^{(r)}]$ yields $\boldsymbol{\varepsilon}_{i,j,p}^{\mathrm{f}}(N)$ and $\boldsymbol{\varepsilon}_{i,j,p}^{\mathrm{b}}(N)$. Then based on (several applications of) (5.125), one can establish the following identity for any i (see [46] for details)

$$\begin{bmatrix} \mathbf{A}_{i,p}^{\mathrm{H}}(N) \\ \mathbf{B}_{i,p}^{\mathrm{H}}(N) \end{bmatrix}^{\mathrm{H}}\begin{bmatrix} \alpha_{i,p}^{-\mathrm{f}}(N) & 0 \\ 0 & -\alpha_{i,p}^{-\mathrm{b}}(N) \end{bmatrix}\begin{bmatrix} \mathbf{A}_{i,p}^{\mathrm{H}}(N) \\ \mathbf{B}_{i,p}^{\mathrm{H}}(N) \end{bmatrix} =$$

$$\begin{bmatrix} \underline{\mathbf{A}}_{i,1,p}^{\mathrm{H}}(N) \\ \vdots \\ \underline{\mathbf{A}}_{i,r,p}^{\mathrm{H}}(N) \\ \underline{\mathbf{B}}_{i,1,p}^{\mathrm{H}}(N) \\ \vdots \\ \underline{\mathbf{B}}_{i,r,p}^{\mathrm{H}}(N) \end{bmatrix}^{\mathrm{H}}\begin{bmatrix} \alpha_{i,1,p}^{-\mathrm{f}}(N) & & & & & \mathrm{O} \\ & \ddots & & & & \\ & & \alpha_{i,r,p}^{-\mathrm{f}}(N) & & & \\ & & & -\alpha_{i,1,p}^{-\mathrm{b}}(N) & & \\ & \mathrm{O} & & & \ddots & \\ & & & & & -\alpha_{i,r,p}^{-\mathrm{b}}(N) \end{bmatrix}\begin{bmatrix} \underline{\mathbf{A}}_{i,1,p}^{\mathrm{H}}(N) \\ \vdots \\ \underline{\mathbf{A}}_{i,r,p}^{\mathrm{H}}(N) \\ \underline{\mathbf{B}}_{i,1,p}^{\mathrm{H}}(N) \\ \vdots \\ \underline{\mathbf{B}}_{i,r,p}^{\mathrm{H}}(N) \end{bmatrix}$$

(5.126)

This identity shows the UDL and LDU factorization of $\alpha_{i,p}^{-\mathrm{f}}(N)$ and $\alpha_{i,p}^{-\mathrm{b}}(N)$, respectively, where the diagonal factors contain the $\alpha_{i,j,p}^{-\mathrm{f}}(N)$ and $\alpha_{i,j,p}^{-\mathrm{b}}(N)$ of the modular FTRLS algorithm, and the triangular factors are absorbed in the modular forward and backward prediction filters. If we put the experiment index $i = q$, then we get the conventional quantities (that we would obtain without sequential processing of the experiments). In [46], one can find a detailed description of how to obtain the triangular factors also, and how to recover the conventional prediction filters from the modular ones.

Similarly, one can exploit the identity $R_{\mathscr{U}}^{-1} = R_{S^{\mathrm{H}}\mathscr{U}}^{-1} - \mathbf{w}_{\mathscr{U}}\alpha_{\mathscr{U}}\mathbf{w}_{\mathscr{U}}^{\mathrm{H}}$ (see T5.2-(5)) in several ways to obtain

$$\mathbf{w}_p(N)\alpha_p^{-1}(N)\mathbf{w}_p^{\mathrm{H}}(N) = \begin{bmatrix} \mathbf{w}_{1,r,p}^{\mathrm{H}}(N) \\ \vdots \\ \mathbf{w}_{q,r,p}^{\mathrm{H}}(N) \end{bmatrix}^{\mathrm{H}}\begin{bmatrix} \alpha_{1,r,p}^{-1}(N) & & \mathrm{O} \\ & \ddots & \\ \mathrm{O} & & \alpha_{q,r,p}^{-1}(N) \end{bmatrix}\begin{bmatrix} \mathbf{w}_{1,r,p}^{\mathrm{H}}(N) \\ \vdots \\ \mathbf{w}_{q,r,p}^{\mathrm{H}}(N) \end{bmatrix} \qquad (5.127)$$

In other words, the modular, multiexperiment, FTRLS algorithm corresponds to an LDU factorization of the $q \times$ q matrix $\alpha_p(N)$. The diagonal factor is propagated and the triangular factor is absorbed into the transformed Kalman gains (and cannot be recovered; this is related to the difficulty of using α^A for the stabilization of the α part of a non-modular, multiexperiment, FTRLS algorithm: how to recover α when only its determinant is given (see (5.78))). Note the duality between this triangular factorization of the likelihood variable $\alpha_p(N)$ in the modularization of the multi-experiment aspect, and the factorizations of the forward and backward prediction error covariances $\alpha_p^{\mathrm{f}}(N)$ and $\alpha_p^{\mathrm{b}}(N)$ in the modularization of the multichannel aspect of the problem. The duality is basically between the two FTRLS time updating strategies.

Notes

1. Superscript H denotes complex conjugate (Hermitain) transpose. In this chapter, we assume that the signals can take on complex values.
2. For notational consistency with the other chapters, we shall denote $\alpha^{-1}(t)$ as $\alpha^*(t)$. Note also that in this chapter λ has been incorporated into the gain vector, leading to a slightly different formulation of the RLS compared to that of Table 2.1 of chapter 2.
3. The fact that the rank of $u(0)\, u^{\mathrm{H}}(0)$ (namely, $\min(q, r)$) is not necessarily equal to r creates extra problems in the general multichannel, multiexperiment case; the general problem does not become fully determined at time $N = p$, but at time $N = pr/q$ (a fractional result means that at the last time instant, one does not require all experiments to make the problem exactly determined). It is conceivable to develop zero soft-constraint initialization procedures for the general case also, especially if one takes the modular approach that we shall discuss later. However, it is clear that soft-constraint initialization is especially convenient for the general multichannel, multiexperiment case.
4. The Kronecker product $\otimes$ of two matrices is defined as $A \otimes B \triangleq [a_{ij}B]$, a block matrix in which block (i, j) is $a_{ij}B$. The direct sum $\oplus$ of matrices is defined as $\oplus_{k+1}^{m} A_k = A_1 \oplus A_2 \oplus \ldots \oplus A_m \triangleq \text{block-diag}\{A_1, A_2, \ldots, A_m\}$.

References

[1] S. Haykin, *Adaptive Filter Theory*, 2nd edn, Prentice Hall: Englewood Cliffs, NJ, 1991.

[2] M.L. Honig and D.G. Messerschmitt, *Adaptive Filters: Structures, Algorithms, and Applications*, Kluwer: Hingham, MA, 1984.

[3] E. Eleftheriou and D. Falconer, 'Tracking properties and steady-state performance of RLS adaptive filter algorithms', *IEEE Trans. Acoust., Speech, Signal Process.*, vol. ASSP-34, pp. 1097–110, 1986.

[4] J.M. Cioffi and T. Kailath 'Fast, recursive least squares transversal filters for adaptive filtering', *IEEE Trans. Acoust., Speech, Signal Process.*, vol. ASSP-32, pp. 304–37, 1984.

[5] B. Widrow, J.M. McCool, M.G. Larimore and C.R. Johnson, Jr. 'Stationary and nonstationary learning characteristics of the LMS adaptive filter', *Proc. IEEE*, vol. 64, pp. 1151–62, 1976.

[6] J.-L. Botto and G.V. Moustakides, 'Stabilizing the fast Kalman Algorithms', *IEEE Trans. Acoust., Speech, Signal Process.*, vol. ASSP-37, pp. 1342–8, 1989.

[7] A. Benallal and A. Gilloire, 'A new method to stabilize fast RLS algorithms based on a first-order model of the propagation of numerical errors', in *Proc. ICASSP 88 Conf.*, vol. 3, pp. 1373–6, 1988.

[8] D.T.M. Slock and T. Kailath, 'Numerically stable fast recursive least-squares transversal filters', in *Proc. ICASSP-88 Conf.*, vol. 3, pp. 1365–8, 1988.

[9] D.T.M. Slock and T. Kailath, 'Numerically stable fast transversal filters for recursive least-squares adaptive filtering', *IEEE Trans. Acoust., Speech, Signal Process.*, vol. ASSP-39, pp. 92–114, 1991.

[10] B. Widrow and S.D. Stearns, *Adaptive Signal Processing*, Prentice Hall: Englewood Cliffs, NJ, 1985.

[11] B. Friedlander, 'Lattice implementation of some recursive parameter-estimation algorithms', *Int. J. Control*, vol. 37, pp. 661–84, 1983.

[12] G.C. Goodwin and K.S. Sin, *Adaptive Filtering Prediction and Control*, Prentice Hall: Englewood Cliffs, NJ, 1984.

[13] D.D. Falconer, and L. Ljung, 'Application of fast Kalman estimation to adaptive equalization', *IEEE Trans. Commun.*, vol. COM-26, pp. 1439–46, 1978.

[14] B. Hätty, 'Recursive least squares algorithms using multirate systems for cancellation of acoustical echoes', in *Proc. ICASSP-90 Conf.*, pp. 1145–8, 1990.

[15] M. Bellanger, *Analyse des Signaux et Filtrage Numérique Adaptatif.*, Masson: CNET-ENST, Paris, 1989.

[16] Y. Boutalis, S. Kollias and G. Carayannis, 'A fast multichannel approach to adaptive estimation and filtering of two dimensional images', in *Proc. ICASSP-87 Conf.*, pp. 1981–4, 1987.

[17] D.T.M. Slock and T. Kailath, 'Multichannel fast transversal filter algorithms for adaptive broadband beamforming', in *Proc. SPIE 89 Conf. – Advanced Algorithms and Architectures for Signal Processing IV*, pp. 22–33, 1989.

[18] J.M. Cioffi, 'Limited-precision effects in adaptive filtering', *IEEE Trans. Circuits Syst.*, vol. CAS-34, pp. 821–33, 1987.

[19] A. Weiss and D. Mitra, 'Digital adaptive filters: Conditions for convergence, rates of convergence, effects of noise and errors arising from the implementation', *IEEE Trans. Inf. Theory*, vol. IT-25, pp. 637–52, 1979.

[20] C. Caraiscos and B. Liu, 'A round-off analysis of the LMS adaptive algorithm', *IEEE Trans. Acoust. Speech, Signal Process.*, vol. ASSP-32, pp. 34–41, 1984.

[21] S. Ljung and L. Ljung, 'Error propagation properties of recursive least-squares adaptation algorithms', *Automatica*, vol. 21, pp. 157–67, 1985.

[22] M.H. Verhaegen and P. Van Dooren, 'Numerical aspects of different Kalman filter implementations', *IEEE Trans. Autom. Control*, vol. AC-31, pp. 907–17, 1986.

[23] S.H. Ardalan and S.T. Alexander, 'Fixed-point roundoff error analysis of the exponentially windowed RLS algorithm for time-varying systems', *IEEE Trans. Acoust. Speech, Signal Process.*, vol. ASSP-35, pp. 770–83, 1987.

[24] C.G. Samson and V.U. Reddy, 'Fixed point error analysis of the normalized ladder algorithm', *IEEE Trans. Acoust. Speech, Signal Process.*, vol. ASSP-31, pp. 1177–91, 1983.

[25] P.A. Regalia, 'Numerical stability properties of a QR-based fast least-squares algorithm', *IEEE Trans.*, submitted.

[26] L. Ljung, M. Morf and D. Falconer, 'Fast calculation of gain matrices for recursive estimation schemes', *Int. J. Control*, vol. 27, pp. 1–19, 1978.

[27] F. Ling and J.G. Proakis, 'Numerical accuracy and stability: Two problems of adaptive estimation algorithms caused by round-off error', in *Proc. ICASSP-84 Conf.*, pp. 30.3.1–4, 1984.

[28] D.W. Lin, 'On digital implementation of the fast Kalman algorithm', *IEEE Trans. Acoust., Speech, Signal Process.*, vol. ASSP-32, pp. 998–1005, 1984.

[29] J.M. Cioffi, Private communication, 1986.

[30] B. Porat, 'Contributions to the theory and applications of lattice filters', PhD Thesis, Stanford University, 1982.

[31] J.-L. Botto and G.V. Moustakides, 'Stabilization of fast recursive least-squares transversal filters', Technical Report 570, INRIA, Rocquencourt, France, 1986.

[32] J.-L. Botto and G.V. Moustakides, 'Stabilization of fast recursive least-squares transversal filters for adaptive filtering', in *Proc. ICASSP-87 Conf.*, vol. 1, pp. 403–7, 1987.

[33] M. Bellanger, 'Engineering aspects of fast least squares algorithms in transversal adaptive filters', in *Proc. ICASSP-87 Conf.*, vol. 4, pp. 2149–52, 1987.

[34] B.D.O. Anderson and J.B. Moore, *Optimal Filtering*, Prentice Hall: Englewood Cliffs, NJ, 1979.

[35] M. Morf and T. Kailath, 'Square-root algorithms for least-squares estimation', *IEEE Trans. Autom. Control*, vol. AC-20, pp. 487–97, 1975.

[36] G.J. Bierman, *Factorization Methods for Discrete Sequential Estimation*, Academic Press: New York, 1977.

[37] G.J. Bierman and C.L. Thornton, 'Numerical comparison of Kalman filter algorithms: Orbit determination case study', *Automatica*, vol. 13, pp. 23–35, 1977.

[38] D.T.M. Slock, 'Backward consistency: A new look at the propagation of numerical errors in the various Kalman filtering algorithms', in *Proc. 25th Information Sciences and Systems Conference, Johns Hopkins University, Baltimore, MD, 1991.*

[39] F. Ling, D. Manolakis and J.G. Proakis, 'A recursive modified Gram–Schmidt algorithm for least-squares estimation', *IEEE Trans. Acoust., Speech, Signal Process.*, vol. ASSP-34, pp. 829–36, 1986.

[40] J.G. McWhirter, 'Recursive least-squares minimization using a systolic array', in *Proc. SPIE (Real-Time Signal Processing IV)*, vol. 431, pp. 105–12, 1983.

[41] J.M. Jover and T. Kailath, 'A parallel architecture for Kalman filter measurement update and parameter estimation', *Automatica*, vol. 22, pp. 43–57, 1986.

[42] H. Lev-Ari, 'Modular architectures for adaptive multichannel lattice algorithms', *IEEE Trans. Acoust., Speech, signal Process.*, vol. ASSP-35, pp. 543–52, 1987.

[43] F. Ling and J.G. Proakis, 'A generalized multichannel least squares lattice algorithm based on sequential processing stages', *IEEE Trans. Acoust., Speech, Signal Process.*, vol. ASSP-32, pp. 381–90, 1984.

[44] H. Sakai, 'Circular lattice filtering using Pagano's method', *IEEE Trans. Acoust., Speech, Signal Process.*, vol. ASSP-30, pp. 279–87, 1982.

[45] P.S. Lewis, 'QR-based algorithms for multichannel adaptive least squares lattice filters', *IEEE Trans. Acoust., Speech, Signal Process.*, vol. ASSP-38, pp. 421–32, 1990.

[46] D.T.M. Slock, L. Chisci, H. Lev-Ari and T. Kailath, 'Modular and numerically stable fast transversal filters for multichannel and multiexperiment RLS', *IEEE Trans. Acoust., Speech, Signal Process*, vol. ASSP-40, pp. 784–802, 1992.

[47] B. Porat, B. Friedlander and M. Morf, 'Square-root covariance ladder algorithms', *IEEE Trans. Autom. Control*, vol. AC-27, pp. 813–29, 1982.

[48] J.M. Cioffi and T. Kailath, 'Windowed fast transversal filters adaptive algorithms with normalization', *IEEE Trans. Acoust., Speech, Signal Process.*, vol. ASSP-33, pp. 607–25, 1985.

[49] H. Lev-Ari, T. Kailath and J. Cioffi, 'Least-squares adaptive lattice and transversal filters: A unified geometric theory', *IEEE Trans. Inf. Theory*, vol. IT-30, pp. 222–36, 1984.

[50] B. Porat, M. Morf and D.R. Morgan, 'On the relationship among several square-root normalized ladder algorithms', in *Proc. CISS 81, Johns Hopkins University, Baltimore, MD*, pp. 496–51, 1981.

[51] D.T.M. Slock and T. Kailath 'A modular prewindowing framework for covariance FTF RLS algorithms', *Signal Process.*, vol. 28, pp. 47–61, 1992.

[52] B. Widrow and E. Walach, 'On the statistical efficiency of the LMS algorithm with nonstationary inputs', *IEEE Trans. Inf. Theory*, vol. IT-30, pp. 211–21, 1984. Special Issue on Adaptive Filtering.

[53] F. Ling and J.G. Proakis, 'Nonstationary learning characteristics of least squares adaptive estimation algorithms', in *Proc. ICASSP-84*, pp. 3.7.1–4, 1984.

[54] H. Kushner, *Approximation and Weak Convergence Methods for Random Processes, with Applications to Stochastic Systems Theory*, MIT Press: Cambridge, MA, 1984.

[55] V. Solo, 'The error variance of LMS with time-varying weights', *IEEE Trans. Signal Process.*, vol. SP-40, pp. 803–13, 1992.

[56] D.T.M. Slock, 'On the convergence behavior of the LMS and NLMS algorithms', in *Proc. EUSIPCO 90, V Eur. Signal Processing conf. Barcelona*, pp. 197–200, 1990.

[57] D.T.M. Slock, 'On the convergence behavior of the LMS and the normalized LMS algorithms', Submitted to *IEEE Trans.*

[58] B. Porat, 'Second-order equivalence of rectangular and exponential windows in least-squares estimation of Gaussian autoregressive processes', *IEEE Trans. Acoust., Speech, Signal Process.*, vol. ASSP-33, pp. 1209–12, 1985.

[59] L. Ljung and T. Söderström, *Theory and Practice of Recursive Identification*, MIT Press: Cambridge, MA, 1983.

[60] A. Benveniste, 'Design of adaptive algorithms for the tracking of time-varying systems', *Int. J. Adaptive Control Signal Process.*, vol. 1, pp. 3–29, 1987.

[61] A. Benveniste and G. Ruget, 'A measure of the tracking capability of recursive stochastic algorithms with constant gains', *IEEE Trans. Autom. Control*, vol. AC-27, pp. 639–49, 1982.

[62] N.J. Bershad and O.M. Macchi, 'Adaptive recovery of a chirped sinusoid in noise, part 2: Performance of the LMS algorithm', *IEEE Trans. Signal Process.*, vol. SP-39, pp. 595–602, 1991.

[63] O.M. Macchi and M.J. Bershad, 'Adaptive Recovery of a chirped sinusoid in noise, part 1: Performance of the RLS algorithm', *IEEE Trans. signal Process.*, vol. SP-39, pp. 583–94, 1991.

[64] N.J. Bershad, 'Analysis of the normalized LMS algorithm with Gaussian inputs', *IEEE Trans. Acoust., Speech, Signal Process.*, vol. ASSP-34, pp. 793–806, 1986.

[65] G. Carayannis, D. Manolakis and N. Kalouptsidis, 'A fast sequential algorithm for least squares filtering and prediction', *IEEE Trans. Acoust., Speech, Signal Process.*, vol. ASSP-31, pp. 1394–402, 1983.

[66] J.D. Markel and A.H. Gray Jr, *Linear Prediction of Speech*, Springer: New York, 1976.

[67] J. Makhoul, 'Linear prediction – A tutorial review', *Proc. IEEE*, vol. 63, pp. 501–80, 1975.

[68] N. Levinson, 'The Wiener r.m.s. (root-mean-square) error criterion in filter design and prediction', *J. Math. Phys.*, vol. 25, pp. 261–78, 1974.

[69] I. Schur, 'Über Potenzreihen die im Innern des Einheitskreises Beschränkt Sind', *Z. Reine Angew. Math.* vol. 147, pp. 205–32, 1917.

[70] T. Kailath, 'Linear estimation for stationary and near-stationary processes', in T. Kailath, (ed.) *Modern Signal Processing*, pp. 59–128, Hemisphere: New York, 1985.

[71] T. Kailath, 'Some new results and insights in linear least-squares estimation theory', in *Proc. 1975 IEEE–USSR Joint Workshop on Information Theory*, pp. 97–104, Moscow, USSR, 1975. Reprinted with corrections in [123].

[72] T. Kailath, L. Ljung and M. Morf, 'Generalized Krein–Levinson equations for efficient calculation of Fredholm resolvents of nondisplacement kernels', in I. Gohberg and M. Kac (eds.) *Topics in Functional Analysis, Dedicated to M.G. Krein on the Occasion of his 70th Birthday*, Academic Press: New York, 1978.

[73] B. Friedlander, T. Kailath, M. Morf and L. Ljung, 'Levinson- and Chandrasekhar-type equations for a general discrete-time linear estimation problem', *IEEE Trans. Autom. Control*, vol. AC-23, pp. 653–9, 1978.

[74] T. Kailath, S.Y. Kung and M. Morf, 'Displacement ranks of matrices and linear equations', *J. Math. Anal. Appl.*, vol. 68, pp. 295–407, 1979. (See also *Bull. Am. Math. Soc.*, vol. 1, pp. 769–73, 1979.)

[75] H. Lev-Ari and T. Kailath, 'Lattice filter parameterization and modeling of nonstationary processes', *IEEE Trans. Inf. Theory*, vol. IT-30, pp. 2–16, 1984.

[76] H. Lev-Ari and T. Kailath, 'Schur and Levinson algorithms for nonstationary processes', in *Proc. ICASSP-81 Conf.*, pp. 860–4, 1981.

[77] M. Morf, 'Fast algorithms for multivariable systems', PhD Thesis, Stanford University, 1974.

[78] N. Kalouptsidis and S. Theodoridis, 'Parallel implementation of efficient algorithms for filtering and prediction', *IEEE Trans. Acoust., Speech, Signal Process.*, vol. ASSP-35, pp. 1465–9, 1987.

[79] M. Morf, B. Dickinson, T. Kailath and A. Vieira, 'Efficient solution of covariance equations for linear prediction', *IEEE Trans. Acoust., Speech, Signal Process.*, vol. ASSP-25, pp. 423–33, 1977.

[80] M. Morf, A. Vieira and D. Lee 'Recursive least-squares ladder estimation algorithms', in *Proc. 1977 Conf. Decision Control, New Orleans, LA, Dec. 1977*, pp. 1074–8.

[81] D.T.L. Lee, M. Morf and B. Friedlander, 'Recursive least-squares ladder estimation algorithms', *IEEE Trans. Acoust., Speech, Signal Process.*, vol. ASSP-29, pp. 627–41, 1981.

[82] J.M. Cioffi, 'A fast QR/frequency-domain RLS adaptive filter', in *Proc. ICASSP-87 Conf.*, vol. 1, pp. 407–10, 1987.

[83] J.M. Cioffi, 'The fast adaptive rotors RLS algorithm', *IEEE Trans. Acoust., Speech, Signal Process.*, vol. ASSP-38, pp. 631–53, 1990.

[84] D.T.M. Slock, 'Reconciling fast RLS lattice and QR algorithms', in *Proc. ICASSP-90 Conf.*, pp. 1591–4, 1990.

[85] M. Morf, T. Kailath and L. Ljung, 'Fast algorithms for recursive identification', in *Proc. 1976 Conf. Decision Control, Florida, Dec. 1976*, pp. 916–21.

[86] G. Demoment and R. Reynaud, 'Fast minimum variance deconvolution', *IEEE Trans. Acoust., Speech, Signal Process.*, vol. ASSP-33, pp. 1324–6, 1985.

[87] G. Demoment, 'Équations de Chandrasekhar et algorithmes rapides pour le traitement du signal et des images', *Traitement du Signal*, vol. 6, pp. 103–115, 1989.

[88] A. Nehorai and M. Morf, 'A new derivation for fast recursive least squares and Levinson algorithms by the conjugate direction method', in *Proc. IEEE Int. Conf. ASSP, Boston, MA, 1983*, pp. 675–9.

[89] A. Houacine and G. Demoment 'Chandrasekhar adaptive regularizer for adaptive filtering', in *Proc. ICASSP-86 Conf.*, pp. 2967–70, 1986.

[90] A. Houacine and G. Demoment, 'Fast adaptive spectrum estimation: Bayesian approach and long AR filters', in *Proc. ICASSP-87 Conf.*, pp. 2085–8, 1987.

[91] A. Houacine and G. Demoment, 'A Bayesian method for adaptive spectrum estimation using high order autoregressive models', in J.G. McWhirter (ed.) *Mathematics in Signal Processing II*, pp. 311–23, Clarendon Press: Oxford, 1990.

[92] A. Houacine, 'Regularized fast recursive least squares algorithms for adaptive filtering', *IEEE Trans. Signal Process.*, vol. SP-39, pp. 860–71, 1991.

[93] X.C. Du, D. Saint-Félix and G. Demoment, 'Comparison between a factorization method and a partitioning method to derive invariant Kalman filters for fast image restoration', in T.S. Duranni *et al.* (eds.) *Mathematics in Signal Processing*, pp. 349–62, Clarendon Press: Oxford, 1987.

[94] D. Saint-Félix, X.C. Du and G. Demoment, 'Image restoration using a non-causal state space model and a fast 2-D Kalman filter', in T.S. Duranni *et al.* (eds.) *Mathematics in Signal Processing*, pp. 362–78, Clarendon Press: Oxford, 1987.

[95] N. Fortier, G. Demoment and Y. Goussard, 'Implementation and practical comparison of two estimators of the smoothing parameter in linear image restoration', in *Proc. ICASSP-88 Conf.*, pp. 1905–8, 1988.

[96] D.T.M. Slock, 'Fast algorithms for fixed-order recursive least-squares parameter estimation', PhD Thesis, Stanford University, 1989.

[97] G.S. Sidhu and T. Kailath, 'Development of new estimation algorithms by innovations analysis and shift-invariance properties', *IEEE Trans. Inf. Theory*, vol. IT-20, pp. 759–62, 1974.

[98] G.V. Moustakides and S. Theodoridis, 'Fast Newton transversal filters – A new class of adaptive estimation algorithms', *IEEE Trans. Signal Process.*, vol. SP-39, pp. 2184–93, 1991.

[99] T. Petillon, A. Gilloire and S. Theodoridis, 'Complexity reduction in fast RLS transversal adaptive filters with application to acoustic echo cancellation', in *Proc. IEEE Int. Conf. ASSP, San Fransisco, CA, March 1992.*

[100] D.T.M. Slock, 'Underdetermined growing and sliding window covariance fast transversal filter RLS algorithms', in *Proc. EUSIPCO 92, VIth Eur. Signal Processing Conf., Brussels, Belgium, 24–27 Aug. 1992.* pp. 1169–72.

[101] D.T.M. Slock, 'The block underdetermined covariance (BUC) fast transversal filter (FTF) algorithm for adaptive filtering', in *Proc. 26th Asilomar Conf. on Signals, Systems and Computers, Pacific Grove, CA, 26–28 Oct. 1992.*

[102] D. Luenberger, *Optimization by Vector Space Methods*, Wiley: New York, 1969.

[103] D.T.M. Slock and T. Kailath, 'Fast transversal filters with data sequence weighting', *IEEE Trans. Acoust., Speech, Signal Process.*, vol. ASSP-37, pp. 346–59, March 1989.

[104] T. Kailath, *Linear Systems*, Prentice Hall: Englewood Cliffs, NJ, 1980.

[105] D.T.M. Slock, 'An overview of some recent advances in fast RLS algorithms', in E. Deprettere and A.-J. van der Veen (eds.) *Algorithms and Parallel VLSI Architectures*, vol. A: Tutorials, Elsevier, Amsterdam, 1991.

[106] D.T.M. Slock, 'Backward consistency concept and round-off error propagation dynamics in recursive least-squares algorithms', *Opt. Eng.*, vol. 31, pp. 1153–69, 1992.

[107] A. Benallal and A. Gilloire, 'Improvements of the tracking capability of the numerically stable fast RLS algorithms for adaptive filtering', in *Proc. ICASSP-89 Conf.*, pp. 1031–4, 1989.

[108] N. Kalouptsidis, G. Carayannis and D. Manolakis, 'A fast covariance type algorithm for sequential LS filtering and prediction', *IEEE Trans. Autom. Control*, vol. AC-29, pp. 752–5, 1984.

[109] A. Dembo and J. Salz, 'On the least squares tap adjustment algorithm in adaptive digital echo cancellers', *IEEE Trans. Commun.*, vol. COM-38, pp. 622–8, 1990.

[110] H. Liu and Z. He, 'The RLS decision feedback equalizer with finite duration exponential windows', in *Proc. EUSIPCO 92, VIth European Signal Processing Conference, Brussels, Belgium, 24–27 Aug. 1992*, pp. 147–50.

[111] D.T.M. Slock and T. Kailath, 'A modular multichannel multiexperiment fast transversal filter RLS algorithm', *Signal Process.*, vol. 28, pp. 25–45, 1992.

[112] S. Karaboyas, N. Kalouptsidis and C. Caroubalos, 'Highly parallel multichannel LS algorithms and application to decision-feedback equalizers', *IEEE Trans. Acoust., Speech, Signal Process.*, vol. ASSP-37, pp. 1380–96, 1989.

[113] S. Karaboyas and N. Kalouptsidis, 'Efficient adaptive algorithms for ARX identification', *IEEE Trans. Signal Process.*, vol. SP-39, pp. 571–82, 1991.

[114] G. Glentis and N. Kalouptsidis, 'Efficient order recursive algorithms for multichannel LS filtering', in *Proc. ICASSP-90 Conf.*, pp. 1429–32, 1990.

[115] G. Glentis and N. Kalouptsidis, 'Fast adaptive algorithms for multichannel LS filtering', in *Proc. ISCAS-90 Conf.*, 1990.

[116] G. Glentis and N. Kalouptsidis, 'Fast algorithms for 2-D least squares FIR filtering', in *Proc. ICASSP-91 Conf.*, pp. 3009–12, 1991.

[117] G. Glentis and N. Kalouptsidis, 'Efficient order recursive algorithms for multichannel least squares filtering', *IEEE Trans. Signal Process.*, vol. SP-40, pp. 1354–74, 1992.

[118] G. Glentis and N. Kalouptsidis, 'Fast adaptive algorithms for multichannel filtering and system identification', *IEEE Trans. Signal Process.*, vol. SP-40, pp. 2433–58, 1992.

[119] G. Glentis and N. Kalouptsidis, 'Efficient adaptive algorithms for multichannel least squares filtering using a channel decomposition technique', *Computers and Electrical Engineering*, Special Issue on Adaptive Signal Processing, to appear.

[120] M.H. Verhaegen, 'Improved understanding of the loss-of-symmetry phenomenon in the conventional Kalman filter', *IEEE Trans. Autom. Control*, vol. AC-34, pp. 331–3, 1989.

[121] W.M. Gentleman, 'Least squares computations by Givens transformations without square roots', *J. Inst. Math. Applic.*, vol. 12, pp. 329–36, 1973.

[122] C.L. Lawson and R.J. Hanson, *Solving Least-Squares Problems*, Prentice Hall: Englewood Cliffs, NJ, 1974.

[123] T. Kailath, *Lectures on Wiener and Kalman filtering*, Springer: Wien–New York, 1981.

6

Lattice Algorithms

Fuyun Ling

Lattice, also called ladder, structures for adaptive estimation and filtering are the result of investigations of linear prediction speech modelling and parametric spectrum estimation [1–3]. Although these early efforts were focused on optimal modelling and estimation of stationary or quasistationary signals using the minimum mean square error (MMSE) criterion, it was soon recognized that the lattice form also provides a natural and computationally efficient structure for exact least-squares (LS) estimation of time series signals [4]. Since then, numerous research papers have been devoted to various lattice algorithms, their applications to adaptive filtering, estimation and parametric modelling and investigation of their properties [5,6]. It is not an exaggeration to say that the research of different aspects of adaptive lattice algorithms has constituted one of the most active areas in the field of adaptive filtering and estimation in the past decade.

Among many attractive properties of lattice algorithms, the most distinctive is *order recursiveness.* By definition, *the order of an estimator based on some optimization criterion is equal to the dimension of the data vector for the estimation.* An order recursive algorithm computes estimation errors of orders from 1 to p, where p is the maximum order of the estimator. In this chapter, we view the lattice algorithms as members of the more general class of order recursive (OR) estimation algorithms.

Consequently, the results obtained from the investigation of lattice algorithms are also, at least in principle, applicable to other order recursive estimation algorithms.

This chapter is organized as follows. In section 6.1, we first establish the statistical foundation of the order recursive estimation by deriving an order recursive relation of MMSE estimation. Then, based on this general relation, we show that the lattice structure is a canonical structure for order recursive estimation of time series data. In section 6.2 we extend the basic order recursive relations to LS estimation and a basic time recursive LS lattice algorithm is derived. In section 6.3, we derive and discuss variations of the basic LS lattice algorithm and extend the result to multichannel LS lattice algorithms. Gradient lattice algorithms are also derived, and thus we establish the relationship between the LMS algorithm and the lattice adaptive algorithm. Numerical, tracking and convergence properties of the LS lattice algorithms are discussed in section 6.4. Their relationships with other ORLS estimation algorithms are also briefly discussed. Finally, section 6.5 summarizes the chapter and gives some final comments.

6.1 The statistical view of order recursive estimation and the lattice structure

As stated above, lattice structures may result from either minimum mean square error (MMSE) estimation or LS estimation of time series signals. We begin our discussion by deriving from a general viewpoint the MMSE lattice estimator of *the order recursive MMSE estimation.*

An MMSE estimator is one way of describing the optimal Wiener filter. Since the basic MMSE estimation equations are obtained from the statistical properties of its input signals, the quantities involved in the MMSE estimation have clear physical, or statistical, interpretations. As a result, the MMSE estimation has special theoretical importance. For practical applications where the statistics of the underlying signal are not known a priori, the MMSE estimator may not be as easily applicable as other data adaptive estimators such as the LS estimator discussed in the next section. Nevertheless, as will become clear during the development of this chapter, the MMSE and the LS estimations have many common properties. The results obtained for the MMSE order recursive estimation and the MMSE lattice filter are also valid for their LS counterparts, and, more generally, are valid for any estimators with a quadratic cost function. We shall use the results obtained in this section to simplify the development of the LS lattice algorithms.

6.1.1 MMSE estimation and its geometric interpretation

Let us consider a desired signal sequence $\{y(k)\}$ and an m-dimensional data vector sequence $\{\mathbf{u}_m(k) \equiv ((u_1(k)\ u_2(k) \ldots u_m(k))^{\mathrm{T}}, k = -\infty, 0, \ldots, k, \ldots, \infty\}$, where $y(k)$ and

$u_i(k)$, $i = 1, 2, \ldots, m$, are samples of complex, jointly stationary stochastic processes, and, thus, are complex random variables. For the MMSE estimations, we only require that they are jointly stationary of second order. Namely, we have that $E[y^*(k)y(k-\tau)]$, $E[y^*(k)u_i(k-\tau)]$ and $E[u_i^*(k)u_j(k-\tau)]$ are functions of τ and are independent of k. The expectation operator $E[\cdot]$ is understood to generate the expectation, or ensemble average value, of the quantity inside the brackets.

MMSE ESTIMATION OF STATIONARY SIGNALS

The general MMSE estimation problem can be described as follows. We first form an estimate $\hat{y}(k)$ of $y(k)$ as a linear combination of $u_i(k)$, $i = 1, 2, \ldots, m$, such that

$$\hat{y}(k) = -\sum_{i=1}^{m} c_i u_i(k) \equiv -\mathbf{c}_m^{\mathrm{T}}\mathbf{u}_m(k)$$

where $\mathbf{c}_m$ is an m-dimensional coefficient vector. The objective of the MMSE estimation is to determine an optimal coefficient vector that minimizes the expectation value of $|e(k)|^2$, i.e. the squared error between $\hat{y}(k)$ and $y(k)$, defined as

$$\sigma_e^2 = E[|e(k)|^2] \equiv E[|y(k) - \hat{y}(k)|^2] = [|y(k) + \mathbf{c}_m^{\mathrm{T}}\mathbf{u}_m(k)|^2] \tag{6.1}$$

and, thus, to generate the optimal MMSE estimate $\hat{y}(k)$ of $y(k)$. We define the coefficient vector that minimizes σ_e^2 as the *MMSE coefficient vector*. It is known that the MMSE coefficient vector, denoted as $\mathbf{c}_{m,\mathrm{MMSE}}$, satisfies

$$\mathbf{c}_{m,\mathrm{MMSE}} = -R_{\mathbf{u}_m\mathbf{u}_m}^{-1}\mathbf{d}_m^{(y)} \tag{6.2}$$

where $R_{\mathbf{u}_m\mathbf{u}_m}$ is the autocorrelation matrix of $\mathbf{u}_m(k)$ defined as

$$R_{\mathbf{u}_m\mathbf{u}_m} = E[\mathbf{u}_m^*(k)\mathbf{u}_m^{\mathrm{T}}(k)] \tag{6.3}$$

and $\mathbf{d}_m^{(y)}$ is the cross-correlation vector between $\mathbf{u}_m(k)$ and $y(k)$, defined as

$$\mathbf{d}_m^{(y)} = E[\mathbf{u}_m^*(k)y(k)]$$

The *m*th-*order MMSE estimate of* $y(k)$ *based on the data vector* $\{\mathbf{u}_m k)\}$, denoted by $\hat{y}_{m,\mathrm{MMSE}}(k)$, is defined as

$$\hat{y}_{m,\mathrm{MMSE}}(k) = -\mathbf{c}_{m,\mathrm{MMSE}}^{\mathrm{T}}\mathbf{u}_m(k) \tag{6.4}$$

Similarly, we define $e_{m,\mathrm{MMSE}}^{(y)}(k)$, the MMSE error between $y(k)$ and $\hat{y}_{m,\mathrm{MMSE}}(k)$, to be *the mth-order MMSE estimation error of* $y(k)$ *based on the data vector* $\mathbf{u}_m(k)$. That is,

$$e_{m,\mathrm{MMSE}}^{(y)}(k) \equiv y(k) - \hat{y}_{m,\mathrm{MMSE}}(k) = y(k) + \mathbf{c}_{m,\mathrm{MMSE}}^{\mathrm{T}}\mathbf{u}_m(k) \tag{6.5}$$

The MMSE estimation is an optimization problem with quadratic cost function that leads to a set of linear equations, i.e. (6.2) given above. The results and conclusions obtained in this section can be generally applied to other optimization problems with an arbitrary quadratic cost function with minor modification. Below we discuss some properties of the MMSE estimation.

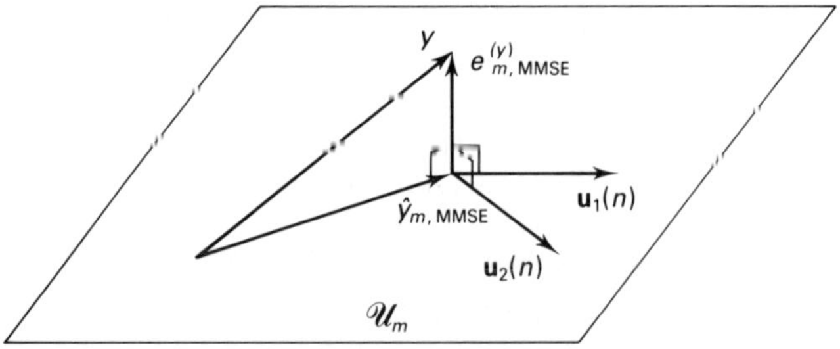

Figure 6.1 Geometric interpretation of orthogonal principle of MMSE estimation

ORTHOGONALITY PRINCIPLE

Premultiplying both sides of (6.5) by $\mathbf{u}_m^*(k)$ and then taking expectation values from (6.2) and 6.3) we obtain

$$E[\mathbf{u}_m^*(k)e^{(y)}_{m,\,\mathrm{MMSE}}(k)] = E[\mathbf{u}_m^*(k)y(k)] + E[\mathbf{u}_m^*(k)\mathbf{u}_m^T(k)]\mathbf{c}_{m,\,\mathrm{MMSE}}$$

$$= \mathbf{d}_m^{(y)} - \mathbf{R}_{\mathbf{u}_m\mathbf{u}_m} R^{-1}_{\mathbf{u}_m\mathbf{u}_m} \mathbf{d}_m^{(y)} = \mathbf{0} \tag{6.6}$$

where we have used the relationship $\mathbf{u}_m^T(k)\mathbf{c}_{m,\,\mathrm{MMSE}} = \mathbf{c}^{\mathrm{T}}_{m,\,\mathrm{MMSE}}\mathbf{u}_m(k)$. Equation (6.6) is called the *orthogonality principle* for MMSE estimation. In words, the orthogonality principle says that *the error from MMSE estimation is orthogonal to every component of the data vector*. This is generally true for any optimization using a quadratic cost funtion with proper interpretation.

GEOMETRIC INTERPRETATION

MMSE estimation or any quadratic minimization has a clear geometric interpretation. For MMSE estimation, the quantities $y(k)$ and $u_i(k)$ are random variables, which can be viewed as vectors in an infinite-dimensional linear space with inner products (Hilbert space), where an inner product is defined to be the expectation value of the product of two random variables (vectors), one of which is the complex conjugated. By definition, the length, or norm, of a vector in such a space is the square root of the inner product of the vector with itself.

The estimate $\hat{y}_{m,\,\mathrm{MMSE}}(k)$ is the projection of $y(k)$ on to the m-dimensional subspace $\mathcal{U}_m$ spanned by $u_i(k)$, $i = 1, 2, \ldots, m$. The MMSE error $e^{(y)}_{m,\,\mathrm{MMSE}}(k)$ is the shortest line from $y(k)$ to the subspace $\mathcal{U}_m$. Thus, the orthogonality principle corresponds to the geometric fact that the shortest line from a point to a subspace is the line that is perpendicular to the subspace. The MMSE error $e^{(y)}_{m,\,\mathrm{MMSE}}(k)$ is called the *orthogonal residual of* $y(k)$ outside $\mathcal{U}_m$. Figure 6.1 depicts the orthogonality principle of MMSE estimation.

EXPRESSION OF ERROR AUTOCORRELATIONS

From (6.1), we can express the minimum mean squared error, or the auto correlation of error $e^{(y)}_{m,\,\mathrm{MMSE}}(k)$, denoted by $\alpha_m^{(yy)}$, as

$$\alpha_m^{(yy)} \equiv \sigma_{m,\mathrm{MMSE}}^{2(y)} = E[\,|\,e_{m,\mathrm{MMSE}}^{(y)}(k)\,|^2] = E[e_{m,\mathrm{MMSE}}^{(y)*}(k)\{y(k) + \mathbf{c}_m^{\mathrm{T}}\mathbf{u}_m(k)\}]$$

It follows from the orthogonality principle that

$$\begin{aligned}\alpha_m^{(yy)} &= E[e_{m,\mathrm{MMSE}}^{(y)*}(k)y(k)] = E[\,|\,y(k)\,|^2] + E[\mathbf{c}_{m,\mathrm{MMSE}}^{*\mathrm{T}}\mathbf{u}_m^*(k)y(k)]\\ &= \rho_0^y + \mathbf{c}_{m,\mathrm{MMSE}}^{\mathrm{H}}\mathbf{d}_m^{(y)} = \rho_0^y - \mathbf{d}_m^{(y)H}R_{\mathbf{u}_m\mathbf{u}_m}^{-1}\mathbf{d}_m^{(y)}\end{aligned} \tag{6.7}$$

where superscript H denotes the complex conjugate and transposing operation, or Hermitian, of a vector or matrix, and $\rho_0^y \equiv E[\,|\,y(k)\,|^2]$ denotes the autocorrelation, or variance, of $y(k)$. Equation (6.7) is a useful expression and will be used in further derivations.

6.1.2 Order recursive MMSE estimation

In section 2.7, it was demonstrated that the Wiener filtering problem can be realized using a lattice structure in an order recursive fashion. In this section, we shall derive the lattice form for MMSE estimation by viewing it as a special case of the more general order recursive MMSE estimation. To begin with, we consider a general order recursive relation of MMSE estimation given by the following theorem.

THEOREM 6.1
(Order recursive relation for MMSE estimation) By partitioning the $(m + 1)$-dimensional data vector $\mathbf{u}_{m+1}(k)$ as

$$\mathbf{u}_{m+1}(k) = (\mathbf{u}_m^{\mathrm{T}}(k),\, u'(k))^{\mathrm{T}}$$

the $(m + 1)$th-order MMSE estimation error of $y(k)$, based on the data vector $\mathbf{u}_{m+1}(k)$, i.e. $e_{m+1,\mathrm{MMSE}}^{(y)}(k)$, can be computed as

$$e_{m+1,\mathrm{MMSE}}^{(y)}(k) = e_{m,\mathrm{MMSE}}^{(y)}(k) - [\alpha_m^{(yu')}/\alpha_m^{(u'u')}]e_{m,\mathrm{MMSE}}^{(u')}(k) \tag{6.8}$$

where $e_{m,\mathrm{MMSE}}^{(y)}(k)$ and $e_{m,\mathrm{MMSE}}^{(u')}(k)$ are the mth-order MMSE estimation errors of $y(k)$ and $u'(k)$ based on the data vector $\mathbf{u}_m(k)$, respectively, and $\alpha_m^{(u'u')}$ denotes the autocorrelation of $e_{m,\mathrm{MMSE}}^{(u')}(k)$ defined by

$$\alpha_m^{(u'u')} \equiv E[\,|\,e_{m,\mathrm{MMSE}}^{(u')}(k)\,|^2]$$

and $\alpha_m^{(yu')}$ denotes the cross-correlation between $e_{m,\mathrm{MMSE}}^{(y)}(k)$ and $e_{m,\mathrm{MMSE}}^{(u')}(k)$, defined as

$$\alpha_m^{(yu')} \equiv E[e_{m,\mathrm{MMSE}}^{(y)}(k)e_{m,\mathrm{MMSE}}^{(u')*}(k)] \tag{6.9}$$

Theorem 6.1 shows that a higher-order MMSE estimation error can be computed from two lower-order MMSE errors. It is therefore called the *order recursive relation for* MMSE estimation.

Proof The mth-order MMSE errors $e_{m,\mathrm{MMSE}}^{(y)}(k)$ and $e_{m,\mathrm{MMSE}}^{(u')}(k)$ are given by (6.5) and

$$e_{m,\mathrm{MMSE}}^{(u')}(k) = u'(k) + \mathbf{c}_{m,\mathrm{MMSE}}^{(u')\mathrm{T}}\mathbf{u}_m(k) \tag{6.10}$$

where $\mathbf{c}^{(u')}_{m,\text{MMSE}}$ is the MMSE coefficient vector that satisfies

$$\mathbf{c}^{(u')}_{m,\text{MMSE}} = -R^{-1}_{\mathbf{u}_m\mathbf{u}_m}\mathbf{d}^{(u')}_m$$

where

$$\mathbf{d}^{(u')}_m = E[\mathbf{u}^*_m(k)u'(k)]$$

By using (6.5) and (6.10) it can be shown that (6.8) can be rewritten as

$$e^{(y)}_{m+1,\text{MMSE}}(k) = y(k) + \begin{pmatrix} \mathbf{c}^{(y)}_{m,\text{MMSE}} - [\alpha^{(yu')}_m/\alpha^{(u'u')}_m]\mathbf{c}^{(u')}_{m,\text{MMSE}} \\ -\alpha^{(yu')}_m/\alpha^{(u'u')}_m \end{pmatrix}^{\text{T}} \begin{pmatrix} \mathbf{u}_m(k) \\ u'(k) \end{pmatrix} \tag{6.11}$$

From (6.11) and the partitioning of $\mathbf{u}_{m+1}(k)$ we observe that $e^{(y)}_{m+1,\text{MMSE}}(k)$ has the form of an estimation error of $y(k)$ based on $\mathbf{u}_{m+1}(k)$. In order to show that $e^{(y)}_{m+1,\text{MMSE}}(k)$ in (6.8) is indeed the $(m+1)$th-order MMSE estimation error of $y(k)$, from the orthogonality principle, it suffices to show that $e^{(y)}_{m+1,\text{MMSE}}(k)$ is orthogonal to $\mathbf{u}_{m+1}(k)$.

From the orthogonality principle, the MMSE errors $e^{(y)}_{m,\text{MMSE}}(k)$ and $e^{(u')}_{m,\text{MMSE}}(k)$ are orthogonal to the data vector $\mathbf{u}_m(k)$. Since $e^{(y)}_{m+1,\text{MMSE}}(k)$ is a linear combination of these two errors, it is orthogonal to $\mathbf{u}_m(k)$. As a result, to prove orthogonality with respect to $\mathbf{u}_{m+1}(k)$, we only need to show that $e^{(y)}_{m+1,\text{MMSE}}(k)$ is also orthogonal to $u'(k)$. Namely,

$$E[e^{(y)*}_{m+1,\text{MMSE}}(k)u'(k)] = 0 \tag{6.12}$$

By replacing $y(k)$ by $u'(k)$ in (6.7) we obtain

$$E[e^{(u')*}_{m,\text{MMSE}}(k)u'(k)] = \alpha^{(u'u')}_m \tag{6.13}$$

and, similarly, we have

$$E[e^{(y)}_{m,\text{MMSE}}(k)u'^*(k)] = \alpha^{(yu')}_m \tag{6.14}$$

From (6.8), (6.13) and (6.14) we obtain

$$\begin{aligned} E[e^{(y)}_{m+1,\text{MMSE}}(k)u'^*(k)] &= E[e^{(y)}_{m,\text{MMSE}}(k)u'^*(k)] - [\alpha^{(yu')}_m/\alpha^{(u'u')}_m]E[e^{(u')}_{m,\text{MMSE}}(k)u'^*(k)] \\ &= \alpha^{(yu')}_m - [\alpha^{(yu')}_m/\alpha^{(u'u')}_m]\alpha^{(u'u')}_m = \alpha^{(yu')}_m - \alpha^{(yu')}_m = 0. \end{aligned}$$

This completes the proof of Theorem 6.1 □

The order recursive relation (6.8) is the basis for order recursive MMSE estimation algorithms. It can be extended to any estimation based on a quadratic cost function. This relation and its extensions shall be repeatedly used in the derivation of MMSE and LS lattice algorithms in this chapter. To emphasize its importance, we first make the following remarks.

Remarks

1. 1By comparing (6.8)–(6.9) with (6.2)–(6.5) we note that (6.8) can be viewed as a '*scalar* MMSE estimation', where $e^{(y)}_{m,\text{MMSE}}(k)$ behaves as the 'desired signal' and $e^{(u')}_{m,\text{MMSE}}(k)$ behaves as the '(scalar) data'. However, we should note that such a scalar MMSE

estimation is only meaningful when the 'desired signal' i.e. $e^{(y)}_{m,\mathrm{MMSE}}(k)$ in (6.8) and the 'scalar data', i.e. $e^{(u')}_{m,\mathrm{MMSE}}(k)$, are both MMSE estimation errors based on the *same* data vector, i.e. $\mathbf{u}_m(k)$, up to a permutation of the elements of the data vector. If this condition is not satisfied, (6.8) will be incorrect or meaningless.

2. The resulting $(m+1)$th-order MMSE error depends on which of the two mth MMSE estimation errors is used as the 'desired signal' and which is used as the 'scalar data'. If we switch the roles of $e^{(y)}_{m,\mathrm{MMSE}}(k)$ and $e^{(u')}_{m,\mathrm{MMSE}}(k)$ in the scalar MMSE estimation described above, we will obtain an $(m+1)$th MMSE estimation error whose desired signal and data vector are $u'(k)$ and $[\mathbf{u}_m(k), y(k)]^{\mathrm{T}}$ respectively.
3. The order recursion defined in Theorem 6.1 is called the *first-order recursion* because the difference in the orders of the MMSE errors $e^{(y)}_{m+1,\mathrm{MMSE}}$ and $e^{(y)}_{m,\mathrm{MMSE}}$ is equal to one. A first-order recursion only involves scalar operations. However, in some applications it is convenient to define higher-order recursions for MMSE estimation. The realization of high-order recursions requires matrix and vector operations. Nevertheless, as will be shown later, it is possible to break down the high-order recursion into multiple, first-order recursions.
4. Although in the partitioning of the data vector $\mathbf{u}_{m+1}(k)$ described in Theorem 6.1 it appears that $u'(k)$ is the last element of $\mathbf{u}_{m+1}(k)$ and $\mathbf{u}_m(k)$ consists of the first m elements of $\mathbf{u}_{m+1}(k)$, the order recursive relation given by (6.8) is valid under more relaxed conditions. Specifically, (6.8) is valid when $u'(k)$ is any one of the elements of $\mathbf{u}_{m+1}(k)$ and $\mathbf{u}_m(k)$ consists of the remaining m elements. Furthermore, the elements of the data vector $\mathbf{u}_m$ may differ by a permutation for the estimation of $e^{(y)}_{m,\mathrm{MMSE}}(k)$ and $e^{(u')}_{m,\mathrm{MMSE}}(k)$. These relaxations can be understood from the proof of Theorem 6.1, where we used only the orthogonality principle, which does not depend on how the elements in the data vector are arranged.
5. It is easy to verify that the autocorrelation $\alpha^{(u'u')}_m$ used above is also the variance, or power of the mth-order MMSE error $e^{(u')}_{m,\mathrm{MMSE}}(k)$. It is called the MMSE variance, or power, of the estimation error. Similar notations will be used to denote the MMSE variance, or power, of the forward/backward estimation errors in the MMSE lattice algorithms discussed later.

GEOMETRIC INTERPRETATION

The order recursive relation for MMSE estimation has a clear geometric meaning. It corresponds to a decomposition of an orthogonal projection on to a linear space into two projections: a projection on to a subspace of the linear space and a projection on to the orthogonal complement of the subspace. Such a decomposition is called the *decomposition of orthogonal projection*. To show this, we first subtract $y(k)$ from both sides of (6.8) and then change their sign. We obtain

$$\hat{y}_{m+1,\mathrm{MMSE}}(k) = \hat{y}_{m,\mathrm{MMSE}}(k) + [\alpha^{(yu')}_m/\alpha^{(u'u')}_m]e^{(u')}_{m,\mathrm{MMSE}}(k) \tag{6.15}$$

Note that, by definition, $\hat{y}_{m+1,\mathrm{MMSE}}(k)$ and $\hat{y}_{m,\mathrm{MMSE}}(k)$ are the orthogonal projections of $y(k)$ on to $\mathscr{U}_{m+1}$ and its subspace $\mathscr{U}_m$ respectively. It can be shown that the last term on the right-hand side of (6.15) represents the orthogonal projection of $y(k)$ on to the orthogonal complement of $\mathscr{U}_m$ spanned by $e^{(u')}_{m,\text{MMSE}}(k)$. Thus, we conclude that (6.15) is the MMSE estimation form of decomposition of an orthogonal projection. It is possible to derive the order recursion of MMSE estimation in a purely geometric manner.

6.1.3 Lattice structure for MMSE estimation

Let us consider the following special kind of estimation problems in which the data vector $\mathbf{u}_m(k)$ is a segment of a time series $\{u(k), k = -\infty, 0, \ldots, k, \ldots, \infty\}$. That is,

$$\mathbf{u}_m(k) = (u(k), u(k-1), \ldots, u(k-m+1))^{\mathrm{T}}$$

Below, we will show that the lattice form is a canonical structure of the order recursive MMSE estimations with such data vectors. To facilitate understanding and the derivation of the MMSE lattice estimation algorithm, we first consider a special class of estimation problems called *linear prediction.*

FORWARD AND BACKWARD LINEAR PREDICTION

Linear prediction is one of the most important problems in time series analysis. It is also widely used in speech coding and high-resolution spectral analysis. Thus, the subject of linear prediction is not only crucial for the development of the MMSE lattice estimation algorithm, but is also important in itself in practical applications.

There are two types of linear prediction: the *forward prediction* and the *backward prediction.* In an mth-order forward prediction we form a linear estimation of a sample $u(k)$ in a time series $\{u(k)\}$ based on its m previous samples in the series. In an MMSE forward linear prediction, we determine a coefficient vector that minimizes the expectation of the squared mth-order foward prediction error $e^{\mathrm{f}}_m(k)$ given by

$$\sigma^{(\mathrm{f})2}_m = E[\,|e^{\mathrm{f}}_m(k)|^2] \equiv E[\,|u(k) - \hat{u}^{\mathrm{f}}_m(k)|^2]$$

where

$$\hat{u}^{\mathrm{f}}_m(k) = -\mathbf{a}^{\mathrm{T}}_m \mathbf{u}_m(k-1)$$

and $\mathbf{a}_m$ is the forward prediction coefficient vector.

From the general MMSE estimation problem discussed in section 6.1.1, it follows that the optimal MMSE forward prediction coefficient vector $\mathbf{a}_{m,\text{MMSE}}$ satisfies

$$\mathbf{a}_{m,\text{MMSE}} = -R^{-1}_{\mathbf{u}_m\mathbf{u}_m}\mathbf{d}^{\mathrm{f}}_m$$

where $R^{\mathrm{T}}_{\mathbf{u}_m\mathbf{u}_m} = E[\mathbf{u}^*_m(k-1)\mathbf{u}^{\mathrm{T}}_m(k-1)] = E[\mathbf{u}^*_m\mathbf{u}^{\mathrm{T}}_m]$ as defined in (6.3) and $\mathbf{d}^{f}_m$ is the correlation vector between $u(k)$ and $\mathbf{u}_m(k-1)$, defined as

$$\mathbf{d}^{\mathrm{f}}_m = E[u(k)\mathbf{u}^*_m(k-1)] = (\rho_1, \rho_2, \cdots, \rho_m)^{\mathrm{T}} \tag{6.16}$$

where $\rho_i \equiv E[u(k)u^*(k-i)]$.The mth-order MMSE forward prediction error is given by

$$e^{\mathrm{f}}_{m,\mathrm{MMSE}}(k) = u(k) - \hat{u}^{\mathrm{f}}_{m,\mathrm{MMSE}}(k) \equiv u(k) + \mathbf{a}^{\mathrm{T}}_{m,\mathrm{MMSE}}\mathbf{u}_m(k-1)$$

In an mth-order backward linear prediction we use a linear combination of m consecutive samples, $u(k), u(k-1), \ldots, u(k-m+1)$, to form an estimate of the sample preceding them, i.e. $u(k-m)$. In an MMSE backward prediction, we determine an MMSE backward prediction coefficient vector $\mathbf{b}_{m,\mathrm{MMSE}}$ that minimizes the expectation value of the squared mth-order backward prediction error, $e^{\mathrm{b}}_m(k)$, given by

$$\sigma^{(\mathrm{b})2}_m = E[\,|e^{\mathrm{b}}_m(k)|^2] \equiv E[\,|u(k-m) - \hat{u}^{\mathrm{b}}_m(k-m)|^2] \equiv E[\,|u(k-m) + \mathbf{b}^{\mathrm{T}}_m\mathbf{u}_m(k)|^2]$$

The MMSE backward prediction coefficient vector $\mathbf{b}_{m,\mathrm{MMSE}}$ satisfies

$$\mathbf{b}_{m,\mathrm{MMSE}} = -R^{-1}_{\mathbf{u}_m\mathbf{u}_m}\mathbf{d}^{\mathrm{b}}_m \tag{6.17}$$

where $\mathbf{d}^{\mathrm{b}}_m$ is the correlation vector between $u(k-m)$ and $\mathbf{u}_m(k)$, defined by

$$\mathbf{d}^{\mathrm{b}}_m = E[u(k-m)\mathbf{u}^*_m(k)] = (\rho^*_m, \rho^*_{m-1}, \cdots, \rho^*_1)^{\mathrm{T}} \tag{6.18}$$

The last equality in (6.18) follows from the relation

$$E[u(k-m)u^*(k-i)] = E[u(k-m+i)u^*(k)] = E[\{u(k)u^*(k-m+i)\}^*] \equiv \rho^*_{m-i}$$

The mth-order MMSE backward prediction error is defined by

$$e^{\mathrm{b}}_{m,\mathrm{MMSE}}(k) = u(k-m) - \hat{u}^{\mathrm{b}}_{m,\mathrm{MMSE}}(k) = u(k-m) + \mathbf{b}^{\mathrm{T}}_{m,\mathrm{MMSE}}\mathbf{u}_m(k)$$

From (6.16) and (6.18), we have

$$\mathbf{d}^{\mathrm{b}}_m = J\mathbf{d}^{\mathrm{f}*}_m \tag{6.19}$$

where J is the exchange matrix defined as

$$J = \begin{pmatrix} 0 & \cdots & 0 & 1 \\ 0 & \cdots & 1 & 0 \\ \vdots & & \vdots & \vdots \\ 1 & \cdots & 0 & 0 \end{pmatrix}$$

It is easy to verify that $JJ = I$.

From the definition of $R_{\mathbf{u}_m\mathbf{u}_m}$ it can be shown that

$$R_{\mathbf{u}_m\mathbf{u}_m} = [JR_{\mathbf{u}_m\mathbf{u}_m}J]^* \tag{6.20}$$

By substituting (6.20) and (6.19) into (6.17), we obtain

$$\mathbf{b}_{m,\mathrm{MMSE}} = -J[R^*_{\mathbf{u}_m\mathbf{u}_m}]^{-1}JJ\mathbf{d}^{\mathrm{f}*}_m = -J[R^{-1}_{\mathbf{u}_m\mathbf{u}_m}\mathbf{d}^{\mathrm{f}}_m]^* = J\mathbf{a}^*_{m,\mathrm{MMSE}}$$

That is, the backward MMSE prediction coefficient vector is equal to the complex conjugate of the forward MMSE prediction coefficient vector in reverse order.

ORDER RECURSIVE LINEAR PREDICTION AND THE LATTICE STRUCTURE

Let us first consider an $(m+1)$th-order MMSE forward linear prediction problem. To apply the general order recursive relation of MMSE estimation given by Theorem 6.1 to the $(m+1)$th-order forward prediction, we first take the upper partition of the data vector $\mathbf{u}_{m+1}(k-1)$ of the forward prediction as

$$\mathbf{u}_{m+1}(k-1) = \begin{pmatrix} \mathbf{u}_m(k-1) \\ u(k-m-1) \end{pmatrix} \tag{6.21}$$

From Theorem 6.1 and using (6.21) it follows that the $(m+1)$th MMSE forward prediction errors $e^{\text{f}}_{m+1,\text{MMSE}}(k)$ can be computed from two sets of MMSE errors: the MMSE errors of $u(k)$ and $u(k-m-1)$ based on the same data vector $\mathbf{u}_m(k-1)$. From the theorem, the first error, which corresponds to $\{e^{(y)}_{m,\text{MMSE}}(k)\}$ in (6.8), is the MMSE error of $u(k)$ based on $\mathbf{u}_m(k-1)$. By definition, it is the mth-order MMSE forward prediction error. The second error, which corresponds to $e^{(u')}_m(k)$ in (6.8), is the MMSE error of $u(k-m-1)$ based on $\mathbf{u}_m(k-1)$. By definition, it is the backward prediction error $e^{\text{b}}_{m,\text{MMSE}}(k-1)$. Specifically, the order recursion of the forward prediction error is expressed as

$$e^{\text{f}}_{m+1,\text{MMSE}}(k) = e^{\text{f}}_{m,\text{MMSE}}(k) - [\beta^{\text{f}}_m/\alpha^{\text{b}}_m]e^{\text{b}}_{m,\text{MMSE}}(k-1) \equiv e^{\text{f}}_{m,\text{MMSE}}(k) + k^{\text{f}}_m e^{\text{b}}_{m,\text{MMSE}}(k-1) \tag{6.22}$$

where

$$\beta^{\text{f}}_m = E[e^{\text{f}}_{m,\text{MMSE}}(k)e^{\text{b}*}_{m,\text{MMSE}}(k-1)] \tag{6.23}$$

and

$$\alpha^{\text{b}}_m = E[|e^{\text{b}}_{m,\text{MMSE}}(k)|^2] = E[|e^{\text{b}}_{m,\text{MMSE}}(k-1)|^2] \tag{6.24}$$

Similarly, we can derive the order recursion for MMSE backward prediction. The desired signal of an $(m+1)$th backward MMSE prediction is $u(k-m-1)$ and its data vector is $\mathbf{u}_{m+1}(k)$, the lower partition of which can be taken as

$$\mathbf{u}_{m+1}(k) = \begin{pmatrix} u(k) \\ \mathbf{u}_m(k-1) \end{pmatrix} \tag{6.25}$$

According to Theorem 6.1, we can compute the $(m+1)$th MMSE backward prediction error $e^{\text{b}}_{m+1,\text{MMSE}}(k)$ from the MMSE errors of $u(k-m-1)$ and $u(k)$ based on the data vector $\mathbf{u}_m(k-1)$. Following similar arguments as above we obtain

$$e^{\text{b}}_{m+1,\text{MMSE}}(k) = e^{\text{b}}_{m,\text{MMSE}}(k-1) - [\beta^{\text{b}}_m/\alpha^{\text{f}}_m]e^{\text{f}}_{m,\text{MMSE}}(k) = e^{\text{b}}_{m,\text{MMSE}}(k-1) + k^{\text{b}}_m e^{\text{f}}_{m,\text{MMSE}}(k) \tag{6.26}$$

where

$$\beta^{\text{b}}_m = E[e^{\text{f}*}_{m,\text{MMSE}}(k)e^{\text{b}}_{m,\text{MMSE}}(k-1)] \tag{6.27}$$

and

$$\alpha^{\text{f}}_m = E[|e^{\text{f}}_{m,\text{MMSE}}(k)|^2]$$

We have shown that both the $(m+1)$th forward and backward prediction errors can be computed through scalar MMSE estimations using the lower-order MMSE forward and backward prediction errors. The results given above are valid for $m \geqslant 1$. When $m = 0$, we can directly use the regular scalar MMSE estimation to compute the first-order prediction errors $e^{\text{f}}_1(k)$ and $e^{\text{b}}_1(k)$, because the desired signal and the

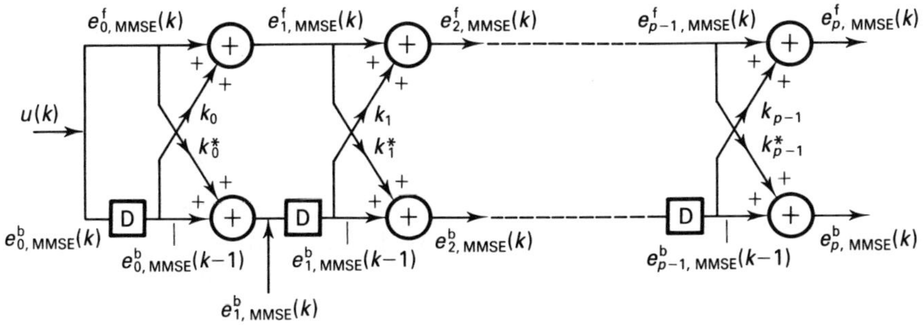

Figure 6.2 Lattice structure for MMSE linear prediction

data – $\{u(k)\}$ and $\{u(k-1)\}$ for forward prediction, and $\{u(k-1)\}$ and $\{u(k)\}$ for backward prediction – are both given. For consistency of notation, we denote

$$e^{f}_{0}(k) \equiv e^{b}_{0}(k) \equiv u(k) \tag{6.28}$$

Based on the above discussion, we conclude that a pth-order forward and/or backward prediction can be implemented in a purely order recursive fashion through a series of order recursive forward and backward predictions for $m = 0$ to $p - 1$. The structure of such a prediction algorithm is depicted in Figure 6.2, and it takes the so-called *lattice form* or *lattice structure*. A predictor implemented in such a way is called a *lattice predictor* which generates both forward and backward prediction errors of orders from 1 to the maximum order p.

In the literature, the scalar estimation coefficients $k^{f}_{m} \equiv -\beta^{f}_{m}/\alpha^{b}_{m}$ in (6.22) and $k^{b}_{m} \equiv -\beta^{b}_{m}/\alpha^{f}_{m}$ in (6.26) are called the forward and backward reflection coefficients respectively.

From (6.7) it follows that

$$\alpha^{f}_{m} = \rho^{u}_{0} + \mathbf{a}^{H}_{m,\text{MMSE}}\mathbf{d}^{f}_{m} = \rho^{u}_{0} - \mathbf{d}^{fH}_{m}R^{-1}_{\mathbf{u}_m\mathbf{u}_m}\mathbf{d}^{f}_{m} \tag{6.29}$$

and

$$\alpha^{b}_{m} = \rho^{u}_{0} + \mathbf{b}^{H}_{m,\text{MMSE}}\mathbf{d}^{b}_{m} = \rho^{u}_{0} - \mathbf{d}^{bH}_{m}R^{-1}_{\mathbf{u}_m\mathbf{u}_m}\mathbf{d}^{b}_{m} \tag{6.30}$$

where we have defined

$$\rho^{u}_{0} \equiv E[\,|u(k)|^{2}] = E[\,|u(k-m)|^{2}]$$

due to the stationarity of $u(k)$. Note that both α^{f}_{m} and α^{b}_{m} are real.

By substituting (6.19) and (6.20) into (6.30) and comparing it with (6.29), we obtain $\alpha^{b}_{m} = \alpha^{f}_{m}$. Thus, we define

$$\alpha_{m} \equiv \alpha^{f}_{m} = \alpha^{b}_{m} \tag{6.31}$$

Furthermore, by comparing (6.27) with (6.23), it is easy to see that β^{b}_{m} is equal to the complex conjugate of β^{f}_{m}, i.e.

$$\beta_m^{\mathrm{b}} = \beta_m^{\mathrm{f}*}$$

It follows that $k_m^{\mathrm{b}} = k_m^{\mathrm{f}*}$, and we define

$$k_m = k_m^{\mathrm{f}} = k_m^{\mathrm{b}*}$$

JOINT MMSE LATTICE ESTIMATION

Now let us return to the general order recursive MMSE estimation problem discussed at the beginning of this section, namely the MMSE estimation of $y(k)$ based on the data vector $\mathbf{u}_m(k)$. It is also called the *joint-process estimation*, because we view $\{y(k)\}$ *as a joint process of* $\{u(k)\}$. The corresponding estimation error is called the *joint estimation error*. Below, we demonstrate how the joint MMSE estimation can be solved using the lattice structure developed above.

From Theorem 6.1 and the upper partitioning of $\mathbf{u}_{m+1}(k)$ the $(m + 1)$th-order joint MMSE estimation $e_{m+1,\mathrm{MMSE}}^{(y)}(k)$ can be computed using two mth-order MMSE estimation errors through a scalar MMSE estimation. They are the MMSE estimation errors of $y(k)$ and $u(k - m)$ based on the data vector $\mathbf{u}_m(k)$, i.e. by definition the mth-order joint MMSE estimation error $e_{m,\mathrm{MMSE}}^{(y)}(k)$ and the MMSE backward prediction error $e_{m,\mathrm{MMSE}}^{\mathrm{b}}(k)$ respectively. The order recursion for the joint MMSE estimation error is given by

$$e_{m+1,\mathrm{MMSE}}^{(y)}(k) = e_{m,\mathrm{MMSE}}^{(y)}(k) - [\beta_m^{(y)}/\alpha_m]e_{m,\mathrm{MMSE}}^{\mathrm{b}}(k) \tag{6.32}$$

where $\beta_m^{(y)}$ is cross-correlation between $e_{m,\mathrm{MMSE}}^{(y)}(k)$ and $e_{m,\mathrm{MMSE}}^{\mathrm{b}}(k)$ given by

$$\beta_m^{(y)} = E[e_{m,\mathrm{MMSE}}^{(y)}(k)e_{m,\mathrm{MMSE}}^{b*}(k)] \tag{6.33}$$

and $\alpha_m \equiv \alpha_m^{\mathrm{b}}$ is defined in (6.24). Similar to the forward and backward reflection coefficients, we define the joint estimation coefficient $k_m^{(y)}$ as

$$k_m^{(y)} = -\beta_m^{(y)}/\alpha_m \tag{6.34}$$

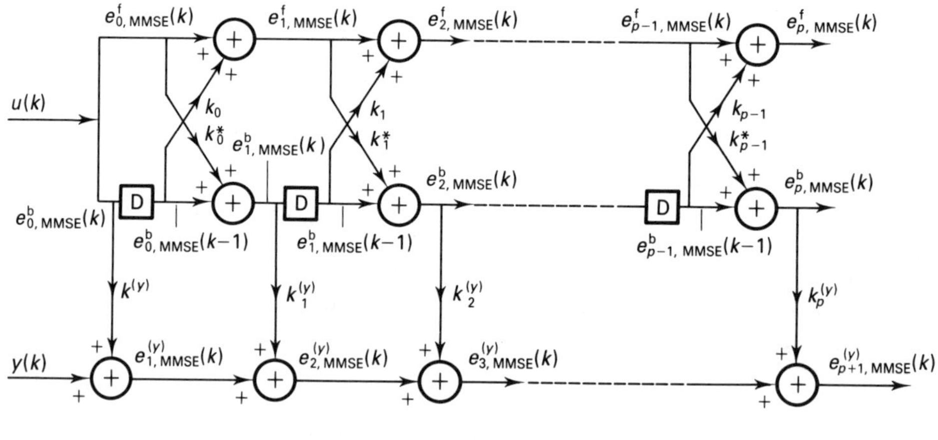

Figure 6.3 MMSE lattice joint-process estimator

Similar to (6.28) we define $e_0^{(y)}(k) \equiv y(k)$. Thus one can realize a pth-order joint estimator using a $(p-1)$th lattice predictor, providing the backward errors, plus p joint estimation stages. By utilizing (6.32)–(6.34) the joint MMSE lattice algorithm results, which are given in Table 6.1 and depicted in Figure 6.3. Note that if only the prediction parameters are of interest then the first five equations of the main loop, corresponding to the lattice part, are used.

Table 6.1 MMSE lattice prediction/joint estimation algorithm

INITIALIZATION

$e^{\mathrm{f}}_{0,\mathrm{MMSE}}(k) = e^{\mathrm{b}}_{0,\mathrm{MMSE}}(k) = u(k)$

$e^{(y)}_{0,\mathrm{MMSE}}(k) = y(k)$

FOR $m = 0$ TO $p-1$ DO

(Prediction)

$\alpha_m = E[\,|e^{\mathrm{f}}_{m,\mathrm{MMSE}}(k)|^2] = E[\,|e^{\mathrm{b}}_{m,\mathrm{MMSE}}(k)|^2]$

$\beta^{\mathrm{f}}_m = E[\,|e^{\mathrm{f}}_{m,\mathrm{MMSE}}(k)e^{\mathrm{b}*}_{m,\mathrm{MMSE}}(k-1)|\,]$

$k_m = -\beta^{\mathrm{f}}_m/\alpha_m$

$e^{\mathrm{f}}_{m+1,\mathrm{MMSE}}(k) = e^{\mathrm{f}}_{m,\mathrm{MMSE}}(k) + k_m e^{\mathrm{b}}_{m,\mathrm{MMSE}}(k-1)$

$e^{\mathrm{b}}_{m+1,\mathrm{MMSE}}(k) = e^{\mathrm{b}}_{m,\mathrm{MMSE}}(k-1) + k^*_m e^{\mathrm{f}}_{m,\mathrm{MMSE}}(k)$

(Joint estimation)

$\beta^{(y)}_m = E[\,|e^{(y)}_{m,\mathrm{MMSE}}(k)e^{\mathrm{b}*}_{m,\mathrm{MMSE}}(k)|\,]$

$k^{(y)}_m = -\beta^{(y)}_m/\alpha_m$

$e^{(y)}_{m+1,\mathrm{MMSE}}(k) = e^{(y)}_{m,\mathrm{MMSE}}(k) + k^{(y)}_m e^{\mathrm{b}}_{m,\mathrm{MMSE}}(k)$

END DO

6.1.4 Properties of the MMSE lattice estimator

In the previous section we have demonstrated that an MMSE estimation of time series signals can be realized using a lattice structure. Compared with the direct form of MMSE estimation using a tapped-delay-line (TDL) structure, the one with lattice structure is more complex and we need to compute more quantities. Nevertheless, the lattice estimator has some important properties that are not provided by the direct structure and, thus, justify its additional complexity.

STAGE-BY-STAGE OPTIMIZATION

One obvious advantage of the lattice structure is that it generates the optimal estimates from the first stage to its maximum order. This property is very important when the optimum order of the estimator is not known *a priori* and has to be determined during estimation. By using the lattice structure, we only need three additional scalar MMSE estimations in order to compute an $(m+1)$th-order MMSE joint-process estimator from the result of an mth-order MMSE lattice estimator. Only two such additional scalar MMSE estimations are required for an MMSE predictor. On the other hand, the direct form of the MMSE estimator does not have such order recursiveness. Assuming that we have computed the coefficient vector in an mth-order, MMSE estimation to compute another MMSE estimator with a different order, we need to recompute the coefficient vector by solving another matrix equation. Thus, the lattice structure is more attractive for such applications.

Commonly, the mean squared error (MSE) of the output error is a good indication of the effectiveness of an estimator. For example, if the MSEs no long decrease for stages m, $m > m_0$, we can conclude that the estimator of order m_0 is the most efficient estimator with the lowest order. For the MMSE linear predictor, the MSEs at different stages are given by α_m^{f} and α_m^{b}, which are readily available from the lattice predictor, and no additional computations are needed. For the joint-process estimators with different orders, additional steps are needed to compute the MSEs of the joint estimation errors at different stages.

CORRELATION PROPERTIES OF LATTICE PREDICTION ERRORS

The correlation properties of the prediction errors and their relationships are discussed in [2]. Below we only discuss the most important property – the orthogonality between backward prediction errors – and list the others in Table 6.2.

Table 6.2 Properties of MMSE prediction errors (all errors are assumed to be MMSE errors)

$E[e_m^{\mathrm{f}}(k)u^*(k-i)] = 0$	for $1 \leqslant i \leqslant m$
$E[e_m^{\mathrm{b}}(k)u^*(k-i)] = 0$	for $0 \leqslant i \leqslant m-1$
$E[e_m^{\mathrm{f}}(k)u^*(k)] = E[e_m^{\mathrm{b}}(k)u^*(k-m)] = c_m$	
$E[e_i^{\mathrm{f}}(k)e_j^{\mathrm{f}*}(k)] = \alpha_{\max(i,j)}$	
$E[e_i^{\mathrm{f}}(k)e_j^{\mathrm{f}*}(k-l) = 0$	for $\begin{cases} 1 \leqslant l \leqslant i-j, & i>j \\ -1 \geqslant l \geqslant i-j, & i<j \end{cases}$
$E[e_i^{\mathrm{b}}(k)e_j^{\mathrm{b}*}(k-l)] = 0$	for $\begin{cases} 0 \leqslant l \leqslant i-j-1, & i>j \\ 0 \geqslant l \geqslant i-j+1, & i<j \end{cases}$
$E[e_i^{\mathrm{f}}(k+i)e_j^{\mathrm{f}*}(k+j)] = 0$	for $i \neq j$
$E[e_i^{\mathrm{b}}(k+i)e_j^{\mathrm{b}*}(k+j)] = \alpha_{\max(i,j)}$	
$E[e_i^{\mathrm{f}}(k)e_j^{\mathrm{b}*}(k)] = \begin{cases} k_{j-1}^*\alpha_i, & i \geqslant j \\ 0, & i<j \end{cases}$	
$E[e_i^{\mathrm{f}}(k)e_j^{\mathrm{b}*}(k-1)] = \begin{cases} -k_j\alpha_i & j \geqslant i \\ 0 & i>j \end{cases}$	
$E[e_i^{\mathrm{b}*}(k-1)u(k) = E[e_i^{\mathrm{f}}(k+1)u^*(k-i)] = \beta_i^{\mathrm{f}} = -k_i\alpha_i$	

One of the most important features of the lattice estimator is that its backward prediction errors of different orders are orthogonal to each other. That is,

$$E[e^{b*}_{i,\mathrm{MMSE}}(k)e^{b}_{j,\mathrm{MMSE}}(k)] = 0 \quad \text{for } i \neq j \tag{6.35}$$

To prove (6.35), assuming that $i > j$, we first note that, from the orthogonality principles, $e^{b}_{i,\mathrm{MMSE}}(k)$ is orthogonal to $\{\mathbf{u}_i(k)\}$. On the other hand, $e^{b}_{j,\mathrm{MMSE}}(k)$ can be expressed as a linear combination of $u(k)$ to $u(k-j)$, which are the components of $\mathbf{u}_i(k)$ for $i > j$. Then (6.35) follows.

As shown in chapter 2, the backward prediction errors result from a Gram–Schmidt type of orthogonalization. This makes the lattice structure advantageous for the adaptive filtering of signals with large eigenvalue spread; that is, when the data autocorrelation matrix $R_{\mathbf{u}_m\mathbf{u}_m}$ has a large maximum to minimum eigenvalue ratio. This will be discussed in section 6.5.

We will not prove the other correlation properties between the prediction errors listed in Table 6.2. Their proof is left as an exercise for the reader.

One of the questions often asked in the study of lattice algorithms or lattice estimators is the following: our starting point for studying MMSE estimation is to find the optimal estimate of $y(k)$ or the optimal estimation coefficients of the transversal filter, e.g. $\mathbf{a}_{m,\mathrm{MMSE}}$, $\mathbf{b}_{m,\mathrm{MMSE}}$ and $\mathbf{c}_{m,\mathrm{MMSE}}$, but why do we end up with a bunch of estimation errors and estimation coefficients? Below we will show that it is possible to obtain the estimate of $y(k)$ directly from the lattice coefficients without resorting to (6.4). As a result, it is not necessary to compute the estimation coefficients explicitly.

ORDER RECURSIVE COMPUTATION OF THE ESTIMATE OF $y(k)$

It follows from (6.32) and 6.34) that

$$e^{(y)}_{m+1,\mathrm{MMSE}}(k) = e^{(y)}_{i,\mathrm{MMSE}}(k) + \sum_{l=i}^{m} k^{(y)}_{l} e^{b}_{l,\mathrm{MMSE}}(k)$$

Since $e^{(y)}_{0,\mathrm{MMSE}}(k) \equiv y(k)$ and letting $i = 0$, we obtain

$$e^{(y)}_{m+1,\mathrm{MMSE}}(k) = y(k) + \sum_{l=0}^{m} k^{(y)}_{l} e^{b}_{l,\mathrm{MMSE}}(k) \tag{6.36}$$

By comparing (6.36) with (6.5) we conclude that

$$\hat{y}_{m+1,\mathrm{MMSE}}(k) = -\sum_{l=0}^{m} k^{(y)}_{l} e^{b}_{l,\mathrm{MMSE}}(k)$$

and we obtain the order recursion for the joint estimate $\hat{y}_{m,\mathrm{MMSE}}(k)$ as

$$\hat{y}_{m+1,\mathrm{MMSE}}(k) = \hat{y}_{m,\mathrm{MMSE}}(k) - k^{(y)}_{m} e^{b}_{m,\mathrm{MMSE}}(k)$$

Similar to the joint estimate, we obtain the order recursion for the forward predictions given by

$$\hat{u}^{f}_{m+1,\mathrm{MMSE}}(k) = \hat{u}^{f}_{m,\mathrm{MMSE}}(k) - k_{m} e^{b}_{m,\mathrm{MMSE}}(k-1)$$

COMPUTATION OF ESTIMATION COEFFICIENTS FROM LATTICE PARAMETERS

We have shown that the estimate of $y(k)$ can be computed without having to solve the MMSE coefficients. On the other hand, once the lattice reflection coefficients and joint estimation coefficients are known, the estimation coefficients of the transversal filter can be obtained by means of the Levinson algorithm if needed. This has been shown in section 2.7.1 and we will not repeat it here.

ORDER RECURSION OF MEAN SQUARED PREDICTION ERRORS α_m

The MSE of the $(m+1)$th-order forward and backward prediction errors can be computed from the mth-order MSE as shown by (2.120). That is,

$$\begin{aligned}\alpha_{m+1} &= \alpha_m + \beta_m^{f*} k_m \\ &= \alpha_m - |\beta_m^f|^2/\alpha_m = \alpha_m - \alpha_m |k_m|^2 = \alpha_m(1 - |k_m|^2) \end{aligned} \tag{6.37}$$

The properties discussed above are not only useful for the MMSE lattice estimator/predictor, they are also valid for other estimators using quadratic cost functions with little or no modification. In the next section, we will show that corresponding properties also exist for the adaptive LS lattice estimator/predictor. By using the results obtained in this section for MMSE estimation, we greatly simplify the derivation of LS lattice algorithm.

6.2 Time-and-order recursive LS estimation and the basic LS lattice algorithm

It has been demonstrated in sections 2.7 and 6.1 that the Wiener filtering problem can be realized using a lattice structure, provided that the statistics of the signal are known. However, in most practical applications, the signal statistics are not known a priori. For such applications, adaptive algorithms are used to estimate the signal statistics directly from the given data while performing estimation or filtering. In this section, we derive adaptive filtering algorithms for LS estimation, which are realized using the same lattice structure as the one described in the previous section. To begin with, we consider the general order recursive LS (ORLS) estimation problem and its time recursive realization.

6.2.1 Basic relations of ORLS estimation

Let us first review the general mth-order LS estimation problem and its properties. Assume that we have a desired signal sequence $\{y(k), k = t_0, \ldots, t\}$ and a data vector sequence $\{\mathbf{u}_m(k), k = t_0, \ldots, t\}$ defined by

$$\mathbf{u}_m(k) = [u_1(k), u_2(k), \ldots, u_m(k)]^{\mathrm{T}}$$

where $\{u_i(k),\ i = 1, 2, \ldots, m\}$ are m complex scalar sequences. These sequences are defined similarly to what we defined in MMSE estimation. However, for practical applications the sequences must start from some point in time. In what follows, we shall, without loss of generality, assume that $t_0 = 0$ and k is understood to run from 0 to t, unless stated otherwise. Then we form an estimate $\hat{y}(k)$ for each $y(k)$ that is a linear combination of $u_i(k)$, $i = 1, 2, \ldots, m$, using a coefficient vector $\mathbf{c}_m$. Hence

$$\hat{y}(k) = -\mathbf{c}_m^{\mathrm{T}}\mathbf{u}_m(k) \tag{6.38}$$

An LS estimation algorithm determines a coefficient vector at each time t, which minimizes the weighted sum of squared errors between $\hat{y}(k)$ and $y(k)$ given by

$$\sigma_e^2(t) = \sum_{k=0}^{t} \lambda(k, t)\,|\,e_m(k)\,|^2 \equiv \sum_{k=0}^{t} \lambda(k, t)\,|\,y(k) + \mathbf{c}_m^{\mathrm{T}}\mathbf{u}_m(k)\,|^2$$

where $\lambda(k, t)$ is a weighting sequence. The most widely used weighting sequence is the so-called *exponential weighting* sequence, i.e. $\lambda(t, k) = \lambda^{t-k}$. The coefficient vector, denoted $\mathbf{c}_m^{(y)}(t)$, that minimizes $\sigma_e^2(t)$ is called the *LS coefficient vector at time t*. It satisfies

$$\mathbf{c}_m^{(y)}(t) = -R_{\mathbf{u}_m\mathbf{u}_m}^{-1}(t)\mathbf{d}_m^{(y)}(t)$$

where $R_{\mathbf{u}_m\mathbf{u}_m}(t)$ and $\mathbf{d}_m^{(y)}(t)$ represent the estimates of the autocorrelation matrix of $\mathbf{u}_m(k)$ and the cross-correlation vector between $\mathbf{u}_m(k)$ and $y(k)$, respectively, defined by

$$R_{\mathbf{u}_m\mathbf{u}_m}(t) = \sum_{k=0}^{t} \lambda^{t-k}\mathbf{u}_m^*(k)\mathbf{u}_m^{\mathrm{T}}(k) \tag{6.39}$$

and

$$\mathbf{d}_m^{(y)}(t) = \sum_{k=0}^{t} \lambda^{t-k}\mathbf{u}_m^*(k)y(k)$$

The optimum estimates $\hat{y}_m(k, t)$ *of* $y(k)$ at t are obtained by substituting the LS coefficient vector $\mathbf{c}_m^{(y)}(t)$ into (6.38). That is,

$$\hat{y}_m(k, t) = -\mathbf{c}_m^{(y)\mathrm{T}}(t)\mathbf{u}_m(k)$$

We call $\hat{y}_m(k, t)$ the *mth-order LS estimate of* $y(k)$ *at time t, based on the data vector sequence* $\{\mathbf{u}_m(\mathrm{k})\}$. Similarly, we call the errors between $y(k)$ and $\hat{y}_m(k, t)$, i.e.

$$e_m^{(y)}(k, t) \equiv y(k) - \hat{y}_m(k, t) = y(k) + \mathbf{c}_m^{(y)\mathrm{T}}(t)\mathbf{u}_m(k) \qquad k = 0, \ldots, t \tag{6.40}$$

the *mth-order LS estimation errors of* $y(k)$ *at time t, based on data vector* $\mathbf{u}_m(k)$. Note that there are two time indices in $\hat{y}_m(k, t)$ and $e_m(k, t)$. The first index k indicates that they are related to the desired signal of time k and the second index t indicates that the estimation uses the optimal LS coefficients at time t. The subscript m and the superscript (y) denote the order and the desired signal $y(k)$ of the estimation respectively.

ORTHOGONALITY PRINCIPLE FOR LS ESTIMATION

Since the LS estimation uses a quadratic cost function, the orthogonality principle also applies. The orthogonality principle for LS estimation is given by

$$\sum_{k=0}^{t} \lambda^{t-k}[\mathbf{u}_m^*(k)e_m^{(y)}(k, t)] = \mathbf{0}$$

which can be proved similar to (6.6)

THEOREM 6.2

(Order recursive relation for LS estimation) By partitioning the $(m + 1)$-dimensional data vector $\mathbf{u}_{m+1}(k)$, $k = 0, 1, \ldots, t$, as

$$\mathbf{u}_{m+1}(k) = (\mathbf{u}_m^{\mathrm{T}}(k), u'(k))^{\mathrm{T}}$$

the $(m + 1)$th-order LS error, $e_{m+1}^{(y)}(k, t)$, of $y(k)$ estimated based on the data vector $\mathbf{u}_{m+1}(k)$ can be computed from the mth LS errors $e_m^{(y)}(k, t)$ and $e_m^{(u')}(k, t)$ such that

$$e_{m+1}^{(y)}(k, t) = e_m^{(y)}(k, t) - [\alpha_m^{(yu')}(t)/\alpha_m^{(u'u')}(t)]e_m^{(u')}(k, t) \equiv e_m^{(y)}(k, t) + k_m^{(yu')}(t)e_m^{(u')}(k, t) \quad (6.41)$$

where $e_m^{(y)}(k, t)$ and $e_m^{(u')}(k, t)$ are the mth-order LS errors of $y(k)$ and $u'(k)$ estimated based on the data vector $\mathbf{u}_m(k)$ at time t, respectively, and $\alpha_m^{(u'u')}(t)$ denotes the autocorrelation of $e_m^{(u')}(k, t)$, defined by

$$\alpha_m^{(u'u')}(t) \equiv \sum_{k=0}^{t} \lambda^{t-k} |e_m^{(u')}(k, t)|^2 \quad (6.42)$$

and $\alpha_m^{(yu')}(t)$ denotes the cross-correlation between $e_m^{(y)}(k, t)$ and $e_m^{(u')}(k, t)$, defined by

$$\alpha_m^{(yu')}(t) \equiv \sum_{k=0}^{t} \lambda^{t-k} e_m^{(y)}(k, t)e_m^{(u')*}(k, t) \quad (6.43)$$

In (6.41) the LS estimation coefficient $k_m^{(yu')}(t)$ is defined by

$$k_m^{(yu')}(t) \equiv -\alpha_m^{(yu')}(t)/\alpha_m^{(u'u')}(t)$$

Theorem 6.2 shows that a higher-order LS error sequence can be computed from two lower-order LS error sequences. It is therefore called the *order recursive relation of LS estimation.*

The proof of Theorem 6.2 follows the proof of Theorem 6.1, almost word for word, by replacing the expectation operator $E[\cdot]$ with the weighted sum operator $\Sigma_{k=0}^{t} \lambda^{t-k}[\cdot]$. It is left as an exercise for the reader.

The order recursive relation (6.41) stated by Theorem 6.2 is the basis of any ORLS estimation algorithm. It will be used repeatedly in the derivation of LS lattice algorithm later in this section. The remarks given in section 6.1.2 for Theorem 6.1 also apply to Theorem 6.2 simply by replacing the 'MMSE' by 'LS'. Below we summarize these remarks for the ORLS estimation.

Remarks

1. The order recursive relation for LS estimation can be interpreted geometrically as the decomposition of an orthogonal projection of $\mathbf{y}(t)$ on to a linear space, spanned by the $(t+1)$-dimensional vectors $\{[u_i(t), \ldots, u_i(0)]^T, \ i = 1, \ldots, m\}$ and $[u'(t), \ldots, u'(0)]^T$, into the projections on to a subspace, spanned by $\{[u_i(t), \ldots, u_i(0)]^T, i = 1, \ldots, m\}$ of the linear space, and on to the orthogonal complement, spanned by $[e_m^{(y)}(t, t), \ldots, e_m^{(y)}(0, t]^T$ of the subspace. It is also possible to prove (6.41) using a purely geometrical approach similar to the MMSE estimation.
2. Equation (6.41) can be viewed as a *scalar LS estimation*, where $\{e_m^{(y)}(k, t)\}$ behaves as the 'desired signal sequence' and $\{e_m^{(u')}(k, t)\}$ behaves as the 'scalar data sequence'. Such a scalar LS estimation is only meaningful when the 'desired signal sequence', i.e. $\{e_m^{(y)}(k, t)\}$ in (6.41), and the 'scalar data sequence', i.e. $\{e_m^{(u')}(k, t)\}$, are both LS estimation error sequences based on the same data vector sequence, i.e. $\{\mathbf{u}_m(k)\}$, up to a permutation of the elements of the data vector.
3. If we use $\{e_m^{(y)}(k, t)\}$ as the 'data sequence' and $e_m^{(u')}(k, t)\}$ as the 'desired signal sequence' in the scalar LS estimation, we obtain an $(m+1)$th-order LS estimation error sequence whose desired signal sequence and data vector sequence are $\{u'(k)\}$ and $\{[\mathbf{u}_m^T(k) : y(k)]^T\}$ respectively.
4. Theorem 6.2 is given in the form of first-order recursion, which only involves scalar operations and is easy to compute. We will discuss a higher-order decomposition later in this chaper.
5. Any of the elements of the data vector $\mathbf{u}_{m+1}(k)$ can be used as $u'(k)$ in the partition, and $\mathbf{u}_m(k)$ consists of the remaining m elements. If the elements of the data vector $\mathbf{u}_m(k)$ differ by a permutation in the estimations of $e_m^{(y)}(k, t)$ and $e_m^{(u')}(k, t)$, (6.41) is still valid.

To facilitate the understanding and derivation of order recursive LS estimation algorithms, we introduce the following more complex but more revealing notations of the LS estimation error sequences $\{e_m^{(y)}(k, t)\}$, $\{e_m^{(u')}(k, t)\}$ and $\{e_{m+1}^{(y)}(k, t)\}$. That is,

$$\begin{aligned} e_{\mathbf{u}_m(k)}^{(y(k))}(t) &\equiv e_m^{(y)}(k, t) \\ e_{\mathbf{u}_m(k)}^{(u'(k))}(t) &\equiv e_m^{(u')}(k, t) \\ e_{[\mathbf{u}_m(k) : u'(k)]}^{(y(k))}(t) &\equiv e_{\mathbf{u}_{m+1}(k)}^{(y(k))}(t) \equiv e_{m+1}^{(y)}(k, t) \end{aligned} \tag{6.44}$$

Note that in the above notation of an LS error, the superscript and subscript denote the desired signal and data vector of the LS error respectively. By using this notation, we can rewrite (6.41) as

$$e_{[\mathbf{u}_m(k) : u'(k)]}^{(y(k))}(t) = e_{\mathbf{u}_m(k)}^{(y(k))}(t) + k_{\mathbf{u}_m}^{(yu')}(t) e_{\mathbf{u}_m(k)}^{(u'(k))}(t) \tag{6.45}$$

where

$$k_{\mathbf{u}_m}^{(yu')}(t) \equiv -\alpha_{\mathbf{u}_m}^{(yu')}(t)/\alpha_{\mathbf{u}_m}^{(u'u')}(t) \equiv -\alpha_m^{(yu')}(t)/\alpha_m^{(u'u')}(t)$$

This order recursive relation can be shown graphically by the block diagram given in Figure 6.4.

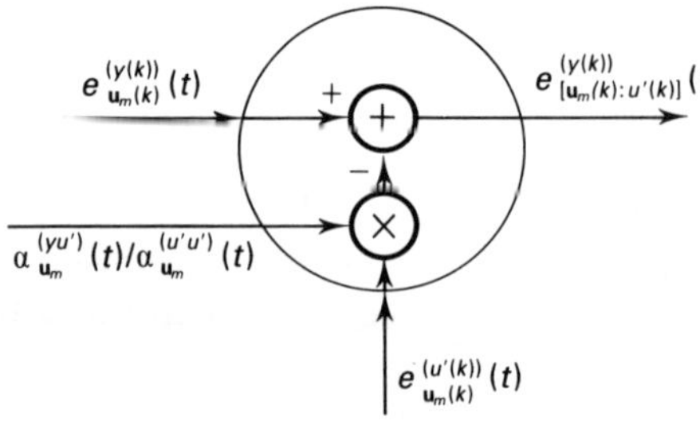

Figure 6.4 Elementary ORLS estimation

We can use these generic relationships of ORLS estimation to derive different ORLS algorithms symbolically, as shown by (6.45), without detailed algebraic manipulation, and, thus greatly simplify the investigation and derivation of such algorithms.

TIME RECURSIVE RELATIONS FOR ORLS ESTIMATION

To obtain the correlations $\alpha_m^{(u'u')}(t)$ and $\alpha_m^{(yu')}(t)$ in (6.41) according to (6.42) and (6.43), we need to compute the entire LS error sequences $\{e_m^{(y)}(k, t)\}$ and $\{e_m^{(u')}(k, t)\}$ for $k = 0, 1, \ldots, t$. Thus at each time instant we must recompute these LS errors. Evidently, such an implementation of the ORLS estimation is not computationally efficient. In many applications of LS estimation, at each time instant t only the most recent error $e_m^{(y)}(t, t)$ or $e_m^{(u')}(t, t)$ is useful. Below, we show that the correlations at time t can be computed from their corresponding values at time $t - 1$, i.e. $\alpha_m^{(u'u')}(t-1)$ and $\alpha_m^{(yu')}(t-1)$, together with the most recent LS errors in the error sequences. By using such an approach, we avoid recomputing the entire error sequences and, hence, greatly reduce the computational burden of the ORLS estimation. This is called the *time recursive relation of the correlations in the ORLS estimation.* It is given by the following theorem.

THEOREM 6.3

(Time recursive relations of correlations in the ORLS estimation) The autocorrelation $\alpha_m^{(u'u')}(t)$ and the cross-correlation $\alpha_m^{(yu')}(t)$ in the ORLS estimation (6.41) can be computed time recursively such that

$$\alpha_m^{(u'u')}(t) = \lambda\alpha_m^{(u'u')}(t-1) + |e_m^{(u')}(t, t)|^2/\alpha_m(t) \tag{6.46}$$

and

$$\alpha_m^{(yu')}(t) = \lambda\alpha_m^{(yu')}(t-1) + e_m^{(u')*}(t, t)e_m^{(y)}(t, t)/\alpha_m(t) \tag{6.47}$$

where $\alpha_m(t)$ is a scalar defined by

$$\alpha_m(t) \equiv 1 - \gamma_m(t) \equiv 1 - \mathbf{u}_m^{\mathrm{T}}(t)R_{\mathbf{u}_m\mathbf{u}_m}^{-1}(t)\mathbf{u}_m^*(t) \tag{6.48}$$

Note that both $\gamma_m(t)$ and $\alpha_m(t)$ in (6.48) are real, positive and with a value between zero and one. As pointed out in Chapter 2 the quantity $\gamma_m(t)$ is related to the maximum likelihood estimation and, therefore, is called the *maximum likelihood factor* in the literature.

Proof The mth-order LS errors $e_m^{(y)}(k, t)$ and $e_m^{(u')}(k, t)$ are computed according to (6.40) and

$$e_m^{(u')}(k, t) = u'(k) + \mathbf{c}_m^{(u')\mathrm{T}}(t)\mathbf{u}_m(k) \tag{6.49}$$

respectively. It is known that the LS coefficient vectors $\mathbf{c}_m^{(y)}(t)$ and $\mathbf{c}_m^{(u')}(t)$ in (6.40) and (6.49) satisfy the time recursive relation for the LS estimation coefficient discussed in Chapter 2. Specifically, the time recursion for $\mathbf{c}_m^{(y)}(t)$ is

$$\mathbf{c}_m^{(y)}(t) = \mathbf{c}_m^{(y)}(t-1) - R_{\mathbf{u}_m\mathbf{u}_m}^{-1}(t)\mathbf{u}^*(t)\varepsilon_m^{(y)}(t)/\alpha_m(t) \tag{6.50}$$

where $\varepsilon_m^{(y)}(t) \equiv e_m^{(y)}(t, t)$ is the most recent LS error of $y(t)$ in the error sequence $\{e_m^{(y)}(k, t)\}$ and is the mth-order a posteriori LS error of $y(t)$. Note that in (6.50) relation (15) of Table 6.4, relating posterior and prior errors, has been used.

By substituting (6.50) into (6.40) the LS errors $e_m^{(y)}(k, t)$, for $k = 0, 1, \ldots, t-1$, can be expressed as

$$\begin{aligned} e_m^{(y)}(k, t) &= y(k) + \mathbf{c}_m^{(y)\mathrm{T}}(t-1)\mathbf{u}_m(k) - \mathbf{u}_m^{\mathrm{T}}(t)R_{\mathbf{u}_m\mathbf{u}_m}^{-1}(t)\mathbf{u}_m^*(k)\varepsilon_m^{(y)}(t)/\alpha_m(t) \\ &= e_m^{(y)}(k, t-1) - \mathbf{u}_m^{\mathrm{T}}(t)R_{\mathbf{u}_m\mathbf{u}_m}^{-1}(t)\mathbf{u}_m^*(k)\varepsilon_m^{(y)}(t)/\alpha_m(t) \end{aligned} \tag{6.51}$$

Similar to (6.50) and (6.51) we have the time update equation for $\mathbf{c}_m^{(u')}(t)$ as

$$\mathbf{c}_m^{(u')}(t) = \mathbf{c}_m^{(u')}(t-1) - R_{\mathbf{u}_m\mathbf{u}_m}^{-1}(t)\mathbf{u}_m^*(t)e_m^{(u')}(t)/\alpha_m(t)$$

and the LS errors $e_m^{(u')}(k, t)$ can be expressed as

$$e_m^{(u')}(k, t) = e_m^{(u')}(k, t-1) - \mathbf{u}_m^{\mathrm{T}}(t)R_{\mathbf{u}_m\mathbf{u}_m}^{-1}(t)\mathbf{u}_m^*(k)\varepsilon_m^{(u')}(t)/\alpha_m(t) \tag{6.52}$$

From the orthogonality principle, it follows that

$$\sum_{k=0}^{t-1} \lambda^{t-1-k} e_m^{(u')}(k, t-1)\mathbf{u}_m^*(k) = 0 \tag{6.53}$$

and

$$\sum_{k=0}^{t-1} \lambda^{t-1-k} e_m^{(y)}(k, t-1)\mathbf{u}_m^*(k) = 0 \tag{6.54}$$

By substituting (6.51) and (6.52) into (6.43) and then using (6.53) and (6.54), we obtain

$$\begin{aligned} \alpha_m^{(yu')}(t) = &\sum_{k=0}^{t-1} \lambda^{t-k} e_m^{(u')*}(k, t-1)e_m^{(y)}(k, t-1) + \mathbf{u}_m^{\mathrm{T}}(t)R_{\mathbf{u}_m\mathbf{u}_m}^{-1}(t) \\ &\sum_{k=0}^{t-1} \lambda^{t-k}[\mathbf{u}_m^*(k)\mathbf{u}_m^{\mathrm{T}}(k)]R_{\mathbf{u}_m\mathbf{u}_m}^{-1}(t)\mathbf{u}_m^*(t)\varepsilon_m^{(y)}(t)\varepsilon_m^{(u')*}(t)/\alpha_m^2(t) + \varepsilon_m^{(y)}(t)\varepsilon_m^{(u')*}(t) \end{aligned} \tag{6.55}$$

To prove (6.47) we first note that

$$\sum_{k=0}^{t-1} \lambda^{t-k} e_m^{(u')*}(k, t-1) e_m^{(y)}(k, t-1) = \lambda \sum_{k=0}^{t-1} \lambda^{t-1-k} e_m^{(u')*}(k, t-1) e_m^{(y)}(k, t-1)$$
$$= \lambda \alpha_m^{(yu')}(t-1) \tag{6.56}$$

and from (6.39) we have

$$\sum_{k=0}^{t-1} \lambda^{t-k} \mathbf{u}_m^*(k) \mathbf{u}_m^{\mathrm{T}}(k) = \sum_{k=0}^{t} \lambda^{t-k} \mathbf{u}_m^*(k) \mathbf{u}_m^{\mathrm{T}}(k) - \mathbf{u}_m^*(k) \mathbf{u}_m^{\mathrm{T}}(k) = R_{\mathbf{u}_m \mathbf{u}_m}(t) - \mathbf{u}_m^*(t) \mathbf{u}_m^{\mathrm{T}}(t) \tag{6.57}$$

From (6.57) and the definitions of $\alpha_m(t)$ and $\gamma_m(t)$, it can be shown that

$$\mathbf{u}_m^{\mathrm{T}}(t) R_{\mathbf{u}_m \mathbf{u}_m}^{-1}(t) \sum_{k=0}^{t-1} \lambda^{t-k} \mathbf{u}_m^*(k) \mathbf{u}_m^{\mathrm{T}}(k) R_{\mathbf{u}_m \mathbf{u}_m}^{-1}(t) \mathbf{u}_m^*(t)$$
$$= \mathbf{u}_m^{\mathrm{T}}(t) R_{\mathbf{u}_m \mathbf{u}_m}^{-1}(t) [R_{\mathbf{u}_m \mathbf{u}_m}(t) - \mathbf{u}_m^*(t) \mathbf{u}_m^{\mathrm{T}}(t)] R_{\mathbf{u}_m \mathbf{u}_m}^{-1}(t) \mathbf{u}_m^*(t)$$
$$= \mathbf{u}_m^{\mathrm{T}}(t) R_{\mathbf{u}_m \mathbf{u}_m}^{-1}(t) \mathbf{u}_m^*(t) - [\mathbf{u}_m^{\mathrm{T}}(t) R_{\mathbf{u}_m \mathbf{u}_m}^{-1}(t) \mathbf{u}_m^*(t)]^2 = \gamma_m(t)[1 - \gamma_m(t)]$$
$$= \gamma_m(t) \alpha_m(t) = [1 - \alpha_m(t)] \alpha_m(t) \tag{6.58}$$

Then, by substituting (6.56) and (6.58) into (6.55) and after some algebra, the time recursion (6.47) of $\alpha_m^{(yu')}(t)$ follows. The time recursion (6.46) of $\alpha_m^{(u'u')}(t)$ can be derived similarly. This completes the proof of Theorem 6.3. □

ORDER RECURSION OF $\gamma_m(t)$ AND $\alpha_m(t)$

The third relation that we need to complete for the ORLS estimation is the order recursion of $\gamma_m(t)$ and $\alpha_m(t)$, which is given by the following theorem.

THEOREM 6.4

(Order recursions of the maximum likelihood factor $\gamma_m(t)$ and $\alpha_m(t) \equiv 1 - \gamma_m(t)$). The maximum likelihood factor $\gamma_m(t)$ satisfies the following order recursive relation:

$$\gamma_{m+1}(t) = \gamma_m(t) + [\varepsilon_m^{(u')}(t)]^2 / \alpha_m^{(u'u')}(t) \tag{6.59}$$

and $\alpha_m(t)$ satisfies

$$\alpha_{m+1}(t) = \alpha_m(t) - [\varepsilon_m^{(u')}(t)]^2 / \alpha_m^{(u'u')}(t) \tag{6.60}$$

Proof By definition, we can write the $(m+1) \times (m+1)$ autocorrelation matrix $R_{\mathbf{u}_{m+1} \mathbf{u}_{m+1}}(t)$ as

$$R_{\mathbf{u}_{m+1} \mathbf{u}_{m+1}}(t) = \sum_{k=0}^{t} \lambda^{t-k} \mathbf{u}_{m+1}^*(k) \mathbf{u}_{m+1}^{\mathrm{T}}(k) = \sum_{k=0}^{t} \lambda^{t-k} \begin{pmatrix} \mathbf{u}_m^*(k) \\ u'^*(k) \end{pmatrix} (\mathbf{u}_m^{\mathrm{T}}(k), u'(k))$$
$$= \begin{pmatrix} \Sigma_{k=0}^t \lambda^{t-k} \mathbf{u}_m^{\mathrm{T}}(k) \mathbf{u}_m^*(k) & \Sigma_{k=0}^t \lambda^{t-k} u'(k) \mathbf{u}_m^*(k) \\ \Sigma_{k=0}^t \lambda^{t-k} \mathbf{u}_m^{\mathrm{T}}(k) u'^*(k) & \Sigma_{k=0}^t \lambda^{t-k} u'(k) u'^*(k) \end{pmatrix} \equiv \begin{pmatrix} R_{\mathbf{u}_m \mathbf{u}_m}(t) & \mathbf{d}_m^{(u')}(t) \\ \mathbf{d}_m^{(u')\mathrm{H}}(t) & \rho_0^{u'}(t) \end{pmatrix} \tag{6.61}$$

Then, by applying the partitioned matrix inversion identities given in [7] and Chapter 8 to the matrix on the right-hand side of (6.61) we obtain

$$R^{-1}_{\mathbf{u}_{m+1}\mathbf{u}_{m+1}}(t) = \begin{pmatrix} R^{-1}_{\mathbf{u}_m\mathbf{u}_m}(t) & \mathbf{0} \\ \mathbf{0}^{\mathrm{T}} & 0 \end{pmatrix} + \begin{pmatrix} -R^{-1}_{\mathbf{u}_m\mathbf{u}_m}(t)\mathbf{d}^{(u')}_m(t) \\ 1 \end{pmatrix}$$

$$\times\, (\rho^{u'}_0(t) - \mathbf{d}^{(u')\mathrm{H}}_m(t)R^{-1}_{\mathbf{u}_m\mathbf{u}_m}(t)\mathbf{d}^{(u')}_m(t))^{-1}\, (-\mathbf{d}^{(u')\mathrm{H}}_m(t)R^{-1}_{\mathbf{u}_m\mathbf{u}_m}(t),\ 1) \tag{6.62}$$

By using the definition $-R^{-1}_{\mathbf{u}_m\mathbf{u}_m}(t)\mathbf{d}^{(u')}_m(t) \equiv \mathbf{c}^{(u')}_m(t)$, and from the orthogonality relation given by (6.53), we obtain the LS counterpart of (6.7) as

$$\alpha^{u'u'}_m(t) = \rho^{u'}_0(t) - \mathbf{d}^{(u')\mathrm{T}}_m(t)R^{-1}_{\mathbf{u}_m\mathbf{u}_m}(t)\mathbf{d}^{(u')}_m(t) \tag{6.63}$$

By substituting the definition of $\mathbf{c}^{(u')}_m(t)$ and (6.63) into (6.62) we obtain

$$R^{-1}_{\mathbf{u}_{m+1}\mathbf{u}_{m+1}}(t) = \begin{pmatrix} R^{-1}_{\mathbf{u}_m\mathbf{u}_m}(t) & \mathbf{0} \\ \mathbf{0}^{\mathrm{T}} & 0 \end{pmatrix} + \begin{pmatrix} \mathbf{c}^{(u')}_m(t) \\ 1 \end{pmatrix}[\alpha^{(u'u')}_m(t)]^{-1}(\mathbf{c}^{(u')H}_m(t),\ 1) \tag{6.64}$$

From the definition of $\gamma_{m+1}(t)$ and (6.64) we have

$$\gamma_{m+1}(t) = \mathbf{u}^{\mathrm{T}}_{m+1}(t)R^{-1}_{m+1m+1}(t)\mathbf{u}^{*}_{m+1}(t) = (\mathbf{u}^{\mathrm{T}}_m(t)\quad u'(t)) \begin{pmatrix} R^{-1}_{\mathbf{u}_m\mathbf{u}_m}(t) & \mathbf{0} \\ \mathbf{0}^{\mathrm{T}} & 0 \end{pmatrix}\begin{pmatrix} \mathbf{u}^{*}_m(t) \\ u'^{*}(t) \end{pmatrix}$$

$$+ (\mathbf{u}^{\mathrm{T}}_m(t)\quad u'(t)) \begin{pmatrix} \mathbf{c}^{(u')}_m(t) \\ 1 \end{pmatrix}[\alpha^{(u'u')}_m(t)]^{-1}(\mathbf{c}^{(u')\mathrm{H}}_m(t)\quad 1)\begin{pmatrix} \mathbf{u}^{*}_m(t) \\ u'^{*}(t) \end{pmatrix}$$

$$= \mathbf{u}^{\mathrm{T}}_m(t)R^{-1}_{\mathbf{u}_m\mathbf{u}_m}(t)\mathbf{u}^{*}_m(t) + [u'(t) + \mathbf{u}^{\mathrm{T}}_m(t)\mathbf{c}^{(u')}_m(t)][\alpha^{(u'u')}_m(t)]^{-1}[u'^{*}(t) + \mathbf{c}^{(u')\mathrm{H}}_m(t)\mathbf{u}^{*}_m(t)]$$

$$= \gamma_m(t) + |\varepsilon^{(u')}_m(t)|^2/\alpha^{(u'u')}_m(t)$$

This proves (6.59) and, since $\alpha_{m+1}(t) = 1 - \gamma_{m+1}(t)$, (6.60) follows. This completes the proof of Theorem 6.4. □

Theorems 6.2–6.4 form the basic equation set of time recursive ORLS estimation. It can be implemented modularly using two types of basic processing cells, as shown in Figure 6.5. A processing cell of the first type, represented by a single circle, implements (6.41), for $k = t$, and (6.47). A processing cell of the second type, represented by a double circle, implements (6.46) and (6.60). ORLS algorithms are well suited for VLSI implementation because of their modular structure.

The basic equation set consisting of (6.41), (6.47), (6.46) and (6.60) forms the basis of any of the ORLS estimation algorithms. Below, we derive the LS lattice algorithms by applying these recursions to the time series estimation problem discussed in section 6.1.3.

6.2.2 Basic LS lattice algorithm with prewindowed data

Similar to the MMSE lattice estimator, or lattice filter, discussed in section 6.1.3, we can implement an LS estimation algorithm of time series signals using a lattice structure. The LS lattice algorithm can be derived in parallel with the derivation of the MMSE lattice algorithm. The simplest form of the LS lattice algorithm is the

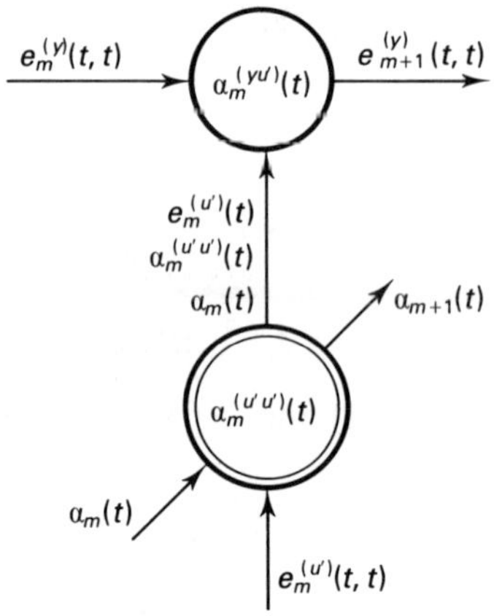

Figure 6.5 Modular realization of an elementary time recursive ORLS estimation

so-called single-channel lattice algorithm with its data vector taken from a *prewindowed* time series, which is defined as follows.

DEFINITION: PREWINDOWED TIME SERIES
A time series $\{u(k), k = 0, 1, \ldots, t\}$ is called prewindowed if $u(k) = 0$ for $k < 0$.

Similar to the derivation of the MMSE lattice algorithm we will first discuss the LS forms of the forward and backward linear prediction problems given in section 6.1.3.

LS FORWARD AND BACKWARD LINEAR PREDICTION
An LS forward linear predictor determines a coefficient vector $\mathbf{a}_m(t)$ that minimizes the exponentially weighted sum of the squared mth-order forward prediction errors. We call $\mathbf{a}_m(t)$ the *LS forward prediction coefficient vector* at t, which satisfies

$$\mathbf{a}_m(t) = -R^{-1}_{\mathbf{u}_m\mathbf{u}_m}(t-1)\mathbf{d}^{\mathrm{f}}_m(t)$$

where $\mathbf{d}^{\mathrm{f}}_m(t)$ is the LS estimate of the correlation vector between $u(k)$ and $\mathbf{u}_m(k-1)$, which is defined by

$$\mathbf{d}^{\mathrm{f}}_m(t) = \sum_{k=0}^{t} \lambda^{t-k} u(k)\mathbf{u}^*_m(k-1)$$

and $R_{\mathbf{u}_m\mathbf{u}_m}(t-1)$ is defined according to (6.39).

The LS estimate of $u(k)$ in the mth-order forward prediction, denoted by $\hat{u}^{\mathrm{f}}_m(k, t)$, and the mth-order LS forward prediction error $e^{\mathrm{f}}_m(t, k)$ at time t resulting from the LS forward prediction is defined as

$$e^{\mathrm{f}}_m(k, t) = u(k) - \hat{u}^{\mathrm{f}}_m(k, t) = u(k) + \mathbf{a}^{\mathrm{T}}_m(t)\mathbf{u}_m(k-1)$$

An LS backward linear predictor determines a coefficient vector $\mathbf{b}_m(t)$ that minimizes the exponentially weighted sum of the squared mth-order backward prediction errors. We call $\mathbf{b}_m(t)$ the *LS backward prediction coefficient vector* at t that satisfies

$$\mathbf{b}_m(t) = -R^{-1}_{\mathbf{u}_m\mathbf{u}_m}(t)\mathbf{d}^{\mathrm{b}}_m(t)$$

where $\mathbf{d}^{\mathrm{b}}_m(t)$ is an estimate of the correlation vector between $u(k-m)$ and $\mathbf{u}_m(k)$ defined as

$$\mathbf{d}^{\mathrm{b}}_m(t) = \sum_{k=0}^{t} \lambda^{t-k} u(k-m)\mathbf{u}^*_m(k)$$

The LS estimate of $u(k-m)$ is the mth-order backward prediction, denoted by $\hat{u}^{\mathrm{b}}_m(k-m, t)$, and the mth-order LS backward prediction error $e^{\mathrm{b}}_m(t, k)$ at time t resulting from the LS backward prediction is defined as

$$e^{\mathrm{b}}_m(k, t) = u(k-m) - \hat{u}^{\mathrm{b}}_m(k-m, t) = u(k-m) + \mathbf{b}^{\mathrm{T}}_m(t)\mathbf{u}_m(k)$$

ORLS LINEAR PREDICTION AND LS LATTICE PREDICTOR

Let us consider an $(m+1)$th-order LS forward linear predictor. By partitioning the data vector $\mathbf{u}_{m+1}(k-1)$ as in (6.21) and applying Theorem 6.2 to the $(m+1)$th LS forward predictor, we can compute the $(m+1)$th prediction errors $\{e^{\mathrm{f}}_{m+1}(k, t), k = 0, \ldots, t\}$ from two sets of LS errors: the LS errors of $u(k)$ and $u(k-m-1)$ based on the same set of data vector $\{\mathbf{u}_m(k-1), k = 0, 1, \ldots, t\}$. By definition, the errors in the first set are the mth-order LS forward prediction errors. By letting $k' = k-1$, the errors in the second set can be viewed as the LS errors of $u(k'-m)$ based on $\mathbf{u}_m(k')$, for $k' = -1, 0, 1, \ldots, t-1$. By definition, the errors in the second set are actually the backward prediction errors $e^{b}_m(k, t-1)$, provided $u(k'-m) = 0$ and $\mathbf{u}_m(k')$ is a zero vector for $k' = -1$. These conditions are satisfied if $\{u(k)\}$ is a *prewindowed* time series.

Thus, if $\{u(k)\}$ is a prewindowed time series, according to Theorem 6.2, the order recursion for the forward prediction errors is given by

$$e^{\mathrm{f}}_{m+1}(k, t) = e^{\mathrm{f}}_m(k, t) - [\beta^{\mathrm{f}}_m(t)/\alpha^{\mathrm{b}}_m(t-1)]e^{\mathrm{b}}_m(k-1, t-1) \tag{6.65}$$

where

$$\beta^{\mathrm{f}}_m(t) = \sum_{k=0}^{t} \lambda^{t-k} e^{\mathrm{f}}_m(k, t)e^{\mathrm{b}*}_m(k-1, t-1) \tag{6.66}$$

and

$$\alpha^{\mathrm{b}}_m(t-1) = \sum_{k=0}^{t-1} \lambda^{t-1-k} |e^{\mathrm{b}}_m(k, t-1)|^2 \tag{6.67}$$

Similarly, we can derive the order recursion for LS backward prediction. The desired signal of an $(m+1)$th backward LS predictor is $u(k-m-1)$ and its data vector is $\mathbf{u}_{m+1}(k)$ which can be partitioned according to (6.25). From Theorem 6.2, we can show that the $(m+1)$th LS backward prediction error $e^{\mathrm{b}}_{m+1}(t)$ can be computed from the LS errors of $u(k-m-1)$ and $u(k)$ based on the data vector $\mathbf{u}_m(k-1)$, $k=0, 1, \ldots, t$. From the discussion above, it is easy to see that they are the mth-order LS forward prediction errors $e_m^{\mathrm{f}}(k, t)$ at time t and the backward prediction errors $e_m^{\mathrm{b}}(k-1, t-1)$ at $t-1$, for prewindowed data. Thus, the order recursion for the backward prediction errors is given by

$$e_{m+1}^{\mathrm{b}}(k, t) = e_m^{\mathrm{b}}(k-1, t-1) - [\beta_m^{\mathrm{b}}(t)/\alpha_m^{\mathrm{f}}(t)]e_m^{\mathrm{f}}(k, t) \tag{6.68}$$

where

$$\beta_m^{\mathrm{b}}(t) = \sum_{k=0}^{t} \lambda^{t-k} e_m^{\mathrm{f}*}(k, t) e_m^{\mathrm{b}}(k-1, t-1) \tag{6.69}$$

and

$$\alpha_m^{\mathrm{f}}(t) = \sum_{k=0}^{t} \lambda^{t-k} |e_m^{\mathrm{f}}(k, t)|^2 \tag{6.70}$$

Thus, we have shown that the $(m+1)$th-order LS backward prediction errors at time t can be computed as the LS estimation errors of the mth-order LS backward prediction errors at time $t-1$ based on the mth-order LS forward prediction errors computed at t through a scalar LS estimation. The $(m+1)$th-order LS forward prediction errors at time t can be computed from the same two sets of the mth-order prediction errors with their roles exchanged. Furthermore, by comparing (6.69) with (6.66) it is easy to see that

$$\beta_m^{\mathrm{b}}(t) = \beta_m^{\mathrm{f}*}(t)$$

By denoting

$$e_0^{\mathrm{f}}(k, t) = e_0^{\mathrm{b}}(k, t) = u(k) \tag{6.71}$$

we can compute the $(m+1)$th forward and backward prediction errors through scalar LS estimations using the mth-order LS forward and backward prediction errors for $m \geqslant 0$. Thus, a pth-order forward and/or backward predictor can be implemented in a purely order recursive fashion through a series of order recursive forward and backward LS predictions for $m = 0$ to $p-1$ using the lattice structure discussed previously for MMSE predictions.

We define *forward and backward LS reflection coefficients*, denoted by $k_m^{\mathrm{f}}(t)$ and $k_m^{\mathrm{b}}(t)$, for LS lattice prediction, as

$$k_m^{\mathrm{f}}(t) = -\beta_m^{\mathrm{f}}(t)/\alpha_m^{\mathrm{b}}(t-1) \quad \text{and} \quad k_m^{\mathrm{b}}(t) = -\beta_m^{\mathrm{f}*}(t)/\alpha_m^{\mathrm{f}}(t) \tag{6.72}$$

respectively. It must be emphasized that the forward and backward reflection coefficients in an LS lattice prediction algorithm are time varying and they are not complex conjugates of each other since $\alpha_m^{\mathrm{f}}(t)$ and $\alpha_m^{\mathrm{b}}(t-1)$ are not equal.

LS LATTICE JOINT-PROCESS ESTIMATION

Now we extend the LS lattice predictor algorithm to the order recursive joint-process LS estimation. Let us denote the $(m+1)$th-order LS estimation of the desired signal sequence $\{y(k)\}$ based on the data vector sequence $\{\mathbf{u}_{m+1}(k)\}$ as $e_{m+1}^{(y)}(k, t)$. By partitioning $\mathbf{u}_{m+1}(k)$ according to (6.25) and following Theorem 6.2, we compute

$e_{m+1}^{(y)}(k, t)$ from the mth-order LS estimation errors of $y(k)$ and $u(k-m)$ based on the data vector $\mathbf{u}_m(k)$, $k = 0, 1, \ldots, t$. By definition, they are the mth-order LS joint estimation errors $e_m^{(y)}(k, t)$ and the mth-order LS backward prediction errors $e_m^{\mathrm{b}}(k, t)$. Thus the order recursion for the joint-process LS estimation error is given by

$$e_{m+1}^{(y)}(k, t) = e_m^{(y)}(k, t) - [\beta_m^{(y)}(t)/\alpha_m^{\mathrm{b}}(t)]e_m^{\mathrm{b}}(k, t) \tag{6.73}$$

where $\beta_m^{(y)}(t)$ is defined by

$$\beta_m^{(y)}(t) = \sum_{k=0}^{t} \lambda^{t-k} e_m^{(y)}(k, t) e_m^{\mathrm{b}*}(k, t)$$

To summarize, the $(m +)$th-order LS estimation errors of $y(k)$ can be computed as the LS estimation errors of the mth-order LS estimation errors of $y(k)$ based on the mth-order LS backward prediction errors at t through a scalar LS estimation.

TIME RECURSIVE IMPLEMENTATION OF THE LS LATTICE PREDICTOR AND JOINT ESTIMATOR

As we have stated above, it is desirable to compute the correlations $\alpha_m^{\mathrm{f}}(t)$, $\alpha_m^{\mathrm{b}}(t)$, $\beta_m^{\mathrm{f}}(t)$ and $\beta_m^{(y)}(t)$ time-recursively to avoid the computation of the entire forward and backward prediction error sequences $\{e_m^{\mathrm{f}}(k, t)\}$ and $\{e_m^{\mathrm{b}}(k, t)\}$. Since the order recursive forward and backward LS predictions are special cases of the ORLS estimations discussed in section 6.1, the time recursive relation given by Theorem 6.3 holds. By comparing (6.66) with (6.43) and replacing $e_m^{(y)}(k, t)$, $e_m^{(u')}(k, t)$ and $\alpha_m^{(yu')}(t)$ in (6.47) with their corresponding quantities $e_m^{\mathrm{f}}(k, t)$, $e_m^{\mathrm{b}}(k-1, t-1)$ and $\beta_m^{\mathrm{f}}(t)$, we obtain

$$\beta_m^{\mathrm{f}}(t) = \lambda\beta_m^{\mathrm{f}}(t-1) + \varepsilon_m^{\mathrm{f}}(t)\varepsilon_m^{\mathrm{b}*}(t-1)/\alpha_m(t-1) \tag{6.74}$$

where, to simplify notation, we have defined

$$\varepsilon_m^{\mathrm{f}}(t) \equiv e_m^{\mathrm{f}}(t, t) \quad \text{and} \quad \varepsilon_m^{\mathrm{b}}(t-1) \equiv e_m^{\mathrm{b}}(t-1, t-1) \tag{6.75}$$

Note that $\alpha_m(t-1)$ appears instead of $\alpha_m(t)$.

We note that the order recursions given by (6.65) and (6.68) are valid for all k including $k = t$ and, thus, are valid for $e_m^{\mathrm{f}}(t)$ and $e_m^{\mathrm{b}}(t)$ in (6.75).

The time recursions for $\alpha_m^{\mathrm{f}}(t)$ and $\alpha_m^{\mathrm{b}}(t)$ are obtained according to (6.46). They are given by

$$\alpha_m^{\mathrm{f}}(t) = \lambda\alpha_m^{\mathrm{f}}(t-1) + |\varepsilon_m^{\mathrm{f}}(t)|^2/\alpha_m(t-1) \tag{6.76}$$

and

$$\alpha_m^{\mathrm{b}}(t) = \lambda\alpha_m^{\mathrm{b}}(t-1) + |\varepsilon_m^{\mathrm{b}}(t)|^2/\alpha_m(t) \tag{6.77}$$

The order recursion of $\alpha_m(t)$ can be obtained by replacing the quantities in (6.60) with the corresponding ones in forward or backward LS linear prediction. If we partition the data vector according to (6.21) and note that the LS estimation error of $u(t-m)$ based on $\mathbf{u}_m(t)$ is $\varepsilon_m^{\mathrm{b}}(t)$, from (6.60) we obtain

$$\alpha_{m+1}(t) = \alpha_m(t) - |\varepsilon_m^{\mathrm{b}}(t)|^2/\alpha_m^{\mathrm{b}}(t) \tag{6.78}$$

Alternatively, if we partition the data vector according to (6.25) and note that the LS estimation error of $u(t)$ based on $\mathbf{u}_m(t-1)$ is $\varepsilon_m^{\mathrm{f}}(t)$, from (6.60) we obtain

$$\alpha_{m+1}(t) = \alpha_m(t-1) - |\varepsilon_m^{\mathrm{f}}(t)|^2/\alpha_m^{\mathrm{f}}(t) \tag{6.79}$$

Either (6.78) or (6.79) can be used for order-updating $\alpha_m(t)$.

The time update equation for $\beta_m^{(y)}(t)$ directly follows from Theorem 6.3 and is given by

$$\beta_m^{(y)}(t) = \lambda\beta_m^{(y)}(t-1) + \varepsilon_m^{(y)}(t)\varepsilon_m^{\mathrm{b}*}(t)/\alpha_m(t) \tag{6.80}$$

where we have used the simplified notation $\varepsilon_m^{(y)}(t) \equiv e_m^{(y)}(t, t)$, which is the most recent error in the error sequence $\{e_m^{(y)}(k, t)\}$ and, thus, obeys the order recursion (6.73).

The time recursive LS lattice and linear joint-process estimation algorithm follows from Equations (6.65), (6.68), (6.74), (6.76), (6.77), (6.78), (6.73) and (6.80). The algorithm is given in Table 6.3. Figure 6.6 depicts a block diagram of the LS lattice predictor/estimator and a typical lattice stage, which uses the symbols of the processing cells defined in Figure 6.5.

Table 6.3 Basic LS lattice algorithm with prewindowed data

INITIALIZATION FOR $t = 0$, $m = 0$, TO $p - 1$:

$\alpha_m^{\mathrm{f}}(0) = \alpha_m^{\mathrm{b}}(0) = \delta$, $\beta_m^{\mathrm{f}}(0) = 0$, $\beta_m^{(y)}(0) = 0$

INITIALIZATION FOR EVERY t:

$\varepsilon_0^{(y)}(t) = y(t)$, $\varepsilon_0^{\mathrm{f}}(t) = \varepsilon_0^{\mathrm{b}}(t) = u(t)$, $\alpha_0(t) = 1$

DO $m = 0$ TO $p - 1$

$\beta_m^{\mathrm{f}}(t) = \lambda\beta_m^{\mathrm{f}}(t-1) + \varepsilon_m^{\mathrm{f}}(t)\varepsilon_m^{\mathrm{b}*}(t-1)/\alpha_m(t-1)$

$\alpha_m^{\mathrm{f}}(t) = \lambda\alpha_m^{\mathrm{f}}(t-1) + |\varepsilon_m^{\mathrm{f}}(t)|^2/\alpha_m(t-1)$

$\alpha_m^{\mathrm{b}}(t) = \lambda\alpha_m^{\mathrm{b}}(t-1) + |\varepsilon_m^{\mathrm{b}}(t)|^2/\alpha_m(t)$

$k_m^{\mathrm{f}}(t) = -\beta_m^{\mathrm{f}}(t)/\alpha_m^{\mathrm{b}}(t-1)$

$k_m^{\mathrm{b}}(t) = -\beta_m^{\mathrm{f}}(t)/\alpha_m^{\mathrm{f}}(t)$

$\varepsilon_{m+1}^{\mathrm{f}}(t) = \varepsilon_m^{\mathrm{f}}(t) + k_m^{\mathrm{f}}(t)\varepsilon_m^{\mathrm{b}}(t-1)$

$\varepsilon_{m+1}^{\mathrm{b}}(t) = \varepsilon_m^{\mathrm{b}}(t-1) + k_m^{\mathrm{b}}(t)\varepsilon_m^{\mathrm{f}}(t)$

$\beta_m^{(y)}(t) = \lambda\beta_m^{(y)}(t-1) + \varepsilon_m^{(y)}(t)\varepsilon_m^{\mathrm{b}*}(t)/\alpha_m(t)$

$k_m^{(y)}(t) = -\beta_m^{(y)}(t)/\alpha_m^{\mathrm{b}}(t)$

$\varepsilon_{m+1}^{(y)}(t) = \varepsilon_m^{(y)}(t) + k_m^{(y)}(t)\varepsilon_m^{\mathrm{b}}(t)$

$\alpha_{m+1}(t) = \alpha_m(t) - |\varepsilon_m^{\mathrm{b}}(t)|^2/\alpha_m^{\mathrm{b}}(t)$

END DO

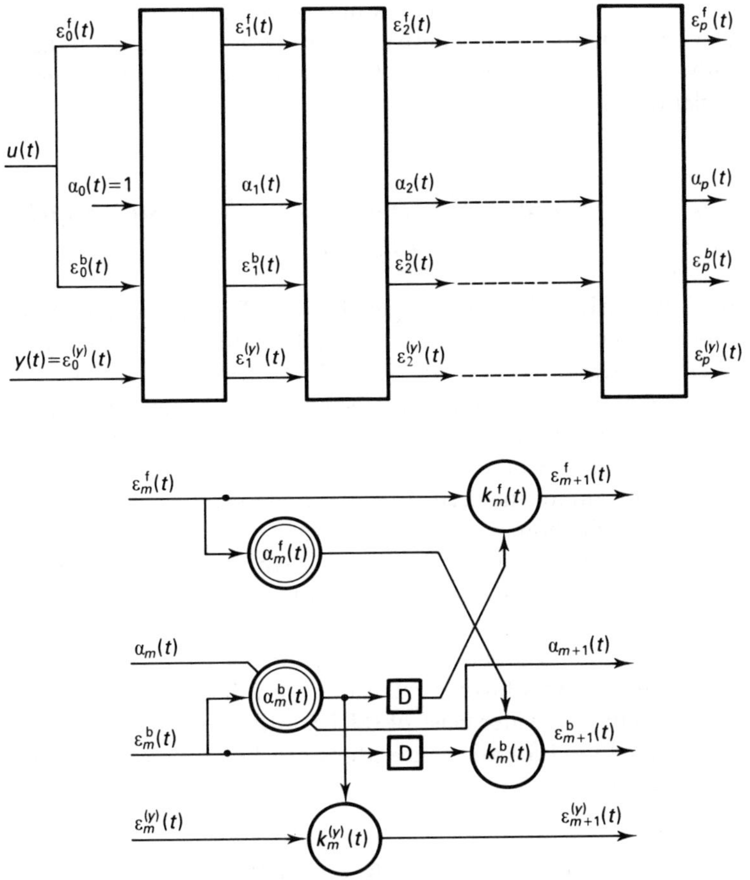

A typical single-channel lattice stage

Figure 6.6 Block diagram of an LS lattice estimator

So far, we have completed the derivation of the basic LS lattice algorithm. We note that the quantities in the LS lattice algorithm can be viewed as the estimates of the corresponding quantities in the MMSE lattice algorithm. Thus, their physical or statistical meaning can also be understood from their counterparts in the MMSE algorithm. It is noteworthy that relations for LS prediction errors similar to the relations of MMSE prediction errors given in Table 6.2 also exist. In the sections following, we will show that the gradient lattice algorithms resulting from the investigation of the MMSE lattice estimator can be viewed as simplified, or non-exact, forms of the variations of the LS lattice algorithm.

In the next section, we will derive a number of alternative forms and extensions of the basic LS lattice algorithm derived in this section. These alternative forms are algebraically equivalent to their basic form. However, they have certain numerical and/or implementational advantages.

6.3 Variations and extensions of the basic LS lattice algorithm

In section 6.2, we derived the basic, single-channel, time recursive LS lattice algorithm for prewindowed data. The basic form of the LS lattice algorithm has two features. First, it uses a posteriori LS errors. Second, the forward and backward reflection coefficients and the joint estimation coefficients are computed as quotients of cross-correlation, and autocorrelations of a posteriori errors. Thus, we say that the basic LS lattice algorithm takes the *a posteriori error correlation quotient form*. This basic form is important for the derivation and understanding of LS lattice algorithms. In this section we will show a number of variations of the basic form that are more appropriate for particular applications and exhibit better numerical behaviour. Furthermore, we will extend the single-channel LS lattice algorithm to a multichannel form. The multichannel lattice algorithm is useful for multichannel spectrum estimation and decision feedback and/or fractionally-spaced equalization (FSE) for data communications. As a further extension, we will discuss the relationship of the LS lattice algorithm to a wider class of ORLS estimation algorithms and show how the results obtained for LS lattice algorithms can be extended to the general ORLS estimation algorithms.

6.3.1 Variations of basic LS lattice algorithm with prewindowed data

From the LS lattice algorithm given in section 6.2.2, we observe that each lattice stage implements three basic order recursive scalar LS estimations, i.e. the forward prediction, the backward prediction and joint estimation. The operations required for these scalar order recursive estimations are identical, but their inputs and outputs are different from each other. These estimations can be viewed as special cases of the general ORLS estimation discussed in section 6.2.1. Thus, in order to derive other forms of the LS lattice algorithm, we only need to investigate alternative ways of implementing the general ORLS estimation and, then, to apply the results to the three basic order recursive estimations in each lattice stage.

Let us first derive a so-called a priori error form of the basic ORLS estimation.

A PRIORI ERROR LS LATTICE ALGORITHM

As can be observed from the ORLS estimation given by (6.41), to compute the $(m+1)$th-order LS estimation error at time t, i.e. $\varepsilon_{m+1}^{(y)}(t) \equiv e_{m+1}^{(y)}(t, t)$, from mth-order

LS errors $\varepsilon_m^{(y)}(t) \equiv e_m^{(y)}(t, t)$ and $\varepsilon_m^{(u')}(t) \equiv e_m^{(u')}(t, t)$, we need first to compute the correlations $\alpha_m^{(yu')}(t)$ and $\alpha_m^{(u'u')}(t)$. It is only possible to compute $\alpha_m^{(yu')}(t)$ using (6.47) if the desired signal $y(t)$ is available (see Table 6.3). In some applications, such as adaptive equalization for data communications, the desired signal $y(t)$ is not available until the LS estimation errors and the estimate of $y(t)$ are computed. As a result, in such applications, instead of using the correlations of time t, we have to use the corresponding correlations at time $t-1$, i.e $\alpha_m^{(yu')}(t-1)$ and $\alpha_m^{(u'u')}(t-1)$, to compute the LS errors and the LS estimate of $y(t)$. The LS errors obtained are those obtained at time t using the optimal coefficients at time $t-1$. They are called a priori errors in the literature and can be expressed as, for example, $e_{m+1}^{(y)}(t, t-1)$ for the $(m+1)$th-order a priori LS error. On the other hand, $\varepsilon_{m+1}^{(y)}(t)$, $\varepsilon_m^{(y)}(t)$ and $\varepsilon_m^{(u')}(t)$ are called a posteriori errors. To simplify the notation in what follows, we denote $e_{m+1}^{(y)}(t) \equiv e_{m+1}^{(y)}(t, t-1)$, $e_m^{(y)}(t) \equiv e_m^{(y)}(t, t-1)$ and $e_m^{(u')}(t) \equiv e_m^{(y)}(t, t-1)$.

The order recursion for the general LS a priori error is given by

$$e_{m+1}^{(y)}(t) = e_m^{(y)}(t) + k_m^{(yu')}(t-1)e_m^{(u')}(t) \equiv e_m^{(y)}(t) - [\alpha_m^{(yu')}(t-1)/\alpha_m^{(u'u')}(t-1)]e_m^{(u')}(t) \tag{6.81}$$

In order to compute the correlations $\alpha_m^{(y'u')}(t)$ and $\alpha_m^{(u'u')}(t)$ time-recursively using a priori errors, we first show the following simple relationship between a priori and a posteriori errors. That is,

$$\varepsilon_m^{(y)}(t) = \alpha_m(t)e_m^{(y)}(t) \tag{6.82}$$

By letting $k = t$ in (6.51) and from the definitions of the a priori error $e_m^{(y)}(t)$, the a posteriori error $\varepsilon_m^{(y)}(t)$ and the maximum likelihood factor $\gamma_m(t)$, we obtain

$$\varepsilon_m^{(y)}(t) = e_m^{(y)}(t) - \gamma_m(t)\varepsilon_m^{(y)}(t)/\alpha_m(t)$$

After some algebra it follows that

$$e_m^{(y)}(t) = [1 + \gamma_m(t)/\alpha_m(t)]\varepsilon_m^{(y)}(t) = \varepsilon_m^{(y)}(t)[\alpha_m(t) + \gamma_m(t)]/\alpha_m(t).$$

By definition, $\alpha_m(t) + \gamma_m(t) = 1$. Thus (6.82) follows.

We note that the relationship given by (6.82) is true for any pair of corresponding a priori and a posteriori LS errors. Thus.

$$\varepsilon_m^{(u')}(t) = \alpha_m(t)e_m^{(u')}(t) \tag{6.83}$$

By using (6.82) and (6.83) we obtain another form of the time recursion (6.47) for the cross-correlation $\alpha_m^{(yu')}(t)$. That is,

$$\alpha_m^{(yu')}(t) = \lambda\alpha_m^{(yu')}(t-1) + \alpha_m(t)e_m^{(u')*}(t)e_m^{(y)}(t) \tag{6.84}$$

Similarly, we obtain

$$\alpha_m^{(u'u')}(t) = \lambda\alpha_m^{(u'u')}(t-1) + \alpha_m(t)\,|\,e_m^{(u')}(t)\,|^2 \tag{6.85}$$

Substituting (6.82) into (6.60) we obtain the order recursion for $\alpha_m(t)$ in terms of a priori LS error as

$$\alpha_{m+1}(t) = \alpha_m(t) - |\,\alpha_m(t)e_m^{(u')}(t)\,|^2/\alpha_m^{(u'u')}(t) = \alpha_m(t)[1 - \alpha_m(t)\,|\,e_m^{(u')}(t)\,|^2/\alpha_m^{(u'u')}(t)] \tag{6.86}$$

Equations (6.81), (6.84), (6.85) and (6.86) constitute the basic equation set for *a priori error ORLS estimation.* Applying these equations to forward and backward predictions as well as to joint estimation results in the *a priori error LS lattice algorithm*, which is given in Table 6.4. This form of LS lattice algorithm is particularly

Table 6.4 A priori error LS lattice alagorithm

INITIALIZATION FOR $t = 0$, $m = 0$ TO $p - 1$:

$$\alpha_m^{\mathrm{f}}(0) = \alpha_m^{\mathrm{b}}(0) = \delta,\ \beta_m^{\mathrm{f}}(0) = 0,\ \beta_m^{(y)}(0) = 0$$

INITIALIZATION FOR EVERY t:

$$e_0^{(y)}(t) = y(t),\ e_0^{\mathrm{f}}(t) = e_0^{\mathrm{b}}(t) = u(t),\ \alpha_0(t) = 1$$

DO $m = 0$ TO $p - 1$

$$e_{m+1}^{\mathrm{f}}(t) = e_m^{\mathrm{f}}(t) + k_m^{\mathrm{f}}(t-1)e_m^{\mathrm{b}}(t-1)$$

$$e_{m+1}^{\mathrm{b}}(t) = e_m^{\mathrm{b}}(t-1) + k_m^{\mathrm{b}}(t-1)e_m^{\mathrm{f}}(t)$$

$$\beta_m^{\mathrm{f}}(t) = \lambda\beta_m^{\mathrm{f}}(t-1) + \alpha_m(t-1)e_m^{\mathrm{f}}(t)e_m^{\mathrm{b}*}(t-1)$$

$$\alpha_m^{\mathrm{f}}(t) = \lambda\alpha_m^{\mathrm{f}}(t-1) + \alpha_m(t-1)|e_m^{\mathrm{f}}(t)|^2$$

$$\alpha_m^{\mathrm{b}}(t) = \lambda\alpha_m^{\mathrm{b}}(t-1) + |e_m^{\mathrm{b}}(t)|^2\alpha_m(t)$$

$$k_m^{\mathrm{f}}(t) = -\beta_m^{\mathrm{f}}(t)/\alpha_m^{\mathrm{b}}(t-1)$$

$$k_m^{\mathrm{b}}(t) = -\beta_m^{\mathrm{f}*}(t)/\alpha_m^{\mathrm{f}}(t)$$

$$e_{m+1}^{(y)}(t) = e_m^{(y)}(t) + k_m^{(y)}(t-1)e_m^{\mathrm{b}}(t)$$

$$\beta_m^{(y)}(t) = \lambda\beta_m^{(y)}(t-1) + \alpha_m(t)e_m^{(y)}(t)e_m^{\mathrm{b}*}(t)$$

$$k_m^{(y)}(t) = -\beta_m^{(y)}(t)/\alpha_m^{\mathrm{b}}(t)$$

$$\alpha_{m+1}(t) = \alpha_m(t) - |\alpha_m(t)e_m^{\mathrm{b}}(t)|^2/\alpha_m^{\mathrm{b}}(t)$$

END DO

suitable for the implementation of adaptive equalizers for data communications and will be discussed in chapter 9.

DIRECT UPDATING OF LATTICE COEFFICIENTS USING AN ERROR FEEDBACK FORMULA

In the LS lattice algorithms given in Tables 6.4 and 6.5, the reflection and joint estimation coefficients are computed as the quotients of the cross- and auto-correlations. It has been found that by modifying some of the equations of the LS lattice algorithms, we can improve their numerical accuracy when a short computer wordlength is used [8]. The modified algorithms are algebraically equivalent to the original algorithm and, thus, still solve the exact LS estimation problem. Below we present a version of the a priori error LS lattice algorithm that uses error feedback to update directly the reflection and joint estimation coefficients instead of computing them as the quotient of the cross- and autocorrelations.

We first consider the general form of the a priori order recursive LS estimation given by (6.81). The estimation coefficient $k_m^{(yu')}(t)$ is computed as the quotient of $\alpha_m^{(yu')}(t)$ and $\alpha_m^{(u'u')}(t)$ according to (6.84) and (6.85). Below, we show that the estimation coefficient $k_m^{(yu')}(t)$ can be directly computed from $k_m^{(yu')}(t-1)$ without computing the cross-correlation $\alpha_m^{(yu')}(t)$. From (6.81), $k_m^{(yu')}(t)$ can be written as

$$\begin{aligned} k_m^{(yu')}(t) &= k_m^{(yu')}(t-1) - \frac{\alpha_m^{(yu')}(t)}{\alpha_m^{(u'u')}(t)} - k_m^{(yu')}(t-1) \\ &= k_m^{(yu')}(t-1) + \frac{-\alpha_m^{(yu')}(t) - \alpha_m^{(u'u')}(t)k_m^{(yu')}(t-1)}{\alpha_m^{(u'u')}(t)} \end{aligned} \qquad (6.87)$$

From (6.84) and (6.85), the numerator in the second term on the right-hand side of (6.87) can be written as

$$\begin{aligned} &-\alpha_m^{(yu')}(t) - \alpha_m^{(u'u')}(t)k_m^{(yu')}(t-1) \\ &= -\lambda\alpha_m^{(yu')}(t-1) - \alpha_m(t)e_m^{(u')*}(t)e_m^{(y)}(t) - [\lambda\alpha_m^{(u'u')}(t-1) + \alpha_m(t)|e_m^{(u')}(t)|^2]k_m^{(yu')}(t-1) \\ &= -\lambda\alpha_m^{(yu')}(t-1) - \alpha_m(t)e_m^{(u')*}(t)e_m^{(y)}(t) + \lambda\alpha_m^{(u'u')}(t-1)\frac{\alpha_m^{(yu')}(t-1)}{\alpha_m^{(u'u')}(t-1)} \\ &\quad - \alpha_m(t)|e_m^{(u')}(t)|^2k_m^{(yu')}(t-1) \\ &= -\alpha_m(t)e_m^{(u')*}(t)e_m^{(y)}(t) - \alpha_m(t)|e_m^{(u')}(t)|^2k_m^{(yu')}(t-1) \\ &= -\alpha_m(t)e_m^{(u')*}(t)[e_m^{(y)}(t) + e_m^{(u')}(t)k_m^{(yu')}(t-1)] \\ &= -\alpha_m(t)e_m^{(u')*}(t)e_{m+1}^{(y)}(t) \end{aligned} \qquad (6.88)$$

The last equality follows from (6.81). By substituting (6.88) into (6.87), we obtain the direct updating equation for $e_{m+1}^{(y)}(t)$ as

$$k_m^{(yu')}(t) = k_m^{(yu')}(t-1) - \frac{\alpha_m(t)e_{m+1}^{(y)}(t)e_m^{(u')*}(t)}{\alpha_m^{(u'u')}(t)} \qquad (6.89)$$

Equations (6.81), (6.85), (6.89) and (6.86) constitute the basic equation set of the a priori error form of the ORLS estimation with direct coefficient updating. Note that the higher-order LS error $e_{m+1}^{(y)}(t)$ is first estimated by using the coefficient $k_m^{(yu')}(t-1)$. Then it is used to time update $k_m^{(yu')}(t-1)$ to $k_m^{(yu')}(t)$ and, thus, forms a negative feedback loop. The error contained in $k_m^{(yu')}(t-1)$ due to numerical precision in computations will be compensated in the feedback loop. This form of LS order recursive estimation is also called the *error feedback* form. Applying the a priori error feedback direct update formula to the LS lattice algorithm, we obtain the a priori error feedback form of the LS lattice algorithm given in Table 6.5. As will be shown in section 6.4, the error feedback LS lattice algorithm has a better numerical behaviour than its correlation-quotient form when a short computer wordlength is used in the implementations.

Table 6.5 A priori error feedback LS lattice algorithm

INITIALIZATION FOR $t = 0$, $m = 0$ TO $p - 1$:

$\alpha_m^{\mathrm{f}}(0) = \alpha_m^{\mathrm{b}}(0) = \delta$, $k_m^{\mathrm{f}}(0) = 0$, $k_m^{\mathrm{b}}(0) = 0$, $k_m^{(y)}(0) = 0$

INITIALIZATION FOR EVERY t:

$e_0^{(y)}(t) = y(t)$, $e_0^{\mathrm{f}}(t) = e_0^{\mathrm{b}}(t) = u(t)$, $\alpha_0(t) = 1$

DO $m = 0$ TO $p - 1$

$e_{m+1}^{\mathrm{f}}(t) = e_m^{\mathrm{f}}(t) + k_m^{\mathrm{f}}(t-1)e_m^{\mathrm{b}}(t-1)$

$e_{m+1}^{\mathrm{b}}(t) = e_m^{\mathrm{b}}(t-1) + k_m^{\mathrm{b}}(t-1)e_m^{\mathrm{f}}(t)$

$\alpha_m^{\mathrm{f}}(t) = \lambda\alpha_m^{\mathrm{f}}(t-1) + \alpha_m(t-1)\,|\,e_m^{\mathrm{f}}(t)\,|^2$

$\alpha_m^{\mathrm{b}}(t) = \lambda\alpha_m^{\mathrm{b}}(t-1) + |\,e_m^{\mathrm{b}}(t)\,|^2\alpha_m(t)$

$k_m^{\mathrm{f}}(t)k_m^{\mathrm{f}}(t-1) - \alpha_m(t-1)e_{m+1}^{\mathrm{f}}(t)e_m^{\mathrm{b}*}(t-1)/\alpha_m^{\mathrm{b}}(t-1)$

$k_m^{\mathrm{b}}(t) = k_m^{\mathrm{b}}(t-1) - \alpha_m(t)e_{m+1}^{\mathrm{b}}(t)e_m^{\mathrm{f}*}(t)/\alpha_m^{\mathrm{f}}(t)$

$e_{m+1}^{(y)}(t) = e_m^{(y)}(t) + k_m^{(y)}(t-1)e_m^{\mathrm{b}}(t)$

$k_m^{(y)}(t) = k_m^{(y)}(t-1) - \alpha_m(t)e_{m+1}^{(y)}(t)e_m^{\mathrm{b}*}(t)/\alpha_m^{\mathrm{b}}(t)$

$\alpha_{m+1}(t) = \alpha_m(t) - |\,\alpha_m(t)e_m^{\mathrm{b}}(t)\,|^2/\alpha_m^{\mathrm{b}}(t)$

END DO

It is not difficult to derive an a posteriori error feedback form of the LS lattice algorithm. The derivation is outlined as follows. From (6.82) we can rewrite (6.81) as

$$e_{m+1}^{(y)}(t) = e_m^{(y)}(t) + k_m^{(yu')}(t-1)e_m^{(u')}(t) = [\varepsilon_m^{(y)}(t) + k_m^{(yu')}(t-1)\varepsilon_m^{(u')}(t)]/\alpha_m(t) \quad (6.90)$$

By substituting (6.90) and (6.82) into (6.89) we obtain

$$k_m^{(yu')}(t) = k_m^{(yu')}(t-1) - \frac{[\varepsilon_m^{(y)}(t) + k_m^{(yu')}(t-1)\varepsilon_m^{(u')}(t)]\varepsilon_m^{(u')*}(t)}{\alpha_m^{(u'u')}(t)\alpha_m(t)} \quad (6.91)$$

This is the a posteriori error form of the direct coefficient updating formula for ORLS estimation. Equations (6.41), (6.91), (6.46) and (6.60) form the basic a posteriori error feedback ORLS estimation. From this general equation set, the derivation of the a posteriori error feedback LS lattice algorithm is straightforward. Simulation has shown that this algorithm has the same numerical robustness as its a priori counterpart [8].

GIVENS-ROTATION-BASED LS LATTICE ALGORITHMS

It is well known that LS estimation problems can be solved using Givens rotation, which is a type of plane rotation known to have good numerical behaviour. Originally, the Givens algorithm was used to solve general LS estimation problems and has a computational complexity proportional to p^2, where p is the order of the estimation

[9]. More recently, LS lattice algorithms based purely on Givens rotations have been suggested for time series LS estimations [10–12]. Below, we derive two types of such Givens lattice algorithms: one with and one without square root operations.

An elementary complex Givens rotation is given by a 2×2 orthogonal matrix called the Givens matrix and defined as

$$G = \begin{pmatrix} c & s \\ -s^* & c \end{pmatrix}$$

where c and s are complex numbers and $|c|^2 + |s|^2 = 1$. It is easy to verify that a Givens matrix performs a plane rotation. For LS estimation, the Givens matrix is commonly used to annihilate an element in a vector such that

$$G\begin{pmatrix} v_1 \\ v_2 \end{pmatrix} = \begin{pmatrix} v_1' \\ 0 \end{pmatrix}$$

To realize this we can let $v_1' = \sqrt{|v_1|^2 + |v_2|^2}$, $c = v_1^*/v_1'$ and $s = v_2^*/v_1'$.

To derive the Givens lattice algorithms, we first introduce another type of LS estimation error, $\tilde{e}_m^{(y)}(t)$, which is defined to be the geometric mean of the corresponding a priori and a posteriori LS errors. That is,

$$\tilde{e}_m^{(y)}(t) = \sqrt{e_m^{(y)}(t)\varepsilon_m^{(y)}(t)} = \sqrt{\alpha_m(t)}e_m^{(y)}(t) = \varepsilon_m^{(y)}(t)/\sqrt{\alpha_m(t)} \tag{6.92}$$

Similarly, we can define $\tilde{e}_m^{(u')}(t) = \sqrt{\alpha_m(t)}e_m^{(u')}(t)$. Since $\sqrt{\alpha_m(t)}$ has the interpretation of the cosine value of the angle between the spaces spanned by the signal up to time $t-1$ and to time t, this new type of LS error is called *angle normalized* LS estimation error [13]. Furthermore, we need to define the following normalized autocorrelation:

$$\tilde{\alpha}_m^{(u'u')}(t) \equiv \sqrt{\alpha_m^{(u'u')}(t)}$$

and cross-correlation:

$$\tilde{k}_m^{(yu')}(t) \equiv -\alpha_m^{(yu')}(t)/\sqrt{\alpha_m^{(u'u')}(t)} = \sqrt{\alpha_m^{(u'u')}(t)}k_m^{(yu')}(t)$$

By using these definitions, we can rewrite (6.81), (6.84) and (6.85) as

$$\tilde{e}_{m+1}^{(y)}(t) = \sqrt{\frac{\alpha_{m+1}(t)}{\alpha_m(t)}}\,\tilde{e}_m^{(y)}(t) - \frac{\alpha_m^{(yu')}(t-1)}{\alpha_m^{(u'u')}(t-1)}\sqrt{\frac{\alpha_{m+1}(t)}{\alpha_m(t)}}\,\tilde{e}_m^{(u')}(t) \tag{6.93}$$

$$\tilde{k}_m^{(yu')}(t) = \sqrt{\frac{\lambda\alpha_m^{(u'u')}(t-1)}{\alpha_m^{(u'u')}(t)}}\sqrt{\lambda}\tilde{k}_m^{(yu')}(t-1) - \frac{\tilde{e}_m^{(u')*}}{\sqrt{\alpha_m^{(u'u')}(t)}}\,\tilde{e}_m^{(y)}(t) \tag{6.94}$$

and

$$\tilde{\alpha}_m^{(u'u')}(t) = \sqrt{\frac{\lambda\alpha_m^{(u'u')}(t-1)}{\alpha_m^{(u'u')}(t)}}\sqrt{\lambda}\tilde{\alpha}_m^{(u'u')}(t-1) + \frac{\tilde{e}_m^{(u')*}}{\sqrt{\alpha_m^{(u'u')}(t)}}\,\tilde{e}_m^{(u')}(t) \tag{6.95}$$

respectively.

From (6.85) and (6.86) we obtain

$$\alpha_{m+1}(t) = \alpha_m(t)\frac{\alpha_m^{(u'u')}(t) - \alpha_m(t)|e_m^{(u')}(t)|^2}{\alpha_m^{(u'u')}(t)} = \alpha_m(t)\frac{\lambda\alpha_m^{(u'u')}(t-1)}{\alpha_m^{(u'u')}(t)}$$

Then we define

$$c_m(t) \equiv \sqrt{\frac{\alpha_{m+1}(t)}{\alpha_m(t)}} = \sqrt{\frac{\lambda\alpha_m^{(u'u')}(t-1)}{\alpha_m^{(u'u')}(t)}} \tag{6.96}$$

and

$$s_m(t) \equiv \frac{\tilde{e}_m^{(u')*}(t)}{\sqrt{\alpha_m^{(u'u')}(t)}} \tag{6.97}$$

Using (6.96) and (6.97) we can write (6.93)–(6.95) as

$$\tilde{e}_{m+1}^{(y)}(t) = c_m(t)\tilde{e}_m^{(y)}(t) + s_m^*(t)\sqrt{\lambda}\tilde{k}_m^{(yu')}(t-1) \tag{6.98}$$

$$\tilde{k}_m^{(yu')}(t) = c_m(t)\sqrt{\lambda}\tilde{k}_m^{(yu')}(t-1) - s_m(t)\tilde{e}_m^{(y)}(t) \tag{6.99}$$

and

$$\tilde{\alpha}_m^{(u'u')}(t) = c_m(t)\sqrt{\lambda}\tilde{\alpha}_m^{(u'u')}(t-1) + s_m(t)\tilde{e}_m^{(u')}(t) \tag{6.100}$$

We can write (6.96)–(6.100) in a compact matrix form as

$$G_m(t)\begin{pmatrix} \sqrt{\lambda}\tilde{\alpha}_m^{(u'u')}(t-1) & -\sqrt{\lambda}\tilde{k}_m^{(yu')}(t-1) \\ \tilde{e}_m^{(u')}(t) & \tilde{e}_m^{(y)}(t) \end{pmatrix} = \begin{pmatrix} \tilde{\alpha}_m^{(u'u')}(t) & -\tilde{k}_m^{(yu')}(t) \\ 0 & \tilde{e}_{m+1}^{(y)}(t) \end{pmatrix} \tag{6.101}$$

where

$$G_m(t) = \begin{pmatrix} c_m(t) & s_m(t) \\ -s_m^*(t) & c_m(t) \end{pmatrix}$$

From (6.84), (6.96) and (6.97) it is easy to see that $|c_m(t)|^2 + |s_m(t)|^2 = 1$. Therefore, the 2×2 matrix $G(t)$ is a Givens matrix, and we conclude that it is possible to use a Givens rotation to perform both time and order updating for general LS order recursive estimation. By applying (6.101) to implement the order recursive forward prediction, backward prediction and joint estimation, we obtain the Givens lattice algorithm with square roots.

The square-root-free Givens lattice algorithm is a variation of the a priori error LS lattice algorithm in error feedback form. By substituting (6.81) into (6.89), we obtain

$$k_m^{(yu')}(t) = k_m^{(yu')}(t-1) - \frac{\alpha_m(t)[e_m^{(y)}(t) + k_m^{(yu')}(t-1)e_m^{(u')}(t)]e_m^{(u')*}(t)}{\alpha_m^{(u'u')}(t)}$$

$$= \left(1 - \frac{\alpha_m(t)\,|\,e_m^{(u')}(t)\,|^2}{\alpha_m^{(u'u')}(t)}\right) k_m^{(yu')}(t-1) - \frac{\alpha_m(t)e_m^{(y)}(t)}{\alpha_m^{(u'u')}(t)}\, e_m^{(u')*}(t) \qquad (6.102)$$

From the result obtained previously, we have

$$1 - \frac{\alpha_m(t)\,|\,e_m^{(u')}(t)\,|^2}{\alpha_m^{(u'u')}(t)} = \frac{\lambda\alpha_m^{(u'u')}(t-1)}{\alpha_m^{(u'u')}(t)} = |\,c_m(t)\,|^2$$

By defining

$$\bar{c}_m(t) = |\,c_m(t)\,|^2 \quad \text{and} \quad \bar{s}_m(t) = \alpha_m(t)e_m^{(y)}(t)/\alpha_m^{(u'u')}(t) \qquad (6.103)$$

we can write (6.102) in the form

$$k_m^{(yu')}(t) = \bar{c}_m(t)k_m^{(yu')}(t-1) - \bar{s}_m(t)e_m^{(u')*}(t) \qquad (6.104)$$

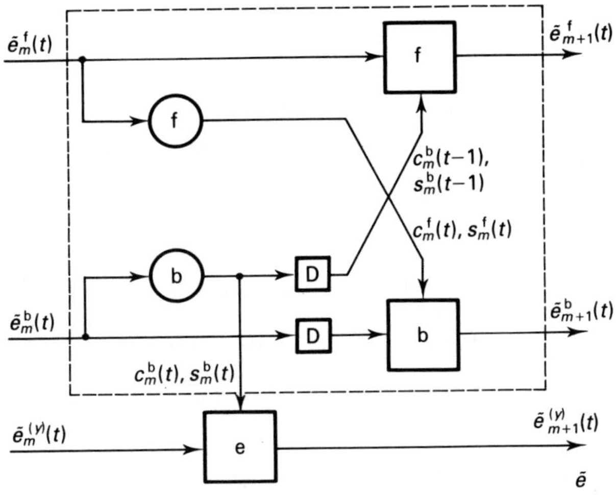

Figure 6.7 An LS lattice estimator

Equations (6.103) and (6.104), together with (6.81), (6.84) and (6.86), form an alternative set of equations that perform a priori error ORLS estimation. By using these results to modify the LS lattice algorithm obtained previously, we can obtain the Givens lattice algorithm without square roots. The complete Givens lattice algorithms with and without square roots are given in Table 6.6. A lattice stage based on the Givens rotations is depicted in Figure 6.7, where a circle denotes the computing element that calculates the rotation parameters and a square represents a computing element that performs rotations.

Table 6.6 The Givens lattice algorithms

Givens lattice algorithm (with square roots)	Givens lattice algorithm (square root free)
Forward rotation parameters:	
$\tilde{\alpha}^f_m(t) = \sqrt{\lambda[\tilde{\alpha}^f_m(t-1)]^2 + \lvert\tilde{e}^f_m(t)\rvert^2}$	$\alpha^f_m(t) = \lambda\alpha^f_m(t-1) + \alpha_m(t-1)\lvert e^f_m(t)\rvert^2$
$c^f_m(t) = \sqrt{\lambda}\tilde{\alpha}^f_m(t-1)/\tilde{\alpha}^f_m(t)$	$\bar{c}^f_m(t) = \lambda\alpha^f_m(t-1)/\alpha^f_m(t)$
$s^f_m(t) = \tilde{e}^{f*}_m(t)/\tilde{\alpha}^f_m(t)$	$\bar{s}^f_m(t) = \alpha_m(t-1)e^{f*}_m(t)/\alpha^f_m(t)$
Backward rotation parameters:	
$\tilde{\alpha}^b_m(t) = \sqrt{\lambda[\tilde{\alpha}^b_m(t-1)]^2 + \lvert\tilde{e}^b_m(t)\rvert^2}$	$\alpha^b_m(t) = \lambda\alpha^b_m(t-1) + \alpha_m(t)\lvert e^b_m(t)\rvert^2$
$c^b_m(t) = \sqrt{\lambda}\tilde{\alpha}^b_m(t-1)/\tilde{\alpha}^b_m(t)$	$\bar{c}^b_m(t) = \lambda\alpha^b_m(t-1)/\alpha^b_m(t)$
$s^b_m(t) = \tilde{e}^{b*}_m(t)/\tilde{\alpha}^b_m(t)$	$\bar{s}^b_m(t) = \alpha_m(t)e^{b*}_m(t)/\alpha^b_m(t)$
Forward rotator:	
$\tilde{e}^f_{m+1}(t) = c^b_m(t-1)\tilde{e}^f_m(t) + \sqrt{\lambda}s^{b*}_m(t-1)\tilde{k}^f_m(t-1)$	$e^f_{m+1}(t) = e^f_m(t) + k^f_m(t-1)e^b_m(t-1)$
$\tilde{k}^f_m(t) = \sqrt{\lambda}c^b_m(t-1)\tilde{k}^f_m(t-1) - s^b_m(t-1)\tilde{e}^f_m(t)$	$k^f_m(t) = \bar{c}^b_m(t-1)k^f_m(t-1) - \bar{s}^b_m(t-1)e^f_m(t)$
Backward rotator:	
$\tilde{e}^b_{m+1}(t) = c^f_m(t)\tilde{e}^b_m(t) + \sqrt{\lambda}s^{f*}_m(t)\tilde{k}^b_m(t-1)$	$e^b_{m+1}(t) = e^b_m(t-1) + k^b_m(t-1)e^f_m(t)$
$\tilde{k}^b_m(t) = \sqrt{\lambda}c^f_m(t)\tilde{\mathbf{k}}^b_m(t-1) - s^f_m(t)\tilde{e}^b_m(t)$	$k^b_m(t) = \bar{c}^f_m(t)k^b_m(t-1) - \bar{s}^f_m(t)e^b_m(t-1)$
Joint estimation rotator:	
$\tilde{e}^{(y)}_{m+1}(t) = c^b_m(t)\tilde{e}^{(y)}_m(t) + \sqrt{\lambda}s^{b*}_m(t)\tilde{k}^{(y)}_m(t-1)$	$e^{(y)}_{m+1}(t) = e^{(y)}_m(t) + k^{(y)}_m(t-1)e^b_m(t)$
$\tilde{k}^{(y)}_m(t) = \sqrt{\lambda}c^b_m(t)\tilde{k}^{(y)}_m(t-1) - s^b_m(t)\tilde{e}^{(y)}_m(t)$	$k^{(y)}_m(t) = \bar{c}^b_m(t)k^{(y)}_m(t-1) - \bar{s}^b_m(t)e^{(y)}_m(t)$

NORMALIZED/SQUARE ROOT LS LATTICE ALGORITHM

The last type of LS lattice algorithm discussed in this section is called the normalized, or square root, LS lattice algorithm. Above we have defined the angle-normalized errors for the Givens lattice algorithm with square roots. The normalized LS lattice algorithm uses full normalized errors, i.e. angle-and-energy-normalized errors, and normalized estimation coefficients. The full normalized error $\tilde{\bar{e}}_m^{(u')}(t)$ is defined as

$$\tilde{\bar{e}}_m^{(u')}(t) = \frac{\varepsilon_m^{(u')}(t)}{\sqrt{\alpha_m^{(u'u')}(t)}\ \sqrt{\alpha_m(t)}} \tag{6.105}$$

Similarly, we define

$$\tilde{\bar{e}}_m^{(y)}(t) = \frac{\varepsilon_m^{(y)}(t)}{\sqrt{\alpha_m^{(yy)}(t)}\ \sqrt{\alpha_m(t)}} \quad \text{and} \quad \tilde{\bar{e}}_{m+1}^{(y)}(t) = \frac{\varepsilon_{m+1}^{(y)}(t)}{\sqrt{\alpha_{m+1}^{(yy)}(t)}\ \sqrt{\alpha_{m+1}(t)}} \tag{6.106}$$

We also define the normalized estimation coefficient $\rho_m^{(yu')}(t)$ as

$$\rho_m^{(yu')}(t) = \frac{\alpha_m^{(yu')}(t)}{\sqrt{\alpha_m^{(yy)}(t)}\ \sqrt{\alpha_m^{(u'u')}(t)}} \tag{6.107}$$

Let us begin by deriving the time update equation of $\rho_m^{(yu')}(t)$. By substituting the above definitions into (6.47), and after some algebra, we obtain

$$\rho_m^{(yu')}(t) = \frac{\sqrt{\lambda\alpha_m^{(yy)}(t-1)}\ \sqrt{\lambda\alpha_m^{(u'u')}(t-1)}}{\sqrt{\alpha_m^{(yy)}(t)}\ \sqrt{\alpha_m^{(u'u')}(t)}}\ \rho_m^{(yu')}(t-1) + \tilde{\bar{e}}(t)\tilde{\bar{e}}_m^{(u')*}(t) \tag{6.108}$$

From (6.85) it can be shown that

$$\frac{\lambda\alpha_m^{(u'u')}(t-1)}{\alpha_m^{(u'u')}(t)} = 1 - \frac{|\varepsilon_m^{(u')}(t)|^2}{\alpha_m(t)\alpha_m^{(u'u')}(t)} = 1 - |\tilde{\bar{e}}_m^{(u')}(t)|^2 \tag{6.109}$$

Similarly, we can show that

$$\frac{\lambda\alpha_m^{(yy)}(t-1)}{\alpha_m^{(yy)}(t)} = 1 - |\tilde{\bar{e}}_m^{(y)}(t)|^2 \tag{6.110}$$

By substituting (6.109) and (6.110) into (6.108), we reach the desired relation

$$\rho_m^{(yu')}(t) = \sqrt{1 - |\tilde{\bar{e}}_m^{(u')}(t)|^2}\ \sqrt{1 - |\tilde{\bar{e}}_m^{(y)}(t)|^2}\ \rho_m^{(yu')}(t-1) + \tilde{\bar{e}}_m^{(y)}(t)\tilde{\bar{e}}_m^{(u')*}(t) \tag{6.111}$$

Another relation needed to complete the normalized order recursive LS estimation is the order update equation of the normalized LS estimation error. To derive this equation, we let $k = t$ in (6.41) and rewrite it in terms of the normalized errors and the normalized estimation coefficient as

$$\tilde{\bar{e}}_{m+1}^{(y)}(t) = \frac{\sqrt{\alpha_m(t)}\ \sqrt{\alpha_m^{(yy)}(t)}}{\sqrt{\alpha_{m+1}(t)}\ \sqrt{\alpha_{m+1}^{(yy)}(t)}}\ (\tilde{\bar{e}}_m^{(y)}(t) - \rho_m^{(yu')}(t)\tilde{\bar{e}}_m^{(u')}(t)) \tag{6.112}$$

By taking exponentially weighted sums of the squared magnitude of both sides of (6.41) and using the orthogonality principle, we obtain

$$\alpha_{m+1}^{(yy)}(t) = \alpha_m^{(yy)}(t) - |\alpha_m^{(yu')}(t)|^2/\alpha_m^{(u'u')}(t)$$

Equivalently,

$$\frac{\alpha_{m+1}^{(yy)}(t)}{\alpha_m^{(yy)}(t)} = 1 - \frac{|\alpha_m^{(yu')}(t)|^2}{\alpha_m^{(yy)}(t)\alpha_m^{(u'u')}(t)} = 1 - |\rho_m^{(yu')}(t)|^2 \tag{6.113}$$

From (6.96) and (6.109) it follows that

$$\frac{\alpha_{m+1}(t)}{\alpha_m(t)} = \frac{\lambda\alpha_m^{(u'u')}(t-1)}{\alpha_m^{(u'u')}(t)} = 1 - |\tilde{\bar{e}}_m^{(u')}(t)|^2 \tag{6.114}$$

By substituting (6.113) and (6.114) into (6.112) we obtain the equation for the order recursion of the normalized error

$$\tilde{\bar{e}}_{m+1}^{(y)}(t) = \sqrt{1 - |\tilde{\bar{e}}_m^{(u')}(t)|^2}\,\sqrt{1 - |\rho_m^{(yu')}(t)|^2}\,(\tilde{\bar{e}}_m^{(y)}(t) - \rho_m^{(yu')}(t)\tilde{\bar{e}}_m^{(u')}(t)) \tag{6.115}$$

Equations (6.115) and (6.111) form the basic equation set of the normalized error ORLS estimation. To obtain the normalized error LS lattice algorithm from them is straightforward. The resulting algorithm is given in Table 6.7. Note that when the desired signal and data for the forward and backward predictions exchange positions, the normalized forward and backward reflection coefficients are complex conjugates of each other. Thus, there are only three equations for a normalized LS lattice prediction stage. However, each of the equations is more complex than other forms of LS lattice algorithms.

Table 6.7 The normalized/square root lattice algorithm

INITIALIZATION FOR $t = 0$:

$\alpha_0^{\mathrm{f}}(0) = \alpha_0^{\mathrm{b}}(0) = \alpha_0^{(y)}(0) = \delta$

INITIALIZATION FOR EVERY t:

$\alpha_0(t) = \lambda\alpha_0(t-1) + |u(t)|^2$

$\alpha_0^{(y)}(t) = \lambda\alpha_0^{(y)}(t-1) + |y(t)|^2$

$\tilde{\bar{e}}_0^{(y)}(t) = y(t)/\sqrt{\alpha_0^{(y)}(t)},\ \tilde{\bar{e}}_0^{\mathrm{f}}(t) = \tilde{\bar{e}}_0^{\mathrm{b}}(t) = u(t)/\sqrt{\alpha_0(t)}$

DO $m = 0$ TO $p - 1$

$\rho_m(t) = \sqrt{1 - |\tilde{\bar{e}}_m^{\mathrm{f}}(t)|^2}\,\sqrt{1 - |\tilde{\bar{e}}_m^{\mathrm{b}}(t) - 1)|^2}\,\rho_m(t-1) + \tilde{\bar{e}}_m^{\mathrm{b}}(t-1)\tilde{\bar{e}}_m^{\mathrm{f}*}(t)$

$\tilde{\bar{e}}_{m+1}^{\mathrm{f}}(t) = (\sqrt{1 - |\tilde{\bar{e}}_m^{\mathrm{b}}(t-1)|^2}\,\sqrt{1 - |\rho_m(t)|^2})^{-1}\,[\tilde{\bar{e}}_m^{\mathrm{f}}(t) - \rho_m(t)\tilde{\bar{e}}_m^{\mathrm{b}}(t-1)]$

$\tilde{\bar{e}}_{m+1}^{\mathrm{b}}(t) = (\sqrt{1 - |\tilde{\bar{e}}_m^{\mathrm{f}}(t)|^2}\,\sqrt{1 - |\rho_m(t)|^2})^{-1}\,[\tilde{\bar{e}}_m^{\mathrm{b}}(t-1) - \rho_m^*(t)\tilde{\bar{e}}_m^{\mathrm{f}}(t)]$

$\rho_m^{(y)}(t) = \sqrt{1 - |\tilde{\bar{e}}_m^{(y)}(t)|^2}\,\sqrt{1 - |\tilde{\bar{e}}_m^{\mathrm{b}}(t)|^2}\,\rho_m^{(y)}(t-1) + \tilde{\bar{e}}_m^{\mathrm{b}}(t)\tilde{\bar{e}}_m^{(y)*}(t)$

$\tilde{\bar{e}}_{m+1}^{(y)}(t) = (\sqrt{1 - |\tilde{\bar{e}}_m^{\mathrm{b}}(t)|^2}\,\sqrt{1 - |\rho_m^{(y)}(t)|^2})^{-1}\,[\tilde{\bar{e}}_m^{(y)}(t) - e_m^{(y)}(t)\tilde{\bar{e}}_m^{\mathrm{b}}(t)]$

END DO

Another important property of this algorithm is that all the quantities involved have a magnitude less than unity. For example, from (6.46) it is easy to show that $|\varepsilon_m^{(u')}(t)|^2/\alpha_m(t) < \alpha_m^{(u'u')}(t)$. Thus it follows that $|\tilde{\varepsilon}_m^{(u')}(t)| < 1$. Similarly, it can be shown that $|\tilde{\varepsilon}_m^{(y)}(t)| < 1$. Finally, $|\rho_m^{(yu')}(t)| \leqslant 1$ follows from (6.107) and the Schwarz inequality. This property greatly simplifies the implementation of the normalized lattice or, in general, normalized ORLS algorithms using fixed point arithmetic.

Both the normalized LS lattice algorithm and the Givens lattice algorithm with square roots need square root operations. As a result, they are not computationally efficient when implemented using conventional hardware. However, they are suitable for implementation using CORDIC VLSI processors [14,15].

6.3.2 Gradient lattice algorithms

In section 6.1, we derived the MMSE lattice estimator. While the MMSE lattice estimator is theoretically optimal, we need to know the signal statistics for its realization. This makes it unrealistic in many practical applications. In section 6.2 we derived the LS lattice algorithm which estimates the statistics of signals directly from the input data. Thus, the LS lattice algorithm, which provides the optimal LS solution, can also be viewed as a realizable approximate implementation of the optimal MMSE lattice estimator. In this section, a family of approximate MMSE lattice algorithms will be presented. These algorithms approach the MMSE lattice estimator asymptotically. Since they are related to the stochastic gradient, or LMS, algorithm discussed in Chapter 4, they are known as *gradient lattice algorithms*. We shall show that these algorithms are actually approximate forms of the LS lattice algorithms. Generally speaking, a gradient algorithm is computationally more efficient than the corresponding LS lattice algorithm but provides a slower initial convergence than the latter. However, as we shall show in the next section, its ability to track time-variant signal characteristics is very close to an LS lattice algorithm.

Several versions of the gradient lattice algorithm exist. We start by deriving an adaptive lattice algorithm using the stochastic gradient method.

STOCHASTIC GRADIENT REALIZATION OF THE MMSE LATTICE ALGORITHM

Assuming that we are able to form an optimal MMSE lattice estimator of order m, the outputs of the mth-stage forward and backward lattice predictors are the mth-order optimal MMSE forward and backward prediction errors. From the discussion in section 6.1, we know that the $(m+1)$th-order MMSE forward prediction error can be computed through scalar MMSE estimations from the mth-order MMSE forward prediction error and the delayed mth-order backward prediction error, which are used as the desired signal and the scalar data respectively. From Chapter 2, we know that such an MMSE estimation can be realized adaptively using the stochastic gradient, or LMS, algorithm. Using this algorithm, the $(m+1)$th-order forward prediction error is computed as

$$e^{\mathrm{f}}_{m+1,\mathrm{grad}}(t) = e^{\mathrm{f}}_{m,\mathrm{grad}}(t) + k^{\mathrm{f}}_{m,\mathrm{grad}}(t-1)e^{\mathrm{b}}_{m,\mathrm{grad}}(t-1) \tag{6.116}$$

The forward prediction coefficient $k^{\mathrm{f}}_{m,\mathrm{grad}}(t-1)$ is updated according to

$$k^{\mathrm{f}}_{m,\mathrm{grad}}(t) = k^{\mathrm{f}}_{m,\mathrm{grad}}(t-1) - \Delta^{\mathrm{f}}_{m}(t)e^{\mathrm{f}}_{m+1,\mathrm{grad}}(t)e^{\mathrm{b}*}_{m,\mathrm{grad}}(t-1) \tag{6.117}$$

Note that $e^{\mathrm{f}}_{m,\mathrm{grad}}(t-1)$ and $e^{\mathrm{b}}_{m,\mathrm{grad}}(t-1)$ are actually MMSE errors according to what we described above. Similarly, the $(m+1)$th-order backward prediction error can be computed as

$$e^{\mathrm{b}}_{m+1,\mathrm{grad}}(t) = e^{\mathrm{b}}_{m,\mathrm{grad}}(t-1) + k^{\mathrm{b}}_{m,\mathrm{grad}}(t-1)e^{\mathrm{f}}_{m,\mathrm{grad}}(t) \tag{6.118}$$

and the backward reflection coefficient is updated according to

$$k^{\mathrm{b}}_{m,\mathrm{grad}}(t) = k^{\mathrm{b}}_{m,\mathrm{grad}}(t-1) - \Delta^{\mathrm{b}}_{m}(t)e^{\mathrm{b}}_{m+1,\mathrm{grad}}(t)e^{\mathrm{f}*}_{m,\mathrm{grad}}(t) \tag{6.119}$$

From the analysis on the LMS algorithm in Chapter 4, we know that the reflection coefficients computed according to (6.117) and (6.119) will be close to their MMSE after certain iterations. The $(m+1)$th-order prediction errors will also approach their MMSE value with some misadjustment error as the reflection coefficients converge. We have denoted the adaptation step sizes for the forward and backward gradient predictors as $\Delta^{\mathrm{f}}_{m}(t)$ and $\Delta^{\mathrm{b}}_{m}(t)$ respectively. This indicates that these step sizes can be either fixed or time varying, as discussed later.

The assumption that the lower-order lattice stages are optimal is neither realistic nor necessary. When starting the gradient lattice algorithm, we do not know the MMSE values of the reflection coefficients. As a result, we have to estimate all the coefficients at all stages from the outlet. Practically, we simply implement the gradient lattice algorithm according to the equations given above without the assumption that the output prediction errors from the lower-order stage already become MMSE prediction errors. The reflection coefficients at all the stages are initialized to zero. Below, we show that the reflection coefficients computed in such a way will approach their MMSE values after many iterations.

To begin with, we consider the forward predictor at the first stage. Since both its desired signal and data $u(t)$ and $u(t-1)$ are given, the first-order forward reflection coefficient will converge towards its MMSE value. For the same reason, the first-order gradient backward predictor will converge towards a first-order MMSE backward predictor. Considering the second stage, at least after the first stage has converged, the second-order forward and backward predictors will start to converge. Similarly, the third stage will converge after the second stage has converged, and so on. Thus, we conclude that at least the gradient lattice algorithm will converge towards an MMSE lattice, in a stage-by-stage fashion. Actually, the higher stages start to converge before the lower-order stages have completely converged [16]. Thus, its convergence is faster than the strict stage-by-stage one.

GRADIENT LATTICE ALGORITHM WITH TIME-VARIANT STEP SIZE

Similar to the LMS algorithm discussed in Chapter 4, we can speed up the convergence of the gradient algorithm using a time-variant step size. Since the convergence rate of the stochastic gradient algorithm is controlled by its step size, we may select the step sizes to make all the stages converge at the same rate. It can be shown from the theory of the LMS adaptive algorithm that the gradient forward predictor at stage $m+1$ converges according to $(1 - \Delta E[e_m^b(t-1)|^2])^n$, where $E[|e_m^b(t-1)|^2]$ is the variance of the scalar data $e_m^b(t-1)$ for the mth-order forward prediction. In order to obtain a uniform convergence for all stages, the step in each stage $m+1$ should be inversely proportional to the variance of $e_m^b(t-1)$. One way to estimate the variance of $e_m^b(t)$ is to form the weighted squared error $\alpha_{m,\mathrm{grad}}^b(t-1)$, such that

$$\alpha_{m,\mathrm{grad}}^b(t) = \lambda\alpha_{m,\mathrm{grad}}^b(t-1) + |e_{m,\mathrm{grad}}^b(t)|^2 \tag{6.120}$$

It can be shown that

$$\alpha_{m,\mathrm{grad}}^b(t) \approx \frac{1}{1-\lambda} E[|e_m^b(t)|^2]$$

Now, if we let

$$\Delta_m^f(t) = 1/\alpha_{m,\mathrm{grad}}^b(t-1) \tag{6.121}$$

for all m, from the discussion above it follows that the output MSE at any stage m converges according to

$$(1 - \Delta_m^f(t)E[|e_m^b(t)|^2])^n \approx [1-(1-\lambda)]^n = \lambda^n$$

Furthermore, if λ is less than but close to unity we have

$$\lambda^n = [1-(1-\lambda)]^n \approx \mathrm{e}^{-n(1-\lambda)}$$

Thus, the MSE of all stages converges uniformly with a time constant of $1/(1-\lambda)$ sampling intervals. In other words, such a gradient lattice algorithm has a uniform converging time constant that is approximately equal to $1/(1-\lambda)$. The backward prediction and joint estimation in each stage are implemented similarly.

SYMMETRIC FORMS OF GRADIENT LATTICE ALGORITHM

It is known from the MMSE lattice theory developed in section 6.1 that $k_{m,\mathrm{MMSE}}^f = k_{m,\mathrm{MMSE}}^{b*}$ if the signal is stationary. Since the gradient lattice algorithm is an adaptive approximation of the MMSE lattice, we can force the estimated forward and backward reflection coefficients in the gradient lattice algorithm to be complex conjugates of each other at each iteration. That is, we define

$$k_{m,\mathrm{grad}}(t) \equiv k_{m,\mathrm{grad}}^f(t) \equiv k_{m,\mathrm{grad}}^{b*}(t)$$

Also, by defining a step size $\Delta_m(t) \equiv \Delta_m^f(t) = \Delta_m^b(t)$, from (6.117) and (6.119) we obtain the time recursion for $k_{m,\mathrm{grad}}(t)$ such that

$$k_{m,\mathrm{grad}}(t) = k_{m,\mathrm{grad}}(t-1) - \frac{\Delta_m(t)}{2}$$
$$\times [e_{m+1,\mathrm{grad}}^{b*}(t)e_{m,\mathrm{grad}}^f(t) + e_{m+1,\mathrm{grad}}^f(t)e_{m,\mathrm{grad}}^{b*}(t-1) \tag{6.122}$$

Because the variances of the MMSE forward and backward prediction errors are equal to each other, we can obtain an estimate of the variance by taking the average of the weighted sums of the forward and backward prediction squared errors. The averaged estimate of the variance can be computed time-recursively as

$$\alpha_{m,\mathrm{grad}}(t) = \lambda\alpha_{m,\mathrm{grad}}(t-1) + [\,|e^{\mathrm{f}}_{m,\mathrm{grad}}(t)|^2 + |e^{\mathrm{b}}_{m,\mathrm{grad}}(t-1)|^2]/2 \qquad (6.123)$$

To obtain a uniform convergence,we select $\Delta_m(t)$ to be equal to

$$\Delta_m(t) = 1/\alpha_{m,\mathrm{grad}}(t) \qquad (6.124)$$

Using (6.122)–(6.124), together with (6.116) and (6.118), we obtain another form of the gradient lattice prediction algorithm. This gradient lattice algorithm has the constraint that its forward and backward reflection coefficients are complex conjugates of each other.

CORRELATION-QUOTIENT FORM OF GRADIENT LATTICE ALGORITHM

Equation (6.8) shows that the optimal MMSE coefficient in a scalar estimation is equal to the quotient of the cross-correlation between the desired signal and the scalar data divided by the autocorrelation of the data. Thus, we may compute the reflection coefficient in a gradient lattice algorithm by first computing the estimate of the auto- and cross-correlations of the forward and backward prediction errors. We have shown that the autocorrelation of the backward prediction error is computed according to (6.120) and that of the forward prediction error is obtained similarly. The estimate of the cross-correlation between forward and delayed backward prediction errors can be computed time-recursively as

$$\beta_{m,\mathrm{grad}}(t) = \lambda\beta_{m,\mathrm{grad}}(t-1) + e^{\mathrm{f}}_{m,\mathrm{grad}}(t)e^{\mathrm{b}*}_{m,\mathrm{grad}}(t-1) \qquad (6.125)$$

Dividing $\beta_{m,\mathrm{grad}}(t)$ by $\alpha^{\mathrm{b}}_{m,\mathrm{grad}}(t-1)$, the autocorrelation of the backward prediction error, we obtain an estimate of the forward prediction coefficient. That is,

$$k^{\mathrm{f}}_{m,\mathrm{grad}}(t) = -\beta_{m,\mathrm{grad}}(t)/\alpha^{\mathrm{b}}_{m,\mathrm{grad}}(t-1) \qquad (6.126)$$

The backward prediction coefficient is computed by dividing $\beta_{m,\mathrm{grad}}(t)$ by the autocorrelation of the forward prediction error. The joint estimation coefficient is computed by dividing the cross-correlation between the joint estimation error and the backward prediction error by the autocorrelation of the backward prediction error.

By using (6.123) we can obtain a symmetric correlation-quotient gradient lattice algorithm. That is, the reflection coefficient is computed as

$$k_{m,\mathrm{grad}}(t) \equiv k^{\mathrm{f}}_{m,\mathrm{grad}}(t) \equiv k^{\mathrm{b}*}_{m,\mathrm{grad}}(t) = -\beta_{m,\mathrm{grad}}(t)/\alpha_{m,\mathrm{grad}}(t) \qquad (6.127)$$

where $\alpha_{m,\mathrm{grad}}(t)$ is computed according to (6.123). It can be shown that the reflection coefficients computed in such a way have magnitudes less than unity. This property is useful for ensuring the stability of a lattice predictor when it is used for AR modelling.

Two symmetric gradient lattice algorithms are summarized in Table 6.8. Asymmetric gradient algorithms are not given because they are very similar to LS lattice algorithms. Actually, they are just the LS lattice algorithms without the factors $\alpha_m(t)$. With the α factors, the LS lattice algorithms perform exact LS estimation and, thus, provide a faster initial convergence.

Table 6.8 Gradient LS lattice algorithms

INITIALIZATION FOR $t = 0$, $m = 0$ TO $p - 1$:

$\alpha_m^f(0) = \alpha_m^b(0) = \delta$, $\beta_{m,\text{grad}}(0) = 0$, $\beta_{m,\text{grad}}^{(y)}(0) = 0$

INITIALIZATION FOR EVERY t

$e_0^{(y)}(t) = y(t)$, $e_0^f(t) = e_0^b(t) = u(t)$, $\alpha_0(t) = 1$

DO $m = 0$ TO $p - 1$ (symmetric correlation quotient form)

$$\beta_{m,\text{grad}}(t) = \lambda\beta_{m,\text{grad}}(t-1) + 2e_{m,\text{grad}}^f(t)e_{m,\text{grad}}^{b*}(t-1)$$

$$\alpha_{m,\text{grad}}(t) = \lambda\alpha_{m,\text{grad}}(t-1) + |e_{m,\text{grad}}^f(t)|^2 + |e_{m,\text{grad}}^b(t-1)|^2$$

$$k_{m,\text{grad}}(t) = -\beta_{m,\text{grad}}(t)/\alpha_{m,\text{grad}}(t)$$

$$e_{m+1,\text{grad}}^f(t) = e_{m,\text{grad}}^f(t) + k_{m,\text{grad}}(t)e_{m,\text{grad}}^b(t-1)$$

$$e_{m+1,\text{grad}}^b(t) = e_{m,\text{grad}}^b(t-1) + k_{m,\text{grad}}^*(t)e_{m,\text{grad}}^f(t)$$

$$\beta_{m,\text{grad}}^{(y)}(t) = \lambda\beta_{m,\text{grad}}^{(y)}(t-1) + 2e_{m,\text{grad}}^{(y)}(t)e_{m,\text{grad}}^b(t)$$

$$k_{m,\text{grad}}^{(y)}(t) = -\beta_{m,\text{grad}}^{(y)}(t)/\alpha_{m,\text{grad}}(t)$$

$$e_{m+1,\text{grad}}^{(y)}(t) = e_{m,\text{grad}}^{(y)}(t) + k_{m,\text{grad}}^{(y)}(t)e_{m,\text{grad}}^b(t)$$

END DO

DO $m = 0$ TO $p - 1$ (symmetric error feedback form)

$$e_{m+1,\text{grad}}^f(t) = e_{m,\text{grad}}^f(t) + k_{m,\text{grad}}(t-1)e_{m,\text{grad}}^b(t-1)$$

$$e_{m+1,\text{grad}}^b(t) = e_{m,\text{grad}}^b(t-1) + k_{m,\text{grad}}^*(t-1)e_{m,\text{grad}}^f(t)$$

$$\alpha_{m,\text{grad}}(t) = \lambda\alpha_{m,\text{grad}}(t-1) + |e_{m,\text{grad}}^f(t)|^2 + |e_{m,\text{grad}}^b(t-1)|^2$$

$$k_{m,\text{grad}}(t) = k_{m,\text{grad}}(t-1) - [e_{m+1,\text{grad}}^f(t)e_{m,\text{grad}}^{b*}(t-1) + e_{m+1,\text{grad}}^b(t)e_{m,\text{grad}}^{f*}(t)]/\alpha_{m,\text{grad}}(t-1)$$

$$e_{m+1,\text{grad}}^{(y)}(t) = e_{m,\text{grad}}^{(y)}(t) + k_{m,\text{grad}}^{(y)}(t-1)e_{m,\text{grad}}^b(t)$$

$$k_{m,\text{grad}}^{(y)}(t) = k_{m,\text{grad}}^{(y)}(t-1) - 2e_{m+1,\text{grad}}^{(y)}(t)e_{m,\text{grad}}^{b*}(t)/\alpha_{m,\text{grad}}(t)$$

END DO

6.3.3 Multichannel LS lattice algorithm

In the previous sections, the data vector in the estimation problem was taken from a single time series, i.e. from a single data channel. Therefore, we call lattice algorithms with such a data vector *single-channel* lattice algorithms. In this section, we extend the LS lattice algorithm to the multichannel case. We consider a multichannel LS estimation problem in which the data are taken from p time series and the desired signals are taken from q time series.

Let us consider an mth-order LS estimation problem in which the data are taken from p time series. The data vector can be written as

$$\mathbf{u}_{lp}(k) \equiv (u_1(k), \ldots, u_p(k), u_1(k-1), \ldots, u_p(k-1), \ldots u_1(k-l+1), \ldots, u_p(k-l+1))^{\mathrm{T}}$$

$$\equiv (u^{\mathrm{T}}(k), u^{\mathrm{T}}(k-1), \ldots, u^{\mathrm{T}}(k-l+1))^{\mathrm{T}} \tag{6.128}$$

where $m = l_p$ and $u(k) \equiv (u_1(k), \ldots u_p(k))^T$ are p-dimensional vectors whose elements $u_i(k)$, $i = 1, \ldots, p$, are samples of p time series $\{u_i(k), k = 0, \ldots, t\}$. Similarly, the desired signal $y(k) \equiv (y_1(k), \ldots, y_q(k))^T$ is a q-dimensional vector whose elements $y_j(k)$, $j = 1, \ldots, q$, are samples of q time series $\{y_i(k), k = 0, \ldots, t\}$.

The mth-order q-channel LS error, which is a q-dimensional vector, of the desired signal $y(k)$ is given by

$$e_m^{(y)}(k, t) = y(k) + \mathbf{c}_m^{(y)T}(t)\mathbf{u}_m(k)$$

where the LS coefficient $\mathbf{c}_m^{(y)}(t)$ is an $m \times q$ matrix that minimizes the trace of the autocorrelation matrix of $e_m^{(y)}(k, t)$. The trace can be written as $\Sigma_{j=1}^{q} \Sigma_{k=0}^{t} \lambda^{t-k} | e_j^{(y)}(k, t) |^2$, where $e_j^{(y)}(k, t)$ is the jth-element of $e_m^{(y)}(k, t)$. It can be shown that the coefficient matrix $\mathbf{c}_m^{(y)}(t)$ satisfies

$$\mathbf{c}_m^{(y)}(t) = -R_{\mathbf{u}_m\mathbf{u}_m}^{-1}(t)\mathbf{d}_m^{(y)}(t)$$

where $R_{\mathbf{u}_m\mathbf{u}_m}^{-1}(t)$ is defined in (6.39) and $\mathbf{d}_m^{(y)}(t)$ is the cross-correlation matrix between $y(k)$ and $\mathbf{u}_m(k)$ defined by

$$\mathbf{d}_m^{(y)}(t) = \sum_{k=0}^{t} \lambda^{t-k}\mathbf{u}_m^*(k)y^T(k)$$

From the definitions given above, it can be seen that each element of the LS error vector $e_m^{(y)}(k, t)$ is simply the scalar LS error of the corresponding element in the desired signal vector $y(k)$ estimated based on the data vector $\mathbf{u}_m(k)$. Thus, a q-channel LS estimation is equivalent to q LS estimations with scalar desired signals. Therefore, we can apply all the results obtained from the LS estimation for scalar desired signals discussed previously. Notably, the elements of the error vector are all orthogonal to the data vector $\mathbf{u}_m(t)$.

To derive the multichannel LS lattice algorithm, we first state the following extension of Theorem 6.2, which is called *the block order recursive relation of LS estimation.*

The order recursive relation for LS estimation given by Theorem 6.2 can be extended to a more general case, as stated below.

THEOREM 6.5

(Block order recursive relation of LS estimation) By partitioning the $(m + p)$-dimensional data vector, $\mathbf{u}_{m+p}(k)$, $k = 0, 1, \ldots, t$, as

$$\mathbf{u}_{m+p}(k) = (\mathbf{u}_m^T(k), u'^T(k))^T \tag{6.129}$$

the $(m + p)$th-order LS error of the scalar desired signal $y(k)$ estimated based on the data vector $\mathbf{u}_{m+p}(k)$, $e_{m+p}^{(y)}(k, t)$, can be computed from the mth-LS errors $e_m^{(y)}(k, t)$ and $e_m^{(u')}(k, t)$ as

$$e_{m+p}^{(y)}(k, t) = e_m^{(y)}(k, t) + k_m^{(yu')T}(t)e_m^{(u')}(k, t)$$

where $e_m^{(u')}(k, t)$, which is a p-dimensional vector, is the mth-order LS error of $u'(k)$ estimated based on $\mathbf{u}_m(k)$ at time t and $k_m^{(yu')\mathrm{T}}(t)$ is the LS estimation coefficient vector, which is defined as

$$k_m^{(yu')}(t) \equiv -[\alpha_m^{(u'u')}(t)]^{-1}\alpha_m^{(yu')}(t)$$

In the above definition, $\alpha_m^{(u'u')}(t)$ is the $p \times p$ autocorrelation matrix of $e_m^{(u')}(k, t)$, defined as

$$\alpha_m^{(u'u')}(t) \equiv \sum_{k=0}^{t} \lambda^{t-k} e_m^{(u')*}(k, t) e_m^{(u')\mathrm{T}}(k, t) \tag{6.130}$$

and $\alpha_m^{(yu')}(t)$ denotes the cross-correlation vector between $e_m^{(y)}(k, t)$ and the vector $e_m^{(u')}(k, t)$, defined as

$$\alpha_m^{(yu')}(t) \equiv \sum_{k=0}^{t} \lambda^{t-k} e_m^{(y)}(k, t) e_m^{(u')*}(k, t) \tag{6.131}$$

Since the difference between the order of $e_{m+p}^{(y)}(k, t)$ and $e_m^{(y)}(k, t)$ is equal to p, which is greater than one, (6.129) is called the *block order recursive relation of LS estimation*.

To prove Theorem 6.5, we only need to note that, as discussed above, each element of the LS error vector $e_m^{(u')}(k, t)$ is the scalar LS error of the corresponding component in the vector $u'(k)$ based on the data vector $\mathbf{u}_m(k)$. Therefore, from the orthogonality principle, $e_m^{(u')}(k, t)$ is orthogonal to $\mathbf{u}_m(k)$ because all of its components are orthogonal to $\mathbf{u}_m(k)$. That is,

$$\sum_{k=0}^{t} \lambda^{t-k} \mathbf{u}_m^*(k) e_m^{(u')\mathrm{T}}(k) = \mathbf{0}$$

Then, Theorem 6.5 can be proved in exactly the same way as Theorems 6.1 and 6.2 using the orthogonality principle. This block recursive relation can be easily extended to the case where the desired signal is a vector by using the result on the decomposition of vector LS estimation discussed at the beginning of this section.

The time update equations for the correlations $\alpha_m^{(u'u')}(t)$ and $\alpha_m^{(yu')}(t)$ can be derived in the same way as their scalar versions given by Theorem 6.3. Because the lower-order LS errors may be vectors, it might be necessary to transpose them to form proper products. It is also possible to obtain direct update equations for the LS estimation coefficient vector $k_m^{(yu')}(t)$. However, the block LS estimation and its time updating need matrix and vector operations. These operations increase the complexity of computation and may cause numerical instability. Below, we show that it is possible to decompose such a block ORLS estimation further. The resulting algorithm requires only scalar LS estimations.

FURTHER DECOMPOSITION OF BLOCK ORLS ESTIMATION

We first define a set of intermediate data vectors $\mathbf{u}_{m+i}(k)$, $i = 1, \ldots, p$, as

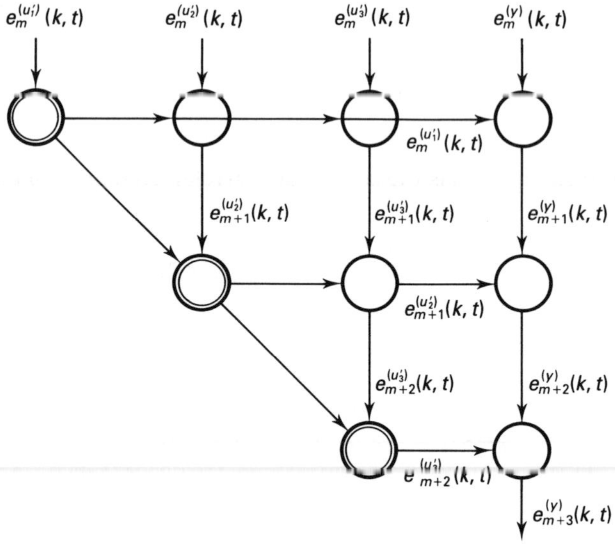

Figure 6.8 Further decomposition of block order recursive LS estimation

$$\mathbf{u}_{m+i}(k) = (\mathbf{u}_m^{\mathrm{T}}(k), u'_1(k), \ldots, u'_i(k))^{\mathrm{T}}$$

where $u'_i(k)$ is the ith element of $u'(k)$. Obviously, $\mathbf{u}_{m+i}(k)_{|i=p} \equiv \mathbf{u}_{m+p}(k)$, as defined previously. To simplify the discussion, we assume $p = 3$ below.

We first compute $e_{m+1}^{(y)}(k, t)$, the $(m + 1)$th-order LS error of $y(k)$ based on the data vector $\mathbf{u}_{m+1}(k, t)$ from the mth-order LS error of $e_m^{(y)}(k, t)$ and $e_m^{(u'_1)}(k, t)$. For later use, we also compute $e_{m+1}^{(u'_2)}(k, t)$ and $e_{m+1}^{(u'_3)}(k, t)$ from $e_m^{(u'_i)}(k, t)$ $i = 1, 2, 3$. Note that $e_m^{(y)}(k, t)$ and $e_m^{(u'_i)}(k, t)$, $i = 1, 2, 3$, the elements of the mth error vector $e_m^{(u')}(k, t)$, are known. These $(m + 1)$th-order LS errors are computed using the ORLS estimation disussed in section 6.2.1. Then we compute the LS errors $e_{m+2}^{(y)}(k, t)$ and $e_{m+2}^{(u'_3)}(k, t)$ from the three $(m + 1)$th LS errors computed from the previous step. Finally, from the two $(m + 2)$th-order LS errors, we compute the desired $(m + 3)$th-order LS error $e_{m+3}^{(y)}(k, t)$.

The computation process described above is depicted in Figure 6.8. The triangular processor actually generates a set of LS errors $e_{m+i-1}^{(u'_i)}(k, t)$, $i = 1, 2, 3$, which are orthogonal to each other, from $e_m^{(u'_i)}(k, t)$, $i = 1, 2, 3$, which are correlated. This processor performs a Gram–Schmidt-type orthogonalization of its inputs. This orthogonalization process enables us to decompose an LS estimation based on the data vector consisting of $e_m^{(u'_i)}(k, t)$ to three scalar LS estimations using $e_{m+i-1}^{(u'_i)}(k, t)$ as the scalar data. If the desired signal is a vector sequence, we can achieve the same order recursive estimation on an element-by-element basis, as discussed above.

MULTICHANNEL LS LATTICE ALGORITHM

By partitioning the data vector according to (6.128), we can define the mth-order p-channel forward and backward predictions, where $m = pl$. The forward LS prediction error is a p-dimensional vector error of $u(k)$ based on the data vector $\mathbf{u}_{lp}(k-1) \equiv (u^{\mathrm{T}}(k-1), \ldots, u^{\mathrm{T}}(k-l))^{\mathrm{T}}$. Similarly, the backward prediction error is the LS error vector of $u(k-l)$ based on the data vector $\mathbf{u}_{lp}(k) \equiv (u^{\mathrm{T}}(k), \ldots, u^{\mathrm{T}}(k-l+1))^{\mathrm{T}}$. Under this definition and using the block order recursive relation given by Theorem 6.5, we can derive the multichannel version of the LS lattice algorithm in the same way as the derivation of the single-channel LS lattice algorithm given in section 6.2. The p-channel lattice algorithm is given in Table 6.9. We note that the forward and backward prediction errors are p-dimensional vectors. The forward and backward reflection coefficients, the cross-correlation between the forward and backward prediction errors and their autocorrelation are all $p \times p$ matrices. The joint estimation coefficients and the cross-correlation between the joint estimation errors and the backward prediction errors are p-dimensional vectors for a scalar desired signal $y(k)$.

Table 6.9 Multichannel LS lattice algorithm

INITIALIZATION FOR $t = 0$, $m = 0$ TO $p - 1$:

$$\alpha_m^{\mathrm{f}}(0) = \alpha_m^{\mathrm{b}}(0) = \delta I,\ k_m(0) = 0,\ k_m^{(y)}(0) = 0$$

INITIALIZATION FOR EVERY t:

$$\varepsilon_0^{(y)}(t) = y(t),\ \varepsilon_0^{\mathrm{f}}(t) = \varepsilon_0^{\mathrm{b}}(t) = u(t),\ \alpha_0(t) = 1$$

DO $m = 0$ TO $p - 1$

$$b_m(t) = \lambda b_m(t-1) + \varepsilon_m^{\mathrm{b}*}(t-1)\varepsilon_m^{\mathrm{fT}}(t)/\alpha_m(t-1)$$

$$\alpha_m^{\mathrm{f}}(t) = \lambda \alpha_m^{\mathrm{f}}(t-1) + \varepsilon_m^{\mathrm{f}*}(t)\varepsilon_m^{\mathrm{fT}}(t)/\alpha_m(t-1)$$

$$\alpha_m^{\mathrm{b}}(t) = \lambda \alpha_m^{\mathrm{b}}(t-1) + \varepsilon_m^{\mathrm{b}*}(t)\varepsilon_m^{\mathrm{bT}}(t)/\alpha_m(t)$$

$$k_m^{\mathrm{f}}(t) = -b_m^{\mathrm{T}}(t)[\alpha_m^{\mathrm{b}}(t-1)]^{-1}$$

$$k_m^{\mathrm{b}}(t) = -b_m(t)[\alpha_m^{\mathrm{f}}(t)]^{-1}$$

$$\varepsilon_{m+1}^{\mathrm{f}}(t) = \varepsilon_m^{\mathrm{f}}(t) + k_m^{\mathrm{f}}(t)\varepsilon_m^{\mathrm{b}}(t-1)$$

$$\varepsilon_{m+1}^{\mathrm{b}}(t) = \varepsilon_m^{\mathrm{b}}(t) + k_m^{\mathrm{b}*}(t)\varepsilon_m^{\mathrm{f}}(t)$$

$$b_m^{(y)}(t) = \lambda b_m^{(y)}(t-1) + \varepsilon_m^{\mathrm{b}*}(t)\varepsilon_m^{(y)\mathrm{T}}(t)/\alpha_m(t)$$

$$k_m^{(y)}(t) = -b_m^{(y)\mathrm{T}}(t)[\alpha_m^{\mathrm{b}}(t)]^{-1}$$

$$\varepsilon_{m+1}^{(y)}(t) = \varepsilon_m^{(y)}(t) + k_m^{(y)}(t)\varepsilon_m^{\mathrm{b}}(t)$$

$$\alpha_{m+1}(t) = \alpha_m(t) - \varepsilon_m^{\mathrm{bT}}(t)[\alpha_m^{\mathrm{b}}(t)]^{-1}\varepsilon_m^{\mathrm{b}*}(t)$$

END DO

It can be seen from Table 6.9 that matrix and vector operations are required to implement the p-channel LS lattice algorithm. Below, we derive two realizations of

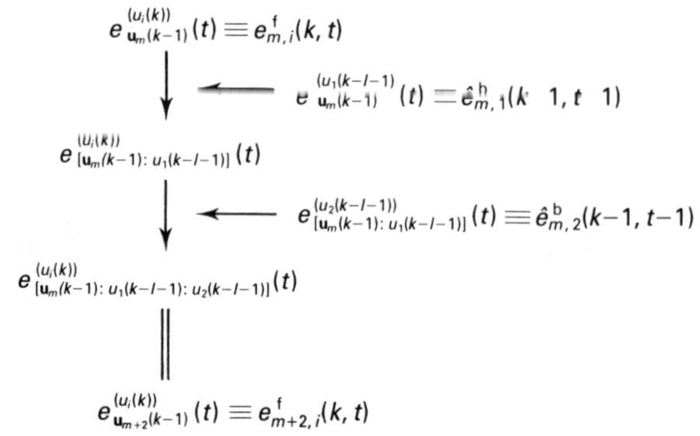

Figure 6.9 Block order update of forward prediction errors using orthogonalized backward prediction errors

the multichannel lattice stages which avoid such operations. Both methods further decompose the block ORLS estimation/predictions involved in the multichannel lattice algorithm into multiple scalar LS estimations. However, they use different decomposition methods and result in two different structures of multichannel lattice stages. The first implementation was given in [17] and called the sequential processing implementation. The second method was proposed in [18]. Detailed derivations of these two algorithms can be found in [18] and [17]. Below, we consider their simplest case where $p = 2$.

SEQUENTIAL PROCESSING IMPLEMENTATION OF MULTICHANNEL LATTICE STAGES

This implementation of the multichannel LS lattice algorithm is a direct consequence of the decomposition of the block ORLS estimation described above. As an example, we consider the forward prediction at stage l of a two-channel LS lattice. The forward predictor computes the $2(l + 1)$th-order forward prediction errr from $2l$th-order forward prediction error and delayed backward prediction error through ORLS estimations. The sequential processing method uses the decomposition approach described above. By employing the triangular processor, we generate a set of two orthogonal errors from the elements of the backward prediction error vector. Then we compute the two elements of the $2(l + 1)$th-order forward prediction error vector through two scalar ORLS estimations.

To see this more clearly, we first examine the elements of the backward and forward prediction errors at the input and output of stage m. By definition, the ith element $e^{\mathrm{f}}_{m,i}(k, t)$ of the forward prediction error vector $e^{\mathrm{f}}_{m}(k, t)$ is the LS error of $u_i(k)$ based on data vector $\mathbf{u}_m(k-1)$, where $m = 2l$ and $i = 1, 2$. By using the more revealing notation given by (6.44) and (6.45) in section 6.2.1, we can denote the ith element as $e^{\mathrm{f}}_{m,i}(k, t) \equiv e^{(u_i(k))}_{\mathbf{u}_m(k-1)}(t)$. Similarly, the ith element of the delayed backward prediction error vector can be denoted as $e^{\mathrm{b}}_{m,i}(k-1, t-1) \equiv e^{(u_i(k-l-1))}_{\mathbf{u}_m(k-1)}(t)$. After processing the

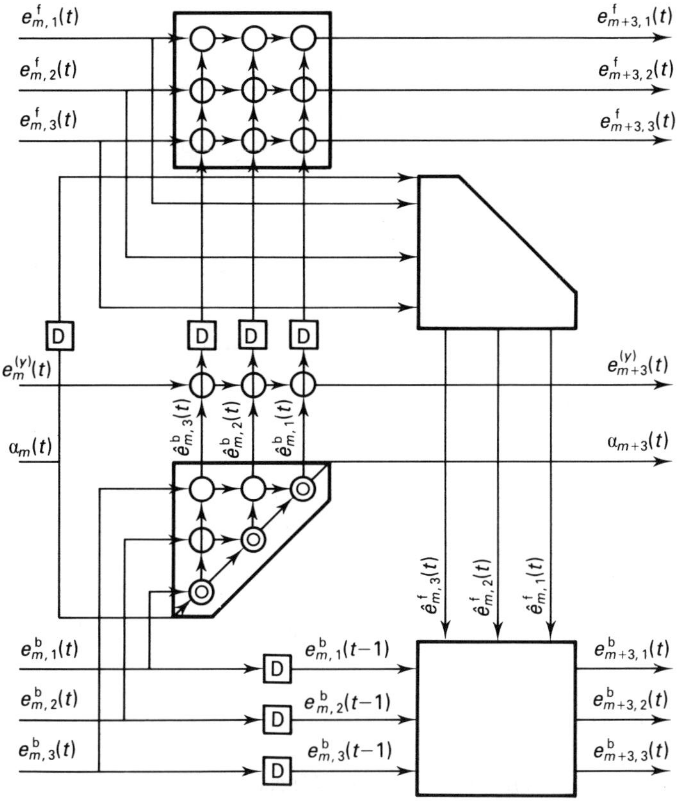

Figure 6.10 A sequential processing three-channel lattice stage

backward prediction errors by the triangular processor, we obtain the orthogonalized mth-order backward prediction errors (delayed by one sample interval) as

$$\hat{e}^{b}_{m,1}(k-1, t-1) \equiv e^{b}_{m,1}(k-1, t-1) \equiv e^{(u_1(k-l-1))}_{\mathbf{u}_m(k-1)}(t)$$

and

$$\hat{e}^{b}_{m,2}(k-1, t-1) \equiv e^{(u_2(k-l-1))}_{[\mathbf{u}_m(k-1):\ u_1(k-l-1)]}(t) \tag{6.132}$$

The elements of the $(m+2)$th-order forward prediction errors, $e^{f}_{m+2}(k, t)$, denoted by $e^{f}_{m+2,i}(k, t) \equiv e^{(u_i(k))}_{\mathbf{u}_{m+2}(k-1)}(t) \equiv e^{(u_i(k))}_{[\mathbf{u}_m(k-1)(t):u_1(k-l-1):u_2(k-l-1)]}(t), i = 1, 2,$ can be computed order-recursively as shown by Figure 6.9. The situations are similar for backward prediction and joint estimation.

The multichannel LS lattice algorithm employing sequential processing can be implemented time-recursively by using the results obtained earlier. Since the elementary operations in the algorithm are scalar LS estimations, any of the variations performing such estimations described in the last two sections can be used for its implementation. The structure of a three-channel, time recursive LS lattice stage using the sequential processing method is depicted in Figure 6.10

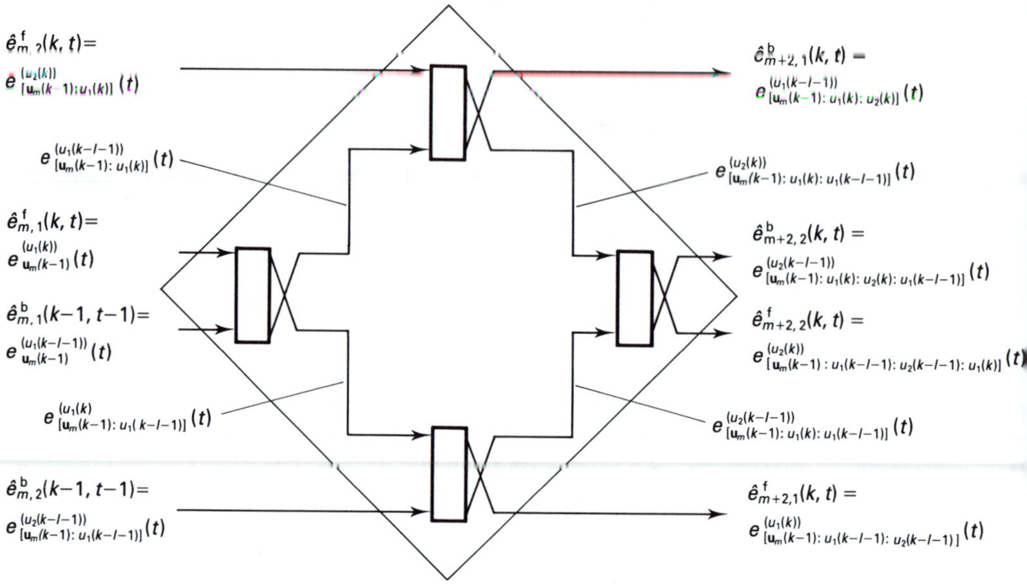

Figure 6.11 Joint order-update of forward and backward prediction errors.

IMPLEMENTATION OF MULTICHANNEL LATTICE STAGES USING JOINT DECOMPOSITION

As described above, in each lattice stage of the sequential processing implementation, we first orthogonalize the mth-order backward and forward prediction errors. After such orthogonalization we reduce the block ORLS estimation to p scalar LS estimations for generating the higher-order prediction errors. The elements of the generated higher-order prediction errors at this stage are orthogonalized again at the next lattice stage. The question is whether we can directly generate higher orthogonalized prediction errors from lower-order orthogonalized prediction errors. It is indeed possible, as shown in [18]. Below, we describe such a two-channel lattice stage.

Similar to (6.132), we express the mth-order orthogonalized forward prediction errors and $(m+2)$th-order orthogonalized forward and backward prediction errors at the output of the $(l+1)$th lattice stage using the notation given above. That is,

$$\hat{e}^{\mathrm{f}}_{m,1}(k,t) \equiv e^{\mathrm{f}}_{m,1}(k,t) \equiv e^{(u_1(k))}_{\mathbf{u}_m(k-1)}(t) \quad \text{and} \quad \hat{e}^{\mathrm{f}}_{m,2}(k,t) \equiv e^{(u_2(k))}_{[\mathbf{u}_m(k-1):u_1(k)]}(t) \tag{6.133}$$

$$\hat{e}^{\mathrm{f}}_{m+2,1}(k,t) \equiv e^{\mathrm{f}}_{m+2,1}(k,t) \equiv e^{(u_1(k))}_{\mathbf{u}_{m+2}(k-1)}(t) \quad \text{and} \quad \hat{e}^{\mathrm{f}}_{m+2,2}(k,t) \equiv e^{(u_2(k))}_{[\mathbf{u}_{m+2}(k-1):u_1(k)]}(t) \tag{6.134}$$

and

$$\hat{e}^{\mathrm{b}}_{m+2,1}(k,t) \equiv e^{\mathrm{b}}_{m+2,1}(k,t) \equiv e^{(u_1(k-l-1))}_{\mathbf{u}_{m+2}(k)}(t) \quad \text{and} \quad \hat{e}^{\mathrm{b}}_{m+2,2}(k,t) \equiv e^{(u_2(k-l-1))}_{[\mathbf{u}_{m+2}(k):u_1(k-1)]}(t) \tag{6.135}$$

respectively.

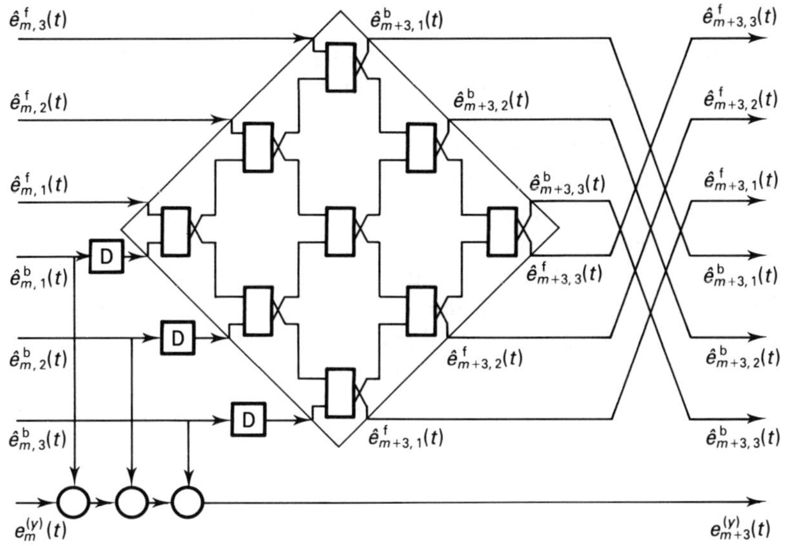

Figure 6.12 A three-channel lattice stage based on single-channel lattice blocks

The processing procedure for generating the orthogonalized $(m + 2)$th prediction errors, given by (6.134) and (6.135), from the orthoganalized mth-order forward prediction errors given by (6.133) and the delayed orthogonalized mth-order backward prediction errors given by (6.132) is shown in Figure 6.11. Each of the rectangular boxes in the figure generates two high-order LS errors from two lower-order LS errors which are inputs to the box. That is, the box performs two scalar LS estimations that use one of the inputs as the desired signal and the other as the scalar data in the estimation, and vice versa. Note that such a box actually performs the same funtion as a single-channel lattice stage. The intermediate quantities generated inside the two-channel lattice stage are also given in the figure. It is not difficult to verify from the figure that, when we use the four elements of the orthogonalized lower-order prediction errors as inputs, such a stage indeed generates the elements of the orthogonalized higher-order prediction errors.

Figure 6.11 shows that both the higher-order orthogonalized backward and forward prediction errors can be jointly generated from the lower-order ones using multiple, single-channel lattice stages. From the time recursive LS estimation theorems derived previously, the scheme of jointly updating forward and backward prediction errors can be easily implemented time-recursively. The result can be extended to the case where $p \geqslant 2$. The structure of such a three-channel lattice stage is depicted in Figure 6.12. It is clear from the figure that such a multichannel LS lattice realization has a more regular structure than the sequentially processing lattice stage. On the other hand, the sequentially processing lattice stage needs less computation than the joint estimation stage. Although the joint orthogonalization multichannel lattice does not perform orthogonalization at each stage, a single orthogonalization processor just like the one used in the sequential processing stage described previously is needed to generate the lowest-order orthogonalized errors [18].

TIME RECURSIVE IMPLEMENTATION

We have shown that a multichannel lattice stage can be implemented based on multiple scalar LS estimations. In particular, the basic processing unit of the multichannel lattice stage implemented using the joint decomposition approach is exactly the same as a single-channel lattice stage. Consequently, we can use any of the variations of implementing scalar LS estimations discussed in section 6.3.1 to implement multichannel lattice algorithms. It is also possible to derive gradient types of multichannel lattice algorithms by using the result obtained from section 6.3.2. In general, there should be no difficulty for readers to derive various multichannel realizations of multichannel lattice algorithms by combining the two structures described in this section and the various time-updating methods derived in previous sections.

MULTICHANNEL LATTICE ALGORITHM FOR ESTIMATION OF UNEQUAL NUMBER OF PARAMETERS IN EACH CHANNEL

In the discussion above, we have assumed that, to form the linear estimate, the multichannel lattice has the same number of data samples, i.e. l, taken from each channel. Such a q-channel lattice algorithm consists of l lattice stages, whose input and output are q-dimensional vectors, which are called lattice stages of order q. In some estimation problems, it may be desirable to take different numbers of data samples from each channel. It has been shown in [17] that it is possible to use a more general form of the multichannel lattice algorithm to perform such an estimation. A generalized q-channel lattice algorithm can be realized by cascading lattice stages having non-descending order up to q. For example, to implement a three-channel LS estimation problem wth p_i data samples taken from the ith channel, $i = 1, 2, 3$, assuming $p_1 \geqslant p_2 \geqslant p_3$, the resulting multichannel lattice algorithm should consist of $p_1 - p_2, p_2 - p_3$ and $p_3 - 1$ lattice stages of order 1, 2 and 3 respectively. Interested readers are referred to [17] for more details. In addition, recent references devoted to multichannel lattice algorithms and their applications are given in [19–25].

6.3.4 Other ORLS estimation algorithms

The single- and multichannel LS lattice algorithms derived above are ORLS estimation algorithms for time series signals for single and multiple channels. It has been shown that, in general, LS estimation of arbitrary signals can also be implemented in an order recursive fashion [27,26]. Such general ORLS algorithms have a triangular structure that is different from the lattice structure for time series signals. These algorithms have a computational complexity proportional to p^2, where p is the dimension of the data vector. Hence, they are computationally more complex than the LS lattice algorithms when p is large. Such triangular ORLS estimation algorithms include the recursive modified Gram–Schmidt algorithm (RMGS) derived in [27] and the Givens-rotation-based LS estimation algorithm [9,28]. It has been shown in [29] that the LS estimation problem of multiple inputs and outputs can be solved order-recursively by using an ORLS algorithm that is implemented as a trapezoidal array.

It has been shown in [30] that all of the ORLS algorithms can be derived using the basic ORLS relations discussed above. As a result, all of the ORLS algorithms share the same basic equation sets of the LS lattice algorithms. Owing to this commonality, the same elementary processors that implement the LS lattice algorithms can be used to implement any of the ORLS algorithms. Furthermore, any variations of the basic equation sets discussed above for the LS lattice algorithm can be used in any of the ORLS algorithms. Recognizing this fact can greatly simplify the derivation and investigation of general ORLS algorithms.

6.4 Convergence and numerical properties of lattice algorithms

The lattice algorithms discussed above are known for their fast initial convergence and superior capability for tracking time-varying signal characteristics. They are also known to have good numerical stability and accuracy. In this section we will discuss these aspects of the LS lattice algorithm. Some common misconceptions are also discussed

6.4.1 Initial convergence of LS lattice algorithms

The LS lattice algorithm is an efficient implementation of LS estimation/filtering algorithms for time series signals. Hence, in general, the *convergence* and *tracking behaviour* of the LS lattice algorithms is no different from other LS algorithms. However, since the most computationally efficient LS lattice algorithm uses prewindowed data, it behaves differently in initial convergence from the LS estimation algorithms without prewindowing. Aspects of convergence and the steady-state mean square error of LS estimation algorithms have been discussed in Chapter 3. There also exists a host of references devoted to this subject. In this section we will only state the main conclusion and discuss some special properties of the lattice algorithms.

IMPACT OF PREWINDOWING AND IMPERFECT INITIALIZATION TO INITIAL CONVERGENCE

As shown previously, in an LS estimation algorithm, we use the time average of signals to approximate their ensemble average. This relies on the assumption that the data vectors and desired signals for all k have the same joint statistics. This assumption is true when the signal is stationary and ergodic, and the corresponding

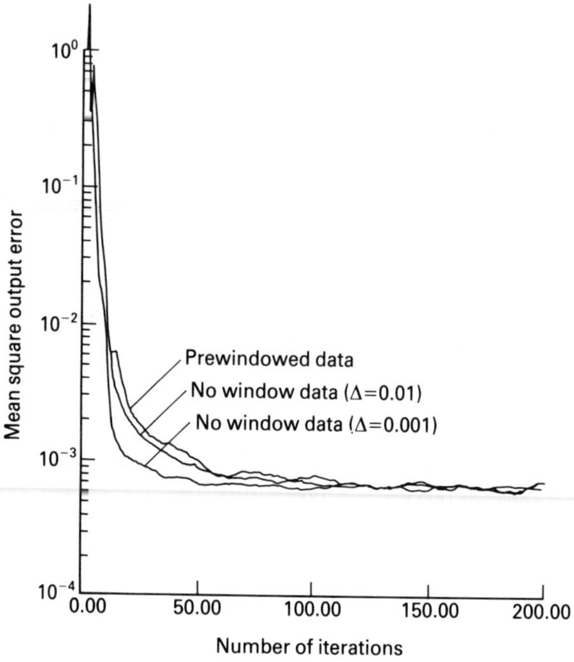

Figure 6.13 Effect of data prewindowing on initial convergence of LS adaptive algorithms

components in the data vectors have the same statistics. However, for prewindowed signals used in the most common form of the LS lattice algorithm, this assumption is obviously false during initialization. This can be illustrated as follows. The first data vector $\mathbf{u}_m(0)$ is of the form

$$\mathbf{u}_m(0) = [u(0), u(-1), \ldots, u(-m+1)]^T$$

By definition, for prewindowed signals, $u(k) = 0$ for $k < 0$. In consequence $\mathbf{u}_m(0)$ has only one non-zero element. The autocorrelation matrix of such a data vector is clearly different from data vectors in steady state, all of whose elements are non-zero. Thus, the time average using such data vectors will create a biased estimate of the autocorrelation matrix. Similarly, the cross-correlation generated from time averaging is also biased. Such a bias also exists for $\mathbf{u}_m(k)$, $k < m$.

Another source of biasing of the LS lattice algorithm is the initial value of the weighted sums, $\alpha^f_m(t)$ and $\alpha^b_m(t)$, of the squared forward and backward prediction errors in each stage. As explained before, to avoid dividing by zero and to control the dynamic range of the quantities in an LS lattice algorithm, it is a common practice to initialize these quantities with a small positive constant Δ. This is equivalent to initializing the autocovariance matrix as a small diagonal matrix. Obviously, such an initialization also introduces bias in estimation.

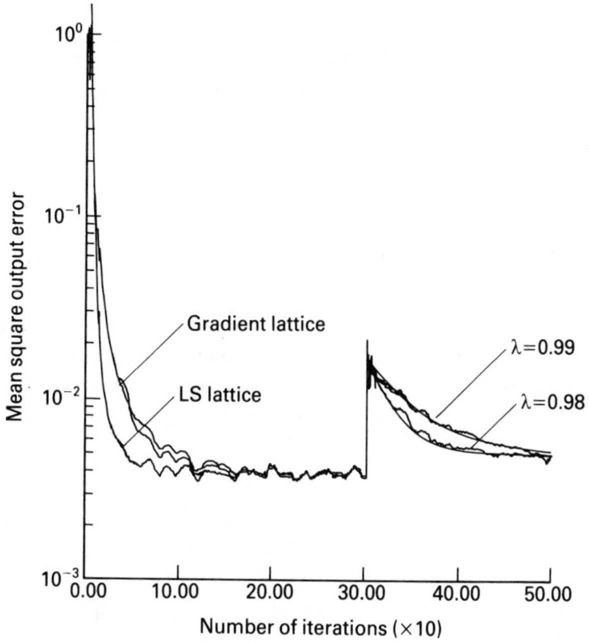

Figure 6.14 Convergence and tracking performance of LS and gradient lattice adaptive algorithms ($\eta_{max}/\eta_{min} = 29$)

As a result of these biases in estimation, the initial covergence of the LS lattice algorithm is slower than unbiased LS algorithms. Figure 6.13 shows plots of the initial convergence curves of LS adaptive algorithms with prewindowing and non-exact initializations but without prewindowing. It can be seen from the curves that the bias created by the prewindowing is much more serious than the non-exact initialization with a small Δ but without prewindowing.

The effect caused by the biases will fade away when exponential windowing is used. The shorter the time constant of the exponential windowing, the faster the bias will fade away. Thus, the estimation bias will not, in general, affect the steady-state performance of LS lattice algorithms.

Figure 6.14 compares the initial convergence of the LS lattice algorithm and the gradient lattice algorithms. It can be seen from the figure that the gradient lattice algorithm takes longer to converge than its LS counterpart. Nevertheless, as shown in [31], it still converges much faster than the LMS algorithm. This figure also shows the tracking behaviour of both algorithms, which will be discussed in the next section.

EFFECT OF EXPONENTIAL WEIGHTING ON CONVERGENCE

The LS estimation algorithms provide optimal or suboptimal initial convergence if the signal statistics are stationary. Thus, unlike the LMS algorithm, in general, an exponential weighting factor $\lambda < 1$ will not accelerate the convergence rate of LS algorithms, provided the LS estimation is unbiased. On the other hand, if a bias was introduced during initialization due to prewindowing and/or the addition of a small diagonal matrix to the autocorrelation matrix as described above, a smaller λ will make the bias fade away in a shorter time. As a result, the algorithm appears to converge faster in some cases when a smaller λ is used.

6.4.2 *Tracking behaviour of lattice algorithms*

When an adaptive algorithm is used in a non-stationary environment, its performance depends on how well it can track the time-varying data characteristics. Thus it is important to investigate the tracking behaviour of an adaptive algorithm.

Although the tracking behaviour of an LMS adaptive algorithm is closely related to its initial convergence characteristics, for the LS estimation algorithms, these are actually two different proportion and should be considered separately. On the other hand, it is a general rule that an algorithm capable of tracking time-varying signal characteristics will introduce excess error, also called self-noise, in a noisy environment. Thus, these two characteristics of adaptive algorithms should always be investigated together. Trade-offs have to be made when we select the adaptation parameters for the use of an adaptive algorithm in a time-varying environment.

As discussed above, after the initial convergence period, an LS lattice algorithm behaves in exactly the same way as any other LS algorithm. Results concerning the tracking behaviour of LS algorithms have been discussed in Chapters 3 and 5. The results also apply to LS lattice algorithms. Interested readers are also referred to [32] and [33] for further information. In this section, we will only provide the main conclusions on the tracking behaviour of exponentially weighted LS algorithms and present simulation results. These results will be compared with the behaviour of gradient lattice algorithms.

An exponentially weighted LS algorithm converges uniformly with a unique time constant. The time constant only depends on the weighting factor λ and is insensitive to the eigenvalue distribution of the data autocorrelation matrix [32]. Assuming that at time t_0 the data characteristics undergo a small step change that results in an increase in the output MSE of the LS estimator, its previously converged coefficients are no longer optimal for the new data characteristics. The LS algorithm will drive the coefficients towards their new optimal values. The difference between the old coefficients and the new optimal coefficients will diminish according to λ^{t-t_0}. When λ is close to unity, we have

$$\lambda^{t-t_0} = [1-(1-\lambda)]^{t-t_0} \approx \mathrm{e}^{-(1-\lambda)(t-t_0)}. \tag{6.136}$$

This equation shows that the LS algorithm has a convergence time constant of $1/(1-\lambda)$ (in units of sampling intervals). Since the output MSE of the adaptive algorithm is a quadratic function of its coefficients, it converges according to $\exp[-2(1-\lambda)(t-t_0)]$.

It has also been shown [31,33] that the steady-state excess MSE $\sigma^2_{e,\mathrm{ex}}$ due to the adaptation of an LS algorithm is given by

$$\sigma^2_{e,\mathrm{ex}} \approx p \frac{1-\lambda}{1+\lambda} \sigma^2_{e,\mathrm{opt}}$$

where p is the order of the estimator and $\sigma^2_{e,\mathrm{opt}}$ is the optimal MSE. From the equations given above, we can see that, as expected, an LS algorithm with a shorter convergence time constant, i.e. a smaller λ, generates a larger steady-state excess MSE.

It is of interest to compare the tracking behaviour of LS and LMS adaptive algorithms. It is known that the convergence behaviour of the LMS algorithm depends on the step size Δ as well as on the eigenvalues of the data autocorrelation matrix. For an eigenvalue η_p, the convergence time constant for the coefficients is $1/\Delta\eta_p$, and $1/2\Delta\eta_p$ for the output MSE. Thus, the LMS algorithm does not converge uniformly when the eigenvalues are not all equal. If the data have some very small eigenvalues relative to the rest, the output MSE corresponding to the small eigenvalues will converge very slowly. On the other hand, if the eigenvalues of the data autocorrelation matrix are all approximately equal to the average eigenvalue $\bar{\eta} \equiv \Sigma_{i=1}^{p} \eta_i/p$, the LMS algorithm will converge uniformly at the time constant of $1/\Delta\bar{\eta}$. By letting $\Delta\bar{\eta} = 1 - \lambda$, both the LMS and the LS algorithms will converge at the same rate. Note that the steady-state excess MSE of the LMS algorithm has been shown to be equal to $(\Delta/2)\ \Sigma_{i=1}^{p} \eta_i$ [34]. It is equal to $(\Delta/2)p\bar{\eta}$ in the case when all the eigenvalues are equal to $\bar{\eta}$. Under this condition, if the step size and the weighting factor satisfy $\Delta\bar{\eta} = 1 - \lambda$, the excess error will be the same for both LMS and LS algorithms, provided all the eigenvalues of the data autocorrelation matrix are approximately equal. In such a case, there is no gain in tracking performance by using the LS algorithm instead of the LMS algorithm. This situation is different from the initial convergence characteristics of the LMS and LS algorithms where the latter is always superior to the former. Actually, the initial convergence of the LS algorithm is optimal for the equal-eigenvalue case provided the additive noise is also white [35].

Now we consider the tracking behaviour of the gradient lattice algorithm. As we discussed in section 6.3.2, by selecting the step sizes in each lattice stage to be inversely proportional to the average of the squared input data signal of the same stage, we can achieve a uniform convergence for the gradient lattice algorithm. If the average of the squared error is computed using exponential weighting, as shown before, the gradient and the LS lattice algorithms will present the same tracking behaviour, provided they both employ the same exponential weighting factor. Thus, we conclude that, although the LS lattice algorithm is always superior in initial convergence to the gradient lattice algorithm, it has little advantage over the latter as far as tracking performance is concerned.

Figures 6.15 and 6.16 show plots of the simulation results that demonstrate and compare the tracking performance of the LMS and the LS algorithms. In the simulations, the signal characteristics undergo a step change at 500 iterations. The average eigenvalue $\bar{\eta}$ is set equal to 1. The coefficients for the LMS algorithms are initialized to their optimum values. Figure 6.15 shows the case where the eigenvalue

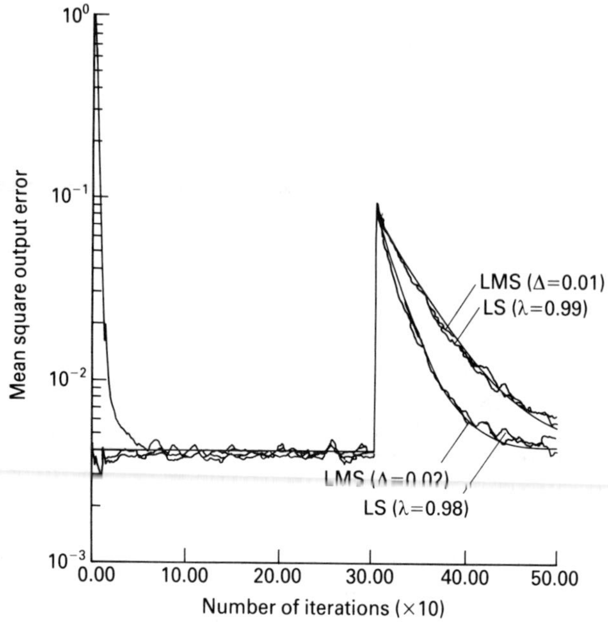

Figure 6.15 Convergence and tracking performance of LS and LMS adaptive algorithms ($\eta_{\max}/\eta_{\min} = 2.6$)

spread, which is the ratio of the largest eigenvalue to the smallest eigenvalue of the data autocorrelation matrix, is small. It can be seen from the plot that both algorithms behave similarly. Simulation results for a signal that has an autocorrelation matrix with a larger eigenvalue spread are depicted in Figure 6.16. It can be seen that the LMS algorithm converges much slower than the LS algorithm, due to the effect of large eigenvalue spread. The tracking behaviour of the gradient lattice, as well as that of the LS lattice algorithms, are shown in Figure 6.14. They behave almost identically for the signal with a large eigenvalue spread. The theoretical tracking characteristics of the LS algorithm given by (6.136) are also shown in the figures by the smooth curves. It can be seen that the simulation and theoretical results are very close.

In a non-stationary environment, the total misadjustment error resulting in the adaptation includes two components: the lag (tracking) error and the self-noise. As we have shown above, a smaller Δ or λ yields a larger self-noise but a smaller lag error. In contrast a larger Δ or λ has the opposite effect. Thus, it is desirable to use an optimal Δ or λ, which minimizes the sum of both errors. This problem was first discussed in [33].

In the discussion of the tracking behaviour of an adaptive algorithm, we used the small-step-change model. Since any slow, time-varying signal characteristics can

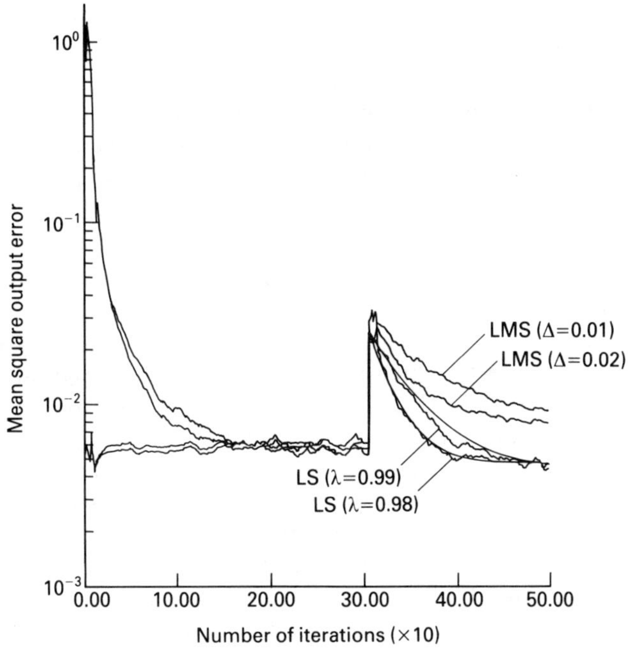

Figure 6.16 Convergence and tracking performance of LS and LMS adaptive algorithms ($\eta_{max}\eta_{min} = 29$)

always be modelled as the result of successive, infinitely small step changes, the results obtained from this model can be extended to more general cases.

Finally, we would like to point out that the above comparison of the tracking behaviour of the LMS and LS algorithms is valid under the assumption that the signal autocorrelation matrix is of full rank, even in the absence of noise, and the noise is white. If these two conditions are not satisfied, the situation becomes more complicated and needs specific consideration. The impact of the relationship between the signal and noise spectra to the total misadjustment error (including self-noise and lag error) has also been considered in Chapter 5. A special case, where the signal is not of full rank, is discussed in [36,37].

6.4.3 Numerical stability and accuracy of lattice algorithms

In this section, we discuss the finite wordlength effects in time recursive LS lattice algorithms. These are also called the numerical properties of LS lattice algorithms. While the different implementations discussed in section 6.3.1 are all equivalent under the assumption that an infinite precision is used in computations, they behave differently in practice because of the finite precision, or roundoff errors. It is especially important to select realizations which are robust to roundoff errors when a short wordlength and/or fixed point arithmetic is used in the computations.

Before we begin the discussion of the numerical properties of LS lattice algorithms, we would like to distinguish between two numerical properties of a general adaptive algorithm: namely, its numerical stability and its numerical accuracy. These numerical properties and their distinctions were originally discussed in [38] and were later used in [39].

DEFINITION OF NUMERICAL STABILITY AND ACCURACY OF ADAPTIVE ALGORITHMS

In numerical analysis, numerical stability has been defined for LS estimation algorithms as the degree to which the accuracy of the estimation is affected by the computational error (e.g. [40] and [41]). More precisely, for the same amount of roundoff error, an algorithm that yields a smaller estimation error is considered to be more stable than one which yields a larger estimation error. This definition has been used to describe the numerical properties of adaptive algorithms such as in [42]. Here, we define this property as the numerical accuracy of adaptive algorithms, since it reflects the estimation accuracy of the algorithms.

There exists another type of stability problem in adaptive algorithms. It occurs when adaptive algorithms operate continuously, in particular in tracking time-varying signal characteristics. Since in such cases the number of iterations may become very large, the stability of adaptive algorithms should be examined through its asymptotic behaviour. The definition of numerical stability, as used in this section, is akin to the definition in system theory. It is defined as the boundness of the estimation error, which can be viewed as the output of a system caused by a finite (bounded) roundoff error at the input. This definition is a natural consequence of the fact that all adaptive algorithms that we have considered are recursive in time. For example, the estimated parameters, i.e. the reflection and the joint estimation coefficients in an LS lattice algorithm, at time t are related to the corresponding parameters at time $t-1$. Hence we may describe the recursive algorithm by using a feedback system model. The stability of the modelling system determines the stability of the adaptive algorithm. Below, we show how this model is used to investigate the numerical stability of lattice algorithms.

A FEEDBACK SYSTEM MODEL FOR THE INVESTIGATION OF NUMERICAL STABILITY OF LS ESTIMATION ALGORITHMS

We use the following general feedback system model to characterize an adaptive algorithm:

$$\boldsymbol{\theta}(t) = f(\boldsymbol{\theta}(t-1), y(t), n(t)) \tag{6.137}$$

where $\boldsymbol{\theta}(t)$ denotes the parameters to be estimated, $y(t)$ is the exact new signal coming in at time t and independent of $\boldsymbol{\theta}(t-1)$, and $n(t)$ denotes roundoff error due to the finite wordlength.

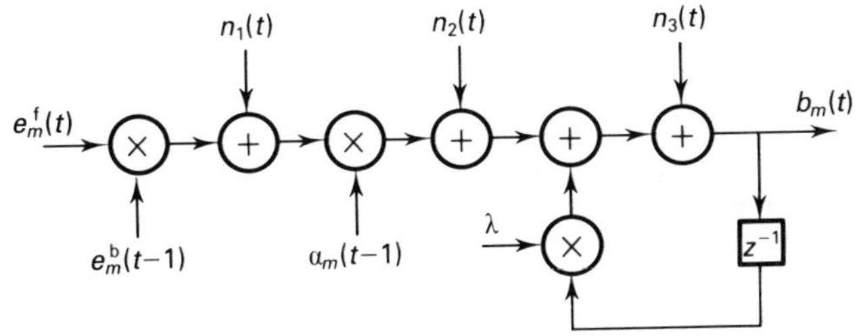

Figure 6.17 A linear feedback system model for LS lattice algorithm parameter estimation

The system may be linear or non-linear. Owing to the presence of the roundoff error $n(t)$, the estimated parameters are not the same as their ideal values obtained using infinite precision. We denote the difference as $\tilde{\theta}(t)$. If this difference for an adaptive algorithm becomes unbounded with t, we say that the algorithm is unstable. Otherwise, the algorithm is stable and we can determine its numerical accuracy at steady state by evaluating the norm of $\tilde{\theta}(t)$.

For a particular adaptive algorithm, (6.137) can be determined from the algorithm itself. Below, we consider the numerical stability of the LS lattice algorithm.

NUMERICAL STABILITY OF LS LATTICE ALGORITHMS

Here, we examine specifically the numerical stability of the correlation-quotient form of the LS lattice algorithm. Since the numerical stability of an adaptive algorithm is related to time recursions, we only need to examine the equations involving time updates. In the algorithm, the cross-correlation between the forward, backward and joint estimation errors and their variances is computed time-recursively. Since all of these update equations have the same form, we consider the time update equation (6.74) of the cross-correlation between the forward and backward prediction errors, $\beta^f_m(t)$, as the representative.

The corresponding system is illustrated in Figure 6.17, where n_1, n_2 and n_3 are the roundoff errors introduced in the computations. The system is a first-order feedback system; its stability is determined by its homogeneous part. It is obvious that this system is stable when $\lambda < 1$. The variances $\alpha^f_m(t)$ and $\alpha^b_m(t)$ in the LS lattice algorithm are also computed time-recursively. Since all these time recursive equations are similar to (6.74), we conclude that the accumulated errors in the correlations and variances are bounded. Hence the LS lattice algorithm is always stable [43].

It can be shown in a similar way that other forms of the LS lattice algorithm are all numerically stable. However, they differ from each other in their numerical accuracy, as shown below.

NUMERICAL ACCURACY OF DIFFERENT FORMS OF LS LATTICE ALGORITHMS

We define the numerical accuracy of an adaptive algorithm in terms of its behaviour in the steady state. In other words, as far as numerical accuracy is concerned, we are interested in the size of the errors in the estimated quantities for a certain amount of roundoff error introduced in the computation, after the algorithm has passed the initial convergence period. If an algorithm is not stable, we can still consider its numerical accuracy before it becomes unstable, as long as the instability occurs long after the algorithm has converged. To determine the numerical accuracy of the LS lattice algorithm, we first provide some simulation results of the LS lattice and other adaptive algorithms.

The simulation results are obtained from a simulated linear equalizer with 11 stages, using different adaptive algorithms, including the correlation-quotient form and error feedback form of the LS lattice algorithm, as well as LMS, fast Kalman and square root Kalman algorithms. The last three algorithms are included for the purpose of comparison. The channel model we use here is a three-tap transversal linear filter with transfer function $H(z) = 0.251 + 0.935z^{-1} + 0.251z^{-2}$. The signal obtained from the channel has a maximum/minimum ratio of the eigenvalues of its correlation matrix equal to 11. An exponential weighting factor $\lambda = 0.975$ is used for the LS algorithms, and a step size $\Delta = 0.025$ is used for the LMS algorithm. The additive noise has a variance of 0.001. The theoretical output MSE is equal to 0.0021, if the precision of the computation is infinite.

Table 6.10 shows the simulation results of the numerical accuracy of adaptive algorithms. Fixed point arithmetic with a precision of 8 bits to 16 bits and floating point arithmetic were used.

Table 6.10 Comparison of numerical accuracy of LS lattice and other adaptive algorithms

Number of bits	Variance of output error ($\times 10^{-3}$)				
	LS lattice algorithm (correlation quotient)	LS lattice algorithm (error feedback)	LMS	RLS (square root)	Fast RLS
Floating point (ideal)	2.10	2.10	2.21	2.10	2.10
16	2.18	2.17	2.30	2.17	2.17
13	3.09	2.22	2.30	2.33	2.21
11	25.2	3.09	19.0	6.14	3.34
9	365.0	31.6	311.0	75.3	*

* Algorithm does not converge to optimal coefficients.

We note that, although it is known that the lattice structure is less sensitive to quantization error than the transversal structure for fixed coefficient filters [44], from the simulation results given, we observe that the numerical accuracy of the correlation-quotient form of the LS lattice algorithm is similar to the accuracy of the LMS algorithm and worse than the accuracy of the square root Kalman algorithm. Both of these algorithms use a transversal structure. Thus we conclude that the lattice structure does not automatically warrant a better numerical accuracy for adaptive algorithms. It is also obvious that the numerical accuracy of the error feedback LS lattice algorithm is superior to that of all other adaptive algorithms given. Below, we explain the cause of the different behaviour between the two forms of the LS algorithm.

We recall that in the correlation-quotient form, the forward and backward reflection coefficients are computed by dividing the cross-correlations between the forward and backward prediction errors by their autocorrelations. Since the correlations and variances cannot be computed accurately when a short wordlength is used in the computation, the computed reflection coefficients are biased. This is the cause of the inaccuracy of the correlation-quotient form. On the other hand, the error feedback form computes, for example, the forward prediction error $e^{\mathrm{f}}_{m+1}(t)$ using the forward reflection coefficient $k^{\mathrm{f}}_m(t-1)$; then the reflection coefficient is time updated using the computed error as

$$k^{\mathrm{f}}_m(t) = k^{\mathrm{f}}_m(t-1) - e^{\mathrm{f}}_{m+1}(t)e^{\mathrm{b}*}_m(t-1)\alpha_m(t-1)/\alpha^{\mathrm{b}}_m(t-1) \tag{6.138}$$

This estimation process is similar to the LMS algorithm if we view $\alpha_m(t-1)/\alpha^{\mathrm{b}}_m(t-1)$ as a variable step size. The better numerical accuracy of the error feedback form of the LS lattice algorithm is mainly due to this similarity, as shown below.

The estimation accuracy of the LS lattice algorithm is determined by how accurately its reflection coefficients are estimated. For the correlation-quotient form, the accuracy of the reflection coefficients is determined by the accuracy of both the cross- and autocorrelation coefficients. On the other hand, in the error feedback form, since the autocorrelations $\alpha^{\mathrm{b}}_m(t-1)$ serve only as variable step sizes, their accuracy does not affect the accuracy of the reflection coefficients. Furthermore, let us assume that the estimated reflection coefficient is biased at some time t. As a result, the correlation between the resulting prediction error $e^{\mathrm{f}}_{m+1}(t)$ and the reference signal $e^{\mathrm{b}}_m(t-1)$ will be non-zero on average, i.e. it also has a bias. Due to the inherent error feedback, the bias in the correlation will drive the bias in the reflection coefficient towards zero. This tendency will only stop when the last term in (6.138) is less than the precision of the wordlength. As shown in [8], the residual bias in the reflection coefficient of the error feedback form is much smaller than that of the correlation-quotient form. The effect of the error feedback on the backward prediction and joint estimation is also similar.

It has been shown experimentally that the Givens lattice algorithm has a numerical accuracy close to the error feedback form. More discussion on this subject can be found in Chapter 7 of this book. On the other hand, although the normalized lattice algorithm is the best suited for fixed point implementation, it also exhibits biasing in its reflection coefficients [45]. The numerical accuracy of the gradient algorithms also varies for different realizations [46]. Those gradient lattice algorithms with inherent error-feedback, given by (6.119) and (6.122), present better numerical accuracy than the correlation-quotient forms due to the same reason given above. It must be emphasized, however, that the behaviour of lattice algorithms with respect to numerical accuracy still requires further investigation.

6.5 Summary and conclusions

In this chapter, we have derived the MMSE and LS lattice algorithms. A number of variations of the lattice algorithms are presented. Their convergence and tracking behaviour as well as their numerical properties are discussed. Typical applications of lattice algorithms in AR modelling and channel equalization can be found in [47–54].

Considering the existing numerous works devoted to the derivation and analysis of adaptive lattice estimation/filtering algorithms, it was the intention of this chapter to put the adaptive lattice algorithm into a new perspective. That is, we view lattice algorithms as a special case of order recursive estimation algorithms, which are discussed in [30] and [55]. From such a perspective, we can apply the results obtained for the lattice algorithms to derive other order recursive estimation algorithms. Conversely, we can use the results obtained for other order recursive algorithms to derive more variations of the lattice algorithm. For example, the Givens lattice algorithm presented in [11] is derived in this way.

Finally, it is important to point out that, in this chapter, we have only discussed the simplest form of the lattice algorithm: the prewindowed lattice algorithm. There are advantages in using a more complex covariance growing window, i.e. unwindowed data, and finite window lattice algorithms including reduced window length and the elimination of initial bias due to prewindowing. These algorithms have been discussed in [56–58]. More recently, an efficient implementation of the finite window lattice algorithm based on Householder transforms was given in [59].

References

[1] F. Itakura and S. Saito, 'Digital filtering techniques for speech analysis and synthesis', *Proc. 7th Int. Conf. on Acoust.*, vol. 3, paper 25C-1, pp. 261–4. 1971.

[2] J. Makhoul and R. Viswanathan, 'Adaptive lattice methods for linear prediction', *Proc. ICASSP–78 Conf.*, pp. 87–90, 1978.

[3] L.J. Griffiths, 'An adaptive lattice structure for noise-canceling applications', *Proc. ICASSP Conf., Tulsa, OK*, pp. 87–90, 1978.

[4] M. Morf and D.T. Lee, 'Recursive least-squares ladder forms for fast parameter tracking'. *Proc. 1979 IEEE Conf. on Decision and Control, San Diego, CA*, pp. 1362–7.

[5] J. Makhoul, 'A class of all-zero lattice digital filters: Properties and applications', *IEEE Trans. Acoust., Speech, Signal Process.*, vol. ASSP-26, pp. 304–14, 1978.

[6] S. Haykin, *Adaptive Filter Theory*, 2nd edn, Prentice Hall, Englewood Cliffs, NJ, 1991.

[7] J.G. Proakis, *Digital Communications*, McGraw-Hill: New York, 1989.

[8] F. Ling, D.G. Monalakis and J.G. Proakis, 'Numerically robust least-squares lattice-ladder algorithms with direct updating of the reflection coefficients', *IEEE Trans. Acoust., Speech, Signal Process.*, vol. ASSP-34, pp. 837–45, 1986.

[9] J.G. McWhirter, 'Recursive least-squares minimization using a systolic array', *Proc. SPIE*, vol. 431, pp. 105–12, 1983.

[10] F. Ling, 'Efficient and robust least-squares lattice algorithms based on Givens-rotation with systolic array implementation', *Proc. ICASSP-89 Conf*, pp. 1290–3, 1989.

[11] F. Ling, 'Givens-rotation based LS lattice algorithm and related algorithms', *IEEE Trans. Acoust., Speech, Signal Process.*, vol SP-39, pp. 1541–51.

[12] I.K. Proudler, J.G. McWhirter and T.J. Shepherd 'QRD-based lattice filter algorithm', *Proc. SPIE* Aug., 1989.

[13] D.T. Lee, M. Morf and B. Friedlander, 'Recursive least squares ladder estimation algorithms', *IEEE Trans. Acoust, Speech, Signal Process.*, vol. ASSP-29, pp. 627–41, 1981.

[14] C.M. Rader and A.O. Steinhardt, 'Hyperbolic Householder transformations', *IEEE Trans. Acoust., Speech, Signal Process.*, vol. ASSP-34, pp. 1589–602, 1986.

[15] H.M. Ahmed, J.M. Delosme and M. Morf, 'Highly concurrent computing structures for matrix arithmetic and signal processing', *IEEE Computer*, Jan., 1982.

[16] M.L. Honig and D.G. Messerschmitt, 'Convergence properties of an adaptive digital lattice filter'. *IEEE Trans. Acoust., Speech, Signal Process.*, vol. ASSP-29, pp. 642–53, 1981.

[17] F. Ling and J.G. Proakis, 'A generalized multichannel least squares lattice algorithm based on sequential processing stages', *IEEE Trans. Acoust., Speech, Signal Process.*, vol. ASSP-32, pp. 381–9, 1984.

[18] H. Lev-Ari, 'Modular architecture for adaptive multichannel lattice algorithms', *IEEE Trans. Acoust., Speech, Signal Process.*, vol. ASSP-87, pp. 543–52, 1987.

[19] G.O. Glentis and N. Kalouptsidis. 'Efficient order recursive algorithms for multichannel least-squares filtering', *IEEE Trans. Signal Process.*, vol. SP-40, pp. 1354–74, July, 1992.

[20] G.O. Glentis and N. Kalouptsidis, 'Fast adaptive algorithms for multichannel filtering and system identification', *IEEE Trans. Signal Process.*, vol SP-40, pp. 2433–59, Oct., 1992.

[21] S. Karaboyas, N. Kalouptsides and C. Caroubalos, 'Higly Parallel transversal batch and block adaptive multichannel LS algorithms and applications to decision feedback equalizers', *IEEE Trans. Acoust., Speech, Signal Process.*, vol. ASSP-37, pp. 1380–96, 1989.

[22] S. Karaboyas and N. Kalouptsidis, 'Efficient adaptive algorithms for ARMA identification', *IEEE Trans. Acoust., Speech, Signal Process.*, vol. ASSP-39, pp. 571–82, 1991.

[23] G. Glentis and N. Kalouptsidis. 'Efficient adaptive algorithms for multichannel least squares filtering using a channel decomposition technique,' *Computer and Electrical Engineering*, Special Issue on Adaptive Signal Processing, 1992.

[24] P.S. Lewis, QR-based algorithms for multichannel adaptive least-squares lattice filters and example applications to magnetoencephalography', *IEEE Trans. Acoust., Speech, Signal Process.*, vol. ASSP-38, pp. 421–32, 1990.

[25] K. Zhao, F. Ling and J.G. Proakis, 'Multichannel Givens lattice adaptive algorithm', *Proc. ICASSP-91 Conf.*, pp. 1849–52, 1991.

[26] S Kalson and K. Yao, 'Geometrical approach to generalized least squares estimation with systolic array processing', *Proc. 22nd Ann. Allerton Conf. on Communications, Control and Computing, Oct., 1984.*

[27] F. Ling, D. Manolakis and J.G. Proakis, 'A recursive modified Gram–Schmidt algorithm for least squares estimation', *IEEE Trans. Acoust., Speech, Signal Process.*, vol. ASSP–34, pp. 829–36, 1986.

[28] H.T. Kung and W.M. Gentleman, 'Matrix triangularization by systolic arrays', *Proc. SPIE (Real-Time Signal Processing IV)*, vol 298, pp. 298–303, 1981.

[29] S. Yuen, K. Abend and R. S. Berkowitz, 'A recursive least-squares algorithm for multiple inputs and outputs and a cyclindrical systolic implementation', *IEEE Trans. Acoust., Speech, Signal Process.*, vol. ASSP-36, pp. 1917–23, 1988.

[30] F. Ling, J. G. Proakis and K. Zhao, 'A systematic treatment of order-recursive LS estimation algorithms', *Proc. SPIE*, July, 1991.

[31] E.H. Satorius and S.T. Alexander, 'Channel equalization using adaptive lattice algo-

rithms', *IEEE Trans. Communications,* vol. COM-27, pp. 899–905, 1979.

[32] F. Ling and J.G. Proakis, Nonstationary learning characteristics of least-squares adaptive algorithms', *Proc. ICASSP-84 Conf.,* pp. 3.7.1–4, 1984.

[33] E. Eleftheriou and D. Falconer, 'Tracking properties and steady-state performance of RLS adaptive filter algorithms', *IEEE Trans. Acoust., Speech, Signal Process.*, vol. ASSP-34, pp. 1094–110, 1986.

[34] B. Widrow, M. McCool , M.G. Larimore and C.R. Johnson, Jr., 'Stationary and nonstationary learning characteristics of the LMS adaptive filter', *IEEE Proc.* vol. 64, pp. 1152–64, 1976.

[35] F. Ling, 'Rapidly convergent adaptive filtering algorithm with application to equalization and channel estimation,' PhD Dissertation, Northeastern University, Boston, 1984.

[36] O. Macchi and N. Bershad, 'Adaptive recovery of a chirped sinusoid in noise. Part I: Performance of the RLS algorithm', *IEEE Trans. Acoust., Speech, Signal Process.*, vol. ASSP-39, pp. 583–95, 1991.

[37] N. Bershad and O. Macchi, 'Adaptive recovery of a chirped sinusoid in noise. Part II: Performance of the LMS algorithm' *IEEE Trans. Acoust., Speech, Signal Process.*, vol. ASSP-39, pp. 583–95 1991.

[38] F. Ling and J.G. Proakis, 'Numerical accuracy and stability: Two problems of adaptive estimation algorithms caused by round-off error', *Proc. ICASSP-84 Conf.*, Mar. 1984.

[39] J.M. Cioffi, 'Limited precision effects in adaptive filtering', *IEEE Trans. Circuits Syst.* CAS-34, pp. 821–33. 1987.

[40] A.S. Householder, *The Theory of Matrixes in Numerical Analysis*, Blaisdell: New York, 1964.

[41] G. Cybenko, 'The numerical stability of the Levension–Durbin algorithm for Toeplitz systems of equations', *SIAM J. Sci. Statist. Comput.*, Sept., 1980.

[42] G.J. Bierman, *Factorization Methods for Discrete Sequential estimation*, Academic Press: New York, 1971.

[43] S. Ljung and L. Ljung, 'Error propagation properties of recursive least squares adaptation algorithms'. Internal Report LITH-ISY-I-0620, Dept of Electrical Engineering. Linköping University 1983.

[44] J.D. Markel and A. H. Gray, 'Roundoff noise characteristics of a class of orthogonal polynomial structures,' *IEEE Trans. Acoust., Speech, Signal Process.,* vol. ASSP-23, pp. 473–86, 1975.

[45] C. Samson and V.U. Reddy, 'Fixed-point error analysis of the normalized ladder algorithm', *IEEE Trans. Acoust., Speech, Signal Process.,* vol. ASSP-31, pp. 1177–191, 1983.

[46] E.H. Satorius, S.W. Larisch, S.C. Lee and J.J. Griffiths, 'Fixed point implementation of adaptive digital filters', *Proc. ICASSP-*83 *Conf.*, pp. 36–6, 1986.

[47] J. Makhoul, 'Linear prediction: A tutorial view', *Proc. IEEE*, vol. 64, pp. 561–80, 1975.

[48] E.H. Satorius and J.D. Park, 'Application of least-squares lattice algorithms to adaptive eqalization', *IEEE Trans. Commum.* vol. COM-29, pp. 136–42, 1981.

[49] F. Ling and J.G. Proakis, 'Generalized multichannel least-squares lattice algorithm and its application to decision-feedback equalization', *Proc. ICASSP-82 Conf.*, 1982.

[50] F. Ling and J.G. Proakis, 'Adaptive lattice decision-feedack equalizers – their performance and application to time-variant multipath channels', *IEEE Trans. Commun.* vol. COM-33, pp. 348–56, 1985.

[51] S. Qureshi and G.D. Forney Jr, 'Performance and properties of a T/2 equalizer', *National Telecommun. Conf. Record*, Los Angeles, CA,. pp. 11.1.1–14, 1977.

[52] R.D. Gitlin and S.B. Weinstein, 'Fractionally-spaced equalization: An improved digital transversal equalizer', *Bell Syst. Tech. J.,* vol. 60, pp. 275–96, 1981.

[53] C.A. Belfiore and J.H. Park Jr, 'Decision-feedback equalization,' *Proc. IEEE*, vol. 67, pp. 1143–56, 1979.

[54] F. Ling and S. Qureshi, 'Lattice predictive decision-feedback equalization', *Proc GlobeCom* '86 Dallas, TX, Dec., 1986.

[55] J.G. Proakis, C. Rader, F. Ling and C. Nikias, *Advanced Digital Signal Processing*, Macmillan: New York, 1992.

[56] B. Porat, B. Friedlander and M. Morf, 'Square-root covariance ladder algorithms', *IEEE Trans. AC*, vol. AC-27, pp. 813–29, 1982.

[57] M. L. Honig and D.G. Messerschmitt, *Adaptive Filters – Structures, Algorithms and Applications*, Kluwer: Boston, MA, 1984.

[58] J.M. Cioffi, 'An unwindowed RLS adaptive lattice algorithm', *IEEE Trans. Acoust., Speech, Signal Process.*, vol. ASSP-36. pp. 365–71, 1988.

[59] K. Zhao, F. Ling, H. Lev-Ari and J. G. Proakis, 'QRD-based sliding window adaptive LS algorithm', *Proc. ICASSP-92 Conf., San Francisco*, 1992.

7

The QR Family*

J.G. McWhirter and I.K. Proudler

Introduction

In this chapter we will show how the method of *QR* decomposition (QRD) may be applied to the adaptive filtering and beamforming problems. QRD is a form of orthogonal triangularization which is particularly useful in least-squares computations and forms the basis of some very stable numerical algorithms.

In section 7.2, we show how the method of QRD by Givens rotations may be applied to the problem of narrowband adaptive beamforming where the data matrix, in general, has no special structure. In particular, it is shown how the least-squares computation may be carried out in parallel using a triangular array of relatively simple processing elements.

In section 7.3, we consider the problem of an adaptive time series filter for which the data vectors exhibit a simple time-shift invariance and the corresponding data matrix is a Toeplitz structure. In this case, the triangular processor array described

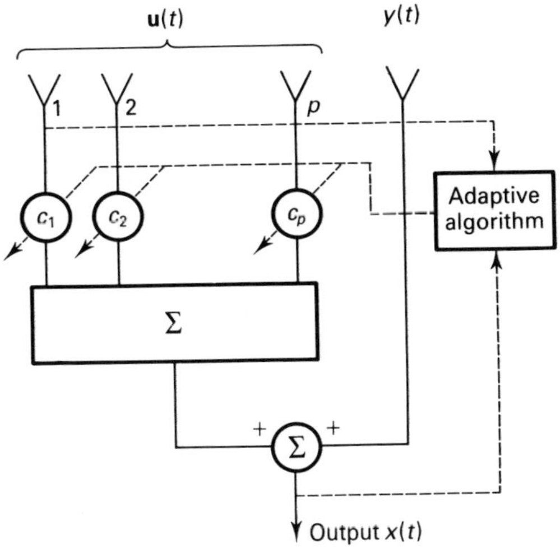

Figure 7.1 Canonical adaptive linear combiner

in section 7.2 is known to be very inefficient. Instead, it is shown how the Toeplitz structure may be used to reduce the computational complexity of the QRD technique. The resulting orthogonal least-squares lattice and fast QRD algorithms, derived in section 7.3, may be implemented using far fewer processing elements. These 'fast' QRD algorithms are very similar to the more conventional ones derived in Chapters 5 and 6 but, in general, are found to have superior numerical properties.

In section 7.4, we consider the more general problem of multichannel adaptive filtering, which arises, for example, in broadband adaptive beamforming. In this case the data matrix has a block Toeplitz structure which may be exploited to generate an efficient multichannel fast QRD or least-squares lattice algorithm. The multichannel least-squares lattice algorithm may be implemented using a lattice of the triangular processor arrays discussed in section 7.2 and so it constitutes a hybrid solution which encompasses the algorithms and architectures of sections 7.2 and 7.3 as special cases.

7.2 Narrowband beamforming

7.2.1 QR decomposition

In Chapter 2 it was shown how an adaptive linear combiner of the type illustrated in Figure 7.1 may be formulated in terms of least-squares minimization and applied directly to the problem of narrowband adaptive beamforming. The combined output at sample time t is denoted by

$$x(t) = \mathbf{u}^{\mathrm{T}}(t)\mathbf{c} + y(t) \tag{7.1}$$

where $\mathbf{u}(t)$ is the p-element (complex) vector of auxiliary signals at time t and $y(t)$ is the corresponding sample of the primary signal. The residual signal power at time N is estimated by the quantity

$$V_N(\mathbf{c}) = \| \boldsymbol{\varepsilon}(N) \|^2 \tag{7.2}$$

where

$$\boldsymbol{\varepsilon}(N) = \Lambda(N)[x(1), x(2), \ldots, x(N)]^{\mathrm{T}} \tag{7.3}$$

and the diagonal matrix

$$\Lambda(N) = \mathrm{diag}[\lambda^{(N-1)/2}, \lambda^{(N-2)/2}, \ldots, 1] \tag{7.4}$$

constitutes an exponential window with $0 < \lambda \leqslant 1$. From Equation (7.1) it follows that the vector of residuals may be written in the form

$$\boldsymbol{\varepsilon}(N) = \mathscr{U}(N)\mathbf{c} + \mathbf{y}(N) \tag{7.5}$$

where

$$\mathscr{U}(N) = \Lambda(N)\begin{bmatrix} \mathbf{u}^{\mathrm{T}}(1) \\ \mathbf{u}^{\mathrm{T}}(2) \\ \vdots \\ \mathbf{u}^{\mathrm{T}}(N) \end{bmatrix} \quad \text{and} \quad \mathbf{y}(N) = \Lambda(N)\begin{bmatrix} y(1) \\ y(2) \\ \vdots \\ y(N) \end{bmatrix} \tag{7.6}$$

$\mathscr{U}(N)$ is simply the matrix of all data samples received by the combiner up to time N, and $\mathbf{y}(N)$ is the corresponding vector of data in the primary or reference channel. For convenience, the matrix $\Lambda(N)$ has simply been absorbed into the definiton of $\boldsymbol{\varepsilon}(N)$, $\mathbf{y}(N)$ and $\mathscr{U}(N)$.

The least-squares weight vector $\mathbf{c}(N)$ is simply the one which minimizes $V_N(\mathbf{c})$. The conventional approach to this problem has already been discussed. It involves the explicit computation of the data coviance matrix $R(N) = \mathscr{U}^{\mathrm{H}}(N)\mathscr{U}(N)$ and, as a result, the condition number of the problem is squared. For a given finite wordlength, this leads to a considerable loss in performance and should be avoided if possible.

An alternative approach to the least-squares estimation problem is the method of QRD which constitutes a form of orthogonal triangularization and has particularly good numerical properties. It will be generalized here to the case of complex data as required, for example, in narrowband adaptive beamforming. An $(N \times N)$ unitary matrix $Q(N)$ is generated such that

$$Q(N)\mathscr{U}(N) = \begin{bmatrix} \tilde{R}(N) \\ \mathrm{O} \end{bmatrix} \tag{7.7}$$

where $\tilde{R}(N)$ is a $p \times p$ upper-triangular matrix and O is the zero matrix. Then, since $Q(N)$ is unitary, we have

$$R(N) = \mathscr{U}^{\mathrm{H}}(N)\mathscr{U}(N) = \mathscr{U}^{\mathrm{H}}(N)Q^{\mathrm{H}}(N)Q(N)\mathscr{U}(N) = \tilde{R}^{\mathrm{H}}(N)\tilde{R}(N) \tag{7.8}$$

so that the triangular matrix $\tilde{R}(N)$ is the Cholesky (square root) factor of the data covariance matrix $R(N)$.

Now, again by virtue of the unitary nature of the matrix $Q(N)$, we have

$$\| \boldsymbol{\varepsilon}(N) \| = \| Q(N)\boldsymbol{\varepsilon}(N) \| = \left\| \begin{bmatrix} \tilde{R}(N) \\ \mathrm{O} \end{bmatrix} \mathbf{c}(N) + \begin{bmatrix} \mathbf{p}(N) \\ \mathbf{v}(N) \end{bmatrix} \right\| \tag{7.9}$$

where

$$\begin{bmatrix} \mathbf{p}(N) \\ \mathbf{v}(N) \end{bmatrix} = Q(N)\mathbf{y}(N) \tag{7.10}$$

It follows that the least-squares weight vector $\mathbf{c}(N)$ must satisfy the equation

$$\tilde{R}(N)\mathbf{c}(N) + \mathbf{p}(N) = \mathbf{0} \tag{7.11}$$

(where $\mathbf{0}$ represents the zero vector) and hence

$$V_N(\mathbf{c}) = \| \mathbf{v}(N) \|^2 \tag{7.12}$$

Since the matrix $\tilde{R}(N)$ is upper triangular, Equation (7.11) is much easier to solve than the Gauss normal equations described earlier. The weight vector $\mathbf{c}(N)$ may be derived quite simply by a process of back-substitution. Equation (7.11) is also much better conditioned since the condition number $\tilde{R}(N)$ is identical to that of the original data matrix $\mathscr{U}(N)$, the two being related by a unitary transformation.

7.2.2 Givens rotations

The triangularization process may be carried out using either Householder transformations [1] or Givens rotations [2]. However, the Givens rotation method is particularly suitable for adaptive filtering since it leads to a very efficient algorithm whereby the triangularization process is recursively updated as each new row of data enters the combiner. Assume that the matrix $\mathscr{U}(N-1)$ has already been reduced to triangular form by the unitary transformation

$$Q(N-1)\mathscr{U}(N-1) = \begin{bmatrix} \tilde{R}(N-1) \\ \mathrm{O} \end{bmatrix} \tag{7.13}$$

and define the unitary matrix

$$\bar{Q}(N-1) = \begin{bmatrix} Q(N-1) & \mathbf{0} \\ \mathbf{0}^{\mathrm{T}} & 1 \end{bmatrix} \tag{7.14}$$

Clearly,

$$\bar{Q}(N-1)\mathcal{U}(N) = \bar{Q}(N-1)\begin{bmatrix} \lambda^{1/2}\mathcal{U}(N-1) \\ \mathbf{u}^{\mathrm{T}}(N) \end{bmatrix} = \begin{bmatrix} \lambda^{1/2}\tilde{R}(N-1) \\ \mathrm{O} \\ \mathbf{u}^{\mathrm{T}}(N) \end{bmatrix} \tag{7.15}$$

and hence the triangularization of $\mathcal{U}(N)$ may be completed by using a sequence of (complex) Givens rotations to eliminate the vector $\mathbf{u}^{\mathrm{T}}(N)$. Each Givens rotation is an elementary unitary transformation of the form

$$\begin{bmatrix} c & s^* \\ -s & c \end{bmatrix}\begin{bmatrix} 0 \ldots 0, & \lambda^{1/2}\tilde{r}_i \ldots & \lambda^{1/2}\tilde{r}_k \ldots \\ 0 \ldots & 0, \; u_i \quad \ldots & u_k \quad \ldots \end{bmatrix} = \begin{bmatrix} 0 \ldots 0, & \tilde{r}'_i \ldots \tilde{r}'_k \ldots \\ 0 \ldots 0, & 0 \ldots u'_k \ldots \end{bmatrix} \tag{7.16}$$

where

$$|c|^2 + |s|^2 = 1 \tag{7.17}$$

and the cosine parameter is assumed to be real without loss of generality. Clearly, we require

$$c = \frac{\lambda^{1/2}\tilde{r}_i}{\sqrt{\lambda\tilde{r}_i^2 + |u_i|^2}} \quad \text{and} \quad s = \frac{u_i}{\sqrt{\lambda\tilde{r}_i^2 + |u_i|^2}} \tag{7.18}$$

The sequence of rotations is applied as follows. The p-element vector $\mathbf{u}^{\mathrm{T}}(N)$ is rotated with the first row $\lambda^{1/2}\tilde{R}(N-1)$ so that the leading element of $\mathbf{u}^{\mathrm{T}}(N)$ is eliminated, and a reduced vector $\mathbf{u}'^{\mathrm{T}}(N)$ is produced. Note that the first row of $\tilde{R}(N-1)$ is modified in the process. The $(p-1)$ element reduced vector $\mathbf{u}'^{\mathrm{T}}(N)$ is then rotated with the second row of $\lambda^{1/2}\tilde{R}(N-1)$ so that the leading element of $\mathbf{u}'^{\mathrm{T}}(N)$ is eliminated, and so on until every element of the data vector has been annihilated. The resulting triangular matrix $\tilde{R}(N)$ then corresponds to a complete triangularization of the matrix $\mathcal{U}(N)$ as defined in (7.7). The corresponding unitary matrix $Q(N)$ is simply given by the recursive expression

$$Q(N) = \hat{Q}(N)\bar{Q}(N-1) \tag{7.19}$$

where $\hat{Q}(N)$ is a unitary matrix representing the sequence of Givens rotation operations described above, i.e.

$$\hat{Q}(N)\begin{bmatrix} \lambda^{1/2}\tilde{R}(N-1) \\ \mathrm{O} \\ \mathbf{u}^{\mathrm{T}}(N) \end{bmatrix} = \begin{bmatrix} \tilde{R}(N) \\ \mathrm{O} \\ \mathbf{0}^{\mathrm{T}} \end{bmatrix} \tag{7.20}$$

It is not difficult to deduce in addition that

$$\hat{Q}(N)\begin{bmatrix} \lambda^{1/2}\mathbf{p}(N-1) \\ \lambda^{1/2}\mathbf{v}(N-1) \\ y(N) \end{bmatrix} = \begin{bmatrix} \mathbf{p}(N) \\ \lambda^{1/2}\mathbf{v}(N-1) \\ \tilde{e}(N) \end{bmatrix} = \begin{bmatrix} \mathbf{p}(N) \\ \mathbf{v}(N) \end{bmatrix} \tag{7.21}$$

and this shows how the vector $\mathbf{p}(N)$ can be updated recursively using the same sequence of Givens rotations. The least-squares weight vector $\mathbf{c}(N)$ may then be derived by solving (7.11). The solution is not defined, of course, if $N < p$, but the

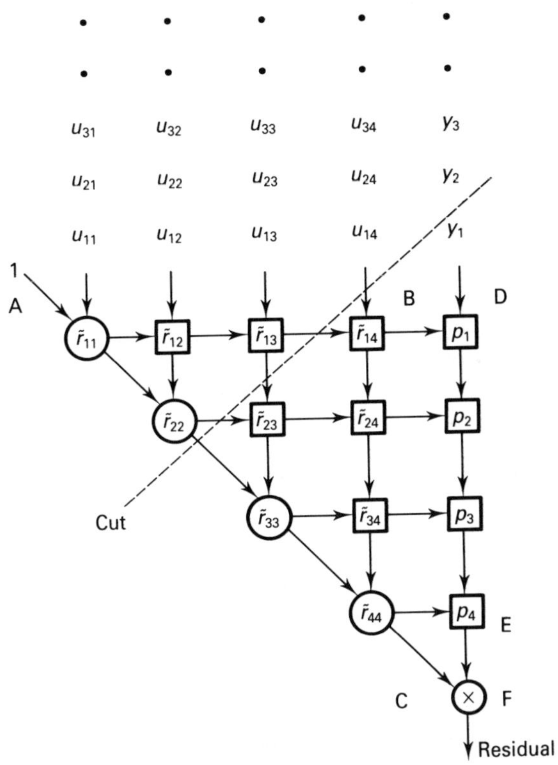

Figure 7.2 Triangular processor array

recursive triangularization process may none the less be initialized by setting $\tilde{R}(0) = O$ and $\mathbf{p}(0) = \mathbf{0}$.

7.2.3 *Parallel implementation*

The Givens rotation algorithm described above may be implemented in parallel using a triangular processor array of the type illustrated in Figure 7.2 for the case $p = 4$. It comprises three distinct sections: the basic triangular array labelled ABC, the right-hand column of cells labelled DE and the final processing cell labelled F.

At time $N - 1$, each cell within the basic triangular array stores one element of the triangular matrix $\tilde{R}(N-1)$ and each cell in the right-hand column DE stores one element of the corresponding vector $\mathbf{p}(N-1)$. On the next clock cycle, the data vector $[\mathbf{u}^T(N), y(N)]$ is input to the top of the array as shown. Each row of cells within the basic triangular array performs a Givens rotation between one row of the stored triangular matrix and a vector of data received from the above so that the

Boundary cell

IF $u_{in}=0$ THEN
$c \leftarrow 1;$ $s \leftarrow 0;$ $\tilde{r} \leftarrow \lambda^{1/2}\tilde{r};$ $\tilde{a}_{out} \leftarrow \tilde{a}_{in}$
ELSE
$r' \leftarrow \sqrt{\lambda\tilde{r}^2+|u_{in}|^2};$
$c \leftarrow ((\lambda^{1/2}\tilde{r})/r');$ $s \leftarrow (u_{in}/r');$
$\tilde{r} \leftarrow r';$ $\tilde{a}_{out} \leftarrow c\tilde{a}_{in};$
ENDIF

Internal cell

$u_{out} \leftarrow cu_{in} - s\lambda^{1/2}\tilde{r}$
$\tilde{r} \leftarrow s^* u_{in} + c\lambda^{1/2}\tilde{r}$

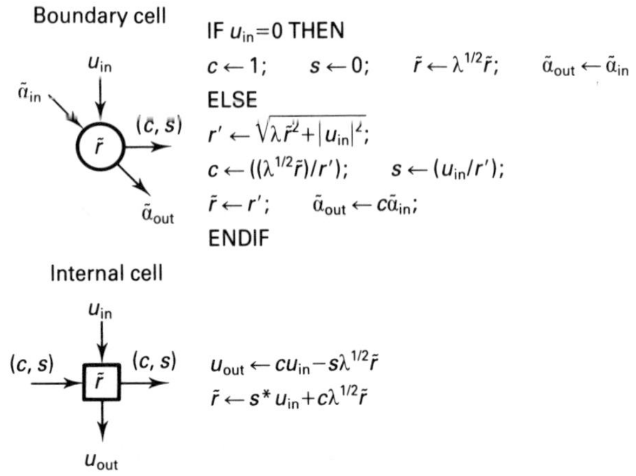

Figure 7.3 Processing elements for triangular QRD array

leading element of the received vector is eliminated. The reduced data vector is then passed downwards through the array. The boundary cell in each row computes the appropriate rotation parameters and passes them on to the right so that the internal cells can apply the same rotation to all other elements in the received data vector. This arrangement ensures that as the input data vector $\mathbf{u}^T(N)$ moves down through the array it interacts with the previously stored triangular matrix $\tilde{R}(N-1)$ and undergoes the sequence of rotations $\hat{Q}(N)$ described in section 7.2.2. All of its elements are thereby eliminated (one on each row of the array) and the updated triangular matrix $\tilde{R}(N)$ is generated and stored in the process.

As the corresponding input sample $y(N)$ moves down through the right-hand column of cells, it undergoes the same sequence of Givens rotations interacting with the stored vector $\mathbf{p}(N-1)$, thereby generating the updated vector $\mathbf{p}(N)$. The resulting output, which emerges from the bottom cell in the right-hand column, is simply the value of the parameter $\tilde{e}(N)$ in (7.21).

The function of the rotation cells is specified in Figure 7.3 and follows immediately from Equations (7.16) and (7.18). The boundary cell includes an additional parameter $\tilde{\alpha}$ which is not required for the basic Givens rotations but will be explained in section 7.2.5 where the function of the final cell F is also made clear. Since the matrix $\tilde{R}$ is initialized to zero, it can be seen that the elements on its leading diagonal (i.e. the values of $\tilde{r}$ within the boundary cells) may be treated as real variables throughout the computation and so the number of real arithmetic operations performed within the boundary cell is, surprisingly, less than that required for an internal cell.

For simplicity and ease of discussion, we have assumed that all cells of the array in Figure 7.2 operate on the same input data vector during a given clock cycle. The critical path for the array therefore involves $2p + 1$ cells and the maximum rate at which data vectors can be input is approximately $1/(2p + 1)T$, where T is the typical processing time for a single cell. When Gentleman and Kung [3] first proposed the triangular array in Figure 7.2, they showed how it could be fully pipelined by introducing latches to store the outputs generated by each cell before they are passed on as inputs to the neighbouring cells. The resulting systolic array can achieve an input clock rate of about $1/T$ and only requires nearest-neighbour cell interconnections, which is highly advantageous for VLSI implementation.

The systolic array may be generated by cutting the diagram in Figure 7.2 along all diagonals parallel to the one indicated and introducing a storage element where each data path crosses a cut line. Note that these cut lines also intersect the input data paths and so each input data vector enters the triangular array in a skewed or staggered manner. Clearly, Figure 7.2 provides a sufficient description of the parallel algorithm without including the detailed pipelining and timing aspects associated with a systolic or wave front array. This relative simplicity means that the parallel processor can be represented more easily in block diagrammatic form later in this chapter.

Table 7.1 SQ/FF narrowband beamformer algorithm

	Add/ subt	Mult	Sqrt/ div
START			
INITIALIZE {all variables := 0};			
FOR t FROM 1 DO			
LET $\tilde{e}(t) := y(t)$; $\tilde{\alpha}(t) := 1$;			
FOR i FROM 1 TO p DO			
LET $\tilde{r}^{(i,i)}(t) := \sqrt{\lambda[\tilde{r}^{(i,i)}(t-1)]^2 + \lvert \mathbf{u}^{(i)}(t) \rvert^2}$;	1	3	1
IF $\tilde{r}^{(i,i)}(t) = 0$ THEN LET $c := 1$; $s := 0$;			
ELSE LET $c := \dfrac{\sqrt{\lambda}\tilde{r}^{(i,i)}(t-1)}{\tilde{r}^{(i,i)}(t)}$; $s := \dfrac{\mathbf{u}^{(i)}(t)}{\tilde{r}^{(i,i)}(t)}$;	–	1	2
END_IF;			
FOR j FROM $i+1$ TO p DO			
LET $u' := (c\mathbf{u}^{(j)}(t) - s\tilde{r}^{(i,j)}(t-1))$;	1	2	–
$\tilde{r}^{(i,j)}(t) := (c\tilde{r}^{(i,j)}(t-1) + s^*\mathbf{u}^{(j)}(t))$;	1	2	–
$\mathbf{u}^{(j)}(t) := u'$			
END_DO;			
LET $\tilde{e}' := (c\tilde{e}(t) - s\mathbf{p}^{(i)}(t-1))$;	1	2	–
$\mathbf{p}^{(i)}(t) := (c\mathbf{p}^{(i)}(t-1) + s^*\tilde{e}(t))$;	1	2	–
$\tilde{e}(t) := \tilde{e}'$;			
$\tilde{\alpha}(t) := c\tilde{\alpha}(t)$	–	1	–
END_DO; {i loop}			
COMMENT pth-order filtered residual COMMENT			
LET $\varepsilon(t) := (\tilde{\alpha}(t)\tilde{e}(t))$	–	1	–
END_DO; {t loop}	$p^2 + 2p$	$2p^2 + 7p + 1$	$3p$
FINISH			

A listing for the $O(p^2)$ QRD-based, least-squares minimization algorithm is shown in Table 7.1 (see section 7.5 for an explanation of the listing.)

7.2.4 Square-root-free version

Gentleman [4] and Hammarling [5] have derived extremely efficient QRD algorithms based on modified Givens rotations which require no square root operation. The essence of these square-root-free algorithms is a factorization of the form

$$\tilde{R}(N) = D^{1/2}(N)\bar{R}(N) \tag{7.22}$$

where

$$D(N) = \text{diag}[\tilde{r}_{11}^2(N), \tilde{r}_{22}^2(N), \ldots, \tilde{r}_{pp}^2(N)] \tag{7.23}$$

and $\bar{R}(N)$ is a unit upper-triangular matrix. The complex Givens rotation in (7.16) then takes the form

$$\begin{bmatrix} c & s^* \\ -s & c \end{bmatrix} \begin{bmatrix} 0 \ldots 0, & \lambda^{1/2}\sqrt{d} & \ldots & \lambda^{1/2}\sqrt{d}\bar{r}_k & \ldots \\ 0 \ldots 0, & \sqrt{\delta\bar{u}_i} & \ldots & \sqrt{\delta\bar{u}_k} & \ldots \end{bmatrix} = \begin{bmatrix} 0 \ldots 0, & \sqrt{d'} & \ldots & \sqrt{d'}\bar{r}'_k & \ldots \\ 0 \ldots 0, & 0 & \ldots & \sqrt{\delta'\bar{u}'_k} & \ldots \end{bmatrix} \tag{7.24}$$

where $\sqrt{d}$ represents an element in the diagonal matrix $D^{1/2}(N)$, and $\bar{r}_k$ denotes an associated off-diagonal element from the matrix $\bar{R}(n)$. Note that each element u_k of the data vector has been expressed in the form $\sqrt{\delta\bar{u}_k}$, where δ is a scaling factor which changes value as a result of the rotation.

The square-root-free algorithm may also be implemented using a type of triangular pocessor array depicted in Figure 7.2 but with the rotation cells as defined in Figure 7.4. In this case, the boundary cells store and update elements of D(N) (i.e. the squares of the diagonal elements of $\tilde{R}(N)$) while the internal cells store and update the off-diagonal elements of $\bar{R}(N)$. The function of these cells may be deduced in a straightforward manner starting from (7.24) and (7.17) and noting that the values of c and s need not be computed explicitly. Note that the scale quantities δ are updated only at the boundary cells and passed diagonally from one row to the next. Note also that the parameters d and δ in each boundary cell may be treated as real variables throughout the computation, assuming that they are assigned real values initially. For the purposes of normal least-squares processing, the δ parameter is initialized to unity so that on input to the array $\bar{\mathbf{u}}(N) = \mathbf{u}(N)$. However, as pointed out by Gentleman [4], the δ parameter associated with each input vector may be assigned an arbitrary initial value which serves to weight that row of data accordingly. The square-root-free array may thus be used to perform a general weighted least-squares computation.

As with the conventional Givens rotation algorithm, cells in the right-hand column perform the same function as those internal to the triangular array. Thus. allowing for the factorization shown in (7.22), these cells update elements of the vector $\bar{\mathbf{p}}(N)$, where

$$\mathbf{p}(N) = D^{1/2}(N)\bar{\mathbf{p}}(N) \tag{7.25}$$

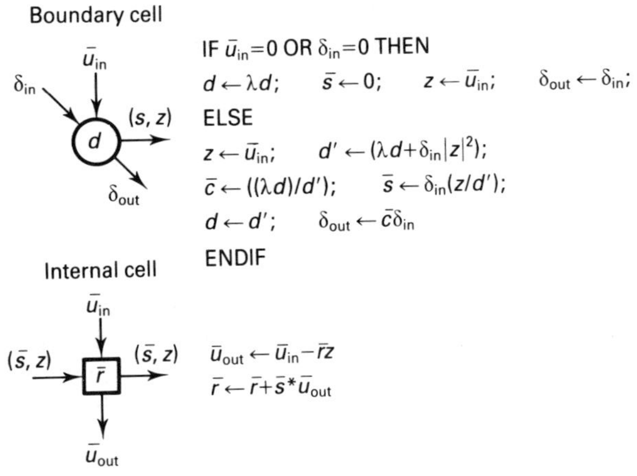

Figure 7.4 Square-root-free processing elements

It follows from (7.22), (7.25) and (7.11) that the weight vector $\mathbf{c}(N)$ is given by

$$\bar{R}(N)\mathbf{c}(N) + \bar{\mathbf{p}}(N) = \mathbf{0} \tag{7.26}$$

and may be obtained, as before, by a simple process of back-substitution. Figure 7.4 specifies the cell functions required for one particular version of the square-root-free algorithm. This version, which requires two multiplications and two additions per cycle to be performed in each internal cell, was first suggested by Golub and reported by Gentleman [4]. We have found it to be particularly stable and accurate throughout an extensive programme of finite wordlength adaptive filtering and beamforming computations [6,7]. These observations are supported by the work of Ling *et al.* [8], who derived an equivalent algorithm (the 'error feedback' algorithm) based on a recursive form of the modified Gram–Schmidt orthogonalization procedure – see section 7.2.8.

The 'square-root-free' algorithm described above is in fact an algorithm for *calculating* the required planar rotation. In the QRD-based least-squares minimization problem these rotations also have to be *applied* to various vectors. In section 7.2.3 the 'natural' implemenation of the Givens rotation (involving the computation of square roots) was applied to various vectors in a feedforward manner: the two components of the rotated vector are calculated independently on the basis of the two input components. The square-root-free algorithm presented above applies the rotation in a feedback mode: one output is now dependent on one input and the other output. It follows that there exists another feedback algorithm where the parameter that is fed back is the stored quantity $\bar{r}$ – see Figure 7.4 – rather than the terms u_{in} or u_{out} which flow through the array.

Table 7.2 Options for implementing a Givens rotation

Method of calculation		*Method of application*
Using square roots (SQ)	×	Feedforward (FF)
Square-root-free (SF)		Feedback of output (FB)
		Feedback of stored parameter (MFB)

Combined with the two methods for calculating the rotation parameters, the feedforward/feedback choice results in six different variations; see Table 7.2. The computer simulations of section 7.3.8 show that, of these four possible variations, the one that is equivalent to the RMGS error feedback algorithm performs the best in terms of numerical stability. It is worth emphasizing that the basic architecture (Figure 7.2) is not affected by the choice of rotation technique so that the only difference between the various options is a change of PEs.

Table 7.3 SF/FB narrowband beamformer algorithm

	Add/subt	Mult	Sqrt/div
START			
INITIALIZE {all variables := 0};			
FOR t FROM 1 DO			
LET $e(t) := y(t)$; $\delta(t) := 1$;			
FOR i FROM 1 TO p DO!?			
LET $\bar{r}^{(i,i)}(t) := (\lambda\bar{r}^{(i,i)}(t-1) + \delta(t)\lvert\mathbf{u}^{(i)}(t)\rvert^2$;	1	3	–
IF $\bar{r}^{(i,i)}(t) = 0$ THEN LET $\bar{c} := 1$; $\bar{s} := 0$;			
ELSE LET $\bar{c} := \dfrac{\lambda\bar{r}^{(i,i)}(t-1)}{\bar{r}^{(i,i)}(t)}$; $\bar{s} := \dfrac{\delta(t)\mathbf{u}^{(i)}(t)}{\bar{r}^{(i,i)}(t)}$;	–	2	2
END_IF;			
FOR j FROM $i+1$ TO p DO			
LET $\mathbf{u}^{(j)}(t) := (\mathbf{u}^{(j)}(t) - \bar{r}^{(i,j)}(t-1)\mathbf{u}^{(i)}(t))$;	1	1	–
$\bar{r}^{(i,j)}(t) := (\bar{r}^{(i,j)}(t-1) + \bar{s}^*\mathbf{u}^{(j)}(t))$	1	1	–
END_DO;			
LET $e(t) := (e(t) - \bar{\mathbf{p}}^{(i)}(t-1)\mathbf{u}^{(i)}(t))$;	1	1	–
$\bar{\mathbf{p}}^{(i)}(t) := (\mathbf{p}^{(i)}(t-1) + \bar{s}^* e(t))$;	1	1	–
$\delta(t) := \bar{c}\delta(t)$	–	1	–
END_DO; {i loop}			
COMMENT pth-order filtered residual COMMENT			
LET $\varepsilon(t) := (\delta(t)e(t))$	–	1	–
END_DO; {t loop}	$p^2 + 2p$	$p^2 + 7p + 1$	$2p$
FINISH			

A listing for the square-root-free with feedback (SF/FB) $O(p^2)$ QRD-based least-squares minimization algorithm is shown in Table 7.3 (see section 7.5 for an explanation of the listing).

7.2.5 Direct residual extraction

In many least-squares problems, and particularly in adaptive noise cancellation, the least-squares weight vector $\mathbf{c}(N)$ is not the principal object of interest. Of more direct concern is the corresponding residual, since this constitutes the noise-reduced output signal from the adaptive combiner [7]. In this section, we show how the a posteriori least-squares residual

$$\varepsilon(N) = \mathbf{u}^{\mathrm{T}}(N)\mathbf{c}(N) + y(N) \tag{7.27}$$

which depends on the most up-to-date weight vector $\mathbf{c}(N)$, may be obtained directly from the type of processor array described in section 7.2.3 without computing $\mathbf{c}(N)$ explicitly [9].

In order to proceed, it is necessary to consider the structure of the $N \times N$ matrix $\hat{Q}(N)$ which, from (7.20), may be expressed in the form

$$\hat{Q}(N) = \begin{bmatrix} \Sigma(N) & \mathrm{O} & \mathbf{q}(N) \\ \mathrm{O} & I & \mathbf{0} \\ \boldsymbol{\sigma}^{\mathrm{T}}(N) & \mathbf{0}^{\mathrm{T}} & \tilde{\alpha}(N) \end{bmatrix} \tag{7.28}$$

where $\Sigma(N)$ is a $p \times p$ matrix, $\mathbf{q}(N)$ and $\boldsymbol{\sigma}(N)$ are $p \times 1$ vectors and $\tilde{\alpha}(N)$ is a scalar. Now $\hat{Q}(N)$ is given by

$$\hat{Q}(N) = \prod_{i=1}^{p} G_i(N) \tag{7.29}$$

where the G_i are elementary $N \times N$ rotation matrices of the form

$$G_i(N) = \begin{bmatrix} 1 & & & & & & \\ & \ddots & & & & & \\ & & 1 & & & & \\ & & & c_i(N) & \cdot\ \cdot\ \cdot & & s_i^*(N) \\ & & & & 1 & & \\ & & & & & \ddots & \\ & & & & & 1 & \\ & & & -s_i(N) & \cdot\ \cdot\ \cdot & & c_i(N) \end{bmatrix} \quad i = 1, 2, 3, \ldots, p \tag{7.30}$$

and all off-diagonal elements are zero except those in the (i,N) and (N,i) locations. It follows directly that

$$\tilde{\alpha}(N) = \prod_{i=1}^{p} c_i(N) \tag{7.31}$$

i.e. $\tilde{\alpha}(N)$ is the product of the cosine terms in the p rotations represented by $\hat{Q}(N)$. Multiplying both sides of (7.20) by $\hat{Q}^{\mathrm{H}}(N)$ and noting that the matrix $\hat{Q}(N)$ is unitary, it is easy to deduce that

$$\mathbf{u}^{\mathrm{T}}(N) = \mathbf{q}^{\mathrm{H}}(N)\tilde{R}(N) \tag{7.32}$$

and similarly, from (7.21), we have

$$y(N) = \mathbf{q}^{\mathrm{H}}(N)\mathbf{p}(N) + \tilde{\alpha}(N)\tilde{e}(N) \tag{7.33}$$

Substituting Equations (7.32) and (7.33) into (7.27) leads to the expression

$$\varepsilon(N) = \mathbf{q}^{\mathrm{H}}(N)\tilde{R}(N)\mathbf{c}(N) + \mathbf{q}^{\mathrm{H}}(N)\mathbf{p}(N) + \tilde{\alpha}(N)\tilde{e}(N) \tag{7.34}$$

and, from (7.11), it follows immediately that

$$\varepsilon(N) = \tilde{\alpha}(N)\tilde{e}(N) \tag{7.35}$$

Now, as noted in section 7.2.3, when the conventional Givens rotation algorithm is employed, the parameter $\tilde{e}(N)$ is generated quite naturally within the triangularization process and simply emerges from the bottom cell of the right-hand column in Figure 7.2; because of its relationship with the a posteriori residual $\varepsilon(N)$, as given in (7.35), the parameter $\tilde{e}(N)$ is known as the angle-normalized residual. The scalar $\tilde{\alpha}(N)$, as given by (7.31), may also be computed very simply. The product of cosine terms is generated recursively by the parameter $\tilde{\alpha}$ as it passes from cell to cell along the chain of boundary processors. The simple product required to form the a posteriori residual as given in (7.35) is computed by the final processing cell F in Figure 7.2.

The direct residual extraction technique obviously avoids a lot of unnecessary computation provided that the weight vector is not required explicitly. As a result, the overall processing architecture is greatly simplified. There is no need for a separate back-substitution processor or any sophisticated control circuitry to ensure that the contents of the triangular array are input to the back-substitution processor in the correct sequence. Consequently, it is much easier to maintain a regular pipelined data flow. A less obvious, but arguably more important, advantage of direct residual extraction is the improved numerical stability which it offers. This derives from the fact that computing the weight vector explicity requires the solution of a linear inverse problem which may be ill-conditioned in circumstances where the optimum weight vector is not well defined. The least-squares residual, however, is always well defined and can be computed reliably. Note, for example, that the correct (zero) residual is obtained even during the first few processing cycles when the data matrix is not of full rank and the corresponding weight vector cannot be uniquely defined. This type of unconditional stability, which may be contrasted with that of the traditional RLS algorithm described in Chapter 2 and avoids the need for 'persistent excitation', is extremely important in the context of real-time signal processing.

The a posteriori residual may be computed in a similar manner when the square-root-free algorithm is employed. The square-root-free algorithm delivers from the bottom cell in the right-hand column a scalar $\bar{e}(N)$ given by

$$\delta^{1/2}(N)\bar{e}(N) = \tilde{e}(N) \tag{7.36}$$

where $\delta^{1/2}(N)$ is the scaling parameter appropriate to the pth row at time N and it has been assumed that $\delta(N)$ is initialized to unity on input to the array. From (7.35) and (7.36) it follows that

$$\varepsilon(N) = \tilde{\alpha}(N)\delta^{1/2}(N)\bar{e}(N) \tag{7.37}$$

where $\tilde{\alpha}(N)$ is the product of cosine terms which arise in the conventional Givens rotation algorithm. However, $\delta(N)$, as computed by the boundary processors for the square-root-free algorithm, is simply the product of all the $\bar{c}$ terms and it can easily be shown that this is equivalent to the product of the conventional cosine parameters squared. Thus,

$$\delta(N) = \tilde{\alpha}^2(N) \tag{7.38}$$

and

$$\varepsilon(N) = \delta(N)\bar{e}(N) \tag{7.39}$$

Hence the a posteriori residual $\varepsilon(N)$ may also be obtained directly from the square-root-free processor array, using a final multiplier cell F, as illustrated in Figure 7.2.

A second form of residual, which occurs naturally in least-squares algorithms such as the least-squares lattice (see Chapter 6), is the a priori residual, denoted $e(N)$. Defined in terms of the previously computed weight vector $\mathbf{c}(N-1)$ and the latest data $[\mathbf{u}^T(N), y(N)]$, it takes the form

$$e(N) = \mathbf{u}^T(N)\mathbf{c}(N-1) + y(N) \tag{7.40}$$

This residual may also be obtained directly from the triangular processor array, as we now show. By substituting the decomposition for $\hat{Q}(N)$ given in (7.28) in the time update relation (Equation (7.20)) we have

$$\boldsymbol{\sigma}^T(N)\lambda^{1/2}\tilde{R}(N-1) + \tilde{\alpha}(N)\mathbf{u}^T(N) = \mathbf{0} \tag{7.41}$$

Eliminating the vector $\mathbf{u}^T(N)$ between (7.40) and (7.41) we find

$$e(N) = [-\boldsymbol{\sigma}^T(N)\lambda^{1/2}\tilde{R}(N-1)\mathbf{c}(N-1) + \tilde{\alpha}(N)y(N)]/\tilde{\alpha}(N) \tag{7.42}$$

Again, by using the decomposition given in (7.28), this time with (7.21), another relationship can be obtained:

$$\boldsymbol{\sigma}^T(N)\lambda^{1/2}\mathbf{p}(N-1) + \tilde{\alpha}(N)y(N) = \tilde{e}(N) \tag{7.43}$$

By eliminating the term $\tilde{\alpha}(N)y(N)$ between (7.42) and (7.43) we have that

$$e(N) = [-\sigma^T(N)\lambda^{1/2}\tilde{R}(N-1)\mathbf{c}(N-1) + \tilde{e}(N) - \sigma^T(N)\lambda^{1/2}\mathbf{p}(N-1)]/\tilde{\alpha}(N) \tag{7.44}$$

and, from (7.11), it follows immediately that

$$e(N) = \tilde{e}(N)/\tilde{\alpha}(N) \tag{7.45}$$

Thus the a priori residual is computed from the same quantities as the a posteriori residual. It follows immediately from (7.35) and (7.45) that they are related by the expression

$$\varepsilon(N) = \tilde{\alpha}^2(N)e(N) \tag{7.46}$$

Note that by eliminating the term $\tilde{\alpha}(N)$ between (7.35) and (7.45) we find that

$$\tilde{e}(N) = \sqrt{\varepsilon(N)e(N)} \tag{7.47}$$

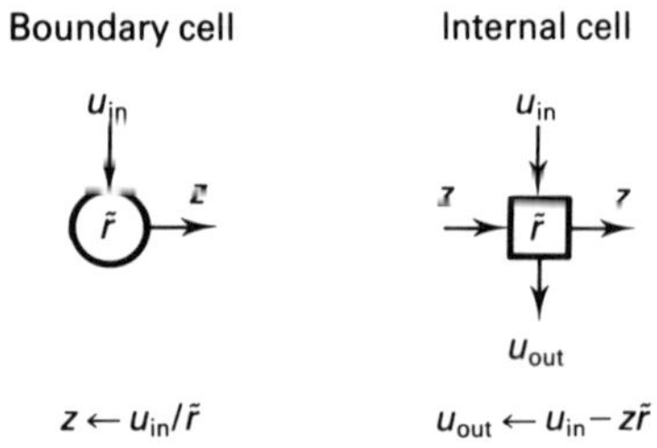

Figure 7.5 Frozen-mode processing elements

so that the angle-normalized residual $\tilde{e}(N)$ can be viewed as the geometric mean of the a priori and the a posteriori residuals.

From (7.38), (7.39) and (7.46) we see that

$$\bar{e}(N) = e(N) \tag{7.48}$$

so that the scalar which emerges naturally from the bottom cell in the right-hand column of the square-root-free processor array is the corresponding a priori residual. Finally, we note that

$$\delta(N) = \tilde{\alpha}^2(N) = \varepsilon(N)/e(N) \tag{7.49}$$

and hence we see that $\delta(N)$ is equal to the angle variable $\alpha(N)$ (see Chapter 6) and

$$\tilde{\alpha}(N) = \sqrt{\alpha(N)} \tag{7.50}$$

7.2.6 Weight freezing and flushing

It has been shown that if a data vector $[\mathbf{u}^T(N), y(N)]$ is input to the triangular processor array in Figure 7.2, the corresponding a posteriori least-squares residual $\varepsilon(N)$ emerges from the final cell F. In order to achieve this result, the array performs two distinct functions:

1. It generates the updated triangular matrix $\tilde{R}(N)$ and corresponding vector $\mathbf{p}(N)$ (or $D(N)$, $\bar{R}(N)$ and $\bar{\mathbf{p}}(N)$ for the square-root-free algorithm) and hence, implicitly, the updated weight vector $\mathbf{c}(N)$.
2. It acts as a simple filter which applies the updated weight vector to the input data according to Equation (7.27).

If the array is subsequently 'frozen' in order to suppress any further update of the stored values, but allowed to function normally in all other respects, it will continue to perform the filtering operation without affecting the implicit weight vector $\mathbf{c}(N)$. Thus, in response to an input of the form $[\mathbf{u}^T, y]$, the frozen network will produce the output residual

$$x = \mathbf{u}^T\mathbf{c}(N) + y \tag{7.51}$$

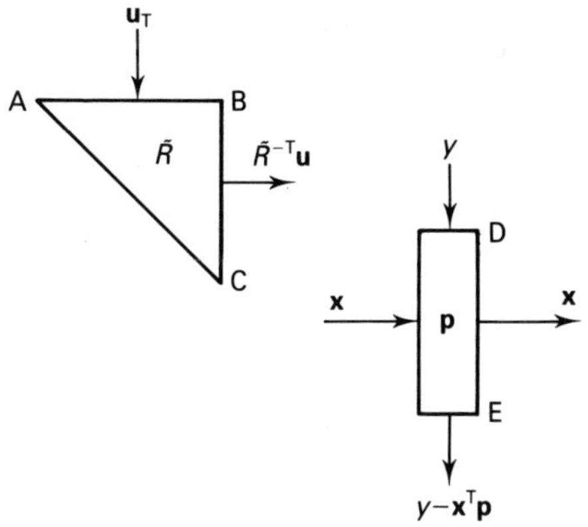

Figure 7.6 Matrix operations

Equation (7.51) may be verified directly by considering the frozen array as a combination of basic matrix operators. Consider first the basic triangular array ABC in Figure 7.2. In frozen mode, the boundary and internal cells perform the reduced processing functions defined in Figure 7.5. Now consider the effect of the simplified network upon a row vector $\mathbf{u}^T$, input from above in the usual manner. This will give rise to a column vector $\mathbf{z}$ which emerges from the right. It is straightforward to verify that the input vector $\mathbf{u}$ is related to the output vector $\mathbf{z}$ by means of matrix transformation

$$\mathbf{u} = \tilde{R}^T\mathbf{z} \tag{7.52}$$

where $\tilde{R}$ is the upper-triangular matrix stored within the array. For example, it is clear that

$$z_1 = u_1/\tilde{r}_{11} \quad \text{and} \quad z_2 = (u_2 - \tilde{r}_{12}/z_1)/\tilde{r}_{22} \tag{7.53}$$

that is,

$$u_1 = \tilde{r}_{11}z_1 \quad \text{and} \quad u_2 = \tilde{r}_{12}z_1 + \tilde{r}_{22}z_2 \tag{7.54}$$

Assuming that $\tilde{R}$ is non-singular (i.e. no diagonal element of $\tilde{R}$ is zero) it follows that

$$\mathbf{z} = \tilde{R}^{-T}\mathbf{u} \tag{7.55}$$

and so the frozen triangular array may be regarded as a $\tilde{R}^{-T}$ matrix operator, as depicted schematically in Figure 7.6.

Now consider the right-hand column of cells DE in Figure 7.2. It is easy to show that, in frozen mode, the effect of this array upon a column vector $\mathbf{x}$ input from the left and a scalar y input from the top is to produce the scalar output $y - \mathbf{x}^T\mathbf{p}$ which emerges from the bottom cell. The vector $\mathbf{x}$ also emerges unchanged from the right, as depicted in Figure 7.6. It follows immediately that if the network in Figure 7.2 is

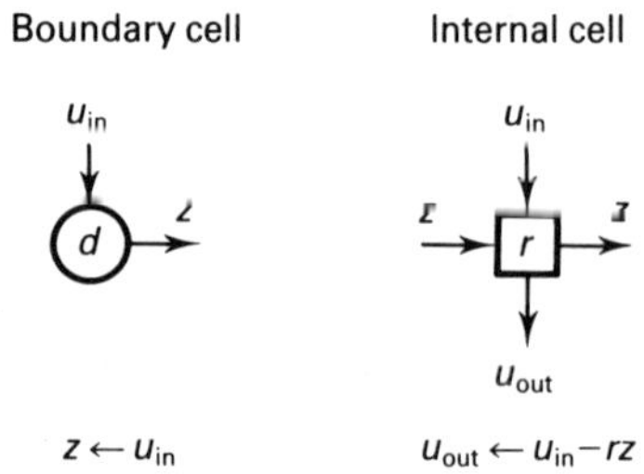

Figure 7.7 Square-root-free frozen processing elements

frozen at time N, its effect upon a vector $[\mathbf{u}^T, y]$ input from the top is to produce the scalar output $y - \mathbf{u}^T R^{-1}(N)\mathbf{p}(N)$, which emerges from the internal cell E. From (7.11), it can be seen that this is precisely the frozen residual defined in (7.51).

When the square-root-free algorithm is employed, the array in Figure 7.2 may be frozen very simply by setting the forgetting factor $\lambda = 1$ and initializing the parameter δ to zero for any input vector which is to be processed in the frozen mode. As pointed out in section 7.2.4, this has the effect of assigning zero weight to that vector within the overall least-squares computation and so the processing does not affect any values stored within the array. This property can be verified quite easily by inspecting the square-root-free cells in Figure 7.4, and the resulting operations for the frozen cells are shown in Figure 7.7. It follows from the discussion above that if a vector $[\mathbf{u}^T, y]$ is input to the top of the frozen square-root-free array at time N, the output which emerges from the internal cell E will be $y - \mathbf{u}^T R^{-1}(N)\bar{\mathbf{p}}(N)$ and from (7.26) it can be seen that this is again the frozen residual defined in (7.51). Note that the parameter δ_{in} for the final processing cell must be set equal to one to avoid suppressing the output residual from the frozen square-root-free network if this technique is used.

Having established that the function of a frozen triangular array is given by (7.51), it is easy to see how the least-squares weight vector, if required, may also be obtained without performing a back-substitution. Let $\mathbf{h}_i$ denote the p-element vector whose only non-zero element is a one, occurring in the *ith* location; it follows that the effect of inputting the sequence of 'impulse' vectors $[\mathbf{h}_i\ 0]$ $(i = 1, 2\ldots, p)$ to the frozen array is to produce the sequence of output values $c_i(N)$ $(i = 1, 2, \ldots, p)$ and so the weight vector $\mathbf{c}(N)$ may be extracted from the array without the need for any additional hardware. This technique, which amounts to measuring the impulse response of the system, is generally referred to as 'weight flushing' [7].

7.2.7 Parallel weight extraction

The technique of weight flushing presented in section 7.2.6 requires that the adaptive filtering be temporarily suspended whilst the sequence of impulse vectors is fed into the array. A pth-order system would therefore have to suspend its data processing for p time instants. Although it is conceivable that this weight-flushing process could be carried out at a higher clock rate, in between the processing of the data, in a high data rate environment this option is unlikely to be viable and an alternative technique is required. As we show below, the weight vector can be produced in parallel with the adaptation process, in several ways, by the addition of extra hardware.

Before describing how the weight vector can be generated, we first consider an important property of the $\hat{Q}$ matrix upon which the parallel weight extraction techniques (and also some other important techniques like MVDR beamforming [10]) depend. We have seen, in section 7.2.2, that the matrix $\hat{Q}(N)$ updates the triangular matrix $\tilde{R}(N-1)$ to $\tilde{R}(N)$ by rotating in the new data vector at time N. Rather surprisingly, the matrix $\hat{Q}(N)$ can also be used to update the triangular matrix $\tilde{R}^{-\mathrm{H}}(n-1)$ to $\tilde{R}^{-\mathrm{H}}(N)$. Consider the following identity:

$$[\lambda^{1/2}\tilde{R}^{\mathrm{H}}(N-1),\ \mathrm{O},\ \mathbf{u}^*(N)]\hat{Q}^{\mathrm{H}}(N)\hat{Q}(N)\begin{bmatrix}\lambda^{-1/2}\tilde{R}^{-\mathrm{H}}(N-1)\\ \mathrm{O}\\ \mathbf{0}^{\mathrm{T}}\end{bmatrix}=I \tag{7.56}$$

Let

$$\hat{Q}(N)\begin{bmatrix}\lambda^{-1/2}\tilde{R}^{-\mathrm{H}}(N-1)\\ \mathrm{O}\\ \mathbf{0}^{\mathrm{T}}\end{bmatrix}=\begin{bmatrix}T(N)\\ \mathrm{O}\\ \mathbf{x}^{\mathrm{T}}(N)\end{bmatrix} \tag{7.57}$$

so that, by means of (7.20) and (7.56), we have:

$$[\tilde{R}^{\mathrm{H}}(N)\ \mathrm{O}\ \mathbf{0}]\begin{bmatrix}T(N)\\ \mathrm{O}\\ \mathbf{x}^{\mathrm{T}}(N)\end{bmatrix}=I \tag{7.58}$$

and clearly,

$$T(N)=\tilde{R}^{-\mathrm{H}}(N) \tag{7.59}$$

Thus (7.57) becomes:

$$\hat{Q}(N)\begin{bmatrix}\lambda^{-1/2}\tilde{R}^{-\mathrm{H}}(N-1)\\ \mathrm{O}\\ \mathbf{0}^{\mathrm{T}}\end{bmatrix}=\begin{bmatrix}\tilde{R}^{-\mathrm{H}}(N)\\ \mathrm{O}\\ \mathbf{x}^{\mathrm{T}}(N)\end{bmatrix} \tag{7.60}$$

and we see that the same rotations that update the matrix $\tilde{R}$ also update the matrix $\tilde{R}^{-\mathrm{H}}$.

Recall from the description of the QRD systolic array in section 7.2.3 that the sine and cosine parameters that define the rotation matrix $\hat{Q}$ are passed from left to right across the array. Thus the matrix $\tilde{R}^{-\mathrm{H}}$ can be calculated by appending an array of (square) processing elements to the right of the basic QRD array: the QRD array stores $\tilde{R}$ and outputs the rotation parameters that specify $\hat{Q}$ as $\tilde{R}$ is updated from time instant to time instant; the rotation parameters are then used by the new array

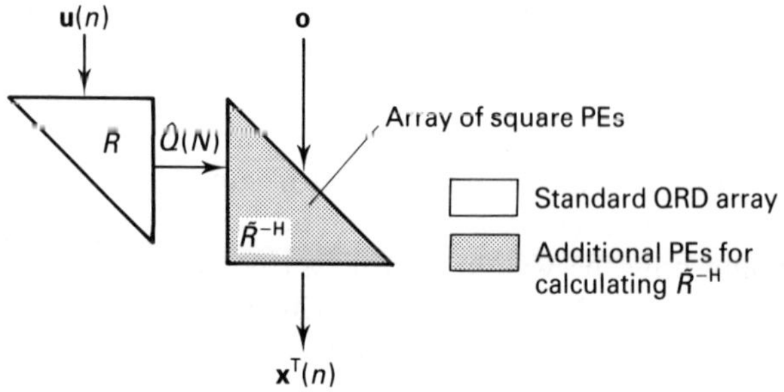

Figure 7.8 Updating the inverse triangular matrix

which stores and updates the matrix $\tilde{R}^{-H}$ by rotating it against a vector of zeros. The matrix $\tilde{R}^{-H}$ is lower triangular and hence the new array will also be lower triangular (see Figure 7.8).

Armed with this property of the $\hat{Q}$ matrix, we can now show how the weight vector can be calculated in parallel with the adaption process [11]. Consider the following augmented data matrix:

$$\mathcal{U}_+(N) = [\mathcal{U}(N) \quad \mathbf{y}(N)] \tag{7.61}$$

The upper-triangular matrix resulting from a QRD will then have the form

$$\tilde{R}_+(N) = \begin{bmatrix} \tilde{R}(N) & \mathbf{p}(N) \\ \mathbf{0}^T & \tilde{\alpha}^y(N) \end{bmatrix} \tag{7.62}$$

where $\tilde{\alpha}^y(N) = \| \mathbf{v}(N) \|$ is the square root of the filtering error power for $y(N)$. It is easy to show that the inverse of an upper-triangular matrix is another upper-triangular matrix. Thus let

$$\tilde{R}_+^{-1}(N) = \begin{bmatrix} T'(N) & \mathbf{s}(N) \\ \mathbf{0}^T & z(N) \end{bmatrix} \tag{7.63}$$

where $T'(N)$ is a $(p+1) \times (p+1)$ upper-triangular matrix, $\mathbf{s}(N)$ is a p-dimensional vector and $z(N)$ is a scalar. Then

$$I = \tilde{R}_+^{-1}(N)\tilde{R}_+(N) = \begin{bmatrix} T'(N) & \mathbf{s}(N) \\ \mathbf{0}^T & z(N) \end{bmatrix} \begin{bmatrix} \tilde{R}(N) & \mathbf{p}(N) \\ \mathbf{0}^T & \tilde{\alpha}^y(N) \end{bmatrix}$$

$$= \begin{bmatrix} T'(N)\tilde{R}(N) & T'(N)\mathbf{p}(N) + \tilde{\alpha}^y(N)\mathbf{s}(N) \\ \mathbf{0}^T & \tilde{\alpha}^y(N)z(N) \end{bmatrix} \tag{7.64}$$

Hence

$$\tilde{R}_+^{-1}(N) = \begin{bmatrix} \tilde{R}^{-1}(N) & -\tilde{\alpha}^{y^{-1}}(N)\tilde{R}^{-1}(N)\mathbf{p}(N) \\ \mathbf{0}^T & \tilde{\alpha}^{y^{-1}}(N) \end{bmatrix} = \begin{bmatrix} \tilde{R}^{-1}(N) & \tilde{\alpha}^{y^{-1}}(N)\mathbf{c}(N) \\ \mathbf{0}^T & \tilde{\alpha}^{y^{-1}}(N) \end{bmatrix} \tag{7.65}$$

where we have used (7.11) to identify the presence of the least-squares weight vector $\mathbf{c}(N)$. Thus we can obtain the weight vector $\mathbf{c}(N)$ directly from the right-hand column of the inverse of the augmented triangular matrix $\tilde{R}_+(N)$ (or, equivalently, the complex conjugate of the bottom row of $\tilde{R}_+^{-H}(N)$).

The matrix $\tilde{R}_+^{-H}(N)$ can be calculated as shown in Figure 7.8 provided we ensure that the QRD array, on the left-hand side of Figure 7.8, is solving the augmented problem. This requires that the data vector being fed into the array, at time N, is $[\mathbf{u}^T(N)\ y(N)]$ and hence that the QRD array is now of dimension $(p+1) \times (p+1)$. The adaptive filtering residual is still available with this architecture since the $(p+1) \times (p+1)$ QRD array can be thought of as consisting of a pth-order adaptive filter processor (as shown in Figure 7.2) with the multiplier cell (F in Figure 7.2) replaced by an additional circular processing element. Clearly, this additional circular processing element can be modified, if necessary, to calculate the adaptive filtering residual.

The parallel weight extraction technique described above is based on the fact that the rotation matrix $\hat{Q}(N)$, as calculated by the basic QRD array on the left-hand side of Figure 7.8, can be used to update the matrix $\tilde{R}^{-H}(N-1)$. An alternative technique for generating $\hat{Q}(N)$ [12–14] leads to a different achitecture for calculating the optimum weight vector. This technique relies on the fact that the rotation matrix $\hat{Q}(N)$ is completely specified by the relevant p rotation angles (see section 7.2.2) and hence may be reconstructed from knowledge of these angles. As we show below, the relevant angles can be recovered from the bottom row of the matrix $\hat{Q}(N)$; furthermore, this row vector can itself be calculated, indirectly, from a knowledge of the matrix $\tilde{R}^{-H}(N-1)$.

From (7.28) and (7.29), we have

$$\hat{Q}^T(N)\boldsymbol{\pi}_N = \begin{bmatrix} \sigma(N) \\ \mathbf{0} \\ \tilde{\alpha}(N) \end{bmatrix} \tag{7.66}$$

or

$$\left[\left(-s_1 \prod_{i=2}^{p} c_i\right), \ldots, (-s_{p-1}c_p), (-s_p), \mathbf{0}^T, \prod_{i=1}^{p} c_i\right]^T \tag{7.67}$$

where the s_i and the c_i are the sines and cosines of the relevant Givens rotations that compose $\hat{Q}(N)$ and the 'pinning' vector $\boldsymbol{\pi}_N$ is an N-dimensional vector such that

$$\boldsymbol{\pi}_N = [0, 0, \ldots, 0, 1]^T \tag{7.68}$$

These sines and cosines uniquely determine the matrix $\hat{Q}(N)$ and can be recovered from $\boldsymbol{\sigma}(N)$ and $\tilde{\alpha}(N)$ in the following orthogonal manner. A sequence of Givens rotations is used successively to annihilate elements of $\boldsymbol{\sigma}(N)$ starting from the top and working down until the pinning vector $\boldsymbol{\pi}_N$ is produced (see (7.66)).

Let $\bar{\alpha} = \Pi_{i=2}^{p} c_i$; then the first rotation takes the form

$$\begin{bmatrix} c & s^* \\ s & c \end{bmatrix}\begin{bmatrix} \bar{\alpha}c_1 \\ \bar{a}s_1 \end{bmatrix} = \begin{bmatrix} g_1 \\ 0 \end{bmatrix} \tag{7.69}$$

and it is easy to show that

$$g_1 = \pm\bar{\alpha} \qquad c = \pm c_1 \qquad s = \mp s_1 \tag{7.70}$$

The ambiguity of sign is immediately resolved by recalling that in the construction of $\hat{Q}_p(N)$ the rotation angle is always chosen such that the cosine is positive. Having applied the first rotation to determine s_1 and c_1, the vector in (7.67) takes the form

$$\left[0, \left(-s_2 \prod_{i=3}^{p} c_i\right), \ldots, (-s_{p-1}c_p), (-s_p), \mathbf{0}^{\mathrm{T}}, \prod_{i=2}^{p} c_i\right]^{\mathrm{T}} \tag{7.71}$$

and so the next Givens rotation serves to compute s_2 and c_2 in a similar manner, and so on. Note that it is also possible to recover the rotation angles that define $\hat{Q}_p(N)$ from its right-hand column. In particular, we have

$$\hat{Q}_p(N)\boldsymbol{\pi}_N = \begin{bmatrix} \mathbf{q}_p(N) \\ \mathbf{0} \\ \tilde{\alpha}_p(N) \end{bmatrix} = \begin{bmatrix} s_1^* \\ c_1 s_2^* \\ \vdots \\ \left(\prod_{i=1}^{p-1} c_i\right)s_p^* \\ \mathbf{0} \\ \prod_{i=1}^{p} c_i \end{bmatrix} \tag{7.72}$$

so that the sines and cosines may be recovered by applying a sequence of Givens rotations which annihilate the elements of the vector $\mathbf{q}_p(N)$ one at a time by rotation against the scalar (initially $\tilde{\alpha}_p(N)$). The order in which these annihilations take place is important and must begin with the bottom element and work upwards: in this case the pair (s_p, c_p) are generated first and (s_1, c_1) last.

Having established that the matrix $\hat{Q}(N)$ can be reconstructed from knowledge of the quantities $\boldsymbol{\sigma}(N)$ and $\tilde{\alpha}(N)$, we now proceed to show how these two quantities can be calculated other than from $\hat{Q}(N)$ itself. Note from (7.41) that

$$\boldsymbol{\sigma}(N) = -\lambda^{-1/2}\tilde{\alpha}(N)\tilde{R}^{-\mathrm{T}}(N-1)\mathbf{u}(N) \tag{7.73}$$

Thus given $\tilde{R}^{\mathrm{T}}(N-1)$ and the new data at time N ($\mathbf{u}(N)$), we may use (7.73) to calculate the vector $\boldsymbol{\sigma}(N)/\tilde{\alpha}(N)$. The value of $\tilde{\alpha}(N)$ can easily be found from the fact that $\hat{Q}(N)$ is orthonormal and hence

$$|\boldsymbol{\sigma}(N)|^2 + \tilde{\alpha}^2(N) = 1 \tag{7.74}$$

or

$$\tilde{\alpha}(N) = (1 + |\boldsymbol{\sigma}(N)/\tilde{\alpha}(N)|^2)^{-1/2} \tag{7.75}$$

Having calculated $\boldsymbol{\sigma}(N)$ and $\tilde{\alpha}(N)$ by this indirect method, we can then calculate $\hat{Q}(N)$ as above and proceed to update the matrix $\tilde{R}^{-H}$, without the need to calculate $\tilde{R}$ explicitly. Clearly, the augmented data matrix $\tilde{R}_{+}^{-H}(N)$ can also be updated in this way (with suitably redefined quantities), thus allowing the weight vector $\mathbf{c}(N)$ to be calculated without the need for the triangular array that performs the QRD of the augmented data matrix. Once the optimum weight vector is known, the adaptive filtering residual could of course be calculated using (7.27) but, since the weight vector can in certain circumstances be undefined, this method is less robust than the direct extraction one (see section 7.2.5).

Another method for extracting the weight vector without interfering with the adaption process also relies on the structure of the $\hat{Q}$ matrix. From (7.73) we see that if a vector $\mathbf{u}^{\mathrm{T}}(N)$ is annihilated by Givens rotations against a triangular matrix $\lambda^{1/2}\tilde{R}(N-1)$, then the bottom row of the $\hat{Q}$ matrix contains the term $\lambda^{-1/2}\tilde{R}^{-\mathrm{T}}(N-1)\mathbf{u}(N)$. Now from (7.11) we have that

$$\mathbf{c}(N) = -\tilde{R}^{-1}(N)\mathbf{p}(N) \tag{7.76}$$

Hence if we use Givens rotations to annihilate the vector $\mathbf{p}(N)$ by rotation against the matrix $\tilde{R}^{\mathrm{T}}(N)$ then the bottom row of the composite rotation matrix will contain the term

$$\tilde{R}^{-1}(N)\mathbf{p}(N) = -\mathbf{c}(N) \tag{7.77}$$

i.e. the negative of the least-squares weight vector. A systolic array for implementing this algorithm can be found in [15].

7.2.8 Comparison with recursive modified Gram–Schmidt algorithm

The square-root-free algorithm and architecture with direct residual extraction as described in sections 7.2.1 to 7.2.5 was obtained independently by Ling *et al.* [8] based on the method of modified Gram–Schmidt (MGS) orthogonalization. The MGS algorithm operates on a fixed block of data and is essentially non-recursive. It solves the least-squares problem by applying a sequence of linear combinations to the columns of the block data matrix $\mathscr{U}(N)$ and transforming it into a matrix $\mathscr{U}'(N)$ whose columns are mutually orthogonal. The complete transformation may be represented by means of an upper-triangular matrix whose elements correspond to the triangular matrix $\tilde{R}(N)$ which would have resulted from applying a QRD to the data matrix $\mathscr{U}(N)$. The process is extended to include the vector of samples $\mathbf{y}(N)$ in the primary channel and hence extract from it the component which is orthogonal to the columns of $\mathscr{U}(N)$. The bottom element of the resulting vector then corresponds to the a posteriori least-squares residual at time N. The QRD and MGS techniques are clearly related except that the former applies an orthogonal transformation to the data matrix $\mathscr{U}(N)$ in order to produce an upper-triangular form, whereas the latter applies an upper-triangular matrix transformation to $\mathcal{U}(N)$ in order to produce a matrix with orthogonal columns. Also, the QRD technique may be applied in a convenient row recursive manner using a sequence of elementary Givens rotations. The most important contribution of Ling *et al.* was to show that the MGS algorithm could also be updated one row at a time, thereby generating the recursive modified Gram–Schmidt (RMGS) algorithm which may be implemented using a triangular processing architecture similar to that in Figure 7.2.

Motivated by the desire to update the stored triangular matrix elements directly rather than as a ratio of other updated terms which occur more naturally in the RMGS approach, they manipulated their algorithm further. The resulting update equations were found to involve an important element of feedback which leads to an improvement in numerical stability. It is interesting to note that the error feedback RMGS algorithm derived by Ling *et al.* [16] is identical to the form of square-root-free Givens rotation algorithm defined in Figures 7.2 and 7.4 (see also Chapter 6). It was this relationship which led us to refer to the basic operation in Figure 7.4 as the square-root-free with feedback (SF/FB) Givens rotation and to identify the underlying feedback mechanism which is inherent to it.

7.2.9 Comparison with Kalman filter algorithms

The QRD algorithms for recursive least-squares processing, as defined in Figures 7.2, 7.3 and 7.4, operate directly on the basic data matrix $\mathcal{U}(N)$ as opposed to the data covariance matrix $R(N)$ and, as noted previously, this has significant numerical advantages. It was also shown in (7.8) that the upper-triangular matrix $\tilde{R}(N)$ which is stored and updated within the processor array is analytically identical to the Cholesky square root factor of the data (or information) covariance matrix. In the nomenclature of Kalman filtering, the QRD algorithm constitutes a numerically stable form of square root information Kalman filter with unit state-space matrix.

It has recently been shown by Chen and Yao [17] how the algorithm and triangular array architecture may, in fact, be extended to the case of a general square root information Kalman filter with arbitrary state-space matrix. Gaston *et al.* [18] have also shown how the triangular QRD array may be used as the core processor in a general square root covariance Kalman filter. Note that in Kalman filtering nomenclature the term covariance does not refer to forming or using the data covariance matrix $R(N)$. Instead, it denotes an algorithm which is based on updating the matrix $R^{-1}(N)$ (or its Cholesky square root factor $\tilde{R}^{-1}(N)$ in the present context) since this specifies the covariance of the weight vector estimate. In the special case of unit state-space matrix the covariance Kalman filter reduces to a least-squares estimation algorithm which makes use of the well-known matrix inversion lemma to perform successive rank 1 updates of the matrix $R^{-1}(N)$. This is often referred to as the ‘recursive least-squares’ (RLS) algorithm, although it is fundamentally different from the QRD-based recursive least-squares technique described in this chapter. For example, as pointed out in Chapter 5, ‘persistent excitation’ is essential if the traditional

RLS algorithm is to retain numerical stability due to the $1/\lambda$ term which occurs in the associated Riccati equation. However, this problem does not arise with the QRD algorithm described in sections 7.2.1 to 7.2.5. It would seem sensible, then, to refer to the traditional RLS and the alternative QRD techniques as the 'covariance' and 'square root information' recursive least-squares algorithms respectively.

Finally, it is worth noting that several authors [12–14] have recently developed stable 'square root covariance' least-squares algorithms and architectures based on the technique illustrated in Figure 7.8 for updating the 'square root covariance' matrix $\tilde{R}^{-1}(N)$. Their technique, however, makes clever use of a pinning vector to avoid storing and updating the 'square root information' matrix $\tilde{R}(N)$ explicitly and only requires a single triangular processor array as described in section 7.2.7. It is not clear how their method would extend to the general 'square root covariance' Kalman filtering problem or how it relates to the 'square root covariance' Kalman filter architectures proposed by Gaston *et al.* [18].

7.3 Adaptive FIR filtering

7.3.1 The QRD approach

In section 7.2 we saw how an adaptive linear combiner could be applied to the problem of narrowband adaptive beamforming. The same linear combiner could be used to construct an adaptive FIR filter. In this case, the combined ouptut at time t is given by the equation

$$x(t) = \mathbf{u}^{\mathrm{T}}(t)\mathbf{c} + y(t) = \sum_{j=0}^{p-1} c_j u(t-j) + y(t) \tag{7.78}$$

which is identical to the narrowband beamformer case (Equation (7.1)) except that the input vector $\mathbf{u}(t)$ now exhibits a high degree of time-shift invariance. This property manifests itself in the fact that the 'data matrix' ($\mathcal{U}_p(N)$ of Equation (7.80)) has a Toeplitz structure, i.e. each row of the matrix is obtained by shifting the previous row one column to the right and introducing one new data sample. Various algorithms have been devised that take advantage of this redundancy and so reduce the computational load, for a pth-order filter, from $O(p^2)$ to $O(p)$ arithmetic operations per sample time (see Chapter 5). The common basis for these fast algorithms is an efficient technique for solving the least-squares linear prediction problem. The concepts of forward and backward linear prediction must both be introduced for this purpose – see Chapters 5 and 6. The adaptive filtering problem can then be solved using quantities already calculated during the linear prediction stages. Unfortunately, the majority of these fast algorithms exhibit some form of numerical instability although much work has been done to overcome the numerical problems, and various rescue procedures have been developed – see Chapter 5.

As we have seen, it is possible to solve a least-squares minimization problem using the technique of QRD. Extensive computer simulations of this algorithm [7] have shown the QRD-based least-squares minimization algorithm to have excellent numerical properties. However, since the recursive QRD algorithm presented in section 7.2 is designed to solve a general recursive least-squares minimization problem, it requires $O(p^2)$ operations per sample time to generate the solution to a pth-order adaptive filter problem. A QRD-based algorithm which is designed for the special case of adaptive filtering and only requires $O(p)$ operations per sample time is thus of considerable interest [19–22].

In a least-squares adaptive filter of order p, the set of p weights $\mathbf{c}_p(N)$ at time N is chosen in order to minimize the sum of the squared differences between a reference signal $y(N)$ and a linear combination of the p samples from a data time series $u(N-i)$ $(0 \leqslant i \leqslant p-1)$. Specifically, the measure to be minimized is $\| \boldsymbol{\varepsilon}_p(N) \|$ where

$$\boldsymbol{\varepsilon}_p(N) = \mathcal{U}_p(N)\mathbf{c}_p(N) + \mathbf{y}(N) \tag{7.79}$$

and

$$\mathcal{U}_p(\mathcal{N}) = \Lambda(N)\begin{bmatrix} u(1) & u(0) & \dots & u(2-p) \\ u(2) & u(1) & \dots & u(3-p) \\ \vdots & \vdots & & \vdots \\ u(N) & u(N-1) & \dots & u(N-p+1) \end{bmatrix} \tag{7.80}$$

and

$$\mathbf{y}(N) = \Lambda(N)[y(1) \dots y(N)]^{\mathrm{T}} \tag{7.81}$$

Note that the computation of $\mathbf{c}_p(N)$ is based on all data received up to time N.

Compared with (7.5) and (7.6), we see that (7.79)–(7.81) constitute a standard least-squares minimization problem except for time-shift invariance of the data. We could therefore proceed to solve this least-squares minimization problem via the QRD technique described in section 7.2.1. In order to do so we must determine an orthogonal matrix $Q_p(N)$ that transforms the matrix $\mathcal{U}_p(N)$ into upper-triangle form[1] and use the same matrix $Q_p(N)$ to rotate the reference vector $y(N)$, i.e.

$$Q_p(N)\mathcal{U}_p(N) = \begin{bmatrix} \tilde{R}_p(N) \\ \mathrm{O} \end{bmatrix} \tag{7.82}$$

and

$$Q_p(N)\mathbf{y}(N) \equiv \begin{bmatrix} \mathbf{p}_p(N) \\ \mathbf{v}_p(N) \end{bmatrix} \tag{7.83}$$

It should be noted that once $Q_p(N)$ has been found, the filtering problem has effectively been solved. Knowledge of $Q_p(N)$ means that $\tilde{\alpha}_p(N)$ is known. It also allows the angle-normalized residual $\tilde{e}_p^y(N)$, i.e. the last component of the vector $\mathbf{v}_p(N)$, to be calculated and thus the least-squares residual may be found (see section 7.2.5). The $O(p^2)$ QRD-based algorithm for the solution of a pth order, least-squares minimization problem as described in section 7.2 has many desirable features. It operates in the 'data domain' and has a time recursive formulation with a time-independent computation requirement and a regular parallel architecture. However, the time shift redundancy in the adaptive filtering problem can be used to improve the method further by reducing the computational load from $O(p^2)$ to $O(p)$. The development of fast QRD algorithms for adaptive filterings is based, almost entirely, on the principle of constructing partially triangularized matrices from known quantities and then finding a set of rotations to complete the process. We recall that this was also a key element in the derivation of the time recursive algorithm in section 7.2.2. The solution at time N was generated by rotating the new input data for time N into the upper triangular matrix associated with the solution at time $(N-1)$.

Recall that the set of rotations $Q_p(N)$, required to solve the adaptive filtering problem are entirely dependent on the matrix $\mathcal{U}_p(N)$. The matrix $\mathcal{U}_p(N)$ can, however, be built up in an order recursive manner by adding extra columns which, because of its Toeplitz structure, consist of one new element and a time-shifted version of the previous column. Consider the following decompositions:

$$\mathcal{U}_p(N) = \Lambda(N)\begin{bmatrix} u(1) & \dots & u(2-p) \\ \vdots & & \vdots \\ u(N) & \dots & u(N-p+1) \end{bmatrix} \tag{7.84}$$

$$\mathcal{U}_p(\mathcal{N}) = [\mathcal{U}_{p-1}(N)\ \mathbf{y}^b_{p-1}(N)] \tag{7.85}$$

$$\mathcal{U}_p(N) = \begin{bmatrix} \lambda^{(N-1)/2}u(1) & \mathbf{z}^{\mathrm{T}} \\ \mathbf{y}^{\mathrm{f}}_{p-1}(N) & \mathcal{U}_{p-1}(N-1) \end{bmatrix} \tag{7.86}$$

where[2]

$$\mathbf{y}^{\mathrm{f}}_{p-1}(N) = \Lambda(N-1)[u(2), \dots, u(N)]^{\mathrm{T}} \tag{7.87}$$

$$\mathbf{y}^{\mathrm{b}}_{p-1}(N) = \Lambda(N)[u(2-p), \dots, u(N-p+1)]^{\mathrm{T}} \tag{7.88}$$

and

$$\mathbf{z} = \lambda^{(N-1)/2}[u(0), \dots, u(2-p)]^{\mathrm{T}} \tag{7.89}$$

Note from (7.85) that if we had already determined the rotation matrix $Q_{p-1}(N)$ which triangularizes the matrix $\mathcal{U}_{p-1}(N)$, we could use it to triangularize the matrix $\mathcal{U}_p(N)$ partially. In doing so we would also have to rotate the vector $\mathbf{y}^{\mathrm{b}}_{p-1}(N)$, but these are exactly the steps required in the QRD-based solution of the $(p-1)$st-order backward linear prediction problem.

In the $(p-1)$st-order backward linear prediction problem at time N, an estimate of $u(N-p+1)$ is formed from a linear combination of the data $\{u(N), \dots, u(N-p+2)\}$. The solution to this problem depends on the triangularization of the matrix $\mathcal{U}_{p-1}(N)$ and the transformation of the reference vector $\mathbf{y}^{\mathrm{b}}_{p-1}(N)$. Hence, knowing the solution of the $(p-1)$st-order backward problem at time N would allow us to construct the partially triangularized matrix $Q_{p-1}(N)\mathcal{U}_p(N)$ from known quantities and thus save a large amount of computation. This partially triangularized

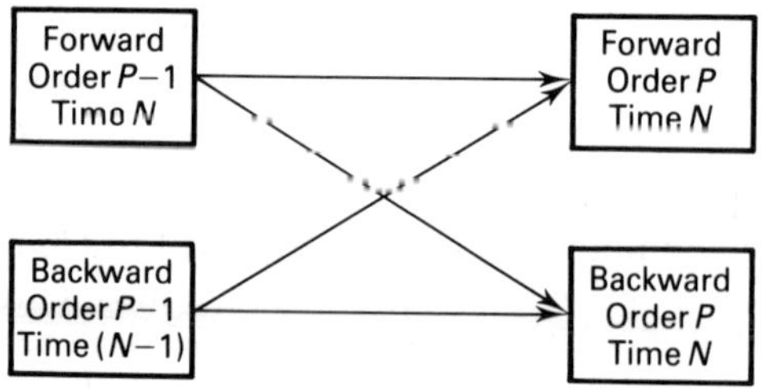

Figure 7.9 Philosophy behind the QRD-based lattice algorithm

matrix could then be transformed into the triangular matrix $\tilde{R}_p(N)$ by a sequence of Givens rotations.

Equation (7.86) allows another partially triangularized version of $\mathscr{U}_p(N)$ to be constructed, this time using quantities from the $(p-1)$st-order forward linear prediction problem. The $(p-1)$st-order (forward) linear prediction problem, at time N, is defined as the estimation of $u(N)$ based upon the data $\{u(N-1), \ldots, u(N-p+1)\}$. This involves the triangularization of the matrix $\mathscr{U}_{p-1}(N-1)$ and the transformation of the relevant reference vector $\mathbf{y}^{\mathrm{f}}_{p-1}(N)$. We could then use the decomposition given in (7.86) to generate a partial triangularization of the matrix $\mathscr{U}_p(N)$ from known quantities. The formation of the triangular matrix $\tilde{R}_p(N)$ could then be achieved very easily using a sequence of Givens rotations.

It is clear that the two linear prediction problems of order $(p-1)$ are intimately connected to the problem of determining a minimum set of rotations which triangularize the data matrix $\mathscr{U}_p(n)$. However, the triangularization of $\mathscr{U}_p(N)$ is central not only to the adaptive filtering problem but also to the pth-order linear prediction problems. The rotations which transform the matrix $\mathscr{U}_p(N)$ into $\tilde{R}_p(N)$ are used to solve the forward problem at time $(N+1)$ and the backward problem at time N. With a suitable time delay, we can therefore construct an order recursive algorithm for linear prediction and adaptive filtering (Figure 7.9).

It is also possible to develop another type of QRD-based fast algorithm. From the discussion above it is clear that we have a fast method for transforming $\tilde{R}_{p-1}(N-1)$ into $\tilde{R}_p(N)$ via (8.86). This transformation allows us to transform the matrix $Q_{p-1}(N-1)$ into $Q_p(N)$ (i.e. we have a time-and-order update). Equation (7.85) constitutes a method for transforming $\tilde{R}_p(N)$ into $\tilde{R}_{p-1}(N)$ and hence $Q_p(N)$ and $Q_{p-1}(N)$ (i.e. an order downdate). Thus, by combining these two transformations we can achieve an overall time update for the rotation matrix Q_p (Figure 7.10). This transformation does not lead directly to the construction of a fast algorithm since having to calculate the various Q matrices explicitly would require $O(p^2)$ operations and does not represent a reduction in the computational requirement. However, we have already seen that the value of this matrix can be inferred from knowledge of its right-hand column (section 7.2.7). Clearly, the transformations that update the matrix Q_p will also perform a time update for this column vector. Then, because we are now

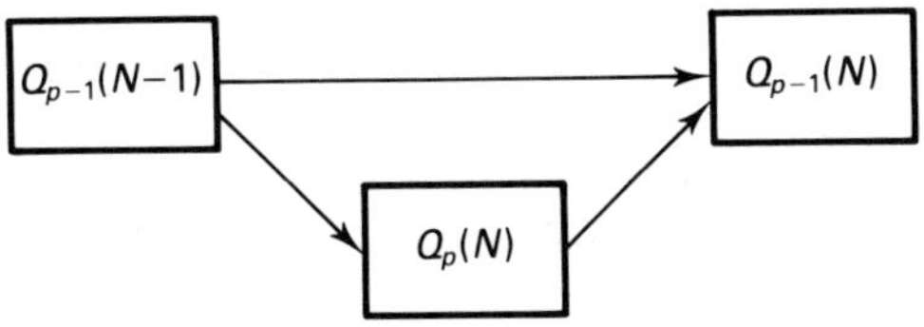

Figure 7.10 Philosophy behind the fast QRD algorithm

dealing with a vector rather than a matrix, the number of computations required to perform this type of update is $O(p)$ and a 'fast' algorithm results.

Traditionally 'fast' adaptive filtering algorithms fall into two classes: the least-squares lattice algorithms and the 'fixed-order' algorithms (e.g. the fast Kalman and fast transversal filters algorithms). Both types of algorithm solve the pth-order linear prediction problem in $O(p)$ operations. The least-squares lattice algorithms do this by solving all of the lower-order problems in sequence, whereas the fixed-order algorithms concentrate on a problem of given order and achieve the reduction in computational load by using the time-and-order update/order downdate technique. Not surprisingly, the two classes of algorithms have tended to be quite different; the fixed-order algorithms calculate the transversal filter coefficients explicitly, whereas the lattice algorithms deal directly with the filter residuals (errors) and calculate reflection coefficients. The fixed-order algorithms also tend to require fewer arithmetic operations than the lattice algorithms, although both are linear in the problem order. The level at which the two different algorithms can be pipelined is different – the lattice algorithms having a higher degree of concurrency. It is also worth noting that the data downdating step which is implicitly required by the fixed-order algorithms gives cause for concern with regard to numerical stability.

The two fast algorithms derived in this chapter are referred to as the QRD-based least-squares lattice algorithm (Figure 7.9) and the fast QRD algorithm (Figure 7.10). The QRD-based least-squares lattice algorithm is clearly a lattice algorithm but the classification of the fast QRD algorithm is somewhat more difficult. Being based on the 'step-up/step-down' technique, it is similar, from a mathematical point of view, to the fixed-order algorithms. It should be noted, however, that it does not calculate the transversal filter coefficients explicitly; instead it generates the required filter output using the QRD-based method of 'direct residual extraction' discussed in section 7.2.5. The fast QRD algorithm also quite naturally produces the solution to all lower-order problems whereas the fixed-order algorithms do not. This property is a natural consequence of using the QRD technique (see section 7.3.6). In these latter respects the fast QRD algorithm has, therefore, more in common with the lattice algorithms than the fixed-order ones.

We begin the detailed derivation of these fast algorithms by considering the problem of determining an efficient method for the solution of the pth-order forward linear prediction problem.

7.3.2 Forward linear prediction

The pth-order forward linear prediction problem, at time N, requires the determination of the vector of filter coefficients $\mathbf{a}_p(N) = [a_p^{(0)}(N), \ldots, a_p^{(p-1)}(N)]^{\mathrm{T}}$ that minimizes the total prediction error $\| \varepsilon_p^{\mathrm{f}}(N) \|$ where

$$\boldsymbol{\varepsilon}_p^{\mathrm{f}}(N) = \mathcal{U}_p(N-1)\mathbf{a}_p(N) + \mathbf{y}_p^{\mathrm{f}}(N)$$

with $\mathcal{U}_p(N-1)$ and $\mathbf{y}_p^{\mathrm{f}}(N)$ as defined in (7.84) and (7.87). In order to solve this least-squares problem via the QRD technique we have to determine the rotation matrix $Q_p(N-1)$ that triangularizes the data matrix $\mathcal{U}_p(N-1)$ and then apply it to the vector $\mathbf{y}_p^{\mathrm{f}}(N)$ in order to calculate the angle-normalized residual $\tilde{e}_p^{\mathrm{f}}(N)$ (cf. Equation (7.21)). We also need to be able to calculate $\tilde{\alpha}_p(N-1)$ in order to generate the a posteriori prediction residual (see Equation (7.35)). Note also that the triangularization of $\mathcal{U}_p(N-1)$ is exactly what we require in the solution of the pth-order adaptive filtering problem at time $(N-1)$ (see Equation (7.79)). Consider, therefore, the following composite matrix:

$$M_1 = [\mathbf{y}_p^{\mathrm{f}}(N), \mathcal{U}_p(N-1), \mathbf{y}(N-1), \boldsymbol{\pi}_{N-1}] \tag{7.91}$$

From (7.14), (7.19) and (7.28), we have that

$$Q_p(N-1)\boldsymbol{\pi}_{N-1} = \hat{Q}_p(N-1)\boldsymbol{\pi}_{N-1} = [\mathbf{q}_p^{\mathrm{T}}(N-1), \mathbf{0}^{\mathrm{T}}, \tilde{\alpha}_p(N-1)]^{\mathrm{T}} \tag{7.92}$$

It should be clear, therefore, that the vector $\boldsymbol{\pi}_{N-1}$ in the above matrix (Equation 7.91)) will enable us to calculate $\tilde{\alpha}_p(N-1)$ just as the vector $\mathbf{y}_p^{\mathrm{f}}(N)$ allows $\tilde{e}_p^{\mathrm{f}}(N)$ to be calculated. Similarly, the presence of the vector $\mathbf{y}(N-1)$ will allow us to calculate $\tilde{e}_p(N-1)$.

Now from (7.85) and (7.87) we have that

$$M_1 = [\mathbf{y}_{p-1}^{\mathrm{f}}(N), \mathcal{U}_{p-1}(N-1), \mathbf{y}_{p-1}^{\mathrm{b}}(N-1), \mathbf{y}(N-1), \boldsymbol{\pi}_{N-1}] \tag{7.93}$$

where we have used the fact that $\mathbf{y}_p^{\mathrm{f}}(N) = \mathbf{y}_{p-1}^{\mathrm{f}}(N)$ in order to emphasize the appearance of the lower-order problem in the decomposition of the matrix M_1. Hence

$$Q_{p-1}(N-1)M_1 = \begin{bmatrix} \mathbf{p}_{p-1}^{\mathrm{f}}(N) & \tilde{R}_{p-1}(N-1) & \mathbf{p}_{p-1}^{\mathrm{b}}(N-1) & \mathbf{p}_{p-1}(N-1) & \mathbf{q}_{p-1}(N-1) \\ \mathbf{v}_{p-1}^{\mathrm{f}}(N) & \mathrm{O} & \mathbf{v}_{p-1}^{\mathrm{b}}(N-1) & \mathbf{v}_{p-1}(N-1) & \mathbf{g}_{p-1}(N-1) \end{bmatrix}$$

$$= M_2 \tag{7.94}$$

where

$$\mathbf{g}_{p-1}(N-1) = [\mathbf{0}^{\mathrm{T}}, \tilde{\alpha}_{p-1}(N-1)]^{\mathrm{T}} \tag{7.95}$$

It is easy to show that $\mathbf{v}_{p-1}^{\mathrm{f}}(N)$ and $\mathbf{v}_{p-1}^{\mathrm{b}}(N-1)$ must have a time recursive decomposition similar to that given in (7.21) for $\mathbf{v}_{p-1}(N-1)$. Hence,

$M_2 =$

$$\begin{bmatrix} \mathbf{p}^{\mathrm{f}}_{p-1}(N) & \tilde{R}_{p-1}(N-1) & \mathbf{p}^{\mathrm{b}}_{p-1}(N-1) & \mathbf{p}_{p-1}(N-1) & \mathbf{q}_{p-1}(N-1) \\ \lambda^{1/2}\mathbf{v}^{\mathrm{f}}_{p-1}(N-1) & \mathrm{O} & \lambda^{1/2}\mathbf{v}^{\mathrm{b}}_{p-1}(N-2) & \lambda^{1/2}\mathbf{v}_{p-1}(N-2) & \mathrm{O} \\ \tilde{e}^{\mathrm{f}}_{p-1}(N) & \mathbf{0}^{\mathrm{T}} & \tilde{e}^{\mathrm{b}}_{p-1}(N-1) & \tilde{e}_{p-1}(N-1) & \tilde{\alpha}_{p-1}(N-1) \end{bmatrix} \tag{7.96}$$

Now suppose that we had already calculated a rotation matrix,[3] $Q^{\mathrm{f}}_p(N-1)$, say, that rotates the vector $\mathbf{v}^{\mathrm{b}}_{p-1}(N-2)$ into a form where only the top element is non-zero, i.e.

$$\begin{bmatrix} Q^{\mathrm{f}}_p(N-1) & \mathbf{0} \\ \mathbf{0}^{\mathrm{T}} & 1 \end{bmatrix} M_2 =$$

$$\begin{bmatrix} \mathbf{p}^{\mathrm{f}}_{p-1}(N) & \tilde{R}_{p-1}(N-1) & \mathbf{p}^{\mathrm{b}}_{p-1}(N-1) & \mathbf{p}_{p-1}(N-1) & \mathbf{q}_{p-1}(N-1) \\ \lambda^{1/2}\tilde{\boldsymbol{\beta}}^{\mathrm{f}}_{p-1}(N-1) & \mathbf{0}^{\mathrm{T}} & \lambda^{1/2}\tilde{\alpha}^{\mathrm{b}}_{p-1}(N-2) & \lambda^{1/2}\tilde{\boldsymbol{\beta}}_{p-1}(N-2) & 0 \\ \lambda^{1/2}\boldsymbol{\varphi}^{\mathrm{f}}_{p-1}(N-1) & \mathrm{O} & \mathbf{0} & \lambda^{1/2}\boldsymbol{\varphi}_{p-1}(N-2) & \mathbf{0} \\ \tilde{e}^{\mathrm{f}}_{p-1}(N) & \mathbf{0}^{\mathrm{T}} & \tilde{e}^{\mathrm{b}}_{p-1}(N-1) & \tilde{e}_{p-1}(N-1) & \tilde{\alpha}_{p-1}(N-1) \end{bmatrix}$$

$$= M_3 \tag{7.97}$$

The new quantities $\tilde{\boldsymbol{\beta}}^{\mathrm{f}}_{p-1}(N-1)$, $\boldsymbol{\varphi}^{\mathrm{f}}_{p-1}(N-1)$, $\tilde{\boldsymbol{\beta}}_{p-1}(N-2)$, $\boldsymbol{\varphi}_{p-1}(N-2)$ and $\tilde{\alpha}^{\mathrm{b}}_{p-1}(N-2)$ are defined by this operation and we note, by analogy with (7.12), that $\tilde{\alpha}^{\mathrm{b}}_{p-1}(N-2)$ is the square root of the $(p-1)$st-order backward prediction energy at time $(N-2)$, i.e.

$$\tilde{\alpha}^{\mathrm{b}}_{p-1}(N-2) = \sqrt{\alpha^{\mathrm{b}}_{p-1}(N-2)} \tag{7.98}$$

Now in order to complete the triangularization of the matrix $\mathcal{U}_p(N-1)$ (see Equation (7.93)) all that is required is the annihilation of the single element $\tilde{e}^{\mathrm{b}}_{p-1}(N-1)$. This can be carried out using a single Givens rotation:

$\hat{Q}^{\mathrm{f}}_p(N)M_3 =$

$$\begin{bmatrix} \mathbf{p}^{\mathrm{f}}_{p-1}(N) & \tilde{R}_{p-1}(N-1) & \mathbf{p}^{\mathrm{b}}_{p-1}(N-1) & \mathbf{p}_{p-1}(N-1) & \mathbf{q}_{p-1}(N-1) \\ \tilde{\boldsymbol{\beta}}^{\mathrm{f}}_{p-1}(N) & \mathbf{0}^{\mathrm{T}} & \tilde{\alpha}^{\mathrm{b}}_{p-1}(N-1) & \tilde{\boldsymbol{\beta}}_{p-1}(N-1) & \bar{q}_p(N-1) \\ \lambda^{1/2}\boldsymbol{\varphi}^{\mathrm{f}}_{p-1}(N-1) & \mathrm{O} & \mathbf{0} & \lambda^{1/2}\boldsymbol{\varphi}_{p-1}(N-2) & \mathbf{0} \\ \tilde{e}^{\mathrm{f}}_p(N) & \mathbf{0}^{\mathrm{T}} & 0 & \tilde{e}_p(N-1) & \tilde{\alpha}_p(N-1) \end{bmatrix} \tag{7.99}$$

$$\equiv \begin{bmatrix} \mathbf{p}^{\mathrm{f}}_p(N) & \tilde{R}_p(N-1) & \mathbf{p}_p(N-1) & \mathbf{q}_p(N-1) \\ \mathbf{v}^{\mathrm{f}}_p(N) & \mathrm{O} & \mathbf{v}_p(N-1) & \mathbf{g}_p(N-1) \end{bmatrix} \tag{7.100}$$

where $\bar{q}_p(N-1)$ is defined by this operation. The identity in (7.100), and hence the labelling of some of the elements in the bottom row of the right-hand matrix in (7.99), follows by definition (see Equation (7.93)). Note from the above that the 'new' quantities $\boldsymbol{\varphi}^{\mathrm{f}}_{p-1}(N-1)$ and $\boldsymbol{\varphi}_{p-1}(N-2)$, introduced in (7.97), are equivalent to existing variables. Indeed, from (7.99) and (7.00) we have that

$$\begin{bmatrix} \lambda^{1/2}\boldsymbol{\varphi}^{\mathrm{f}}_{p-1}(N-1) & \lambda^{1/2}\boldsymbol{\varphi}_{p-1}(N-2) \\ \tilde{e}^{\mathrm{f}}_{p}(N) & \tilde{e}_{p}(N-1) \end{bmatrix} = [\mathbf{v}^{\mathrm{f}}_{p}(N) \quad \mathbf{v}_{p}(N-1)] \tag{7.101}$$

so that (see Equation (7.21))

$$\boldsymbol{\varphi}^{\mathrm{f}}_{p-1}(N) = \mathbf{v}^{\mathrm{f}}_{p}(N) \tag{7.102}$$

and

$$\boldsymbol{\varphi}_{p-1}(N) = \mathbf{v}_{p}(N) \tag{7.103}$$

From the above we see that the sequence of orthogonal transformation shown in (7.94), (7.97) and (7.99) solves the pth-order forward linear prediction problem Note, however, that the matrix operated upon by $\hat{Q}^{\mathrm{f}}_{p}(N)$ in (7.99) consists entirely of quantities that would be available if the $(p-1)$st-order forward and backward problems had already been solved at times N and $N-1$ respectively. If this assumption were true then we could have constructed this intermediate matrix directly, thereby circumventing the need for the operations as outlined in (7.94) and (7.97). Only the single Givens rotation of (7.99) would actually need to be performed and so the number of arithmetic operations required would be independent of p: only eight elements, one of which is zero, of the left-hand matrix in (7.99) are affected by the required rotation. Having derived a fast method for solving the forward linear prediction problem, we now consider a fast update method for the auxiliary (backward) problem.

7.3.3 Backward linear prediction

The pth-order backward linear prediction problem, at time N, requires the determination of the vector of filter coefficients $\mathbf{b}_p(N) = [b_p^{(0)}(N), \ldots, b_p^{p-1}(N)]^{\mathrm{T}}$ that minimizes the total prediction error $\| \boldsymbol{\varepsilon}^{\mathrm{b}}_p(N) \|$ where

$$\boldsymbol{\varepsilon}^{\mathrm{b}}_p(N) = \mathscr{U}_p(N)\mathbf{b}_p(N) + \mathbf{y}^{\mathrm{b}}_p(N) \tag{7.104}$$

Again, the least-squares solution to this problem can be found by the QRD method. It is necessary to determine the rotation matrix $Q_p(N)$ that triangularizes the data matrix $\mathscr{U}_p(N)$ and then apply it to the vector $\mathbf{y}^{\mathrm{b}}_p(N)$ in order to calculate $\tilde{e}^{\mathrm{b}}_p(N)$ (cf. Equation (7.21)). We also need to be able to calculate $\tilde{\alpha}_p(N)$ (see Equation (7.35)) in order to generate the a posteriori prediction residual. Consider, therefore, the following composite matrix and the illustrated decomposition which is a simple extension of (7.86):

$$[\mathscr{U}_p(N) \quad \mathbf{y}^{\mathrm{b}}_p(N) \quad \boldsymbol{\pi}_N] = \begin{bmatrix} \lambda^{(N-1)/2}u(1) & \mathbf{0}^{\mathrm{T}} & 0 & 0 \\ \mathbf{y}^{\mathrm{f}}_{p-1}(N) & \mathscr{U}_{p-1}(N-1) & \mathbf{y}^{\mathrm{b}}_{p-1}(N-1) & \boldsymbol{\pi}_{N-1} \end{bmatrix} = M_4 \tag{7.105}$$

In (7.105), it has been assumed that the data sequence $u(N)$ is prewindowed (i.e. $u(N) = 0$ for $N \leqslant 0$). Note that this is the only place in the analysis where we require this assumption.[4] Consider the effect of the rotation matrix $Q_{p-1}(N-1)$ on the lower part of the matrix in (7.105):

$$\begin{bmatrix} 1 & \mathbf{0}^T \\ \mathbf{0} & Q_{p-1}(N-1) \end{bmatrix} M_4 = \begin{bmatrix} \lambda^{(N-1)/2}u(1) & \mathbf{0}^T & 0 & 0 \\ \mathbf{p}^f_{p-1}(N) & \tilde{R}_{p-1}(N-1) & \mathbf{p}^b_{p-1}(N-1) & \mathbf{q}_{p-1}(N-1) \\ \mathbf{v}^f_{p-1}(N) & O & \mathbf{v}^b_{p-1}(N-1) & \mathbf{g}_{p-1}(N-1) \end{bmatrix}$$

$$= M_5 \tag{7.106}$$

As before, all the vectors on the bottom row of the matrix M_5 may be written in terms of their underlying time recursion and thus we obtain the expression

$$M_5 = \begin{bmatrix} \lambda^{(N-1)/2}u(1) & \mathbf{0}^T & 0 & 0 \\ \mathbf{p}^f_{p-1}(N) & \tilde{R}_{p-1}(N-1) & \mathbf{p}^b_{p-1}(N-1) & \mathbf{q}_{p-1}(N-1) \\ \lambda^{1/2}\mathbf{v}^f_{p-1}(N-1) & O & \lambda^{1/2}\mathbf{v}^b_{p-1}(N-2) & \mathbf{0} \\ \tilde{e}^f_{p-1}(N) & \mathbf{0}^T & \tilde{e}^b_{p-1}(N-1) & \tilde{\alpha}_{p-1}(N-1) \end{bmatrix} \tag{7.107}$$

Now suppose that we have already constructed a rotation matrix $Q^b_p(N-1)$ that annihilates the vector $\mathbf{v}^f_{p-1}(N-1)$ by rotation against the element $\lambda^{(N-1)/2}u(1)$, i.e.

$$\begin{bmatrix} Q^b_p(N-1) & \mathbf{0} \\ \mathbf{0}^T & 1 \end{bmatrix} M_5$$

$$= \begin{bmatrix} \lambda^{1/2}\tilde{\alpha}^f_{p-1}(N-1) & \mathbf{0}^T & \lambda^{1/2}\tilde{\beta}^b_{p-1}(N-1) & 0 \\ \mathbf{p}^f_{p-1}(N) & \tilde{R}_{p-1}(N-1) & \mathbf{p}^b_{p-1}(N-1) & \mathbf{q}_{p-1}(N-1) \\ \mathbf{0} & O & \lambda^{1/2}\boldsymbol{\varphi}^b_{p-1}(N-2) & \mathbf{0} \\ \tilde{e}^f_{p-1}(N) & \mathbf{0}^T & \tilde{e}^b_{p-1}(N-1) & \tilde{\alpha}_{p-1}(N-1) \end{bmatrix} = M_6 \tag{7.108}$$

Now let $\hat{Q}^b_p(N)$ be the rotation matrix that annihilates the element $\tilde{e}^f_{p-1}(N)$ by rotation against the element $\lambda^{1/2}\tilde{\alpha}^f_{p-1}(N-1)$. Application of the transformation $\hat{Q}^b_p(N)$ to the above matrix yields the result

$$\hat{Q}^b_p(N)M_6 = \begin{bmatrix} \tilde{\alpha}^f_{p-1}(N) & \mathbf{0}^T & \tilde{\beta}^b_{p-1}(N) & \tilde{q}_p(N) \\ \mathbf{p}^f_{p-1}(N) & \tilde{R}_{p-1}(N-1) & \mathbf{p}^b_{p-1}(N-1) & \mathbf{q}_{p-1}(N-1) \\ \mathbf{0} & O & \lambda^{1/2}\boldsymbol{\varphi}^b_{p-1}(N-2) & \mathbf{0} \\ 0 & \mathbf{0}^T & \hat{e}^b_{p-1}(N) & \hat{\alpha}_{p-1}(N) \end{bmatrix} = M_7 \tag{7.109}$$

where the new quantities $\tilde{q}_p(N)$, $\hat{e}^b_{p-1}(N)$ and $\hat{\alpha}_{p-1}(N)$ are defined by this equation. Bearing in mind the underlying data matrix (see Equation (7.105)), recall that we are attempting to create an upper-triangular $p \times p$ matrix in the upper left-hand corner of the matrix in (7.109). At present this submatrix is not quite triangular but a little thought shows that it is easy to construct a matrix ($\tilde{Q}^b_p(N)$, say) which will complete the required triangularization. Specifically, let $\tilde{Q}^b_p(N)$ be constructed from a sequence of Givens rotations such that each rotation annihilates one element of the vector $\mathbf{p}^f_{p-1}(N)$ in turn by rotation against the element in the top row (initially $\tilde{\alpha}^f_{p-1}(N)$) of matrix M_7 above. Provided we start with the last elements $\mathbf{p}^f_{p-1}(N)$ and work upwards, the sequence of rotations will not destroy the triangular structure of the matrix $\tilde{R}_{p-1}(N-1)$, although its value will be changed as a result. Thus

$$\tilde{Q}^b_p(N)M_7 = \begin{bmatrix} \tilde{R}_p(N) & \mathbf{p}^b_p(N) & \mathbf{q}_p(N) \\ O & \mathbf{v}^b_p(N) & \mathbf{g}_p(N) \end{bmatrix} \tag{7.110}$$

Note that the matrix $\tilde{Q}_p^b(N)$ only affects the upper part of the partitioned matrix on the left-hand side of (7.110). It should therefore be clear that

$$\hat{e}_{p-1}^b(N) = \tilde{e}_p^b(N) \quad \text{and} \quad \hat{\alpha}_{p-1}(N) = \tilde{\alpha}_p(N) \tag{7.111}$$

Also note that

$$\begin{bmatrix} \lambda^{1/2}\boldsymbol{\varphi}_{p-1}^b(N-2) \\ \tilde{e}_p^b(N) \end{bmatrix} = \mathbf{v}_p^b(N) \tag{7.112}$$

and so the 'new' quantity $\boldsymbol{\varphi}_{p-1}^b(N-2)$, introduced in (7.108), is equivalent to an existing variable. Indeed, from (7.21) we have

$$\boldsymbol{\varphi}_{p-1}^b(N-1) = \mathbf{v}_p^b(N) \tag{7.113}$$

Hence the sequence of orthogonal rotations given in (7.106), (7.108), (7.109) and (7.110) solve the pth-order backward linear prediction problem. Following the development of the solution to the forward problem in section 7.3.2, note that the matrix on the left-hand side of (7.109) could be constructed directly given the solutions to the $(p-1)$st-order forward and backward linear prediction problems at time N and $N-1$ respectively. Thus the transformation shown in (7.106) and (7.108) could be bypassed. Furthermore, assuming that we are only interested in the prediction residuals, the transformation shown in (7.110) is not required either, since $\tilde{e}_p^b(N)$ and $\tilde{\alpha}_p(N)$ are both available in the matrix M_7. Thus only the Givens rotation summarized in (7.109) need actually be performed and the number of arithmetic operations required is independent of p; only six elements, one of which is zero, of the matrix M_6 are affected by the rotation.

7.3.4 The QRD least-squares lattice algorithm

Gathering together the results of sections 7.3.2 and 7.3.3 we see that it is possible to utilize various terms from the solution to the $(p-1)$st-order forward and backward linear prediction problems, at time N and $(N-1)$, respectively, to generate corresponding terms for the solution to the pth-order problems at time N (see Figure 7.11). Note that the processing elements shown in Figure 7.11 are the same as those used in the triangular systolic array described in section 7.2.3 (i.e. as shown in Figure 7.3). It is possible to show that the corresponding square-root-free implementation may be obtained very simply by substituting the processing elements shown in Figure 7.4.

Given that zeroth-order linear prediction is trivial, we can thus generate the solution to the pth-order problem by iteration in order using a cascade of the sections shown in Figure 7.11. The resultant architecture (Figure 7.12) has a lattice structure and, since the number of operations per stage is independent of p, $O(p)$ operations are required to solve the pth-order problem (see Table 7.4). Note that by including the adaptive filtering vector $\mathbf{y}(N-1)$ in the calculation of the pth-order forward linear prediction problem (section 7.3.2) we automatically solve the pth-order adaptive filtering problem for time $(N-1)$. The solution to the adaptive filtering problem for

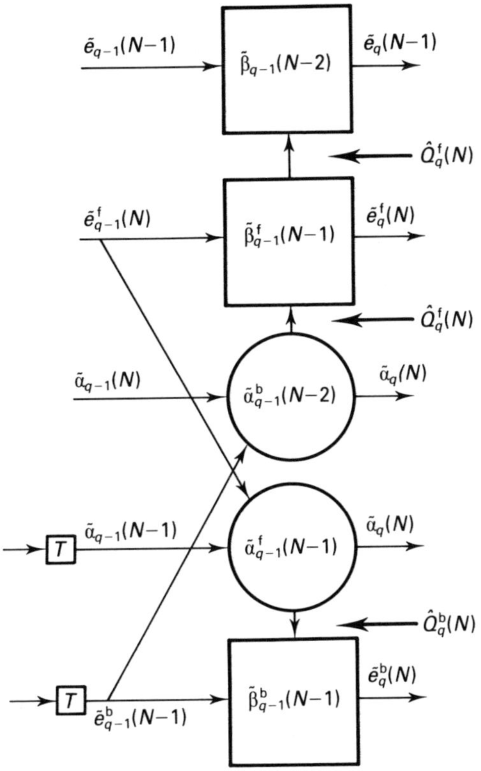

Figure 7.11 QRD-based least-squares lattice section

time N can easily be derived from the above. Note that in Figure 7.11 the rotation parameters used (in the square processors) to calculate each order update of the delayed joint-process residual are evaluated (in the round processors) from the delayed backward residual. Thus, by using the undelayed backward residual, we can calculate the rotations required to update the undelayed joint-process residual – see Figure 7.13. In this form, a stage of the QRD-based lattice can be seen to be very reminiscent of that of the standard least-squares lattice; the only difference is that the former structure has rotation processors instead of the (reflection coefficient) multipliers of the conventional form (see Figure 6.2). Although conceptually appealing, the QRD-based lattice filter stage shown in Figure 7.13(b) is inefficient because there is duplication in the calculatons of the rotation parameters of the forward prediction and joint estimation residuals. Any practical implementation would clearly calculate the rotation parameters only once in the joint-process estimation channel and store them for use in the next time instant in the forward prediction channel. Indeed, the algorithm shown in Table 7.4 is written in just such a manner (see section 7.5 for an explanation of the listing).

Table 7.4 SQ/FF lattice algorithm

	Add/subt	Mult	Sqrt/div
START			
INITIALIZE {all variables := 0};			
FOR q FROM 1 TO p DO			
LET $c^{\mathrm{f}}_q(0) := 1;\ \ s^{\mathrm{f}}_q(0) := 0;$			
END_DO			
FOR t FROM 1 DO			
LET $\tilde{e}^{\mathrm{f}}_0(t) := u(t);\ \tilde{e}^{\mathrm{b}}_0(t) := u(t);\ \tilde{e}_0(t) := y(t);\ \tilde{\alpha}_0(t) := 1.0;$	–	–	–
FOR q FROM 1 TO p DO			
LET $\tilde{\alpha}^{\mathrm{b}}_{q-1}(t) := \sqrt{\lambda(\tilde{\alpha}^{\mathrm{b}}_{q-1}(t-1))^2 + \lvert \tilde{e}^{\mathrm{b}}_{q-1}(t) \rvert^2};$	1	3	1
IF $\tilde{\alpha}^{\mathrm{b}}_{q-1}(t) = 0$ THEN LET $c^{\mathrm{f}}_q(t) := 1;\ \ s^{\mathrm{f}}_q(t) := 0;$	–	–	–
ELSE LET $c^{\mathrm{f}}_q(t) := \dfrac{\lambda^{1/2}\tilde{\alpha}^{\mathrm{b}}_{q-1}(t-1)}{\tilde{\alpha}^{\mathrm{b}}_{q-1}(t)};\ \ s^{\mathrm{f}}_q(t) := \dfrac{\tilde{e}^{\mathrm{b}}_{q-1}(t)}{\tilde{\alpha}^{\mathrm{b}}_{q-1}(t)};$	–	1	2
END_IF;			
LET $\tilde{\beta}^{\mathrm{f}}_{q-1}(t) := c^{\mathrm{f}}_q(t-1)\lambda^{1/2}\tilde{\beta}^{\mathrm{f}}_{q-1}(t-1) + s^{\mathrm{f}*}_q(t-1)\tilde{e}^{\mathrm{f}}_{q-1}(t);$	1	3	–
$\tilde{e}^{\mathrm{f}}_q(t) := c^{\mathrm{f}}_q(t-1)\tilde{e}^{\mathrm{f}}_{q-1}(t) - s^{\mathrm{f}}_q(t-1)\lambda^{1/2}\tilde{\beta}^{\mathrm{f}}_{q-1}(t-1);$	1	3	–
$\tilde{\beta}_{q-1}(t) := c^{\mathrm{f}}_q(t)\lambda^{1/2}\tilde{\beta}_{q-1}(t-1) + s^{\mathrm{f}*}_q(t)\tilde{e}_{q-1}(t);$	1	3	–
$\tilde{e}_q(t) := c^{\mathrm{f}}_q(t)\tilde{e}_{q-1}(t) - s^{\mathrm{f}}_q(t)\lambda^{1/2}\tilde{\beta}_{q-1}(t-1);$	1	3	–
$\tilde{\alpha}_q(t) := c^{\mathrm{f}}_q(t)\tilde{\alpha}_{q-1}(t);$	–	1	–
COMMENT qth-order forward prediction residual COMMENT			
$\varepsilon^{\mathrm{f}}_q(t) := \tilde{\alpha}_q(t-1)\tilde{e}^{\mathrm{f}}_q(t);$			
COMMENT qth-order filtered residual COMMENT			
$\varepsilon_q(t) := \tilde{\alpha}_q(t)\tilde{e}_q(t);$			
LET $\tilde{\alpha}^{\mathrm{f}}_{q-1}(t) := \sqrt{\lambda(\tilde{\alpha}^{\mathrm{f}}_{q-1}(t-1))^2 + \lvert \tilde{e}^{\mathrm{f}}_{q-1}(t) \rvert^2};$	1	3	1
IF $\tilde{\alpha}^{\mathrm{f}}_{q-1}(t) = 0$ THEN LET $c^{\mathrm{b}}_q(t) := 1;\ \ s^{\mathrm{b}}_q(t) := 0;$	–	–	–
ELSE LET $c^{\mathrm{b}}_q(t) := \dfrac{\lambda^{1/2}\tilde{\alpha}^{\mathrm{f}}_{q-1}(t-1)}{\tilde{\alpha}^{\mathrm{f}}_{q-1}(t)};\ \ s^{\mathrm{b}}_q(t) := \dfrac{\tilde{e}^{\mathrm{f}}_{q-1}(t)}{\tilde{\alpha}^{\mathrm{f}}_{q-1}(t)};$	–	1	2
END_IF;			
LET $\tilde{\beta}^{\mathrm{b}}_{q-1}(t-1) := c^{\mathrm{b}}_q(t)\lambda^{1/2}\tilde{\beta}^{\mathrm{b}}_{q-1}(t-2) + s^{\mathrm{b}*}_q(t)\tilde{e}^{\mathrm{b}}_{q-1}(t-1);$	1	3	–
$\tilde{e}^{\mathrm{b}}_q(t) := c^{\mathrm{b}}_q(t)\tilde{e}^{\mathrm{b}}_{q-1}(t-1) - s^{\mathrm{b}}_q(t)\lambda^{1/2}\tilde{\beta}^{\mathrm{b}}_{q-1}(t-2);$	1	3	–
COMMENT qth-order backward prediction residual COMMENT			
$\varepsilon^{\mathrm{b}}_q(t) := \tilde{\alpha}_q(t)\tilde{e}^{\mathrm{b}}_q(t);$			
END_DO	8	27	6
COMMENT pth-order filtered residual COMMENT			
$\varepsilon_p(t) := \tilde{\alpha}_p(t)\tilde{e}_p(t);$	–	1	–
END_DO			
FINISH	$8p$	$27p + 1$	$6p$

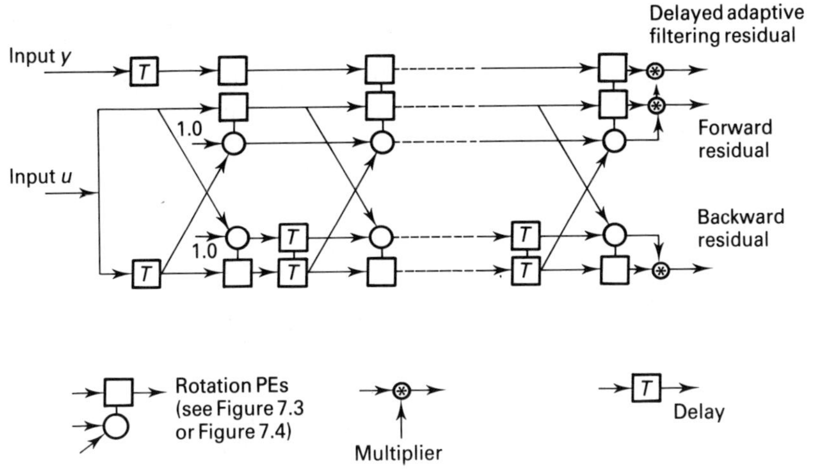

Figure 7.12 'Delayed' adaptive filtering lattice

In section 7.2.8 we discussed the recursive modified Gram-Schmidt (RMGS) algorithm with error feedback. This was proposed by Ling *et al.* [8] for the general narrowband adaptive beamforming problem and leads to a triangular array processor equivalent to the one described in Figures 7.2 and 7.4. Ling and Proakis [24] subsequently developed the technique to produce an efficient RMGS algorithm with error feedback which requires $O(p)$ arithmetic operations per sample time to solve a pth-order adaptive FIR filtering problem (as discussed in Chapter 6). Their algorithm (Table 6.5) also has a lattice structure and, in view of the equivalence referred to above, it is not surprising to find that it corresponds exactly to the QRD-based least-squares lattice algorithm derived in this chapter. In terms of the notation

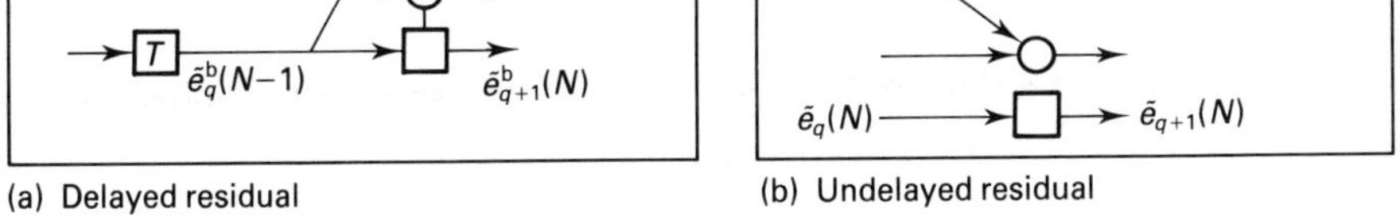

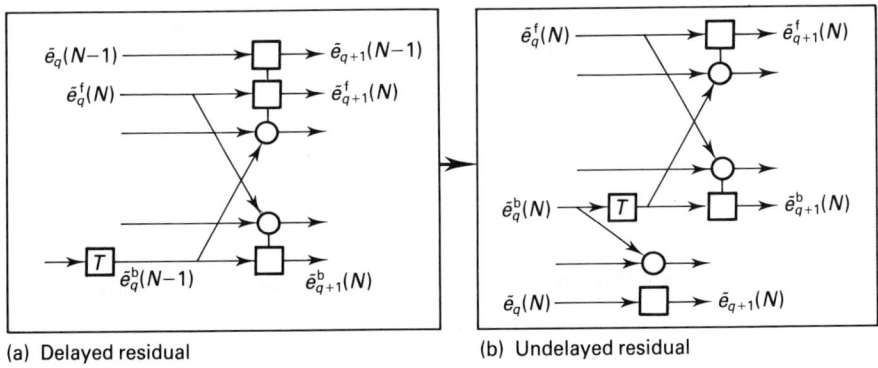

Figure 7.13 Equivalent order updates for joint process residual

introduced in this chapter, the RMGS lattice algorithm with error feedback (Table 6.5) is the SF/FB QRD lattice algorithm (i.e. using the square-root-free Givens rotations shown in Figure 7.4), whereas the two algorithms shown in Table 6.6 are the SQ/FF and SF/FF algorithms. The RMGS lattice algorithm with error feedback was the first numerically robust least-squares lattice algorithm to be developed and it is interesting, therefore, to note this correspondence to an algorithm based entirely on orthogonal rotations.

7.3.5 The 'fast QRD' algorithm

In this section we expand on the remarks made in section 7.3.1 about the connection between the forward and backward linear prediction problems for a fixed-order algorithm and develop the fast QRD algorithm. We take as our starting point the situation where we have solved the pth-order forward linear prediction problem for time N and are attempting to update this solution to the next time instant. Specifically, if we could generate the matrix $\hat{Q}_p(N)$ using $O(p)$ arithmetic operations then the linear prediction solution could be updated efficiently.

The original derivation of a QRD-based fast QRD algorithm was presented by Cioffi [23] although his algorithm differs somewhat from that presented here. The material presented in this section follows that of Proudler *et al.* [25] and leads to the same algorithm as derived by Regalia and Bellanger [26]. Both of these derivations were based on Cioffi's work and the approach presented here actually follows the same sequence of ideas used in Cioffi's original paper. The difference in the resulting algorithms is due to the choice of triangular matrix used in the QRD. Here we use an upper right-hand triangular matrix to conform with the work of Gentleman and Kung [3]. Cioffi, on the other hand, chose to use an upper left-hand triangular matrix. This choice results in an algorithm which is slightly more complex than the one derived here. We refer interested readers to the original references for further details. As in the case of the lattice algorithm, the solution to the adaptive filtering problem is updated along with the linear prediction one. In fact, since we will be explicitly calculating the matrix $\hat{Q}_p(N)$, there is clearly no need to include the adaptive filtering problem explicitly in the following analysis.

Note from (7.106), (7.108), (7.109) and (7.110) that

$$Q_{p+1}(N) = \tilde{Q}^{\mathrm{b}}_{p+1}(N)\hat{Q}^{\mathrm{b}}_{p+1}(N)\begin{bmatrix} Q^{\mathrm{b}}_{p+1}(N-1) & \mathbf{0} \\ \mathbf{0}^{\mathrm{T}} & 1 \end{bmatrix}\begin{bmatrix} 1 & \mathbf{0}^{\mathrm{T}} \\ \mathbf{0} & Q_p(N-1) \end{bmatrix} \tag{7.114}$$

which can be viewed as a time-and-order update relationship for the matrix $Q_p(N-1)$. Also from (7.94), (7.97) and (7.99) we have

$$Q_{p+1}(N) = \hat{Q}^{\mathrm{f}}_{p+1}(N+1)\begin{bmatrix} Q^{\mathrm{f}}_{p+1}(N) & \mathbf{0} \\ \mathbf{0}^{\mathrm{T}} & 1 \end{bmatrix} Q_p(N) \tag{7.115}$$

or, given that the inverse of an orthogonal matrix is just its transpose,

$$Q_p(N) = \begin{bmatrix} Q^{\rm f}_{p+1}(N) & \mathbf{0} \\ \mathbf{0}^{\rm T} & 1 \end{bmatrix}^{\rm T} [\hat{Q}^{\rm f}_{p+1}(N+1)]^{\rm T} Q_{p+1}(N) \tag{7.116}$$

which is an order downdate relationship for $Q_{p+1}(N)$. Taken together, (7.114) and (7.116) represent, at least in principle, a means of updating the matrix Q_p from one time instant to the next. However, we are really interested in obtaining a time update relationship for the matrix $\hat{Q}_p(N)$, since the matrix Q_p operates on the entire data matrix and would therefore lead to an algorithm which is not time recursive and requires an ever increasing amount of storage. Furthermore, since explicit evaluation of the relationships derived above would require $O(p^3)$ multiplications, this approach could not lead to a 'fast' algorithm.

The observant reader may also have noticed that the matrix $\hat{Q}^{\rm f}_{p+1}(N+1)$ depends on the solution to the pth-order backward problem at time N. This is somewhat paradoxical given that one of the purposes of trying to calculate $Q_p(N)$ is to calculate the solution! However, as we shall see, it is possible to avoid this paradox by doing the matrix multiplications required by (7.114) and (7.116) implicitly, thus computing the matrix $\hat{Q}_p(N)$ efficiently and generating a fast algorithm. We do this by constructing only the right-hand column of the matrix $\hat{Q}_p(N)$ and then inferring the whole matrix from this vector (as in section 7.2.7). This technique reduces the dimension of the problem (from matrices to vectors) and thus scales down the computational requirement; in fact, the evaluation of (7.114) and (7.116) is actually reduced to $O(p)$ arithmetic operations in this case. Note that the right-hand column of the matrix $\hat{Q}_p(N)$ can be considered to be the result of applying the rotation matrix $\hat{Q}_p(N)$ to the pinning vector $\boldsymbol{\pi}_N$ defined in (7.68). The occurrence of the pinning vector in the calculations is another point of contact between the fast QRD algorithm and the fixed-order algorithms (see Chapter 5).

Returning to (7.114), we see that

$$\begin{bmatrix} \mathbf{q}_{p+1}(N) \\ \mathbf{0} \\ \tilde{\alpha}_{p+1}(N) \end{bmatrix} = Q_{p+1}(N)\boldsymbol{\pi}_N = \tilde{Q}^{\rm b}_{p+1}(N)\hat{Q}^{\rm b}_{p+1}(N) \begin{bmatrix} Q^{\rm b}_{p+1}(N-1) & \mathbf{0} \\ \mathbf{0}^{\rm T} & 1 \end{bmatrix} \begin{bmatrix} 0 \\ \mathbf{q}_p(N-1) \\ \mathbf{0} \\ \tilde{\alpha}_p(N-1) \end{bmatrix} \tag{7.117}$$

and, from (7.115),

$$\begin{bmatrix} \mathbf{q}_{p+1}(N) \\ \mathbf{0} \\ \tilde{a}_{p+1}(N) \end{bmatrix} = \hat{Q}^{\rm f}_{p+1}(N+1) \begin{bmatrix} Q^{\rm f}_{p+1}(N) & \mathbf{0} \\ \mathbf{0}^{\rm T} & 1 \end{bmatrix} \begin{bmatrix} \mathbf{q}_p(N) \\ \mathbf{0} \\ \tilde{a}_p(N) \end{bmatrix} \tag{7.118}$$

Now from (7.108), it is clear that the vector $[0, \mathbf{q}_p^{\mathrm{T}}(N-1), \mathbf{0}^{\mathrm{T}}, \tilde{a}_p(N-1)]^{\mathrm{T}}$ is unaffected by the matrix $Q_{p+1}^{\mathrm{b}}(N-1)$. Similarily, the vector $[\mathbf{q}_p^{\mathrm{T}}(N), \mathbf{0}^{\mathrm{T}}, \tilde{\alpha}_p(N)]^{\mathrm{T}}$ is invariant under the action of the matrix $Q_{p+1}^{\mathrm{f}}(N)$ (see Equation (7.97)). Hence, we deduce that

$$\begin{bmatrix} \mathbf{q}_{p+1}(N) \\ \mathbf{0} \\ \tilde{\alpha}_{p+1}(N) \end{bmatrix} = \tilde{Q}_{p+1}^{\mathrm{b}}(N)\hat{Q}_{p+1}^{\mathrm{b}}(N) \begin{bmatrix} 0 \\ \mathbf{q}_p(N-1) \\ \mathbf{0} \\ \tilde{\alpha}_p(N-1) \end{bmatrix} \tag{7.119}$$

and

$$\begin{bmatrix} \mathbf{q}_{p+1}(N) \\ \mathbf{0} \\ \tilde{\alpha}_{p+1}(N) \end{bmatrix} = \hat{Q}_{p+1}^{\mathrm{f}}(N+1) \begin{bmatrix} \mathbf{q}_p(N) \\ \mathbf{0} \\ \tilde{\alpha}_p(N) \end{bmatrix} \tag{7.120}$$

Finally, with reference to (7.99), we note that

$$\hat{Q}_{p+1}^{\mathrm{f}}(N+1) \begin{bmatrix} \mathbf{q}_p(N) \\ \mathbf{0} \\ \tilde{\alpha}_p(N) \end{bmatrix} = \begin{bmatrix} \mathbf{q}_p(N) \\ \bar{q}_{p+1}(N) \\ \mathbf{0} \\ \tilde{\alpha}_{p+1}(N) \end{bmatrix} \tag{7.121}$$

Clearly, if the rotation parameters corresponding to $\hat{Q}_{p+1}^{\mathrm{f}}(N+1)$ are $(c_{p+1}^{\mathrm{f}}, s_{p+1}^{\mathrm{f}})$ then

$$\begin{bmatrix} \bar{q}_{p+1}(N) \\ \tilde{\alpha}_{p+1}(N) \end{bmatrix} = \begin{bmatrix} c_{p+1}^{\mathrm{f}} & s_{p+1}^{\mathrm{f}*} \\ -s_{p+1}^{\mathrm{f}} & c_{p+1}^{\mathrm{f}} \end{bmatrix} \begin{bmatrix} 0 \\ \tilde{\alpha}_p(N) \end{bmatrix} \tag{7.122}$$

and hence the rotation matrix $\hat{Q}_{p+1}^{\mathrm{f}}(N+1)$ can be calculated indirectly from the known quantities $\bar{q}_{p+1}(N)$ and $\tilde{\alpha}_{p+1}(N)$.

Thus by means of (7.119) we can transform the vector $[\mathbf{q}_p^{\mathrm{T}}(N-1), \mathbf{0}^{\mathrm{T}}, \tilde{\alpha}_p(N-1)]^{\mathrm{T}}$ into the vector $[\mathbf{q}_{p+1}^{\mathrm{T}}(N), \mathbf{0}^{\mathrm{T}}, \tilde{\alpha}_{p+1}(N)]^{\mathrm{T}}$ in $O(p)$ orthogonal operations. Equation (7.120) then provides the basis for an $O(p)$ method for transforming this latter vector into $[\mathbf{q}_p^{\mathrm{T}}(N), \mathbf{0}^{\mathrm{T}}, \tilde{\alpha}_p(N)]^{\mathrm{T}}$, and finally $\hat{Q}_p(N)$ can be calculated, again in $O(p)$ operations, from $[\mathbf{q}_p^{\mathrm{T}}(N), \mathbf{0}^{\mathrm{T}}, \tilde{\alpha}_p(N)]^{\mathrm{T}}$ as shown in section 7.2.7. The resulting algorithm may be implemented using the parallel computing architecture shown in Figure 7.14 with rotation processors as defined in Figure 7.15. Note that these rotation processors are essentially the same as those used in the triangular array and lattice algorithms. The main difference is that the processing elements in Figure 7.15 do not store any internal variables: all variables are passed either into or out of the processing element. In fact, if these processing elements are equipped with a storage element and the correct output variable fed back to the relevant input (as shown in the left-hand column of cells in Figure 7.14), then they are essentially identical to the processing elements shown in Figure 7.3. A listing of the SQ/FF fast QRD algorithm is shown in Table 7.5 (see section 7.5 for an explanation of the listing).

Table 7.5 SQ/FF fast QRD algorithm

	Add/ subt	Mult	Sqrt/ div
START			
INITIALIZE {all variables := 0};			
FOR q FROM 1 TO p DO			
LET $c_q := 1.0$;	–	–	–
END_DO;			
LET $\tilde{\alpha}_p := 1.0$; –	–	–	
FOR t FROM 1 DO			
LET $\tilde{e}^{\mathrm{f}}_0(t) := u(t)$;	–	–	–
FOR q FROM 1 TO p DO			
LET $\mathbf{p}^{f(q)}_p(t) := c_q \lambda^{1/2} \mathbf{p}^{f(q)}_p(t-1) + s^*_q \tilde{e}^{\mathrm{f}}_{q-1}(t)$;	p	$3p$	–
$\tilde{e}^{\mathrm{f}}_q(t) := c_q \tilde{e}^{\mathrm{f}}_{q-1}(t) - s_q \lambda^{1/2} \mathbf{p}^{f(q)}_p(t-1)$;	p	$3p$	–
END_DO;			
COMMENT pth-order forward prediction residual COMMENT			
$\varepsilon^{\mathrm{f}}_p(t) := \tilde{\alpha}_p \tilde{e}^{\mathrm{f}}_p(t)$;			
LET $\tilde{\alpha}^{\mathrm{f}}_p(t) := \sqrt{\lambda(\tilde{\alpha}^{\mathrm{f}}_p(t-1))^2 + \lvert \tilde{e}^{\mathrm{f}}_p(t) \rvert^2}$;	1	3	1
IF $\tilde{\alpha}^{\mathrm{f}}_p(t) = 0$ THEN LET $c^{\mathrm{b}}_{p+1} := 1.0$; $s^{\mathrm{b}}_{p+1} := 0.0$;	–	–	–
ELSE LET$c^{\mathrm{b}}_{p+1} := \dfrac{\lambda^{1/2} \tilde{\alpha}^{\mathrm{f}}_p(t-1)}{\tilde{\alpha}^{\mathrm{f}}_p(t)}$; $s^{\mathrm{b}}_{p+1} := \dfrac{\tilde{e}^{\mathrm{f}}_p(t)}{\tilde{\alpha}^{\mathrm{f}}_p(t)}$;	–	1	2
END_IF;			
LET $\tilde{\alpha}_{p+1} := c^{\mathrm{b}}_{p+1} \tilde{\alpha}_p$; $\bar{q}_p := s^{\mathrm{b}}_{p+1} \tilde{\alpha}_p$;	–	2	–
FOR q FROM p TO 1 DO			
LET $\tilde{\alpha}^{\mathrm{f}}_{q-1}(t) := \sqrt{(\tilde{\alpha}^{\mathrm{f}}_q(t))^2 + \lvert \mathbf{p}^{f(q)}_p(t) \rvert^2}$;	p	$2p$	p
IF $\tilde{\alpha}^{\mathrm{f}}_{q-1}(t) = 0$ THEN LET $\tilde{c}^{\mathrm{b}}_q := 1.0$; $\tilde{s}^{\mathrm{b}}_q := 0.0$;	–	–	–
ELSE LET $\tilde{c}^{\mathrm{b}}_q := \dfrac{\tilde{\alpha}^{\mathrm{f}}_q(t)}{\tilde{\alpha}^{\mathrm{f}}_{q-1}(t)}$; $\tilde{s}^{\mathrm{b}}_q := \dfrac{\mathbf{p}^{\mathrm{f}(q)}_q(t)}{\tilde{\alpha}^{\mathrm{f}}_{q-1}(t)}$;	–	–	$2p$
END_IF;			
LET $\bar{q}_{q-1} := \tilde{c}^{\mathrm{b}}_q \bar{q}_q + \tilde{s}^{\mathrm{b}*}_q \mathbf{q}^{(q)}(t-1)$;	p	$2p$	–
$\mathbf{q}^{(q+1)}(t) := \tilde{c}^{\mathrm{b}}_q(t) \mathbf{q}^{(q)}(t-1) - \tilde{s}^{\mathrm{b}}_q(t) \bar{q}_q$;	p	$2p$	–
END_DO			
LET $\mathbf{q}^{(1)}(t) := \bar{q}_0$;	–	–	–
LET $\tilde{\alpha}_p := \sqrt{(\tilde{\alpha}_{p+1})^2 + \lvert \mathbf{q}^{(q+1)}(t) \rvert^2}$;	1	2	1
FOR q FROM p TO 1 DO			
LET $\tilde{\alpha}_{q-1} := \sqrt{(\tilde{\alpha}_q)^2 + \lvert \mathbf{q}^{(q)}(t) \rvert^2}$;	p	$2p$	p
IF $\tilde{\alpha}_{q-1} = 0$ THEN LET $c_q := 1$; $s_q := 0$;	–	–	–
ELSE LET $c_q := \dfrac{\tilde{\alpha}_q}{\tilde{\alpha}_{q-1}}$; $s_q := \dfrac{\mathbf{q}^{(q)*}(t)}{\tilde{\alpha}_{q-1}}$;	–	–	$2p$
END_IF;			
END_DO;			

LET $\tilde{e}_0(t) := y(t)$;	–	–	–
FOR q FROM 1 TO p DO			
LET $\mathbf{p}^{(q)}(t) := c_q\lambda^{1/2}\mathbf{p}^{(q)}(t-1) + s_q^*\tilde{e}_{q-1}(t)$;	p	$3p$	
$\tilde{e}_q(t) := c_q\tilde{e}_{q-1}(t) - s_q\lambda^{1/2}\mathbf{p}^{(q)}(t-1)$;	p	$3p$	–
END_DO			
COMMENT pth-order filtered residual COMMENT			
$\varepsilon_p(t) := \tilde{\alpha}_p(t)\tilde{e}_p(t)$;	–	1	–
END_DO			
FINISH	$8p+4$	$20p+7$	$5p+3$

—

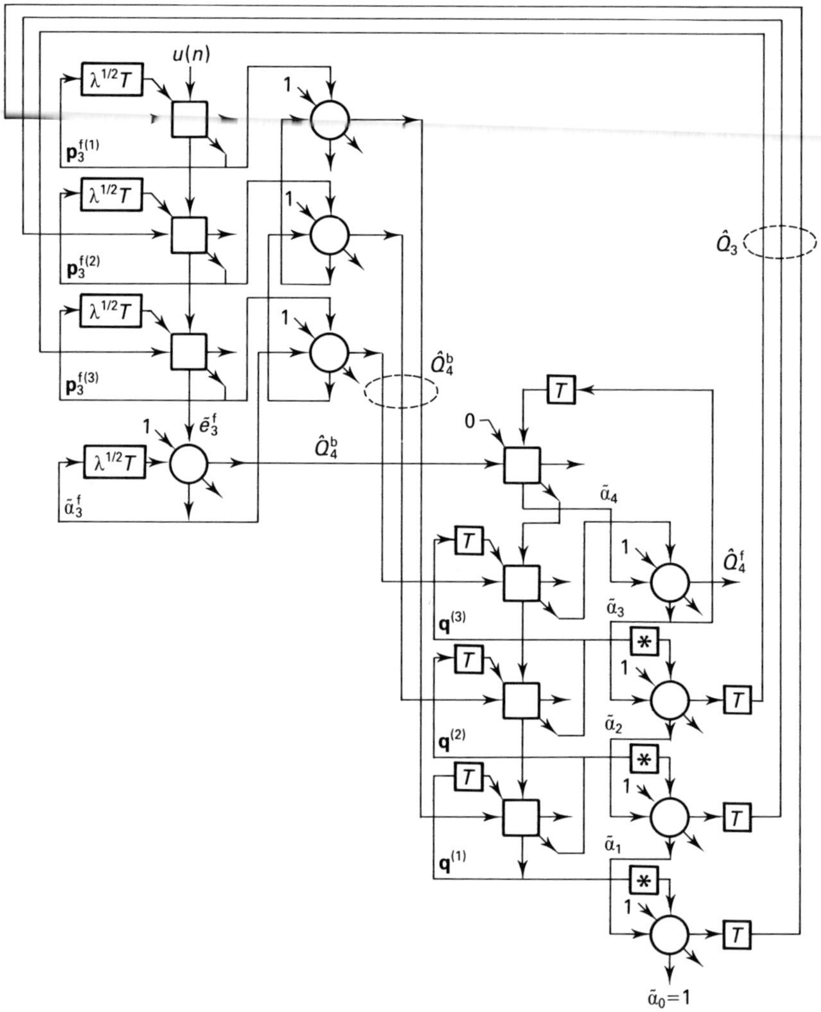

Figure 7.14 Fast QRD architecture

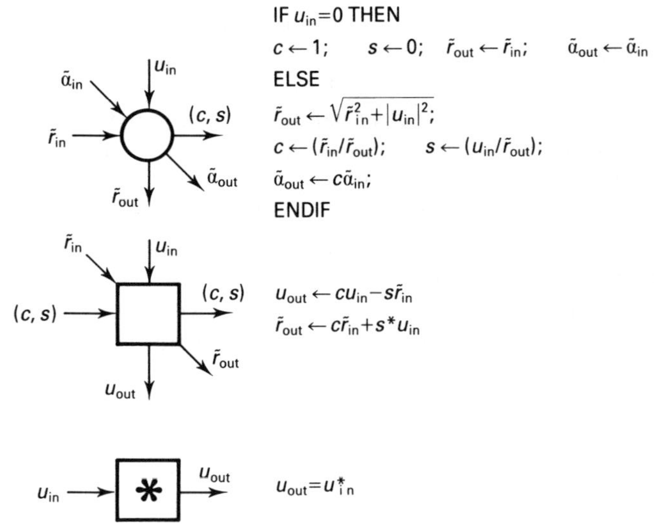

Figure 7.15 Fast QRD processors

An interesting consequence of using the QRD approach is that the fast QRD algorithm not only produces the solution to a given order problem but also that for all lower-order problems. This is because the QRD-based approach to least-squares minimization is inherently an order recursive process (see section 7.3.6). For instance, consider the triangular processor array described in section 7.2.3. It should be clear that the quantity being passed down to the boundary cells (the product of the various cosine terms for earlier rotations) is the quantity $\tilde{\alpha}(N)$ required by the lower-order problems. A little thought will also reveal that the angle-normalized residual, $\tilde{e}(N)$, for a given lower-order problem is the quantity passed down to the boundary processor from the last internal processor of that particular column (see Figure 7.16).

It is interesting to note that, at first glance, the algorithm pictured in Figure 7.14 appears not to include any quantities related to the backward linear prediction problem; however, this is not true. It is possible to show (see section 7.3.6) that the vector $\mathbf{q}_p(N)$ consists of the backward prediction residuals for orders 0 to $p-1$ normalized by the respective prediction error energy. Indeed, Regalia and Bellanger [26] derived their algorithm using these quantities explicitly.

7.3.6 Physical interpretation of fast algorithm parameters

The QRD-based approach to least-squares minimization is just one of many different ways in which the problem can be solved. However, the quantities used in a QRD-based algorithm appear to be radically different from those to be found in the more familiar approaches. Clearly, because the underlying problem is the same, the variables in a QRD-based algorithm must be related to more conventional quantities. In this section we point out some of these relationships along with some other interesting inter relationships between quantities found in the QRD based approach. Indeed, the QRD-based approach to least-squares minimization, and linear prediction in particular, offers many useful insights.

We have already seen that the forward and backward prediction residual powers (α_p^{f} and α_p^{b}) appear quite naturally in the QRD-based lattice algorithm – albeit in terms of their square roots: $\tilde{\alpha}_p^{\mathrm{f}}$ and $\tilde{\alpha}_p^{\mathrm{b}}$ (see Equations (7.99) and (7.109)). The lattice algorithm also calculates estimates of the partial correlation coefficients. Consider the term $\tilde{\beta}_{p-1}^{\mathrm{f}}$ in (7.99): if the rotation parameters corresponding to $\hat{Q}_p^{\mathrm{f}}(N)$ are (c_p^{f}, s_p^{f}) then

$$\begin{bmatrix} c_p^{\mathrm{f}} & s_p^{\mathrm{f}*} \\ -s_p^{\mathrm{f}} & c_p^{\mathrm{f}} \end{bmatrix} \begin{bmatrix} \lambda^{1/2}\tilde{\beta}_{p-1}^{\mathrm{f}}(N-1) & \lambda^{1/2}\tilde{\alpha}_{p-1}^{\mathrm{b}}(N-2) \\ \tilde{e}_{p-1}^{\mathrm{f}}(N) & \tilde{e}_{p-1}^{\mathrm{b}}(N-1) \end{bmatrix} = \begin{bmatrix} \tilde{\beta}_{p-1}^{\mathrm{f}}(N) & \tilde{\alpha}_{p-1}^{\mathrm{b}}(N-1) \\ \tilde{e}_p^{\mathrm{f}}(N) & 0 \end{bmatrix} \tag{7.123}$$

that is,

$$\tilde{\beta}_{p-1}^{\mathrm{f}}(N) = c_p^{\mathrm{f}}\lambda^{1/2}\tilde{\beta}_{p-1}^{\mathrm{f}}(N-1) + s_p^{\mathrm{f}*}\tilde{e}_{p-1}^{\mathrm{f}}(N) \tag{7.124}$$

where

$$c_p^{\mathrm{f}} = \frac{\lambda^{1/2}\tilde{\alpha}_{p-1}^{\mathrm{b}}(N-2)}{\tilde{\alpha}_{p-1}^{\mathrm{b}}(N-1)} \quad \text{and} \quad s_p^{\mathrm{f}} = \frac{\tilde{e}_{p-1}^{\mathrm{b}}(N-1)}{\tilde{\alpha}_{p-1}^{\mathrm{b}}(N-1)} \tag{7.125}$$

Combining (7.124) and (7.125), and setting

$$b_{p-1}(N) = \tilde{\alpha}_{p-1}^{\mathrm{b}}(N-1)\tilde{\beta}_{p-1}^{\mathrm{f}}(N) \tag{7.126}$$

we find, after some algebra, that

$$b_{p-1}(N) = \lambda b_{p-1}(N-1) + \tilde{e}_{p-1}^{\mathrm{f}}(N)\tilde{e}_{p-1}^{\mathrm{b}*}(N-1) \tag{7.127}$$

$$= \sum_{n=1}^{N} \lambda^{(N-n)}\tilde{e}_{p-1}^{\mathrm{f}}(n)\tilde{e}_{p-1}^{\mathrm{b}*}(n-1) \tag{7.128}$$

Now, by comparison with (7.47), we have

$$\tilde{e}_{p-1}^{\mathrm{f}}(N) = \sqrt{\varepsilon_{p-1}^{\mathrm{f}}(N)e_{p-1}^{\mathrm{f}}(N)} \quad \text{and} \quad \tilde{e}_{p-1}^{\mathrm{b}}(N-1) = \sqrt{\varepsilon_{p-1}^{\mathrm{b}}(N-1)e_{p-1}^{\mathrm{b}}(N-1)} \tag{7.129}$$

Substituting these relations into (7.128) gives

$$b_{p-1}(N) = \sum_{n=1}^{N} \lambda^{(N-n)}e_{p-1}^{\mathrm{f}}(n)\varepsilon_{p-1}^{\mathrm{b}*}(n-1) \tag{7.130}$$

Thus we see that $b_{p-1}(N)$ is a (weighted) estimate of the cross-correlation between the a priori forward residual and the complex conjugate of the a posteriori backward residual. For stationary signal statistics, once a recursive least-squares prediction

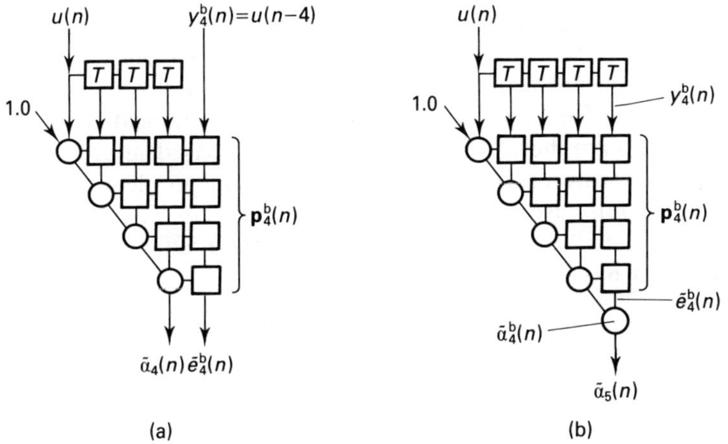

Figure 7.16 Order recursion within the QRD-based LS algorithm: (a) fourth-order backward linear prediction; (b) fifth order triangular array

algorithm has converged, the a priori and a posteriori residuals are identical; hence $b_{p-1}(N)$ can be viewed as an estimate of the partial correlation coefficient for the forward linear prediction problem. Therefore,

$$\tilde{\beta}^{\mathrm{f}}_{p-1}(N) \approx \frac{\beta^{\mathrm{f}}_{p-1}(N)}{\tilde{\alpha}^{\mathrm{b}}_{p-1}(N-1)} \tag{7.131}$$

is a normalized version of the partial correlation coefficient. In a similar manner, it is possible to show, again from (7.99), that

$$\tilde{\beta}_{p-1}(N-1) \approx \frac{\beta_{p-1}(N-1)}{\tilde{\alpha}^{\mathrm{b}}_{p-1}(N-1)} \tag{7.132}$$

and, from (7.109), that

$$\tilde{\beta}^{\mathrm{b}}_{p-1}(N) \approx \frac{\beta^{\mathrm{b}}_{p-1}(N)}{\tilde{\alpha}^{\mathrm{f}}_{p-1}(N)} \tag{7.133}$$

Next we consider the order recursive nature of the QRD approach to linear prediction. Note that (7.99) and (7.100) provide a recursive decomposition of the matrix $\tilde{R}_p(N-1)$. Specifically, and for time N rather $(N-1)$,

$$\tilde{R}_p(N) = \begin{bmatrix} \tilde{R}_{p-1}(N) & \mathbf{p}^{\mathrm{b}}_{p-1}(N) \\ \mathbf{0}^{\mathrm{T}} & \tilde{\alpha}^{\mathrm{b}}_{p-1}(N) \end{bmatrix} \tag{7.134}$$

This shows that the diagonal elements of the matrix $\tilde{R}_p(N)$ are in fact the square roots of the backward prediction residual energy terms for each of the suborder problems. It also indicates why it is sensible for the Givens rotations used in QRD-based linear prediction to ensure that the diagonal elements of the $\tilde{R}(N)$ matrix are always positive. Equation (7.134) also shows that the off-diagonal parts of the triangular matrix are the various '**p**' vectors. This is not surprising considering that the linear prediction problem could be solved using a full $p \times p$ systolic array with the time series $u(N)$ fed in via a tapped delay line, as indicated in Figure 7.16. It therefore serves to emphasize the fact that a QRD-based approach generates the solution to all suborder problems as well as the target problem of a given order.

Order recursion also plays an important part in any least-squares lattice algorithm. Traditionally, the order recursion for the prediction residuals takes the form (see Chapter 6)

$$\begin{aligned} \varepsilon_p^{\mathrm{f}}(N) &= \varepsilon_{p-1}^{\mathrm{f}}(N) + k_{p-1}^{\mathrm{f}}(N)\varepsilon_{p-1}^{\mathrm{b}}(N-1) \\ \varepsilon_p^{\mathrm{b}}(N) &= \varepsilon_{p-1}^{\mathrm{b}}(N-1) + k_{p-1}^{\mathrm{b}}(N)\varepsilon_{p-1}^{\mathrm{f}}(N) \end{aligned} \tag{7.135}$$

where $k_{p-1}^{\mathrm{f}}(N)$ and $k_{p-1}^{\mathrm{b}}(N)$ are the $(p-1)$th-order reflection coefficients. Unlike the conventional lattice algorithms, the QRD-based one derived in section 7.3.4 does not calculate reflection coefficients explicitly; instead the order update for the (angle-normalized) residuals takes the form (see Equations (7.99) and (7.109))

$$\begin{aligned} \tilde{e}_p^{\mathrm{f}}(N) &= c_p^{\mathrm{f}}(N)\tilde{e}_{p-1}^{\mathrm{f}}(N) - s_p^{\mathrm{f}}(N)\lambda^{1/2}\tilde{\beta}_{p-1}^{\mathrm{f}}(N-1) \\ \tilde{e}_p^{\mathrm{b}}(N) &= c_p^{\mathrm{b}}(N)\tilde{e}_{p-1}^{\mathrm{b}}(N-1) - s_p^{\mathrm{b}}(N)\lambda^{1/2}\tilde{\beta}_{p-1}^{\mathrm{b}}(N-1) \end{aligned} \tag{7.136}$$

where $c_p^{\mathrm{f}}(N)$, $s_p^{\mathrm{f}}(N)$, $c_p^{\mathrm{b}}(N)$ and $s_p^{\mathrm{b}}(N)$ are the sines and cosines of the transformations $\hat{Q}_p^{\mathrm{f}}(N)$ and $\hat{Q}_p^{\mathrm{b}}(N)$ respectively. Equation (7.136) appears quite different from the usual lattice equations shown in (7.135); however, note that (7.136) is written in terms of angle-normalized residuals and not a posteriori ones, and that the sines and cosines are functions of the residuals (see equations (7.99) and (7.109)). Converting from the angle-normalized residuals to the a posteriori ones using (7.35) and evaluating the sines and cosines, we obtain, after some manipulation,

$$\begin{aligned} \varepsilon_p^{\mathrm{f}}(N) &= \varepsilon_{p-1}^{\mathrm{f}}(N) - \frac{\tilde{\beta}_{p-1}^{\mathrm{f}}(N)}{\tilde{\alpha}_{p-1}^{\mathrm{b}}(N-1)}\,\varepsilon_{p-1}^{\mathrm{b}}(N-1) \\ \varepsilon_p^{\mathrm{b}}(N) &= \varepsilon_{p-1}^{\mathrm{b}}(N-1) - \frac{\tilde{\beta}_{p-1}^{\mathrm{b}}(N)}{\tilde{\alpha}_{p-1}^{\mathrm{f}}(N)}\,\varepsilon_{p-1}^{\mathrm{f}}(N) \end{aligned} \tag{7.137}$$

from which it follows that

$$\begin{aligned} k_{p-1}^{\mathrm{f}}(N) &= -\frac{\tilde{\beta}_{p-1}^{\mathrm{f}}(N)}{\tilde{\alpha}_{p-1}^{\mathrm{b}}(N-1)} \\ k_{p-1}^{\mathrm{b}}(N) &= -\frac{\tilde{\beta}_{p-1}^{\mathrm{b}}(N)}{\tilde{\alpha}_{p-1}^{\mathrm{f}}(N)} \end{aligned} \tag{7.138}$$

This result is not surprising; the reflection coefficients in (7.135) are defined to be those values that minimize the terms $\Sigma_{n=1}^{N} |\varepsilon_p^{\mathrm{f}}(n)|^2$ and $\Sigma_{n=1}^{N} |\varepsilon_p^{\mathrm{b}}(n)|^2$. In other words, the reflection coefficients are the coefficients in a first-order least-squares minimization problem. From section 7.2.1 we know that, when using the QRD technique, the least-squares coefficients are given by

$$\mathbf{c} = -\tilde{R}^{-1}\mathbf{p} \tag{7.139}$$

In the case of a first-order problem, both the matrix $\tilde{R}$ and the vector $\mathbf{p}$ are scalars and for the least-squares minimization problem shown in (7.135) these quantities equate to those shown in (7.138). This equality can most easily be seen with reference to Figure 7.11. If the least-squares minimization problem is one of first order, then the triangular QRD array (cf. Figure 7.2) will be a single circular processing element and the right-hand column will reduce to a one square processing element. These structures can be identified in Figure 7.11 from which the relationship shown in (7.135) can readily be deduced.

It was remarked earlier that the quantities $\tilde{\beta}^{\mathrm{f}}_{p-1}$ and $\tilde{\beta}^{\mathrm{b}}_{p-1}$ were normalized partial correlation coefficients; however, it is clear from (7.138) that they can also be thought of as scaled reflection coefficients. We have adopted the former view in this chapter since they quite naturally appear as correlation variables; whereas in Chapter 6 the latter interpretation was used – where they were referred to as $\tilde{k}^{\mathrm{f}}_{p-1}$ and $\tilde{k}^{\mathrm{b}}_{p-1}$ respectively – due to the emphasis on lattice structures. This difference in notation explains the minor differences between the algorithms summarized in Table 7.4 and their counterpart in Table 6.6.

As remarked earlier, the fast QRD algorithm does not calculate conventional quantities (neither the optimum coefficients nor the reflection coefficients). As described in section 7.3.5, the fast QRD algorithm calculates the rotation matrix $\hat{Q}_p(N)$. This matrix is then used as shown in (7.21) to compute the vector $\mathbf{p}_p(N)$ and produce the adaptive filtering residual. However, we note that the vector $\mathbf{p}_p(N)$ is merely a transformed version of the optimum coefficients $\mathbf{c}_p(N)$, as discussed above.

Another unusual feature of the algorithm derived in section 7.3.5 is that it does not appear to make use of any quantities related to the backward prediction problem. This is not true, however, since the vector $\mathbf{q}_p(N)$ may be interpreted in terms of the backward prediction residuals. To see the equivalence, consider (7.121), (7.122) and (7.125); it is clear that

$$\mathbf{q}_p(N) = \begin{bmatrix} \mathbf{q}_{p-1}(N) \\ s^{\mathrm{f}*}_p \tilde{\alpha}_{p-1}(N) \end{bmatrix} = \begin{bmatrix} \mathbf{q}_{p-1}(N) \\ \dfrac{\tilde{\alpha}_{p-1}(N)\tilde{e}^{\mathrm{b}*}_{p-1}(N)}{\tilde{\alpha}^{\mathrm{b}}_{p-1}(N)} \end{bmatrix} = \begin{bmatrix} \mathbf{q}_{p-1}(N) \\ \dfrac{\varepsilon^{\mathrm{b}*}_{p-1}(N)}{\tilde{\alpha}^{\mathrm{b}}_{p-1}(N)} \end{bmatrix} \tag{7.140}$$

and we see that the vector $\mathbf{q}_p(N)$ consists of the energy-normalized backward prediction residuals of order $(p-1)$ and below.

7.3.7 Weight extraction from fast algorithms

The QRD-based least-squares lattice and fast QRD algorithms presented above are based on the 'direct residual extraction' technique and as such produce the adaptive filtering residual without explicitly calculating the optimum weight vector. This is highly desirable in an adaptive filtering context since the residual is the quantity of

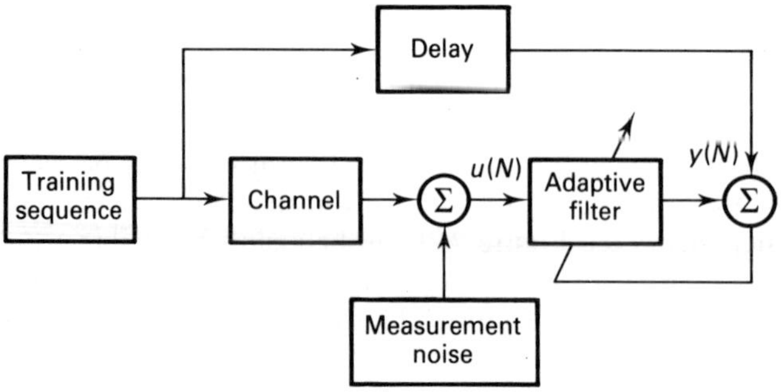

Figure 7.17 Channel equalizer experiment

interest; however, in system identification the primary goal is the calculation of the weight vector. The two weight extraction techniques presented for the full QRD-based algorithm – weight flushing (section 7.2.6) and parallel weight extraction (secton 7.2.7) – are equally applicable to the fast algorithms. Note, however, that neither of the fast algorithms calculates the triangular matrix $\tilde{R}$ explicitly so that the back-substitution method (Equation (7.11)) is not available.

Both the lattice and fast QRD algorithms use the same processing elements as the full triangular QRD array and can therefore be operated in the frozen mode. If a unit impulse is fed into the frozen filter, its output will clearly be the impulse response of the system, i.e. the set of filter weights. Note, however, that unlike the narrowband beamformer of section 7.2, the adaptive filters presented here are not memoryless systems when operated in their frozen mode: the (implicit) tapped delay line must continue to operate. It is therefore necessary to ensure that the (implicit) tapped delay line is 'full of zeros' before applying the unit impulse. Thus it would require an input series of length $2p + 1$ consisting of a single sample of value unity sandwiched between two set of p consecutive zero samples.

Although this operation will produce the filter's impulse response, it is an invasive procedure – again because of the fact that the system has memory. Having frozen the filter and passed the impulse sequence through it, the state of the filter (the contents of the implicit tapped delay line) will have been altered compared with the point in time when the filter was frozen. Thus it is not now possible to unfreeze the filter and continue with the adaption process as if nothing had happened. If the system is allowed to adapt from this incorrect filter state then the output will be in error – until sufficient data have been processed so that the error in the filter's state has decayed away. This may well be acceptable in certain situations since this process of converging to the required solution is exactly what happens when the filter is first started. One way in which weight flushing can be made non-invasive is for the contents of the filter delay elements to be stored before the weight flushing begins and then restored before the adaption continues.

Recall from section 7.2.7 that in order to extract the filter weights in parallel with the adaption process, it is necessary to have available the rotation matrix $\hat{Q}(N)$. This matrix is explicitly calculated in the fast QRD algorithm and, indeed, is also available in the lattice algorithm in a disguised form: it can be shown that the rotation matrices $\hat{Q}_i^{\mathrm{f}}(N)$ $(1 \leqslant i \leqslant p)$ are equal to the elementary Givens rotations $G_i(N)$ $(1 \leq i \leq p)$ that make up the matrix $\hat{Q}(N)$ – see (7.30). Thus it is possible to use the additional hardware shown in Figure 7.8 in conjunction with the fast algorithm (instead of the triangular processor array) to generate the weights. However, the utility of this approach is questionable since the addition of the extra processing means that the computational load of the complete algorithm rises from $O(p)$ to $O(p^2)$ and the advantage of the fast algorithm is lost.

7.3.8 *Computer simulations*

The main concern with 'fast' algorithms is that they are potentially more sensitive to numerical errors than their generic counterparts. This is because the fast algorithms exploit exact mathematical relationships between various quantities in the generic algorithm in order to reduce the computational load. In the case of the least-squares lattice algorithms, for example, the assumption that the problem has already been solved for the one order allows the solution to the next order problem to be generated efficiently. Now in practice the calculations can only be done to a finite accuracy, so, strictly speaking, the assumptions upon which the fast algorithms are based (e.g. the existence of the solution to the lower-order problems) are only approximately true. We can therefore expect to pay some penalty, in terms of numerical stability, for the reduced computational load. The perceived advantage with the fast QRD-based algorithms is that they should be more robust in the presence of numerical errors than other fast algorithms owing to the orthogonal nature of the QRD algorithm.

In order to investigate the effects of finite precision on the fast QRD-based algorithms we consider the application of an adaptive filter to a typical channel equalization problem (Figure 7.17). In particular, we consider[5] the case of an adaptive equalizer applied to a data channel with a 'raised cosine' impulse response (Equation (7.141)):

$$h(n) = \begin{cases} \dfrac{1}{2}\left[1 + \cos\left(\dfrac{2\pi}{W}(n-2)\right)\right] & n = 1,\ 2,\ 3 \\ \mathrm{O} & \text{otherwise} \end{cases} \tag{7.141}$$

By varying the parameter W, the amount of interference between a given symbol and the two either side of it can be changed. This, in effect, controls the eigenvalue spread of the data covariance matrix (see Table 7.6).

Table 7.6 Eigenvalue spread

W	$\|\lambda_{max}/\lambda_{min}\|$
2.9	6.1
3.1	11.2
3.3	21.9
3.5	47.5

An 11th-order QRD-based least-squares adaptive filter is used to equalize the channel response. In the simplest of situations, the equalizer would be trained periodically by transmitting a known sequence and adapting the equalizer with a stored version of this signal as the 'reference' signal (see section 7.3.1). In between these training sessions, with the adaption frozen, the channel can be used for the transmission of data, hopefully with the intersymbol interference much reduced.

In our computer experiment, the transmission channel is fed with a polar (± 1) pseudorandom training sequence. This sequence, delayed by seven time instants (see chapter 9), is used as the reference signal for the adaptive filtering algorithm. A small quantity of 'measurement' noise, in the form of a pseudorandom sequence with an approximately Gaussian probability distribution function, is added to the channel output. The noise sequence used has zero mean and variance of 0.001. The 'forgetting factor' λ (see Equation (7.4)) was fixed at a value of 0.992, which implies an effective data window (i.e. the duration for which any data vector has an effect on the filter adaption) of 250 time samples.

All calculations within the algorithm were performed using limited-precision floating point arithmetic. Only the number of bits in the mantissa is varied during the experiments: the number of bits in the exponent is fixed at eight. No quantities internal to the adaptive filtering algorithm are held to a greater precision than the outputs: the results of *all* arithmetic operations are immediately reduced to the required precision. The numerical performance of most algorithms can be improved by using higher precision for some internal calculations; however, it is necessary to have a good understanding of the algorithm and, in particular, to identify the critical intermediate quantities for this method to be used effectively.

The performance of the equalizer is monitored by recording the ensemble-averaged, squared a priori equalization error (see Equation (7.40)). This has the advantage that it shows how close to convergence the algorithm is whilst showing, asymptotically, the least-squares equalization error. The ensemble average is taken over 100 realizations of the experiment. We have carried out several experiments using various combinations of parameters and algorithms but only the main results are discussed below. Care should be taken in the interpretation of any computer simulation experiment. In particular, when the numerical stability of an algorithm is being investigated it should be noted [27] that instability is often preceded by a period of apparent stability (e.g. Figure 7.19, 8 bit mantissa plot). Thus simulation

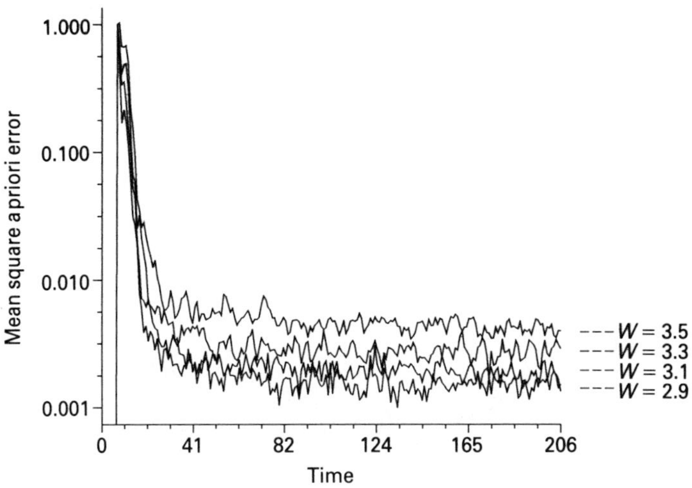

(a)

Figure 7.18(a) SQ/FF QRD lattice: effect of eigenvalue spread

experiments can only confirm lower bounds on the sequence length required to cause instability and do not 'prove' that a system is stable.

Figures 7.18 and 7.19 show the basic performance of the fast QRD-based equalizer algorithms, using the square-root feedforward Givens rotations (SQ/FF) for different values of wordlength and eigenvalue spread. Figure 7.18 shows that, with double-precision arithmetic, the rate of convergence is more or less insensitive to the different eigenvalue spread settings, as would be expected from a recursive least-squares minimization process. Figure 7.19 illustrates how the wordlength affects the perform-

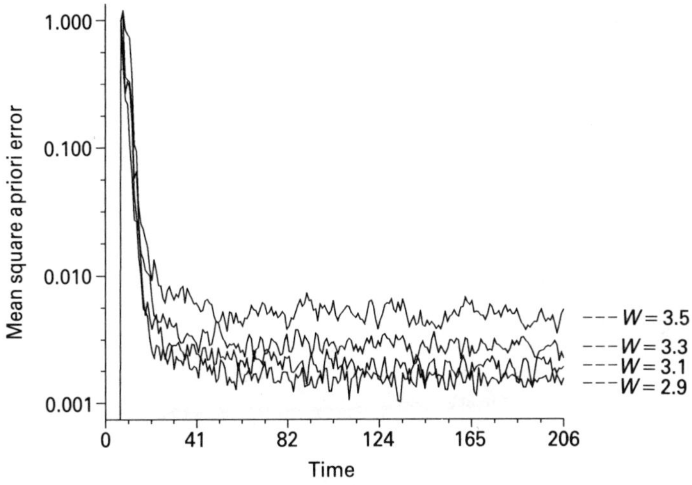

(b)

Figure 7.18(b) SQ/FF fast QRD: effect of eigenvalue spread

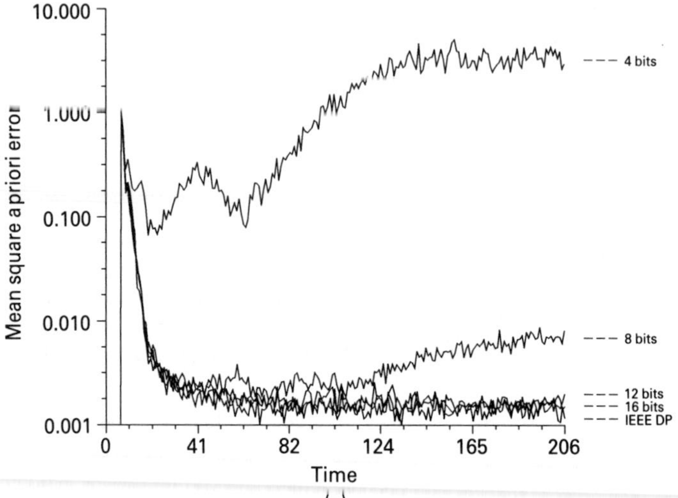

(a)

Figure 7.19(a) SQ/FF QRD lattice: effect of wordlength

ance for a fixed eigenvalue spread ($W = 2.9$). There is very little discernible difference between the filters using 12, 16 and 56 (IEEE double-precision) bit mantissas and these filters appear to be well behaved for data sequences of length up to 200. The first sign of any instability appears with a mantissa of 8 bits. Here the filters initially show signs of converging to a stable state but then begin to diverge, producing an ever increasing error. The 4 bit systems clearly do not behave in a sensible manner as would be expected from such a short wordlength. Note that there is very little difference between the two fast algorithms (the lattice and the fast QRD algorithm).

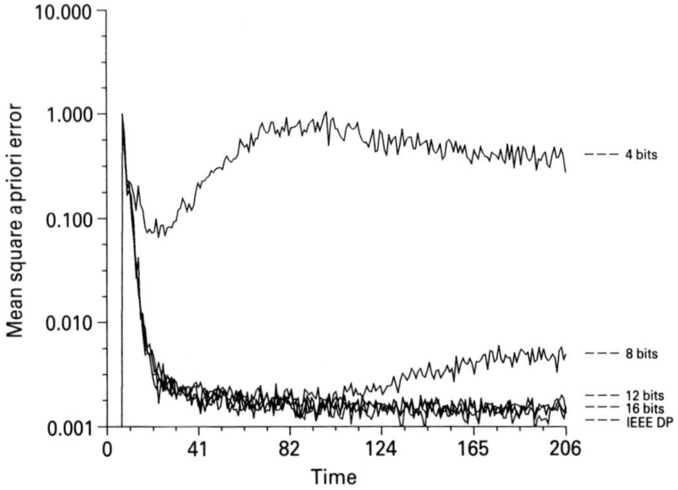

(b)

Figure 7.19(b) SQ/FF fast QRD: effect of wordlength

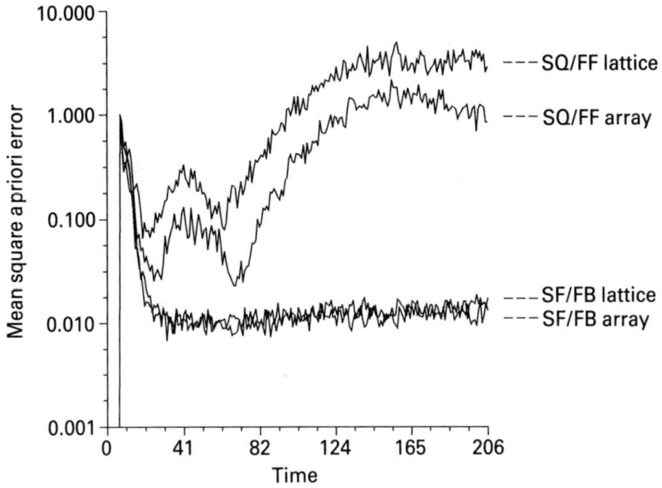

Figure 7.20 Comparison of lattice and array

Figure 7.20 shows a comparison of the QRD-based lattice algorithm with the full QRD-based triangular systolic array version. Four systems are also shown in this figure: they are the 'square-root-free with feedback' (SF/FB) forms of the lattice algorithm (Table 6.5) and the array algorithm, along with the SQ/FF versions of both algorithms. This shows the case of a fixed eigenvalue spread setting ($W = 2.9$) and 4 bit mantissas. This may be considered to be an excessively short wordlength. The reason for this choice is that finite precision effects are often only manifested after the roundoff errors have had time to accumulate [27]. By using a small wordlength, the appearance of such effects occur sooner, thus reducing the time necessary to perform the simulation.

In most cases, a pth-order RLS adaptive filter will converge within $2p$ time instants. At this point the a priori residual will have reached a value primarily determined by the eigenvalue spread and not the wordlength. As roundoff errors accumulate, the a priori error will increase, indicating a loss of accuracy in the algorithm. Up to a run length of 10 000, the longest simulation run to date by the authors, the SQ/FF lattice algorithm remains stable with 12 bit mantissas. In the case of the SF/FB lattice, the same behaviour is seen using only 4 bit mantissas.

It can be seen, in Figure 7.20, that in the SQ/FF mode the faster lattice algorithm is only marginally worse than the full triangular array version, thus demonstrating that little penalty has been paid in reducing the computational load. As expected, the square-root-free with feedback versions of the algorithms perform better than the basic versions. In this case, there was no discernible difference between the lattice version and the array version in any of the simulations run so far. This would seem to demonstrate the power of the feedback technique in improving the numerical accuracy of these algorithms.

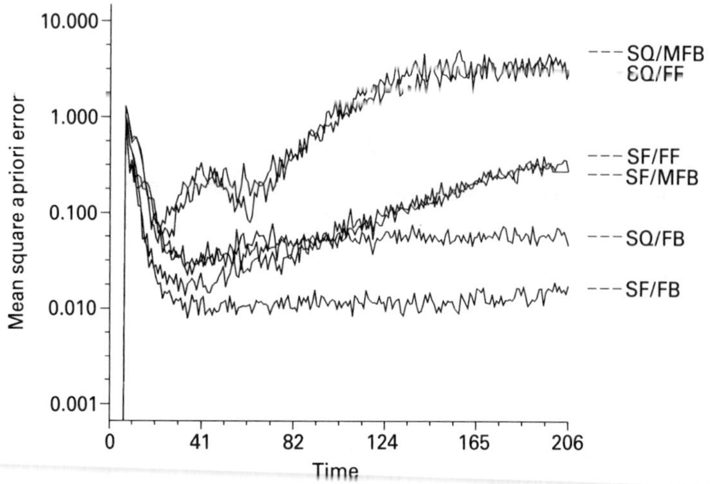

Figure 7.21 Comparison of Givens algorithms

The relative effect of the square-root-free and the feedback techniques can be seen in Figure 7.21. This shows the performance of the lattice algorithms with 4 bit mantissas and fixed eigenvalue spread ($W = 2.9$) for six possible Givens rotation algorithms: SQ/FF (Table 7.4), SF/FB (Table 6.5), square-root Givens rotations with feedback (SQ/FB), square-root-free feedforward rotations (SF/FF), square root rotations with the stored parameter fed back (SF/MFB), and square-root-free rotations with the stored parameter fed back (SF/MFB) – see section 7.2.4 for more details of the various ways of implementing a Givens rotation. From this it can be seen that there is indeed a numerical advantage in avoiding the square-root operation but that the most significant improvement comes about by introducing the 'error feedback' – provided that the feedback is applied properly. Figure 7.21 shows that there is little difference in performance between the 'FF' and 'MFB' versions, whereas the 'FB' versions (i.e. the 'error feedback' algorithms) are significantly better.

In conclusion, we have found that for both the triangular QRD processor in Figure 7.2 and the least-squares lattice filter in Figure 7.12, the best numerical performance over a wide range of computer simulations was obtained using the SF/FB Givens rotations defined in Figure 7.4. The fact that a square-root-free algorithm performs best is not entirely surprising. A closer analysis of the conventional Givens rotation algorithm defined in Figure 7.3 shows that the square root operation is only required in situations where we sum the squares of two numbers and then compute their square root. Avoiding this process could certainly improve the numerical performance. The fact that the feedback form of square-root-free Givens rotation in Figure 7.4 performs best is contrary to the initial expectation of numerical analysts [5]. However, it is entirely consistent with the fact that the corresponding error feedback RMGS lattice algorithm has also been found to exhibit robust numerical stability. Since Ling *et al.* were first inspired to introduce the error feedback mechanism by analogy with the stability techniques used in control, it would appear that this provides a better way of analyzing the numerical stability of other signal processing algorithms.

We have repeated the above experiments with values of the W parameter other than 2.9 and all of the above observations appear to hold, essentially independently of the eigenvalue spread.

7.4 Wideband beamforming

7.4.1 Multichannel adaptive filters

In a multichannel least-squares adaptive filtering problem at time N, a set of r p-dimensional weight vectors, $\mathbf{c}_p^{(i)}(N)$ $(0 \leqslant i \leqslant r-1)$, are to be found that minimize the sum of the squared differences between a reference signal $y(N)$ and a linear combination of r samples from each of p data time series $u_i(N-j)$ $(1 \leqslant i \leqslant p,\ 0 \leqslant j \leqslant r-1)$. This is equivalent to filtering adaptively p separate time series in order to form the best estimate of the reference signal (see Figure 7.22). If the p data sequences come from spatially separate sensors then we have a spatial as well as a temporal filtering problem. In this sense, the multichannel adaptive filtering problem subsumes both the narrowband beamforming problem and the (single-channel) adaptive filtering problem.

To be specific, the measure[6] $V_N(\mathbf{c}_r') = \| \boldsymbol{\varepsilon}_r(N) \|^2$ is to be minimized, where

$$\boldsymbol{\varepsilon}_r(N) = \mathcal{U}_r(N)\mathbf{c}_r' + \mathbf{y}(N) \tag{7.142}$$

$$\mathcal{U}_r(N) = \Lambda(N)\begin{bmatrix} u^{\mathrm{T}}(1) & \dots & u^{\mathrm{T}}(2-r) \\ \vdots & & \vdots \\ u^{\mathrm{T}}(N) & \dots & u^{\mathrm{T}}(N-r+1) \end{bmatrix} \tag{7.143}$$

$$u^{\mathrm{T}}(N) = [u_1(N) \quad u_2(N) \quad \dots \quad u_p(N)] \tag{7.144}$$

$$c_r'(N) = \begin{bmatrix} c_p^{(0)}(N) \\ c_p^{(1)}(N) \\ \vdots \\ c_p^{(r-1)}(N) \end{bmatrix} \tag{7.145}$$

and

$$\mathbf{y}(N) = \Lambda(N)[y(1), \dots, y(N)]^{\mathrm{T}} \tag{7.146}$$

Equation (7.144) serves to define the new quantity $u(N)$. Note that, apart from the change from scalar to vector quantities, (7.143) is identical to (7.80). It is possible to generalize the system shown in Figure 7.22 and to consider the case where the reference signal $y(N)$ is replaced by several such signals. In this case we would have

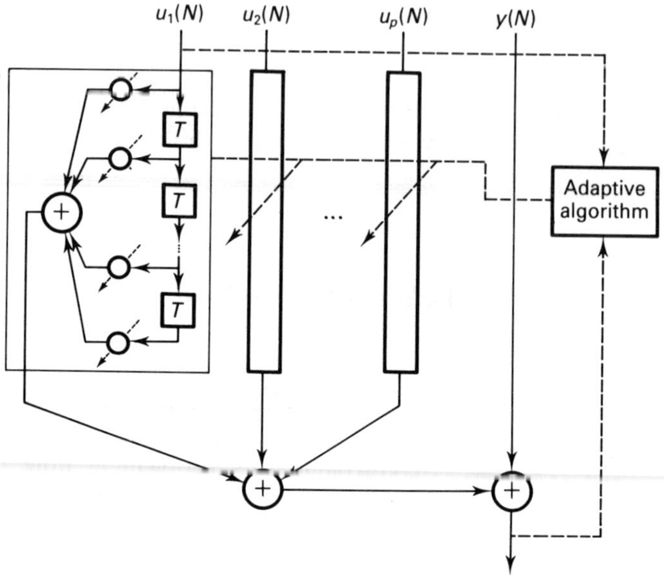

Figure 7.22 Multichannel adaptive filter

to replace the vector $y(N)$ by a matrix. We will not persue the idea of multichannel joint-process estimation any further here but note that a similar situation naturally arises in section 7.4.2 when we consider multichannel linear prediction.

The solution of this vector least-squares minimization problem via QRD follows the usual pattern and requires the determination of an orthogonal matrix $Q_r(N)$ that transforms the matrix $\mathscr{U}_r(N)$ into upper-triangular form. The fact that the matrix $\mathscr{U}_r(N)$ is block-Toeplitz allows us to use the ideas developed in section 7.3 to construct 'fast' algorithms. As one might expect, it is possible to derive both a fast QRD and a lattice algorithm.

7.4.2 *Multichannel lattice*

The extension of the lattice algorithm presented in section 7.3.4 to the wideband beamforming problem is relatively straightforward: the only change required is that certain scalar quantities be replaced by vectors and some vectors be replaced by matrices [28]. The essential features of the derivation presented in section 7.3.4 carry over exactly. The only point where the derivation of the multichannel case deviates in any appreciable way from that given in section 7.3.4 is in the extension to p dimensions of operations on one-dimensional objects.

In the solution of the pth-order, single-channel, forward linear prediction problem it was necessary to determine the rotation matrix $Q_p^{\mathrm{f}}(N-1)$ that annihilated all but one component of the vector $\mathbf{v}_{p-1}(N-2)$ (see Equation (7.97)). In the rth-order multichannel forward linear prediction problem we have to determine a similar rotation matrix. However, in this case, the vector $\mathbf{v}_{p-1}(N-2)$ is replaced by a suitably defined matrix $\mathbf{V}_{r-1}(N-2)$ (with p columns, one for each channel). The equivalent operation to that of $Q_p^{\mathrm{f}}(N-1)$ is to convert $\mathbf{V}_{r-1}(N-2)$ into an upper-triangular matrix, i.e. to perform a QRD decomposition! Indeed, the operation in the single-channel algorithm can be considered to be a QRD on a vector and the resulting, single, non-zero component to be a 1×1 triangular matrix. Similarly, the next step in the derivation (cf. Equation (7.99)) consists of calculating the p rotations necessary to perform the recursive QRD update on the above triangular matrix, instead of just one rotation.

The resultant architecture (see Figure 7.23) has a lattice structure where each stage of the lattice contains two triangular systolic arrays. The total number of operations necessary to solve an rth-order multichannel adaptive filtering problem, with p channels, is thus $O(rp^2)$. Note that in Figure 7.23 some of the vector data lines have been 'twisted'. This merely signifies the fact that data are fed into the triangular arrays in the order specified by the mathematics.[7]

The architecture shown in Figure 7.23 is intuitively satisfying for the following reasons. It was well known (see Chapter 6 or [29]) that the lattice structure of linear prediction algorithms is inherent to the problem. Indeed, it is easy to show in more general terms that a pth-order linear prediction problem can be solved using a lattice structure which, in effect, solves two least-squares minimization problems at each stage. These least-squares minimization problems relate to the determination of the forward and backward reflection coefficients and calculate the order-updated residuals. Specifically, for a multichannel problem,

$$
\begin{aligned}
\varepsilon_r^{\mathrm{f}}(N) &= \varepsilon_{r-1}^{\mathrm{f}}(N) - k_{r-1}^{\mathrm{f}}(N)\varepsilon_{r-1}^{\mathrm{b}}(N-1) \\
\varepsilon_r^{\mathrm{b}}(N) &= \varepsilon_{r-1}^{\mathrm{b}}(N-1) - k_{r-1}^{\mathrm{b}}(N)\varepsilon_{r-1}^{\mathrm{f}}(N)
\end{aligned}
\tag{7.147}
$$

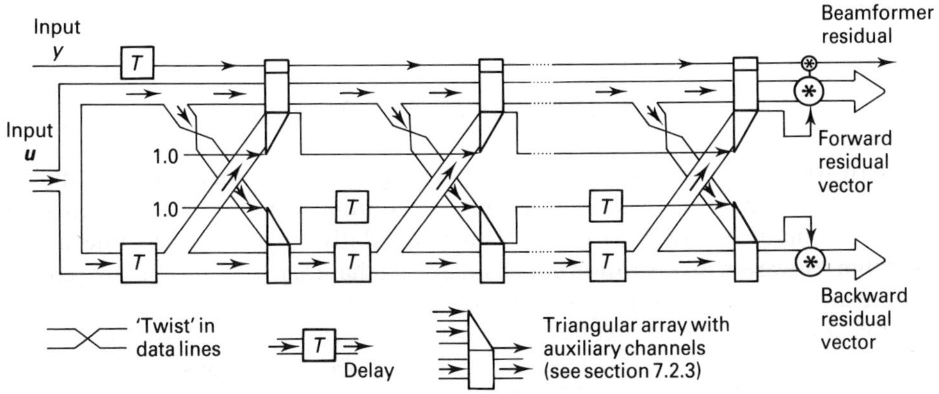

Figure 7.23 Lattice of triangular arrays

where $\varepsilon_r^{\mathrm{f}}$ and $\varepsilon_r^{\mathrm{b}}$ are the rth-order forward and backward a posteriori residual vectors and k_{r-1}^{f} and k_{r-1}^{b} are $p \times p$ reflection coefficient matrices respectively. As we have shown, each of the triangular arrays depicted in Figure 7.23 is capable of performing a recursive least-squares minimization and calculating the residual directly. A close look at Figure 7.23 will show that these triangular arrays operate on the forward and backward residual vectors in precisely the manner required to solve these problems (cf. discussion following Equation (7.138)).

An alternative method [22] for deriving the multichannel QRD-based lattice algorithm actually begins with the standard multichannel lattice algorithm and transforms it into a purely orthogonal square root information algorithm. In the 'standard' RLS lattice algorithm, the forward prediction residuals for order r (say) are found by subtracting linear combinations of the $(r-1)$st-order backward prediction residuals from the $(r-1)$st-order forward residuals; the coefficients involved in this linear combination are, of course, just the reflection coefficients. A similar relationship holds for the rth-order backward residuals. The calculation of the reflection coefficients requires an inversion of the data covariance matrix (see Chapter 6), which is computationally expensive and often ill-conditioned. Using a Cholesky decomposition of the data covariance matrix, Lewis [19] showed how this part of the algorithm could be transformed into a recursive square root information process. In this algorithm the reflection coefficients are no longer calculated explicitly: quantities analogous to the vector **p** in (7.11) and the vector **q** in (7.92) are calculated instead. Nevertheless, the rth-order residuals are still calculated in terms of the difference between the $(r-1)$st-order residuals and another term (now a function of the quantities analogous to **p** and **q**). As the bulk of the calculation is exactly the computation of the reflection coefficients, Lewis proceeded no further with this reformulation and apparently failed to notice that the 'non-orthogonal' part of his algorithm is in fact redundant. Yang and Böhme [22] observed that the adaptive filtering residuals were effectively being produced along with the computation of the vectors **p** and **q** as shown in section 7.2.5: this observation results in the construction of a purely orthogonal algorithm and is equivalent to that derived from the first principles here.

7.4.3 Multichannel fast QRD algorithm

The derivation of the multichannel fast QRD algorithm is somewhat more difficult than the lattice equivalent. If the various substitutions of scalars for vectors and vectors for matrices are carried out then it is relatively easy to generate a 'fast' algorithm. However, an operation count will reveal that $O(rp^3)$ operations are required to solve a p-channel rth-order problem. Assuming that $r > p$, this is 'fast' compared with the $O(r^2p^2)$ operations required when using a generic triangular systolic array fed by tapped delay lines, but falls short of the $O(rp^2)$ operations required by the lattice algorithm. It is possible [30] to generate an $O(rp^2)$ multichannel fast QRD algorithm and, as in the single-channel case, the pinning vector is used to allow the inference of a rotation matrix from vector quantities. As before, this technique reduces the problem by one dimension and produces an $O(rp^2)$ algorithm.

The 'bottleneck' in the naïve generalization of the single-channel algorithm (with $O(rp^3)$ operations) is the calculation equivalent to that shown in (7.110). For convenience we reproduce this equation, in part, below:

$$\tilde{Q}^{\mathrm{b}}_{p+1}(N)\begin{bmatrix} \tilde{\alpha}^{\mathrm{f}}_p(N) & \mathbf{0} \\ \mathbf{p}^{\mathrm{f}}_p(N) & \tilde{R}_p(N-1) \\ \mathbf{0} & \mathrm{O} \\ 0 & \mathbf{0}^{\mathrm{T}} \end{bmatrix} = \begin{bmatrix} R_{p+1}(N) \\ \mathrm{O} \end{bmatrix} \qquad (7.148)$$

In the multichannel equivalent, the vector $\mathbf{p}^{\mathrm{f}}_p(N)$ is replaced by an $rp \times p$ matrix $\tilde{\mathbf{P}}^{\mathrm{f}}_r(N)$ (say) and the scalar $\tilde{\alpha}^{\mathrm{f}}_p(N)$ by a $p \times p$ triangular matrix $(\tilde{A}^{\mathrm{f}}_r(N))$. The operation of annihilating this $rp \times p$ matrix – equivalent to a block recursive QRD update – requires $O(rp^3)$ Givens rotations: it requires $O(p)$ operations to annihilate a p-dimensional vector by rotation against a $p \times p$ triangular matrix, and the matrix $\tilde{\mathbf{P}}^{\mathrm{f}}_r(N)$ has rp rows.

An alternative procedure for annihilating the matrix $\tilde{\mathbf{P}}^{\mathrm{f}}_r(N)$ is to eliminate it one column at a time rather than one row at a time as in the recursive QRD update derived in section 7.2.2. Each column of $\tilde{\mathbf{P}}^{\mathrm{f}}_r(N)$ has rp components and will therefore require $O(rp)$ operations to annihilate it. If this was all that was required we would have a fast algorithm: $\tilde{\mathbf{P}}^{\mathrm{f}}_r(N)$ has p columns making a total of $O(rp^2)$ operations. However, it is not sufficient for each column of the matrix $\tilde{\mathbf{P}}^{\mathrm{f}}_r(N)$ just to be annihilated individually: the rotations that annihilate a given column must also be applied to the other columns. Columns that have previously been annihilated clearly will not be affected by this operation but the non-zero ones will be. Thus the rotations that annihilate the first column must be applied to the $(p-1)$ other columns; the rotations that annihilate the second column will have to be applied to $(p-2)$ other columns, and so on. This sequence of steps clearly requires $O(rp^3)$ operations.

In the above scheme, the column vectors of $\tilde{\mathbf{P}}^{\mathrm{f}}_r(N)$ are subject to, in general, several rotations. This is exactly the situation we faced in the construction of the single-channel fast QRD algorithm (see Equations (7.114) and (7.115)). The solution to this problem was to use the pinning vector effectively to condense the rotations down to a single one. Using this idea in the multichannel case leads to an $O(rp^2)$ algorithm. The sequence of operations is then as follows. The left-hand column of the matrix $\tilde{\mathbf{P}}^{\mathrm{f}}_r(N)$ is annihilated ($O(rp)$ operations) and this rotation is applied to the second column and a pinning vector ($O(rp)$ operations). The transformed second column is then annihilated ($O(rp)$ operations) and the rotations applied to the transformed pinning vector ($O(rp)$ operations). The doubly transformed pinning vector can then be 'unrotated' ($O(rp)$ operations) to generate a rotation that is equivalent to the combined effect of the earlier pair of rotations. This combined rotation is then applied to the third column and another pinning vector and the process continues. At each stage only $O(rp)$ operations are required and thus all p columns of $\tilde{\mathbf{P}}^{\mathrm{f}}_r(N)$ can be annihilated in $O(rp^2)$ operations.

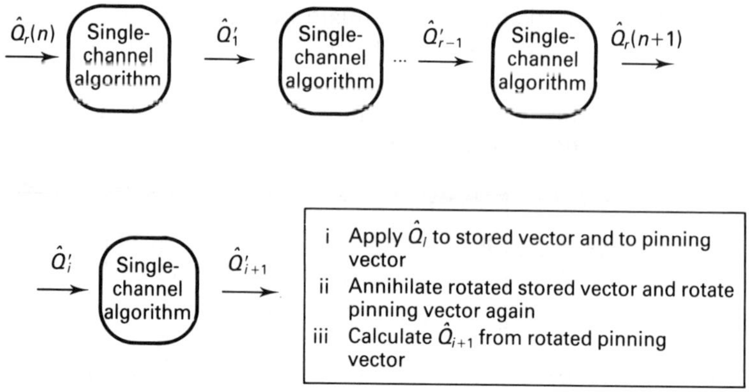

Figure 7.24 Multichannel fast QRD algorithm

A more detailed study of the multichannel fast QRD algorithm shows that it is, in effect, just the application of the single-channel algorithm many times. It should be clear from the above that the annihilation of one of the columns of the matrix $\tilde{\mathbf{P}}_r^{\mathrm{f}}(N)$ involves the three steps: the application of a known rotation, annihilation of a vector and subsequent calculation of the combined rotation based on the rotated pinning vector. In the single-channel algorithm, exactly the same type of operations are required: the known rotation is $\hat{Q}_p(N)$ and the resultant rotation is $\hat{Q}_p(N+1)$. The multichannel algorithm thus has the structure shown in Figure 7.24. Each pass of the 'single-channel' algorithm requires $O(rp)$ operations (since the matrix $\tilde{\mathbf{P}}_r^{\mathrm{f}}(N)$ has columns of dimension rp) and p such passes are required (because the matrix $\tilde{\mathbf{P}}_r^{\mathrm{f}}(N)$ has p columns), so the final operation count is $O(rp^2)$.

7.5 Algorithm listings

The algorithm listings given in this chapter are written in pseudo-ALGOL, and are for a narrowband beamformer and a single-channel adaptive filter only. Two narrowband beamformer algorithms are presented: the first algorithm uses the obvious, feedforward implementation of Givens rotations using square roots (Table 7.1); the second one avoids taking square roots by calculating transformed quantities and implements the rotations via the feedback algorithm (Table 7.3). Both the least-squares lattice and fast QRD algorithms are given but only using the conventional Givens rotations using square roots (Tables 7.4 and 7.5 respectively). Based on the diagrammatic representations of the algorithms and the interchangeability of the processing elements, it should be possible for the reader to generate any of the 'missing' algorithms. In the same spirit, it should be clear how to modify these algorithms to include such aspects as parallel weight extraction.

The computations count given in the right-hand column of each table assumes that the signals being processed are real – although the mathematics is written assuming complex quantities. The complexity of division has been equated to that of square-rooting and multiplication by the exponential weighting factor λ is counted as one general-purpose multiplication – which may not be the case if λ is chosen to be of a simple form such as $1 - 2^n$. Note that the algorithms are not optimized: the computational load could be reduced by rewriting the algorithms to take advantage of intermediate quantities common to two or more calculations; such compact forms of the algorithms are not presented here, in order that the regularity of the calculation should not be obscured.

The narrowband beamformer algorithms take as inputs a p-dimensional vector of auxiliary signals $\mathbf{u}(t)$ and a reference sequence $y(t)$ and calculate the beamformer residual. The adaptive filtering algorithms calculate the filter residual for a pth-order system fed with a prewindowed data sequence $u(t)$ and a reference sequence $y(t)$.

Note that $r^{(i,j)}(t)$ is the (i, j)th component of the matrix $R(t)$; and $\mathbf{u}^{(i)}(t)$ is the ith component of the vector $\mathbf{u}(t)$.

Notes

1. In the following, we use shading in order to emphasize the structure of the non-zero elements of a matrix.
2. Note that the subscript p attached to the vector $\mathbf{y}_p^{\mathrm{f}}$ is superfluous and that $\mathbf{y}_p^{\mathrm{f}} = \mathbf{y}_{p-1}^{\mathrm{f}}$, etc. Its use is merely to preserve symmetry with the vector $\mathbf{y}_p^{\mathrm{b}}$ for which the subscript is necessary.
3. The notation for the rotation matrices introduced in this section is somewhat arbitrary: in order to solve the pth-order *forward* linear prediction problem, we must annihilate quantities from the $(p - 1)$st-order *backward* prediction problem The rotation matrices used here are labelled according to the problem to which they relate rather than the quantities they annihilate.
4. It is possible to develop a fast QRD algorithm in which $u(N) \neq 0$ for $N \leqslant 0$. The resulting algorithm is based on much of the prewindowed version presented here but with some extra computation – see Cioffi [23] for further details. The authors are not aware of any similar work for the QRD-based lattice algorithm.
5. Suggested by S. Haykin.
6. To be rigorous, we should label quantities involved with a p-channel, r-tap, multichannel least-squares minimization problem with two indices (p and r). However, in the following we will be considering only iterations in the number of taps and not the number of channels. Thus for notational simplicity we will indicate explicitly only the number of taps being considered – the number of channels being assumed to be fixed.
7. It is easy to show that this data twist is not necessary since a permutation of the data does not affect the value of the residuals; however, we do not pursue this idea any further in this book.

References

[1] G.H. Golub, 'Numerical methods for solving linear least-squares problems', *Numer. Math.*, vol. 7, pp. 206–16, 1965.

[2] W. Givens, 'Computation of plane unitary rotations transforming a general matrix to triangular form', *J. Soc. Ind. Appl. Math.*, vol. 6, pp. 26–50, 1958.

[3] W.M. Gentleman and H.T. Kung, 'Matrix triangularisation by systolic array', *Proc. SPIE (Real-Time Signal Processing IV)*, pp. 298–303, 1981.

[4] W.M. Gentleman, 'Least-squares computations by Givens transformations without square-roots', *J. Inst. Maths. Applic.*, vol. 12, pp. 329–69, 1973.

[5] S. Hammarling, 'A note on modifications to the Givens plane rotation', *J. Inst. Math. Applics.*, vol. 13, pp. 215–18, 1974.

[6] I.K. Proudler, J.G. McWhirter and T.J. Shepherd 'The QRD-based least squares lattice algorithm: Some computer simulations using finite wordlengths', *Proc. IEEE Int. Symp. on Circuits and Systems, New Orleans, LA, May, 1990,* pp. 258–61.

[7] C.R. Ward, P.J. Hargrave and J.G. McWhirter, 'A novel algorithm and architecture for adaptive digital beamforming', IEEE Trans. *Antennas Propag.*, vol. AP-34, pp. 338–46, 1986.

[8] F. Ling, D. Manolakis and J.G. Proakis, 'A recursive modified Gram–Schmidt algorithm for least-squares estimation', *IEEE Trans. Acoust., Speech, Signal Process.*, vol. ASSP-34, pp. 829–36, 1986.

[9] J.G. McWhirter, 'Recursive least squares minimisation using a systolic array', *Proc. SPIE (Real-Time Signal Processing IV)*, vol. 431, pp. 105–12, 1983.

[10] J.G. McWhirter and T.J. Shepherd, 'Systolic array processor for MVDR beamforming', *IEE Proc.,* pt F, vol. 130, no. 2, pp. 75–80, 1989.

[11] T.J. Shepherd and J. Hudson, 'Parallel weight extraction from a systolic adaptive beamformer', *Proc. IMA Conf. on Mathematics in Signal Processing, Warwick, England, Dec. 1988.*

[12] S.T. Alexander and A.L. Ghirnikar, 'A method for recursive least squares filtering based upon an inverse QR decomposition', Submitted to *IEE trans. Signal Process.* SP, 1991. (See A.L. Ghirnikar and S.T. Alexander, 'Stable recursive least squares filtering using an inverse QR decomposition', *Proc. IEE Int. Conf. on ASSP*, Albuquerque, NM, April, pp. 1623–26, 1990.)

[13] M. Moonen and J. Vandewalle, 'Recursive least squares with stabilized inverse factorization', *IEEE trans. Signal Process.*, vol. SP–21, pp. 1–15, 1990.

[14] C.T. Pan and R.J. Plemmons, 'Least squares modifications with inverse factorization: Parallel implementations', *Comput Appl. Math.*, vol. 27, pp. 109–27, 1989.

[15] A.P. Varvitsiotis and S. Theodoridis, 'A pipelined structure for QR adaptive LS system identification, *IEEE Trans. Signal Process.*, vol. SP–39, pp. 1920–23, 1991.

[16] F. Ling, D. Manolakis and J.G. Proakis, 'A flexible, numerically robust array processing algorithm and its relationship to the Givens transformation' *Proc. IEEE Int. Conf. on ASSP, Tokyo, Japan, Apr., 1986.*

[17] M.J. Chen and K. Yao, 'On realizations of least squares estimation and Kalman filtering', *Proc. 1st Int. Workshop on Systolic Arrays, Oxford*, pp. 161–70, 1986.

[18] F.M.F. Gaston, G.W. Irwin and J.G. McWhirter, 'Systolic square-root covariance Kalman filtering', *J. VLSI Signal Processing*. vol. 2, pp. 37–49, 1990.

[19] P.S. Lewis, 'QR-based algorithms for multichannel adaptive least squares lattice filters', *IEEE trans. Acoust., Speech, Signal Process.*, vol ASSP-38, pp. 421–32, 1990.

[20] F. Ling, 'Givens rotation based least squares lattice and related algorithms', *IEEE trans. Signal Process.*, vol. SP-39, pp. 1541–52, 1991.

[21] I.K. Proudler, J.G. McWhirter and T.J. Shepherd, 'Computationally efficient, *QR decomposition approach to least squares adaptive filtering', IEE proc.*, pt. F, no 4, vol. 138, no. 4, pp. 341–53, 1991.

[22] B. Yang and J.F. Böhme, 'On a parallel implementation of the adaptive multichannel least-squares lattice filter', *Proc. Int. Symp, on Signals and Electronics, Erlangen, FRG, Sept., 1989.*

[23] J.M. Cioffi, The fast adaptive rotors RLS algorithm', *IEEE trans. Acoust., Speech, Signal Process.* vol. ASSP-38, pp. 631–51, 1990.

[24] F. Ling and J.G. Proakis, 'A generalized multichannel least squares lattice algorithm based on sequential processing stages', *IEEE Trans. Acoust., Speech, Signal Process.* vol. ASSP-32, pp. 381–9, 1987.

[25] I.K. Proudler, J.G. McWhirter and T.J. Shepherd, 'Fast QRD-based algorithms for least squares linear prediction', *Proc. IMA Conf. on Mathematics in Signal Processing, Warwick, England Dec., 1988.*

[26] P.A. Regalia and M.G. Bellanger, 'On the duality between fast QR methods and lattice methods in least squares adaptive filtering', *IEEE Trans. Signal Process.* vol. SP-39, pp. 879–92, 1991.

[27] J.M. Cioffi, 'Finite precision effects in adaptive filtering', *IEEE Trans. Circuits Syst.* vol. CAS-34 pp. 821–33, 1987.

[28] I.K. Proudler, J.G. McWhirter and T.J. Shepherd, 'Computationally efficient QRD-based wideband beamforming', *Proc. IEEE Int. Conf. on ASSP, Albuquerque, NM,* paper no. 348 A13.12,

[29] M.J. Shensa, 'Recursive least squares lattice algorithms – A geometric approach', *IEEE trans.*, vol. 26, pp. 696–702, 1981.

[30] M.G. Bellanger and P.A. Regalia, 'The FLS-QR algorithm for adaptive filtering The case of multichannel signals', *Signal Process.*, vol. 22, no 2, Feb. 1991.

8

Spectral Analysis

S. Theodoridis and N. Kalouptsidis

8.1 Introduction

The power spectral density (PSD) or power spectrum of a wide-sense stationary (WSS) discrete time random process is defined as the Fourier transform of its autocorrelation function, i.e.

$$S(\omega) = \sum_{t=-\infty}^{+\infty} \rho(t)e^{-j\omega t} \tag{8.1}$$

where the autocorrelation function is

$$\rho(t) = E[x(t+\tau)x(\tau)] \tag{8.2}$$

Spectral analysis is concerned with the estimation of the power spectrum based on a limited set of observation points of the corresponding WSS process. The availability of short data records is imposed either by the physical problem, i.e. lack of data (economics), changing statistics of the pertinent process for longer observation periods (speech) or by excessive computational burdens required for large data sets. Typical application areas of spectral analysis are encountered in communications, signal detection, speech processing, transient analysis phenomena and seismic signal processing, to name but a few. The intriguing task of estimating the second-order statistics from a short observation interval of a realization of a WSS process has given rise to a number of different techniques. These methods are required to produce power spectral estimates that meet the demands of high resolution and good statistical performance. Such requirements are often conflicting.

Resolution is the ability of a method to estimate the unknown power spectrum with sufficient spectral detail [1]. This is in general an empirical figure of merit and there are no means of quantifying the resolution capability of a method against an optimal value, or even among different methods. The resolution of a method is usually tested against known signals for its ability to identify closely located spectral peaks as a function of data record length and/or observation noise. This involves a great deal of subjective judgement and thus has to be carried out with great caution. These issues are discussed in detail in [1,2].

The other major issue in spectral analysis is that of the statistical properties of the resulting estimate as expressed by its bias and variance. Obviously large variance estimators enjoy little confidence. Again, the mathematical analysis of these issues is rather involved and is usually carried out under strong assumptions. In practice, one has again to resort to experimental verifications using Monte Carlo techniques.

The classical methods of power spectral estimation are direct outcomes of the definition of the power spectrum and are classified as the correlogram and periodogram methods.

Given a limited set of input samples, correlogram methods estimate first the autocorrelation sequence and then the Fourier transform to obtain the power spectrum estimate. Periodogram methods are based on an equivalent to definition (8.1) of the PSD. Indeed, it can be shown that under certain weak limiting conditions of the input autocorrelation sequence, the power spectral density $S(\omega)$ is given by

$$S(\omega) = \lim_{N \to \infty} E\left[\frac{1}{2N+1}\left|\sum_{t=-N}^{N} x(t)\mathrm{e}^{-j\omega t}\right|^2\right] \tag{8.3}$$

Periodogram methods compute the Fourier transform of the input sequence and then square it, and the PSD estimate results through some type of averaging. These methods became very popular after the advent of Fast Fourier Transform (FFT).

Both methods assume that unknown input samples or autocorrelation lags are implicitly zero. This windowing results in spectral estimates which are smeared versions of the true one, thus introducing bias and reducing resolution. Spectral leakage is another undesirable effect introduced by windowing. It describes the extent to which one spectral component contaminates nearby spectral estimates due to the sidelobes associated with the discontinuity of the imposed window. As a result, lower-power spectral peaks may be hidden in the sidelobes of a nearby high-power peak. Different types of windows have been introduced trading off resolution with spectral leakage. In general, windowing is artificially induced into the problem by assuming unavailable data to be zero and changing perfectly good data or lag estimates by a weighting function.

An alternative approach is based on parametric methods as introduced in Chapter 2. Improved performance can be achieved *provided the WSS process is adequately described by the chosen parametric model.* This is reasonable since modelling the process gives us the means implicitly to extrapolate information beyond the known interval where Fourier-based methods assume zero values.

8.2 Basic guidelines from the theory of spectral analysis of stochastic processes

As stated in the book's introduction, the major concern of this chapter is the development of efficient spectral estimation algorithms. The majority of the schemes we are about to describe deal with the computation of the power spectral density of signals produced by AR sources. This is because AR modelling leads to a linear set of equations for which efficient computational schemes can be derived. Another reason is that the more general ARMA modelling can also be treated by means of approximate AR models, as we shall see later. To help the reader form a perspective of how representative the class of AR models is, we collect and highlight some important results from the theory of spectral analysis of WSS stochastic processes in this section. For a detailed presentation, see [3,4]. The reader primarily interested in spectral estimation algorithms may skip this section.

Let us consider a wide-sense stationary discrete time process $x(t)$ with autocorrelation sequence $\rho(t)$. The discrete time Fourier transform pair

$$S(\omega) = \sum_{t=-\infty}^{\infty} \rho(t) \mathrm{e}^{-\mathrm{j}\omega t} \tag{8.4}$$

$$\rho(t) = \frac{1}{2\pi} \int_{-\pi}^{\pi} \mathrm{e}^{\mathrm{j}\omega t} S(\omega)\, \mathrm{d}\omega \tag{8.5}$$

is not well defined in the ordinary sense, unless the autocorrelation function is properly confined. Let us take, for instance, the sinusoidal signal

$$x(t) = A \cos(\omega t + \phi) \tag{8.6}$$

where A and ω are constants, while the phase ϕ is a random variable uniformly distributed on the interval $-\pi \leqslant \phi \leqslant \pi$; $x(t)$ is WSS. Indeed,

$$E[x(t)] = A \int_{-\pi}^{\pi} \cos(\omega t + \phi) \frac{1}{2\pi} \, d\phi = 0 \tag{8.7}$$

and

$$\rho(t) = E[x(t+\tau)x(\tau)] = \frac{A^2}{2\pi} \int_{-\pi}^{\pi} \cos[\omega(t+\tau) + \phi] \cos(\omega\tau + \phi) \, d\phi$$

$$= \frac{A^2}{2\pi} \int_{-\pi}^{\pi} [\cos(2\omega\tau + \omega t + 2\phi) + \cos \omega t) \, d\phi$$

As in the computation of the mean, the first term is zero. Hence,

$$\rho(t) = \frac{1}{2} A^2 \cos \omega t \tag{8.8}$$

More generally, the so-called *harmonic process*

$$x(t) = \sum_{k=1}^{p} A_k \cos(\omega_k t + \phi_k) \tag{8.9}$$

where A_k and ω_k are constants and ϕ_k are independent random variables uniformly distributed over $[-\pi, \pi]$, is a WSS process with autocorrelation function

$$\rho(t) = \frac{1}{2} \sum_{k=1}^{p} A_k^2 \cos \omega_k t \tag{8.10}$$

This equation follows from (8.8) and independence.

We observe that $\rho(t)$ is bounded, yet it is not an l_1 sequence, as it does not die away as t approaches infinity. Moreover, the Fourier transform does not exist if (8.5) is interpreted as an ordinary integral. On the other hand, harmonic processes exhibit a periodic-like pattern and should be amenable to a spectral characterization. To assign a spectral content to harmonic processes we must broaden our definition of (8.5). One approach is to use generalized functions. Let $\delta(t)$ denote the Dirac function. As is well known, $\delta(t)$ renders the formal expression

$$\int_{-\infty}^{+\infty} f(t-\tau)\delta(\tau) \, d\tau = f(t) \tag{8.11}$$

meaningful. It then follows that the Fourier transform of (8.10) for $p = 1$ is

$$S(\omega) = \frac{A^2\pi}{2} \sum_{k=-\infty}^{\infty} [\delta(\omega + \omega_1 - 2\pi k) + \delta(\omega - \omega_1 - 2\pi k)] \tag{8.12}$$

As Equation (8.11) suggests, formal expressions involving Dirac functions become legitimate once they are integrated. Suppose $-\pi < \omega_1 < \pi$. Integration of (8.12) gives

$$F(\omega) = \int_{-\pi}^{\omega} S(v) \, dv = \begin{cases} 0 & -\pi < \omega < -\omega_1 \\ A^2\pi/2 & -\omega_1 < \omega < \omega_1 \\ A^2\pi & \omega_1 < \omega < \pi \end{cases} \tag{8.13}$$

$F(\omega)$ is a perfectly defined ordinary step function. It is called the *integrated spectrum* of $x(t)$.

$F(\omega)$ and $S(\omega)$ are related via the formal expression

$$\frac{\mathrm{d}F(\omega)}{\mathrm{d}\omega} = S(\omega) \tag{8.14}$$

The above analysis indicates that spectral analysis of a signal like (8.9) could be carried out by means of the integrated spectrum. Indeed, if (8.14) is substituted into (8.5) we obtain

$$\rho(t) = \frac{1}{2\pi} \int_{-\pi}^{\pi} \mathrm{e}^{\mathrm{j}\omega t} \, \mathrm{d}F(\omega) \tag{8.15}$$

This is a Riemann–Stieltjes integral defined by the limit

$$\lim \sum_{n} \mathrm{e}^{\mathrm{j}\omega_n t}(F(\omega_n) - F(\omega_{n-1})) \quad \text{as} \quad \max(\omega_n - \omega_{n-1}) \to 0$$

If $F(\omega)$ is differentiable, the Riemann–Stieltjes integral converts into an ordinary Riemann integral. On the other hand the integral (8.15) is defined for functions $F(\omega)$ that are not differentiable, such as the integrated spectrum of a harmonic process, which is a step function.

The validity of the spectral representation of the autocorrelation function based on the integrated spectrum and the Riemann–Stieltjes integral (8.15) is the content of the discrete version of the celebrated Wiener–Khintchine theorem established by Wold. A precise statement follows.

DISCRETE WIENER–KHINTCHINE THEOREM

If $\rho(t)$ is the autocorrelation function of a WSS process $x(t)$, there exists a function $F(\omega)$ such that (8.15) holds.

The converse statement is also true. Indeed, the integrated spectrum is bounded on $[-\pi, \pi]$, $F(-\pi) = 0$, and is non-decreasing, i.e. $F(\omega_1) \leqslant F(\omega_2)$, if $\omega_1 < \omega_2$. If these properties accompany a function $F(\omega)$, then the sequence defined by (8.15) is the autocorrelation function of a WSS process.

The *normalized integrated spectrum*

$$F_n(\omega) = \frac{1}{F(\pi)} F(\omega) \tag{8.16}$$

has the properties of a probability distribution function $F_X(x)$ associated with a random variable X,

$$F_X(x) = P[X \leqslant x]$$

namely

$$F_n(-\pi) = 0 \qquad F_n(\pi) = 1 \tag{8.17a}$$

$$F_n(\omega_1) \leqslant F_n(\omega_2) \qquad \omega_1 < \omega_2 \tag{8.17b}$$

Therefore if $F_n(\omega)$ is differentiable, the *normalized spectral density*

$$S_n(\omega) = \frac{1}{2\pi\rho(0)} S(\omega) \tag{8.18}$$

becomes a probability density function.

The interpretation of the normalized integrated spectrum as a probability distribution function enables us to make use of the Lebesgue decomposition theorem and to shed new light on the spectral characteristics of a WSS process.

LEBESGUE DECOMPOSITION THEOREM

Any distribution $F(\omega)$ can be written as a convex combination of three distributions

$$F(\omega) = a_1 F_1(\omega) + a_2 F_2(\omega) + a_3 F_3(\omega) \qquad \sum_{i=1}^{3} a_i = 1 \qquad a_i \geqslant 0 \tag{8.19}$$

where $F_2(\omega)$ is a discontinuous function formed from the discontinuity jumps of $F(\omega)$, which are at most countable. $F_1(\omega)$ and $F_3(\omega)$ are both continuous. $F_1(\omega)$ is differentiable almost everywhere, while $F_3(\omega)$ has zero derivative almost everywhere.

Since distribution $F_1(\omega)$ is differentiable, it has a probability density function

$$S_1(\omega) = \frac{dF_1(\omega)}{d\omega} \tag{8.20}$$

The distribution $F_2(\omega)$ is a step function. It can be written as

$$F_2(\omega) = \sum_{i:\omega_i \leqslant \omega} p_i \tag{8.21}$$

where ω_i denote the frequencies where jumps occur and p_i denote the jump heights. Finally, $F_3(\omega)$ exhibits a pathological behaviour and in practice we ignore it.

The application of the Lebesgue decomposition theorem on the normalized integrated spectrum of the WSS process $x(t)$ with autocorrelation function $\rho(t)$ leads to the decomposition

$$\rho(t) = a_1\rho_1(t) + a_2\rho_2(t) + a_3\rho_3(t) \tag{8.22}$$

$$\rho_1(t) = \frac{1}{2\pi}\int_{-\pi}^{\pi} e^{j\omega t}\, dF_1(\omega) = \frac{1}{2\pi}\int_{-\pi}^{\pi} e^{j\omega t} S_1(\omega)\, d\omega \tag{8.23}$$

$$\rho_2(t) = \frac{1}{2\pi}\int_{-\pi}^{\pi} e^{j\omega t}\, dF_2(\omega) = \frac{1}{2\pi}\int_{-\pi}^{\pi} e^{j\omega t} \sum_i p_i \delta(\omega - \omega_i)\, d\omega = \frac{1}{2\pi}\sum_i p_i e^{j\omega_i t}$$

or

$$\rho_2(t) = \sum (A_i \cos \omega_i t + B_i \sin \omega_i t) \tag{8.24}$$

As pointed out above, the third term $\rho_3(t)$ is usually disregarded.

EXAMPLES

A. LTI sources driven by white noise Let us consider a process $x(t)$ produced by a linear time-invariant (LTI) source excited by white noise of variance σ^2; $x(t)$ admits the representation

$$x(t) = \sum_{k=-\infty}^{\infty} h(k)\eta(t-k) \tag{8.25}$$

The filter with impulse response $h(t)$ represents the source. If it is BIBO stable, i.e.

$$\sum_{t=-\infty}^{\infty} |\, h(t)\,| < \infty$$

then the process $x(t)$ is WSS with autocorrelation function

$$\rho(t) = h(t) * h(-t) * \delta(t)\sigma^2 \tag{8.26}$$

The power spectral density exists and is given by

$$S(\omega) = |\, H(\omega)\,|^2\sigma^2 \tag{8.27}$$

where $H(\omega)$ is the filter frequency response, i.e. the discrete time Fourier transform of $h(t)$.

With reference to the spectral decomposition (8.19) or (8.22) the output spectrum of an LTI source excited by white noise corresponds to the case $a_1 = 1$, $a_2 = a_3 = 0$.

B. ARMA models ARMA models were introduced in Chapter 2 and have the form

$$y(t) + \sum_{i=1}^{n_a} a_i y(t-i) = \sum_{j=0}^{n_c} c_j \eta(t-j) \tag{8.28}$$

with $c_0 = 1$, without loss of generality. A process $y(t)$ generated by the above model is known as an ARMA (n_a, n_c) signal. ARMA models are special cases of LTI sources driven by white noise. Hence, they admit the representation (8.25) with the further restriction that the impulse response $h(t)$ is a rational signal, i.e. its z-transform is a ratio of polynomials. In accordance with the discussion of the preceding example the power spectrum of $y(t)$ is

$$S(\omega) = H(\mathrm{e}^{\mathrm{j}\omega})H(\mathrm{e}^{-\mathrm{j}\omega})\sigma^2 \tag{8.29}$$

where

$$H(z) = \frac{1 + c_1 z^{-1} + \ldots + c_{n_c} z^{-n_c}}{1 + a_1 z^{-1} + \ldots + a_{n_a} z^{-n_a}} = \frac{C(z)}{A(z)} \tag{8.30}$$

The poles of $H(z)$ are located inside the unit circle.

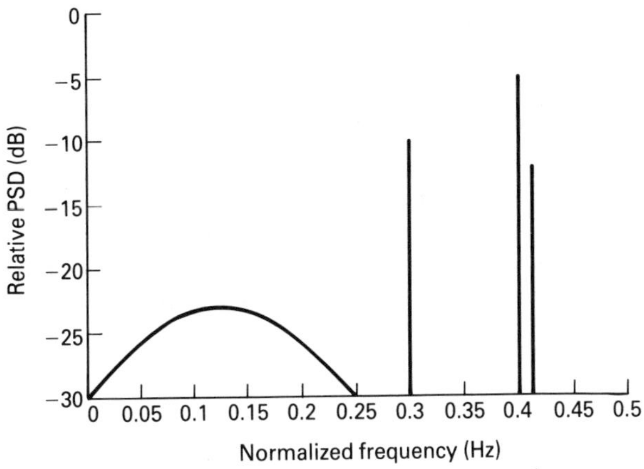

Figure 8.1 Power spectral density of a mixed process

C. Harmonic models The harmonic process was defined (see Equation (8.9)) as a sum of sinusoids

$$x(t) = \sum_{k=1}^{p} A_k \cos(\omega_k t + \phi_k) \tag{8.31}$$

The autocorrelation function is given by (8.10). The spectral density is a sum of impulses and the integrated spectrum is a step function. Thus, the harmonic process corresponds to the case $a_1 = 0$, $a_2 = 1$ and $a_3 = 0$.

D. Mixed processes By definition, the spectrum of a mixed process involves both a continuous part and spectral lines (discrete part). It results if the output of an LTI source excited by white noise is added to a harmonic process

$$z(t) = x(t) + y(t) \tag{8.32}$$

$$y(t) = \sum_{k=-\infty}^{\infty} h(k)\eta(t-k) \tag{8.33}$$

$$x(t) = \sum_{k=1}^{p} A_k \cos(\omega_k t + \phi_k) \tag{8.34}$$

As an example, consider the process of Figure 8.1. It is a mixed process with a power spectrum whose continuous part is obtained by passing white noise through a bandpass filter and whose discrete part is a harmonic signal with frequencies located outside the filter passband.

SPECTRAL REPRESENTATION OF STOCHASTIC PROCESSES

Mixed processes are more or less typical representatives of WSS processes. This statement is made precise using the spectral representation of WSS processes. Let us start with the remark that a single realization of a WSS process cannot be represented by a Fourier integral

$$x(t) = \int e^{j\omega t}\, dA(\omega) = \int e^{j\omega t} a(\omega)\, d\omega$$

Indeed, if this were true, $x(t)$ would decay to zero as time went to infinity, contradicting the stationarity behaviour of the process which requires a more or less steady-state pattern in time. Although such a direct attack fails, the process $x(t)$ can be decomposed into a continuous range of exponential signals with random magnitudes and phases. The precise statement is given in the following theorem.

SPECTRAL REPRESENTATION THEOREM

Let $x(t)$ be a WSS process. Then there exists a stochastic process $z(\omega)$ defined on the interval $[-\pi, \pi]$ such that

$$x(t) = \int_{-\pi}^{\pi} e^{j\omega t}\, dz(\omega) \quad \forall t \tag{8.35}$$

The process $z(\omega)$ has the following properties:

(i)

$$E[dz(\omega)] = 0 \tag{8.36}$$

(ii)

$$E[|dz(\omega)|^2] = dH(\omega) \quad \forall\omega \tag{8.37}$$

where $H(\omega)$ is the integrated spectrum of $x(t)$

(iii) $dz(\omega)$ is an orthogonal process, i.e. $\forall\omega_1 \neq \omega_2]$

$$E[dz(\omega_1)d^*z(\omega_2)] = 0 \tag{8.38}$$

The spectral representation theorem in conjunction with the Lebesgue decomposition of the integrated spectrum provides important information about the basic consti-tuents of a WSS process. Indeed, let us consider the decomposition of the integrated spectrum $F(\omega)$ in (8.19), and let us disregard the pathological part $F_3(\omega)$. Let Ω_1 be the continuity points of $F(\omega)$ and Ω_2 the discontinuity points of $F(\omega)$; Ω_1 and Ω_2 are disjoint sets. We set

$$z_i(\omega) = \begin{cases} z(\omega) & \text{if } \omega \in \Omega_i \\ 0 & \text{otherwise} \end{cases} \qquad i = 1, 2 \tag{8.39}$$

It then follows that

$$z(\omega) = z_1(\omega) + z_2(\omega) \tag{8.40}$$

and

$$x(t) = x_1(t) + x_2(t) = \int_{\omega=-\pi}^{\pi} e^{j\omega t}\, dz_1(\omega) + \int_{\omega=-\pi}^{\pi} e^{j\omega t}\, dz_2(\omega) \tag{8.41}$$

Moreover,

$$E[\,|\,dz_1(\omega)\,|^2] = dH_1(\omega) = h(\omega)\,d\omega \tag{8.42}$$

In an analogous way we can treat the general case and arrive at the decomposition

$$x(t) = x_1(t) + x_2(t) + x_3(t) \tag{8.43}$$

where $x_1(t)$ has a power spectral density $h(\omega)$, $x_2(t)$ has a line spectrum, and $x_3(t)$ is the pathological part.

WSS SIGNALS HAVING POWER SPECTRAL DENSITY ARE OBTAINED FROM LTI SYSTEMS DRIVEN BY WHITE NOISE

Let us consider the $x_1(t)$ component of $x(t)$ giving rise to the continuous part $h(\omega)$ of the spectrum of $x(t)$. We will show that $x_1(t)$ can be obtained at the output of a non-causal LTI source driven by white noise (see Example A).

We factorize $h(\omega)$ as follows:

$$h(\omega) = \phi(\omega)\phi^*(\omega) = |\,\phi(\omega)\,|^2 \tag{8.44}$$

This is allowed because the power spectral density is non-negative. We represent $\phi(\omega)$ by its Fourier series

$$\phi(\omega) = \sum_{k=-\infty}^{\infty} \Phi_k e^{jk\omega} \tag{8.45}$$

Let

$$dv(\omega) = \frac{1}{\phi^*(\omega)}\,dz_1(\omega) \tag{8.46}$$

where we assume $\phi(\omega) \neq 0$, an assumption which can be removed. Now we substitute (8.46) into (8.41):

$$x_1(t) = \int_{-\pi}^{\pi} e^{j\omega t}\,dz_1(\omega) = \int_{-\pi}^{\pi} e^{j\omega t}\phi^*(\omega)\,dv(\omega)$$

$$= \int_{-\pi}^{\pi} e^{j\omega t} \sum_{k=-\infty}^{+\infty} \Phi_k^* e^{-jk\omega}\,dv(\omega) = \sum \Phi_k^*\left(\int e^{j(t-k)\omega}\,dv(\omega)\right)$$

Thus $x_1(t)$ has the form

$$x_1(t) = \sum_{k=-\infty}^{\infty} \Phi_k^*\eta(t-k) \tag{8.47}$$

where

$$\eta(t) = \int_{-\pi}^{\pi} e^{j\omega t}\,dv(\omega) \tag{8.48a}$$

The process $\eta(t)$ is white noise. Indeed, using (8.36)

$$E[\eta(t)] = \int_{-\pi}^{\pi} e^{j\omega t} E[dv(\omega)] = 0 \tag{8.48b}$$

Let $H_\eta(\omega)$ be the integrated spectrum of $\eta(t)$. Then by (8.37) we have

$$dH_\eta(\omega) = E[dv^2(\omega)] = \frac{1}{|\phi(\omega)|^2} E[|dz_1(\omega)|^2] = \frac{1}{|\phi(\omega)|^2} dH_1(\omega)$$

$$= \frac{1}{|\phi(\omega)|^2} h(\omega)\, d\omega = \frac{1}{|\phi(\omega)|^2} |\phi(\omega)|^2\, d\omega = d\omega$$

Therefore,

$$h_\eta(\omega) = \frac{dH\eta(\omega)}{d\omega} = 1 \tag{8.48c}$$

and $\eta(t)$ is a white-noise signal.

Equation (8.47) states that $x_1(t)$ is obtained at the output of an LTI system with impulse response $\Phi^*(k)$ and excited by white noise.

CAUSALITY AND THE DISCRETE PALEY–WIENER CONDITION

The system (8.47) is not causal in general. To determine a causal representation, some additional assumptions must be imposed. One such assumption is the discrete version of the **Paley–Wiener condition:**

$$-\infty < P \equiv \int_{-\pi}^{\pi} \ln h(\omega)\, d\omega \tag{8.49}$$

If the latter condition holds there exists a factorization of the form (8.44) such that the Fourier series (8.45) involves positive terms only. Consequently, the impulse response of (8.47) is causal. This is the content of the following theorem.

SPECTRAL FACTORIZATION THEOREM

If condition (8.49) holds then there exists a unique signal $g(t)$, $t = 0, 1, 2, \ldots$, with $g(0)$ real, positive, having finite energy and such that

$$h(\omega) = C\,|G(e^{-j\omega})|^2 \tag{8.50}$$

with C being a constant. The z-transform of $g(t)$, $G(z)$, is analytic outside the unit circle.

It then follows that $x(t)$ can be obtained at the output of the causal filter with impulse response $g(t)$ and with input white noise.

$$x(t) = \sum_{n=0}^{t} g(n)\eta(t-n) \tag{8.51}$$

Important insight into the above theorem is obtained if (8.49) is strengthened, using the following method of Whittle [4]. Consider the z-transform of the autocorrelation sequence

$$h(z) = \sum_{t=-\infty}^{\infty} \rho(t) z^{-t} \tag{8.52}$$

and suppose that ln $h(z)$ exists with a region of convergence that contains the unit circle. This assumption implies (8.49). Indeed, log $h(z)$ is the z-transform of a unique signal $c(t)$

$$\ln h(z) = \sum_{t=-\infty}^{+\infty} c(t)z^{-t} \tag{8.53}$$

Since the unit circle is contained in the region of convergence of ln $h(z)$, $c(t)$ can be recovered by the inverse Fourier transform

$$c(t) = \frac{1}{2\pi}\int_{-\pi}^{\pi} \ln h(\omega)e^{j\omega t}\, d\omega \tag{8.54}$$

and in particular

$$c(0) = \frac{1}{2\pi}\int_{-\pi}^{\pi} \ln h(\omega)\, d\omega \tag{8.55}$$

Thus (8.49) holds. Notice now that

$$h(z) = \exp\left(\sum_{t=-\infty}^{+\infty} c(t)z^{-t}\right) = e^{c(0)}\exp\left(\sum_{t=-\infty}^{-1} c(t)z^{-t}\right)\exp\left(\sum_{t=1}^{+\infty} c(t)z^{-t}\right) \tag{8.56}$$

Since $\rho(t)$ is real and even, $h(\omega)$ is even, and thus we have

$$c(t) = c(-t) \tag{8.57}$$

Therefore, if we define

$$G(z) = \exp\left(\sum_{t=1}^{+\infty} c(t)z^{-t}\right) \tag{8.58}$$

Equation (8.56) becomes

$$h(z) = e^{c(0)}G(z)G(z^{-1}) \tag{8.59}$$

and on the unit circle

$$h(\omega) = e^{c(0)}\,|\,G(e^{j\omega})\,|^2 \tag{8.60}$$

G gives the required spectral factorization. Indeed, by (8.58) $G(z)$ is analytic on the exterior of a disk and in particular it is analytic on the unit circle. Therefore, it involves one-sided terms only. Moreover, it has no zeros outside the unit circle because $1/G(z)$ is analytic on the same region. In other words, $G(z)$ is a *minimum phase* polynomial. If we repeat the arguments leading to (8.47) we obtain the causal representation (8.51). As with equation (8.48c) we find

$$h_\eta(\omega) = \frac{1}{|\,G(e^{-j\omega})\,|^2}\, h(\omega) = e^{c(0)} \tag{8.61}$$

and thus

$$E[\eta^2(t)] = \frac{1}{2\pi}\int_{-\pi}^{\pi} h_\eta(\omega)\, d\omega = e^{c(0)} \tag{8.62}$$

The stability of $G^{-1}(z)$, established above, implies that $\eta(t)$ can be obtained from $x(t)$. Indeed, from (8.51)

$$\eta(t) = G^{-1}(q)x(t)$$

where q denotes the time-shift operator. The filter $G^{-1}(q)$ is also known as the whitening filter and $\eta(t)$ the *innovations series* of $x(t)$. Furthermore, $\eta(t)$ is the optimum prediction error when $x(t)$ is predicted from its infinite past. Define

$$\hat{x}(t) = \sum_{n=1}^{\infty} g(n)\eta(t-n)$$

$\hat{x}(t)$ is the optimum predictor since $x(t) - \hat{x}(t) = g_0\eta(t)$ and this is orthogonal to $\hat{x}(t)$, due to the white nature of $\eta(t)$. Note that $g_0 = 1$. This is readily seen from (8.58) if we let $z \to \infty$. In other words, (8.62) provides the corresponding prediction error power.

We have completed our treatment of the term $x_1(t)$. Next we turn to the term $x_2(t)$. Since $x_2(t)$ has a discrete spectrum the following representation can be established:

$$x_2(t) = \sum_i A_i e^{j\omega_i t} \tag{8.63}$$

where the spectral amplitudes A_i are uncorrelated.

We recapitulate our findings as follows:

> Every WSS process $x(t)$ can be decomposed in three terms (8.43). $x_1(t)$ has a power spectral density and can be obtained at the output of an LTI filter fed with white noise. This filter becomes causal iff the discrete Paley–Wiener condition holds. The part $x_2(t)$ is a harmonic process of the form (8.63) and $x_3(t)$ is the ill-conditioned part with continuous integrated spectrum and derivative zero almost everywhere.

LINEAR PREDICTION

Further important insight into the structure of WSS processes can be obtained by the study of the linear prediction problem. Let $x(t)$ be a WSS process. For each t, let $\mathscr{H}_t$ designate the smallest linear subspace spanned by finite linear combinations of the random variables $\{x(t), x(t-1), \ldots,\}$ and closed under the limits of sequences in this space. For each t, $\mathscr{H}_t$ is a Hilbert space with inner product

$$\langle x, y\rangle = E[xy] \tag{8.64}$$

Let $\hat{x}(t)$ denote the projection of $x(t)$ on to $\mathscr{H}_{t-1}$. Then $\hat{x}(t)$ is expressed as

$$\hat{x}(t) = \sum_{n=1}^{\infty} a_n x(t-n) \tag{8.65}$$

Since $\hat{x}(t)$ is the projection of $x(t)$ on to $\mathscr{H}_{t-1}$, the error

$$e^f(t) = x(t) - \hat{x}(t) \tag{8.66}$$

is orthogonal to $\mathscr{H}_{t-1}$, i.e.

$$\langle x(t) - \hat{x}(t), x(n) \rangle = 0 \qquad n = t-1, t-2, \ldots \tag{8.67}$$

and the distance of $x(t)$ to $\mathscr{H}_{t-1}$ is minimized.

Note that since projection of $x(t)$ is conducted over the entire infinite past history of $x(t-1), x(t-2), \ldots$, the predictor model parameterization is not finite as in the setup of Chapter 2.

Let α^{f} denote the minimum prediction error

$$\alpha^{\mathrm{f}} = E[|e^{\mathrm{f}}(t)|^2] \tag{8.68}$$

Clearly, $\alpha^{\mathrm{f}} \geqslant 0$. If $\alpha^{\mathrm{f}} = 0$, then

$$x(t) = \hat{x}(t) = \sum_{n=1}^{\infty} a_n x(t-n) \tag{8.69}$$

almost everywhere. In this case $x(t)$ is linearly predicted from its past and hence is called a singular or *deterministic* or *totally predictable* signal. If, on the other hand, $\alpha^{\mathrm{f}} > 0$, $x(t)$ is called a *non-deterministic* or *regular process.* The following theorem gives the necessary and sufficient conditions for regularity [5].

THEOREM

A WSS process with PSD $S(\omega)$ is regular if and only if (iff) $S(\omega)$ is positive almost everywhere and satisfies the Paley–Wiener condition, i.e.

$$S(\omega) > 0 \qquad \int_{-\pi}^{\pi} \ln S(\omega)\, d\omega > -\infty$$

The predictive characteristics of a WSS stationary process are illuminated by the following important theorem, i.e. [5].

WOLD DECOMPOSITION

A WSS process $x(t)$ can be decomposed into a regular process $r(t)$ with power spectral density and a deterministic process $s(t)$ as

$$x(t) = r(t) + s(t) \tag{8.70}$$

where (a) $r(t)$ and $s(t)$ are uncorrelated processes, and (b) combining the above theorem, the discussion on causality and the Paley–Wiener condition indicates that $r(t)$ is obtained by filtering white noise $\eta(t)$, which is uncorrelated with $s(t)$

$$r(t) = \sum_{k=0}^{\infty} g(k)\eta(t-k) \qquad E[\eta(t)s(\tau)] = 0 \quad \forall t, \tau \tag{8.71}$$

Moreover, the filter impulse response $g(t)$ and the white-noise signal $\eta(t)$ are uniquely determined from the signal $x(t)$.

Let us now combine the above decomposition with the spectral representation (8.43). It follows from (8.63) that $x_2(t)$ is predictable. Moreover, $x_3(t)$ is singular because the derivative of its integrated spectrum vanishes almost everywhere. The part $x_1(t)$ may or may not be predictable, depending on the Paley–Wiener condition. If (8.49) holds, the total prediction error with the aid of (8.62) turns out to be

$$\alpha^{\mathrm{f}} = \mathrm{e}^{P/2\pi} \tag{8.72}$$

Hence $\alpha^{\mathrm{f}} > 0$ and $x_1(t)$ is regular. The opposite is also true; that is, if $P = -\infty$, then $\alpha^{\mathrm{f}} = 0$ and $x_1(t)$ is predictable. In summary,

$$r(t) = x_1(t) \qquad s(t) = x_2(t) + x_3(t) \qquad \text{if } P > -\infty \tag{8.73a}$$

$$r(t) = 0 \qquad s(t) = x(t) \qquad \text{if } P = -\infty \tag{8.73b}$$

EXAMPLE: BANDLIMITED SIGNALS

A continuous bandlimited signal has a power spectral density which vanishes outside a frequency interval. Consequently, the integral (8.49) tends to minus infinity and the signal is singular. In a similar fashion, a mixed bandlimited signal (see Figure 8.1) is singular as a consequence of (8.73b).

EXAMPLE: AR, MA AND ARMA MODELS

Let us consider the ARMA process (8.28). Since the z-transform of the logarithm of the autocorrelation sequence ln $S(z)$ is analytic on a region containing the unit circle, the discrete Paley–Wiener condition holds. Therefore, the ARMA process is a special case of a regular process with rational spectrum. In this case it is also easy to derive explicit expressions for the spectral factors of the power spectral density. Thus we have

$$S(z) = \frac{N(z)N(z^{-1})}{D(z)D(z^{-1})} \qquad z = \mathrm{e}^{\mathrm{j}\omega} \tag{8.74}$$

where $N(z)$ and $D(z)$ are minimum phase polynomials, i.e. they have their zeros inside the unit circle.

Indeed, it follows from (8.74) that if z_0 is a zero of the numerator polynomial, so is z_0^{-1}, and similarly for the denominator. Hence we may factor $S(z)$ as

$$S(z) = C\,\frac{\Pi_k(1-\mu_k z)(1-\mu_k z^{-1})}{\Pi_k(1-\lambda_k z)(1-\lambda_k z^{-1})} \tag{8.75}$$

where $|\mu_k| < 1$, $|\lambda_k| < 1$ for all k and C is a constant. The factor $G(z)$ is then

$$G(z) = \frac{\Pi_k(1-\mu_k z)}{\Pi_k(1-\lambda_k z)}\,C^{1/2} \tag{8.76}$$

AR and MA processes, being special cases of ARMA signals, are also regular processes.

8.3 Parametric models and autoregressive spectral estimation

The basic parametric model we shall be concerned with is the autoregressive model. The AR(p) model with parameters (p, σ^2) has the form

$$y(t) + \sum_{k=1}^{p} a_k y(t-k) = \eta(t) \tag{8.77}$$

$\eta(t)$ is a white-noise signal with variance σ^2. AR models appear to be the most popular among parametric models in spectral estimation. A major reason for their popularity is that they lead to linear regressions (see Chapter 2) and hence to linear system solvers. Once model (8.77) is specified the power spectral density is computed by the formula

$$S_y(\omega) = \frac{\sigma^2}{|A(e^{j\omega})|^2} \tag{8.78}$$

where

$$A(z) = 1 + a_1 z^{-1} + \ldots + a_p z^{-p} \tag{8.79}$$

AR processes are special cases of ARMA processes. They are obtained by setting the moving average part of the ARMA model equal to zero. We recall from Chapter 2 that ARMA models do not lead to linear equations. A possibility is to determine the ARMA coefficients and then the power spectral density (8.29) through some variant of the prediction error algorithm, presented in Chapter 2. In what follows we will establish the relationship between the parameters of an ARMA model and its equivalent AR(∞) model. This is of interest to suboptimal schemes, where the AR and MA parts of the ARMA model are estimated *separately*. Furthermore it allows the approximation of the ARMA parameters by an equivalent 'long enough' AR model.

Consider the ARMA model (8.28) and the infinite-order AR model

$$y(t) + \sum_{k=1}^{\infty} g_k y(t-k) = \eta(t) \tag{8.80}$$

or

$$G(q)y(t) = \eta(t) \tag{8.81}$$

where q^{-1} is the time-shift operator and

$$G(q) = 1 + g_1 q^{-1} + g_2 q^{-2} + \ldots \tag{8.82}$$

The systems (8.28) and (8.80) produce the same input–output map if

$$C(q)G(q) = A(q) \tag{8.83}$$

Equating coefficients, we obtain

$$g_n + \sum_{k=1}^{n_c} c_k g_{n-k} = a_n \qquad n \leqslant n_a \tag{8.84}$$

$$g_n + \sum_{k=1}^{n_c} c_k g_{n-k} = 0 \quad n > n_a \tag{8.85}$$

Thus, knowing the ARMA coefficients a_k and c_k, we can compute the AR coefficients g_k from the above difference equations. Initial values $g_{-nc}, \ldots, g_{-1}$ are set equal to zero.

Conversely, if the coefficients g_k of an AR(∞) model are given, the equivalent ARMA (n_a, n_c) parameters are specified from the following set of linear equations, as (8.85) suggests:

$$\begin{pmatrix} g_{n_a} & g_{n_a-1} & \cdots & g_{n_a-n_c+1} \\ g_{n_a+1} & g_{n_a} & \cdots & g_{n_a-n_c+2} \\ \vdots & \vdots & & \vdots \\ g_{n_a+n_c-1} & g_{n_a+n_c-2} & \cdots & g_{n_a} \end{pmatrix} \begin{pmatrix} c_1 \\ c_2 \\ \vdots \\ c_{n_c} \end{pmatrix} = - \begin{pmatrix} g_{n_a+1} \\ g_{n_a+2} \\ \vdots \\ g_{n_a+n_c} \end{pmatrix} \tag{8.86}$$

and the AR part of the ARMA model is in turn provided from

$$\begin{pmatrix} g_1 & 1 & 0 & \cdots & 0 \\ g_2 & g_1 & 1 & \cdots & 0 \\ \vdots & \vdots & \vdots & & \vdots \\ g_{n_a} & g_{n_a-1} & g_{n_a-2} & \cdots & g_{n_a-n_c} \end{pmatrix} \begin{pmatrix} 1 \\ c_1 \\ \vdots \\ c_{n_c} \end{pmatrix} = \begin{pmatrix} a_1 \\ a_2 \\ \vdots \\ a_{n_a} \end{pmatrix} \tag{8.87}$$

which is the expansion of (8.84). We shall return to ARMA modelling in a later section. Next we will focus on AR spectral analysis.

Let us consider an AR model of order p, generating the WSS process $y(t)$. If we multiply both sides of (8.77) with $y(t-m)$ and take expectations we obtain

$$\rho(m) = - \sum_{k=1}^{p} a_k \rho(m-k) \qquad m > 0 \tag{8.88}$$

Equation (8.88) written in matrix form yields the Yule–Walker equations

$$\begin{pmatrix} \rho(0) & \rho(1) & \cdots & \rho(p-1) \\ \rho(1) & \rho(0) & \cdots & \rho(p-2) \\ \vdots & \vdots & & \vdots \\ \rho(p-1) & \rho(p-2) & \cdots & \rho(0) \end{pmatrix} \begin{pmatrix} a_1 \\ a_2 \\ \vdots \\ a_p \end{pmatrix} = \begin{pmatrix} \rho(1) \\ \rho(2) \\ \vdots \\ \rho(p) \end{pmatrix} \tag{8.89}$$

This equation set coincides with the normal equations for the Wiener solution discussed in Chapter 2; see Equation (2.100).

A straightforward outcome of the Levinson algorithm (Equations (2.106), (2.118) and (2.119)) is that the sequence of prediction powers is non-increasing:

$$\alpha_m^{\mathrm{f}} \geqslant \alpha_{m+1}^{\mathrm{f}} > 0 \qquad m = 1, 2, \ldots \tag{8.90}$$

This is reasonable since longer models accommodate longer time history into the prediction process. Clearly, the prediction power sequence converges:

$$\alpha_m^{\mathrm{f}} \to \alpha^{\mathrm{f}} \qquad m \to \infty \tag{8.91}$$

If the predicted sequence $y(t)$ obeys an AR model of order p, the above sequence stabilizes at α_p^{f}, i.e.

$$\alpha_k^{\mathrm{f}} = \alpha_p^{\mathrm{f}} \qquad k = p+1, p+2, \ldots \qquad \alpha^{\mathrm{f}} = \alpha_p^{\mathrm{f}} \tag{8.92}$$

This is a consequence of (8.88) and (8.89). Indeed, the following set of equations can be readily verified:

$$\begin{pmatrix} \rho(0) & \rho(1) & \cdots & \rho(p-1) & \rho(p) \\ \rho(1) & \rho(0) & \cdots & \rho(p-2) & \rho(p-1) \\ \vdots & \vdots & & \vdots & \vdots \\ \rho(p-1) & \rho(p-2) & \cdots & \rho(0) & \rho(1) \\ \rho(p) & \rho(p-1) & \cdots & \rho(1) & \rho(0) \end{pmatrix} \begin{pmatrix} a_1 \\ a_2 \\ \vdots \\ a_p \\ 0 \end{pmatrix} = \begin{pmatrix} \rho(1) \\ \rho(2) \\ \vdots \\ \rho(p) \\ \rho(p+1) \end{pmatrix} \tag{8.93}$$

Note, however, that (8.93) is the Yule–Walker equation for the predictor of order $p+1$. Thus we conclude that the predictor of order $p+1$ is formed by appending an extra zero to the pth-order predictor. Generalization to higher orders is easily established. Going back to the Levinson algorithm, we can show that the above is equivalent to having

$$k_m = 0 \quad m = p, p+1, \ldots \tag{8.94}$$

This equation, together with (2.120) of Chapter 2, leads to (8.92).

The preceding discussion indicates that all the predictable information regarding an AR process of order p is enclosed in p previous samples and no improvement results if higher-order predictors are called. This conforms with the property that such a process is a WSS Markov process of order p, namely

$$E[y(t) \mid y(t-1), y(t-2), \ldots] = E[y(t) \mid y(t-1), \ldots, y(t-p)]$$

Next we exploit the above results in a spectral analysis context. We are given $p+1$ values $\rho(0), \ldots \rho(p)$ of the autocorrelation sequence of a process $y(t)$, and we seek to estimate the power spectrum $S(\omega)$ of $y(t)$. Conventional Fourier methods solve the problem by extending the autocorrelation sequence with zeros. However, this leads to various windowing effects, as discussed in the introduction, and their implicit limitations, as demonstrated in a number of cases [2]. Another alternative is to extrapolate the unknown values of the autocorrelation, i.e.. $\rho(\tau)$, $\tau > p$, by assuming a model describing the stochastic process. Again, such a model is not uniquely defined and there are different ways of achieving this goal. Going back to the spectral representation theorem, one may assume that the process consists of spectral lines implying a certain type of autocorrelation extrapolation of the form (8.24) [6, page 489]. Another possibility is to assume that the process is a regular one with a further restriction that the spectrum is rational. This leads to the three alternatives, namely ARMA, MA and AR modelling. The extent to which the resulting spectra will resemble the true one depends on how well the adopted assumption represents the generation mechanism of the stochastic process. If the latter is a mixed process, it further complicates the problem and different methods may be more appropriate for different parts of the spectrum, thus making the decision of the best method overall not a straightforward task. This has been demonstrated with an example in [7]. Below, we will concentrate on AR modelling; it is quite popular as a result of the linearity of the underlying equations and its satisfactory description of a wide class of signals encountered in practice.

The AR assumption is equivalent to extending the autocorrelation sequence according to the difference equation (8.88) and with the known autocorrelation lags as initial conditions. An equivalent assumption is that the prediction power, $\alpha_m^{\mathrm{f}} > 0$, is not reduced by assuming prediction models greater than p, i.e. $\alpha_m^{\mathrm{f}} = \alpha_p^{\mathrm{f}} \equiv \alpha^{\mathrm{f}}$, for $m = p+1, p+2, \ldots$. This has an interesting implication. For a general *regular* process the whole infinite past is required to obtain the predictable part of the process and minimum prediction error is the power of the equivalent innovation series [6, page 297]. For such a general process, imposing (for practical reasons) the AR assumption (i.e. prediction based on the finite past) is equivalent to maximizing the minimum attainable prediction error power, subject to the constraint imposed by the known autocorrelation lags. Indeed, the pth-order optimum predictor is uniquely determined by (8.89) and the minimum prediction error power α_p^{f} is determined, also uniquely, by the Levinson algorithm. The particular choice of the unknown autocorrelation lags will determine the values of the non-increasing sequence of error powers

$$\alpha_{p-1}^{\mathrm{f}} \geqslant \alpha_p^{\mathrm{f}} \geqslant \ldots \geqslant \rightarrow \alpha^{\mathrm{f}} \qquad p \rightarrow \infty$$

The choice (8.88), corresponding to AR modelling of order p, implicitly assigns all values of the above sequence equal to α_p^{f} (Equation (8.92)). In other words, adopting an AR model is equivalent to maximizing the minimum prediction error power. This in turn has another implication. According to (8.62) the minimum attainable error power is given by

$$E[\eta^2(t)] = \exp\left\{\frac{1}{2\pi}\int_{-\pi}^{\pi} \ln h(\omega)\, \mathrm{d}\omega\right.$$

Hence, maximizing the minimum error power is equivalent to maximizing

$$H = \int_{-\pi}^{\pi} \ln h(\omega)\, \mathrm{d}\omega$$

For a zero-mean Gaussian process, H reduces to the entropy rate of the process. Hence, extrapolation of the autocorrelation sequence according to the AR assumption is equivalent to estimating the power spectral density $h(\omega)$ so that the entropy rate of the Gaussian signal is maximized. A further interesting consequence is that each of the unknown lags is specified so as to maximize the determinant of the autocorrelation matrix, subject to the known lags constraint [8]. Recently, the above arguments have been extended to the non-stationary case, where the autocorrelation matrix is not Toeplitz. This has led to a new class of adaptive estimation algorithms [9].

To conclude, we recapitulate that given the set of $p+1$ first autocorrelation lags, $\rho(0), \ldots, \rho(p)$, the AR spectral estimate is provided by (8.78) and (8.79) with the parameters provided by (8.89). σ^2 is the prediction power $\alpha_p^{\rm f}$ and is a byproduct of the Levinson algorithm for solving (8.89).

8.4 AR spectral estimation based on observation data

In the previous section we assumed that the first $p + 1$ autocorrelation lags were known and that the AR model parameters resulted from them by the Levinson/Schur algorithms. In practice, however, the correlation lags are unknown. What is available is a finite set of observation samples from a single realization of the signal. The aim of this section is to describe the methods and algorithms used to estimate the AR spectra from this limited set of data. Both block and time recursive schemes are considered.

8.4.1 Schemes based on the Toeplitz structure of the autocorrelation matrix

Algorithms exploiting the Toeplitz structure of the autocorrelation matrix relating the first $p + 1$ autocorrelation lags and the AR parameters in a one-to-one correspondence were presented in Chapter 2. The following two types of algorithmic methods are naturally associated with these schemes.

AUTOCORRELATION METHOD

According to this scheme, sometimes known as the Yule–Walker approach, the autocorrelation lags are computed from the available data set by means of the estimates

$$\hat{\rho}(m) = \frac{1}{N} \sum_{n=0}^{N-m} y(n)y(n+m) \qquad m = 1, 2, \ldots p \tag{8.95}$$

These estimates ensure that the autocorrelation matrix is positive definite, a necessary requirement. However, this method inherits the problems of windowing, since data beyond N are assumed to be zero. This affects the resolution which, for small N, can be particularly poor with respect to the other AR methods described below [1,2].

BURG'S METHOD

Burg exploited the Levinson algorithm in a reverse fashion. He first computed the reflection coefficients directly from the data set in the interval $[M, N]$. Then he used the order recursion of the Levinson algorithm to pass to the AR tap weights. In more precise terms, the algorithm proceeds along the following steps:

- Assume that $k_1, k_2, \ldots, k_{m-2}$ are known from previous recursions.
- Use the lattice realization of the AR model at stage m:

$$e_m^{\rm f}(n) = e_{m-1}^{\rm f}(n) + k_{m-1}e_{m-1}^{\rm b}(n-1) \tag{8.96a}$$

$$e_m^{\rm b}(n) = e_{m-1}^{\rm b}(n-1) + k_{m-1}e_{m-1}^{\rm f}(n) \tag{8.96b}$$

• Compute k_{m-1} from an LS problem by minimizing the total squared error of order m over the given interval. The total squared error is the sum of the forward plus backward errors. A point that has to be emphasized here is that the resulting minimum is not the global minimum obtained if minimization is carried out with respect to all reflection coefficients $k_1, k_2, \ldots, k_{m-2}$. The latter problem is non-linear and can be addressed using iterative techniques [10].

The total squared error at stage m is

$$\mathscr{E}_m = \frac{1}{2}\left(\sum_{n=M+m}^{N} (e_m^{\mathrm{f}}(n))^2 + \sum_{n=M+m}^{N} (e_m^{\mathrm{b}}(n))^2 \right) \tag{8.97}$$

The lower limit in the summation ensures that only known data in the interval $[M, N]$ are used; thus no assumption is made about unavailable data samples. Thus, Burg's method frees itself of any windowing assumptions. This is also within the spirit of the maximum entropy approach where the resulting estimate is maximally non-committal with regard to the unavailable information [11]. Another point worth stressing is that the sum of forward and backward errors is minimized. In a stationary environment, the corresponding mean squared errors are equal. However, in the finite data case under consideration, this is not true. Minimization of (8.97) with respect to k_{m-1} gives the following values for the reflection coefficient:

$$\hat{k}_{m-1} = \frac{-2\Sigma_{n=M+m}^{N} e_{m-1}^{\mathrm{f}}(n)e_{m-1}^{\mathrm{b}}(n-1)}{\Sigma_{n=M+m}^{N}(e_{m-1}^{\mathrm{f}}(n))^2 + \Sigma_{n=M+m}^{N}(e_{m-1}^{\mathrm{b}}(n-1))^2} \tag{8.98}$$

It is not difficult to see that $\hat{k}_{m-1} \leqslant 1$.

Burg's method offers improved resolution properties with respect to the autocorrelation method. On the other hand, Burg's method suffers from two disadvantages:

1. *Spectral line splitting:* For sinusoidal plus noise signals the spectrum may exhibit two spectral lines instead of the true one. Spectral line splitting is more severe when (a) the SNR is high, (b) AR orders are relatively long with respect to the number of data samples used, (c) the initial phase of sinusoids is an odd multiple of 45° and (d) data window length is an odd multiple of quarter sinusoidal cycles. A justification for line splitting is provided in [12].
2. *Bias sensitivity with respect to phase:* Burg's method is very sensitive in detecting the position of a frequency peak. Experiments in [13] have shown that as the phase of the sinusoid varies, the bias in frequency position exhibits a sinusoidal-like variation and errors of as much as 16% of resolution cells Δf (defined as $\Delta f \approx 1/(N - M + 1)$) can occur. A justification for this variation can be found in [14]. Table 8.1 summarizes Burg's algorithm. $w_m(n)$ is a window function incorporated in (8.98). It has been suggested, i.e. [15], that the use of a window function is effective in reducing the occurrence of line splitting and also reducing the frequency bias. If no windowing is used the denominator (D_{m-1}) in (8.98) can be updated recursively as

Table 8.1 Burg's algorithm

INITIALIZATION

$e_0^{\mathrm{f}}(n) = e_0^{\mathrm{b}}(n) = y(n),\ n = M, \ldots, N$

END

FOR $m = 1, 2, \cdots, p$

(1) $\hat{k}_{m-1} = \dfrac{-\sum_{n=M+m}^{N} e_{m-1}^{\mathrm{f}}(n) e_{m-1}^{\mathrm{b}}(n-1)}{\sum_{n=M+m}^{N} [(e_{m-1}^{\mathrm{f}}(n))^2 + (e_{m-1}^{\mathrm{b}}(n-1))^2]}$

(2) $\mathbf{a}_m = \begin{bmatrix} \mathbf{a}_{m-1} \\ \mathbf{0} \end{bmatrix} + \begin{bmatrix} \mathbf{1} \\ J\mathbf{a}_{m-1} \end{bmatrix} \hat{k}_{m-1}$

FOR $n = M + m, \ldots, N$

(3) $e_m^{\mathrm{f}}(n) = e_{m-1}^{\mathrm{f}}(n) + \hat{k}_{m-1} e_{m-1}^{\mathrm{b}}(n-1)$

(4) $e_m^{\mathrm{b}}(n) = e_{m-1}^{\mathrm{b}}(n-1) + \hat{k}_{m-1} e_{m-1}^{\mathrm{f}}(n)$

END

END

$$D_m = (1 - k_{m-1}^2) D_{m-1} - (e_m^{\mathrm{f}}(M+m))^2 - (e_m^{\mathrm{b}}(N))^2 \qquad m = 1, 2, \ldots$$

This is a direct consequence of the definition in (8.97), the lattice recursions (8.96) and Burg's formula (8.98).

8.4.2 Least-squares method

The setup of the previous section utilized mean squared optimality and the associated lattice recursions (8.96a) and (8.96b). Least-squares optimization was applied locally for each reflection coefficient k_m. In this section we will adopt the total squared error over the finite data interval $[M, N]$ as the performance index. In accordance with the notation of Chapter 2 we consider the cost function

$$V_N(\mathbf{a}_p) = \sum_{t=M+p}^{N} e_p^2(t) \tag{8.99}$$

where for AR(p) model fitting we have

$$e_p(t) = y(t) + \sum_{k=1}^{p} a_k y(t-k) \tag{8.100}$$

The range of summation in (8.99) is chosen so that only available samples enter the minimization process. Equation (2.151) of Chapter 2 now becomes

$$R_p(M, N-1)\mathbf{a}_p(M, N) = -\mathbf{r}_p^{\mathrm{f}}(M, N) \tag{8.101}$$

where both M and N are used to indicate the observation interval $[M, N]$. The pertinent parameters are

$$R_p(M, N-1) = \sum_{t=M+p-1}^{N-1} \mathbf{y}_p(t)\mathbf{y}_p^{\mathrm{T}}(t) \tag{8.102}$$

$$\mathbf{r}_p^{\mathrm{f}}(M, N) = \sum_{t=M+p-1}^{N-1} \mathbf{y}_p(t)y(t+1) \tag{8.103}$$

In what follows we will make use of the definitions of $R_m(M, N)$ and $\mathbf{r}_m^{\mathrm{f}}(M, N)$ in (8.102) and (8.103) for $m = 1, 2, \ldots, p$. Note that $R_m(M, N)$ are symmetric and non-negative definite. We shall assume that they are positive definite, and hence invertible, and we shall derive a Levinson-type scheme for the solution of (8.101). Owing to the shift invariance of the regressor vector the following properties hold:

$$\begin{aligned} R_m(M, N) &= \sum_{t=M+m-1}^{N} \begin{pmatrix} \mathbf{y}_{m-1}(t) \\ y(t-m+1) \end{pmatrix} (\mathbf{y}_{m-1}^{\mathrm{T}}(t) \; y(t-m+1)) \\ &= \sum_{t=M+m-1}^{N} \begin{pmatrix} y(t) \\ \mathbf{y}_{m-1}(t-1) \end{pmatrix} (y(t) \; \mathbf{y}_{m-1}(t-1)) \end{aligned} \tag{8.104}$$

or for order $m+1$

$$R_{m+1}(M, N) = \begin{pmatrix} R_m(M+1, N) & \mathbf{r}_m^{\mathrm{b}}(M, N) \\ \mathbf{r}_m^{\mathrm{bT}}(M, N) & r_{0m}^{\mathrm{b}}(M, N) \end{pmatrix} \tag{8.105a}$$

$$R_{m+1}(M, N) = \begin{pmatrix} r_{0m}^{\mathrm{f}}(M, N) & \mathbf{r}_m^{\mathrm{fT}}(M, N) \\ \mathbf{r}_m^{\mathrm{f}}(M, N) & R_m(M, N-1) \end{pmatrix} \tag{8.105b}$$

with

$$\mathbf{r}_m^{\mathrm{b}}(M, N) = \sum_{t=M+m}^{N} \mathbf{y}_m(t)y(t-m) \tag{8.106a}$$

$$r_{0m}^{\mathrm{b}}(M, N) = \sum_{t=M+m}^{N} y(t-m)y(t-m) \tag{8.106b}$$

$$r_{\mathrm{om}}^{\mathrm{f}}(M, N) = \sum_{t=M+m}^{N} y(t)y(t) \tag{8.106c}$$

These expressions are generalizations of the formulae presented in Chapter 2 for the prewindowed case. From (8.102) the following time updates are readily derived:

$$R_m(M, N) = R_m(M, N-1) + \mathbf{y}_m(N)\mathbf{y}_m^{\mathrm{T}}(N) \tag{8.107a}$$

$$R_m(M+1, N) = R_m(M, N) - \mathbf{y}_m(M+m-1)\mathbf{y}_m^{\mathrm{T}}(M+m-1) \tag{8.107b}$$

We have collected all the necessary ingredients to derive efficient order recursive schemes for the solution of (8.101). The key point is the relationship between the upper and lower partitionings of $R_{m+1}(M, N)$ in (8.105). For Toeplitz matrices these are equal. In (8.105) the two resulting matrices are different, however, and are related by (8.107). Depending on the way the nesting properties of (8.105) are exploited, respective algorithmic variants result. One possibility is that both $R_m(M+1, N)$ and $R_m(M, N-1)$ are corrected back to $R_m(M, N)$ through the rank 1 corrections, as (8.107) suggests. A second alternative is based on the update

$$R_m(M, N-1) = R_m(M+1, N) + \mathbf{y}_m(M+m-1)\mathbf{y}_m^{\mathrm{T}}(M+m-1) - \mathbf{y}_m(N)\mathbf{y}_m^{\mathrm{T}}(N) \tag{8.108}$$

This is different: the lower partitioning is corrected back to the upper partitioning through a rank 2 correction. This leads to a different algorithmic structure where successive partitionings are now expressed in terms of lower-order parameters defined on successively smaller intervals, i.e. $R_m(M+1, N)$, $R_{m-1}(M+2, N)$, $R_{m-2}(M+3, N)$, etc.

Relations (8.107), as well as the correction form (8.108), are special cases of a more general type of well-structured matrices for which the difference between the upper submatrix R_m and the lower submatrix S_m is a low-rank matrix, that is,

$$S_m - R_m = \sum_{k=1}^{\rho} \mathbf{v}_k \mathbf{v}_k^{\mathrm{T}} \qquad m = 1, 2, \ldots, p-1 \tag{8.109}$$

where $\mathbf{v}_k$ are m-dimensional vectors and ρ is small compared with p; ρ is called the Toeplitz distance of the near-to-Toeplitz matrix R_p. It follows from (8.108) that $R_m(M, N)$ is near-to-Toeplitz with distance 2. It is not difficult to see that this definition coincides with that of α_0 in Equation (5.39) of Chapter 5.

In the rest of the section we will deal with algorithms belonging to the former type. We will see that these schemes provide the predictors of all orders $m = 1, 2, \ldots, p$ that are optimal in the same interval $[M, N]$. This is of importance when the exact order of the system is not known; hence monitoring prediction power provides a criterion for determining a good predictor order. On the other hand, the second near-to-Toeplitz approach provides optimal estimates in different intervals, thus making such a criterion meaningless. However, the overall complexity of such schemes is lower [16].

LEVINSON-TYPE (COVARIANCE) ALGORITHM

A Levinson-type algorithm for the solution of (8.101) first appeared in [17,18] and was later improved in [19]. This improved version is presented in Table 8.2. The proof is supplied in Appendix A and consists of repeated applications of the matrix inversion lemma on (8.104)–(8.107). To gain insight into the structure of the algorithm's parameters we extract the following expressions from Table 8.2:

Table 8.2 Covariance algorithm

INITIALIZATION

$S = \Sigma_{t=M+1}^{N-1} y(t)y(t)$

$R_1(M, N-1) = y(M)y(M) + S$

$R_1(M, N) = R_1(M, N-1) + y(N)y(N)$

$R_1(M+1, N) = S + y(N)y(N)$

$r_1^{\mathrm{f}}(M, N) = \Sigma_{t=M}^{N-1} y(t)y(t+1)$

$r_1^{\mathrm{b}}(M, N) = \Sigma_{t=M+1}^{N} y(t)y(t-1)$

$q_1(M, N) = -R_1^{-1}(M, N)y(M)$

$\alpha_1^q(M, N) = 1 + y(M)q_1(M, N)$

$w_1(M, N) = -R_1^{-1}(M, N-1)y(N)$

$\alpha_1^*(M, N) = 1 - w_1(M, N)y(N)$

$a_1(M, N) = -R_1^{-1}(M, N-1)r_1^{\mathrm{f}}(M, N)$

$b_1(M, N) = -R_1^{-1}(M+1, N)r_1^{\mathrm{b}}(M, N)$

$\alpha^{\mathrm{b}}(M, N) = y(M)y(M) + S + b_1(M, N)r_1^{\mathrm{b}}(M, N)$

$\alpha^{\mathrm{f}}(M, N) = S + y(N)y(N) + a_1(M, N)r_1^{\mathrm{f}}(M, N)$

$\star r_1(M, N) = \Sigma_{t=M}^{N} y(t)z(t)$

$\star c_1(M, N) = -R_1^{-1}(M, N)r_1(M, N)$

$\star \alpha_1^z = \Sigma_{t=M}^{N} z(t)z(t) + c_1(M, N)r_1(M, N)$

END

FOR $m = 1$ TO p DO

(1) $\mathbf{r}_m^{\mathrm{b}}(M, N-1) = \mathbf{r}_m^{\mathrm{b}}(M, N) - \mathbf{y}_m(N)y(N-m)$

(★2) $\tilde{\beta}_m(M, N) = \mathbf{y}_m^{\mathrm{T}}(M+m-1)\mathbf{c}_m(M, N) + z(M+m-1)$

(★3) $\tilde{k}_m(M, N) = -\alpha_m^{-q}(M, N)\tilde{\beta}_m(M, N)$

(★4) $\mathbf{c}_m(M+1, N) = \mathbf{c}_m(M, N) + \mathbf{q}_m(M, N)\tilde{k}_m(M, N)$

(★5) $\alpha_m^z(M+1, N) = \alpha_m^z(M, N) + \tilde{\beta}_m(M, N)\tilde{k}_m(M, N)$

(6) $\tilde{\beta}_m^w(M, N) = \mathbf{w}_m^{\mathrm{T}}(M, N)\mathbf{y}_m(M+m-1)$

(7) $\tilde{k}_m^q(M, N) = -\alpha_m^{*-1}(M, N)\tilde{\beta}_m^w(M, N)$

(8) $\mathbf{q}_m(M, N-1) = \mathbf{q}_m(M, N) + \mathbf{w}_m(M, N)\tilde{k}_m^q(M, N)$

(9) $\alpha_m^q(M, N-1) = \alpha_m^q(M, N) + \tilde{\beta}_m^w(M, N)\tilde{k}_m^q(M, N)$

(10) $\tilde{k}_m^w(M, N) = -\alpha_m^{-q}(M, N-1)\tilde{\beta}_m^w(M, N)$

(11) $\mathbf{w}_m(M+1, N) = \mathbf{w}_m(M, N) + \mathbf{q}_m(M, N-1)\tilde{k}_m^w(M, N)$

(12) $\tilde{\beta}_m^{\mathrm{f}}(M, N) = \mathbf{a}_m^{\mathrm{T}}(M, N)\mathbf{y}_m(M+m-1) + y(M+m)$

(13) $\tilde{k}_m^{\mathrm{f}}(M, N) = -\alpha_m^{-q}(M, N-1)\tilde{\beta}_m^{\mathrm{f}}(M, N)$

(14) $\mathbf{a}_m(M+1, N) = \mathbf{a}_m(M, N) + \mathbf{q}_m(M, N-1)\tilde{k}_m^{\mathrm{f}}(M, N)$

(15) $\alpha_m^*(M+1, N) = \alpha_m^*(M, N) - \tilde{\beta}_m^w(M, N)\tilde{k}_m^w(M, N)$

(16) $\tilde{\beta}_m^{\mathrm{b}}(M, N) = \mathbf{w}_m^{\mathrm{T}}(M+1, N)\mathbf{r}_m^{\mathrm{b}}(M, N-1) + y(N-m)$

(17) $\tilde{k}^{\mathrm{b}}_m(M, N) = -\alpha^{*-1}_m(M+1, N)\tilde{\beta}^{\mathrm{b}}_m(M, N)$

(18) $\mathbf{b}_m(M, N-1) = \mathbf{b}_m(M, N) + \mathbf{w}_m(M+1, N)\tilde{k}^{\mathrm{b}}_m(M, N)$

(19) $\alpha^{\mathrm{b}}_m(M, N-1) = \alpha^{\mathrm{b}}_m(M, N) + \tilde{\beta}^{\mathrm{b}}_m(M, N)\tilde{k}^{\mathrm{b}}_m(M, N)$

(20) $\alpha^{\mathrm{f}}_m(M+1, N) = \alpha^{\mathrm{f}}_m(M, N) + \tilde{\beta}^{\mathrm{f}}_m(M, N)\tilde{k}^{\mathrm{f}}_m(M, N)$

(21) $r^{\mathrm{b}}_{m+1}(M, N) = \Sigma^N_{t=M+m+1} y(t)y(t-m-1)$

(22) $\beta^{\mathrm{b}}_m(M, N) = \mathbf{a}^{\mathrm{T}}_m(M+1, N)\mathbf{r}^{\mathrm{b}}_m(M, N-1) + r^{\mathrm{b}}_{m+1}(M, N)$

(23) $k^{\mathrm{b}}_m(M, N) = -\alpha^{-\mathrm{f}}_m(M+1, N)\beta^{\mathrm{b}}_m(M, N)$

(24) $k^{\mathrm{f}}_m(M, N) = -\alpha^{-\mathrm{b}}_m(M, N-1)\beta^{\mathrm{b}}_m(M, N)$

(25) $\mathbf{b}_{m+1}(M, N) = \begin{bmatrix} 0 \\ \mathbf{b}_m(M, N-1) \end{bmatrix} + \begin{bmatrix} 1 \\ \mathbf{a}_m(M+1, N) \end{bmatrix} k^{\mathrm{b}}_m(M, N)$

(26) $\mathbf{a}_{m+1}(M, N) = \begin{bmatrix} \mathbf{a}_m(M+1, N) \\ 0 \end{bmatrix} + \begin{bmatrix} \mathbf{b}_m(M, N-1) \\ 1 \end{bmatrix} k^{\mathrm{f}}_m(M, N)$

(27) $\mathbf{r}^{\mathrm{b}}_{m+1}(M, N) = \begin{bmatrix} r^{\mathrm{b}}_m(M, N) \\ \mathbf{r}^{\mathrm{b}}_m(M, N-1) \end{bmatrix}$

(28) $k^{q}_m(M, N) = -\alpha^{-\mathrm{f}}_m(M, N)\tilde{\beta}^{\mathrm{f}}_m(M, N)$

(29) $k^{w}_m(M, N) = -\alpha^{-\mathrm{b}}_m(M, N-1)\tilde{\beta}^{\mathrm{b}}_m(M, N)$

(30) $\mathbf{q}_{m+1}(M, N) = \begin{bmatrix} 0 \\ \mathbf{q}_m(M, N-1) \end{bmatrix} + \begin{bmatrix} 1 \\ \mathbf{a}_m(M, N) \end{bmatrix} k^{q}_m(M, N)$

(31) $\mathbf{w}_{m+1}(M, N) = \begin{bmatrix} \mathbf{w}_m(M+1, N) \\ 0 \end{bmatrix} + \begin{bmatrix} \mathbf{b}_m(M, N-1) \\ 1 \end{bmatrix} k^{w}_m(M, N)$

(★32) $r_{m+1}(M, N) = \Sigma^N_{t=M+m} y(t-m)z(t)$

(★33) $\beta_m(M, N) = \mathbf{c}^{\mathrm{T}}_m(M+1, N)\mathbf{r}^{\mathrm{b}}_m(M, N) + r_{m+1}(M, N)$

(★34) $k_m(M, N) = -\alpha^{-\mathrm{b}}_m(M, N)\beta_m(M, N)$

(★35) $\mathbf{c}_{m+1}(M, N) = \begin{bmatrix} \mathbf{c}_m(M+1, N) \\ 0 \end{bmatrix} + \begin{bmatrix} \mathbf{b}_m(M, N) \\ 1 \end{bmatrix} k_m(M, N)$

(★36) $\alpha^{z}_{m+1}(M, N) = \alpha^{z}_m(M+1, N) + \beta_m(M, N)k_m(M, N)$

(37) $\alpha^{*}_{m+1}(M, N) = \alpha^{*}_m(M+1, N) - \tilde{\beta}^{\mathrm{b}}_m(M, N)k^{w}_m(M, N)$

(38) $\alpha^{q}_{m+1}(M, N) = \alpha^{q}_m(M, N-1) + \tilde{\beta}^{\mathrm{f}}_m(M, N)k^{q}_m(M, N)$

(39) $\alpha^{\mathrm{b}}_{m+1}(M, N) = \alpha^{\mathrm{b}}_m(M, N-1) + \beta^{\mathrm{b}}_m(M, N)k^{\mathrm{b}}_m(M, N)$

(40) $\alpha^{\mathrm{f}}_{m+1}(M, N) = \alpha^{\mathrm{f}}_m(M+1, N) + \beta^{\mathrm{b}}_m(M, N)k^{\mathrm{f}}_m(M, N)$

$$\mathbf{a}_{m+1}(M, N) = \begin{pmatrix} \mathbf{a}_m(M+1, N \\ 0 \end{pmatrix} + \begin{pmatrix} \mathbf{b}_m(M, N-1) \\ 1 \end{pmatrix} k^{\mathrm{f}}_m(M, N) \qquad (8.110\mathrm{a})$$

$$\mathbf{b}_{m+1}(M, N) = \begin{pmatrix} 0 \\ \mathbf{b}_m(M, N-1) \end{pmatrix} + \begin{pmatrix} 1 \\ \mathbf{a}_m(M+1, N) \end{pmatrix} k^{\mathrm{b}}_m(M, N) \qquad (8.110\mathrm{b})$$

$$k^{\mathrm{f}}_m(M, N) = -\alpha^{-\mathrm{b}}_m(M, N-1)\beta^{\mathrm{b}}_m(M, N) \qquad (8.110\mathrm{c})$$

$$k^{\mathrm{b}}_m(M, N) = -\alpha^{-\mathrm{f}}_m(M+1, N)\beta^{\mathrm{b}}_m(M, N) \qquad (8.110\mathrm{d})$$

$$\beta_m^{\mathrm{b}}(M, N) = \mathbf{a}_m^{\mathrm{T}}(M+1, N)\mathbf{r}_m^{\mathrm{b}}(M, N-1) + r_{m+1}^{\mathrm{b}}(M, N) \tag{8.110e}$$

Exact definitions of the quantities appearing above are provided in Table 8.3. The variables $\mathbf{b}_m(M, N)$ are the optimum backward predictors over $[M, N]$, $\alpha_m^{\mathrm{f}}(M, N)$ and $\alpha_m^{\mathrm{b}}(M, N)$ are the error powers of the forward and backward predictors, respectively, $\mathbf{r}_m^{\mathrm{b}}(M, N)$ is the cross-covariance vector between $y(t-m)$ and $\mathbf{y}_m(t)$ and $r_{m+1}^{\mathrm{b}}(M, N)$ is the top element of the vector $\mathbf{r}_{m+1}^{\mathrm{b}}(M, N)$. The similarity of the above algorithm to the Levinson algorithm is apparent. There are, however, the following distinct differences:

Table 8.3 Internal variables involved in the covariance algorithm

Correlations:	
$\mathbf{r}_m^{\mathrm{f}}(M, N) = \Sigma_{t=M+m-1}^{N-1}\mathbf{y}_m(t)y(t+1)$	$\mathbf{r}_m^{\mathrm{b}}(M, N) = \Sigma_{t=M+m}^{N}\mathbf{y}_m(t)y(t-m)$
$r_{m+1}^{\mathrm{f}}(M, N) = \Sigma_{t=M+m}^{N-1}y(t-m)y(t+1)$	$r_{m+1}^{\mathrm{b}}(M, N) = \Sigma_{t=M+m+1}^{N}y(t-m-1)y(t)$ $= r_{m+1}^{\mathrm{f}}(M, N)$
$r_{0m}^{\mathrm{f}}(M, N) = \Sigma_{t=M+m}^{N}y(t)y(t)$	$r_{0m}^{\mathrm{b}}(M, N) = \Sigma_{t=M+m}^{N}y(t-m)y(t-m)$
$\mathbf{r}_m(M, N) = \Sigma_{t=M+m-1}^{N}\mathbf{y}_m(t)z(t)$	$r_m = \Sigma_{t=M+m}^{N}y(t-m)z(t)$
Filters:	
$\mathbf{a}_m(M, N) = -R_m^{-1}(M, N-1)\mathbf{r}_m^{\mathrm{f}}(M, N)$	$\mathbf{b}_m(M, N) = -R_m^{-1}(M+1, N)\mathbf{r}_m^{\mathrm{b}}(M, N)$
$\mathbf{c}_m(M, N) = -R_m^{-1}(M, N)\mathbf{r}_m(M, N)$	
$\mathbf{q}_m(M, N) = -R_m^{-1}(M, N)\mathbf{y}_m(M+m-1)$	$\mathbf{w}_m(M, N) = -R_m^{-1}(M, N-1)\mathbf{y}_m(N)$
Power-related variables:	
$\alpha_m^{\mathrm{f}}(M, N) = r_{0m}^{\mathrm{f}}(M, N) + \mathbf{r}_m^{\mathrm{fT}}(M, N)\mathbf{a}_m(M, N)$	$\alpha_m^{\mathrm{b}}(M, N) = r_{0m}^{\mathrm{b}}(M, N) + \mathbf{r}_m^{\mathrm{bT}}(M, N)\mathbf{b}_m(M, N)$
$\alpha_m^{q}(M, N) = 1 + \mathbf{q}_m^{\mathrm{T}}(M, N)\mathbf{y}_m(M+m-1)$	$\alpha_m^{z}(M, N) = \Sigma_{t=M+m-1}^{N}z(t)z(t) + \mathbf{r}_m^{\mathrm{T}}(M, N)\mathbf{c}_m(M, N)$
$\alpha_m^{*}(M, N) = 1 - \mathbf{w}_m^{\mathrm{T}}(M, N)\mathbf{y}_m(N)$	

1. The backward predictor is not simply the reverse of the forward predictor. As a consequence an extra recursive update is required.
2. Prediction powers and associated reflection coefficients are different.
3. The order interleaving between the backward and forward prediction takes place through time-shifted versions. These in turn introduce the extra variables $\mathbf{a}_m(M+1, N)$ and $\mathbf{b}_m(M, N-1)$, which must be translated into the interval $[M, N]$. The computational complexity of the algorithm is $O(12m + L)$ MADs per order recursion where $L = N + 1 - M$. In practical applications where the exact order is not known a priori, monitoring the non-increasing forward prediction power $\alpha_m^{\mathrm{f}}(M, N)$ as a function of m can provide a good order estimate. In Table 8.2 the part of the algorithm denoted by stars corresponds to the extra equations required if the goal is to compute the optimal LS filter $\mathbf{c}_p(M, N)$ which transforms the input series $y(t)$ into a desired response signal $z(t)$. The overall complexity for

this case is $O(15m + 2L)$ per recursion. $\alpha^z(M, N)$ is the error power of the filter estimate operating in the interval $[M, N]$.

SCHUR-TYPE ALGORITHM

We have pointed out in Chapter 2 that in a parallel processing environment, recursions of the type (25) and (26) of Table 8.2 can be performed in parallel in a time equivalent to 1 MAD, i.e. independent of m. This is not possible with the inner product operation of type (22) of Table 8.2, for which the best one can achieve is to reduce the time to $O(\log_2 m)$ units. This bottleneck can be circumvented by resorting to an extension of the Schur algorithm for Toeplitz systems, discussed in Chapter 2. Such an algorithm was first presented in [20,21].

Let us concentrate on one of these variables requiring the computation of an inner product, i.e. $\beta^{\mathrm{b}}_m(M, N)$ in (22) of Table 8.2. The reason that such a variable cannot be computed in an order recursive manner by one addition and one multiplication, as is the case with α^*_m for example ((37), Table 8.2), is due to the partition properties of the vectors involved. Indeed, the forward predictor accepts an upper partition in terms of quantities known from previous recursion ((26), Table 8.2). On the other hand $\mathbf{r}^{\mathrm{b}}_m$ accepts a lower partitioning in terms of known quantities ((27), Table 8.2). Hence the inner product cannot be expressed in a recursive manner in terms of its lower-order counterparts (even if time shifted). To remedy this drawback we observe that the parameter $\beta^{\mathrm{b}}_m(M, N)$ could be viewed as the error at the output of the filter $\mathbf{a}_m(M + 1, N)$ at a particular time instant. The input vector and the desired response signal, at this specific time instant, are provided by $\mathbf{r}^{\mathrm{b}}_m(M, N - 1)$ and $r^{\mathrm{b}}_{m+1}(M, N)$ respectively. Motivated by this remark we introduce a 'virtual time' variable n, in which the above filter operates and the following definitions are made:

$$\mathbf{r}^{\mathrm{b}}_m(n \mid M, N - 1) = \sum_{t=M+n}^{N-1} \mathbf{y}_m(t)y(t - n) \qquad n \geqslant m \tag{8.111a}$$

$$r^{\mathrm{b}}(n \mid M, N) = \sum_{t=M+n}^{N} y(t)y(t - n) \tag{8.111b}$$

For fixed M, N the output error of the above filter $\mathbf{a}_m(M + 1, N)$ at different 'virtual time' instants n is determined by the above signals and is given by

$$\tilde{e}^{\mathrm{f}}_m(n \mid M, N) = \mathbf{a}^{\mathrm{T}}_m(M + 1, N)\mathbf{r}^{\mathrm{b}}_m(n - 1 \mid M, N - 1) + r^{\mathrm{b}}(n \mid M, N) \qquad n \geqslant m \tag{8.112}$$

It is not difficult to observe that at 'virtual time' $n = m + 1$

$$\tilde{e}^{\mathrm{f}}_m(m + 1 \mid M, N) = \beta^{\mathrm{b}}_m(M, N) \tag{8.113}$$

The inequality $n \geqslant m$ ensures that data are not running outside the known interval $[M, N]$. Having introduced the extra free parameter of the 'virtual time n' the following partitions are valid:

$$\mathbf{r}^{\mathrm{b}}_{m+1}(n \mid M, N) = \begin{bmatrix} \mathbf{r}^{\mathrm{b}}_m(n \mid M, N) \\ r^{\mathrm{b}}(n - m \mid M, N - m) \end{bmatrix} \tag{8.114a}$$

$$= \begin{bmatrix} r^{\mathrm{b}}(n \mid M, N) \\ \mathbf{r}_m^{\mathrm{b}}(n-1 \mid M, N-1) \end{bmatrix} \tag{8.114b}$$

That is, the new vector quantity introduced in (8.111a) accepts both upper and lower partitions in terms of the same vector of lower order. Note that time shifts in both real time and 'virtual time' are present. The above partitions constitute the basis of the algorithm and lead to lattice-type propagation of the error variable defined in (8.112). Specifically, if we combine (8.114a) with the order recursion (26) and the time update (14) of Table 8.2, the following update pattern results.

$$\tilde{e}_m^{\mathrm{f}}(n \mid M, N) = e_m^{\mathrm{f}}(n \mid M, N) + \tilde{k}_m^{\mathrm{f}} e_m^q(n-1 \mid M, N-1) \tag{8.115a}$$

$$e_{m+1}^{\mathrm{f}}(n \mid M, N) = \tilde{e}_m^{\mathrm{f}}(n \mid M, N) + k_m^{\mathrm{f}}(M, N) e_m^{\mathrm{b}}(n-1 \mid M, N-1) \tag{8.115b}$$

Table 8.4 Internal variables for the Schur-type algorithm

$$
\begin{aligned}
e_m^{\mathrm{f}}(n \mid M, N) &= \mathbf{a}_m^{\mathrm{T}}(M, N)\mathbf{r}_m^{\mathrm{b}}(n-1 \mid M, N-1) + r^{\mathrm{b}}(n \mid M, N) \\
\tilde{e}^{\mathrm{f}}(n \mid M, N) &= \mathbf{a}_m^{\mathrm{T}}(M+1, N)\mathbf{r}_m^{\mathrm{b}}(n-1 \mid M, N-1) + r^{\mathrm{b}}(n \mid M, N) \\
e_m^{\mathrm{b}}(n \mid M, N) &= \mathbf{b}_m^{\mathrm{T}}(M, N)\mathbf{r}_m^{\mathrm{b}}(n \mid M, N) + r^{\mathrm{b}}(n-m \mid M, N-m) \\
e_m^{q}(n \mid M, N) &= \mathbf{q}_m^{\mathrm{T}}(M, N)\mathbf{r}_m^{\mathrm{b}}(n \mid M, N) \\
e_m^{\mathrm{w}}(n \mid M, N) &= \mathbf{w}_m^{\mathrm{T}}(M, N)\mathbf{r}_m^{\mathrm{b}}(n \mid M, N-1) + y(N-n) \\
\tilde{e}_m^{\mathrm{w}}(n \mid M, N) &= \mathbf{w}_m^{\mathrm{T}}(M+1, N)\mathbf{r}_m^{\mathrm{b}}(n \mid M, N-1) + y(N-n) \\
\varepsilon_m^{\mathrm{f}}(n \mid M, N) &= \mathbf{a}_m^{\mathrm{T}}(M, N)\mathbf{y}_m(n-1 \mid M) + y(M+n-1) \\
\tilde{\varepsilon}_m^{\mathrm{f}}(n \mid M, N) &= \mathbf{a}_m^{\mathrm{T}}(M+1, N)\mathbf{y}_m(n-1) \mid M) + y(M+n-1) \\
\varepsilon_m^{\mathrm{b}}(n \mid M, N) &= \mathbf{b}_m^{\mathrm{T}}(M, N)\mathbf{y}_m(n \mid M) + y(M+n-m-1) \\
\varepsilon_m^{q}(n \mid M, N) &= \mathbf{q}_m^{\mathrm{T}}(M, N)\mathbf{y}_m(n \mid M) \\
\varepsilon_m^{\mathrm{w}}(n \mid M, N) &= \mathbf{w}_m^{\mathrm{T}}(M, N)\mathbf{y}_m(n \mid M) \\
\tilde{\varepsilon}_m^{\mathrm{w}}(n \mid M, N) &= \mathbf{w}_m^{\mathrm{T}}(M+1, N)\mathbf{y}_m(n \mid M)
\end{aligned}
$$

$$
\begin{aligned}
e_m(n \mid M, N) &= \mathbf{c}_m^{\mathrm{T}}(M, N)\mathbf{r}_m^{\mathrm{b}}(n \mid M, N) + r(n \mid M, N) \\
\tilde{e}_m(n \mid M, N) &= \mathbf{c}_m^{\mathrm{T}}(M+1, N)\mathbf{r}_m^{\mathrm{b}}(n \mid M, N) + r(n \mid M, N) \\
\varepsilon(n \mid M, N) &= \mathbf{c}_m^{\mathrm{T}}(M, N)\mathbf{y}_m(n \mid M) + z(M+n-1) \\
\tilde{\varepsilon}(n \mid M, N) &= \mathbf{c}_m^{\mathrm{T}}(M+1, N)\mathbf{y}_m(n \mid M) + z(M+n-1)
\end{aligned}
$$

The exact definition of the variables is provided in Table 8.4. What is important in (8.115) is that $\tilde{e}_m^{\mathrm{f}}(n \mid M, N)$ is obtained from lower-order quantities with complexity equal to two MADs. It must be pointed out of course that at a particular order $m+1$ the error variable has to be computed at different 'virtual time' instants n, to be used later at higher-order recursions. Thus, in a uniprocessor environment this way of inner product computation leads to similar complexity as that of the covariance algorithm. However, in a multiprocessor environment the computations in (8.115) for the different 'virtual time' instants can take place concurrently, thus resulting in an order-of-magnitude reduction in computational time. A careful observation of the definitions in Table 8.4 and the computations of the inner products in Table 8.2 reveals the following identities:

$$\varepsilon_m^{\mathrm{w}}(m \mid M, N) = \tilde{\beta}_m^{\mathrm{w}}(M, N) \quad \tilde{e}_m^{\mathrm{w}}(m \mid M, N) = \tilde{\beta}_m^{\mathrm{b}}(M, N)$$

$$\varepsilon_m^{\mathrm{f}}(m+1 \mid M, N) = \tilde{\beta}_m^{\mathrm{f}}(M, N) \quad \tilde{e}_m^{\mathrm{f}}(m+1 \mid M, N) = \beta_m^{\mathrm{b}}(M, N)$$

$$\varepsilon_m(m \mid M, N) = \tilde{\beta}_m(M, N) \quad \tilde{e}_m(m \mid M, N) = \beta_m(M, N)$$

Table 8.5 summarizes the algorithm. A close observation of Tables 8.4 and 8.5 reveals the presence of two types of error variables denoted by e and ε respectively. The former are associated with the vector defined in (8.114) and the latter with the vector of the initial samples $\mathbf{y}_m(M+m-1)$. The proof of the order propagation of these variables (Table 8.5) is obtained from the respective definitions, the filter order updates of Table 8.2, the partitioning (8.114) and the following definitions:

Table 8.5 Schur-type algorithm

FOR $m = 0, p-1$
  IN PARALLEL DO
  BEGIN
    $\tilde{k}_m^q = -\alpha_m^{*-1}(M, N)\varepsilon_m^{\mathrm{w}}(m \mid M, N)$
    $k_m^q = -\alpha_m^{-\mathrm{f}}(M, N)\varepsilon_m^{\mathrm{f}}(m+1 \mid M, N)$
    $\tilde{k}_m = -\alpha_m^{-q}(M, N)\varepsilon_m(m \mid M, N)$
  END
  IN PARALLEL DO
  BEGIN
    FOR $n = m, p-1$ IN PARALLEL DO
      $e_m^q(n \mid M, N-1) = e_m^q(n \mid M, N) + \tilde{k}_m^q e_m^{\mathrm{w}}(n \mid M, N)$
      $\varepsilon_m^q(n \mid M, N-1) = \varepsilon_m^q(n \mid M, N) + \tilde{k}_m^q \varepsilon_m^{\mathrm{w}}(n \mid M, N)$
      $\tilde{e}_m(n \mid M, N) = e_m(n \mid M, N) + \tilde{k}_m e_m^q(n \mid M, N)$
    END
    $\alpha_m^q(M, N-1) = \alpha_m^q(M, N) + \tilde{k}_m^q \varepsilon_m^{\mathrm{w}}(m \mid M, N)$
    FOR $i = 1, m$ IN PARALLEL DO
      $q_m^i(M, N-1) = q_m^i(M, N) + w_m^i(M, N)\tilde{k}_m^q$
      $c_m^i(M+1, N) = c_m^i(M, N) + q_m^i(M, N)\tilde{k}_m$
    END
  END
  IN PARALLEL DO
  BEGIN
    $\tilde{k}_m^w = -\alpha_m^{-q}(M, N-1)\varepsilon_m^{\mathrm{w}}(m \mid M, N)$
    $k_m = -\alpha_m^{-\mathrm{b}}(M, N)\tilde{e}_m(m \mid M, N)$
  END
  IN PARALLEL DO
  BEGIN
    FOR $n = m, p-1$ IN PARALLEL DO

$\tilde{e}_m^w(n-1\,|\,M, N) = e_m^w(n-1\,|\,M, N) + \tilde{k}_m^w e_m^q(n-1\,|\,M, N-1)$

$\tilde{\varepsilon}_m^w(n-1\,|\,M, N) = \varepsilon_m^w(n-1\,|\,M, N) + \tilde{k}_m^w \varepsilon_m^q(n-1\,|\,M, N-1)$

$\varepsilon(n-1\,|\,M, N) = \varepsilon_m(n-1)\,|\,M, N) + \tilde{k}_m \varepsilon_m^q(n \quad 1\,|\,M, N)$

END

$\alpha_m^*(M+1, N) = \alpha_m^*(M, N) - \tilde{k}_m^w \varepsilon_m^w(m\,|\,M, N)$

FOR $i = 1, m$ IN PARALLEL DO

$w_m^i(M+1, N) = w_m^i(M, N) + q_m^i(M, N-1)\tilde{k}_m^w$

END

END

IN PARALLEL DO

BEGIN

$\tilde{k}_m^b = -\alpha_m^{*-1}(M+1, N)\tilde{e}_m^w(m\,|\,M, N)$

$\tilde{k}_m^f = -\alpha_m^{-q}(M, N-1)\varepsilon_m^f(m+1\,|\,M, N)$

END

IN PARALLEL DO

BEGIN

FOR $n = m+1, p$ IN PARALLEL DO

$\tilde{e}_m^f(n\,|\,M, N) = e_m^f(n\,|\,M, N) + \tilde{k}_m^f e_m^q(n-1\,|\,M, N-1)$

$\tilde{\varepsilon}_m^f(n\,|\,M, N) = \varepsilon_m^f(n\,|\,M, N) + \tilde{k}_m^f \varepsilon_m^q(n-1\,|\,M, N-1)$

$e_m^b(n-1\,|\,M, N-1) = e_m^b(n-1\,|\,M, N) + \tilde{k}_m^b \tilde{e}_m^w(n-1\,|\,M, N)$

$\varepsilon_m^b(n-1\,|\,M, N-1) = \varepsilon_m^b(n-1\,|\,M, N) + \tilde{k}_m^b \tilde{\varepsilon}_m^w(n-1\,|\,M, N)$

$e_{m+1}^q(n\,|\,M, N) = e_m^q(n-1\,|\,M, N-1) + k_m^q e_m^f(n\,|\,M, N)$

$\varepsilon_{m+1}^q(n\,|\,M, N) = \varepsilon_m^q(n-1\,|\,M, N-1) + k_m^q \varepsilon_m^f(n\,|\,M, N)$

$\star e_{m+1}(n\,|\,M, N) = \tilde{e}_m(n\,|\,M, N) + k_m e_m^b(n\,|\,M, N)$

$\star \varepsilon_{m+1}(n\,|\,M, N) = \tilde{\varepsilon}_m(n\,|\,M, N) + k_m \varepsilon_m^b(n\,|\,M, N)$

END

$\alpha_m^b(M, N-1) = \alpha_m^b(M, N) + \tilde{k}_m^b \tilde{e}_m^w(m\,|\,M, N)$

$\alpha_m^f(M+1, N) = \alpha_m^f(M, N) + \tilde{k}_m^f \varepsilon_m^f(m+1\,|\,M, N)$

FOR $i = 1, m$ IN PARALLEL DO

$b_m^i(M, N-1) = b_m^i(M, N) + w_m^i(M+1, N)\tilde{k}_m^b$

$a_m^i(M+1, N) = a_m^i(M, N) + q_m^i(M, N-1)\tilde{k}_m^f$

$q_{m+1}^{i+1}(M, N) = q_m^i(M, N-1) + a_m^i(M, N)k_m^q$

END

IN PARALLEL DO

BEGIN

$k_m^f = -\alpha_m^{-b}(M, N-1)\tilde{e}_m^f(m+1\,|\,M, N)$

$k_m^b = -\alpha_m^{-f}(M+1, N)\tilde{e}_m^f(m+1\,|\,M, N)$

$k_m^w = -\alpha_m^{-b}(M, N-1)\tilde{e}_m^w(m\,|\,M, N)$

END

IN PARALLEL DO

BEGIN

FOR $n = m+1, p$ IN PARALLEL DO

$e_{m+1}^f(n\,|\,M, N) = \tilde{e}_m^f(n\,|\,M, N) + k_m^f e_m^b(n-1\,|\,M, N-1)$

$\varepsilon_{m+1}^f(n\,|\,M, N) = \tilde{\varepsilon}_m^f(n\,|\,M, N) + k_m^f \varepsilon_m^b(n-1\,|\,M, N-1)$

$e^{\mathrm{b}}_{m+1}(n \mid M, N) = e^{\mathrm{b}}_m(n-1 \mid M, N-1) + k^{\mathrm{b}}_m \tilde{e}^{\mathrm{f}}_m(n \mid M, N)$
$\varepsilon^{\mathrm{b}}_{m+1}(n \mid M, N) = \varepsilon^{\mathrm{b}}_m(n-1 \mid M, N-1) + k^{\mathrm{b}}_m \tilde{\varepsilon}^{\mathrm{f}}_m(n \mid M, N)$
$e^{\mathrm{w}}_{m+1}(n-1 \mid M, N) = \tilde{e}^{\mathrm{w}}_m(n-1 \mid M, N) + k^{\mathrm{w}}_m e^{\mathrm{b}}_m(n-1 \mid M, N-1)$
$\varepsilon^{\mathrm{w}}_{m+1}(n-1 \mid M, N) = \tilde{\varepsilon}^{\mathrm{w}}_m(n-1 \mid M, N) + k^{\mathrm{w}}_m \varepsilon^{\mathrm{b}}_m(n-1 \mid M, N-1)$
END
$\alpha^{*}_{m+1}(M, N) = \alpha^{*}_m(M+1, N) - k^{\mathrm{w}}_m \tilde{e}^{\mathrm{w}}_m(m \mid M, N)$
$\alpha^{q}_{m+1}(M, N) = \alpha^{q}_m(M, N-1) + k^{q}_m \varepsilon^{\mathrm{f}}_m(m+1 \mid M, N)$
$\alpha^{\mathrm{b}}_{m+1}(M, N) = \alpha^{\mathrm{b}}_m(M, N-1) + k^{\mathrm{b}}_m \tilde{e}^{\mathrm{f}}_m(m+1 \mid M, N)$
$\alpha^{\mathrm{f}}_{m+1}(M, N) = \alpha^{\mathrm{f}}_m(M+1, N) + k^{\mathrm{f}}_m \tilde{e}^{\mathrm{f}}_m(m+1 \mid M, N)$
FOR $i = 1, m$ IN PARALLEL DO
$w^{i}_{m+1}(M, N) = w^{i}_m(M+1, N) + b^{i}_m(M, N-1)k^{\mathrm{w}}_m$
$b^{i+1}_{m+1}(M, N) = b^{i}_m(M, N-1) + a^{i}_m(M+1, N)k^{\mathrm{b}}_m$
$a^{i}_{m+1}(M, N) = a^{i}_m(M+1, N) + \mathrm{b}^{i}_m(M, N-1)k^{\mathrm{f}}_m$
$\star c^{i}_{m+1}(M, N) = c^{i}_m(M+1, N) + b^{i}_m(M, N)k_m$
END
$w^{m+1}_{m+1}(M, N) = k^{\mathrm{w}}_m$
$b^{1}_{m+1}(M, N) = k^{\mathrm{b}}_m$
$a^{m+1}_{m+1}(M, N) = k^{\mathrm{f}}_m$
$q^{1}_{m+1}(M, N) = k^{q}_m$
$\star c^{m+1}_{m+1}(M, N) = k_m$
END
END

$$r(n \mid M, N) \equiv \sum_{t=M+n}^{N} z(t)y(t-n) \tag{8.116a}$$

$$\mathbf{y}_{m+1}(n \mid M) \equiv \begin{bmatrix} y(M+n-1) \\ \vdots \\ y(M+n-m-1) \end{bmatrix} = \begin{bmatrix} \mathbf{y}_m(n \mid M) \\ y(M+n-m-1) \end{bmatrix} = \begin{bmatrix} y(M+n-1) \\ \mathbf{y}_m(n-1 \mid M) \end{bmatrix} \tag{8.116b}$$

The proof for the time shifts is slightly more tricky, and is explained at the end of Appendix A. Figure 8.2 describes the evolution, with time, of the computations involved in the algorithm of Table 8.5. The time interval indices M, N have been suppressed for simplicity. The number of processors involved is p, and it is assumed that recursions are currently at step m. We have focused our attention on the e variables. The ε variables follow exactly the same evolutionary pattern. The variables resting at each processor at the beginning of the recursion are indicated on the right-hand side. After time τ_1 (division) has elapsed reflection coefficients $\tilde{k}^{q}_m$, k^{q}_m have been computed in the processor $pr \neq 1$ and they are propagated to the rest of the

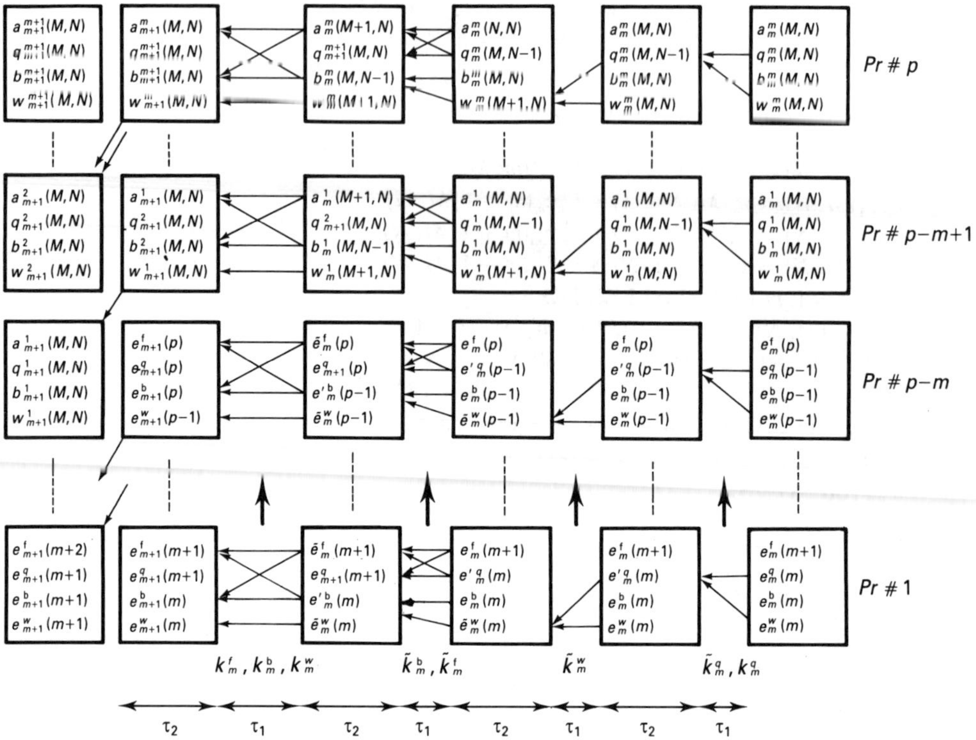

Figure 8.2 Evolution of the computed variables in the LS Schur-type algorithm

processors. Subsequently, $e_m^q(n \mid M, N-1) \equiv \acute{e}_m^q(n)$ are produced in the lower group of processors and $q_m^i(M, N-1)$ in the top group. The subsequent computations of the rest of the variables, as time goes by are illustrated in Figure 8.2. At the end of the recursions, after $4(\tau_1 + \tau_2)$ time units (τ_2 time for add–mult), each processor contains the new variables. Before the next recursion starts, processors must interchange certain variables. Specifically, $pr \# 1$ receives $e_{m+1}^{w}(m+1 \mid M, N)$, $e_{m+1}^{f}(m+1 \mid M, N)$ from $pr \# 2$, and so on. Note that $e_{m+1}^{w}(m+1 \mid M, N)$, $e_{m+1}^{f}(m+1 \mid M, N)$ will not be used and may not be computed. Processor $pr \# p-m$ passes the variables $e_{m+1}^{w}(p-1 \mid M, N)$, $e_{m+1}^{f}(p \mid M, N)$ to processor $pr \# p-m-1$. Note that variables $e_{m+1}^{q}(p \mid M, N)$, $e_{m+1}^{b}(p \mid M, N)$ will not be used further and may not be computed. Processor $pr \# p-m$ receives $w_{m+1}^{1}(M, N)$, $a_{m+1}^{1}(M, N)$ from processor $pr \# p-m+1$ and at the same time substitutions for $q_{m+1}^{1}(M, N)$, $b_{m+1}^{1}(M, N)$ are completed. In the topmost processor the substitutions $q_{m+1}^{m+1}(M, N)$, $a_{m+1}^{m+1}(M, N)$ are made. It is clear, therefore, that the processors required for the order updates of the e are decreasing with order while the processors required for the transversal parameter updates are increasing. At the end of the recursions the processors will contain the elements of the transversal parameters' vectors. This combined mode of

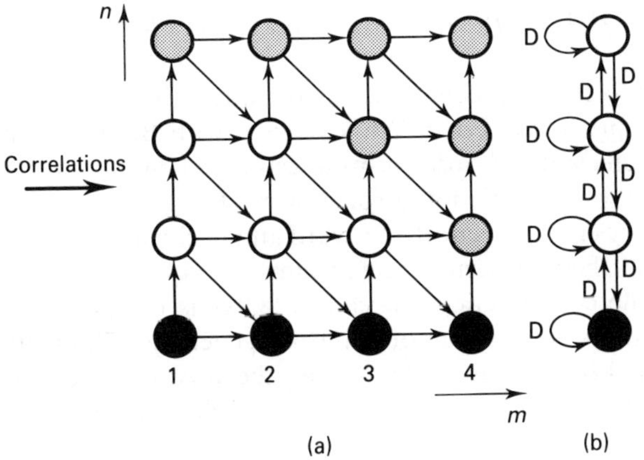

Figure 8.3 (a) DG and (b) SFG of the LS Schur type algorithm

operation was first suggested in [22]. The computational data dependencies of the above description associated with the single assignment code of Table 8.5 are shown by Figure 8.3(a). Shaded nodes correspond to transversal parameter computations. Note that only multiply–add operations are performed, with the exception of the lower (dark) nodes where divisions are also required. Figure 8.3(b) shows the signal flow graph resulting from projection across the default schedule. It is apparent that only local processor communications are required, a most desirable feature for VLSI integration. Furthermore, if the global communication of the reflection coefficients among the processing cells is undesirable, a systolic/wavefront mode of operation is possible, using systolization techniques [23].

SIMPLIFICATIONS

The Schur-type algorithm in Table 8.5 corresponds to the unwindowed case. Obviously, if prewindowing is assumed the algorithm is substantially simplified. Indeed under this assumption the initial data vector $\mathbf{y}_m(M + m - 1)$ becomes identically zero. Hence all the ε variables, as well as $\mathbf{q}_m(M, N)$, also become zero. The computational time per recursion becomes $2(\tau_1 + \tau_2)$. In [22] a different version of the algorithm is presented, well suited for the block adaptive mode of operation. The final solution is the same as that computed by the algorithm in Table 8.5. However, the intermediate-order quantities are different. The resulting algorithm requires $2\tau_1 + 3\tau_2$ time units per recursion and it is also structurally simpler. It is also suggested that if a two-dimensional processor array is used a fully pipelined operation is possible for block adaptation. Furthermore, it is shown that all error variables involved are well bounded, a property which can be exploited for fixed point implementations of the algorithm. A Shur-type algorithm for the more general ρ-Toeplitz case is provided in [24]. Square root variants of the Schur algorithm have also appeared [25].

8.4.3 The forward backward LS method

In the preceding section the forward predictor was estimated by minimizing the total squared forward prediction error. While dealing with Burg's technique we saw that the reflection coefficients of order m were estimated by minimizing the sum of the forward and backward prediction error powers. The key concept behind this idea is that in a stationary environment the forward and backward predictors are the reverse of each other and their respective power sums are equal. Thus adopting the sum of these errors as the cost function, instead of (8.99), is reasonable and is expected to lead to better results since more error points are generated. This method was suggested independently in [26,27] and is known as the forward backward LS (FBLS) or modified covariance method [2]. The new cost function is now

$$\sum_{t=M+p}^{N} \{[e_p^{\mathrm{f}}(t)]^2 + [e_p^{\mathrm{b}}(t)]^2\} \tag{8.117a}$$

where

$$e_p^{\mathrm{f}}(t) = y(t) + \mathbf{y}_p^{\mathrm{T}}(t-1)J\mathbf{b}_p \tag{8.117b}$$

$$e_p^{\mathrm{b}}(t) = y(t-p) + \mathbf{y}_p^{\mathrm{T}}(t)\mathbf{b}_p(t) \tag{8.117c}$$

and where $\mathbf{b}$ has been chosen to represent the FBLS optimum backward predictor resulting from (8.117). In effect this is a constrained minimization problem with respect to the forward and backward predictors under the constraint that they are the reverse of each other. Minimization of (8.117) with respect to $\mathbf{b}_p$ results in the following set of linear equations [28,29]:

$$S_p(M, N)\mathbf{b}_p(M, N) = -\mathbf{s}_p(M, N) \tag{8.118a}$$

where

$$S_p(M, N) = R_p(M+1, N) + JR_p(M, N-1)J \tag{8.118b}$$

$$\mathbf{s}_p(M, N) = \mathbf{r}_p^{\mathrm{b}}(M, N) + J\mathbf{r}_p^{\mathrm{f}}(M, N) \tag{8.118c}$$

with $R_p(M, N)$ being the covariance matrix and $\mathbf{r}_p^{\mathrm{b}}(M, N)$, $\mathbf{r}_p^{\mathrm{f}}(M, N)$ the backward and forward cross-correlations whose exact definitions are provided in Table 8.6, which summarizes the definitions of the various correlation quantities involved in the algorithm. If only the backward error term was present in (8.117) then only the first terms on the right-hand side of (8.118b) and (8.118c) would be present (resulting in the LS backward predictor). Matrix $S_p(M, N)$ is symmetric, positive definite and has a special rich structure inherited from $R_p(M, N)$. This structure was exploited first in [28] for the derivation of a fast recursive scheme of order $O(p^2) + O(p(L))$ MADs. A highly parallel Schur-type algorithm was given in [30] and can be performed in $O(p) + O(L)$ time units on a linear array of $O(p)$ processors.

Table 8.6 Definitions of the correlation variables involved in the FBLS algorithms

$$Q_m(M, N) = R_m(M, N) + JR_m(M, N)J$$
$$S_m(M, N) = R_m(M + 1, N) + JR_m(M, N - 1)J$$
$$\mathbf{r}_m^{\mathrm{b}}(M,k\ N) = \Sigma_{n=M+m}^{N} \mathbf{y}_m(n)y(n - m)$$
$$\mathbf{r}_m^{\mathrm{f}}(M, N) = \Sigma_{n=M+m-1}^{N-1} \mathbf{y}_m(n)y(n + 1)$$
$$\mathbf{s}_m(M, N) = J\mathbf{r}_m^{\mathrm{f}}(M, N) + \mathbf{r}_m^{\mathrm{b}}(M, N)$$
$$\mathbf{q}_m(M, N) = \mathbf{r}_m^{\mathrm{f}}(M + 1, N) + J\mathbf{r}_m^{\mathrm{b}}(M, N - 1)$$
$$r_{0m}^{\mathrm{b}}(M, N) = \Sigma_{n=M+m}^{N} y^2(n - m)$$
$$r_{0m}^{\mathrm{f}}(M, N) = \Sigma_{n=M+m}^{N} y^2(n)$$
$$s_m^0(M, N) = r_{0m}^{\mathrm{f}}(M, N) + r_{0m}^{\mathrm{b}}(M, N)$$
$$q_m^0(M, N) = r_{0m}^{\mathrm{f}}(M + 1, N) + r_{0m}^{\mathrm{b}}(M, N - 1)$$
$$r_{m+1}^{\mathrm{f}}(M, N) = \Sigma_{n=M+m+1}^{N} y(n - m - 1)y(n) = r_{m+1}^{\mathrm{b}}(M, N)$$
$$s_{m+1} = 2r_{m+1}^{\mathrm{f}}(M, N)$$

In this section we focus our attention on a recently developed order recursive algorithm which belongs to the split Levinson algorithm family. A Schur-type version of this algorithm is still to be derived.

In Chapter 2 we saw how the split Levinson algorithm overcame the computational redundancy of the Levinson algorithm by imposing symmetry on to the problem and exploiting its computational merits. The same philosophy will be adopted here and we shall see that the algorithm in [28] is computationally redundant. To this end we define the following symmetric vectors [31]:

$$\mathbf{b}_{m+1}^{s1}(M, N) = \begin{bmatrix} \mathbf{b}_m(M, N) \\ 1 \end{bmatrix} + \begin{bmatrix} 1 \\ J\mathbf{b}_m(M, N) \end{bmatrix} \tag{8.119a}$$

$$\mathbf{b}_m^s(M, N) = \mathbf{b}_m(M, N) + J\mathbf{b}_m(M, N) \tag{8.119b}$$

A first observation is that if $\mathbf{b}_m^{s1}(M, N)$ and $\mathbf{b}_m^s(M, N)$ are known, $\mathbf{b}_m(M, N)$ can be obtained without multiplications via additions only. Table 8.7 summarizes this simple algorithmic scheme. Hence the effort is focused on obtaining an efficient algorithm for the update of the above symmetric vectors. Our purpose here is to point out the basic features of the derived algorithm and its distinctive traits compared with previously derived algorithms.

Table 8.7 Scheme for computing the FBLS backward predictor in terms of the introduced auxiliary symmetric vectors

$b_m^m(M, N) = b_{m+1}^{s1,1}(M, N) - 1$

FOR $i = 1, m/2$ (m even) or $(m - 1)/2$ (m odd)

  $b_m^i(M, N) = b_m^{s,i}(M, N) - b_m^{m+1-i}(M, N)$

  $b_m^{m-i}(M, N) = b_{m+1}^{s1,i+1}(M, N) - b_m^i(M, N)$

END

From the definition in (8.118) it is straightforward to see that

$$Q_{m+1}(M, N)\begin{bmatrix}\mathbf{b}_{m+1}(M, N)\\ 1\end{bmatrix} = \begin{bmatrix}\mathbf{0}\\ \alpha^{\mathrm{b}}_{m+1}(M, N)\end{bmatrix} \tag{8.120a}$$

where

$$Q_{m+2}(M, N) \equiv \begin{bmatrix} S_{m+1}(M, N) & \mathbf{s}_{m+1}(M, N)\\ \mathbf{s}^{\mathrm{T}}_{m+1}(M, N) & s^{0}_{m+1}(M, N)\end{bmatrix} \tag{8.120b}$$

This equation is the counterpart of (2.113). The definition of $\alpha^{\mathrm{b}}_{m+1}$ is readily seen from (8.120a) and Table 8.6:

$$\alpha^{\mathrm{b}}_{m+1}(M, N) = s^{0}_{m+1}(M, N) + \mathbf{b}^{\mathrm{T}}_{m+1}(M, N)(J\mathbf{r}^{\mathrm{f}}_{m+1}(M, N) + \mathbf{r}^{\mathrm{b}}_{m+1}(M, N)) \tag{8.121}$$

It is not difficult to see that the above quantity $\alpha^{\mathrm{b}}_{m+1}(M, N)$ is the minimum total forward plus backward squared error in (8.117a). From (8.120b) we can show that

$$Q_{m+2}(M, N) = R_{m+2}(M, N) + JR_{m+2}(M, N)J \tag{8.122}$$

This matrix has been extensively studied in the context of linear phase filtering [32–35]. It is near-to-Toeplitz, positive definite, symmetric and persymmetric, that is,

$$JQ_m(M, N)J = Q_m(M, N) \tag{8.123a}$$

$$Q_m(M, N)J = JQ_m(M, N) \tag{8.123b}$$

The upper partitioning of $Q_m(M, N)$ is available from the definition in (8.120b). In order to derive a fast order recursive scheme, one more partitioning is necessary. So far, a lower partitioning has been used. In contrast, a third type of partitioning will be introduced here. We will call it middle partitioning and it is the direct outcome of the following:

$$\mathbf{y}_{m+2}(t) = [y(t) \quad \mathbf{y}^{\mathrm{T}}_m(t-1) \quad y(t-m-1)]^{\mathrm{T}} \tag{8.124}$$

Introducing (8.124) into the definition (8.122) results in

$$Q_{m+2}(M, N) = \begin{bmatrix} q^{0}_m(M, N) & \mathbf{q}^{\mathrm{T}}_m(M, N) & s_{m+1}(M, N)\\ \mathbf{q}_m(M, N) & Q_m(M+1, N-1) & J\mathbf{q}_m(M, N)\\ s_{m+1}(M\ N) & \mathbf{q}^{\mathrm{T}}_m(M, N)J & q^{0}_m(M, N)\end{bmatrix} \tag{8.125}$$

The various quantities are defined in Table 8.6, although the exact definitions are of no interest to us for the time being. The importance of (8.125) lies in establishing the relation of matrix $Q_{m+2}(M, N)$ to its two-orders-lower predecessor time shifted with respect to both M and N. The advantage of the above is that the matrix Q_m resulting from the partitioning is also persymmetric, thus providing the means to exploit the symmetries imposed on the problem (Equations (8.119)). This would not be possible if a standard low partitioning of $Q_{m+2}(M, N)$ was used, which would involve $JS_{m+1}(M, N)J$, as can easily be checked. Combining (8.123) and (8.120a) results in

$$Q_{m+2}(M, N)\mathbf{b}^{s1}_{m+2}(M, N) = \begin{bmatrix}\mathbf{0}\\ \alpha^{\mathrm{b}}_{m+1}(M, N)\end{bmatrix} + \begin{bmatrix}\alpha^{\mathrm{b}}_{m+1}(M, N)\\ \mathbf{0}\end{bmatrix} \tag{8.126}$$

From (8.125) and (8.126) and using the form of the matrix inversion lemma, given in Appendix B, it can be shown that

$$\mathbf{b}^{s1}_{m+2}(M, N) = \begin{bmatrix} 1 \\ \mathbf{g}^{s}_{m}(M, N) \\ 1 \end{bmatrix} \alpha^{-1}_{m}(M, N)\alpha^{b}_{m+1} M, N) \tag{8.127}$$

$\mathbf{g}^{s}_{m}(M, N)$ is a symmetric vector with a special meaning: it is the LS optimum symmetric predictor predicting simultaneously in the forward and backward directions; we will call it the *bidirectional predictor*. That is, from the data set $y(N-1)$, $y(N-2), \ldots,$ $y(N-m)$, it predicts the sum $y(N) + y(N-m-1)$ and is defined by

$$Q_m(M+1, N-1)\mathbf{g}^{s}_{m}(M, N) = -[\mathbf{q}_m(M, N) + J\mathbf{q}_m(M, N)] \tag{8.128}$$

$\alpha_m(M, N)$ is the respective error power in the simultaneous prediction, explicitly defined in Table 8.8. Note that Table 8.8 provides the definitions of all the internally used vectors. As is the case with $\mathbf{g}^{s}_{m}(M, N)$, each of these vectors is LS optimal for an appropriate combination of input and desired response signal. The defining equations are the respective normal equations. The table also provides, together with the filters, the definitions of the respective errors and LS minimum error power related variables. Predictor $\mathbf{g}^{s}_{m}(M, N)$ is the result of the middle partitioning in (8.125). The matrix inversion lemma forces $Q^{-1}_{m}(M+1, N-1)$ to be combined with the sum of vectors to the left and right of $Q_m(M+1, N-1)$ in the partitioning of (8.125). This is analogous to the covariance case where the matrix inversion lemma forces the upper partitioning $R_m(M+1, N)$ in (8.105a) to be combined with the vector to its right, resulting in the LS backward predictor.

Table 8.8 Filter, error, power definitions

	Error- and/or power-related variables
Non-symmetric filters:	
$S_m(M, N)\mathbf{b}_m(M, N) = -\mathbf{s}_m(M, N)$	$\alpha^{b}_{m}(M, N) = r^{f0}_{m}(M, N) + r^{b0}_{m}(M, N) + \mathbf{b}^{T}_{m}(M, N)\mathbf{s}_m(M, N)$ $\varepsilon^{b}_{m}(M, N) = h_m(M, N) + \mathbf{H}^{T}_{m}(M, N)\mathbf{J}\mathbf{b}_m(M, N)$ $\varepsilon^{br}_{m}(M, N) = h_0(M, N) + \mathbf{H}^{T}_{m}(M+1, N-1)\mathbf{b}_m(M, N)$
$Q_m(M+1, N-1)\mathbf{u}_m(M, N) = -J\mathbf{H}_m(M, N)$	$\alpha^{u}_{m}(M, N) = I_2 - \mathbf{H}^{T}_{m}(M, N)J\mathbf{u}_m(M, N)$ $\varepsilon^{ur}_{m}(M, N) = \mathbf{H}^{T}_{m}(M, N)\mathbf{u}_m(M, N)$
$Q_m(M, N)\mathbf{u}^{*}_{m}(M, N) = -\mathbf{H}_m(M, N)$	$\alpha^{u*}_{m}(M, N) = I_2 + \mathbf{H}^{T}_{m}(M, N)\mathbf{u}^{*}_{m}(M, N)$ $\varepsilon^{u*}_{m}(M, N) = \mathbf{H}^{T}_{m}(M, N)J\mathbf{u}^{*}_{m}(M, N)$
$S_m(M, N)\mathbf{v}^{*}_{m}(M, N) = -J\mathbf{H}_m(M, N)$	$\alpha^{v*}_{m}(M, N) = I_2 + \mathbf{H}^{T}_{m}(M, N)J\mathbf{v}^{*}_{m}(M, N)$

$$\varepsilon_m^{v*}(M, N) = \mathbf{H}_m^{\mathrm{T}}(M+1, N-1)\mathbf{v}_m^*(M, N)$$

$$S_m(M, N)\mathbf{w}_m(M, N) = -\mathbf{H}_m(M, N) \qquad \alpha_m^w(M, N) = I_2 - \mathbf{H}_m^{\mathrm{T}}(M, N)\mathbf{w}_m(M, N)$$

$$\varepsilon_m^w(M, N) = \mathbf{H}_m^{\mathrm{T}}(M, N)J\mathbf{w}_m(M, N)$$

$$S_m(M+1, N-1)\mathbf{v}_m(M, N) = -J\mathbf{H}_m(M, N) \qquad \alpha_m^v(M, N) = I_2 - \mathbf{H}_m^{\mathrm{T}}(M, N)J\mathbf{v}_m(M, N)$$

$$Q_m(M+1, N-1)\mathbf{g}_m(M, N) = -J\mathbf{q}_m(M, N) \qquad \varepsilon_m^g(M, N) = h_m(M, N) + \mathbf{H}_m^{\mathrm{T}}(M, N)J\mathbf{g}_m(M, N)$$

$$\varepsilon_m^{gr}(M, N) = h_0(M, N) + \mathbf{H}_m^{\mathrm{T}}(M+1, N-1)\mathbf{g}_m(M, N)$$

$$\varepsilon_m^g(M+1, N-1/M, N) = h_{m+1}(M, N) + \mathbf{H}_m^{\mathrm{T}}(M+1, N-1)J\mathbf{g}_m(M, N)$$

$$\alpha_m^g(M, N) = q_m^0(M, N) + \mathbf{q}_m^{\mathrm{T}}(M, N)J\mathbf{g}_m(M, N)$$

Symmetric filters;

$$Q_m(M+1, N-1)\mathbf{u}_m^s(M, N) = -(\mathbf{H}_m(M, N) + J\mathbf{H}_m(M, N)) \qquad \alpha_m^{us}(M, N) = I_2 - \mathbf{H}_m^{\mathrm{T}}(M, N)\mathbf{u}_m^s(M, N)$$

$$Q_m(M, N)\mathbf{u}_m^{*s}(M, N) = -(\mathbf{H}_m(M, N) + J\mathbf{H}_m(M, N)) \qquad \alpha_m^{u*s}(M, N) = I_2 + \mathbf{H}_m^{\mathrm{T}}(M, N)\mathbf{u}_m^{*s}(M, N)$$

$$Q_m(M+1, N-1)\mathbf{g}_m^s(M, N) = -(\mathbf{q}_m(M, N) + J\mathbf{q}_m(M, N)) \qquad \varepsilon_m^{gs}(M, N) = \tilde{h}_{m+1}(M, N) + \mathbf{H}_m^{\mathrm{T}}(M+1, N-1)\mathbf{g}_m^s(M, N)$$

$$e_m^{gs}(M, N) = \tilde{h}_{m+1}(M, N) + \mathbf{H}_m^{\mathrm{T}}(M+1, N-1)\mathbf{g}_m^s(M+1, N-1)$$

$$\alpha_m(M, N) = q_m^0(M, N) + s_{m+1}(M, N) + \mathbf{q}_m^{\mathrm{T}}(M, N)\mathbf{g}_m^s(M, N)$$

where

$$\mathbf{H}_m(M, N) = (\mathbf{y}_m(M+m-1)\ J\mathbf{y}_m(N)) \qquad h_m(M, N) = \begin{pmatrix} y(M+m) \\ y(N-m) \end{pmatrix}$$

$$\tilde{h}_m(M, N) = h_m(M, N) + h_0(M, N) \qquad h_0(M, N) = \begin{pmatrix} y(M) \\ y(N) \end{pmatrix}$$

Having expressed the vector $\mathbf{b}_m^{s1}(M, N)$ in terms of a symmetric vector, we will attempt the same for $\mathbf{b}_m^s(M, N)$. In Appendix B it is shown that

$$\mathbf{b}_m^s(M, N) = \mathbf{g}_m^s(M, N) + \mathbf{u}_m^s(M, N)\varepsilon_m^{\mathrm{b}}(M, N) \tag{8.129}$$

$\varepsilon_m^{\mathrm{b}}(M, N)$ is an error 2×1 vector. From its definition in Table 8.8 it is apparent that its top element is the forward error at time $M + m$ with predictor $J\mathbf{b}_m(M, N)$ and its lower element the backward error at time N with predictor $\mathbf{b}_m(M, N)$. In other words, this vector summarizes the constraint nature of $\mathbf{b}_m(M, N)$. $\mathbf{u}_m^s(M, N)$ is an $m \times 2$ J invariant matrix ($J\mathbf{u}_m^s(M, N) = \mathbf{u}_m^s(M, N)$) defined in Table 8.8. A careful study of this quantity reveals its equivalence with the vectors $[\mathbf{w}_m, \mathbf{q}_m]$ appearing in the covariance algorithm of Table 8.2. We will see later that under the prewindowed assumption this is the (prior) Kalman gain associated with the time update of the $\mathbf{g}_m^s(N)$. A closely related quantity is $\mathbf{u}_m^{*s}(M, N)$, defined in Table 8.8. It is not difficult to show that

$$\mathbf{u}_m^s(M, N) = \mathbf{u}_m^{*s}(M, N)\alpha_m^{-u*s}(M, N) \tag{8.130}$$

with the latter quantity also given in Table 8.8. Equations (8.127) and (8.129) consist of two types of variables:

1. Those which are either J symmetric ($\mathbf{g}_m^s$, $\mathbf{u}_m^{*s}$, $\mathbf{u}_m^s$) or computed in terms of J symmetric variables (α_m, Table 8.8).
2. Those which are not explicitly expressed in terms of J symmetric quantities (α_m^{b}, $\varepsilon_m^{\mathrm{b}}$).

The above distinction leads to an algorithm consisting of the following three steps:

Table 8.9 Unwindowed order recursive symmetric algorithm for the bidirectional predictor

		Add	Mult
	FOR $q = 1, 2, \ldots, p$		
	Loop for the prediction part		
	FOR $m = 2q - 1, 2q$		
(1)	$\mathbf{H}_{m+2}^s(M, N) = \mathbf{H}_{m+2}(M, N) + J\mathbf{H}_{m+2}(M, N)$	$1m$	
(2)	$\mathbf{H}_{m+2}^s(M+1, N-1) = \mathbf{H}_{m+2}(M+1, N-1) + J\mathbf{H}_{m+2}(M+1, N-1)$	$1m$	
	Covariance computations:		
(3)	$r_{m+2}^{\mathrm{f}}(M+1, N-1) = \Sigma_{j=M+m+2}^{N-2} y(j-m-1)y(j+1)$	$L-m$	$L-m$
(4)	$s_{m+2}(M+1, N-1) = r_{m+2}^{\mathrm{f}}(M+1, N-1) + r_{m+2}^{\mathrm{f}}(M+1, N-1)$		
(5)	$q_{m+1}^0(M+1, N-1) = q_m^0(M+1, N-1) - h_{m+2}^{\mathrm{T}}(M, N)h_{m+2}(M, N)$		
(6)	$\mathbf{s}_{m+2}^{\mathrm{T}}(M+1, N-1) = [s_{m+2}(M+1, N-1), \mathbf{q}_{m+1}^{\mathrm{T}}(M+1, N-1)J]$		
(7)	$\mathbf{q}_{m+2}(M+1, N-1) = \mathbf{s}_{m+2}(M+1, N-1)$ $- \mathbf{H}_{m+2}(M+1, N-1)h_{m+2}(M+1, N-1)$	$2m$	$2m$
(8)	$\mathbf{q}_{m+2}^s(M+1, N-1) = \mathbf{q}_{m+2}(M+1, N-1)$ $+ J\mathbf{q}_{m+2}(M+1, N-1)$	$0.5m$	
(9)	$\mathbf{q}_{m+2}^2(M, N) = \mathbf{q}_{m+2}^s(M+1, N-1) + \mathbf{H}_{m+2}^s(M+1, N-1)\tilde{h}_{m+3}(M, N)$	$1m$	$1m$
(10)	$\alpha_{m+1}(M+1, N-1) = q_{m+1}^0(M+1, N-1)$ $+ s_{m+2}(M+1, N-1)$ $+ 0.5\mathbf{q}_{m+1}^{s\mathrm{T}}(M+1, N-1)\mathbf{g}_{m+1}^s(M+1, N-1)$	$0.5m$	$0.5m$
(11)	$\beta_{m+2}^{gs}(M, N) = \alpha_{m+1}(M+1, N-1)$ $+ (e_{m+1}^{gs\mathrm{T}}(M, N) - e_m^{gs\mathrm{T}}(M, N))\varepsilon_{m+1}^{gs}(M, N)$		

(12) $\beta^{u*s}_{m+2}(M, N) = \varepsilon^{gs\mathrm{T}}_{m+1}(M, N) - e^{gs\mathrm{T}}_{m+1}(M, N)\alpha^{u*s}_{m+1}(M + 1, N - 1)$

Filters' order updates:

(13) $k^{gs}_{m+2}(M, N) = -\alpha^{-1}_{m}(M + 1, N - 1)\beta^{gs}_{m+2}(M, N)$

(14) $$\mathbf{g}^{s}_{m+2}(M, N) = \begin{pmatrix} \mathbf{g}^{s}_{m+1}(M, N) \\ 1 \end{pmatrix} + \begin{pmatrix} 1 \\ \mathbf{g}^{s}_{m+1}(M, N) \end{pmatrix} - \begin{pmatrix} 0 \\ \mathbf{u}^{*}_{m}(M + 1, N - 1) \\ 0 \end{pmatrix} \alpha^{-u*s}_{m}(M + 1, N - 1)\varepsilon^{gs}_{m+1}(M, N) + \begin{pmatrix} 1 \\ \mathbf{g}^{s}_{m}(M + 1, N - 1) \\ 1 \end{pmatrix} k^{gs}_{m+2}(M, N) \qquad 2m \qquad 1.5m$$

(15) $k^{u*s}_{m+2}(M, N) = -\alpha^{-1}_{m}(M + 1, N - 1)\beta^{u*s}_{m+2}(M, N)$

(16) $$\mathbf{u}^{*s}_{m+2}(M + 1, N - 1) = \begin{pmatrix} \mathbf{u}^{*s}_{m+1}(M + 1, N - 1) \\ 0 \end{pmatrix} + \begin{pmatrix} 0 \\ \mathbf{u}^{*s}_{m+1}(M + 1, N - 1) \end{pmatrix} - \begin{pmatrix} 0 \\ \mathbf{u}^{*s}_{m}(M + 1, N - 1) \\ 0 \end{pmatrix} \alpha^{-u*s}_{m}(M + 1, N - 1)\alpha^{u*s}_{m+1}(M + 1, N - 1) + \begin{pmatrix} 1 \\ \mathbf{g}^{s}_{m}(M + 1, N - 1) \\ 0 \end{pmatrix} k^{u*s}_{m+2}(M, N) \qquad 4m \qquad 3m$$

Errors and error powers:

(17) $\alpha_{m+1}(M, N) = \alpha_{m+1}(M + 1, N - 1) + e^{gs\mathrm{T}}_{m+1}(M, N)\varepsilon^{gs}_{m+1}(M, N)$

(18) $\alpha^{u*s}_{m+2}(M + 1, N - 1) = I_2 + 0.5\mathbf{H}^{s\mathrm{T}}_{m+2}(M + 1, N - 1)\mathbf{u}^{*s}_{m+2}(M + 1, N - 1) \qquad 2m \qquad 2m$

(19) $\tilde{k}^{u*s}_{m+2}(M, N) = -\alpha^{-1}_{m}(M, N)\varepsilon^{gs\mathrm{T}}_{m}(M, N)$

(20) $\alpha^{u*s}_{m+2}(M, N) = \alpha^{u*s}_{m}(M + 1, N - 1) + \varepsilon^{gs}_{m}(M, N)\tilde{k}^{u*s}_{m+2}(M, N)$

(21) $\varepsilon^{gs}_{m+2}(M, N) = \tilde{h}_{m+3}(M, N) + 0.5\mathbf{H}^{s\mathrm{T}}_{m+2}(M + 1, N - 1)\mathbf{g}^{s}_{m+2}(M, N) \qquad 1m \qquad 1m$

(22) $e^{gs}_{m+2}(M, N) = \alpha^{-u*s}_{m+2}(M + 1, N - 1)\varepsilon^{gs}_{m+2}(M, N)$

Filters' time updates:

(23) $\mathbf{g}^{s}_{m+2}(M + 1, N - 1) = \mathbf{g}^{s}_{m+2}(M, N) - \mathbf{u}^{*s}_{m+2}(M + 1, N - 1)e^{gs}_{m+2}(M, N) \qquad 1m \qquad 1m$

(24) $$\mathbf{u}^{*s}_{m+2}(M, N) = \begin{pmatrix} 0 \\ \mathbf{u}^{*s}_{m}(M + 1, N - 1) \\ 0 \end{pmatrix} + \begin{pmatrix} 1 \\ \mathbf{g}^{s}_{m}(M, N) \\ 1 \end{pmatrix} \tilde{k}^{u*s}_{m+2}(M, N) \qquad 1m \qquad 1m$$

END

Linear phase filtering part:

★(25) $d_{2q+1}(M, N) = \Sigma^{N}_{j=M+2q}\,[y(j) + y(j-2q)]z^{\mathrm{T}}(j-q)$	$2L - 2m$	$L - m$
★(26) $\beta^{0}_{2q+1}(M, N) = d_{2q+1}(M, N) + 0.5\mathbf{q}^{s\mathrm{T}}_{2q-1}(M, N)\mathbf{c}^{0}_{2q-1}(M+1, N-1)$	$0.5m$	$0.5m$
★(27) $k^{0}_{2q+1}(M, N) = -\alpha^{-1}_{2q-1}(M, N)\beta^{0}_{2q+1}(M, N)$		
★(28) $\mathbf{c}^{0}_{2q+1}(M, N) = \begin{pmatrix} 0 \\ \mathbf{c}^{0}_{2q-1}(M+1,\ N-1) \\ 0 \end{pmatrix} + \begin{pmatrix} 1 \\ \mathbf{g}^{s}_{2q-1}(M,\ N) \\ 1 \end{pmatrix} k^{0}_{2q+1}(M, N)$	$0.5m$	$0.5m$
★(29) $\varepsilon^{0}_{2q+1}(M, N) = h^{z}_{q}(M, N) + 0.5\mathbf{H}^{s\mathrm{T}}_{2q+1}(M, N)\mathbf{c}^{0}_{2q+1}(M, N)$	$1m$	$1m$
★(30) $\mathbf{c}^{0}_{2q+1}(M+1, N-1) = \mathbf{c}^{0}_{2q+1}(M, N)$		
$\quad - \mathbf{u}^{*s}_{2q+1}(M, N)\alpha^{-u*s}_{2q+1}(M, N)\varepsilon^{0}_{2q+1}(M, N)$	$1m$	$1m$

END

where

$$h^{z}_{q}(M, N) = \begin{pmatrix} z(M+q) \\ z(N-q) \end{pmatrix},\ h_{m}(M, N) = \begin{pmatrix} y(M+m) \\ y(N-m) \end{pmatrix},\ \tilde{h}_{m}(M, N) = \begin{pmatrix} y(M+m) + y(M) \\ y(N) + y(N-m) \end{pmatrix}$$

Remarks

(a) For each recursion of the linear phase filter (i.e. relations (25)–(30)) two recursions of the prediction part of the algorithm are required.

(b) When the optimum LS smoother is required then the above algorithm can also be used, with only slight modifications. Indeed, by setting $d_{2q+1}(M, N) = 0$ and $h^{z}_{q}(M, N) = 0$ the filter $\mathbf{c}^{0}_{2q+1}(M, N)$ equals the optimum symmetric smoother.

Step 1. Update variables of type 1). This part of the algorithm is given in Table 8.9 and is the equivalent to the prediction part of the covariance algorithm of Table 8.2. A structural comparison between the two algorithms reveals the following differences:

(a) Two filters $\mathbf{g}^{s}_{m}$ and $\mathbf{u}^{*s}_{m}$ are present instead of the four ($\mathbf{a}_m$, $\mathbf{b}_m$, $\mathbf{w}_m$, $\mathbf{q}_m$) in the covariance prediction part. Recall, however, that $\mathbf{u}^{*s}_{m}$ is an $m \times 2$ matrix. It must be emphasized here that the use of $\mathbf{u}^{*s}_{m}$ instead of $\mathbf{u}^{s}_{m}$ was dictated by the partition properties of the involved quantities, as explained in Appendix B.

(b) The order updates of these filters, i.e. (13)–(16) of Table 8.9, involve three-term recursions (three successive orders are involved) instead of the two-term recursions of Table 8.2.

(c) Since both of these filters are J symmetric, only half of the computations need to be performed in the operations where they are involved.

The proof of the algorithm is provided in Appendix B. Its complexity amounts to $L + 12m$ multiplications and $L + 16m$ additions, where $L = N - M + 1$. The complexity of each recursion is provided on the right of Table 8.9.

It has been shown that in a system identification task where the system is constrained to be linear phase, the LS estimate of the symmetric impulse response is efficiently computed by the algorithm of Table 8.9. It needs only six extra recursions, indicated by ★, to obtain the unknown impulse response estimate [34,36].

Step 2. Obtain variables $\alpha_m^{\mathrm{b}}, \varepsilon_m^{\mathrm{b}}$. These are the power and error variables associated with the non-symmetric vector $\mathbf{b}_m$. Their respective definitions are given in Table 8.8. The goal of this algorithmic step, summarized in Table 8.10, is to compute the above quantities in terms of those of Table 8.9 which are computed by means of symmetric quantities.

The scheme of Table 8.10 consists of two parts. In the first one, i.e. Equations (1)–(12), quantities α_m^{b}, $\varepsilon_m^{\mathrm{b}}$ are computed. This is achieved in an order recursive lattice-type operation springing from the quantities $\varepsilon_1^{\mathrm{b}}$, ε_2^{u*}, α_1^{b}, α_2^{b}, k_2^{b}, k_2^{u*}, k_1^{g} appearing at the input of the lattice. The computation of these seeds is the direct result of the respective definitions provided in the tables. Subsequently, the different stages are fed with the quantities ε_m^{gs}, α_m, α_m^{u*s}, $m = 1, 2, \ldots, p$, computed from Table 8.9. The second part consists of the computation of the symmetric vectors $\mathbf{b}_m^{s1}$, $\mathbf{b}_m^{s}$. The overall contribution of this step to the algorithm is 1 MAD per order recursion.

Table 8.10 Order recursive scheme for step 2 for the symmetric FBLS algorithm

(1) $\varepsilon_m^{\mathrm{br}}(M, N) = \varepsilon_{m-1}^{gs}(M, N)\alpha_{m-1}^{-1}(M, N)\alpha_m^{\mathrm{b}}(M, N) - \varepsilon_m^{\mathrm{b}}(M, N)$

(2) $\varepsilon_m^{v*}(M, N) = \varepsilon_{m+1}^{u*}(M, N) - \varepsilon_m^{\mathrm{br}}(M, N)k_{m+1}^{u*}(M, N)$

(3) $\varepsilon_m^{gr}(M, N) = \varepsilon_m^{\mathrm{br}}(M, N) + \varepsilon_m^{v*}(M, N)k_m^{g}(M, N)$

(4) $\varepsilon_{m+1}^{\mathrm{b}}(M, N) = \varepsilon_m^{gs}(M, N) - \varepsilon_m^{gr}(M, N)[1 - k_{m+1}^{\mathrm{b}}(M, N)]$

(5) $k_{m+2}^{u*}(M, N) = -\alpha_{m+1}^{-\mathrm{b}}(M, N)\varepsilon_{m+1}^{\mathrm{bT}}(M, N)$

(6) $\alpha_{m+1}^{v*}(M, N) = \alpha_{m+1}^{u*}(M, N) - \varepsilon_{m+1}^{u*}(M, N)\alpha_{m+1}^{-u*}(M, N)\varepsilon_{m+1}^{u*}(M, N)$

(7) $k_{m+1}^{g}(M, N) = -\alpha_{m+1}^{-v*}(M, N)\varepsilon_{m+1}^{\mathrm{b}}(M, N)$

(8) $\alpha_{m+1}^{g}(M, N) = \alpha_{m+1}^{\mathrm{b}}(M, N) + \varepsilon_{m+1}^{\mathrm{bT}}(M, N)k_{m+1}^{g}(M, N)$

(9) $k_{m+2}^{\mathrm{b}}(M, N) = 1 - \alpha_{m+1}^{-g}(M, N)\alpha_{m+1}(M, N)$

(10) $\alpha_{m+2}^{\mathrm{b}}(M, N) = \alpha_{m+1}^{g}(M, N)[1 - (k_{m+2}^{\mathrm{b}}(M, N))^2]$

(11) $\alpha_{m+2}^{u*}(M, N) = \alpha_{m+1}^{v*}(M, N) + \varepsilon_{m+1}^{\mathrm{b}}(M, N)k_{m+2}^{u*}(M, N)$

(12) $\varepsilon_{m+2}^{u*}(M, N) = \alpha_{m+2}^{u*s}(M, N) - \alpha_{m+2}^{u*}(M, N)$

(13) $\mathbf{b}_{m+2}^{s1}(M, N) = \begin{pmatrix} 1 \\ \mathbf{g}_m^{s}(M, N) \\ 1 \end{pmatrix} \alpha_m^{-1}(M, N)\alpha_{m+1}^{\mathrm{b}}(M, N)$

(14) $\mathbf{b}_{m+1}^{s}(M, N) = \mathbf{g}_{m+1}^{s}(M, N) + \mathbf{u}_{m+1}^{*s}(M, N)\alpha_{m+1}^{-u*s}(M, N)\varepsilon_{m+1}^{\mathrm{b}}(M, N)$

Remark: Vector $\mathbf{u}_{m+1}^{s}(M, N)$ involved in (8.129) is replaced, in the final scheme, with the equivalent $u_{m+1}^{*s}(M, N)a_{m+1}^{-u*s}(M, N)$ (see (14) above). This is because the latter expression consists of quantities explicitly involved in the first part of the algorithm of Table 8.9.

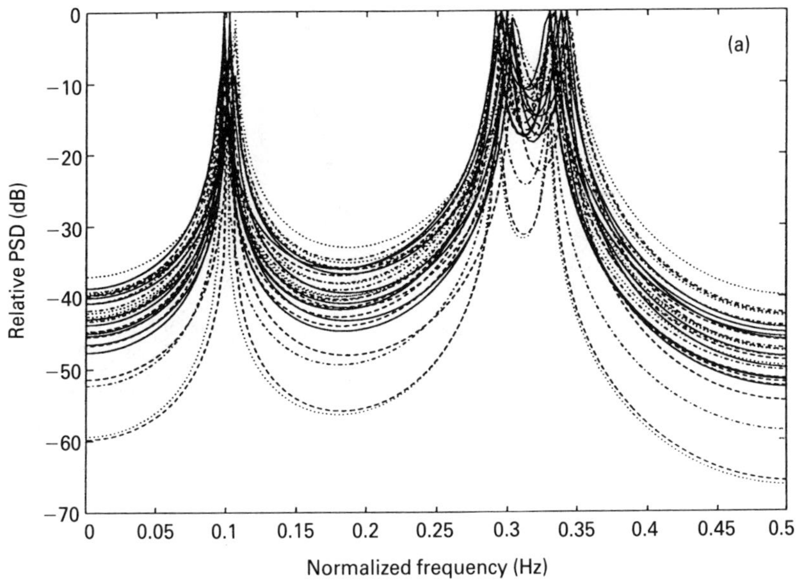

Figure 8.4 Power spectral densities obtained with (a) the Burg algorithm; (b) the FBLS technique

Step 3. Obtain the non-symmetric $\mathbf{b}_m$ from $\mathbf{b}^{s1}$, $\mathbf{b}^s_m$. This is achieved by means of Table 8.7; it requires p additions and only needs to be computed at the final order.

Thus the overall complexity of the algorithm is $L + 12m$ multiplications and $L + 17m$ additions. Hence a saving of $3m$ multiplications is achieved with respect to the algorithm presented in [2,28].

SOME SIMULATION RESULTS AND DISCUSSION

The comparative performance of the FBLS, the covariance and Burg's methods have been studied and reported extensively in [26, 37–39]. The FBLS method has overcome the two drawbacks of Burg's method, that is, spectral line splitting and bias sensitivity with respect to phase, with little extra computational complexity. From a statistical point of view, both methods exhibit the least spectral variance, compared with other AR methods, when tested with non-sinusoidal processes [26]. In [2] it is pointed out that Burg's technique exhibits less amplitude variance and the FBLS method less frequency variance.

Figures 8.4(a) and (b) show the resulting spectra for the Burg and the FBLS methods respectively. The time series used consisted of three sinusoids at the normalized frequencies of 0.1, 0.3 and 0.325 Hz. The number of data points was 32, the AR order 6 and the signal-to-noise ratio 30 dB. All sinusoids were of the same

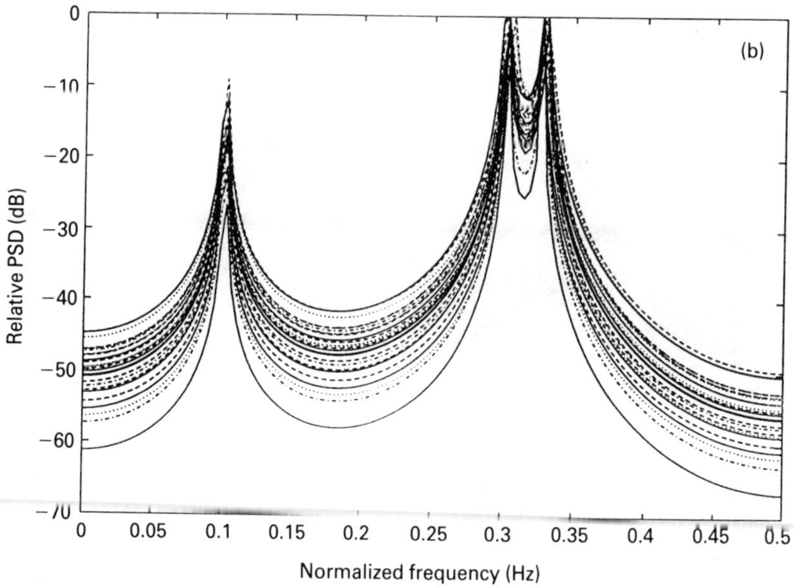

Figure 8.4 (*cont.*) (b) the FBLS technique

amplitude. Each figure is the result of overlaying (the same) 40 realizations with different initial phases and noise samples. The figures demonstrate in a clear way the sensitivity of Burg's technique to variation in phase. For smaller data lengths the results were even more catastrophic for Burg's method. It must be emphasized that the covariance method always resulted in inferior-quality spectra, from the resolution point of view, and sensitivity to phase dependence, compared with the FBLS method. The difference in performance becomes more noticeable as the data length decreases, as expected. It must also be emphasized that the improved performance of the symmetric FBLS algorithm, presented in this section, compared with the covariance algorithm of Table 8.2, is achieved at approximately the same computational cost.

An important issue in AR spectral analysis is that of selecting the right order p for the model. A number of different criteria have been suggested [40,41], but none of these seems to work well in practice, especially in the case of short data lengths [39,42]. As suggested in [2], the choice of model order is more an art than a science and is still based on subjective judgement.

Observation noise is another critical factor whose presence degrades the performance of an AR(p) estimator. Assume $\hat{\eta}(t)$ to be the additive noise corrupting the signal which then becomes

$$\tilde{y}(t) = y(t) + \hat{\eta}(t)$$

It is easy to see that $\tilde{y}(t)$ obeys an ARMA model of order (p, p) [2]. Thus for low SNRs, AR(p) is not adequate for modelling and ARMA techniques should be utilized.

For high-to-medium-level noise components the effects of noise can be compensated by an increase in the order of the AR model. However, an excessive increase in order results in spurious peaks and may also lead to ill-conditioned matrices. A theoretical limit, for example in the case of the covariance method, is $p \leqslant L/2$, to ensure that the number of summands in the sum (8.104) is greater than p, which is a sufficient condition for $R_p(M, N)$ to be positive definite. For the FBLS method thc theoretical limit is obviously $p \leqslant 2L/3$.

8.4.4 Adaptive techniques

All the methods presented so far are of the block processing type, that is, a block of data has been assumed to be available. For real-time applications, time sequential techniques are desirable, which update the AR parameters as each new sample is received. Assuming that the statistics of the signals are slowly time varying, the incorporation of an appropriate window, as explained in Chapter 2, allows the algorithm to adapt to the changing dynamics of the environment. If the covariance PSD is to be estimated, any of the algorithms described in the previous chapters can be used to adapt the AR parameters. The performance and properties of these algorithms have already been discussed and we will not consider them any further here.

Time recursive schemes for the FBLS method (and the related linear phase filtering case) have been suggested in a series of articles [29,43,44,45,46], covering the transversal, sliding-window transversal, lattice and normalized lattice cases. Here we shall concern ourselves with the recently [31] proposed transversal algorithm. This scheme evolves around symmetric internal variables and it is the time adaptive version of the algorithm presented in section 8.4.3. The prewindowed case will be considered, that is, the initial set of samples $\mathbf{y}_p(M + p - 1)$ is identically zero. Thus notation will be relaxed from the initial time M.

The issue of windowing is slightly tricky when dealing with the FBLS method. As is common practice with the adaptive algorithms, an exponential window is usually adopted to weight the error power samples entering the minimized cost function. If this approach is followed with the cost function in (8.117a) it destroys those matrix structural properties which lead to fast computational schemes. In particular, partition (8.120b) will not be valid. A way out of this difficulty is to weight directly the time series samples [47,48]. Thus at time N the signal appears to be

$$y(N)\ \lambda y(N-1)\ \lambda^2 y(N-2)\ \ldots$$

and at time $N + 1$

$$y(N+1)\ \lambda y(N)\ \lambda^2 y(N-1)\ \ldots$$

Hence, a second index N is needed to describe the signal for each time instant

$$\begin{aligned}\mathbf{y}^{\mathrm{T}}_{m,N}(t) &= [\lambda^{N-t}y(t), \lambda^{N-t-1}y(t-1), \ldots, \lambda^{N-m-t+1}y(t-m+1)] \\ &= \lambda^{N-t}\boldsymbol{\Lambda}\mathbf{y}_m(t)\end{aligned}$$

where

$$\Lambda = \text{diag}[1, \lambda^{-1}, \ldots, \lambda^{-m+1}]$$

As a consequence all the signal related quantities should be modified accordingly. Table 8.11 is the resulting counterpart of Table 8.6. The defining equations for the various filters, errors and error powers can be obtained from those of Table 8.8 if we add the second index N to the subscript of the matrices and vectors involved. Also, the signal samples should be weighted accordingly. The corresponding equation (1) of Table 8.8, for example, becomes

$$S_{m,N}(N)\mathbf{b}_m(N) = -\mathbf{s}_{m,N}(N)$$

$$\varepsilon_m^{\mathrm{b}}(N) = \lambda^m y(N-m) + \mathbf{y}_{m,N}^{\mathrm{T}}(N)\mathbf{b}_m(N)$$

where all vector quantities involving the initial set of samples have been set to zero as a result of the prewindowing assumption. As was the case with the block symmetric algorithm, its adaptive version also consists of three parts:

1. Adaptive computation of the symmetric bidirectional predictor $\mathbf{g}_p^s(N)$. This is summarized in Table 8.12 [49]. The algorithm exhibits certain distinct characteristics in its philosophy compared with the non-symmetric adaptive schemes of Chapter 5. It consists of two parts. In the first one (Equations (1) to (8)) the step-up update is completed involving four filters of successive orders, i.e. p, $p+1$, $p+2$, $p+3$. The combined time–order update of the gain (Equation (4)) is of the form $(p, N-1) \to (p+2, N)$ and $(p+1, N-1) \to (p+3, N)$. The second part completes the step down (Equation (10)) by means of three-term recursions in the following philosophy: $\{(p+3, N), (p+2, N)\} \to (p+1, N)$ and then $\{(p+2, N), (p+1, N)\} \to (p, N)$. If the aim is not the FBLS predictor $\mathbf{b}_m$ but a linear phase filter, three extra recursions are needed [34]. The algorithm of Table 8.12 is basically a rearrangement of the block processing algorithm of Table 8.9, after the prewindowing simplification and the incorporation of λ. In particular, Equation (10) of Table 8.12 is the same as (16) of Table 8.9 where (8.130) has been exploited together with the easily proven relationship.

$$\alpha^{u*s}(N) = \alpha^{-us}(N)$$

   Equation (9) of Table 8.12 is the direct outcome of the particular structure of (10) of Table 8.12. Equations (4) and (7) of Table 8.12 are the counterparts of (24) and (20) of Table 8.9 and are derived in an analogous way. Equation (13) of Table 8.12 results from the respective definitions of the variables involved [31].
2. Computation of the non-symmetric, vector-related quantities (Table 8.13). The proof of the formulae is the result of the respective definitions and the order–time updates of Table 8.14, which can be shown from the respective definitions. The detailed steps of the proof (for $\lambda = 1$) can be found in [31]. Note that the 'prior error' variables, denoted by e, are defined as their 'posterior' counterparts (ε) by replacing the time index N with $N-1$ in the respective filter.
3. Computation of the FBLS backward predictor (Table 8.7).

The overall complexity of the algorithm is $8.5p$ multiplications and $9p$ additions. To initialize the algorithm we assume that prior to time $N = 0$, when the first sample

Table 8.11 Definitions related to the adaptive FBLS algorithm

$\mathbf{y}_{m,N}(n) = [\lambda^{N-n}y(n) \ldots \lambda^{N-n+m-1}y(n-m+1)]^{\mathrm{T}}$	$R_{m,N}(n) = \Sigma_{t=0}^{n}\mathbf{y}_{m,N}(t)\mathbf{y}_{m,N}^{\mathrm{T}}(t)$
$Q_{m,N}(n) = R_{m,N}(n) + JR_{m,N}(n)J$	$S_{m,N}(n) = R_{m,N}(n) + JR_{m,N}(n-1)J$
$\mathbf{r}_{m,N}^{\mathrm{b}}(n) = \Sigma_{t=0}^{n}\lambda^{N-t+m}\mathbf{y}_{m,N}(t)y(t-m)$	$\mathbf{r}_{m,N}^{\mathrm{f}}(n) = \Sigma_{t=0}^{n-1}\lambda^{N-t-1}\mathbf{y}_{m,N}(t)y(t+1)$
$\mathbf{s}_{m,N}(n) = \mathbf{r}_{m,N}^{b}(n-1) + J\mathbf{r}_{m,N}^{\mathrm{f}}(n-1)$	$\mathbf{q}_{m,N}(n) = \mathbf{r}_{m,N}^{\mathrm{f}}(n) + J\mathbf{r}_{m,N}^{\mathrm{b}}(n-1)$
$r_{0m,N}^{\mathrm{b}}(n) = \Sigma_{t=0}^{n}\lambda^{2(N-t+m)}y^2(t-m)$	$r_{0m,N}^{\mathrm{f}}(n) = \Sigma_{t=0}^{n}\lambda^{2(N-t)}y^2(t)$
$s_{m,N}^{0}(n) = r_{0m,N}^{\mathrm{b}}(n) + r_{0m,N}^{\mathrm{f}}(n)$	$q_{m,N}^{0}(n) = r_{0m,N}^{\mathrm{f}}(n) + r_{0m,N}^{\mathrm{b}}(n-1)$
$r_{m+1,N}^{\mathrm{f}}(n) = \Sigma_{t=0}^{n}\lambda^{2(N-t)+m+1}y(t)y(t-m-1) = r_{m+1,N}^{\mathrm{b}}(n)$	$s_{m+1,N}(n) = 2r_{m+1,N}^{\mathrm{f}}(n)$
$\mathbf{y}_{m,N}^{s}(n) = \mathbf{y}_{m,N}(n) + J\mathbf{y}_{m,N}(n)$	$\mathbf{q}_{m,N}^{s}(n) = \mathbf{q}_{m,N}(n) + J\mathbf{q}_{m,N}(n)$

Table 8.12 Adaptive algorithm for the bidirectional predictor

		Mults	Adds
	FOR $m = p, p+1$		
(1)	$e_m^{gs}(N) = y(N) + \beta_m^{u}(N-1)$		
(2)	$\varepsilon_m^{gs}(N) = e_m^{gs}(N)\alpha_m^{-us}(N-1)$		
(3)	$k_{m+2}^{us}(N) = -\lambda^{-2}e_m^{gs}(N)\alpha_m^{-1}(N-1)$		
(4)	$\mathbf{u}_{m+2}^{s}(N) = \lambda^{-1}\begin{bmatrix} 0 \\ \mathbf{u}_m^{s}(N-1) \\ 0 \end{bmatrix} + \begin{bmatrix} 1 \\ \mathbf{g}_m^{s}(N-1) \\ 1 \end{bmatrix} k_{m+2}^{us}$	$2p$	$1p$
(5)	$\alpha_m(N) = \lambda^2\alpha_m(N-1) + e_m^{gs}(N)\varepsilon_m^{gs}(N)$		
(6)	$\mathbf{g}_m^{s}(N) = \mathbf{g}_m^{s}(N-1) + \lambda^{-1}\mathbf{u}_m^{s}(N-1)\varepsilon_m^{gs}(N)$	$1p$	$1p$
(7)	$\alpha_{m+2}^{us}(N) = \alpha_m^{us}(N-1) - e_m^{gs}(N)k_{m+2}^{us}(N)$		
(8)	$\mathbf{y}_{m,N}^{s}(N) = \mathbf{y}_{m,N}(N) + J\mathbf{y}_{m,N}(N)$		$1p$
	END		
	FOR $m = p+1, p$		
(9)	$k_{m+2}^{u*s}(N) = u_{m+2}^{s,1}(N)\alpha_{m+2}^{-us} - u_{m+1}^{s,1}(N)\alpha_{m+1}^{-us}(N)$		
(10)	$\begin{bmatrix} 0 \\ \mathbf{u}_m^{s}(N) \\ 0 \end{bmatrix} = \begin{bmatrix} \mathbf{u}_{m+1}^{s}(N) \\ 0 \end{bmatrix} + \begin{bmatrix} 0 \\ \mathbf{u}_{m+1}^{s}(N) \end{bmatrix} + \begin{bmatrix} 1 \\ \mathbf{g}_m^{s}(N) \\ 1 \end{bmatrix}$ $\times\, k_{m+2}^{u*s}\alpha_{m+1}^{us}(N) - \mathbf{u}_{m+2}^{s}(N)\alpha_{m+1}^{us}(N)\alpha_{m+2}^{-us}(N)$	$2p$	$3p$
(11)	$\alpha_m^{us}(N) = 1 - 0.5\mathbf{y}_{m,N}^{s\mathrm{T}}(N)\mathbf{u}_m^{s}(N)$	$1p$	$1p$
	END		
(12)	$\beta_{p-1}^{u}(N) = \lambda^{p}y(N-p+1) + 0.5\mathbf{y}_{p-1,N+1}^{s\mathrm{T}}(N)\mathbf{g}_{p-1}^{s}(N)$	$0.5p$	$0.5p$
(13)	$\beta_p^{u}(N) = \beta_{p-1}^{u}(N) - \lambda\alpha_{p-1}(N)k_{p+1}^{u*s}(N)\alpha_p^{us}(N)$		
(14)	$\mathbf{y}_{p,N+1}(N+1) = [y(N+1)\ \lambda\mathbf{y}_{p,N}^{\mathrm{T}}(N)]^{\mathrm{T}}$	p	

is received, all data are zero except $y(-p-2)=\sigma$, where σ is a small quantity. This is necessary in order to make $\alpha_p^{\mathrm{b}}(-1)$, $\alpha_p(-1)$, required by the algorithm at $N=0$, invertible. It is easy to show that all vector quantities at time -1 are zero. Indeed, under the above assumption all cross-correlation sums from $-\infty$ to -1 become zero, leading the corresponding filters to have zero values. Note that $Q_{p,-1}(-1)$, $S_{p,-1}(-1)$ are diagonal and invertible. All error quantities are also zero. All power sums, i.e. the αs, at time -1 are unity except

$$\alpha_p^{\mathrm{b}}(-1)=\alpha_p(-1)=\alpha_{p-1}(-1)=2\sigma^2\lambda^{2(p+1)}\equiv c$$

Figure 8.5 demonstrates the effect of λ in the direct data windowing. Originally the signal consisted of seven sinusoids at the normalized frequencies of 0.1, 0.15, 0.2, 0.25, 0.3, 0.35, 0.4 Hz of equal amplitude and arbitrary (random) relative phases. At time $t=130$ s the frequency of the last sinusoid changed to 0.45 Hz. The spectra on the left of the figure corresponded to $\lambda=0.98$ and on the right to $\lambda=1$. The spectra also corresponded to time instants $t=130$, $t=200$, $t=240$, and the order of the AR model was 21. The SNR was 25 dB in all cases. Owing to the non-unity value of λ the spectral line at 0.4 Hz has been forgotten at time $t=240$, contrary to the spectrum on the right.

Table 8.13 Computation of non-symmetric quantiles via symmetric variables for the adaptive FBLS algorithm

(1) $e_p^{gr}(N)=e_p^{gs}(N)-\lambda\varepsilon_p^{g}(N-1)$

(2) $e_p^{\mathrm{b}r}(N)=e_p^{gr}(N)+\lambda\alpha_p^{u}(N-1)\varepsilon_p^{w}(N-1)e_p^{\mathrm{b}}(N-1)$

(3) $e_p^{\mathrm{b}}(N)=\alpha_p^{\mathrm{b}}(N-1)e_{p-1}^{gs}(N)\alpha_{p-1}^{-1}(N-1)-e_p^{\mathrm{b}r}(N)$

(4) $\varepsilon_p^{gr}(N)=e_p^{\mathrm{b}r}(N)\alpha_p^{-w}(N-1)$

(5) $k_{p+1}^{ur}(N)=-\lambda^{-2}e_p^{\mathrm{b}r}(N)\alpha_p^{-\mathrm{b}}(N-1)$

(6) $k_{p+1}^{u}(N)=-\lambda^{-2}e_p^{\mathrm{b}}(N)\alpha_p^{-\mathrm{b}}(N-1)$

(7) $\alpha_p^{g}(N)=\lambda^2\alpha_p^{\mathrm{b}}(N-1)+\varepsilon_p^{gr}(N)e_p^{\mathrm{b}r}(N)$

(8) $\alpha_{p+1}^{u}(N)=\alpha_p^{w}(N-1)-e_p^{\mathrm{b}r}(N)k_{p+1}^{ur}(N)$

(9) $\alpha_p^{v}(N)=\alpha_{p+1}^{u}(N)+e_p^{b}(N)k_{p+1}^{u}(N)$

(10) $e_p^{w}(N)=\lambda\alpha_p^{v}(N)-\lambda\alpha_{p+1}^{us}(N)-\lambda[e_p^{\mathrm{b}}(N)+e_p^{\mathrm{b}r}(N)]k_{p+1}^{u}(N)$

(11) $\alpha_p^{u}(N)=\alpha_p^{v}(N)-\lambda^{-2}(e_p^{w}(N))^2\alpha_p^{-w}(N-1)$

(12) $\varepsilon_p^{w}(N)=1-\alpha_p^{us}(N)\alpha_p^{-u}(N)$

(13) $\alpha_p^{w}(N)=\alpha_p^{us}(N)[1+\varepsilon_p^{w}(N)]$

(14) $\varepsilon_p^{g}(N)=e_p^{\mathrm{b}}(N)+\lambda^{-1}e_p^{w}(N)\varepsilon_p^{gr}(N)$

(15) $\varepsilon_p^{\mathrm{b}}(N)=\varepsilon_p^{g}(N)\alpha_p^{-u}(N)$

(16) $\alpha_p^{\mathrm{b}}(N)=\alpha_p^{g}(N)+\varepsilon_p^{\mathrm{b}}(N)\varepsilon_p^{g}(N)$

(17) $\mathbf{b}_{p+1}^{s1}(N)=\begin{bmatrix}1\\ \mathbf{g}_{p-1}^{s}(N)\\ 1\end{bmatrix}\alpha_p^{\mathrm{b}}(N)\alpha_{p-1}^{-1}(N)$

(18) $\mathbf{b}_p^{s}(N)=\mathbf{g}_p^{s}(N)+\mathbf{u}_p^{s}(N)\varepsilon_p^{\mathrm{b}}(N)$

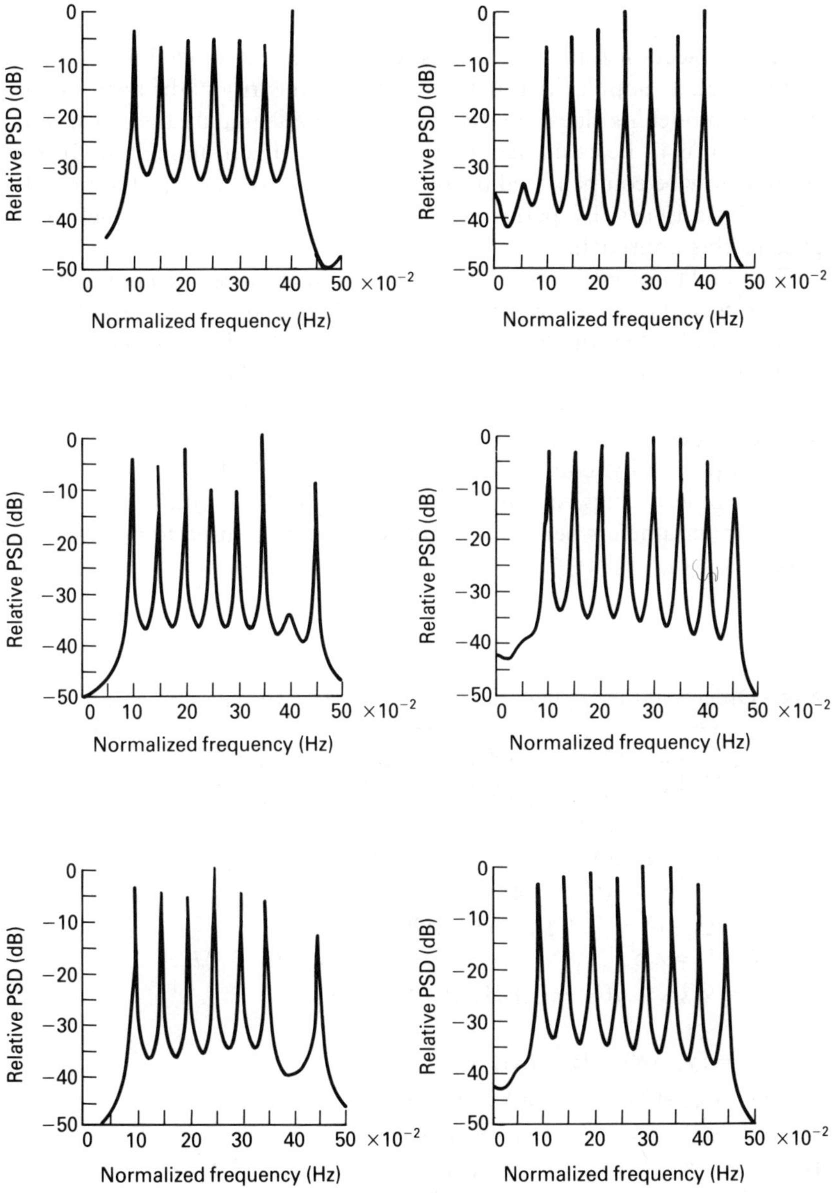

Figure 8.5 Power spectral densities obtained with the adaptive symmetric algorithm for $\lambda = 0.98$ (left) and $\lambda = 1$ (right)

8.5 ARMA spectral analysis: basic directions

In the previous section our interest was focused on AR modelling. As has already been pointed out, its popularity is partly due to the linearity of the resulting equations. However, AR modelling does not always describe adequately the stochastic process whose spectrum is to be estimated, due to the lack of zero modelling. It is true that this can somehow be compensated by allowing longer AR models. On the other hand, spectral analysis is usually performed with short data lengths which is a serious hindrance to this approach.

In Chapter 2 the prediction error method was introduced as a popular, non-linear optimization technique and applied to ARMAX modelling. The algorithm of Table 2.2 is readily applicable for the ARMA model if we let the exogenous input $u(t)$ be identically zero. Furthermore, the fast computational schemes, discussed in Chapter 5, can be employed to update the gain with a complexity linearly dependent on the order of the model. This can be done by embedding the problem in a multichannel formulation. Specifically, consider $y^{\mathrm{f}}(N)$ and $e^{\mathrm{f}}(N)$, of Table 2.2, as the two channels of an input sample, as suggested in Chapter 2. Any fast algorithm, with the provision of different tap parameters per input channel, can now be used to update the Kalman gain [50].

Table 8.14 Order-and-time updates for the derivation of the adaptive FBLS algorithm

(1) $$Q_{m+1,N}(N) = \begin{bmatrix} S_{m,N}(N) & [\mathbf{s}_{m,N}(N)] \\ \mathbf{s}^{\mathrm{T}}_{m,N}(N) & [s^{0}_{m,N}(N)] \end{bmatrix} = \begin{bmatrix} s^{0}_{m,N}(N) & [\mathbf{s}^{\mathrm{T}}_{m,N}(N)] \\ J\mathbf{s}_{m,N}(N) & [S_{m,N}(N)] \end{bmatrix}$$

(2) $$Q_{m+2,N}(N) = \begin{bmatrix} q^{0}_{m,N}(N) & \mathbf{q}^{\mathrm{T}}_{m,N}(N) & s_{m+1,N}(N) \\ \mathbf{q}_{m,N}(N) & Q_{m,N}(N-1) & J\mathbf{q}_{m,N}(N) \\ s_{m+1,N}(N) & \mathbf{q}^{\mathrm{T}}_{m,N}(N)J & q^{0}_{m,N}(N) \end{bmatrix}$$

(3) $$S_{m,N}(N) = Q_{m,N}(N-1) + \mathbf{y}_{m,N}(N)\mathbf{y}^{\mathrm{T}}_{m,N}(N)$$

(4) $$Q_{m,N}(N-1) = \lambda^2 S_{m,N-1}(N-1) + J\mathbf{y}_{m,N}(N-1)\mathbf{y}^{\mathrm{T}}_{m,N}(N-1)J$$

(5) $$Q_{m,N}(N) = \lambda^2 Q_{m,N-1}(N-1) + \tilde{\mathbf{H}}_{m,N}(N)\tilde{\mathbf{H}}^{\mathrm{T}}_{m,N}(N)$$

where

$$\tilde{\mathbf{H}}_{m,N}(N) = [\mathbf{y}_{m,n}(N),\quad J\mathbf{y}_{m,N}(N)]$$

(6) $$\mathbf{s}_{m,N}(N) = J\mathbf{q}_{m,N}(N) + \lambda^{m}\mathbf{y}_{m,N}(N)y(N-m)$$

(7) $$J\mathbf{q}_{m,N}(N) = \lambda^2 \mathbf{s}_{m,N-1}(N-1) + J\mathbf{y}_{m,N}(N-1)y(N)$$

(8) $$\mathbf{q}_{m,N}(N) = \lambda^2 \mathbf{q}_{m,N-1}(N-1) + \mathbf{y}_{m,N-1}(N-1)\,[\lambda y(N) + \lambda^{m+2}y(N-m-1)]$$

(9) $$s_{m+1,N}(N) = \lambda^2 s_{m+1,N-1}(N-1) + 2\lambda^{m+1}y(N-m-1)y(N)$$

(10) $$q^{0}_{m,N}(N) = \lambda^2 q^{0}_{m,N-1}(N-1) + y^2(N) + \lambda^{2(m+1)}y^2(N-m+1)$$

(11) $$\mathbf{y}_{m+1,N}(n) = \begin{bmatrix} \lambda^{N-n}y(n) \\ \mathbf{y}_{m,N}(n-1) \end{bmatrix} = \begin{bmatrix} \mathbf{y}_{m,N}(n) \\ \lambda^{N+m-n}y(n-m) \end{bmatrix} = \begin{bmatrix} \lambda^{N-n}y(n) \\ \mathbf{y}_{m-1,N}(n-1) \\ \lambda^{N+m-n}y(n-m) \end{bmatrix}$$

A different type of approach is to estimate the AR and MA parts of the ARMA model separately. This suboptimum class of techniques exploits the specific nature of the ARMA model. Multiplying both sides of (8.28) by $y(t-m)$ and taking expected values results in the following set of equations:

$$\rho(m) = -\sum_{k=1}^{n_a} a_k \rho(m-k) \qquad m \geqslant n_c + 1 \tag{8.131}$$

where the white-noise nature of the model $\eta(t)$ has been taken into consideration. Assuming that the autocorrelation lags are known the AR parameters are obtained as the solution of a linear system of equations which results from the above equation for $m = n_c + 1, n_c + 2, \ldots, n_c + n_a$. The resulting system of equations is known as the *modified Yule–Walker equations.* In practice the autocorrelation lags are not known and their estimates $\hat{\rho}(m)$ are used instead and computed from the input sequence.

A refinement of the above technique stems from the observation that (8.131) can be interpreted as a prediction operation, where the correlation estimates $\hat{\rho}(k)$ constitute the time series. Let us assume that the data length of the input sequence is long enough so that 'good' estimates can be obtained for correlation lags greater than $n_a + n_c$. This gives the possibility of collecting at least $n_a + 1$ equations of the type (8.131), for $m = n_c + 1, n_c + 2, \ldots$. The predictor (AR) parameters can then be obtained as the least-squares estimates. This can be achieved using the covariance algorithm of section 8.4.2 and using $\hat{\rho}(k)$ as the input samples [51,52].

The next step is to estimate the MA parameters. Using the shift operator q^{-1}, (8.28) can be written as

$$y(t) = \frac{C(q)}{A(q)} \eta(t)$$

with

$$C(q) = 1 + c_1 q^{-1} + \ldots + c_{n_c} q^{-n_c}$$

$$A(q) = 1 + a_1 q^{-1} + \ldots + a_{n_a} q^{-n_a}$$

Let $\hat{A}(q)$ be the corresponding polynomial involving the obtained estimates $\hat{a}_i$ of a_i, $i = 1, 2, \ldots, n_a$. If we pass $y(t)$ through the linear filter $\hat{A}(q)$ the resulting time series $\hat{y}(t)$ is

$$\hat{y}(t) = \frac{C(q)}{A(q)} \hat{A}(q)\eta(t) \approx C(q)\eta(t)$$

which is an approximate MA process of order n_c. To compute the estimates of the moving average parameters we recall the equivalence of an MA model with an AR(∞) one, as discussed in section 8.3. A long enough AR model is adapted to fit the filtered time series $\hat{y}(t)$. The MA parameters result from (8.87), where the right-hand side is now identically zero. Variations of this basic scheme are also available. The interested reader may consult [2] and the references therein.

8.6 Modified Prony's technique for sinusoidal modelling

So far we have been concerned with modelling the regular part resulting from the spectral decomposition of a WSS process. In this section we will focus on the predictable part assuming that it is a harmonic process. The original Prony's technique was first developed for modelling sampled data in terms of exponentials. Here we will present the LS version of Prony's technique as applied to sinusoidal modelling, known as the modified Prony's method.

Assume that our input signal known in the interval $[M, N]$ consists of p sinusoids, that is,

$$y(n) \equiv y(nT) = \sum_{i=1}^{p} A_i \cos(2\pi f_i nT + \phi_i) \tag{8.132}$$

The task is to estimate the optimum (in the least-squares sense) parameters A_i, f_i, ϕ_i, $i = 1, 2, \ldots, p$. This is a non-linear optimization problem. However, following Prony's idea, a suboptimal solution is possible which drastically simplifies the procedure. Its philosophy lies in the decoupling of the above parameters during the optimization process. This is achieved as follows.

Using Euler's formulae (8.132) becomes

$$y(n) = \sum_{i=1}^{p} (c_i z_i^n + c_i^* z_i^{*n}) \tag{8.133a}$$

with

$$c_i = \frac{A_i}{2} e^{j\phi_i} \tag{8.133b}$$

$$z_i = e^{j2\pi f_i} \tag{8.133c}$$

Equation (8.133a) can now be thought of as the solution of a linear difference equation whose characteristic polynomial $P(z)$ has z_i, z_i^* as its roots. That is,

$$P(z) = \prod_{i=1}^{p} (z - z_i)(z - z_i^*) \tag{8.134}$$

Since for each z_i its complex conjugate z_i^* is also a root and z_i are of unit modulus, $P(z)$ is a symmetric real polynomial and is written as

$$P(z) = \sum_{i=1}^{2p} g_{|i-p|} z^i \tag{8.135}$$

Obviously, $g_0 = g_{2p} = \Pi_{i=1}^{p}(z_i z_i^*) = 1$. Thus the difference equation obeyed by $y(n)$ is

$$y(n) + y(n - 2p) = -\sum_{i=1}^{p} g_i[y(n - i) + y(n - 2p + i)] \tag{8.136}$$

In the least-squares version of Prony's technique, where the number of data points exceeds the number of equations and observation noise as well as modelling approximation errors are present, g_i are estimated so as to minimize the total power in $[M, N]$ of the error

$$\varepsilon(n) \equiv \hat{y}(n) + \hat{y}(n-2p) - [y(n) + y(n-2p)]$$

where $\hat{y}(n) + \hat{y}(n-2p)$ is now the LS optimal estimate

$$\hat{y}(n) + \hat{y}(n-2p) = -\sum_{i=1}^{p} g_i[y(n-i) + y(n-2p+i)]$$

The symmetric predictor parameterized in terms of g_i is readily recognized as the bidirectional predictor introduced in section 8.4.3. Using the above methodology, computations of f_i, A_i, ϕ_i have been decoupled to the following stages:

Stage 1. Estimate the symmetric bidirectional predictor of order $2p$. This can be done using the algorithm of Table 8.9. Thus a saving of $6m + 0.5L$ multiplications is achieved, compared with the algorithms of [2]. When a parallel processing environment is available, a Schur-type scheme for the computation of the inner products of Table 8.9 is possible and has been presented in [36].

Stage 2. Compute the roots of polynomials $P(z)$ in (8.135). This is carried out by any root-finding algorithm. It must be emphasized that the resulting roots may not necessarily have unit magnitude although our experience with the bidirectional predictor has shown that this very rarely happens in practice. Once the roots are computed the frequencies f_i are estimated.

Stage 3. When the estimates $\hat{f}_i$ of the frequencies have been obtained the respective amplitudes can be estimated using another LS problem. The original equation (8.132) now becomes

$$\hat{y}(n) = \sum_{i=1}^{p} A_i \cos(2\pi \hat{f}_i nT)$$

Collecting the above equation for all η in the interval $[M, N]$ we get

$$V\mathbf{A} = \hat{\mathbf{Y}}$$

where

$$V = \begin{bmatrix} 1 & 1 & \cdots & 1 \\ \cos(2\pi\hat{f}_1 T) & \cos(2\pi\hat{f}_2 T) & \cdots & \cos(2\pi\hat{f}_p T) \\ \vdots & \vdots & & \vdots \\ \cos(2\pi\hat{f}_1 pT) & \cos(2\pi\hat{f}_2 pT) & \cdots & \cos(2\pi\hat{f}_p \mathrm{pT}) \end{bmatrix}$$

and

$$\mathbf{A} = [A_1, A_2, \ldots, A_p]^{\mathrm{T}}$$
$$\hat{\mathbf{Y}} = [\hat{y}(M), \hat{y}(M+1), \ldots, \hat{y}(N)]^{\mathrm{T}}$$

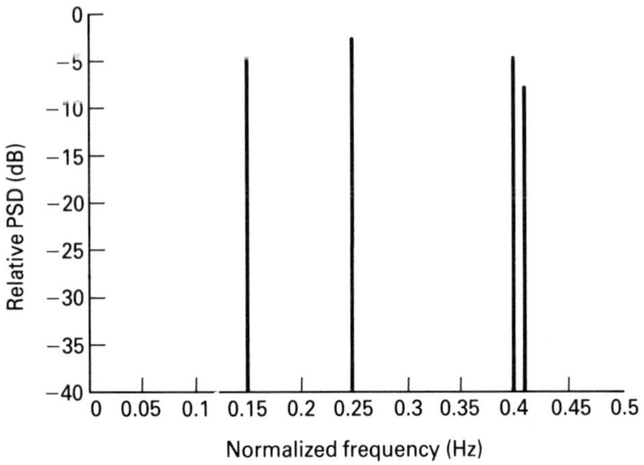

Figure 8.6 Spectral lines resulting from Prony's technique

The matrix V is known as the Vandermode matrix. The LS estimate of **A** minimizes the sum $\Sigma_{i=M}^{N}(y(i) - \hat{y}(i))^2$ and is given by

$$\mathbf{A} = [V^{\mathrm{T}}V]^{-1}V^{\mathrm{T}}\mathbf{Y}$$

where

$$\mathbf{Y} = [y(M)\ y(M+1) \ldots y(N)]$$

Efficient schemes for the inversion of the Grammian of the Vandermode matrix have been presented in [53].

Further discussion on Prony's method treating the issues of additive observation noise can be found in [2].

Figure 8.6 shows the Prony spectrum resulting from the above procedure. The input series consisted of four sinusoids at the normalized frequencies of 0.15, 0.25, 0.4 and 0.41 Hz with amplitudes of 0.7, 1.0, 0.7 and 0.4 respectively.

Appendix A

This appendix summarizes the basic steps for the derivation of the algorithm given Table 8.2. A full proof is given in [19].

A close look at Table 8.2 reveals the following four types of recursions:

1. Vector order updates of the form (26).
2. Vector time updates of the form (14).
3. Scalar time updates of the form (20).
4. Scalar order updates of the form (37).

The first type of order update is the direct result of applying the following lemma to the respective definitions of the involved quantities.

INVERSION LEMMA I
Given the partitioned linear system of equations

$$\begin{bmatrix} A & B \\ C & D \end{bmatrix}\begin{bmatrix} X \\ Y \end{bmatrix} = -\begin{bmatrix} E \\ F \end{bmatrix} \tag{A.1}$$

the solution can be expressed as

$$\begin{bmatrix} X \\ Y \end{bmatrix} = \begin{bmatrix} -A^{-1}E \\ 0 \end{bmatrix} + \begin{bmatrix} -A^{-1}B \\ 1 \end{bmatrix} Y \tag{A.2a}$$

$$Y = -(D - CA^{-1}B)^{-1}(F - CA^{-1}E) \tag{A.2b}$$

or alternatively

$$\begin{bmatrix} X \\ Y \end{bmatrix} = \begin{bmatrix} 0 \\ -D^{-1}F \end{bmatrix} + \begin{bmatrix} 1 \\ -D^{-1}C \end{bmatrix} X \tag{A.3a}$$

$$X = -(A - BD^{-1}C)^{-1}(E - BD^{-1}F) \tag{A.3b}$$

The proof is easily established by inspection. Inverses are assumed to exist. Equation (26) of Table 8.2 for the forward predictor is the result of applying the above lemma to (8.105a), after the following substitutions:

$$A \to R_m(M+1, N) \quad B \to \mathbf{r}_m^{\mathrm{b}}(M, N) \quad C \to \mathbf{r}_m^{\mathrm{bT}}(M, N)$$

$$Y \to k_m^{\mathrm{f}}(M, N) \quad E \to \mathbf{r}_m^{\mathrm{f}}(M, N) \quad F \to r_{m+1}^{\mathrm{f}}(M, N)$$

$$D \to r_{0m}^{\mathrm{b}}(M, N)$$

The element $r_{m+1}^{\mathrm{f}}(M, N)$ results from the partitioning of the right-hand side of (8.101), that is,

$$\mathbf{r}_{m+1}^{\mathrm{fT}}(M, N) = [\mathbf{r}_m^{\mathrm{fT}}(M+1, N), r_{m+1}^{\mathrm{f}}(M, N)] \tag{A.4a}$$

$$r_{m+1}^{\mathrm{f}}(M, N) = \sum_{t=M+m}^{N-1} y(t-m)y(t+1) \tag{A.4b}$$

Quantity $D - CA^{-1}B \equiv r_{0m}^{\mathrm{b}}(M, N) + \mathbf{r}_m^{\mathrm{bT}}(M, N)\mathbf{b}_m(M, N)$ is the LS optimum backward error power. This is easily established after substituting the LS optimum backward predictor in the defining equation $\alpha_m^{\mathrm{b}}(M, N) = \Sigma_{t=M+m}^{N}(y(t-m) + \mathbf{b}_m^{\mathrm{T}}(M, N)\mathbf{y}_m(t))^2$. The order recursions for the rest of the quantities, i.e. $\mathbf{b}_m(M, N)$, $\mathbf{w}_m(M, N)$, $\mathbf{q}_m(M, N)$, are derived similarly. These vectors, whose definitions are provided in Table 8.3, are the result of the following lemma, underlying the second type of update, that is, the vector time updates.

INVERSION LEMMA II

Given the linear system of equations

$$(A \pm BB^{\mathrm{T}})X = C + BD \tag{A.5}$$

then

$$X = A^{-1}C - A^{-1}BE \tag{A.6a}$$

$$E = -(I \pm B^{\mathrm{T}}A^{-1}B)^{-1}(D \mp B^{\mathrm{T}}A^{-1}C) \tag{A.6b}$$

As with the previous lemma the proof is readily checked by substitution and the matrix inverses are assumed to exist. For the evolution of the algorithm it is necessary to time-shift the quantities $\mathbf{a}_m(M+1, N)$, $\mathbf{b}_m(M, N-1)$ in the interval $[M,N]$, where they are assumed to be known from the previous order recursion. We shall concentrate on the first of the two quantities. The respective equation (14) (Table 8.2) is the result of applying Lemma II in (8.107b) after the following substitutions:

$$A \to R_m(M, N-1) \quad B \to \mathbf{y}_m(M+m-1)$$

$$C \to -\mathbf{r}^{\mathrm{f}}_m(M, N) \quad D \to y(M+m)$$

The last two equations are easily understood from the defining equation (8.103). It is interesting to highlight the physical meaning of the two quantities introduced, i.e. $\mathbf{q}_m(M, N)$, $\alpha^q_m(M, N)$ (Table 8.2). The vector $\mathbf{q}_m(M, N)$ can be thought of as the LS optimum filter whose input $y(t)$ is forced to track, in the LS sense, the desired response sequence $\delta(t - M - m + 1)$ which is zero everywhere except at time $M + m - 1$, where it is one. $\alpha^q_m(M, N)$ is the corresponding error power. A similar procedure is followed to show time corrections with respect to N. These include the vector $\mathbf{w}_m(M, N)$ and the scalar $\alpha^*_m(M, N)$; $\mathbf{w}_m(M, N)$ is identical to the gain vector present in the RLS and its fast versions. The quantity $\alpha^*_m(M, N)$ is directly related to the angle variable appearing in the RLS and lattice schemes. Indeed, under the prewindowing assumption

$$\alpha^*_m(M, N) = 1 + \mathbf{y}^{\mathrm{T}}_m(N)R^{-1}_m(N-1)\mathbf{y}_m(N)$$

The angle variable is defined as

$$\alpha_m(N) = 1 - \mathbf{y}^{\mathrm{T}}_m(N)R^{-1}_m(N)\mathbf{y}_m(N)$$

Substituting $R^{-1}_m(N)$ in terms of $R^{-1}_m(N-1)$ and using the inversion lemma II, it is straightforward to show that

$$\alpha_m(N) = \frac{1}{\alpha^*_m(N)}$$

$\alpha_m(N)$ has an interesting interpretation. It is the error power of the optimum LS filter, forced to track the desired response $\delta(t - N)$, which is zero everywhere except at time $t = N$, where it is one. Being an error power, hence positive, and recalling the positive definiteness of $R_m(N)$, it is readily observed that

$$0 \leqslant \alpha_m(N) \leqslant 1$$

This justifies its name as an angle variable. Obviously, this is also true for the likelihood variable, i.e.

$$0 \leqslant \gamma_m(N) \leqslant 1$$

From the above relationships it is not difficult to show that the posterior Kalman gain, introduced in Chapter 2, is equal to

$$\mathbf{w}_m(N) = -R_m^{-1}(N)\mathbf{y}_m(N)$$

The third type of update, that is, scalar time updates (i.e. Equation (20)), results from the respective definitions and the corresponding vector time updates. For example, the defining equation for the forward error power is

$$\alpha_m^{\mathrm{f}}(M+1, N) = r_{0m}^{\mathrm{f}}(M+1, N) + \mathbf{r}_m^{\mathrm{fT}}(M+1, N)\mathbf{a}_m(M+1, N) \tag{A.7}$$

However,

$$r_{0m}^{\mathrm{f}}(M+1, N) = r_{0m}^{\mathrm{f}}(M, N) - y(M+m-1)y(M+m-1) \tag{A.8a}$$

$$\mathbf{r}_m^{\mathrm{f}}(M+1, N) = \mathbf{r}_m^{\mathrm{f}}(M, N) - \mathbf{y}_m(M+m-1)y(M+m) \tag{A.8b}$$

$$\mathbf{a}_m(M+1, N) = \mathbf{a}_m(M, N) + \mathbf{q}_m(M, N-1)\tilde{k}_m^{\mathrm{f}}(M, N) \tag{A.8c}$$

The combination of (A.8) and (A.7) leads to (20). The scalar order updates are the result of the vector order updates of Table 8.2 and the related partitioning properties of the vectors involved in the respective definitions (i.e. Equation (27) of Table 8.2, (A.4a), as well as the partitions appearing in (8.104)).

In what follows we will concentrate on the proof of the time shift for $e_m^q(n \mid M, N)$. From the respective definition in Table 8.4 we have

$$e_m^q(n \mid M, N-1) = \mathbf{q}_m^{\mathrm{T}}(M, N-1)\mathbf{r}_m^{\mathrm{b}}(n \mid M, N-1) \tag{A.9}$$

Recalling (8) of Table 8.2 and the definition in (8.111) we have the following:

$$\begin{aligned} e_m^q(n \mid M, N-1) = {} & \mathbf{q}_m^{\mathrm{T}}(M, N)\mathbf{r}_m^{\mathrm{b}}(n \mid M, N) + \tilde{k}_m^q(M, N)\mathbf{w}_m^{\mathrm{T}}(M, N)\mathbf{r}_m^{\mathrm{b}}(n \mid M, N-1) \\ & + \tilde{k}_m^q(M, N)\mathbf{w}_m^{\mathrm{T}}(M, N)\mathbf{y}_m(N)y(N-n) \\ & + \mathbf{q}_m^{\mathrm{T}}(M, N-1)\mathbf{y}_m(N)y(N-n) \end{aligned} \tag{A.10}$$

After substituting $\mathbf{q}_m(M, N-1)$ in the last term in (A.10) from its definition in Table 8.3 and the definition of $\alpha_m^*(M, N)$ in the same table, we obtain

$$\begin{aligned} e_m^q(n \mid M, N-1) = {} & e_m^q(n \mid M, N) + \tilde{k}_m^q(M, N)\mathbf{w}_m^{\mathrm{T}}(M, N)\mathbf{r}_m^{\mathrm{b}}(n \mid M, N-1) \\ & + \tilde{k}_m^q(M, N)[1 - \alpha_m^*(M, N)]y(N-n) \\ & - \mathbf{w}_m^{\mathrm{T}}(M, N)\mathbf{y}_m(M+m-1)y(N-n) \end{aligned} \tag{A.11}$$

Inserting (6) and (7) of Table 8.2 into (A.11) finally results in

$$e_m^q(n \mid M, N-1) = e_m^q(n \mid M, N) + \tilde{k}_m^q(M, N)e_m^w(n \mid M, N)$$

A similar approach is followed for the formula $e^b_m(n \mid M, N-1)$.

Appendix B

This appendix will detail the derivation of the algorithm of Table 8.9 and is complementary to section 8.4.3. It consists of three parts. In the first one, formulae (8.128) and (8.129) are derived. In the second one the order recursive algorithm for the efficient computation of $\mathbf{g}^s_m(M, N)$ (Table 8.9) is developed. The third part outlines the proof for the lattice formulae of Table 8.10.

PART 1

In Appendix A we gave a set of basic formulae related to the matrix inversion lemma for (upper and lower) partitioned matrices. Here we provide their counterpart for the middle partitioning case. Consider the following linear set of equations:

$$A\mathbf{y} = -\mathbf{H} \tag{B.1}$$

where A and $\mathbf{H}$ are partitioned as follows:

$$A = \begin{bmatrix} a^0 & \mathbf{w}^{\mathrm{T}} & b^0 \\ \mathbf{w} & \tilde{A} & J\mathbf{w} \\ b^0 & \mathbf{w}^{\mathrm{T}}J & a^0 \end{bmatrix} \qquad H = \begin{bmatrix} h_1 \\ \mathbf{h} \\ h_1 \end{bmatrix} \tag{B.2}$$

and $\tilde{A}$ is an invertible matrix. Then it is easily checked by substitution that the solution of (B.1) satisfies the following [45]:

$$\mathbf{y} = \begin{bmatrix} \mathbf{0} \\ \tilde{\mathbf{y}} \\ 0 \end{bmatrix} + \begin{bmatrix} 1 \\ \mathbf{z} \\ 1 \end{bmatrix} k \qquad k = -a^{-1}\beta \tag{B.3a}$$

where

$$\tilde{A}\tilde{\mathbf{y}} = -\mathbf{h} \qquad \tilde{A}\mathbf{z} = -(\mathbf{w} + J\mathbf{w}) \tag{B.3b}$$

$$a = a^0 + b^0 + \mathbf{w}^{\mathrm{T}}\mathbf{z} \tag{B.3c}$$

$$\beta = h_1 + \mathbf{w}^{\mathrm{T}}\tilde{\mathbf{y}} \tag{B.3d}$$

Inserting (B.3) into (8.125) and (8.126) and making the following substitutions

$$Q_m(M+1, N-1) \leftarrow \tilde{A} \qquad \mathbf{q}_m(M, N) \leftarrow \mathbf{w}$$
$$q^0_m(M, N) \leftarrow a^0 \qquad s_{m+1}(M, N) \leftarrow b^0$$
$$\mathbf{0} \leftarrow \mathbf{h} \qquad \alpha^{\mathrm{b}}_{m+1}(M, N) \leftarrow h_1$$

we obtain (8.127) with

$$a \to \alpha_m(M, N) = q^0_m(M, N) + s_{m+1}(M, N) + \mathbf{q}^{\mathrm{T}}_m(M, N)\mathbf{g}^s_m(M, N) \tag{B.4a}$$

$$\mathbf{z} \rightarrow \mathbf{g}^s_m(M, N) = -Q_m^{-1}(M+1, N-1)[\mathbf{q}_m(M, N) + J\mathbf{q}_m(M, N)] \quad \text{(B.4b)}$$

To establish (8.129) recall the definition (8.118a) for $\mathbf{b}_m(M, N)$ and relations (1) and (3) of Table B.1, summarizing basic relations. The following results:

Table B.1 Basic time-and-order relations for the derivation of the symmetric FBLS algorithm

(1)	$S_m(M, N) = Q_m(M+1, N-1) + J\mathbf{H}_m(M, N)\mathbf{H}^T_m(M, N)J$
(2)	$S_m(M, N) = Q_m(M, N) - \mathbf{H}_m(M, N)\mathbf{H}^T_m(M, N)$
(3)	$\mathbf{s}_m(M, N) = J\mathbf{q}_m(M, N) + J\mathbf{H}_m(M, N)h_m(M, N)$
(4)	$Q_m(M+1, N-1) = Q_m(M, N) - [\mathbf{H}_m(M, N)\ J\mathbf{H}_m(M, N)]\begin{bmatrix}\mathbf{H}^T_m(M, N)\\ \mathbf{H}_m(M, N)J\end{bmatrix}$
(5)	$\mathbf{r}^f_m(M, N-1) = \mathbf{r}^f_m(M, N) - J\mathbf{H}_m(M, N-1)\begin{bmatrix}0\\ y(N)\end{bmatrix}$
(6)	$\mathbf{r}^b_m(M+1, N) = \mathbf{r}^b_m(M, N) - \mathbf{H}_m(M+1, N)\begin{bmatrix}y(M)\\ 0\end{bmatrix}$
(7)	$\mathbf{r}^f_{m+1}(M, N) = \begin{bmatrix}\mathbf{r}^f_m(M, N)\\ r^f_{m+1}(M, N)\end{bmatrix} - \begin{bmatrix}\mathbf{y}_m(M+m-1)y(M+m)\\ 0\end{bmatrix}$
(8)	$\mathbf{r}^b_{m+1}(M, N) = \begin{bmatrix}r^b_{m+1}(M, N)\\ \mathbf{r}^b_m(M, N)\end{bmatrix} - \begin{bmatrix}0\\ \mathbf{y}_m(N)y(N-m)\end{bmatrix}$
(9)	$\mathbf{H}_{m+1}(M, N) = \begin{bmatrix}h^T_m(M, N)\\ \mathbf{H}_m(M, N)\end{bmatrix} = \begin{bmatrix}\mathbf{H}_m(M+1, N-1)\\ h^T_0(M, N)\end{bmatrix} = \begin{bmatrix}h^T_{m+1}(M, N)\\ \mathbf{H}_{m-1}(M+1, N-1)\\ h^T_0(M, N)\end{bmatrix}$
(10)	$\mathbf{g}^s_m(M, N) = \mathbf{g}_m(M, N) + J\mathbf{g}_m(M, N)$
(11)	$\mathbf{u}^s_m(M, N) = \mathbf{u}_m(M, N) + J\mathbf{u}_m(M, N)$

$$\mathbf{b}_m(M, N) = \mathbf{g}_m(M, N) + \mathbf{u}_m(M, N)\varepsilon^b_m(M, N) \quad \text{(B.5)}$$

Equation (8.129) is now the direct result of (B.5) and (8.119b) using (10) and (11) of Table B.1. In the above derivations the symmetric and persymmetric properties of Q have been exploited.

PART 2

We shall first show that $\mathbf{g}^s_m(M, N)$ is the LS optimal symmetric filter predicting simultaneously in the forward and backward directions, as stated in section 8.4.3. That is, given the samples $y(n-m), \ldots, y(n-1)$ it predicts the sum of $y(n) + y(n-m-1)$. It is already well established [2,28] that the normal equations of a filter, constrained to be symmetric, are of the form $(R + JRJ)\mathbf{g}^s = -(\mathbf{r} + J\mathbf{r})$, where R, $\mathbf{r}$ are the covariance matrix and the cross-covariance vector, respectively, of the unconstrained problem. For the particular case of interest here the covariance matrix is defined in the interval $[M+1, N-1]$

$$\sum_{n=M+m}^{N-1} \mathbf{y}_m(n)\mathbf{y}_m^{\mathrm{T}}(n) = R_m(M+1, N-1)$$

and the input desired output cross-correlation vector is

$$\sum_{n=M+m}^{N-1} \mathbf{y}_m(n)[y(n+1) + y(n-m)] = -\mathbf{r}_m^{\mathrm{f}}(M+1, N) - \mathbf{r}_m^{\mathrm{b}}(M, N-1)$$

where the definitions of Table 8.6 have been used. Thus the optimum symmetric filter predicting the sum $y(n) + y(n-m-1)$ is given by

$$\begin{aligned} Q_m(M+1, N-1)\mathbf{g}_m^s(M, N) &= -\mathbf{r}_m^{\mathrm{f}}(M+1, N) - \mathbf{r}_m^{\mathrm{b}}(M, N-1) \\ &\quad - J\mathbf{r}^{\mathrm{f}}(M+1, N) - J\mathbf{r}_m^{\mathrm{b}}(M, N-1) \\ &\equiv -\mathbf{q}_m(M, N) - J\mathbf{q}_m(M, N) \end{aligned}$$

according to the definitions given in Table 8.6. From a closer look at the right-hand side of the above equation and combining the two $\mathbf{r}^{\mathrm{f}}$ and $\mathbf{r}^{b}$ terms, it can be seen that

$$\mathbf{g}_m^s(M, N) = \mathbf{a}_m^s(M+1, N) + \mathbf{b}_m^s(M, N-1) \tag{B.6}$$

where $\mathbf{a}_m^s$ and $\mathbf{b}_m^s$ are the LS symmetric forward and backward predictors respectively. We will use the above relationship to derive the order recursion of $\mathbf{g}_m^s$. The proof is quite tricky and we will describe the philosophy behind the procedure in some detail.

From the definition of $\mathbf{a}_m^s$ we have

$$Q_m(M+1, N-1)\mathbf{a}_m^s(M+1, N) = -[\mathbf{r}_m^{\mathrm{f}}(M+1, N) + J\mathbf{r}_m^{\mathrm{f}}(M+1, N)] \tag{B.7}$$

We consider the product

$$Q_{m+2}(M+1, N-1)\begin{bmatrix} \mathbf{a}_{m+1}^s(M+1, N) \\ 1 \end{bmatrix} \equiv X$$

First we substitute Q_{m+2} by its partition in terms of S_{m+1} (8.120b) and use relations (2) and (5) of Table B.1 and the definition of $\mathbf{s}_m$ in Table 8.6. Then we carry out the multiplication between the resulting matrix and the vector to show that the above product is equal to

$$\begin{aligned} X = \begin{bmatrix} -\mathbf{r}_{m+1}^{\mathrm{f}}(M+1, N) \\ \varepsilon_{m+1}^{s,a}(M+1, N) \end{bmatrix} + \begin{bmatrix} \mathbf{r}_{m+1}^{\mathrm{b}}(M+1, N-1) \\ 0 \end{bmatrix} \\ - \begin{bmatrix} \mathbf{H}_{m+1}(M+1, N-1) \\ 0 \end{bmatrix} \varepsilon_{m+1}^{a}(M+1, N) \end{aligned} \tag{B.8}$$

where

$$\begin{aligned} \varepsilon_{m+1}^{s,a}(M+1, N) &= \mathbf{s}_{m+1}^{\mathrm{T}}(M+1, N-1)\mathbf{a}_{m+1}^s(M+1, N) \\ &\quad + s_{m+1}^{0}(M+1, N-1) \end{aligned} \tag{B.9a}$$

$$\varepsilon^{a}_{m+1}(M+1,\ N) = \begin{bmatrix} 0 \\ y(N) \end{bmatrix} + \mathbf{H}^{\mathrm{T}}_{m+1}(M+1,\ N-1)\mathbf{a}^{s}_{m+1}(M+1,\ N) \qquad \text{(B.9b)}$$

Next we consider the definition (B.7) for order $m+2$. We apply on the right-hand side the partition (7) of Table B.1 in the form

$$\begin{bmatrix} \mathbf{r}^{\mathrm{f}}_{m+1}(M+1,\ N) \\ -\varepsilon^{s,a}_{m+1}(M+1,\ N) \end{bmatrix} + \begin{bmatrix} \mathbf{0} \\ r^{\mathrm{f}}_{m+2}(M+1,\ N) + \varepsilon^{s,a}_{m+1}(M+1,\ N) \end{bmatrix}^{\mathrm{T}}$$

where $\varepsilon^{s,a}$ has been added and subtracted. In what follows we substitute the first summand above by its equivalent from (B.8) and multiply both sides by $Q^{-1}_{m+2}(M+1,\ N-1)$. The following results:

$$\mathbf{a}^{s}_{m+2}(M+1,\ N) = \begin{bmatrix} \mathbf{a}^{s}_{m+1}(M+1,\ N) \\ 1 \end{bmatrix} + \begin{bmatrix} 1 \\ \mathbf{a}^{s}_{m+1}(M+1,\ N) \end{bmatrix}$$
$$+ \begin{bmatrix} 0 \\ \mathbf{b}^{s}_{m}(M+1,\ N-2) \\ 0 \end{bmatrix} - \begin{bmatrix} 0 \\ \mathbf{u}^{s}_{m}(M+1,\ N-1) \\ 0 \end{bmatrix} \varepsilon^{a}_{m+1}(M+1,\ N)$$
$$+ \begin{bmatrix} 1 \\ \mathbf{g}^{s}_{m}(M+1,\ N-1) \\ 1 \end{bmatrix} k^{a}_{m+2}(M+1,\ N) \qquad \text{(B.10a)}$$

where

$$k^{a}_{m+2}(M+1,\ N) = -\alpha^{-1}_{m}(M+1,\ N-1)\beta^{a}_{m+2}(M+1,\ N) \qquad \text{(8.11b)}$$
$$\begin{aligned}\beta^{a}_{m+2}(M+1,\ N) = r^{\mathrm{b}}_{m+1}(M+1,\ N-1) &+ r^{\mathrm{f}}_{m+2}(M+1,\ N) \\ &+ \mathbf{q}^{\mathrm{T}}_{m}(M+1,\ N-1)\mathbf{b}^{s}_{m}(M+1,\ N-2) \\ &- \beta^{u}_{m}(M+1,\ N-1)\varepsilon^{a}_{m+1}(M+1,\ N) \\ &+ \varepsilon^{s,a}_{m+1}(M+1,\ N)\end{aligned} \qquad \text{(8.11c)}$$
$$\begin{aligned}\beta^{us}_{m}(M+1,\ N-1) = h_{m}(M+1,\ N-1)& \\ + \mathbf{q}^{\mathrm{T}}_{m}(M+1,\ N-1)&\mathbf{u}^{s}_{m}(M+1,\ N-1)\end{aligned} \qquad \text{(B.10d)}$$

In the above derivation the matrix inversion lemma (B.1)–(B.3) has been used repeatedly. A similar procedure is followed for the order recursive computation of $\mathbf{b}^{s}_{m+2}(M,\ N-1)$. Then, using the definition (B.6) and after some algebra recursion (14) for $\mathbf{g}^{s}_{m+2}$ in Table 8.9 is obtained, with the involved variables defined as

$$\beta^{gs}_{m+2}(M,\ N) = \beta^{0}_{m+2}(M,\ N) - \beta^{us}_{m}(M+1,\ N-1)\varepsilon^{gs}_{m+1}(M,\ N) \qquad \text{(B.11a)}$$
$$\beta^{0}_{m+2}(M,\ N) = r^{\mathrm{b}}_{m+2}(M,\ N-1) + r^{\mathrm{f}}_{m+2}(M+1,\ N) + \varepsilon^{sg}_{m+1}(M,\ N) \qquad \text{(B.11b)}$$
$$\varepsilon^{sg}_{m+1}(M,\ N) = \mathbf{s}^{\mathrm{T}}_{m+1}(M+1,\ N-1)\mathbf{g}^{s}_{m+1}(M,\ N) + s^{0}_{m+1}(M+1,\ N-1) \qquad \text{(B.11c)}$$
$$\varepsilon^{gs}_{m+1}(M,\ N) = h_{0}(M,\ N) + h_{m+1}(M+1,\ N-1) + \mathbf{H}^{\mathrm{T}}_{m+1}(M+1,\ N)\mathbf{g}^{s}_{m+1}(M,\ N) \qquad \text{(B.11d)}$$

In the algorithm β^{gs} is computed not via its definition in (B.11a) but indirectly, i.e. (11) of Table 8.9, which is shown by the definitions of the involved variables ((21), (22) of Table 8.9). Note also that $\mathbf{u}_m^{*s}$ has been used instead of $\mathbf{u}_m^s$, as pointed out in section 8.4.3 (Equation (8.130)). The derivation of (16) of Table 8.9 follows similar arguments as above. As was the case with β^{gs}, β^{u*s} defined as

$$\begin{aligned}\beta_{m+1}^{u*s}(M, N) = &-[h_m^{\mathrm{T}}(M+1, N-1) + \mathbf{q}_m^{\mathrm{T}}(M+1, N-1)\mathbf{u}_m^{*s}(M+1, N-1)] \\ &\times \alpha_{m+1}^{u*}(M+1, N-1) \\ &+ h_{m+1}^{\mathrm{T}}(M+1, N-1) + \mathbf{s}_{m+1}^{\mathrm{T}}(M+1, N-1)\mathbf{u}_{m+1}^{*s}(M+1, N-1)\end{aligned}$$

is also computed indirectly, in order to reduce complexity. The rest of the formulae involved in Table 8.9 result from the respective definitions and identities in Table B.1 [31, 34].

PART 3

The formulae of Table 8.10, relating errors and error powers, result from the equivalent relations among the different filters which are summarized in Table B.2, which results from the respective definitions in Table 8.8 [31]. For example, (2) of 8.10 results from (3a) of Table B.2, (3) of 8.10 from (2a) of B.2, (4) of 8.10 from (1a) of B.2 and (9) of B.1, (1) of 8.10 from (8.119a), (8.127) and (9) of B.2. Similar arguments hold for the rest of the formulae of Table 8.10.

Table B.2 Filter relations for the derivation of the algorithmic part of Table 8.10.

(1a)	$\mathbf{b}_{m+1}(M, N) = \begin{bmatrix} 0 \\ \mathbf{g}_m(M, N) \end{bmatrix} + \begin{bmatrix} 1 \\ J\mathbf{g}_m(M, N) \end{bmatrix} k_{m+1}^{\mathrm{b}}(M, N)$
(1b)	$k_{m+1}^{\mathrm{b}}(M, N) = -\alpha_m^{-g}(M, N)\beta_{m+1}^{\mathrm{b}}(M, N)$
(1c)	$\beta_{m+1}^{\mathrm{b}} = s_{m+1}(M, N) + \mathbf{q}_m^{\mathrm{T}}(M, N)\mathbf{g}_m(M, N)$
(2a)	$\mathbf{g}_m(M, N) = \mathbf{b}_m(M, N) + \mathbf{v}_m^*(M, N)k_m^g(M, N)$
(2b)	$k_m^{\mathrm{g}}(M, N) = -\alpha_m^{-v*}(M, N)\varepsilon_m^{\mathrm{b}}(M, N)$
(3a)	$\begin{bmatrix} 0 \\ J\mathbf{v}_m^*(M, N) \end{bmatrix} = \mathbf{u}_{m+1}^*(M, N) - \begin{bmatrix} 1 \\ J\mathbf{b}_m(M, N) \end{bmatrix} k_{m+1}^{u*}(M, N)$
(3b)	$k_{m+1}^{u*}(M, N) = -\alpha_m^{-\mathrm{b}}(M, N)\varepsilon_m^{\mathrm{b}}(M, N)$
(4a)	$\mathbf{v}_m^*(M, N) = J\mathbf{u}_m^*(M, N) + \mathbf{u}_m^*(M, N)k_m^{v*}(M, N)$
(5b)	$k_m^{v*}(M, N) = -\alpha_m^{-u*}(M, N)\varepsilon_m^{u*}(M, N)$

References

[1] S.M. Kay, *Modern Spectral Esimation*, Prentice Hall: Englewood Cliffs, NJ, 1987.

[2] S.L. Marple, *Digital Spectral Analysis with Applications*, Prentice Hall: Englewood Cliffs, NJ, 1987.

[3] M.B. Priestley, *Spectral Analysis and Time Series*, Academic Press: London, 1981.

[4] P. Whittle, *Prediction and Regulation*, English Universities Press: London, 1963.

[5] J.L. Doob, *Stochastic Processes*, Wiley: New York, 1953.

[6] A. Papoulis, *Probability, Random Variables and Stochastic Processes*, 2nd edn, McGraw-Hill: 1984.

[7] S.L. Marple, 'A tutorial overview of modern spectral estimation', *Proc ICASSP-89 Conf.*, pp. 2152–7, 1989.

[8] A. Van Den Bos, 'An alternative interpretation of the maximum entropy spectral analysis', *IEEE Trans. Inf. Theory*, vol. IT–17, pp. 493–4, 1971.

[9] G. Moustakides and S. Theodoridis, 'Fast Newton transversal filters – A new class of adaptive estimation algorithms', *IEEE Trans. Acoust., Speech, Signal Process.*, vol. ASSP-39, pp. 2184–94, 1991.

[10] P.F. Fougere, 'A solution to the problem of spontaneous line splitting in maximum entropy spectral analysis', *J. Geophys. Res.*, vol. 82, pp. 1051–4, 1977.

[11] E.T. Jaynes, 'Prior probabilities', *IEEE, Trans.*, vol. System Science and Cybernetics–4 pp. 227–41, 1968.

[12] R.W. Herring, 'The cause of line splitting in Burg maximum entropy spectral analysis', *IEEE, Trans. Acoust., Speech, Signal Process.*, vol. ASSP-28, pp. 692–701. 1980.

[13] W.Y. Chen and G.R. Stegen, 'Experiments with maximum entropy power spectra of sinusoids', *J. Geophys. Res.*, vol. 79, pp. 3019–22, 1974.

[14] D.N. Swingler, 'Frequency errors in MEM processing', *IEEE Trans. Acoust., Speech, Signal Process.*, vol ASSP-28, pp. 257–9, 1980

[15] B. Helme and C.L. Nikias, 'Improved spectrum performance via data adaptive weighted Burg Technique', *IEEE Trans. Acoust., Speech, Signal Process.*, vol. ASSP-33, pp. 903–10, 1985.

[16] N. Kalouptsidis, D. Manolakis and G. Karayannis, 'Efficient recursive triangularization, inversion and system solution of near-to-Toeplitz matrices and applications to signal processing', *Signal Process.*, vol. 6, pp. 235–59, 1984.

[17] M. Morf, B. Dickinson, T. Kailath and A. Vieira, 'Efficient solution of covariance equations for linear prediction', *IEEE Trans. Acoust., Speech, Signal Process.*, vol. ASSP-25, pp. 429–33, 1977.

[18] S.L. Marple, 'Efficient least squares FIR system identification', *IEEE Trans. Acoust., Speech, Signal Process.*, vol. ASSP-29, pp. 62–73, 1978.

[19] N. Kalouptsidis, G. Karayannis, D. Manolakis and E. Koukoutsis, 'Efficient recursive in order LS FIR filtering and prediction', *IEEE Trans., Acoust., Speech, Signal Process.*, vol. ASSP–10, pp. 1175–87, 1985.

[20] N. Kalouptsidis and S. Theodoridis, 'Parallel implementation of efficient LS algorithms for filtering and prediction', *IEEE Trans. Acoust., Speech, Signal Process.*, vol. ASSP-35, pp. 1565–9, 1987.

[21] S. Theodoridis and N. Kalouptsidis, 'A highly parallel block processing algorithm for LS FIR filtering', *Proc. Int. Symp. on Circuits and Systems, Helsinki*, pp. 293–6, 1988.

[22] S. Theodoridis, 'Parallel architecture for block apadtive LS FIR filtering and prediction', *IEEE Trans. Acoust., Speech, Signal Process.*, vol. ASSP-38. pp. 81–90, 1990.

[23] S.Y. Kung, *VLSI Array Processors*, Prentice Hall: Engelwood Cliffs, NJ, 1988

[24] S. Theodoridis and A. Liavas, 'Highly concurrent algorithm for the solution of ρ-Toeplitz system of equations', *Signal Process.*, vol. 24, pp. 165–76, 1991.

[25] P. Strobach, 'The square root Schur RLS adaptive filter', *Proc. ICASSP-91*, pp. 1845–8, 1991.

[26] A.H. Nuttall, 'Spectral analysis of a univariate process with bad data points via maximum entropy and linear predictive techniques', Naval Underwater Systems Center Technical Report. TR-5303. New London, CT, Mar., 1976.

[27] T.J. Ulrych and T.N. Bishop, 'Maximum entropy spectral analysis and autoregressive decomposition', *Rev. Geophys. Space Phys.*, vol. 13, pp. 183–200, 1975.

[28] S.L. Marple, 'A new autoregressive spectrum analysis algorithm', *IEEE Trans., Acoust., Speech, Signal Process.*, vol. ASSP-28, pp. 441–54, 1980.

[29] N. Kalouptsidis and S. Theodoridis, 'Fast adaptive least squares algorithms for power spectral estimation', *IEEE Trans. Acoust., Speech, Signal Process.*, vol. ASSP-35, pp. 661–70, 1987.

[30] S. Theodoridis, N. Kalouptsidis and D. Bakirtzis, 'Pipelined algorithm for LS FIR filters with symmetric impulse response', *IEEE Trans.*, vol. ASSP-38, pp. 260–71, 1990.

[31] K. Berberidis and S. Theodoridis, 'Efficient symmetric algorithms for the modified covariance method for autoregressive spectral analysis', to appear *IEEE Trans Signal Process.*, Jan., 1993.

[32] N. Kalouptsidis and G. Koyas 'Efficient block LS design of FIR filters with linear phase', *IEEE Trans., Acoust., Speech, Signal Process.*, vol. ASSP-33, pp. 1435–45, 1985.

[33] S. Theodoridis and N. Kalouptsidis, 'A fast algorithm for block LS design of FIR filters with linear phase and optimum lag', *IEEE Trans., Acoust., Speech, Signal Process.*, vol. ASSP-35, pp. 1079–82, 1987.

[34] N. Kalouptsidis and S. Theodoridis, 'Efficient structurally symmetric algorithms for least squares FIR filters with linear phase', *IEEE Trans., Acoust., Speech, Signal Process.*, vol. ASSP-36, pp. 1454–65, 1988.

[35] S.L. Marple, 'Fast algorithms for linear prediction and system identification filters with linear phase', *IEEE Trans. Acoust., Speech, Signal Process.*, vol. ASSP-30, pp. 942–53, 1982.

[36] K. Berberidis and S. Theodoridis, 'New Levinson, Schur, and lattice type algorithms for linear phase filtering'. *IEEE Trans., Acoust., Speech, Signal Process.*, vol. ASSP-38, pp. 1879–93, 1990.

[37] P.F. Fougere, E.J. Zawalick, H.R. Radoski, 'Spontaneous line splitting in MEM spectrum analysis', *Phys. Earth Planet. Inter.*, vol. 12, pp. 201–7, Aug. 1976.

[38] S. Shon and K. Mehrotra, 'Performance comparison of autoregressive estimation methods', *Proc. IEEE Int. Conf. on Speech, Acoustics and Signal Processing*, 1984.

[39] S. Theodoridis and D.C. Cooper, 'Application of the maximum entropy spectrum analysis technique to signals with spectral peaks with finite width.' *Signal Process.* vol. 3, pp. 109–22, 1981.

[40] H. A. Akaike, 'A new look at the statistical model identification.' *IEEE Trans. Auto. Control*, vol. AC-19, pp. 716–23, 1974.

[41] J.A. Rissanen, 'A universal prior for the integers and estimation by minimum description length', *Ann. Stat.*, vol. 11, pp. 417–31, 1983.

[42] T.J. Ulrych and R. W. Clayton, 'Time series modelling and maximum entropy', *Phys. Earth Planet. Inter.*, vol. 12., pp. 188–200, 1976.

[43] N. Kalouptsidis, 'Efficient lattice and transversal algorithms for linear phase multichannel filtering', *IEEE Trans. Circuits Syst.,* vol. CAS-35 pp. 425–31, 1988.

[44] K. Berberidis and S. Theodoridis, 'A normalized lattice algorithm for AR spectral analysis and system identification filters with symmetric impulse response', *IEEE Trans., Acoust., Speech, Signal Process.*, vol. ASSP-38, pp. 397–405, 1990.

[45] S. Theodoridis, K. Berberidis and N. Kalouptsidis, 'A new adaptive covariance symmetric algorithm and a fast initialization scheme for LS FIR filters with symmetric impulse response', *Signal Process.*, vol. 18, pp. 153–67, 1989.

[46] S. Theodoridis and K. Berberidis, 'Adaptive LS internally symmetric algorithms for linear phase filtering', in *Algorithms and VLSI Architectures*, ed. E. Deprettere, North-Holland Amsterdam, 1990.

[47] D.T.M. Slock and T. Kailath, 'Fast transversal filters with data sequence weighting', *IEEE Trans., Acoust., Speech, Signal Process.,* vol. ASSP-37, pp. 346–60, 1989.

[48] S. Pasupathy and B. Toplis, 'Tracking improvements in fast RLS algorithms using a variable forgetting factor', *IEEE Trans. Acoust., Speech, Signal Process.*, vol. ASSP-36, pp. 206–27, 1988.

[49] I. Marantidis, 'Efficient symmetric algorithms for adaptive spectral estimation', Final year project report, Dept of Computer Engineering, University of Patras, 1990.

[50] G. Glentis and N. Kalouptsidis, 'Efficient adaptive algorithms for multichannel least squares filtering', IEEE SP–40, NO. IV, pp. 2433–58, 1992.

[51] B. Porat and B. Friedlander, 'Asymptotic analysis of the bias of the modified Yule–Walker estimator', *IEEE Trans. Authom. Control.* vol. AC-30. pp. 765–67, 1985.

[52] B. Friedlander and K. Sharman. 'Performance evaluation of the modified Yule–Walker Estimator', *IEEE Trans., Acoust., Speech, Signal Process.*, vol. ASSP-33, pp. 719–25, 1985.

[53] C.J. Demeure and L.S. Scharf, 'Fast least squares solution of Vandermonde systems of equations', *Proc ICASSP-89 Conf.*, pp. 2198–201, 1989.

9

Channel Equalization

C.F.N. Cowan

9.1 Introduction

The process of channel equalization [1] in digital communications systems is perhaps the most heavily exploited area of application for adaptive filtering algorithms. The purpose of this chapter is to review the subject of adaptive equalization and highlight some of the problems endemic to modern communications systems which are resulting in ongoing efforts to improve and extend the performance of adaptive filters.

Section 9.2 of this chapter introduces equalization in its simplest form, that of a finite impulse response (FIR) equalizer. This is restricted to the case of purely real signals in the first instance, with the development of complex equalizers for use on modulated channels considered in section 9.6. The concept of decision-directed training is also developed to show how real-time tracking of slowly time-variant channel responses may be achieved.

Section 9.3 examines the effect of the choice of adaptive algorithm on equalizer performance. This section describes the effects of the various algorithm variants, which have already been developed in previous chapters, within the equalizer environment. The self-orthogonalizing algorithm variant is also briefly discussed in this section.

The preceding sections are all developed with the idea of symbol-spaced filters in mind. However, a better structure may often be provided by sub-symbol sampling. Such equalizers are discussed in section 9.4, which highlights the advantages and disadvantages of the structure, including a brief description of the tap leakage algorithm.

Section 9.5 is devoted to a description of the decision feedback equalizer. These equalizers provide a much more efficient form of filter than the simple FIR structures described in section 9.2 but can suffer from problems of error extension and instability.

In section 9.6 the structure of complex equalizers is described. All the filters discussed in the previous sections were designed to operate with real, or baseband, signals. However, in practice, modems deal with complex, passband, signals. The only significant difference is in the increased complexity of the hardware to deal with complex numbers and a slight modification to the adaptation algorithm.

The final section deals with a description of the use of non-linear filtering processes in equalization. It is shown here that optimum equalization is, in fact, a non-linear process and a number of possible architectures are considered in order to solve this problem.

9.2 Linear FIR Equalizers

The basic problem which is introduced in a digital communications system is illustrated in Figure 9.1. Here, a transmitter codes and modulates a digital information sequence $\{y(t)\}$ in a manner suitable for the transmission channel in question. This signal is then transmitted through the channel which will introduce both time dispersion and additive noise. If the communication channel is a cable then this dispersion will have a continuous impulse response which may spread over many symbol intervals, thus causing intersymbol interference (ISI). In the case of a radio channel the dispersion is more likely to be discrete in nature and caused by multipath effects. If the multipath spread exceeds the symbol spacing then ISI is once more introduced. For the moment we shall assume that the channel under consideration is time invariant and the additive noise level is negligible. In this case, the task of the equalizer is to reconstruct an estimate $\{\hat{y}(t)\}$ of the original information sequence $\{y(t)\}$ by using observations of the channel output $\{u(t)\}$. It will do this by attempting to remove the ISI. This process is what is classically referred to as channel equalization.

It should be noted here that it is only necessary to equalize the channel at the symbol-sampling instants and not over all time. Thus, if we can define the sampled transfer function of the channel as $H(z)$ then clearly the transfer function of the optimum equalizer is given by

$$G(z) = H^{-1}(z) \tag{9.1}$$

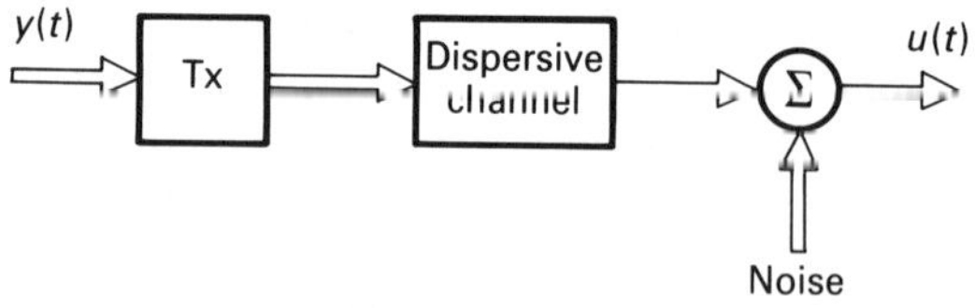

Figure 9.1 Basic communications channel mode

This rather simplistic view appears to offer a perfectly satisfactory solution which only requires the measurement of $H(z)$ and its inversion. However, such an inverse will be unstable in the situation where $H(z)$ possesses zeros outside the unit circle in the z-plane (non-minimum phase channels). This situation is fairly common in practical systems so, clearly, the simple solution of (9.1) is not a good one. A common way of getting round this problem is to approximate the exact inverse in (9.1) by a finite, all-zero filter which is achieved by performing the division in (9.1) and then truncating the series. The problem of forming an equalizer for the non-minimum phase case is resolved by introducing finite lag (time delay) into the equalizer. The following simple example demonstrates how this operates. First assume that the transmission channel has the following, non-minimum phase transfer function:

$$H(z) = 0.5 + z^{-1} \tag{9.2}$$

The simple inverse for this channel is given by

$$G(z) = \frac{z}{1 + 0.5z} \tag{9.3}$$

This function has a pole at $z = -2$ and therefore, clearly, may not be implemented directly. However, if (9.3) is divided out we achieve the infinite series

$$G(z) = z - 0.5z^2 + 0.25z^3 - \dots \tag{9.4}$$

This sequence (which is convergent in positive time) may be truncated after a finite number of samples which may be determined by the required residual ISI allowable in the receiver. As an example we will assume this number of samples to be three in this case (i.e. truncation after the z^3 term). This leaves us with a non-causal filter which may be made realizable by introducing a delay of three samples, i.e. multiplying (9.4) by z^{-3}. This yields a final realizable equalizer of the form

$$\begin{aligned} G'(z) &= z^{-3}G(z) \\ &= 0.25 - 0.5z^{-1} + z^{-2} \end{aligned} \tag{9.5}$$

This provides us with an acceptable, if simplistic, solution to the equalization problem provided the channel transfer function is known. However, in practical systems function $H(z)$ will not be known and may indeed be time variant. If we, initially, retain our assumption of time invariance then a simple adaptive filter structure may

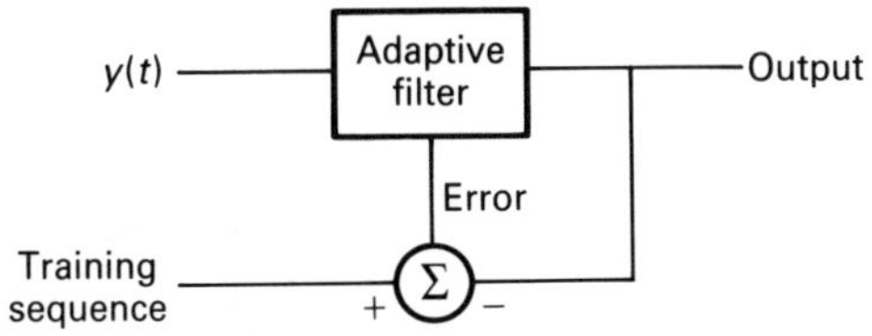

Figure 9.2 Simple equalizer structure using explicit training

be deployed to resolve the problem of initially acquiring the inverse. This is done by using a training period during which a known data sequence is transmitted over the channel. Since the sequence is known and the channel output is observable it is clearly possible to use this information to generate an estimate of the channel transfer function using an adaptive filter. However, a simpler solution is to acquire the inverse (equalizer) directly using the structure in Figure 9.2. This structure may be run using the training sequence until such time as the adaptive filter has converged and then freezing the coefficients so that real data may be transmitted.

If we now introduce the more realistic situation of time-variant channels the simplistic solution of Figure 9.2 no longer applies. Provided the time variations may be guaranteed to be sufficiently slow it is still possible to use this structure by implementing periodic retraining. However, a frequently used procedure is that of decision-directed training [1]. Here we may use a simple example where the transmitted sequence is of random bipolar form 1, -1, i.e. the symbol alphabet has only two terms. The structure illustrated in Figure 9.3 may then be applied, with the a priori assumption that the initial state of the equalizer coefficients is near optimum. What then happens is that the output of the equalizing filter is passed through a non-linear device, a slicer, which returns the noisy estimate to the noise-free, correct symbol values. This noise-free signal is then used as a training sequence for the adaptive equalizer. This is a robust solution to the tracking problem provided that no rapid changes in the transmission channel occur. Basically this operates because slow variations in the channel transfer function will result in a gradual degradation of the filter output away from the correct symbol values. However, the slicer will

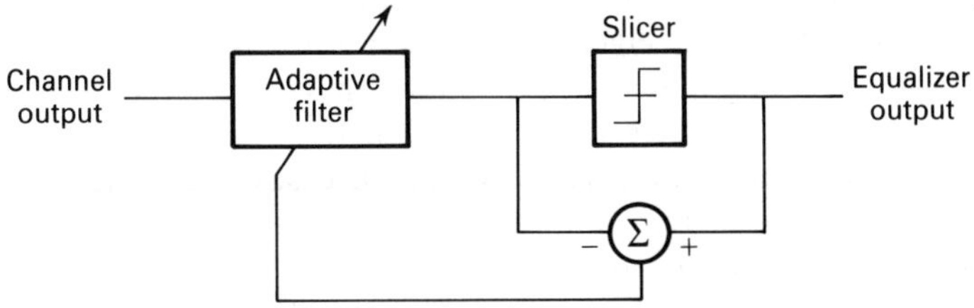

Figure 9.3 Decision-directed adaptive equalizer structure

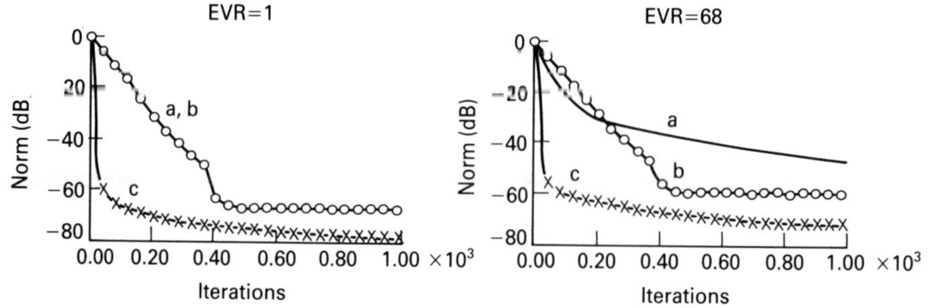

Figure 9.4 Convergence of the coefficient error norm for a system identification experiment at input eigenvalue ratios of 1 and 68 for: (a) LMS; (b) self-orthogonalized; (c) RLS adaptive algorithms

restore the correct values which are fed back as a training signal, thus 'bootstrapping' the filter coefficients back towards the correct values.

In many systems a combination of a priori training, retraining and decision-directed feedback is used in order to maintain the correct equalization. This is currently done in rapidly time-variant systems such as HF radio [2] and proposals for mobile digital cellular systems.

9.3 Adaptive algorithm performance

Thus far nothing has been said about the actual adaptive algorithms which would be used in these equalizers. This is a complex issue in the case of equalizers where the choice of algorithm is affected by a number of issues:

1. The initial convergence time of the adaptive filter. This dictates the initial setup time of the equalizer as it determines the time taken from call initiation to the ability to transmit data. Although the originally transmitted data sequence will have clearly defined statistical properties the introduction of the transmission channel will introduce distortion which can have a serious effect on the signal conditioning at the equalizer input. Therefore, the eigenvalue spread of the input autocorrelation matrix can be quite considerable. The set of simulation results in Figure 9.4 shows the differences in convergence rates between LMS, RLS and self-orthogonalizing [3] algorithms (in a system-modelling scenario) for the cases where the ratio of maximum to minimum eigenvalues of the input autocorrelation matrix is either 1 or 68. The simulated equalizer is of order 16 and is explicitly trained with correct symbols. It should be noted here that the eigenvalue ratio of 1 corresponds to a channel introducing no distortion and an eigenvalue ratio of 68 is not a particularly severely distorting channel.

The algorithms which have just been mentioned include the LS (transversal or lattice) and LMS, which have been discussed in detail in earlier chapters. However, self-orthogonalizing (SO) algorithms [4] have not been explicitly discussed as a generic set. There is, indeed, often confusion between the SO and RLS algorithms which may be avoided if we examine the rationale behind the SO algorithm from the equalization viewpoint.

The convergence rate of the LMS algorithm deteriorates in direct relationship to the spread of eigenvalues in the input autocorrelation matrix to the equalizer. This effect is illustrated in the simulation studies in Figure 9.4 and has already been discussed in Chapter 4. The overall effect of this is to make the convergence time of the equalizer, based on the LMS algorithm, unpredictable.

One way in which this problem may be alleviated is by applying an orthonormal transformation to the input signal and then equalizing the power at every output from this transform; this is the basic self-orthogonalizing adaption algorithm. The effect of performing this operation is first to diagonalize the autocorrelation matrix via the transform and then to reduce this matrix to the identity by the use of power denormalization. The problem, of course, is that the orthonormal transformation required to do this is non-trivial and indeed data dependent. The optimal procedure for accomplishing this is the Karhunen–Loeve transform (KLT) [5] which may be achieved by implementing an adaptive lattice filter (Chapter 6) [6]. This approach does have the disadvantage of incorporating two adaptive structures in order to implement the overall equalizer, the decorrelation lattice and the output combiner.

Approximations which make use of fixed transformations such as the discrete Fourier transform (DFT) discrete cosine transform (DCT) and Walsh transform [7] have been suggested. These techniques do have the effect of reducing the sensitivity to input signal conditioning but do not remove it entirely.

An alternative strategy for implementation of the SO algorithm is to effect the orthogonalization of the various convergence modes within the equalizer adaptation loop [8] rather than by orthogonalizing the input observation space. This may be done by using the following recursion:

$$\mathbf{c}(t+1) = \mathbf{c}(t) + \alpha R^{-1}\mathbf{u}(t)[y(t) - \hat{y}(t)] \tag{9.6}$$

where α is a constant gain term and R is the input autocorrelation matrix. This form is clearly rather simpler than the procedure described above but does have the problem of requiring exact knowledge of the input autocorrelation (which is time variant in any case). Therefore, in practice, R must be estimated in real time and periodically updated. The confusion with RLS arises because this algorithm is equivalent to RLS once the RLS algorithm has passed the initial convergence transient.

Another subtle point which has to be highlighted is related to the application of lattice structures for equalization. During the training period the desired response sequence $y(t-l)$, where l is the delay introduced for non-minumum phase channels, is available at each time instant. Hence, any form of the lattice structures presented in Chapter 6 can be used. However, during the decision-directed period the desired response is not available until an estimate has been made. Hence, only a priori lattice forms can be used. From Table 6.5 of Chapter 6, once the backward errors at current time t have been computed, the estimate $\tilde{y}(t - l)$ at stage m can be computed recursively as

$$\tilde{y}_{m+1}(t - l) = \tilde{y}_m(t - l) + k_m^y(t - 1)e_m^{\mathrm{b}}(t) \tag{9.7a}$$

The estimate from the final stage p

$$\tilde{y}_p(t - l) = \sum_{m=0}^{p-1} k_m^y(t - 1)e_m^{\mathrm{b}}(t) \tag{9.7b}$$

is sent to the slicer to become the decision $\hat{y}(t - 1)$ and it is in turn used to compute the prior errors

$$e_m^y(t) = \hat{y}(t - l) - \tilde{y}_m(t - l) \tag{9.7c}$$

Equations (9.7a) and (9.7c) substitute the update equation for $e_{m+1}^y(t)$ in Table 6.5. $\tilde{y}_0(t - l)$ is initialized to zero.

The salient facts which may be elicited from the results in Figure 9.4 are that, first, the LMS converges more slowly than its alternatives, even in the perfect channel case. Second, the convergence rate of the LMS degrades dramatically with channel degradation, which does not happen with the alternative algorithms This second point is clearly a severe disadvantage in the case of the LMS at it must be possible to place an upper bound on the setup time for the equalizer which may be inordinately long for LMS equalizers. The RLS variant has the fastest and most consistent convergence rate, but this must be balanced against the increased complexity of the algorithm and the accompanying numerical problems, which have all been discussed in earlier chapters of this book. The SO algorithm is slower than the RLS but retains the advantage of having convergence which is robust in relation to channel condition. However, the algorithm simulated here [3] is not in fact practical since it uses perfect knowledge of the autocorrelation matrix in a time-variant situation. Compromises between the LMS and SO algorithm may be used which scale the LMS convergence factor according to estimates of the input signal power, but this still leaves some considerable variation in convergence rates. Alternative, approximate, self-orthogonalizing, algorithms are possible by using approximate orthogonal transforms to precondition the signal [9]. These are really only practical when the equalizer order is large.

2. The second issue of importance in choice of algorithms is the tracking ability of various algorithms. The importance of this issue is dependent on the channel involved. For instance, cable channels tend to change their characteristics very slowly relative to the transmission data rates and there is no real problem involved in tracking such channels. However, the case of some radio channels is not quite so straightforward. Two particular instances of channels which introduce rapidly changing channels (fast fading [10]) are high-frequency (HF) radio and mobile digital.

Both these channel types introduce time variations which can cause serious problems in terms of tracking. It has often been assumed that RLS algorithms will outperform LMS algorithms in terms of tracking since their initial convergence is faster. Recent work has, however, indicated that this may not be entirely true [11,12]. The reasons for this are dominated by the way in which we look at misadjustment in an adaptive filter. Initial convergence time is a transient effect which determines the time taken to translate the filter from an initial state to the region of the correct operating point. If this operating point is stationary then the accuracy of the result obtained at this stage is dominated by the additive noise and the approximations made in determining the filter length. However, if this operating point moves continuously then there is an additional error introduced as a result of the lag involved in the adaptive algorithm acquiring sufficient information to follow the correct solution. This is a significantly different effect from the initial transient problem and there is evidence to indicate that the RLS algorithm will not always outperform the LMS in this mode of operation [12]. However, this is an issue which is still being actively pursued by a number of research teams and it is certainly too early to draw any conclusions on this matter at present.

9.4 Subsymbol-spaced equalizers

The foregoing discussion has centred on the consideration of symbol-spaced equalization only. The ideal equalizer is in fact symbol spaced (i.e. sampling at the symbol rate) with a channel matched filter before it [1]. The channel-matched filter may not be implemented by a symbol-spaced filter since its sampling rate is too low. However, both these operations may be subsumed into a single adaptive filter if the sampling rate is increased beyond the symbol rate. This form of adaptive filter is commonly termed a fractionally spaced equalizer. For digital implementations sampling intervals are constrained to NT/M, where T is the symbol spacing (in seconds) and N, M are the positive integers.

In order to understand the process involved here we can examine the simplified schematic diagram of the transmission system shown in Figure 9.5. Here the input data are present at a rate of $1/T$ symbols per second. This is applied to a low-pass filter which limits the spectrum going into the channel. The output of the channel is passed through an identical low-pass filter and the data are then recovered at the original rate of $1/T$ symbols per second. However, the low-pass filters involved cannot be perfect, that is, they must have finite roll-off rates at the band edge. The transfer

Figure 9.5 Simple data transmission system

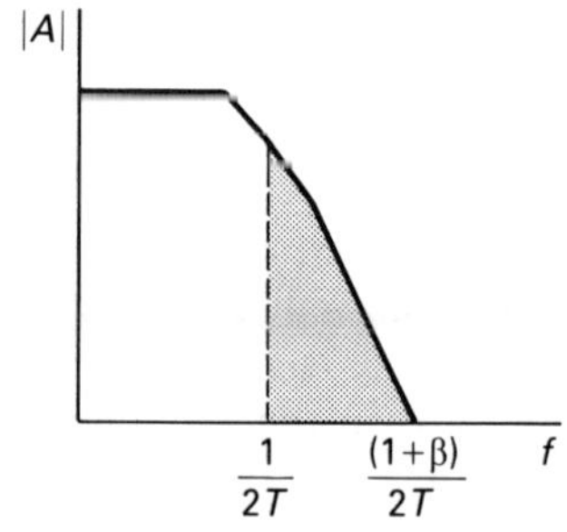

Figure 9.6 Transfer function of the low-pass filter in Figure 9.5

function of such a filter is illustrated in Figure 9.6, which shows that the signal spectrum extends beyond $1/2T$ Hz, where β is the roll-off factor and is limited to values between 0 and 1.

Clearly, if the output of this arrangement is sampled at rate $1/T$ then the Nyquist frequency occurs at $1/2T$ and thus spectral folding occurs (the shaded part in Figure 9.6). Given properly designed transmit and receive filters, correct sampling phase and no distortion, this folding will result in both a flat amplitude spectrum and a flat group delay response [13]. However, if the group delay distortion introduced by the channel is sufficiently severe in the region around $1/2T$ then the spectral folding can introduce deep nulls in the sampled frequency response. The T-spaced equalizer will attempt to equalize (i.e. amplify) these parts of the spectrum with resulting noise enhancement. The obvious way to combat this problem is to raise the output sample rate to above $1/T$ [14].

Such equalizers also have the property of insensitivity to sampling phase (timing). Symbol-spaced equalizers achieve optimum performance when sampling is ideally synchronized to the point of maximum signal power in the received pulse. Any deviation from this timing phase results in degraded signal-to-noise ratio and consequent noise enhancement by the equalizer [15]. A fractionally spaced equalizer, however, is insensitive to timing phase and thus allows the possibility of less complex timing recovery circuits.

A very frequently used fractional equalizer is one which samples at twice the symbol rate (spacing $= T/2$). The output of such a filter is then sampled at the symbol rate to calculate the error. The update process then takes place at half the filter sampling rate.

Significant points about the performance of such equalizers are, first, that the time span required to implement the equalizer is reduced such that the same number of delay stages is used in the fractional equalizer as for the symbol-spaced system. It should be noted, however, that the required sampling rate is obviously higher (this is particularly important at high data rates such as those found in microwave links). However, this is counteracted by the fact that there is no noise enhancement and a lack of sensitivity to timing phase acquisition [14].

A problem which does occur is that, owing to the high sampling rate, the input signal spectrum will fall towards zero, for vanishingly small noise, within the Nyquist limits (Figure 9.6). It has been shown in [16] that for finite-length equalizers a unique solution is still guaranteed, provided there is less than 100% excess bandwidth ($\beta < 1$). However, the input correlation matrix can be ill-conditioned with large eigenvalue spread. This causes the elliptical contours of equal mean squared error (MSE) around the optimal solution, in the parameter space, to have large eccentricity involving parameters with large values. Hence, as the adaptive algorithm converges towards the optimal solution, following a path of decreasing MSE, some of the tap parameters can become excessively large. A solution to this is the so-called tap leakage algorithm [16] (see also Chapter 4)

$$\mathbf{c}(t+1) = (1-\alpha)\mathbf{c}(t) + \boldsymbol{\phi}(t) \tag{9.8}$$

where $\boldsymbol{\phi}(t)$ is the adaptation algorithm increment. This algorithm introduces a decay window on the parameter vector controlled by the term $\alpha(0 < \alpha \ll 1)$. This tends to constrain the parameter solution to the one closest to zero, thus avoiding the coefficient growth problem. Some bias does result from adoption of this strategy and has been analyzed in [17].

To exploit the LS algorithms presented in the previous chapters for the case of fractionally spaced equalizers, one has to adopt the multichannel formulations. Indeed, for the $T/2$ case the received signal is sampled twice per symbol interval. These two samples can be grouped together forming a two-dimensional input data vector, and a two-channel transversal or lattice algorithm can be used.

9.5 Decision feedback equalizers

The foregoing discussion has all centred on the use of simple feedforward (transversal) equalizers. We will now consider ways in which an alternative structure may be employed which leads to a more efficient equalizer realization and also avoids the noise enhancement evidenced in the previous symbol-spaced filters.

If we first consider the case of a minimum phase distorting channel then decisions can be made on the channel output with zero lag (see section 9.2). The distortion introduced into this symbol is then due to the convolution of the remaining part of the channel response with the data sequence, i.e. the impulse response following the main peak. Thus the ISI may be removed by convolving the output of the decision circuit with an adaptively acquired model of this residual response and subtracting the result from the input signal, i.e. the residual ISI is cancelled. Such a structure is shown in Figure 9.7 and is known as a decision feedback equalizer (DFE) [18], with the assumption that the transmitted sequence is from the bipolar set -1, 1.

This type of equalizer has two main advantages relative to the transversal equalizer. The first is that one does not require as many coefficients to realize the filter since the all-zero approximation (Equation (9.4)) is not used. Therefore, it may

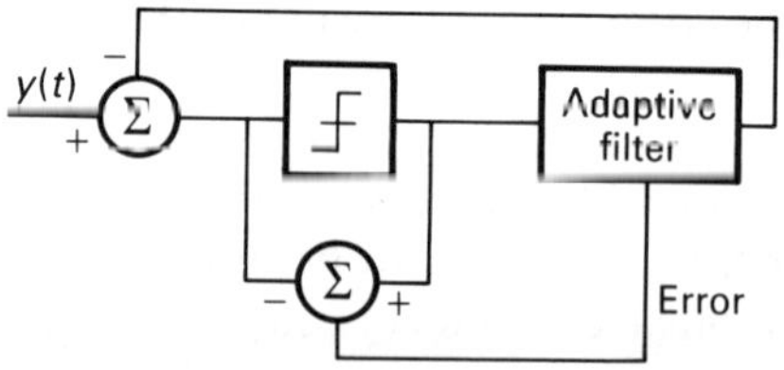

Figure 9.7 Configuration of the standard decision feedback equalizer

be assumed that the DFE will in fact provide a more accurate inverse for the channel. The second point is that no noise enhancement takes place in the DFE since the coefficients are operating on the noise-free output from the decision circuit.

There are, however, some disadvantages associated with this structure. The first may be observed if we consider what happens if an incorrect decision is made. In the transversal equalizer such an error could be treated in isolation but in the DFE the error is fed back to the input again and may result in additional errors, the so-called error extension [19]. This error extension is extremely difficult to quantify as the filter is operating in a pseudorecursive fashion with a strong non-linearity in the feedback path. However, there has been some recent work on applying bounds to the error probabilities resulting from these effects [20].

The other problem which arises is what happens when the channel in question is non-minimum phase in nature, that is, the impulse response of the channel has low-amplitude precursors before the main peak. This is not as severe a problem as one might first assume, since the DFE is not a genuine infinite impulse response filter and therefore the fact that the channel zeros are outside the unit circle does not automatically imply instability. Indeed, provided the precursor amplitude is sufficiently above the additive noise so that it does not adversely affect the operation of the decision circuit, the DFE of Figure 9.7 will still operate correctly. However, if the precursor signal-to-noise ratio falls sufficiently to start introducing errors in the decision device then the structure fails catastrophically due to error extension effects.

The solution to this problem is to introduce an additional stage of equalization using a feedforward equalizer before the decision circuit (see Figure 9.8). This has the effect of introducing lag into the equalizer and attempts to restore near minimum phase characteristics at the output of the pre-equalizer by removing (equalizing) the precursor part of the channel response. This structure may obviously deal with more complex mixed-phase responses, with the prefilter dealing with precursor elements of the impulse response and the feedback filter cancelling postcursor parts of the response. The general form of a decision feedback equalizer is given by

$$\tilde{y}(t-l) = \sum_{i=0}^{l} c_i u(t-i) + \sum_{i=1}^{p} c_{l+i}\hat{y}(t-l-i) \tag{9.9}$$

The first sum involves the received samples (feedforward) and the second sum the previous decisions (feedback part). The error extension effects of the original structure

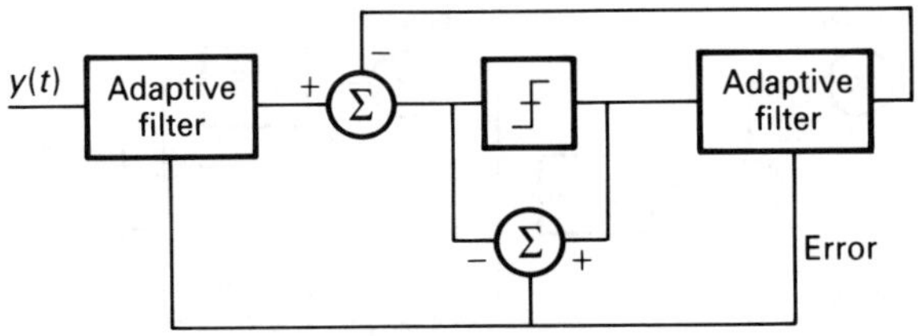

Figure 9.8 Full DFE structure with feedforward section

will of course still apply here and some noise enhancement will be introduced by the feedforward filter.

However, in many situations these structures provide a much more efficient filter than their simpler linear counterparts. Using a two-channel interpretation, with the feedforward section as one channel and the feedback as another, the multichannel algorithms of previous chapters, as well as recently developed schemes, i.e. [21], [22], can be employed for the efficient adaptation of a DFE. It should be emphasized here that the number of coefficients in each of the feedforward and feedback sections is not necessarily the same. Finally, the feedforward section could be fractionally spaced. For a $T/2$ feedforward section a three-channel formulation is required for the efficient adaptation of the resulting DFE.

9.6 Complex equalizers

The foregoing discussion has concentrated exclusively on equalizers processing purely real signals. There are some applications where this form of processing is applied, e.g. equalizers in local subscriber loops for ISDN access. However, the majority of equalizers are deployed in modem applications on radio channels of one type or another. The principal modulation technique used in these applications tends to be some combination of amplitude and phase-shift keying. This means that the baseband signal is in fact complex rather than real valued. As a result the equalizer structures already discussed have to be modified to deal with complex-valued signals.

The overall structure of a transmission system which utilizes quadrature amplitude modulation (QAM) is illustrated in Figure 9.9. Here it can be seen that a two-dimensional (complex) signal is used as input and therefore the required equalizer is also of dimension 2. The actual structure of such an equalizer is shown in Figure 9.10 which, as we can see, in fact requires four real filters in order to implement the complex filter. This in turn requires that the adaptation algorithm be modified to a complex form. This modification is, in fact, quite simple and in the case of the LMS algorithm the coefficient update recursion becomes [23]

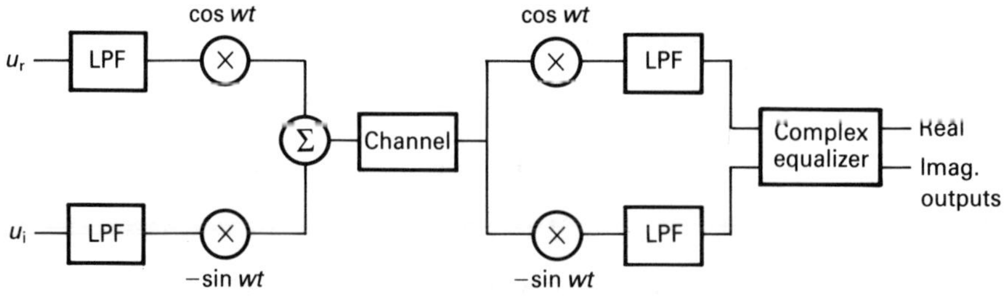

Figure 9.9 Configuration of a QAM transmission system configured with a complex baseband equalizer

$$\mathbf{c}(t+1) = \mathbf{c}(t) + \mu e(t)\mathbf{u}^*(t) \tag{9.10}$$

where $\mathbf{u}^*(t)$ is the complex conjugate of $\mathbf{u}(t)$, and $e(t)$ is the estimation error at time t.

An alternative to the complex baseband equalizer of Figure 9.10 is the passband equalizer. This form performs the equalization in the passband as shown in Figure 9.11. This structure clearly requires more complex hardware as there is a requirement for two complex modulators first to generate the error signal and then to translate the error back to the passband. Despite this additional complexity the passband form is frequently preferred since it does not introduce the delay through the equalizer filter in the phase-tracking loop. It is therefore easier to track rapid phase jitter in this case.

9.7 Non-linear equalizers

Perhaps the best known form of non-linear detection systems is the maximum likelihood sequence estimator (MLSE), depicted in Figure 9.12 [24]. In simple terms

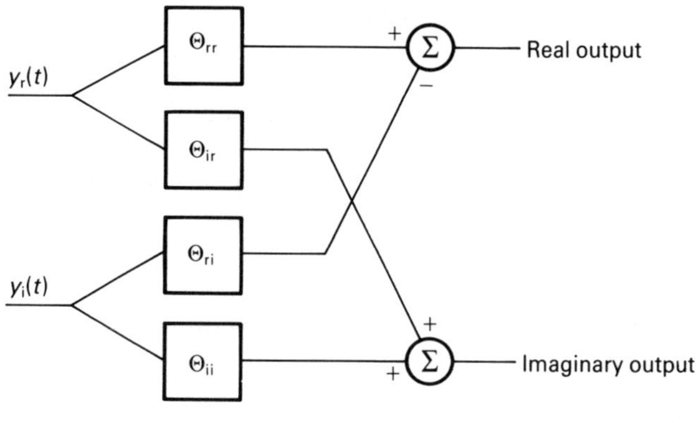

Figure 9.10 Configuration of a complex baseband equalizer

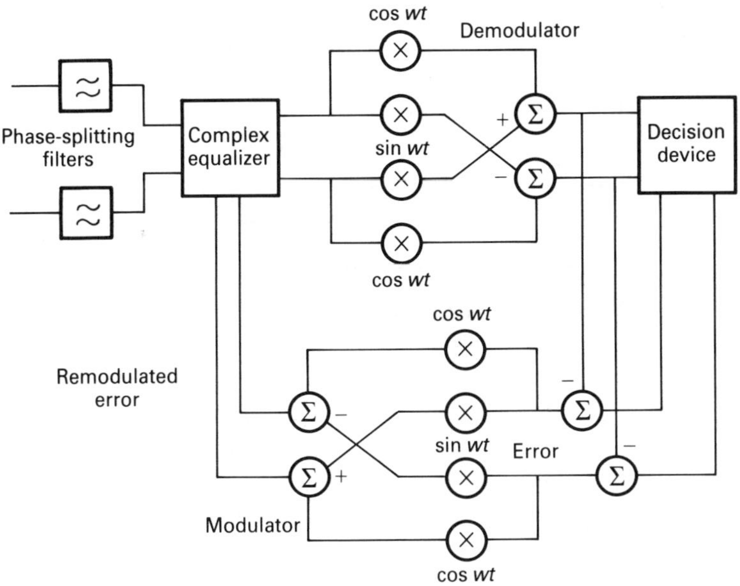

Figure 9.11 Passband equalizer

this processor has an adaptive prefilter [25] which ensures that the input to the detector is minimum phase in nature. This is equivalent to the necessary assumption that the noise component is white, which is dictated by the use of Euclidean distance classification in the MLSE detector. A practical way to achieve this is by means of a DFE. The output of the feedforward section is approximately minimum phase (at moderately high SNRs) and is used as the input to the non-linear detector. The feedback section provides an estimate of the equivalent causal channel parameters, for lags greater than zero. The zero-lag parameter is assumed to be one. In a time-varying environment the adaptive schemes discussed in the previous sections may be used. The detector then makes use of this observation sequence plus the estimate of the overall (causal) received impulse response in order to arrive at an estimate of the transmitted sequence. In practice such a detector can be very complex in structure if the span of the ISI is long. A number of suggestions have been put forward involving approximate, near maximum likelihood detectors to ease the complexity problem [26].

Something of the relationship between MLSE and the equalizers already considered may be clarified by examining the operation of non-linear equalizers. For the purposes of the current explanation we will assume a very simple transmission model where the input is a binary baseband data set from -1, 1. The distorting channel involved has transfer function

$$H(z) = 1 + 0.5z^{-1} \tag{9.11}$$

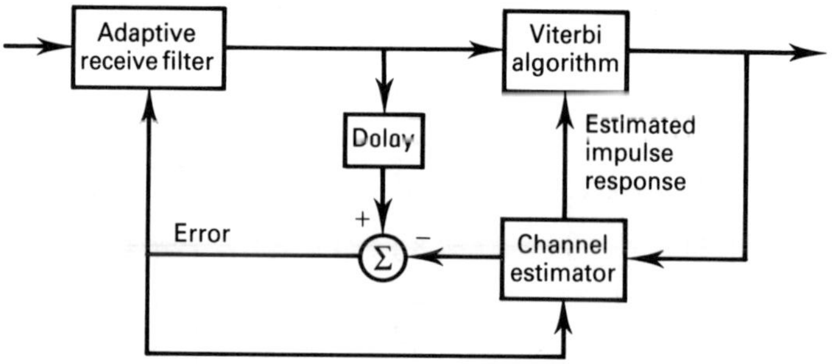

Figure 9.12 Structure of the general maximum likelihood sequence estimator

and the receiver equalizer considers only the last two received observation samples (symbol spaced). Given this scenario the set of all possible received coordinates is shown in Table 9.1. This is then plotted in the form of a two-dimensional diagram in Figure 9.13, where the crosses denote that the symbol $y(t)$ is a -1 and the circles indicate the positions for $y(t)$ to be $+1$.

Table 9.1 Set of all possible channel output pairs

$u(t)$	$u(t-1)$	$u(t-2)$	$y(t)$	$y(t-1)$
−1	−1	−1	−1.5	−1.5
−1	−1	1	−1.5	−0.5
−1	1	−1	−0.5	0.5
−1	1	1	−0.5	1.5
1	−1	−1	0.5	−1.5
1	−1	1	0.5	−0.5
1	1	−1	1.5	0.5
1	1	1	1.5	1.5

If the two-coefficient equalizer uses a simple linear combiner followed by a binary decision circuit then the decision boundary formed is that shown as a solid line in Figure 9.13 [27]. This is, in fact, the boundary formed when the combiner realizes the optimum Wiener filter for this application. The major point which will be noticed here is that the solid boundary is not equidistant from adjacent observation points. Therefore, in the case when noise is present there are unequal probabilities of detection of various states. For instance, it is more likely that the symbol at (1.5, 0.5) will be correctly detected than the symbol at (−0.5, 0.5). This will inevitably lead to a degradation in performance as channel noise levels increase.

The optimum decision boundary in this case, using a minimum a posteriori error criterion, is in fact that shown as the dashed line in Figure 9.13, which is clearly non-linear. This boundary may be achieved by replacing the linear combiner by a non-linear adaptive network. A number of such structures have been suggested,

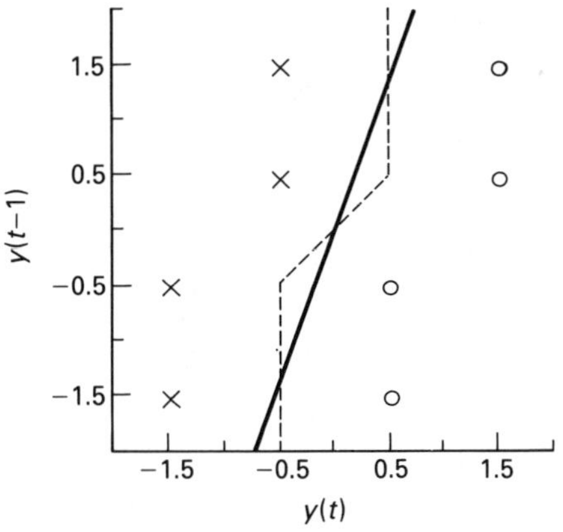

Figure 9.13 Two-dimensional representation of the received observation vector formed by two successive received samples

including the multilayer perceptron [27], Volterra series [28] and radial basis functions [29]. All of these techniques produce good estimates of the non-linear boundary, as shown for the MLP case in Figure 9.14. The resultant performance improvement is illustrated in terms of bit error rate plotted against signal-to-noise ratio in Figure 9.15. It is clear that substantial error rate improvement results from this strategy.

The remaining step in linking the equalizer to MLSE may be clarified by examining Figure 9.16. Here it is clear that even the non-linear equalizer does not make optimum use of the information available to it since it is possible to identify the two previous states in the transmitted sequence and not just the present one. If

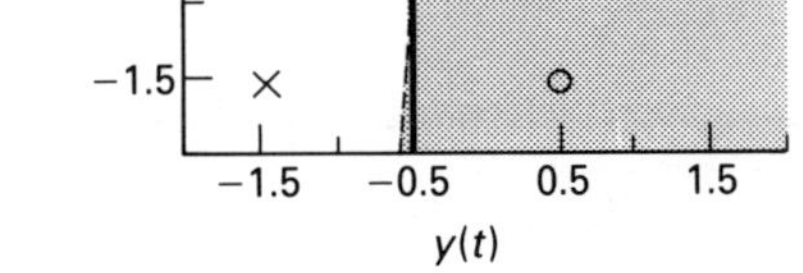

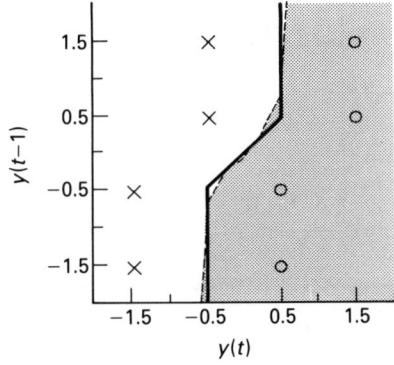

Figure 9.14 Actual decision boundary converged on by the MLP equalizer

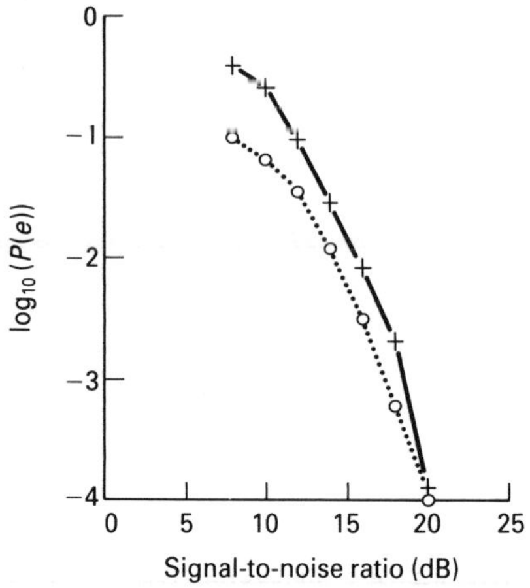

Figure 9.15 Simulation results showing relative bit error rate performance for: o, MLP-based equalizer with input dimension 5; ×, LMS linear equalizer with the same input dimension

this is done, possibly by the use of clustering analysis [30], then the time-shifting nature of the sequence may be exploited in order to deploy the Viterbi algorithm to provide a degree of error protection in the detection process. This is still an active research area but it does indicate that there may be a larger range of possible equalizer structures which fill the gap in performance between linear equalizers and MLSE.

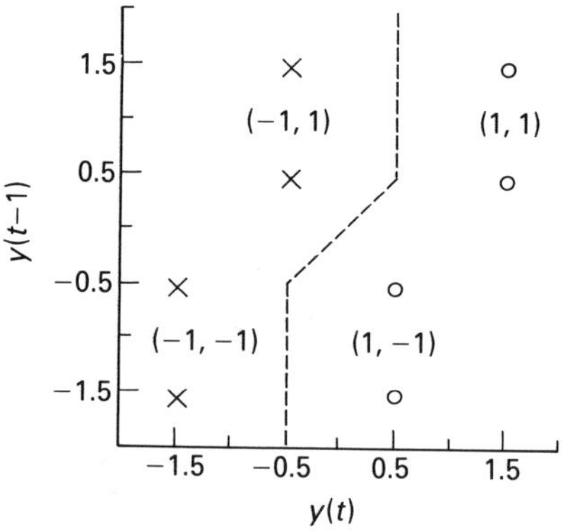

Figure 9.16 Indication of the two previous transmitted symbols given two successive received observations

9.8 Conclusions

This chapter has attempted to provide a concise overview of the key elements involved in the design of adaptive equalizers as they are currently understood. The mathematics associated with the control of the structures has not been discussed here as this has been adequately covered in preceding chapters. Despite the fact that equalization has been an active field of research for many years and may be viewed as a mature area, many problems remain to be solved in order to extend the useful performance of communications systems further.

Two particular areas of interest are the tracking of highly time-variant channels [11,12] and the use of non-linear structures in equalization [24–30]. The first idea has already been mentioned. The use of non-linear structures, however, is an active research area with two main focuses. The first is that equalization is in fact inherently a non-linear process since the exact linear inverse for a channel may never be implemented. It is indeed this which inevitably leads to techniques such as maximum likelihood sequence estimation, as already discussed in section 9.7. The second focus is on the potential non-linearity of many components within the transmission electronics. This can arise owing to the accuracy of voltage-level setting at the modulation stage and to amplifier saturation at the transmitter output. Both effects introduce non-linearity with memory into the overall channel response. The best means of tackling these problems is currently unclear because of the immature nature of non-linear systems theory. These problems have much in common with other application areas and in particular work in image and speech classification is producing significant advances which may be useful in future developments of adaptive equalizers.

References

[1] S.U.H. Qureshi 'Adaptive equalization', *Proc. IEEE*, vol. 73, pp. 1349–87, 1985.

[2] K. Brayer, *Data Communications via Fading Channels*, IEEE Press reprint vol., New York, 1975.

[3] C.F.N. Cowan, 'Performance comparison of finite linear adaptive filters', *IEEE Proc.*, vol. 13, pp. 211–16, 1978.

[4] R.D. Gitlin and F.R. Magee, 'Self-orthogonalizing adaptive equalizer algorithms', *IEEE Trans. Commun.*, vol. COM-25, pp. 666–72, 1977.

[5] R.W. Chang 'A new equalizer structure for fast start-up digital communications', *Bell Syst. Tech. J.*, vol 50, pp. 1969–2014, 1971,

[6] E.H. Satorius and S.T. Alexander, 'Channel equalization using adaptive lattice algorithms' *IEEE Trans. Commun.*, vol. COM-27, pp. 899–905, 1979.

[7] B. Mulgrew and C.F.N. Cowan, *Adaptive Filters and Equalizers*, Kluwer: Boston, 1988.

[8] K.H. Mueller 'A new fast-converging mean-square algorithm for adaptive equalizers with partial-response signaling', *Bell Syst. Tech. J.*, vol. 54, pp. 143–53, 1975.

[9] R.P. Bitmead and B.D.O. Anderson, 'Adaptive frequency sampling filters', *IEEE Trans. Circuits Syst.*, vol CAS-28 pp. 524–34, 1981.

[10] P. Monsen, 'Fading channel communications', *IEEE Commun. Soc. Mag.*, vol. 18, pp. 27–36, 1980.

[11] S. McLaughlin, B. Mulgrew and C.F.N. Cowan, 'Performance comparison of least squares and least mean squares algorithms as HF channel estimators', *Proc. ICASSP-87 Conf.* pp. 2105–8, 1987.

[12] A.P. Clark, and R. Harun, 'Assessment of Kalman filter channel estimators for an HF radio link', *Proc. IEE*, pt F, vol. 133, pp. 513–21, 1986.

[13] J.W. Smith, 'The joint optimization of transmitted signal and receiving filter for data transmission systems', *Bell Syst. Tech J.*, vol. 44, pp. 2362–92, 1965.

[14] R.D. Gitlin, and S.B. Weinstein, 'Fractional-spaced equalization: An improved digital transversal equalizer', *Bell Syst. Tech. J.*, vol. 60, pp. 275–96, 1961.

[15] P.E. Adams, 'Adaptive filters in telecommunications', in *Adaptive Filters*, eds. C.F.N. Cowan and P.M. Grant, Prentice-Hall: Englewood Cliffs, NJ, 1985.

[16] R.D. Gitlin, H.C. Meadors and S.B. Weinstein, 'The tap leakage algorithm: An algorithm for the stable operation of a digitally implemented fractionally-spaced adaptive equalizer', *Bell Syst. Tech. J.*, vol. 61, pp. 1817–939, 1982.

[17] J.C. Pesquet, O. Macchi and G. Tziritas, 'Second-order statistical analysis of two constrained LMS algorithms', *Proc. EUSIPCO, Barcelona, Sept, 1990*, pp. 193–96.

[18] C.A. Belfiore and J.H. Park, 'Decision feedback equalization', *Proc. IEEE*, vol. 67, pp. 1143–56, 1979.

[19] J.J. O'Reilly and A.M. de Oliveira Duarte, 'Error propagation in decision feedback receivers', *IEE Proc.*, pt, F, vol. 132, pp. 561–66, 1985.

[20] R.A. Kennedy, B.D.O. Anderson and R.R. Bitmead, 'Tight bounds on the error probabilities of decision feedback equalisers', *IEEE Trans. Commun*, vol. COM-35, pp. 1022–28, 1987.

[21] F. Ling and J.G. Proakis, 'Adaptive lattice DFE equalizers. Their performance and application to time varying multipath channels', *IEEE Trans. Commun.*, vol. COM-33, pp. 348–56, 1985.

[22] S. Theodoridis and G. Moustakides 'A novel class of fast adaptive algorithms for multichannel filtering', *Proc. EUSIPCO 90, Barcelona, Sept., 1990*, pp. 409–12.

[23] B. Widrow, J. McCool and M. Ball, 'The complex LMS algorithm', *Proc. IEEE*, vol. 63, pp. 719–20, 1975.

[24] G.D. Forney, 'Maximum-likelihood sequence estimation of digital sequences in the presence of intersymbol interference', *IEEE Trans. Inf. Theory*, vol. IT-18, pp. 363–77, 1972.

[25] B. Mulgrew, 'Adaptive prefilter for maximum likelihood sequence estimation', *Proc. EUSIPCO, Barcelona, Sept., 1990*. pp. 245–8.

[26] A.P. Clark and S.F. Hau, 'Adaptive adjustment of the receiver for distorted digital signals', *IEE Proc.* pt F, vol. 131, pp. 526–36, 1984.

[27] S. Chen, G.J. Gibson, C.F.N. Cowan and P.M. Grant, 'Adaptive equalisation of finite non-linear channels using multilayer perceptrons', *Signal Process.*, vol. 20, pp. 107–20, 1990.

[28] S. Chen, G.J. Gibson and C.F.N. Cowan, 'Adaptive channel equalization using a polynomial-perceptron structure', *IEE Proc.*, pt I, vol. 137, pp. 257–64, 1990.

[29] S. Chen, G.J. Gibson, C.F.N. Cowan and P.M. Grant, 'Reconstruction of binary signals pusing an adaptive radial basis function equaliser', *Signal Process.*, vol. 22, pp. 77–93, 1991.

[30] C.F.M. Cowan, P.M. Grant, S. Chen, and G.J. Gibson, 'Nonlinear classification and adaptive structures', *Proc. SPIE, San Diego, July, 1990*, pp. 62–72.

10

Echo Cancellation

Fuyun Ling

Echo cancellation is a suitable area for the application of adaptive filtering. An adaptive echo canceller estimates the responses of an underlying echo-generating system in real time, generates a synthesized echo based on the estimate, and cancels the echo by subtracting the synthesized echo from the received signal. Echoes may be generated due to reflections from objects surrounding an acoustic source and are called acoustic echoes, or they may be generated in a telephone network, e.g. in a public switched telephone network (PSTN), and are called telephone echoes. Echo cancellers for cancelling such echoes are called acoustic cancellers and telephone echo cancellers respectively. Compared with acoustic echo cancellers, various types of telephone echo cancellers have been studied more thoroughly theoretically and for practical applications. In this chapter, our discussion will mainly focus on telephone echo cancellers, in particular a special type of telephone canceller, the modem data echo canceller.

This chapter is organized as follows.

In section 10.1, we will discuss how echoes are generated in telephone networks and how they affect the quality of voice and data communications. Three types of telephone echo cancellers, voice echo cancellers and data echo cancellers for modem signal and digital subscriber loop data transmission signals are introduced. We will also discuss the cancellation requirements for these echo cancellers.

In sections 10.2 and 10.3 various aspects of telephone echo cancellers will be discussed. The modem data echo canceller, or simply modem echo canceller, will be the focus of discussion. Also in section 10.2, we will present different structures of modem echo cancellers. Their converging and tracking characteristics will be discussed in detail.

Echo cancellers are usually implemented using high-speed digital signal processors (DSPs). DSPs, either general-purpose DSPs or custom-designed ones for echo cancellation, use a finite wordlength in the computation, which may degrade the echo canceller performance. In section 10.3, we will analyze the relationship between the wordlength and achievable echo cancellation. The results given there can serve as guidelines in echo canceller design. Algorithms to improve echo canceller performance when using a DSP with short wordlength are presented.

Owing to the widespread applications of echo cancellers, many advances have been made to improve their performance. In section 10.4, we briefly describe some of the subjects related to different echo cancellers, such as digital rate conversion, echo phase-roll compensation and schemes for the fast training of modem echo cancellers. Although, the topics discussed in sections 10.2 and 10.3 are described in the context of modem echo cancellers, they can be applied to other kinds of echo cancellers with some modifications. On the other hand, other types of echo cancellers have their own special requirements that are different from those of modem echo cancellers. Section 10.4 also provides a brief discussion of such special requirements for ISDN echo cancellers, the most important echo canceller for subscriber-loop data transmission and voice telephone echo cancellers and their implementations.

10.1 Echoes in telephone networks and their cancellati

In order to understand how an echo canceller works and what the req for different kinds of echo cancellers, we first take a close look at the e mechanism on telephone networks.

10.1.1 Echoes in telephone networks and their impac oice and data transmission

The existence of echoes on telephones has been well known for a long time [1]. It is an easily recognizable phenomenon whenever we make a telephone call. When we speak into the microphone of a telephone set, we can almost immediately hear our own voice from the earphone. This phenomenon shows that part of the signal generated by the microphone comes back to the earphone, and can be viewed as a type of echo. For the telephone conversation, this type of echo is actually desirable. Experiments show that while speaking, if the speaker does not hear his or her own voice, the speaker will unconsciously raise his or her voice because he or she would assume that the person on the other end does not hear the speaker's voice either. On the other hand, during a long-distance call, we may hear a delayed form of our own voice. This type of echo is usually annoying during a telephone conversation, and it has been shown that the longer the delay, the more annoying the echo becomes. We call these two types of echoes the near echo, meaning that it has a short delay, and the far echo, meaning that it has a long delay. To ensure the quality of telephone conversation, it is necessary to control the far echo. As shown later, both the near and far echoes are harmful for data transmission over telephone networks.

In order effectively to control echoes on telephone networks, it is necessary to model the echo signals accurately. We now describe how echoes are generated on a telephone network.

ECHO GENERATION ON A TELEPHONE NETWORK

In a typical telephone connection, a telephone is connected to one end of a two-wire telephone line. The other end of the telephone line connects to a local telephone central office (CO). Such a two-wire telephone circuit is called a *local loop*.

In a telephone handset there is a microphone and an earphone. The microphone is a transmitting device that converts acoustic speech signals into electrical signals and sends them out through the local loop. On the other hand, the earphone is a receiving device that receives the speech signal in electrical form from the same local loop and converts it into sounds that we can hear. Obviously, the local loop carries the signals travelling in both directions simultaneously. In other words, the local loop is a *bidirectional two-wire circuit*.

To connect the microphone and earphone to the two-wire local loop, we need a four-wire to two-wire coupling device called a *hybrid coupler*, or simply a *hybrid*. If the listener is located far away, four-wire trunk circuits are usually used for connection. In this case, at the CO, another hybrid coupler is used to connect the other end of the local loop with the four-wire circuit. The four-wire circuit, which may be realized using microwave cables, fiber optics cables, microwave relay links or satellite links, consists of a pair of two-wire circuits. The transmission may be achieved in analog or digital form. In any case, however, modulation and demodulation are involved during the transmission process. Unlike the local loop, each of the two-wire circuits only allows a signal to travel in one direction. The four-wire circuit connects the local CO to a remote CO, which connects to the remote telephone through the remote local loop and completes the telephone connection. Such a typical telephone connection is depicted in Figure 10.1.

A hybrid coupler is a two-wire-to-four-wire converter. The single port on its two-wire sides connects to a bidirectional transmission line, e.g. the local loop. There are two ports on the four-wire side: the transmitting port, or the input port, and the receiving port, or the output port. The signals applied to the transmitting port are sent out through the bidirectional transmission line. The signals from the transmission

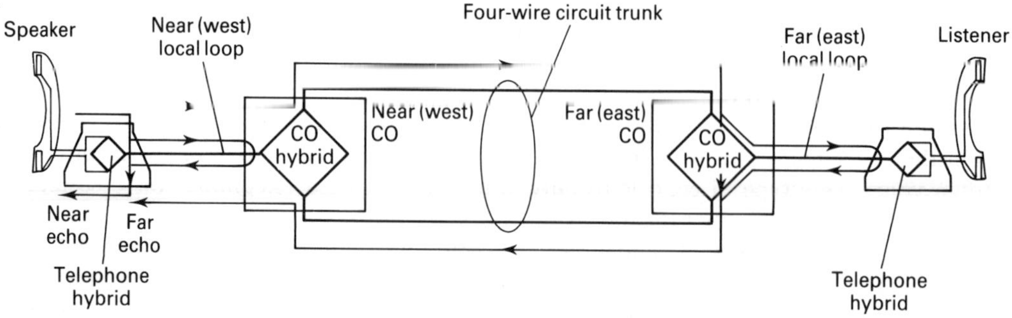

Figure 10.1 A typical telephone connection

line appear at the receiving port of the hybrid coupler. The function of the hybrid is to separate the transmitted and the received signals, i.e. to prevent the signal applied to its transmitted port from leaking to its receiving port. A hybrid may be realized using passive or active elements, and is called a *passive hybrid* or an *active hybrid* respectively. In order to eliminate completely the signal leaking from the transmitting port to the receiving port, the hybrid must exactly match the impedance of the bidirectional transmission line. Practically, such an exact matching is not realistic. As a result, part of the transmitted signals will always leak back to the receiving port. This is the main source of echoes generated in telephone networks.

Now let us use Figure 10.1 as an example to show how echoes are generated on a telephone network. Let us assume that a person on the left (west) side is speaking into the microphone of the telephone set. The speech signal is transmitted through a hybrid in the telephone and the local loop to the near (west) CO. Meanwhile, this signal will leak back to the receiver side of the hybrid as a result of the impedance mismatching of the hybrid. Moreover, any impedence mismatch along the local loop and at the hybrid in the near CO will also cause reflection of the transmission signal. All these echoes have short delays relative to the transmitted signal. The combination of these echoes is called the *near echo*. As stated above, an echo with a short delay may be desirable for telephone conversation, and therefore the hybrid in the telephone set is often deliberately mismatched. However, for data transmission over telephone lines, the near echo must be completely eliminated because it is usually much stronger than the useful received signal from the remote side.

No echoes are generated along the four-wire transmission circuits in which signals can only travel in one direction along each pair of wires. The transmitted signal travels from west to east along the top pair of wires. It arrives at one port of the hybrid in the remote (east) CO. Owing to the mismatch of the hybrid, part of the signal will leak back to the other port of the hybrid. Most of the signal will reach the remote telephone set through the remote local loop. The signal leaking through the hybrid in the east CO, together with the signals reflected from the hybrid in the remote telephone and along the remote local loop, will then travel back from east to west along the bottom pair of wires and arrive at the west, or near, CO. Finally, these signals arrive at the earphone of the local telephone after passing through the hybrid in the near CO, the near local loop and the hybrid in the west telephone. The signal reflected back from the remote side is called the *far echo*. Since the far echo travels through the whole four-wire circuit loop, it has a delay equal to approximately twice the time required for the signal travelling from one telephone handset to the other. This delay could be as large as 150 ms for microwave relay or cable transmission. It will be about 600 ms for single-satellite hop transmission and about 1200 ms for double-satellite hops. The far echo is usually weaker than the useful received signal from the remote side. This is because, as can be seen from Figure 10.1, the far echo is attenuated twice on the four-wire circuit, while the direct speech or data signal is only attenuated once. Nevertheless, the far echo is very annoying for speech telephone communications and must be controlled for speech transmission. For data communication, the far echo is normally easier to cancel than the near echo because it is weaker than the received signal. However, the far echo may be frequency translated when it passes through frequency converters (modulators and demodulators) on the four-wire circuits [3]. This phenomenon is called phase roll or frequency offset of the far echo, and is a non-linear impairment. Non-linear compensation techniques must be used for effective far echo cancellation if the phase roll exists. More information on the characteristics of echoes on telephone networks can be found in [2] and [3].

The near echo and the far echo without phase roll can be viewed as the output of the transmitted signal passing through a linear system. If we can accurately estimate the linear system, we can precisely synthesize these echoes using the transmitted signal and the estimated system parameters. Thus, in order to remove echoes from the received signal, we need only to subtract the synthesized echoes from the received signal. This is the principle of echo cancellation. Below we will examine the structure and requirements for speech and data echo cancellers.

10.1.2 Voice echo control and cancellation

As shown above, the near echo need not be cancelled for speech telephone transmission. However, noticeable far echoes with a delay of more than 10 to 20 ms constitute one of the most serious forms of impairment in telephone channels. Their subjective annoyance increases with both echo amplitude and propagation delay. It has been found that if there is no delay, one can tolerate an echo that is only 1.4 dB below the remote speech signal. However, if the echo delay increased to 20, 60 and 100 ms, the far echo must be reduced to 11.1, 22.7 and 30.9 dB below the speech signal, respectively, to avoid annoying the listener [4]. A number of echo control methods, including echo cancellation and other methods, are currently used in telephone networks.

ECHO SUPPRESSOR

For telephone connections with small propagation delay, for instance for a delay less than 40 ms or for telephone circuits shorter than 2000 miles (3220 km) in length, transmission losses proportional to line length are inserted in the four-wire circuits.

As a result, far echoes are attenuated twice as much as the desired signal. On longer circuits, devices called *echo suppressors* are used instead to prevent the useful signal from being overattenuated.

An echo suppressor is essentially a voice-operated electronic switch that monitors and controls the voice signals travelling in both directions during a long-distance conversation. The suppressor detects which person is talking and blocks the signal travelling in the opposite direction. This prevents the echo from looping back to the speaker through the listener's transmission path. The main problem with echo suppressors is that they tend to 'chop' the telephone conversation. For example, when two persons talk quickly back and forth, the suppressor must quickly insert and remove loss in the talking paths. Thus, some speech may be blocked partially, causing it to be 'choppy'. Furthermore, during double talk, i.e. when both parties talk simultaneously, echo suppressors cannot really block any of the paths. As a result, echo control becomes ineffective. Despite these disadvantages, it has been shown that echo suppressors can provide adequate performance for terrestrial circuits. However, their performance on satellite circuits is not satisfactory, as was demonstrated in [5]. A more effective method is *echo cancellation*, as was shown in [6] and [7].

VOICE ECHO CANCELLER

The arrangement of echo cancellers in a telephone network is shown in Figure 10.2. As we can see from the figure, a pair of echo cancellers are needed per telephone conversation. The details of a voice echo canceller are depicted in Figure 10.3. From Figure 10.2, we note that the echo canceller is placed in the four-wire path, near the origin of the echo, i.e. near the CO hybrid. Let us assume that the person on the left (west) side is talking. The speech travels along the top pair of wires from west to east until it reaches the hybrid in the east CO. Parts of the signal are reflected back from the hybrid and from the east local loop, which form the far echo for the talker. To cancel this echo, an echo canceller is placed near the east CO hybrid. As stated above, the echo can be modelled as a linearly transformed transmitted signal. The transmitted signal is also called the *reference signal* for the voice echo canceller. For the example discussed here, the reference signal is the signal on the input port side of the hybrid and the signal on its port side contains the echo. As shown by Figure 10.3, the echo canceller is a *finite impulse response* (FIR) filter. The input to the delay line of the filter is the sampled reference signal. If the FIR filter coefficients are selected such that its output is a synthesized copy of the sampled echo signal, we can eliminate the echo by subtracting the synthesized echo from the signal at the output port of the hybrid. It is easy to see that the echo canceller can effectively eliminate echo even under double-talk conditions. During double talk, the signal at the output port of the hybrid contains both the far echo and the speech signal from the east talker. After the synthesized echo is subtracted from it, the remaining signal is basically the speech signal.

The echo canceller in the west CO operates exactly in the same way as described above. It cancels the echo generated from the speech of the east talker. It is noteworthy that a voice echo canceller cancels the echo for the remote telephone user. This is

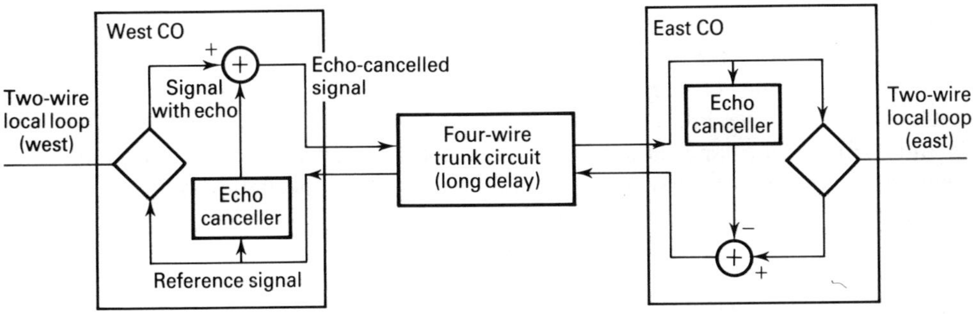

Figure 10.2 Telephone echo canceller arrangement in telephone network

different from a data echo canceller, as we will see later. The advantage of such an arrangement is that, because the canceller is close to the place where the echo is generated, the delay of the echo from the reference signal is short. Therefore, the delay line of the echo canceller can be short also. This simplifies the implementation of the echo canceller[1].

In order for the echo canceller to be able to model the echo response precisely, it must be trained before effective echo cancellation can be performed. Any of the adaptive algorithms discussed in the previous chapters can be used for training the voice echo canceller. In practice, the LMS algorithm is most widely used because of its computational efficiency. Other adaptive algorithms such as various least-squares

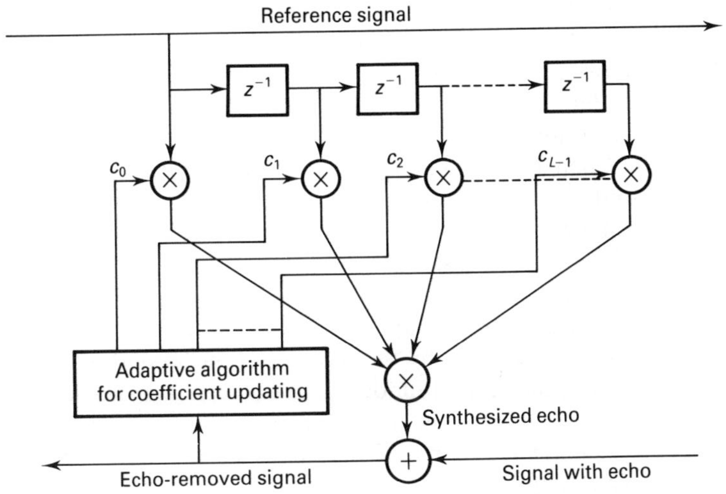

Figure 10.3 Structure of a voice telephone echo canceller

(LS) and lattice algorithms may be desirable in some cases because of their faster convergence and better tracking capability for such applications.

In general, the echo rejection requirements for voice echo cancellers are not as high as those for the data echo canceller discussed below. Usually one requires only about 30 dB of echo cancellation for the voice echo canceller [8]. However, practically, there are many factors that make effective voice echo cancellation difficult. We will briefly discuss these special problems in section 10.4.

10.1.3 Modem data echo cancellation

Owing to the widespread applications of digital computers today, there are increasing demands for exchanges of data. As a result, telephone networks are not only used to carry speech signals, but also to act as channels for data transmission. Data transmission over the telephone network is achieved through devices called *modems*, which is a contraction of 'modulators–demodulators'. Data transmission can be classified as half-duplex, for which data can only flow in one direction at any given time, and full duplex, for which data will flow in opposite directions simultaneously. Full-duplex data transmission provides a higher data throughput than half-duplex transmission, because it eliminates modem turnaround time. The easiest way to realize full-duplex data transmission is to use a pair (four-wire) of telephone circuits so that each telephone circuit is only responsible for the data flow in one direction. Obviously, it would be desirable to use only one (two-wire) telephone circuit to perform full-duplex data transmission. However, as shown previously, when transmitted over a two-wire telephone circuit the signal will partially return back to the receiver section of the modem in the form of an echo that interferes with its operation. Thus, the main obstacle for full-duplex data transmission over a two-wire telephone circuit is the echo.

Classically, time-division multiplexing (TDM) and frequency-division multiplexing (FDM) are two approaches for separating the received signal from the echo of the transmitted signal in a modem. A TDM system is a form of half-duplex transmission system; namely, time is divided for the modems on each end during data transmission. When a modem is transmitting, a second modem is only receiving and will not transmit. Thus, there is no echo for the receiver of the second modem. The situation is similar for the transmission of the second modem. On the other hand, in an FDM system, the usable bandwidth of telephone channels, which is typically from 300 Hz to 3400 Hz for voiceband telephone connections, is divided between the two modems. For example, the first modem uses the high band for transmitting its data to be received by the receiver of the second modem at the other end. On the other hand, the second modem is transmitting in the low band and the signal is received by the receiver of the first modem. Because the echo and the received signal for the same modem are in different frequency bands, they can be easily separated by proper filtering.

A common problem for TDM and FDM systems is that the modems at both ends of a telephone connection cannot use all the available time and bandwidth

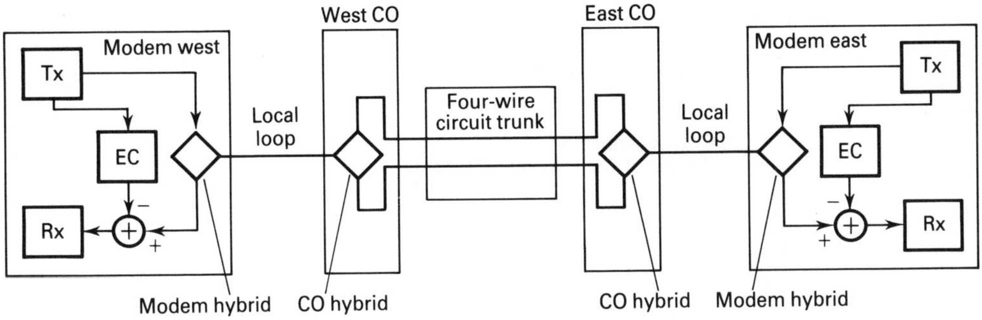

Figure 10.4 Data communication with echo cancellation modems

simultaneously. In other words, they do not fully use all the available resources provided by a telephone channel. To achieve more efficient full-duplex data transmission on two-wire telephone circuits, thanks to the advances in digital signal processing, we can use echo cancellation to separate the echo and the received signal in a modem. Such a modem is called an *echo cancellation* modem. Their application enables us to implement two-wire true full-duplex service with full bandwidth available in both directions for data transmisssion.

ECHO CANCELLATION FOR DATA TRANSMISSION

Unlike voice echo cancellers, which are installed on telephone networks, data echo cancellers are resident inside modems. Figure 10.4 shows two echo cancellation modems connected through local loops to a telephone network. There is a transmitter and a receiver in each of the modems. The transmitter converts data to be sent into the transmitter signal or Tx signal, suitable for transmission over the telephone network. The Tx signal is transmitted over the telephone network via a hybrid coupler inside the modem. The receiver receives the useful signal, also called the received signal or Rx signal, from the remote modem via the receiver port of the hybrid. The Rx signal is corrupted by echoes and channel noise. The echo canceller estimates the echo channel response and synthesizes the echo using the transmitted signals and the emulated echo channel. The echoes in the Rx signal can be eliminated by simply subtracting the synthesized echoes from the signal, as long as the echo channel is estimated and emulated accurately.

Since there exist a near echo and a far echo, the total echo channel response may become very long, e.g. up to 600 ms or longer. It would be unrealistic from the viewpoint of computational complexity to emulate such a long channel. Fortunately, as can be seen from the way in which echoes are generated, as described in section 10.1.1, the near and far echoes are separated by a long silence, as shown in Figure 10.5. Thus, an echo canceller can be implemented as two shorter parts, each of which has a span of about 10 to 20 ms. These two parts of the echo canceller, called near and far echo cancellers, respectively, are separated by a long time delay equal to the

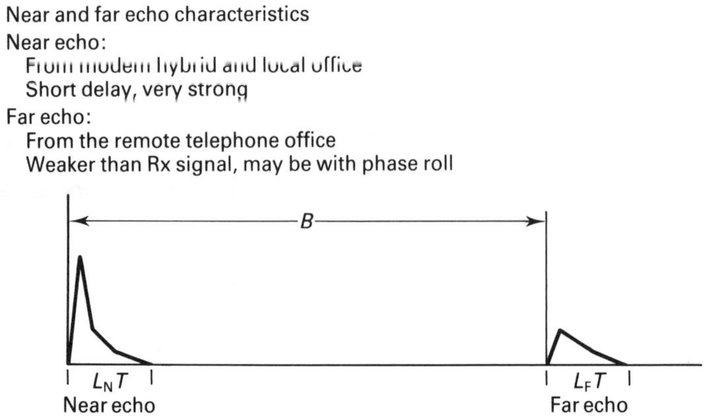

Figure 10.5 Echo profile and characteristics

time delay between the near and far echoes. The structure of such an echo canceller is depicted in Figure 10.6.

Figure 10.6 shows the general structure of a modem echo canceller. Practically, echo cancellers can be implemented in various forms. These forms of echo cancellers and their convergence and tracking characteristics will be discussed in section 10.2. Below we first describe the echo rejection requirements of modem echo cancellers.

ECHO REJECTION REQUIREMENTS

Let us use the CCITT V.32[9] echo cancellation modem standard as an example. For the 2400 baud, 9600 bits per second (bps) signal employing trellis-coded modulation (TCM) specified by the V.32 standard, a signal-to-noise ratio (SNR) equal to 18 db or higher is required to ensure normal operation. In order for the residual echo not to affect the receiver performance noticeably, we would like the residual echo to be 6 dB below the allowable noise level. Typically, the nominal level of the transmitted signal delivered to the local loop is equal to -9 dBm[2], and the range of the signal level received from the loop is from -43 dBm to -9 dBm. For the lowest signal level (-43 dBm), which imposes the most severe echo cancellation requirement, a residual echo level of -67 ($= -43 - 18 - 6$) dBm, or lower, is required. On the other hand, the typical insertion loss of a hybrid, i.e. the attenuation from the transmitter port to the receiver port, is equal to 9 dB, and the minimum insertion loss is equal to 6 dB. Hence, the highest echo level at the hybrid receiver port is -15 dBm. Therefore, the echo canceller is required to provide a 52 dB echo rejection to attenuate the maximum echo level (-15 dBm) to the required residual echo level (-67 dBm). The ratio of the echoes before and after echo cancellation is called the *echo return loss enhancement* (ERLE) of the echo canceller. The signalling defined by the CCITT standard V.32bis requires an SNR that is 6 dB higher than the signal for the V.32 standard. As a consequence, we would like the echo canceller in a V.32bis

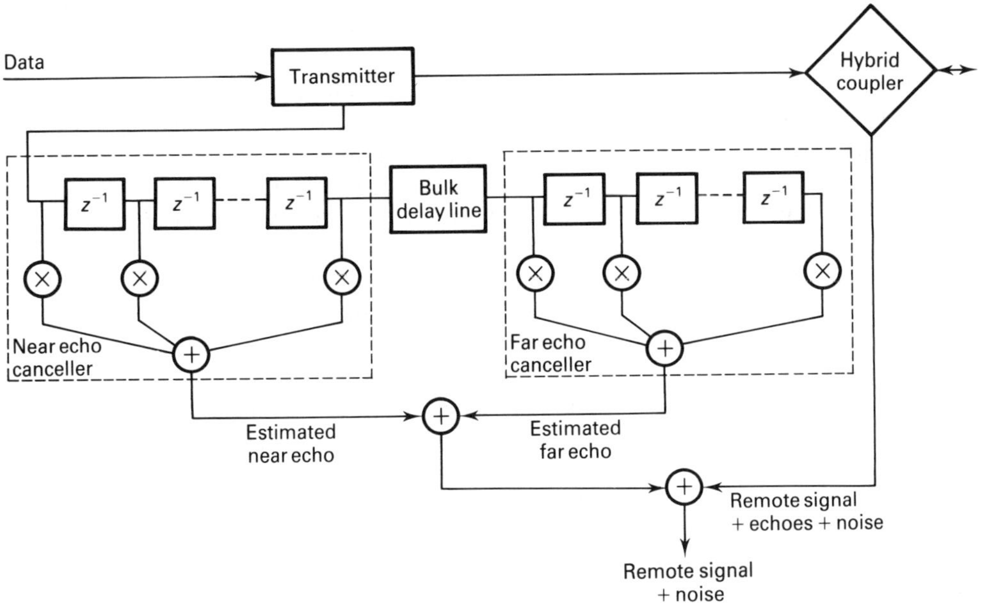

Figure 10.6 Structure of a modem echo canceller

modem to provide an ERLE of 58 dB. In other words, an echo canceller for V.32 or V.32bis modems must reduce the echo power by a factor of 160 000 or 640 000 respectively. Obviously, an echo canceller must estimate the echo channel very accurately to achieve such a required ERLE.

Assuming that the echoes are indeed a perfectly linear transformed transmitted signal, it is possible, at least in theory, to achieve an infinite ERLE. Practically, there are two major factors that limit our ability to achieve such an infinite ERLE. First, there are always some non-linearities in the echo channel. Thus, our linear model is just an ideal case. Obviously, any non-linear echo cannot be cancelled by a linear echo canceller. Second, a modem echo canceller is normally implemented using a digital signal processing device. The accuracy in modelling the echo channel is always limited by finite precision effects in the computation, which also limits the ERLE of an echo canceller. We will discuss the finite effects for echo cancellation in section 10.3.

The strongest echo is that part of the transmitted signal leaking through the hybrid which constitutes a part of the near echo and is usually much stronger than the received signal. Although the far echo is usually much weaker, it may become modulated as a result of the imperfect matching of the modulators and demodulators in the carrier system. Because modulation is a non-linear operation, non-linear compensation schemes must be used to achieve the effective rejection of such a far echo. The generation mechanism of phase roll in far echo and its compensation are discussed in section 10.4.2.

10.1.4 Digital subscriber-loop echo cancellation

Although almost all data transmission over new telephone networks is realized in digital form, it is still more economical to use analog signals for high-speed data transmission over the local loop, i.e. from the end user to the near central office. It has been shown that the echo cancellation can provide a greater range for full-duplex transmission than alternative methods such as TDM and FDM. Echo cancellers used for high-speed data transmission over local loops are called subscriber-loop echo cancellers. Since the subscriber-loop data transmission for the integrated service digital network (ISDN) is the only type of such services specified today [10], we will discuss subscriber-loop data transmission and echo cancellation in what follows only in the context of the ISDN.

Unlike modem data transmission, in which data are transmitted in analog form (including digitized analog form) from one user to another, ISDN data transmission is only in analog form from the user to the near CO. Between COs data are transmitted in digital form. Thus, a higher data rate can be achieved due to the shorter transmission distance. For the ISDN subscriber loop, the data rate is 144 kbps (kilobits per second) as opposed to 9600 bps for the V.32 standard. Even higher data rates up to 800 kbps for new services are also under consideration.

Because analog data transmission is only implemented between the user and the local CO, echo cancellers must be employed at both ends, i.e. at the end user premises and the local CO. ISDN echo cancellers only need to eliminate near echoes since there is no far echo. Furthermore, there is only one master clock source located at the CO that controls data transmission in both directions. Thus, the transmitter clock and receiver clock are always synchronous. These properties of ISDN data transmission make the design of an ISDN echo canceller simpler. On the other hand, transmission at such high data rates requires the echo canceller to have a higher echo rejection and entails more computation. An echo reduction ratio of 70 dB is usually required for such echo cancellers. As a result, in addition to using linear FIR echo cancellers, IIR and non-linear echo cancellers are often used for realizing ISDN echo cancellers. Synchronization of the transmitter and receiver clock rates also creates new problems for both the receiver and the echo canceller. Other differences between the modem and ISDN data transmission include data symbol sets and baud rates. Roughly speaking, transmitting data symbols are simpler but the baud rate is much higher for ISDN data transmission than that of modem data transmission. The features of ISDN echo cancellers must be considered in the design and will be briefly discussed in section 10.4.

10.2 Data-driven Nyquist echo cancellers and their converging and tracking characteristics

Echo cancellers are successfully used in today's echo cancellation modems. Modem echo cancellers are probably the most investigated and best understood echo cancellers among the many different types of echo cancellers. Although many different types of echo cancellers have been proposed, today's echo cancellation modems almost exclusively employ the so-called *data-driven echo cancellers* [11,12]. In this section, we first discuss the structures and operations of two types of data-driven echo cancellers, the real passband echo cancellers and the (baseband or passband) analytic (complex) echo cancellers, and establish their mathematical models. Then their convergence and tracking characteristics are derived. The results obtained for the modem data echo cancellers are also, at least in principle, applicable to other types of echo cancellers, Finally, the different aspects and properties of the two types of echo cancellers are compared.

10.2.1 Structures of modem data echo cancellers

The basic principle in echo cancellation is to synthesize an accurate replica of the echo contained in the received signal by using a linear combination of samples of the transmitted signal. The synthesized echo is subtracted from the received signal to achieve echo cancellation. To form the synthesized echo, we may use either the transmitted (Tx) data symbols or passband Tx signals. An echo canceller that uses the Tx data symbols to form the synthesized echo is called a data-driven echo canceller. Because the data contained in the input delay line of a data-driven echo canceller are usually randomized, they have a flat (white) spectrum. Thus, the eigenvalues of the data autocorrelation matrix are all equal. As a consequence, for whatever adaptive algorithm used, data-driven echo cancellers always provide the best possible convergence and tracking performance.

Echo subtraction can be performed at the received data symbol rate that is synchronous with the remote transmitter rate [11, 13]. It can also be performed at any sampling rate of the received signal that satisfies the Nyquist criterion, i.e. the sampling rate is at least twice that of the highest frequency in the signal [12,14]. It is usually selected to be an integer multiple of the Tx data symbol rate. These two types of echo cancellations are called *symbol rate echo cancellation* and Nyquist echo cancellation respectively. Historically, symbol rate echo cancellation was the first one proposed. However, while symbol rate echo cancellers are used in ISDN echo cancellers, Nyquist echo cancellers are almost exclusively used in modems. This is because Nyquist echo cancellers eliminate the need for complicated symbol synchronization between the local and remote transmitter symbol rates that are usually not synchronous to each other for data modems. Nyquist echo cancellers are also suitable for use in conjunction with fractionally spaced equalizers that are widely used in modems. We will use such data-driven Nyquist echo cancellers to illustrate the operation and characteristics of data echo cancellers.

STRUCTURES AND OPERATION OF DATA-DRIVEN NYQUIST ECHO CANCELLATION MODEMS

The Nyquist echo canceller was first proposed in [12] and is now widely used in most commercial modem products. Several variations have been proposed since then.

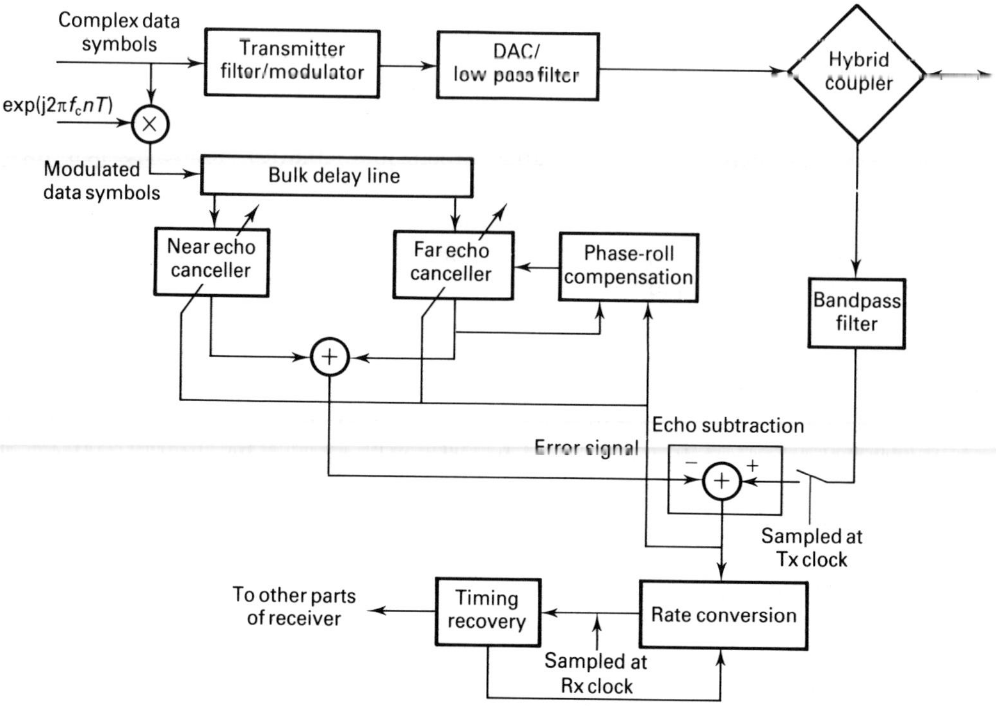

Figure 10.7 Structure of an echo cancellation modem

Depending on where echo subtraction is performed, Nyquist echo cancellers can be classified as real passband echo cancellers and analytic echo cancellers, which, in turn, can be classified as passband-analytic and baseband-analytic echo cancellers. To begin with, we first describe the structure operation of a typical echo cancellation modem.

A block diagram of a modem with a data-driven echo canceller is given in Figure 10.7. A modem interfaces to a piece of digital terminal equipment (DTE), such as a computer data terminal. The binary data from the DTE are first randomized by a scrambler and then encoded by a channel encoder, which ensures more reliable transmission over telephone channels, and/or mapped into a sequence of complex-valued data symbols, which are also random. As a result of the randomization, the autocorrelation of the symbol sequence is an impulse function. Such a sequence is called white because it has a flat, or white, power spectrum which, by definition, is the Fourier transform of the autocorrelation function of the sequence. The purpose of the randomization is to facilitate effective transmission of the data over the telephone channel. Moreover, it is known from adaptive filtering theory that an adaptive filter has the best convergence when the input signal is white. For adaptive echo cancellers discussed in this chapter, their input signal is the data symbol sequence. As a result, randomization also improves the convergence of the echo canceller.

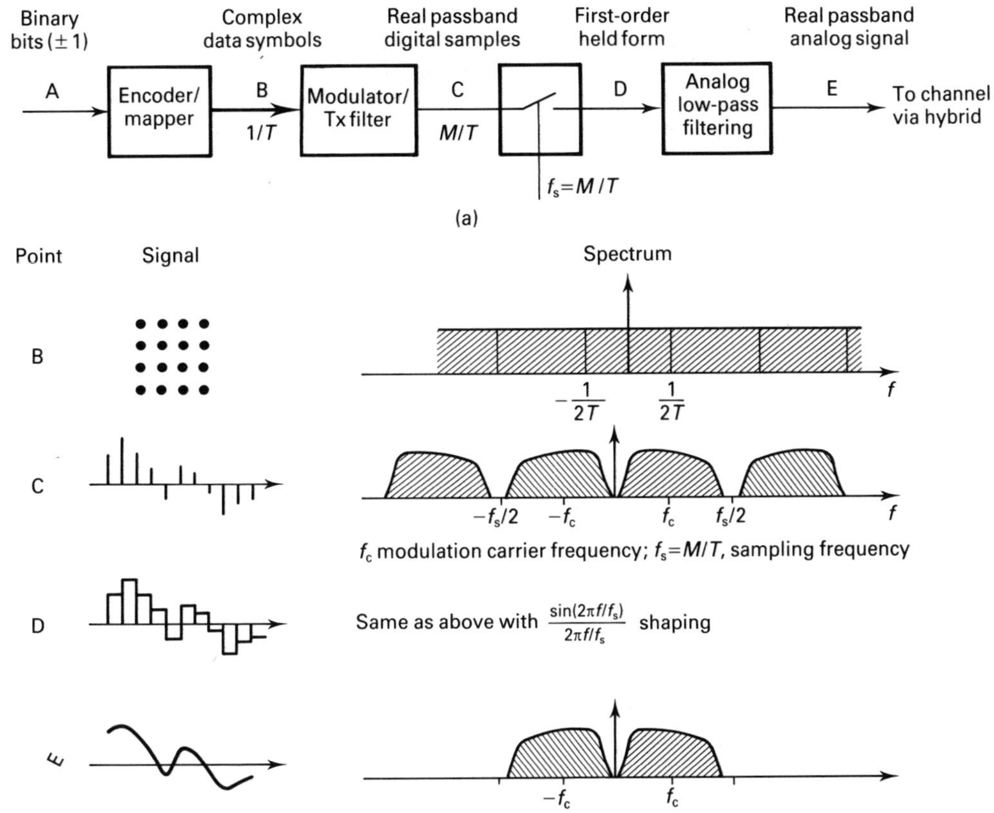

Figure 10.8 Generation of modem transmitter signal: (a) Tx signal generation (T, symbol interval, $M > 1$); (b) signals and their spectra at different stages

The complex-valued, information-bearing data symbols, denoted by u_m at a *symbol rate*, also called *baud rate*, of $1/T$, where T is the *symbol interval*, are then modulated by a modulator at the modulating carrier frequency f_c and spectrum shaped by a filter called a transmitter (Tx) filter. After these operations, the spectrum of the resulting signal is determined by the frequency response of the Tx filter and is suitable for transmission over telephone channels. It is important to note that both the modulation and filtering operations are implemented in digital form. As a result, for complex digital data, modulation is simply a sample-by-sample rotation. The real part of the modulated and filtered digital signal, which is called the *real passband* signa, usually has a sample frequency of M/T, where M is an integer and $M \geqslant 3$. The real passband signal is then converted to analog form by a digital-to-analog converter (DAC). The analog real passband Tx signal, after it is filtered by an analog bandpass filter, is applied to the transmitter port of a hybrid, which sends the signal to the remote modem through a telephone channel, at least a part of which is a bidirectional two-wire circuit. The process of generating the transmitted signal described above is

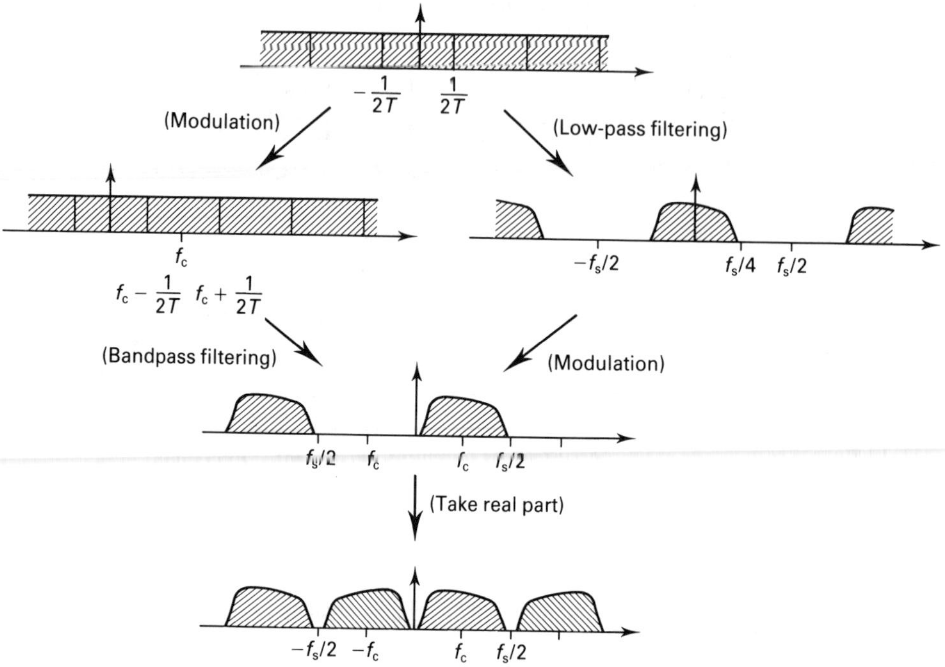

Figure 10.9 Two arrangements of modulation and Tx filtering

depicted in Figure 10.8. On the figure, we have also shown the time signal and its frequency domain representation (spectrum).

The modulating/Tx filtering of the transmitter signal can be implemented in two ways, as shown in Figure 10.9. We may first modulate the complex data symbols and then spectrally shape the modulated signal using a bandpass filter. The resulting signal is a modulated complex signal and is called a *passband-analytic* signal. This process is depicted by the left path in Figure 10.9. Alternatively, we can first shape the unmodulated complex symbol using a low-pass filter. The filtered signal, which is called a *baseband-analytic* signal, is then modulated to generate the passband-analytic signal. The process is shown by the right path in Figure 10.9. The digital real passband signal, which is the real part of the passband-analytic signal, is then sent to the DAC to be converted to an analog signal.

We note that the Tx filter/modulator is a multirate system. Specifically, the output sample rate must be higher than the Nyquist rate of the passband signal to be transmitted. Furthermore, because only the real part of the Tx filter output is used, the transmitter filter has to ensure that no significant, undesired, negative frequency components will alias back to the useful signal band. As mentioned above, the output sampling rate of the Tx filter is usually selected to be an integer multiple of the input Tx symbol rate $1/T$. We illustrate these operations and requirements with the following example.

EXAMPLE 1

The Tx symbol rate $1/T$ and the modulation carrier frequency f_c specified by the CCITT V.32 echo cancellation modem standard are 2400 symbols per second and 1800 Hz respectively. In order to preserve all the Tx information, the baseband signal bandwidth must be at least equal to the Tx symbol rate, i.e. 2400 Hz. For a practical implementation, some excess bandwidth is usually required. Assuming that the excess bandwidth is 15%, the total signal bandwidth is then equal to 2400 Hz $\times$ $(1 + 0.15) = 2760$ Hz. First let us consider the generating process specified by the right path of Figure 10.9. The frequency response of a baseband Tx filter for generating such a complex (analytic) signal usually has the shape of a square root of a raised cosine between ± 1380 Hz. After modulation, the analytic passband Tx signal will be located between 1800 ± 1380 Hz, i.e. from 420 Hz to 3180 Hz. Such a signal will be adequate for transmission over telephone channels which have a usable bandwidth from 300 Hz to 3400 Hz. The real part of the analytic passband signal is sent to the DAC. To satisfy the Nyquist criterion, the sampling frequency f_s of the signal at the input to the DAC must be higher than $2 \times 3180 = 6360$ samples per second. Such a criterion can be satisfied by selecting the sampling frequency f_s to be greater than or equal to $3/T = 7200$ samples per second. With such a sampling frequency, we only need to consider the signal in the frequency band from -3600 Hz to 3600 Hz.

To avoid aliasing, the negative frequency component of the complex passband signal must be eliminated before taking its real part. Specifically for this example, the residual signal in the band from -3600 Hz to 0 Hz, in particular from -3180 Hz to -420 Hz, must be far below the positive frequency passband signal to be sent. It is not difficult to verify that the baseband Tx filter must eliminate the signal in the frequency band from -3600 Hz to -2220 Hz (-420 Hz to -1800 Hz), and its symmetric frequency band from 220 Hz to 3600 Hz. If the modulation is performed first, the Tx filter should simply reject the signal in the frequency range from -3180 Hz to -420 Hz, as shown by the left path in Figure 10.9.

The signal from the remote modem, which may be corrupted by additive noise and echoes of the locally transmitted signal, appears at the receiver port of the hybrid. The signal is bandpass filtered and converted to digital samples. The receiver sampling rate must also satisfy the Nyquist criterion. For an echo cancellation modem, it is selected to be an integer multiple M_r, usually equal to three or four times the Tx symbol rate. The real received signal samples are denoted by $x(kT/M_r)$.

As has been stated above, the synthesized echo can be subtracted from the real received signal samples or from the baseband or passband-analytic received signal samples. The passband-analytic signal samples may be generated by passing the real samples through a Hilbert transformer to eliminate the negative components in the received signal. After the negative frequency components are eliminated, the sampling rate of the analytic signal samples can be reduced to one-half of the sampling rate of the real samples without aliasing. The baseband-analytic signal samples can be generated by demodulating the passband-analytic signal samples. Practically, the sampling rate of the analytic signal samples must also be an integer multiple of the

Tx symbol rate. The analytic received signal samples are denoted by $\tilde{x}(kT/M_c)$, with M_c usually chosen to be equal to 2 with $M_r = 4$.

In a Nyquist real passband echo cancellation modem, the echo canceller, consisting of a near echo and a far echo canceller, generates real T/M_r-spaced digital samples of a synthesized echo that is an accurate replica of the true echo in the received signal samples. By subtracting the synthesized echo samples from the real received signal samples $x(kT/M_r)$, we obtain real received signal samples containing mainly the desired, remotely transmitted, signal that is substantially echo free but may be corrupted by additive noise. On the other hand, in a Nyquist analytic echo cancellation modem, the echo canceller generates T/M_c-spaced analytic signal samples of a synthesized echo that is subtracted from the analytic received signal samples $\tilde{x}(kT/M_c)$.

The echo-free received signal is then converted to samples at a different rate that is controlled by a timing generator circuit. The timing circuit recovers information from the remote transmitter clock to maintain the converted samples in synchronism with the remote transmitter clock. The converted samples are then processed by a demodulator and an equalizer to recover the information bearing remote transmitted symbols. Information data bits are recovered from the symbols and output to the DTE.

Based on the description given above, we now establish mathematical models for the data-driven Nyquist echo cancellers. Although the real passband Nyquist echo canceller has some implementation advantages over its analytic counterpart, the analytic Nyquist echo canceller is conceptually simpler. We will discuss this one first.

MODEL OF A DATA-DRIVEN ANALYTIC NYQUIST ECHO CANCELLER

As is known from the data communication theory (see e.g. [15]) the transmitted line quadrature amplitude modulation (QAM) signal $x_t(t)$ can be expressed as

$$x_t(t) = \mathrm{Re}\left(e^{j\omega t} \sum_{m=-\infty}^{\infty} u_m \tilde{h}_t(t - mT) \right) \cong \mathrm{Re}[\tilde{x}_t(t)] \tag{10.1}$$

where u_m are the (complex) QAM data symbols, $\tilde{h}_t(t - mT)$ is the baseband time response of the transmitter channel (from the input of the transmitter filter to the hybrid output) and $\tilde{x}_t(t)$ is the analytic form of $x_t(t)$. We note that $x_t(t)$ and $\tilde{x}_t(t)$ are cyclostationary signals with a period of T. The received signal $x(t)$ at the receiver port of the hybrid can be expressed as

$$x(t) = r(t) + e_N(t) + e_F(t) + w(t) \cong \mathrm{Re}[\tilde{r}(t)] + \mathrm{Re}[\tilde{e}_N(t)] + \mathrm{Re}[\tilde{e}_F(t)] + \mathrm{Re}[\tilde{w}(t)] \tag{10.2}$$

where $r(t)$ is the signal from the remote modem, $e_N(t)$ is the near echo, $e_F(t)$ is the far echo and $w(t)$ is the additive noise which is assumed to be Gaussian and white. The symbols with a tilde are the analytic forms of the corresponding real quantities. As discussed above, we assume that the received signal $x(t)$ is sampled at M_r/T. The lth sample $x(l(T/M_r))$ can be expressed as

$$x\left(l\frac{T}{M_r}\right)=r\left(l\frac{T}{M_r}\right)+e_N\left(l\frac{T}{M_r}\right)+e_F\left(l\frac{T}{M_r}\right)+w\left(l\frac{T}{M_r}\right) \tag{10.3}$$

After passing $x(l(T/M_r))$ through a Hilbert transformer, we obtain its analytic form. If the Hilbert transformer is ideal, its output is indeed $\tilde{x}(l(T/M_r))$. Practically, the Hilbert transformer will always introduce delay and linear distortion. To simplify the analysis, we shall assume that the Hilbert transformer is ideal and the signal at its output is the analytic form of the real signal $x(l(T/M_r))$. Similar to (10.3), the analytic received signal samples can be expressed as

$$\tilde{x}\left(l\frac{T}{M_c}\right)=\tilde{r}\left(l\frac{T}{M_c}\right)+\tilde{e}_N\left(l\frac{T}{M_c}\right)+\tilde{e}_F\left(l\frac{T}{M_c}\right)+\tilde{w}\left(l\frac{T}{M_c}\right) \tag{10.4}$$

where $\tilde{x}(l(T/M_c))$, $\tilde{r}(l(T/M_c))$, $\tilde{e}_N(l(T/M_c))$, $\tilde{e}_F(l(T/M_c))$ and $\tilde{w}(l(T/M_c))$ are the analytic forms of their corresponding real quantities. It has been assumed that the sampling rate of the output samples of the Hilbert transformer is M_c/T, and M_c may be different from and usually is an integer submultiple of M_r. The sampled analytic near echo signal can be expressed as

$$\tilde{e}_N\left(l\frac{T}{M_c}\right)=e^{j\omega t}\sum_{m=-\infty}^{\infty}u_m\tilde{h}_N(t-mT)\Big|_{t=l\frac{T}{M_c}}=e^{j\omega l\frac{T}{M_c}}\sum_{m=-\infty}^{\infty}u_m\tilde{h}_N\left(l\frac{T}{M_c}-mT\right) \tag{10.5}$$

where $\tilde{h}_N(t)$ is the baseband time response of the total near echo channel (from the input of the transmitter filter to the hybrid receiver port output). The far echo can be expressed similarly by using the baseband far echo channel response $\tilde{h}_F(t)$ instead of $\tilde{h}_N(t)$.

By writing $l(T/M_c)$ as $nT+(i(T/M_c))$, $i=0,\ldots,M_c-1$, we can separate (10.5) into M_c equations. That is,

$$\tilde{e}_{N,i}(n)\cong\tilde{e}_N\left(nT+i\frac{T}{M_c}\right)=\exp\left[j\omega\left(nT+i\frac{T}{M_c}\right)\right]\sum_{m=-\infty}^{\infty}u_m\tilde{h}_N\left(nT+i\frac{T}{M_c}-mT\right)$$

$$i=0,\ldots,M_c-1 \tag{10.6}$$

It can be seen from (10.6) that the near echo can be viewed as being generated by M_c independent and interleaved complex linear subsystems, each of which is sampled at every T seconds and spaced by T/M_c seconds from the adjacent samples.

Since the near echo has a finite time response, i.e. $\tilde{h}_N(t)\neq 0$, for $0\leqslant t\leqslant T_N\leqslant(L_N-1)T$, we can rewrite (10.6) as

$$\tilde{e}_{N,i}(n)=\exp\left[j\omega\left(nT+i\frac{T}{M_c}\right)\right]\sum_{m=n-L_N+1}^{n}u_m\tilde{h}_N\left(nT+i\frac{T}{M_c}-mT\right)$$

We say such a near echo spans L_N symbol intervals, or bauds. By letting $k=n-m$, and after some algebra, we obtain

$$\begin{aligned}\tilde{e}_{\mathrm{N},i}(n) &= \sum_{k=0}^{L_{\mathrm{N}}-1} \exp[\mathrm{j}\omega(n-k)T]u_{n-k} \exp\left[\mathrm{j}\omega\left(kT + i\frac{T}{M_{\mathrm{c}}}\right)\right]\tilde{h}_{\mathrm{N}}\left(kT + i\frac{T}{M_{\mathrm{c}}}\right) \\ &\cong \sum_{k=0}^{L_{\mathrm{N}}-1} \exp[\mathrm{j}\omega(n-k)T]u_{n-k}\tilde{h}_{\mathrm{N},i}(k) \end{aligned} \tag{10.7}$$

where

$$\tilde{h}_{\mathrm{N},i}(k) \cong \exp\left[\mathrm{j}\omega\left(kT + i\frac{T}{M_{\mathrm{c}}}\right)\right]\tilde{h}_{\mathrm{N}}\left(kT + i\frac{T}{M_{\mathrm{c}}}\right) \quad i = 0, \ldots, M_{\mathrm{c}} - 1$$

is called the ith (passband) response of the near echo.

The generation and modelling of the far echo are similar to the near echo. The difference is that we must take into account the long delay usually associated with the far echo. Using the model presented above, the far echo channel can also be separated into M_{c} subchannels. Assuming that the frequency offset does not exist, the far echo samples generated by the ith echo subchannel can be written as

$$\begin{aligned}\tilde{e}_{\mathrm{F},i}(n) \cong \tilde{e}_{\mathrm{F}}\left(nT + i\frac{T}{M_{\mathrm{c}}}\right) &= \exp\left[j\omega\left(nT + i\frac{T}{M_{\mathrm{c}}}\right)\right] \\ &\times \sum_{m=-\infty}^{\infty} u_m \tilde{h}_{\mathrm{F}}\left(nT + i\frac{T}{M_{\mathrm{c}}} - mT\right) \quad i = 0, \ldots, M_{\mathrm{c}} - 1 \end{aligned} \tag{10.8}$$

For a far echo with a delay of BT and span of L_{F} bauds, we have that $h_{\mathrm{F}}(t) \neq 0$, only for $BT \leqslant t \leqslant BT + (L_{\mathrm{F}} - 1)T$. Similar to the development of (10.7), (10.8) can be rewritten as

$$\begin{aligned}\tilde{e}_{\mathrm{F},i}(n) &= \exp\left[\mathrm{j}\omega\left(nT + i\frac{T}{M_{\mathrm{c}}}\right)\right]\sum_{m=-\infty}^{\infty} u_m\tilde{h}_{\mathrm{F}}\left(nT + i\frac{T}{M_{\mathrm{c}}} - mT\right) \\ &= \sum_{k=0}^{L_{\mathrm{F}}-1} \exp[\mathrm{j}\omega(n-B-k)T]u_{n-B-k} \exp\left[\mathrm{j}\omega\left(kT + i\frac{T}{M_{\mathrm{c}}}\right)\right]\tilde{h}_{\mathrm{F}}\left(kT + i\frac{T}{M_{\mathrm{c}}}\right) \\ &\cong \sum_{k=0}^{L_{\mathrm{F}}-1} \exp[\mathrm{j}\omega(n-B-k)T]u_{n-B-k}\tilde{h}_{\mathrm{F},i}(k) \end{aligned} \tag{10.9}$$

where

$$\tilde{h}_{\mathrm{F},i}(k) \cong \exp\left[\mathrm{j}\omega\left(kT + i\frac{T}{M_{\mathrm{c}}}\right)\right]\tilde{h}_{\mathrm{F}}\left(kT + i\frac{T}{M_{\mathrm{c}}}\right)$$

is the ith (passband) far echo response.

From the above discussion, we conclude that the near echo canceller can be implemented as M_{c} independent and interleaved sub-echo-cancellers, each of which is an FIR filter with a sampling rate of $1/T$. The M_{c} subcancellers share a common input delay line of modulated data symbols $\exp[\mathrm{j}\omega(n-k)T]u_{n-k}$, $k = 0, \ldots, L_{\mathrm{N}} - 1$. The realization of the near echo canceller thus reduces to accurately determining the echo channel coefficients $\tilde{h}_{\mathrm{N},i}(k)$, $i = 0, \ldots, M_{\mathrm{c}}$ and $k = 0,1, \ldots, L_{\mathrm{N}} - 1$, and to implement the FIR filters. The implementation of the far echo canceller is similar to the implementation of the near echo canceller. Owing to the delay BT of the far echo, at time nT, the symbols used to synthesize the far echo are the modulated symbols $e^{j\omega lT}u_l$, for $l = n - B$ to $n - B - L_F + 1$. Thus, a 'bulk' delay line with a delay of B symbols is used to generate the delayed data.

An alternative implementation of the analytic echo canceller is to demodulate the received signal before echo subtraction. In such a case, we can use the unmodulated transmitted data symbols to form synthesized echoes. Such an echo canceller is called a *baseband analytic* echo canceller. Similarly, the analytic echo canceller described above is often called a passband-analytic echo canceller. There is no substantial difference between these two types of analytic echo cancellers and their performance is similar. However, a modem designer may prefer to use one form over another solely as a result of specific implementation considerations. Moreover, there is no difference in analyzing these two types of analytic echo cancellers. The analysis of the passband-analytic echo given later in this section also applies to the baseband analytic echo canceller.

MODEL OF A DATA-DRIVEN REAL PASSBAND NYQUIST ECHO CANCELLER

The data-driven real passband echo canceller is very similar to the data-driven analytic one discussed above. The real near echo $e_N(l(T/M_r))$ in (10.3) can be written as the real part of a hypothetical analytic near echo $\tilde{e}_N(l(T/M_r))$, which is the same as the anlytic near echo given by (10.5) using an ideal Hilbert transformer, i.e.

$$e_N\left(l\frac{T}{M_r}\right) = \mathrm{Re}\left[\tilde{e}_N\left(l\frac{T}{M_r}\right)\right]$$

Following the same development for $\tilde{e}_N(l(T/M_r))$, it can be shown that the real near echo can be viewed as the real parts of M_r independent and interleaved complex linear subsystems, each of which is sampled at every T seconds and spaced by T/M_r seconds from the adjacent samples. The ith echo sample can be written as

$$e_{N,i}(n) = \mathrm{Re}\left(\sum_{k=0}^{L_N-1} \exp[j\omega(n-k)T]u_{n-k}\tilde{h}_{N,i}(k)\right) \tag{10.10}$$

Similarly, the ith real far echo sample at nT can be expressed as

$$e_{F,i}(n) = \mathrm{Re}\left(\sum_{k=0}^{L_F-1} \exp[j\omega(n-B-k)T]u_{n-B-k}\tilde{h}_{F,i}(k)\right) \tag{10.11}$$

To summarize, the real passband echo canceller can be implemented as Mr independent and interleaved sub-echo-cancellers, each of which is an FIR filter with a sampling rate of $1/T$ and shares a common input delay line of modulated complex data symbols $\exp[j\omega(n-k)T]u_{n-k}$. Only the real part of the echo canceller output needs to be computed to perform echo cancellation. Each of the subcancellers consists of a near and far echo canceller section separated by a bulk delay line of BT symbols.

The determination of the echo canceller coefficients is equivalent to the linear system indentification problem discussed in Chapter 2. Furthermore, the time responses of telephone channels are slowly time varying due to, for example, temperature changes. It will be desirable for the echo canceller to track the time varying echo channel. Because of these requirements, today's high-speed echo cancellation modems almost exclusively employ adaptive echo cancellers. Various adaptive algorithms can be used to implement echo cancellers. However, because of the large number of coefficients involved, the LMS algorithm discussed in Chapter 4 is most commonly used for echo channel estimation because of its computational efficiency. It is also suitable for tracking slow echo channel changes because the input data to the echo cancellers are white and the LMS algorithm yields the same tracking performance as the LS algorithms under such circumstances.

The determination of the echo canceller coefficients, called the training period of the echo canceller, is performed during half-duplex modem operations. That is, only the transmitter of the modem whose echo canceller is under training sends signals, while the modem at the other end is silent. Thus, the term $r(l(T/M_r))$ is zero. On the other hand, an echo canceller must track slow time variations of the echo channel under full-duplex modem operation. In the latter case, the signal $r(l(T/M_r))$ from the remote modem is not zero and is much stronger than the residual echo. The adaptation parameters of the LMS echo canceller have to be selected accordingly and are very different in these two cases. These problems are discussed in the next two sections.

10.2.2 Convergence characteristics of LMS Nyquist echo cancellers

In this section, we derive the convergence characteristics of Nyquist LMS echo cancellers. We begin by considering the analytic Nyquist echo canceller.

CONVERGENCE CHARACTERISTICS OF LMS ANALYTIC ECHO CANCELLERS

Recall that a Nyquist analytic echo canceller comprises M_c subcancellers. Each subcanceller, in turn, has a near echo canceller section and a far echo canceller section. At nT, the ith subcanceller output $\hat{\tilde{e}}_i(n)$ is given by

$$\hat{\tilde{e}}_i(n) = \hat{\tilde{e}}_{\mathrm{N},i}(n) + \hat{\tilde{e}}_{\mathrm{F},i}(n) = \sum_{k=0}^{L_{\mathrm{N}}-1} \tilde{u}_{n-k} c_{\mathrm{N},i,k}(n) + \sum_{l=0}^{L_{\mathrm{F}}-1} \tilde{u}_{n-B-l} c_{\mathrm{F},i,l}(n) \qquad (10.12)$$

where $c_{\mathrm{N},i,k}(n)$ and $c_{\mathrm{F},i,l}(n)$ are the near and far echo coefficients of the ith subcanceller at time nT. Let us assume that the durations of the near echo and the far echo are both about 15 ms and the symbol rate of the modem is 2400 symbols per second. The time span of an echo canceller must be longer than the duration of the echoes to be cancelled. Thus, we may choose the number of coefficients of each of the near echo subcancellers as $L_{\mathrm{N}} = 40$, since $40 \times 1/2400 = 16.7$ ms > 15 ms. Similarly, we may also choose $L_{\mathrm{F}} = 40$. To simplify notation, we combine the two summations in (10.12) into one. That is,

$$\hat{\tilde{e}}_i(n) = \sum_{k=0}^{L-1} \alpha_{n-k} c_{i,k}(n) \tag{10.13}$$

where $L = L_N + L_F$ and, depending on the index k, α_{n-k} and $c_{i,k}(n)$ represent $\tilde{u}_{n-k}$ or $\tilde{u}_{n-B-l}$ and $c_{N,i,k}(n)$ or $c_{F,i,l}(n)$, respectively. We can rewrite (10.13) using matrix notation as

$$\hat{\tilde{e}}_i(n) = \mathbf{c}_i^T(n)\mathbf{u}_n \tag{10.14}$$

where $\mathbf{c}_i(n) = [c_{i,0}(n), \ldots, c_{i,L-1}(N)]^T$ and $\mathbf{u}_n = [\alpha_n, \ldots, \alpha_{n-L+1}]^T$.

To estimate the echo canceller coefficients adaptively, the LMS echo canceller first computes the error signal $\tilde{\varepsilon}_i(n)$ between the ith received signal $\tilde{x}_i(n)$ and the ith estimated echo using the coefficient vector $\mathbf{c}_i(n)$. That is,

$$\tilde{\varepsilon}_i(n) = \tilde{x}_i(n) - \hat{\tilde{e}}_i(n) \tag{10.15}$$

The coefficient vector $\mathbf{c}_i(n)$ is then time updated according to

$$\mathbf{c}_i(n+1) = \mathbf{c}_i(n) + \mu\tilde{\varepsilon}_i(n)\mathbf{u}_n^*. \tag{10.16}$$

The new coefficient vector $\mathbf{c}_i(n+1)$ is then used in (10.14) at time $n+1$. The analysis of the convergence of the LMS echo canceller given below follows the approach used in [11] and [12].

Since the training is performed during the half-duplex operation of the modem, the analytic received signal sample $\tilde{x}_i(n)$ can be written as

$$\tilde{x}_i(n) = \tilde{e}_{N,i}(n) + \tilde{e}_{F,i}(n) + \tilde{w}_i(n) = \mathbf{h}_i^T\mathbf{u}_n + \tilde{z}_i(n) \tag{10.17}$$

where $\mathbf{h}_i = [\tilde{h}_{i,N,0}, \ldots, \tilde{h}_{i,N,L_N-1}, \tilde{h}_{i,F,0}, \ldots, \tilde{h}_{i,F,L_F-1}]^T$ are the linear echo channel coefficients. In (10.17), $\mathbf{h}_i^T\mathbf{u}_n$ represents the linear echo, and $\tilde{z}_i(n)$ represents the sum of the noise $\tilde{w}_i(n)$ and the non-cancellable non-linear echo. Ideally, when $\mathbf{c}_i(n) = \mathbf{h}_i$, we can completely eliminate all the linear echo.

By defining the coefficient error vector $\xi_i(n) \equiv \mathbf{c}_i(n) - \mathbf{h}_i$, and after simplification, the variance of the error $\tilde{\varepsilon}_i(n)$ can be expressed as

$$\begin{aligned}\sigma_{\varepsilon_i}^2(n) &= E[|\tilde{\varepsilon}_i(n)|^2] = E[|\xi_i^T(n)\mathbf{u}_n + \tilde{z}_i(n)|^2] \\ &= E[\xi_i^T(n)\mathbf{u}_n\mathbf{u}_n^H\xi_i^*(n)] - 2\mathrm{Re}\{E[\tilde{z}_i^*(n)\xi_i^T(n)\mathbf{u}_n]\} + E[|\tilde{z}_i(n)|^2]\end{aligned} \tag{10.18}$$

The total residual error $\tilde{z}_i(n)$ is uncorrelated with $\xi_i^T(n)$ and $\mathbf{u}_n$. By using the so-called first *independence assumption* for the analysis of the LMS algorithm [16], we assume that they are also independent. So, the second term in (10.118) is zero. Furthermore, we note that $\xi_i^T(n)\mathbf{u}_n\mathbf{u}_n^H\xi_i^*(n) = \mathrm{tr}\{\mathbf{u}_n\mathbf{u}_n^H\xi_i^*\xi_i^T(n)]$. By adopting the second *independence assumption*,[3] i.e. $\mathbf{u}_i$ and $\mathbf{u}_j$ are independent for $i \neq j$, it follows that $\mathbf{u}_n$ and $\xi_i(n)$ are independent. Therefore, we can take the ensemble average of $\mathbf{u}_n$ and $\xi_i(n)$ independently. Thus, it follows that

$$\begin{aligned}\sigma_{\varepsilon_i}^2(n) &= \mathrm{tr}\{E[\mathbf{u}_n\mathbf{u}_n^H]E[\xi_i^*(n)\xi_i^T(n)]\} + E[|\tilde{z}_i(n)|^2] \\ &= \sigma_u^2 E[\xi_i^T(n)\xi_i^*(n)] + E[|\tilde{z}_i(n)|^2] = \sigma_{\varepsilon_i,\mathrm{ex}}^2(n) + \sigma_{\varepsilon_i,\mathrm{opt}}^2\end{aligned} \tag{10.19}$$

where $\sigma^2_{\varepsilon_i,\mathrm{opt}}$ is the optimal MSE at the ith echo canceller output which is equal to the variance of $\tilde{z}(n)$ and $\sigma^2_{\varepsilon_i,\mathrm{ex}}(n)$ is excess MSE at time nT. In addition, we have used the relation $E[\mathbf{u}_n\mathbf{u}_n^{\mathrm{H}}] = \sigma_u^2 I$, because $\{u_n\}$ is a white sequence.

From (10.16) and the definition of $\xi_i(n)$ we have

$$\xi_i(n+1) = \xi_i(n) + \mu\mathbf{u}_n[-\mathbf{u}_n^{\mathrm{H}}\xi_i(n) + \tilde{z}_i(n)] = [I - \mu\mathbf{u}_n\mathbf{u}_n^{\mathrm{H}}]\xi_i(n) + \mu\tilde{z}_i(n)\mathbf{u}_n \tag{10.20}$$

By substituting (10.20) into (10.19) and again using the first independence assumption that $\tilde{z}_i(n)$ is independent of $\mathbf{u}_i$ and $\xi_i(n)$, we obtain

$$\begin{aligned}\sigma^2_{\varepsilon_i,\mathrm{ex}}(n+1) &= \sigma_u^2 E[\xi_i^{\mathrm{T}}(n+1)\xi_i^*(n+1)]\\ &= \mathrm{tr}\{E[(I-\mu\mathbf{u}_n\mathbf{u}_n^{\mathrm{H}})^2]E[\xi_i(n)\xi_i^{\mathrm{H}}(n)]\sigma_u^2\} + \mu^2 E[|\tilde{z}_i(n)|^2]E[\mathbf{u}_n^{\mathrm{H}}\mathbf{u}_n]\sigma_u^2 \end{aligned} \tag{10.21}$$

The term $E[(I - \mu\mathbf{u}_n\mathbf{u}_n^{\mathrm{H}})^2]$ can be evaluated as

$$E[(I-\mu\mathbf{u}_n\mathbf{u}_n^{\mathrm{H}})^2] = I - 2\mu E[\mathbf{u}_n\mathbf{u}_n^{\mathrm{H}}] + \mu^2 E[\mathbf{u}_n\mathbf{u}_n^{\mathrm{H}}\mathbf{u}_n\mathbf{u}_n^{\mathrm{H}}] \tag{10.22}$$

It can be shown that $E[\mathbf{u}_n\mathbf{u}_n^{\mathrm{H}}\mathbf{u}_n\mathbf{u}_n^{\mathrm{H}}] = (L-1+\kappa)](\sigma_u^2)^2 I$, where κ is the kurtosis of u_n defined by $\kappa = E[|u_n|^4]/(\sigma_u^2)^2$. By substituting these relations and (10.22) into (10.21), and after simplification, we obtain the desired recursive relation of $\sigma^2_{\varepsilon_i,\mathrm{ex}}(n)$ as

$$\sigma^2_{\varepsilon_i,\mathrm{ex}}(n+1) = [1 - 2\mu\sigma_u^2 + \mu^2(L-1+\kappa)(\sigma_u^2)^2]\sigma^2_{\varepsilon_i,\mathrm{ex}}(n) + \mu^2 L(\sigma_u^2)^2\sigma^2_{\varepsilon_i,\mathrm{opt}} \tag{10.23}$$

The iterative relation given above is a first-order difference equation. The excess MSE $\sigma^2_{\varepsilon_i,\mathrm{ex}}(n)$ is a decreasing function if and only if $1 - \mu 2\sigma_u^2 + \mu^2(L-1+\kappa)(\sigma_u^2)^2 < 1$, or

$$\mu < \frac{2}{(L-1+\kappa)\sigma_u^2} \cong \mu_{\max} \tag{10.24}$$

This is the convergence condition for $\sigma^2_{\varepsilon_i,\mathrm{ex}}(n)$. When the step size μ satisfies (10.24) we say that the LMS algorithm *converges in variance.*

For an arbitrary $\mu < \mu_{\max}$, the convergence characteristic of $\sigma^2_{\varepsilon_i,\mathrm{ex}}(n)$ is given by

$$\begin{aligned}\sigma^2_{\varepsilon_i,\mathrm{ex},\mu}(n+1) &= [1 - 2\mu\sigma_u^2 + \mu^2(L-1+\kappa)(\sigma_u^2)^2]^n\sigma^2_{\varepsilon_i,\mathrm{ex}}(0)\\ &\quad + \frac{1 - [1 - 2\mu\sigma_u^2 + \mu^2(L-1+\kappa)(\sigma_u^2)^2]^n}{2 - \mu(L-1+\kappa)\sigma_u^2}\mu\sigma_u^2\sigma^2_{\varepsilon_i,\mathrm{opt}}\end{aligned} \tag{10.25}$$

From (10.23) or (10.25), it can be shown that the most rapid convergence occurs if

$$\mu = \frac{1}{(L-1+\kappa)\sigma_u^2} = \frac{\mu_{\max}}{2} = \mu_{\mathrm{opt}}. \tag{10.26}$$

When $\mu = \mu_{\mathrm{opt}}$, the convergence characteristic of the LMS canceller is given by

$$\begin{aligned}\sigma^2_{\varepsilon_i,\mathrm{ex}}(n+1) &= \left(1 - \frac{1}{L-1+\kappa}\right)^n\sigma^2_{\varepsilon_i,\mathrm{ex}}(0)\\ &\quad + \left[1 - \left(1 - \frac{1}{L-1+\kappa}\right)^n\right]\frac{L}{L-1+\kappa}\sigma^2_{\varepsilon_i,\mathrm{opt}}.\end{aligned} \tag{10.27}$$

The coefficient vector $\mathbf{c}_i(n)$ is usually initialized to be a zero vector, i.e. $\mathbf{c}_i(0) = \mathbf{0}$. It can be shown from (10.13) and (10.15), when $\mathbf{c}_i(0) = \mathbf{0}$, that $\hat{\tilde{e}}_i(0) = 0$ and $\tilde{\varepsilon}_i(0) = \tilde{x}_i(0)$. Hence, $\sigma^2_{\varepsilon_i,\text{ex}}(0) = E[\,|\tilde{x}_i(0)|^2]$, which is equal to the maximum level of the total echo. When $L \gg 1$, $\sigma^2_{\varepsilon_i,\text{ex}}(n)$ converges exponentially with a convergence time constant equal to $(L - 1 + \kappa)$ bauds starting from the maximum echo level. Furthermore, for a practical distribution of data symbols, we have $\kappa - 1 \ll L$. We can approximate the excess error $\varepsilon_{i,\text{ex}}(n)$ as

$$\sigma^2_{\varepsilon_i,\text{ex}}(n+1) = \left(1 - \frac{1}{L}\right)^n \sigma^2_{\varepsilon_i,\text{ex}}(0) + \left[1 - \left(1 - \frac{1}{L}\right)^n\right] \sigma^2_{\varepsilon_i,\text{opt}} \tag{10.28}$$

This equation shows that, when the optimal step size μ_{opt} is used, the LMS analytic equalizer converges according to $(1 - 1/L)^n$, approximately. From the relation that $(1 - 1/L)^n \approx e^{n/L}$, for $1/L \ll 1$, it follows that the convergence time constant is approximately equal to L symbol intervals, i.e. the span of the echo canceller, for an LMS analytic echo canceller with optimal step size. If we choose $\mu \ll \mu_{\text{opt}}$, in (10.25) we have $\mu^2(L - 1 + \kappa)(\sigma_u^2)^2 \ll 2\mu\sigma_u^2$. As a result, the excess residual error converges according to

$$(1 - 2\mu\sigma_u^2)^n \approx \left(1 - \frac{2\mu}{\mu_{\text{opt}}}\frac{1}{L}\right)^n \approx \exp\left(\frac{2\mu}{\mu_{\text{opt}}}\frac{n}{L}\right)$$

Thus, we have shown that when the step size μ is small, the convergence time constant of the echo canceller is equal to $(\mu_{\text{opt}}/2\mu)L$.

During the initial convergence period $\sigma^2_{\varepsilon_i,\text{ex}}(0) \gg \sigma^2_{\varepsilon_i,\text{opt}}$. We need only consider the first term in (10.28). In order for the residual echo power to be reduced 100 times, i.e. 20 dB, we need to satisfy the condition $(1 - 1/L)^n = 0.01$. After some simple calculation, we find that $n \approx 4.6L$. In words, the maximum rate of echo reduction is inversely proportional to the length of the echo canceller. The number of bauds required to reduce the echo by 20 dB is equal to 4.6 times the length of an LMS analytic echo canceller.

EXAMPLE 2

The total span of an analytic echo canceller is 80 bauds. Assuming $\kappa = 1$, the echo decays according to $(1 - 1/80)^n$. To reduce the echo 50 dB from its maximum value, the minimum required number of bauds can be computed as $(1 - 1/80)^n = 10^{-5}$, or $n = -5/\log(79/80) = 915$ bauds. Using the approximate formula, we have $4.6 \times 80 \times (50/20) = 920$ bauds.

For an echo canceller with M subcancellers, each sub-echo-canceller is operated independently and the above analysis is valid for all of them. Therefore, the convergence time does not depend on the number of subcancellers, if all of the subcancellers are updated every baud.

CONVERGENCE OF THE LMS REAL PASSBAND ECHO CANCELLER

A derivation similar to the one given above can be performed for the real passband echo canceller as was given in [12] and [14]. Below we provide a simpler derivation by directly using the result obtained for the analytic echo canceller.

We first write the complex coefficient vector in terms of its real part $\mathbf{c}_{r,i}(n)$ and imaginary part $\mathbf{c}_{i,i}(n)$, such that

$$\mathbf{c}_i(n) = \mathbf{c}_{i,r}(n) + j\mathbf{c}_{i,i}(n) \tag{10.29}$$

Similarly, the data vector $\mathbf{u}_n$ can also be written as

$$\mathbf{u}_n = \mathbf{u}_{n,r} + j\mathbf{u}_{n,i} \tag{10.30}$$

The real estimated echo can then be written as

$$\hat{e}_i(n) = \mathrm{Re}[\mathbf{c}_i^{\mathrm{T}}(n)\mathbf{u}_n] = \mathbf{c}_{i,r}^{\mathrm{T}}(n)\mathbf{u}_{n,r} - \mathbf{c}_{i,i}^{\mathrm{T}}(n)\mathbf{u}_{n,i} \tag{10.31}$$

By forming a real augmented data vector $\bar{\mathbf{u}}_n$ and a real augmented coefficient vector $\bar{\mathbf{c}}_i(n)$, such that

$$\bar{\mathbf{u}}_n = \begin{bmatrix} \mathbf{u}_{n,r} \\ \mathbf{u}_{n,i} \end{bmatrix} \quad \text{and} \quad \bar{\mathbf{c}}_i(n) = \begin{bmatrix} \mathbf{c}_{i,r}(n) \\ -\mathbf{c}_{i,i}(n) \end{bmatrix}$$

we can write the real estimated echo as

$$\hat{e}_i(n) = \bar{\mathbf{c}}_i^{\mathrm{T}}(n)\bar{\mathbf{u}}_n \tag{10.32}$$

The LMS real passband echo canceller is implemented by first computing the real error as

$$\varepsilon(n) = x_i(n) - \hat{e}_i(n) \tag{10.33}$$

and then performing the LMS updating such that

$$\bar{\mathbf{c}}_i(n+1) = \bar{\mathbf{c}}_i(n) + \mu\varepsilon(n)\bar{\mathbf{u}}_n \tag{10.34}$$

Because $\bar{\mathbf{c}}_i(n)$ and $\bar{\mathbf{u}}_n$ are $2L$-dimensional vectors, by following exactly the same derivation as for the analytic echo canceller given above, we can show that the convergence condition for the real passband echo canceller is

$$\mu < \frac{2}{(2L - 1 + \kappa)\sigma_{u,r}^2} \cong \mu_{\max} \tag{10.35}$$

where $\sigma_{u,r}^2$ is the variance of the real part of α_n. It is equal to the variance of the imaginary part of α_n and equal to a half of σ_u^2. That is,

$$\sigma_{u,r}^2 = \sigma_{u,i}^2 = \sigma_u^2/2 \tag{10.36}$$

By substituting (10.36) into (10.35), it can be shown that $\mu_{\max}$ for the real passband echo canceller is the same as $\mu_{\max}$ for the analytic echo canceller. The optimal step size for convergence, μ_{opt}, is a half of $\mu_{\max}$, which is the same as μ_{opt} for the analytic echo canceller, when $\kappa - 1 \ll L$. We can express (10.35) approximately as

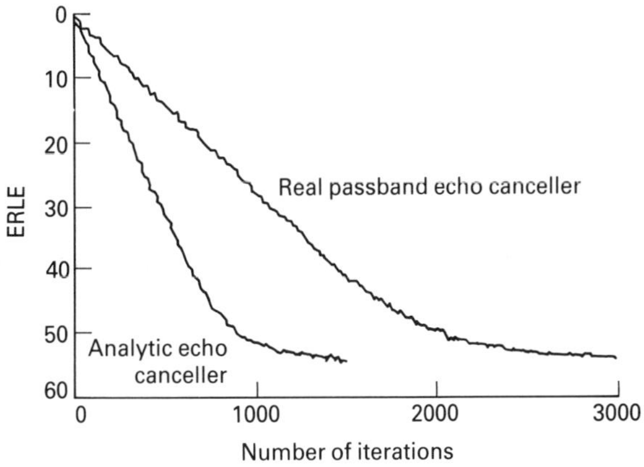

Figure 10.10 Convergence characteristics of real passband and analytic echo cancellers

$$\sigma^2_{\varepsilon_i,\,\mathrm{ex}}(n+1) = \left(1 - \frac{1}{2L}\right)^2 \sigma^2_{\varepsilon_i,\,\mathrm{ex}}(0) + \left[1 - \left(1 - \frac{1}{2L}\right)^n\right]\sigma^2_{\varepsilon_i,\,\mathrm{opt}}. \tag{10.37}$$

Similar to the analytic echo canceller above, in order for the residual echo power of a real passband echo canceller to be reduced by 20 dB, we need $1 - 1/2L)^n = 0.01$, or $n \approx 9.2L$.

EXAMPLE 3

The total span of a real passband echo canceller is 80 bauds. Assuming $\kappa = 1$, the echo decays according to $(1 - 1/160)^n$. To reduce the echo by 50 dB from its maximum value, the minimum required number of bauds can be computed as $(1 - 1/160)^n = 10^{-5}$, or $n = -5/\log(159/160) = 1836$ bauds. Using the approximate formula, we have $9.2 \times 80 \times (50/20) = 1840$ bauds.

Figure 10.10 shows the convergence curves of simulated real passband and analytic echo cancellers with the parameters described in the examples given above. It can be seen from the figure that the simulation results agree well with the analytical formulae obtained above.

When $\kappa - 1 \ll L$ and using (10.36) it can be shown that for the same Tx data symbols, μ_{opt} and $\mu_{\max}$ are the same for the real and complex echo cancellers. However, the convergence time of a real echo canceller will be approximately twice as long as that of a complex echo canceller to achieve the same ERLE.

10.2.3 Steady-state excess MSE and selection of step sizes

By using the expressions that describe the convergence characteristics of the real passband and analytic Nyquist echo cancellers derived in section 10.2.2, we can also obtain the steady-state excess mean square error of each of the Nyquist cancellers as a function of the adaptation step-sizes as follows.

Let us first consider the analytic echo canceller, whose excess MSE at time $(n + 1)T$ is described by (10.25). By definition, the steady-state excess MSE is the asymptotic excess MSE when n goes to infinity. For $n = \infty$, and the step size satisfies (10.24), we have

$$[1 - 2\mu\sigma_u^2 + \mu^2(L - 1 + \kappa)(\sigma_u^2)^2]^\infty = 0$$

Hence

$$\sigma^2_{\varepsilon_i,\mathrm{ex},\mu}(\infty) = \frac{1}{2 - \mu(L - 1 + \kappa)\sigma_u^2}\,\mu L\sigma_u^2\sigma^2_{\varepsilon_i,\mathrm{opt}} = \left(\frac{2}{\mu L\sigma_u^2} - \frac{L - 1 + \kappa}{L}\right)^{-1}\sigma^2_{\varepsilon_i,\mathrm{opt}} \qquad (10.38)$$

The above expression can be further simplified by making the approximation that $L - 1 + k \approx L$ as before. That is,

$$\sigma^2_{\varepsilon_i,\mathrm{ex},\mu}(\infty) \approx \left(\frac{2}{\mu L\sigma_u^2} - 1\right)^{-1}\sigma^2_{\varepsilon_i,\mathrm{opt}} = \left(\frac{2}{L\sigma_u^2}\Big/\mu - 1\right)^{-1}\sigma^2_{\varepsilon_i,\mathrm{opt}} \approx \frac{1}{\mu_{\max}/\mu - 1}\,\sigma^2_{\varepsilon_i,\mathrm{opt}} \qquad (10.39)$$

This is the desired expression of the steady-state excess MSE for an arbitrary μ. It also expresses the steady-state excess MSE of the real passband Nyquist echo canceller.

Since $\mu_{\mathrm{opt}} = \mu_{\max}/2$, from (10.39) we have

$$\sigma^2_{\varepsilon_i,\mathrm{ex},\mu_{opt}}(\infty) \approx \frac{1}{\mu_{\max}/\mu_{\mathrm{opt}} - 1}\,\sigma^2_{\varepsilon_i,\mathrm{opt}} = \sigma^2_{\varepsilon_i,\mathrm{opt}} \qquad (10.40)$$

Thus, when $\mu = \mu_{\mathrm{opt}}$, the steady-state excess MSE is as strong as the optimal MSE. In other words, the total output MSE of an echo canceller is 3 dB higher than the optimal MSE.

STEP-SIZE SELECTION ECHO CANCELLER TRAINING

Although by selecting $\mu = \mu_{\mathrm{opt}}$ we can obtain the most rapid initial convergence for LMS echo cacellers, such a step size results in a total MSE 3 dB higher than the output MSE generated by an optimal echo canceller. Using a smaller μ will generate a smaller steady-state MSE. However, a smaller μ will slow down the initial convergence of the echo canceller. A common technique used in designing echo cancellers is to use different step sizes at different stages of training. That is, at the beginning of training, we use μ_{opt} to achieve the fastest convergence. When the echo canceller is close to convergence, it switches to a smaller μ to reduce the steady-state MSE, i.e. increase the ERLE. Since the echo canceller has almost converged, at the second stage, we only need to reduce the MSE by a few more decibels and it will not increase training time significantly even though a smaller step size is used. This technique is sometimes called *gear shifting* for echo canceller training. More than one stage of gear shifting may be used if necessary. We provide the following example to illustrate how this technique works.

EXAMPLE 4

For the analytic echo canceller described in the previous example and if the optimal residual MSE is -55 dB below the total echo level, it will take 915 bauds to achieve 50 dB echo cancellation by using μ_{opt}. At the second stage of training we reduce the step size by a factor of 5. As described above, such a step size will have a convergence time constant that is approximately 2.5 times longer than μ_{opt}, i.e., $2.5 \times 4.6 \times 80$ bauds per 20 dB. Hence, we need approximately an additional 230 bauds to reduce the MSE by another 5 dB using such a step size. The final excess MSE for such a step size will be

$$\frac{1}{10-1}\epsilon_{i,\text{opt}} = 0.111\epsilon_{i,\text{opt}}$$

The final total MSE will be 0.46 dB above the optimal MSE, which is practically acceptable, and the increased training time is insignificant.

STEP-SIZE SELECTION DURING FULL-DUPLEX MODE

When an echo cancellation modem is in full-duplex operation, the signal from the remote modem appears at the output of the echo canceller. For the echo canceller, the remote signal behaves like a very strong noise. As a result, it will cause significant excess error that interferes with the receiver operation unless a very small step size is used. From the expression for the excess error given by (10.40), we note that the excess is proportional to the optimal MSE for a certain step size. Since the remote signal power is much stronger than other errors, the optimal output MSE is approximately equal to the remote signal power. Hence, the excess error is mainly determined by the received remote signal power. When the received remote signal level is high, the excess error could be the main source of the interference for the receiver. Thus, in order to ensure reliable operation of the modem, we must limit the echo canceller excess error.

From (10.39) we obtain

$$\frac{\sigma^2_{\varepsilon_i,\text{opt}}}{\sigma^2_{\varepsilon_i,\text{ex},\mu}} \approx \mu_{\max}/\mu_{\text{full}} - 1 \tag{10.41}$$

where μ_{full} is the step size in full-duplex mode. To achieve a signal to residual echo ratio of $D = 30$ dB, we need $\sigma^2_{\varepsilon_i,\text{opt}}/\sigma^2_{\varepsilon_i,\text{ex},\mu} = 10^{D/10}$. From (10.41) we obtain

$$\mu_{\text{full}} = \mu_{\max}/(10^{D/10} + 1) \approx \mu_{\max}/10^{D/10}. \tag{10.42}$$

EXAMPLE 5

For a CCITT standard V.32bis modem, a signal to residual echo ratio of $D = 30$ dB is required to ensure its normal operation. To achieve such a signal for an analytic echo canceller that spans 80 bauds, we have the full-duplex step size equal to $\mu_{\text{full}} = 2/(80 \times 1000) \times \sigma_u^2$. Assuming $\sigma_u^2 = 1$, we have $\mu_{\text{full}} = 2.5 \times 10^{-5}$.

From the above example, it can be seen that the full-duplex step size should be very small to achieve a high signal to residual echo level. This will cause a serious numerical precision problem when the echo canceller is implemented using digital signal processors with a short wordlength. The finite wordlength effects in digital data echo cancellers are discussed in section 10.3.

10.2.4 Comparison of analytic and real passband echo cancellers

Theoretically, the analytic and real passband echo cancellers are equivalent. However, they are different in many practical aspects. A modem designer must compare these aspects to decide which structure is more suitable for the particular hardware and software platform under consideration. In this section, we will discuss and compare these differences.

CONVERGENCE AND TRACKING CHARACTERISTICS

The convergence characteristics of the analytic and real passband echo cancellers have been discussed in the last section. It was shown that the optimal convergence speed of the analytic echo canceller is twice as fast as the optimal speed of the real passband echo canceller. Therefore, the analytic echo canceller is preferable when a faster convergence speed is necessary for the echo cancellation modem.

For the tracking characteristics in steady state, the conclusion is the same. In order to generate the same amount of excess MSE, or the same signal to residual echo ratio, the analytic echo canceller can track echo channel change twice as fast as the real passband echo canceller. Thus, the analytic echo canceller is again in favour, although the tracking speed in full-duplex mode is usually not an important factor for echo cancellation modems because the echo channels are essentially time invariant or very slowly time varying.

COMPUTATIONAL COMPLEXITY

The implementation of an echo canceller includes two steps: the computation of the estimated echo by performing convolution of the data symbols with the coefficients and the coefficient updating. For an analytic echo canceller that has M_c subcancellers and spans L bauds, its convolution requires LM_c complex multiplication and accumulations (MACs), or equivalently, $4LM_c$ real MACs. The coefficient updating also needs LM_c complex MACs, or $4LM_c$ real MACs. Thus, in total $8LM_c$ real MACs are needed. On the other hand, a real passband echo canceller, which spans L bauds, with M_r subcancellers needs $2LM_r$ real MACs each for echo computation and coefficient updating. In total, $4LM_r$ real MACs are needed Assuming that $M_r = 2M_c$, for instance $M_c = 2$ and $M_r = 4$, their computational complexities are roughly the same. However, practically, M_c and M_r must be integer numbers, so the smallest number of subcancellers of an analytic echo canceller is equal to two. On the other hand, the smallest number of subcancellers of a real passband echo canceller can be chosen to be equal to three. As a result, the complexity of the real passband echo canceller is equal to, or can be lower than, the complexity of the analytic echo canceller, depending on the selection of the sampling rate.

The analytic echo canceller requires a Hilbert transformer to generate the analytic received signal samples. To evaluate the impact of the Hilbert transformer on the complexity of the echo cancellers, we must also consider the structure of the receiver. The analytic echo canceller is usually used in conjunction with a complex fractionally spaced equalizer (FSE) in the receiver. No additional computational complexity is introduced by the Hilbert transformer, because a Hilbert transformer is needed for a complex FSE anyway. On the other hand, it is possible to use the so-called phase splitting FSE (PSFSE) described and analyzed in [17] and [18] that does not need a separate Hilbert transformer. This design reduces the computational complexity of modems with real passband echo cancellation.

In conclusion, the implementation of a real passband echo canceller is the same as or more efficient than the implementation of an analytic echo canceller in computational complexity, at the expense of a slower convergence and tracking speed.

MEMORY REQUIREMENTS

Memory is required to form the input data symbol delay line and to store the echo canceller coefficients. The required delay line memory is the same for both types of echo canceller. Because each subcanceller needs $2L$ memory locations for storing the coefficients for either type of echo canceller, the analytic echo canceller, always requires less memory than the real passband echo canceller which always has more subcancellers than the former. The extra memory required for the Hilbert transformer for an analytic echo canceller can be compensated by the memory required for the Hilbert transformer in the receiver or the extra memory required by PSFSE relative to that for a complex FSE. Thus, the analytic echo canceller is more efficient in the use of memory than the real passband echo canceller.

THE IMPACT OF A NON-IDEAL TX FILTER ON THE PERFORMANCE OF ECHO CANCELLERS

Perfect echo cancellation can only be achieved when the linear echo assumption is satisfied. For an analytic echo canceller, non-linear effects are introduced not only by the non-linear elements in the echo path, but also by non-perfect negative frequency rejection of the Tx filter and the Hilbert transformer. We will analyze such non-linear effects and show why the performance of the real passband echo canceller is not affected by the non-linearity due to the non-ideal Tx filter.

We can write the spectrum of the received signal at the input of the Hilbert transformer as

$$S(\omega) = A_r(\omega)T_r(\omega)Ch(\omega) - A_i(\omega)T_i(\omega)Ch(\omega) \tag{10.43}$$

where $A_r(\omega)$, $A_i(\omega)$, $T_r(\omega)$, $T_i(\omega)$ and $Ch(\omega)$ are the frequency responses of the real and imaginary parts of the Tx data symbol sequence, the real and imaginary parts of the Tx filter and the channel respectively. The Fourier transforms of the real and imaginary outputs of the Hilbert transformer can be written as

$$\begin{aligned} E_r(\omega) &= H_r(\omega)[A_r(\omega)T_r(\omega)Ch(\omega) - A_i(\omega)T_i(\omega)Ch(\omega)] \\ E_i(\omega) &= H_i(\omega)[A_r(\omega)T_r(\omega)Ch(\omega) - A_i(\omega)T_i(\omega)Ch(\omega)] \end{aligned} \tag{10.44}$$

respectively, where $H_r(\omega)$ and $H_i(\omega)$ are the frequency responses of the real and imaginary part of the Hilbert transformer.

The Fourier transforms of the real and imaginary parts of the analytic echo canceller output can be expressed as

$$\hat{E}_r(\omega) = A_r(\omega)C_r(\omega) - A_i(\omega)C_i(\omega) \quad \text{and} \quad \hat{E}_i(\omega) = A_r(\omega)C_i(\omega) + A_i(\omega)C_r(\omega) \tag{10.45}$$

In order to eliminate echo completely we must have

$$\hat{E}_r(\omega) = E_r(\omega) \quad \text{and} \quad \hat{E}_i(\omega) = E_i(\omega) \tag{10.46}$$

From (10.44) and (10.45), in order to satisfy (10.46), we need

$$\begin{aligned} C_r(\omega) &= H_r(\omega)T_r(\omega)Ch(\omega) = H_i(\omega)T_i(\omega)Ch(\omega) \\ C_i(\omega) &= H_r(\omega)T_i(\omega)Ch(\omega) = H_i(\omega)T_r(\omega)Ch(\omega) \end{aligned} \tag{10.47}$$

It follows from (10.47) that, in the region where the involved quantities are not zero,

$$[H_r(\omega)]^2 = -[H_i(\omega)]^2 \quad \text{or} \quad H_r(\omega) = \pm jH_i(\omega) \tag{10.48}$$

and

$$[T_r(\omega)]^2 = -[T_i(\omega)]^2 \quad \text{or} \quad T_r(\omega) = \pm jT_i(\omega) \tag{10.49}$$

We can completely eliminate the linear echo using the anlytic echo canceller if and only if (10.48) and (10.49) are satisfied.

Let us first consider the requirement for $T(\omega)$, the frequency response of the transmitter filter. Since the real and imaginary portions of the transmitter filter are both real filters, their frequency responses are conjugate symmetric. That is, $T_r(\omega) = T_r^*(-\omega)$ and $T_i(\omega) = T_i^*(-\omega)$. By denoting the positive and negative frequency portions of $T_r(\omega)$ by $T_r^+(\omega)$ and $T_r^-(\omega)$, respectively, we have $T_r^-(\omega) = T_r^{+*}(-\omega)$. Similarly we have $T_i^-(\omega) = T_i^{+*}(-\omega)$. By assuming that[4].

$$T_i^+(\omega) = jT_r^+(\omega) \tag{10.50}$$

we obtain

$$T_i^-(\omega) = T_i^{+*}(-\omega) = [jT_r^+(-\omega)]^* = -jT_r^{+*}(-\omega) = -jT_r^-(\omega) \tag{10.51}$$

The relations given by (10.50) and (10.51) are the conditions for a perfect Hilbert transformer that completely rejects the negative frequency components. Another possibility is to have $T_i^+(\omega) = -jT_r^+(\omega)$. Then we must have $T_i^-(\omega) = -jT_r^-(\omega)$, which also forms a perfect Hilbert transformer. It can be shown similarly that $H_r(\omega)$ and $H_i(\omega)$ must also satisfy the same condition. Thus, we conclude that in order to eliminate completely the linear echo introduced by the channel, the transmitter filter and the Hilbert transformer must completely reject the negative frequency components in the transmitted signal and in the received signal. Any imperfection in these filters will cause non-cancellable echo and degrade the performance of the analytic echo canceller.

Using a non-ideal transmitter filter will also, in general, impair the performance of the receiver in any modem, independent of whether or not an echo canceller is used. The difference is that a receiver only needs a signal-to-noise ratio of about 30–40 dB to operate while an echo canceller needs to reduce the echo by more than 50 dB. Thus, these filters used in a modem employing an analytic echo canceller must meet tighter requirements that are not necessary for the receiver.

As far as the real-passband echo canceller is concerned, it can be shown that the requirement would be such that

$$C_r(\omega) = T_r(\omega)Ch(\omega) \quad \text{and} \quad C_i(\omega) = T_i(\omega)Ch(\omega) \tag{10.52}$$

Thus, the real passband echo canceller will operate satisfactorily for any frequency response of the transmitter filter. Therefore, the real passband echo canceller does not impose any special requirements on the transmitter filter. Of course, the transmitter filter must still sufficiently reject the negative frequencies for the receiver operation. For the real passband echo canceller the only non-cancellable echo would be the echo introduced by the non-linearities of analog system components, channel and digital quantization error discussed in the next section.

10.3 Finite wordlength effects in echo cancellation

Almost all of today's echo cancellers are implemented using high-speed digital processors (DSPs) based on VLSI technology. They may be either general-purpose DSPs or specifically tailored ones for echo cancellation applications. Since implementation of an echo canceller is a computationally intensive task, in either case, it is desirable to use a shorter computer wordlength in computation to increase the speed of computation and/or reduce hardware complexity. On the other hand, a shorter wordlength will degrade the performance of the echo canceller. In this section, we analyze how a short wordlength affects this performance and the achievable echo cancellation for a given wordlength used in computation. Finally, methods for improving echo cancellation performance using DSPs with a certain wordlength are discussed.

10.3.1 How finite wordlength affects echo canceller operation

Echo cancellers are usually implemented using general-purpose DSPs or specially designed VLSI circuits with a digital processing unit. In either case, an arithmetic unit, which usually contains a fast hardware multiplier and an accumulator, is used to perform the arithmetic operations required to implement the echo canceller. Although some of the existing DSPs support floating point arithmetic and/or have a long wordlength of 24–32 bits, in order to obtain fast computation speed and to reduce hardware complexity, the majority of the available DSPs only support fixed point operations with a computer wordlength of 16 bits. The analysis of echo canceller performance given in the previous sections is based on the assumption that the computations are performed using infinite arithmetic. For the design and analysis of practical echo cancellers based on a practical DSP implementation, we must also consider finite wordlength effects.

BASICS OF FINITE WORDLENGTH EFFECTS

The general effects of a finite wordlength in digital signal processing have been discussed in the literature, e.g. [19]. Here we shall review some of the basics and only consider fixed point arithmetic operations.

For fixed point arithmetic, numbers are assumed to have a magnitude that is less than unity. When the computer wordlength used in computation is b bits, 1 bit is used to represent the sign of the number and the remaining $b-1$ bits are used to represent its magnitude. The range of the numbers for such a fixed point representation is from -1.0 to $+1.0-2^{-(b-1)}$. The smallest non-zero magnitude of a number represented by such a fixed number system is $2^{-(b-1)}$. A number that has a magnitude less than 2^{-b}, if rounding is used as explained below, is represented by zero and is called a machine zero.

When a fixed point number is added to or subtracted from another number, the result may have a magnitude greater than that of either of the operands. If the result is larger than unity, we say an *overflow* occurs. Thus, proper scaling must be performed on the operands before the operations to ensure that the result will have a magnitude less than one to avoid overflow. On the other hand, when two such numbers are multiplied, the product has a magnitude less than one. As a result, we don't have to worry about the problem of overflow. However, we may have to prescale the operands to prevent the magnitude of the product from becoming too small. Otherwise, the produce will be a machine zero and we call such a situation *underflow*.

When multiplication is performed by the multiplier in a DSP unit, the immediate result usually has more bits than the wordlength. For the number to be stored or used for further operations, we must eliminate the extra bits in the product. There are two ways of eliminating these extra bits: truncation and rounding. To truncate a number, we simply set the bits below $2^{-(b-1)}$ to zero. On the other hand, to round a number, we set these bits to zero if the residual bits have a magnitude less than 2^{-b}. Otherwise, the magnitude of the number will increase by $2^{-(b-1)}$. The easiest way to perform rounding is to add 2^{-b} to the mantissa part of the product without the sign (i.e. the mantissa is viewed as a positive number) before performing a truncation on it.

It can be easily seen that the maximum error caused by truncation is $2^{-(b-1)}$, while the maximum error caused by rounding is 2^{-b}. In addition to providing a more accurate result, rounding also reduces bias in the result. It can be shown that the bias after truncation is equal to 2^{-b}, assuming the product before rounding has $2b - 1$ bits. This bias reduces to 2^{-2b+1} after rounding using the method described above. The reduction of bias is much more important than the reduction in absolute error for echo cancellation [20]. Below we always assume that rounding is performed unless specified otherwise.

The error that results from rounding is called roundoff error[5]. For a b bit fixed point number, the smallest absolute value of a non-zero number is $2^{-(b-1)}$. The roundoff error e_{round} has a uniform distribution and $-2^{-b} < e_{\text{round}} < 2^{-b}$. It can be shown that the variance of the roundoff error is equal to $2^{-2b}/3$. Since the peak value of the numbers that can be represented by such a fixed point number is approximately one, the peak signal to roundoff noise ratio (PSRNR) is equal to 3×2^{2b}, or $6b + 4.8$ dB. For example, PSRNR = 100.8 dB for a 16 bit fixed point number representation.

Finally, it should be noted that in most of today's DSPs the multipliers and accumulators usually have more bits than the default wordlength. Numbers only round to the default wordlength when they are moved out from the accumulator for other operations or for storing into memory. Thus we can obtain better numerical results if it is possible to perform a series of many multiplications and accumulations (MACs) without removing the result out from the accumulator before completion.

FINITE WORDLENGTH EFFECTS IN ECHO COMPUTATION

Adaptive echo cancellation is realized in two steps: computing the synthesized echo and updating the echo canceller coefficients. We first examine the error introduced in echo computation.

The synthesized echo is computed by convolving the data symbol sequence and the echo canceller coefficients. Because the convolution can be implemented using a series of MACs, as stated above we can ignore the error introduced in the convolution. The error due to the finite wordlength is mainly the result of the error in representing the coefficients and the data symbols. Let us consider the variance of the data symbols σ_u^2. It is a value less than one and the square root of $1/\sigma_u^2$ is called the peak-to-RMS value of the data symbols. We usually assume that σ_u^2 for complex modem symbols is less than 1/8. Then, the ratio of peak to mean square value of the real and imaginary parts of the complex symbols is equal to 16, i.e. they have a peak-to-RMS value of 4. Thus, in the sequel, we shall asume the RMS values of the real and imaginary parts of the complex symbols to be equal to 0.25. The error in the digital representations of the data symbols and the coefficients is equal to e_{round}[6]. Below, we first derive the expression of the variance of the total error introduced in computing the synthesized echo of a real passband echo canceller.

The synthesized echo at nT is computed according to

$$\hat{e}(\mathrm{n}) = \mathrm{Re}\left(\sum_{k=0}^{L-1} c_k(n)u_{n-k}\right) = \sum_{k=0}^{L-1} [c_{k,r}(n)u_{n-k,r} - c_{k,\mathrm{i}}(n)u_{n-k,\mathrm{i}}] \tag{10.53}$$

Since the symbol sequence $\{u_n\}$ is white and its real and imaginary parts are uncorrelated and have the same power, we can express the total echo power as

$$P_e = E[\hat{e}^2(n)] = (\sigma_u^2/2) \sum_{k=0}^{L-1} [\{c_{k,r}(n)\}^2 + \{c_{k,i}(n)\}^2] \tag{10.54}$$

In order for $\hat{e}^2(n)$ not to overflow while fully utilizing the available precision of the coefficients, it is desirable to scale the quantities in echo computation such that $\sum_{k=0}^{L-1} [\{c_{k,r}(n)\}^2 + \{c_{k,i}(n)\}^2] \approx 1$ and we shall assume this is true in what follows. By denoting the numerical errors of $c_{k,r/i}(n)$ and $u_{n-k,r/i}$ as $\delta c_{k,r/i}(n)$ and $\delta u_{n-k,r/i}$, respectively, and ignoring high-order errors, the roundoff error $\hat{e}(n)$ can be expressed as

$$\delta\hat{e}(n) = \sum_{k=0}^{L-1} [c_{k,r}(n)\delta u_{n-k,r} + u_{n-k,r}\delta c_{k,r}(n) - c_{k,i}(n)\delta u_{n-k,i} - u_{n-k,i}\delta c_{k,i}(n)] \tag{10.55}$$

Similar to (10.54), we can express the power of the roundoff error $\delta\hat{e}(n)$ as

$$P_{\delta e} = E[(\delta\hat{e}(n))^2] = (\sigma_u^2/2) \sum_{k=0}^{L-1} [\{\delta c_{k,r}(n)\}^2 + \{\delta c_{k,i}(n)\}^2]$$

$$+ \sum_{k=0}^{L-1} E[\{\delta u_{n-k,r}(n)\}^2]\{c_{k,r}(n)\}^2 + \sum_{k=0}^{L-1} E[\{\delta u_{n-k,i}(n)\}^2]\{c_{k,i}(n)\}^2 \tag{10.56}$$

Since the variance of the roundoff errors is equal to $2^{-2b}/3$ as shown above, we can express $P_{\delta e}$ as

$$P_{\delta e} \approx 2L \times \sigma_u^2/2 \times 2^{-2b}/3 + (2^{-2b}/3) \sum_{k=0}^{L-1} [\{c_{k,r}(n)\}^2 + \{c_{k,i}(n)\}^2] \tag{10.57}$$

From (10.54) and (10.57) we obtain the roundoff noise-to-signal ratio as

$$P_e/P_{\delta e} = \frac{\sigma_u^2}{2^{-2b+1}}/3 \times \left(1 + L\sigma_u^2 / \sum_{k=0}^{L-1} [\{c_{k,r}(n)\}^2 + \{c_{k,i}(n)\}^2]\right)^{-1} \tag{10.58}$$

By using the assumption that $\sum_{k=0}^{L-1} [\{c_{k,r}(n)\}^2 + \{c_{k,i}(n)\}^2] \approx 1$, we obtain the signal to roundoff error ratio of the synthesized echo as

$$P_e/P_{\delta e} = \frac{\sigma_u^2}{2^{-2b+1}/3} \times \frac{1}{1 + L\sigma_u^2} \tag{10.59}$$

Furthermore, we assume that the variance of the data symbols σ_u^2 is approximately equal to 1/8. Thus, we obtain

$$P_e/P_{\delta e} = \frac{1}{2^{-2b+4}/3} \times \frac{1}{1 + L/8} \tag{10.60}$$

For $L = 80$ and $b = 16$, we have that $P_e/P_{\delta e} = 3 \times 2^{28}/11 = 7.32 \times 10^7$, or 78 dB. Thus, a DSP with a 16 bit wordlength is sufficient to implement a modem data echo canceller for echo computation. However, the situation changes when we consider the wordlength requirements for adaptively updating the echo canceller coefficients, as shown below.

WORDLENGTH REQUIREMENTS FOR ECHO CANCELLER COEFFICIENT UPDATING

Commonly, the LMS algorithm is used constantly to adapt the coefficients of a modem echo canceller. The purpose of the adaptation is twofold: in the initialization stage, i.e, during the half-duplex echo canceller training , adaptation is used quickly to learn the optimal echo canceller coefficients; during the data mode, i.e. full-duplex or double-talk mode, it is used for the echo canceller to track the slow time variations of the echo channel characteristics. Wordlength requirements are different in these two cases. We begin by analyzing the requirements for the wordlength in the training stage.

RELATIONSHIP OF WORDLENGTH AND ACHIEVABLE ECHO CANCELLATION AFTER TRAINING

Let us recall that the LMS coefficient updating of each subcanceller is given by first computing the error as

$$\varepsilon(n) = x(n) - \hat{e}(n) \tag{10.61}$$

where $\hat{e}(n)$ is given by (10.31) and we have omitted the subscript i because this equation applies to any of the subcancellers. The coefficient vector is then updated according to

$$\mathbf{c}(n+1) = \mathbf{c}(n) + \mu\varepsilon(n)\mathbf{u}_n^* \tag{10.62}$$

Unlike the echo computation described in the previous section where the finite wordlength effect introduces additive roundoff noise in the estimated echo, the main problem caused by finite wordlength for the coefficient updating is underflow. That is, when the error $\varepsilon(n)$ is reduced to a certain level, the second term on the right side of (10.62) becomes a vector of machine zeros. No adaptation occurs in such a case. This prevents the error from diminishing further and limits the possible echo canceller performance.

Assuming that rounding is used, the condition for the echo canceller coefficients to keep updating is

$$|\mu\varepsilon(n)u_{n-k}^*| = \mu|\varepsilon(n)|\,|u_{n-k}^*| \geqslant 2^{-b} \tag{10.63}$$

Because the magnitudes of $|\varepsilon(n)|$ and $|u_{n-k}^*|$ vary from sample to sample, it is very difficult to describe accurately when the updating has completely stopped without knowing the exact distribution of these quantities. However, practically, we can use the RMS values of $\varepsilon(n)$ and u_{n-k}^* in place of $|\varepsilon(n)|$ and $|u_{n-k}^*|$ in (10.63)[7]. Thus, the minimum of $|\varepsilon(n)|$ is given by

$$|\varepsilon_{\min}(n)| = \frac{2^{-b}}{\mu|u_{n-k}|}$$

or, equivalently, the minimum energy of $\varepsilon(n)$ can be expressed as

$$\sigma_{\varepsilon,\min}^2 = E[\varepsilon_{\min}^2(n)] \approx \frac{2^{-2b}}{\mu^2 E[|u_{n-k}|^2]} = \frac{2^{-2b}}{\mu^2\sigma_u^2} \tag{10.64}$$

From (10.64) we observe that we can improve echo cancellation by 6 dB if the wordlength increases by 1 bit. Since the total echo before calculation can be computed according to (10.54) as

$$\sigma_{\mathrm{e}}^2 = \sigma_u^2/2 \sum_{k=0}^{L-1} [\{c_{k,r}(n)\}^2 + \{c_{k,i}^2(n)\}^2] \approx \frac{\sigma_u^2}{2}$$

the maximum ERLE can be expressed as

$$\sigma_{\mathrm{e}}^2/\sigma_{\varepsilon,\min}^2 = \frac{\sigma_u^2}{2} \Bigg/ \frac{2^{-2b}}{\mu^2\sigma_u^2} = \mu^2(\sigma_u^2)^2 2^{2b-1} \qquad (10.65)$$

EXAMPLE 6

For an echo canceller that uses a wordlength of 16 bits for computations and assuming that $\sigma_u^2 = 0.125$ and $L = 80$, let us compute the achievable maximum echo cancellation after training using the optimal step size for convergence.

From (10.26) we know that the optimal step size for convergence is equal to $\mu_{\mathrm{opt}} = 1/L\sigma_u^2 = 0.1$. By using (10.65), we have

$$\sigma_{\mathrm{e}}^2/\sigma_{\varepsilon,\min}^2 \approx \mu_{\mathrm{opt}}^2(\sigma_u^2)^2 2^{2b-1} = (0.1)^2 \times (0.125)^2 \times 2^{31} = 3.36 \times 10^5$$

or equivalently, 55 dB.

The above example shows that a 16 bit long wordlength is marginally sufficient for the implementation of modem echo cancellers during training.

DATA (FULL-DUPLEX) MODE

In full-duplex mode, the received remote signal dominates the echo canceller output $\varepsilon(n)$. The signal-to-noise ratio of the echo canceller output is determined by the ratio of the received remote signal to the excess MSE of $\varepsilon(n)$. As a consequence, a small step size should be used to reduce the excess error to a level that is far below the remote signal. To facilitate our discussion, we first define a quantity as the ratio of the remote signal power to excess MSE. According to (10.39), in order for the ratio to be equal to ρ, the step size should be equal to

$$\mu_\rho \approx 2 \times \mu_{\mathrm{opt}}/\rho \qquad (10.66)$$

By substituting (10.66) into (10.63), the required wordlength b should satisfy

$$2^{-b} \approx \mu_\rho |\varepsilon(n)|\, |u_{n-k}^*| = |\varepsilon(n)|\, |u_{n-k}^*| \times 2 \times \mu_{\mathrm{opt}}/\rho \qquad (10.67)$$

Obviously, the required wordlength b is determined by the magnitude of $\varepsilon(n)$. If we select the scaling of the quantities such that the RMS value of the maximum echo before cancellation is equal to 0.25, and the ratio of the maximum echo to received signal is equal to ρ_1, then $|\varepsilon(n)|$ is

$$|\varepsilon(n)| = 0.25/\sqrt{\rho_1} \qquad (10.68)$$

By substituting (10.68) into (10.67) and assuming that $|u_{n-k}| = 0.25$ and $\sigma_u^2 = 0.125$, as before,we obtain

$$2^{-b} \approx 1/(L \times \rho \times \sqrt{\rho_1})$$

or, equivalcntly,

$$b \approx \log_2(L) + \log_2\rho + \tfrac{1}{2}\log_2\rho_1$$

That is,

$$10\log_{10}\rho = 3b - 10\log_{10}L - 10\log_{10}\rho - \tfrac{1}{2} \times 10\log_{10}\rho_1 \qquad (10.69)$$

From (10.69), we observe that by increasing the wordlength by 1 bit, we can only gain 3 dB in the ratio of the remote signal to residual MSE as opposed to 6 dB in half-duplex training. On the other hand, by decreasing the maximum echo power by A dB, we can gain $A/2$ dB in this ratio.

EXAMPLE 7
Let us assume that $L = 80$, the maximum echo level is -15 dBm, and the lowest received signal level is -43 dBm, i.e. $\rho_1 = 630$. To achieve a ratio of the remote signal to excess MSE equal to 27 dB, i.e. $\rho = 500$, the required wordlength b is $6.32 + 8.96 + 4.64 = 19.92 \approx 20$ bits.

The above example shows that a DSP unit with a wordlength of 16 bits is not adequate to implement echo cancellers if we require the echo canceller to track changes in channel characteristic as well as generate low excess MSE that does not interfere with the operation of the modem receiver. In the next section, we show how to remedy this problem by using a technique called *gradient averaging.*

10.3.2 Gradient averaging algorithm for echo canceller implementation

In the previous sections, we have shown that in order to reduce excess error at the echo canceller output, which behaves like an additive noise to the received remote signal, it is necessary to decrease the adaptation step size to a very small number. On the other hand, the small step size causes numerical problems for the echo canceller in full-duplex mode. One method to avoid this problem is simply to use a DSP unit with a longer wordlength. However, this may not be proper for some practical designs due to hardware limitations. In this section, we discuss the *gradient averaging algorithm* that effectively extends the wordlength of a DSP unit [21].

As is known from the theory of the LMS adaptive algorithm, the term $\varepsilon(n)\mathbf{u}_n^*$ in (10.62) is an unbiased but noisy estimate of the true gradient. We can obtain a better estimate of the true gradient by using an average of the gradient estimates, i.e. $(1/M)$ $\sum_{k=n}^{n+M-1} \varepsilon(k)\mathbf{u}_k^*$. Thus, we obtain a modified form of (10.63) as

$$\mathbf{c}[(n+1)M] = \mathbf{c}(nM) + \frac{\mu}{M} \sum_{k=nM}^{(n+1)M-1} \varepsilon(k)\mathbf{u}_k^* \qquad (10.70)$$

In other words, the coefficient vector is updated every M symbol intervals by using the averaged gradient. By using a derivation similar to that used in section 10.2.1, it can be shown that the excess MSE generated using (10.70) is M times smaller than the conventional LMS algorithm for the same step size μ. On the other hand, because the LMS algorithm employs gradient averaging and updates its coefficient vector every M bauds, its time constant is M times longer than the conventional LMS algorithm. We can achieve the same excess MSE and the same converging/tracking behaviour as the gradient averaging technique by using a conventional LMS algorithm with a step size equal to μ/M. Thus, from this point of view, we do not gain anything from using the gradient technique if infinite precision is used in the computation. However, this technique improves the numerical behaviour of an LMS echo canceller when finite precision is used in the implementation.

It follows from the above discussion that, for a small μ, a gradient averaging LMS echo canceller that updates the coefficient vector according to

$$\mathbf{c}[(n+1)M] = \mathbf{c}(nM) + \mu \sum_{k=nM}^{(n+1)M-1} \varepsilon(k)\mathbf{u}_k^* \qquad (10.71)$$

has almost the same excess MSE and tracking/convergence characteristics as the LMS echo canceller given by (10.62). Assuming that we use the gradient averaging algorithm after half-duplex training, the echo canceller has approximately reached its optimum state and $\varepsilon(k)$ has become uncorrelated with the elements of u_k^*. Furthermore $\varepsilon(k)\mathbf{u}_k^*$ with different time index k are uncorrelated to each other. Thus, the variance of $\Sigma_{k=nM}^{(n+1)M-1} \varepsilon(k)\mathbf{u}_k^*$ is M times larger than $\varepsilon(n)\mathbf{u}_n^*$ in (10.63). It follows that their magnitudes differ by $\sqrt{M}$ times. The condition (10.67) for continuing to update in the data mode becomes

$$2^{-b} \approx \mu_\rho \sqrt{\mathrm{M}}\, |\varepsilon(n)|\, |u_{n-k}^*| = \sqrt{M}\, |\varepsilon(n)|\, |u_{n-k}^*| \times 2 \times \mu_{\text{opt}}/\rho \qquad (10.72)$$

for the gradient averaging echo canceller. In other words, the required wordlength reduces $\frac{1}{2}\log_2 M$ bits.

EXAMPLE 8

Under the same conditions given in Example 7 for an echo canceller which employs the gradient averaging technique and uses a wordlength of 16 bits to track channel changes, let us determine the required averaging length.

As shown above, without gradient averaging we need a DSP unit with a 20 bit wordlength. To reduce the wordlength to 16 bits, we need $\frac{1}{2}\log_2 M = 4$, i.e. $M = 256$ bauds.

IMPLEMENTATIONAL CONSIDERATIONS OF THE GRADIENT AVERAGING ALGORITHM

The gradient averaging LMS algorithm can be implemented in two stages. First, we compute the *estimated gradient* according to

$$\hat{\mathbf{g}}[(n+1)M] = \mu_1 \sum_{k=nM}^{(n+1)M-1} \varepsilon(k)\mathbf{u}_k^* \tag{10.73}$$

The estimated gradient is computed time-recursively for M bauds, such that

$$\hat{\mathbf{g}}(k) = \hat{\mathbf{g}}(k-1) + \mu_1 \varepsilon(k)\mathbf{u}_k^* \quad \text{for } k = nM, \ldots, (n+1)M \tag{10.74}$$

Then, at the Mth baud, we update the coefficient vector accordng to

$$\mathbf{c}[(n+1)M] = \mathbf{c}(nM) + \mu_2 \hat{\mathbf{g}}[(n+1)M] \tag{10.75}$$

It is necessary to select the constants μ_1 and μ_2 properly such that $\mu_1\mu_2 = \mu$ for the echo canceller to have the desired excess MSE, and μ_1 is selected to be as large as possible while keeping $\hat{\mathbf{g}}(k+1)$ from overflowing.

It is important to note that the gradient averaging algorithm is usually used in the data or full-duplex mode and, maybe, also in the last stage of half-duplex training, when higher precision is needed. If it is used at the beginning of training, the estimated gradient may overflow because at that stage the step size μ and the error $\varepsilon(k)$ both have large magnitudes. In order to prevent overflow, we must scale other quantities in the algorithm, but this will cause other numerical problems, as was analyzed in [22]. Practically, it is not necessary to use gradient averaging during that stage, anyway, as was shown in Example 6.

AN ALTERNATIVE IMPLEMENTATION OF THE GRADIENT AVERAGING ALGORITHM

An alternative implementation of the gradient averaging algorithm was described in [23]. Instead of computing the gradient of the whole coefficient vector, the alternative method computes the gradient of each component of the vector at a time as described below.

Instead of computing the gradient vector estimate recursively every baud, and then updating the whole coefficient vector as given by (10.74) and (10.75), we may compute one component of the gradient vector estimate at a time and then immediately update the corresponding coefficient. To do so, the errors $\varepsilon(k)$ and the data symbols a_k are stored in buffers each M bauds long. The ith component $c_i(n)$ of the coefficient vector is updated according to

$$c_i(n+M) = c_i(n) + \sum_{k=n}^{n+M-1} \varepsilon(k)u_{k-i} \quad \text{for } i = 0, 1, \ldots \tag{10.76}$$

Since the averaging form is implemented as a series of MACs, as stated above, it has a high precision of computation. Its numerical properties will be as good as or better than the previously described form. This form of the gradient averaging algorithm is also more computationally efficient for an implementation using the majority of commercially available DSPs, because such DSPs are usually cycle efficient to perform many MACs sequentially. On the other hand, the algorithm given by (10.74) and (10.75) needs to load the estimated gradient and store the result before and after each MAC operation. As far as memory is concerned, to implement this alternative form, the memory required for storing the errors is proportional to the block size. As a consequence, it will use more memory when a long averaging block size M is needed. The storage required by the original gradient algorithm is proportional to the number of ccoefficients.

10.4 Related topics and references

In the previous sections, we described the generation mechanism of the echoes in telephone networks and discussed different aspects of adaptive echo cancellation. In particular, we derived the convergence characteristics and tracking behaviour of LMS adaptive echo cancellers. We have also investigated finite wordlength effects in implementing echo cancellers using digital signal processors. These subjects are fundamental and important to the understanding of adaptive echo cancellation. However, to improve echo canceller performance further and to address the requirements for various practical applications, many related subjects have to be addressed. Although these subjects may or may not directly relate to adaptive filtering, they are important for the implementation of effective and efficient echo cancellation in practical applications. In this section, we will briefly discuss some of the problems and provide references on these topics for further reading.

10.4.1 Far echo frequency offset compensation

Let us recall from section 10.1 that there exist two types of echoes on the telephone network: the near echo and the far echo. The near echo consists of the signal leaking through the hybrid inside the modem and the signal reflected back along the local loop and from the local CO. The near echo is usually much stronger than the received signal, but it is essentially linear and can be cancelled almost completely. The far echo consists of the portions of the transmitted signal reflected back from the remote CO, remote local loop and remote modem. The far echo is usually weaker than the received signal. However, since the far echo travels through the telephone trunk carrier system, it is affected by impairments in the carrier system. The most damaging impairment in the far echo is phase roll, or frequency offset. Below we describe the generation of the phase roll in the far echo and briefly discuss its compensation.

GENERATION OF PHASE ROLL IN FAR ECHO

Figure 10.11 shows a telephone connection that has a four-wire circuit including a carrier system. When the modem signal passes through the trunk carrier portion of the four-wire circuit, it is first modulated to a carrier frequency f_{c1} that is usually very high, e.g. in the microwave frequency range. At the other end of the carrier system, the modulated signal is demodulated by a demodulator with frequency f_{d2} to become a baseband signal again that can be received by the remote modem. On

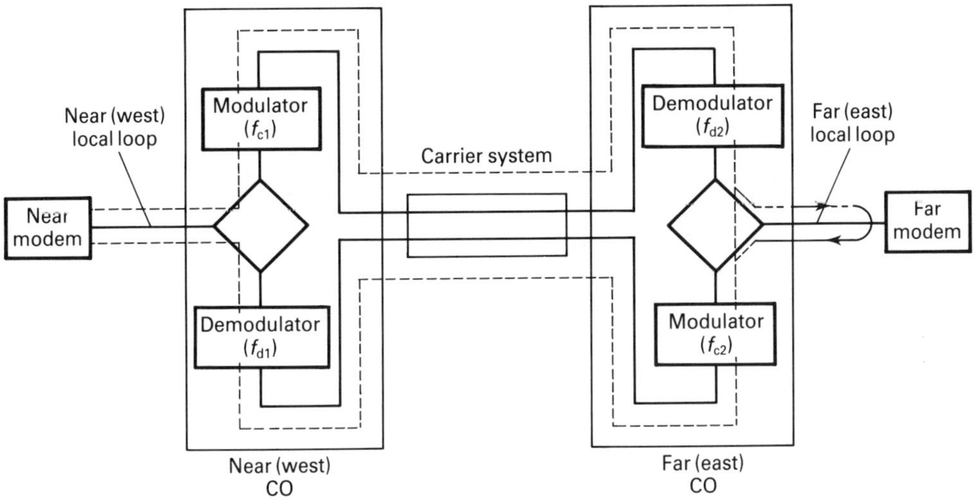

Figure 10.11 Generation of phase roll in far echo

the other direction of the carrier system, the signal from the remote modem is modulated by another modulator with frequency f_{c2} and then demodulated by a demodulator with frequency f_{d1} as shown in Figure 10.11. Theoretically, it is desirable that $f_{c1} = f_{d2}$ and $f_{c2} = f_{d1}$. However, practically, it is very difficult to achieve such a perfect frequency matching. Thus there is often a frequency offset between the transmitted signal and the signal received at the other end. As the far echo travels through the complete four-wire loop, it is modulated at least twice and demodulated at least twice by both modulators and demodulators. Thus, the offset frequency between the transmitted signal and the far echo is equal to $\Delta f = f_{c1} - f_{d2} + f_{c2} - f_{d1}$. It is easy to see that $\delta f = 0$ if either $f_{c1} = f_{d2}$ and $f_{c2} = f_{d1}$, or $f_{c1} = f_{d1}$ and $f_{c2} = f_{d2}$. The second condition is easier to satisfy, because the modulator and demodulator at each end are usually collocated and could use the same frequency reference source. In most US telephone networks, especially in telephone networks within one telephone company, phase roll is usually not a problem. However, it is a different story for international telephone communications. Nevertheless, the offset frequency of the phase roll in far echo, whose value rarely exceeds 1 Hz [2], is often smaller than the offset frequency in the modem received signal. The latter is only affected by the carrier system in one direction and may have an offset frequency as large as 10 Hz.

COMPENSATION OF PHASE ROLL IN FAR ECHO

When the phase roll exists, according to (10.9), we can express the ith analytic far echo at time nT that has phase roll as

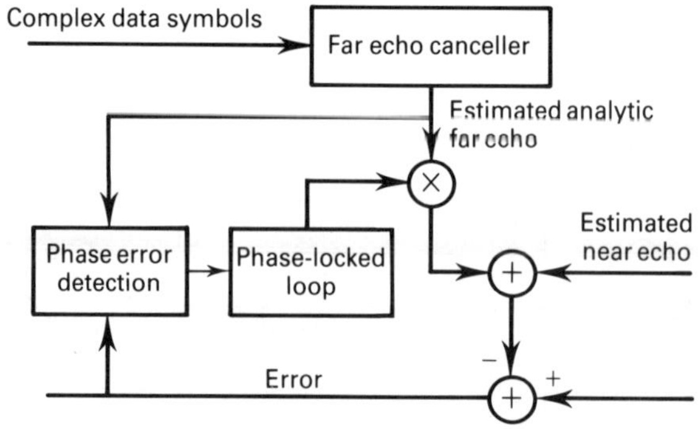

Figure 10.12 Analytic far echo canceller with phase-roll compensation

$$\tilde{e}_{F,i}^{(\omega_p)}(n) = \exp[j(\omega_p nT + \phi)]\tilde{e}_{F,i}^{(0)}(n) \cong \exp[j(\omega_p nT + \phi)] \sum_{k=0}^{L_F - 1} \exp[j\omega(n - B - k)T]u_{n-B-k}\tilde{h}_{F,i}(k) \quad (10.77)$$

where ω_p is the offset frequency of the phase roll in the far echo, ϕ is a constant offset phase angle and $\tilde{e}_{F,i}^{(0)}(n)$ and $\tilde{h}_{F,i}(k)$ are the far echo and the far echo channel coefficients under the assumption that there is no phase roll. If the offset angle and offset frequency can be estimated accurately, we may first estimate $\tilde{h}_{F,i}(k)$ and then compute $\tilde{e}_{F,i}^{(0)}(n)$. The true analytic far echo with phase roll can be obtained by multiplying $\tilde{e}_{F,i}^{(0)}(n)$ by $\exp[j\omega_p nT + \phi)]$. As is well known in communication and control theory, the offset frequency and angle can be computed by using a phase locked loop (PLL). Such

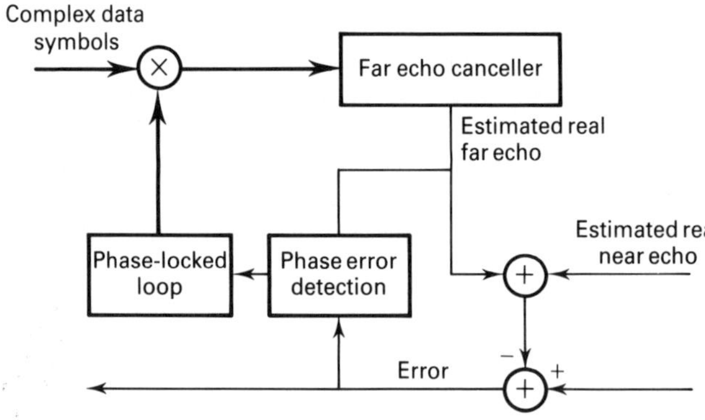

Figure 10.13 Real passband far echo canceller with phase-roll compensation

an arrangement for far echo cancellation with phase roll compensation is given in Figure 10.12.

In order to compensate far echo phase roll using the method described above, we must compute both the real and the imaginary parts of the analytic far echo. This does not require any additional computation for an analytic Nyquist echo canceller because these two parts of the echo would be computed anyway. However, for a real passband echo canceller, only the real part of the analytic echo is used in echo cancellation. To compute the imaginary part means doubling the computations of echo estimation and is not desirable. Assuming that the offset frequency is small relative to the baud rate, the offset angle is approximately constant in a time period that is equal to the far echo time span. Thus, we can rewrite (10.77) as

$$\begin{aligned}\tilde{e}_{\mathrm{F},i}^{(\omega_\mathrm{p})}(n) &= \exp[\mathrm{j}(\omega_\mathrm{p} nT + \phi)]\tilde{e}_{\mathrm{F},i}^{(0)}(n)\\ &\approx \sum_{k=0}^{L_\mathrm{F}-1} \{\exp(\mathrm{j}\phi')\exp[\mathrm{j}(\omega+\omega_\mathrm{p})(n-B-k)T]u_{n-B-k}\}\tilde{h}_{\mathrm{F},i}(k) \qquad (10.78)\end{aligned}$$

where ϕ' is some constant phase angle.

The quantity in braces in (10.78) is the modulated form of the transmitted data symbol u_{n-B-k}. Thus, the compensation of the far echo phase roll can also be realized by modulating the data symbols in the echo canceller delay line instead of modulating the estimated echo. This method is computationally more efficient for the real passband echo canceller. A block diagram of a real passband far echo canceller with such a compensation approach is depicted in Figure 10.13. It has been shown experimentally that such a far echo canceller is effective for far echoes with a low offset frequency, e.g. $\omega_\mathrm{p}/2\pi < 10$ Hz. As shown in [2], in practice, the far echo offset frequency rarely exceeds 1.0 Hz. Therefore, both of the compensation techniques are effective for cancelling the far echo with phase roll and are being used in echo cancellation modems.

Above we have presented two basic types of far echo phase-roll compensation methods. It is possible to derive other variations of the basic methods. Interested readers are referred to [12], [24–26] and [27] for further analytical and implementation details of different echo phase-roll compensation techniques.

10.4.2 Analog and digital sampling rate conversion

As discussed in sections 10.1 and 10.2, in an echo cancellation modem employing a Nyquist echo canceller, it is desirable first to sample the received analog signal at an integer multiple of the local transmitted symbol rate for performing echo subtraction. The echo-removed signal must then be resampled at a new rate that is synchronous with the remote transmitter symbol rate, required by the modem receiver. Classically, such a rate conversion is performed by analog means. There are few possible arrangements for performing echo subtraction and rate conversion as described below.

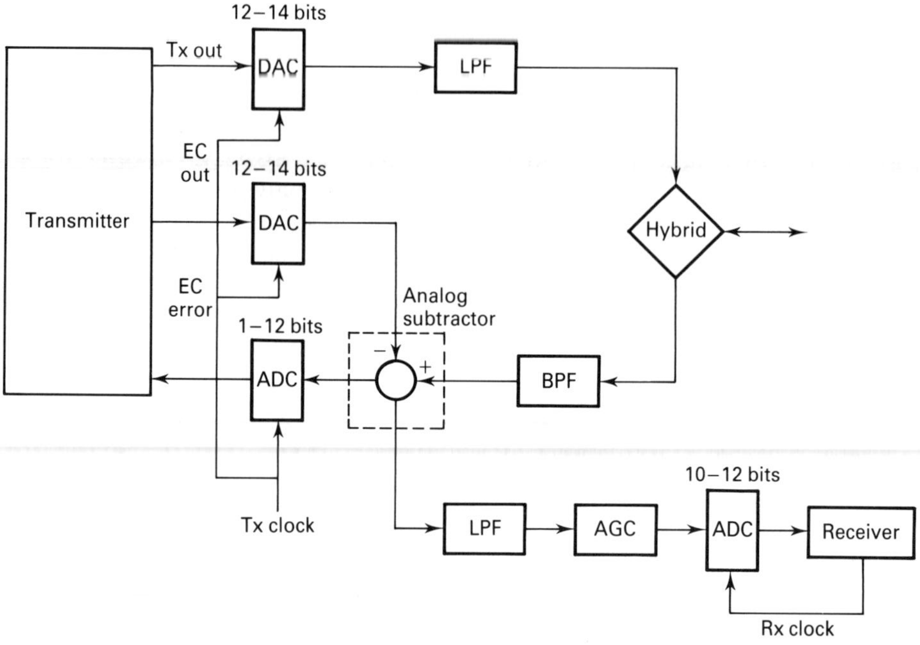

Figure 10.14 An echo cancellation modem with analog subtraction/analog rate conversion

ANALOG SUBTRACTION/ANALOG RATE CONVERSION

The first scheme uses analog subtraction to subtract the echo from the received signal, as depicted in Figure 10.14. The synthesized echo samples are generated at a rate that is an integer multiple of the local transmitter symbol rate. They are then converted to analog form by using an analog-to-digital converter (ADC) and holding that value for one sample time interval. The received signal is also sampled at the same rate and the analog samples also hold for one sample interval. These sampled/held analog values are applied to an analog subtracter, which is usually realized using a conventional operational amplifier (OP-AMP). After the input values are stabilized, the difference (error) signal at the subtracter output is then converted to digital values using a digital-to-analog converter (DAC). The digital error signals are used by the echo canceller to update its coefficients. On the other hand, the subtracter output, which is an analog signal in staircase form, passes through a low-pass analog filter to remove the high-frequency components due to the sampling and holding operations. The filtered analog signal, after being power adjusted by an automatic gain control (AGC) circuit, is then sampled by another ADC at a rate synchronous with the remote transmitter clock. The second ADC is controlled by a timing recovery circuit that recovers the remote clock information.

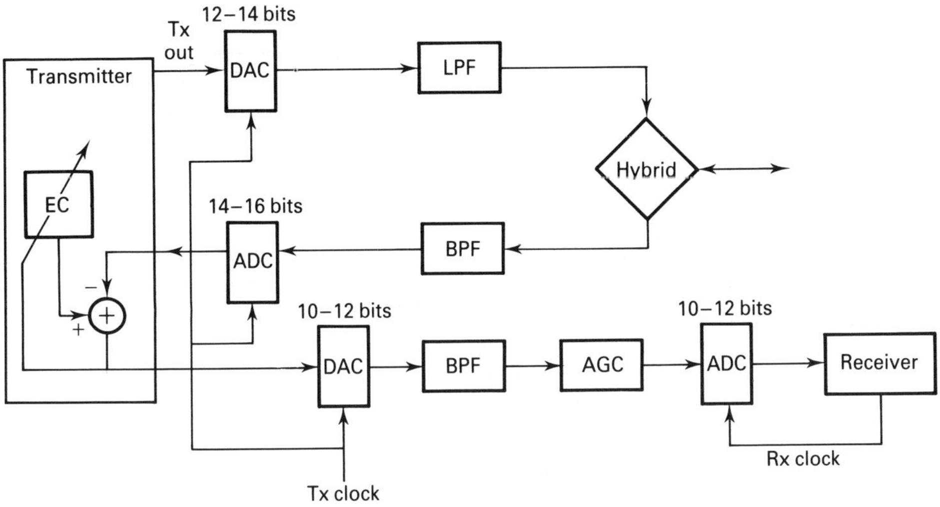

Figure 10.15 An echo cancellation modem with digital subtraction/analog rate conversion

DIGITAL SUBTRACTION ANALOG RATE CONVERSION

An alternative arrangement that uses digital subtraction/analog interpolation is depicted in Figure 10.15. Using such a scheme, the received signal is first converted to digital samples synchronous with the local transmitter clock by a first ADC. A digitally synthesized echo is subtracted from the converted digital received signal samples by the DSP in the modem. The difference (error) digital signal is directly used to update the echo canceller coefficient. It is also sent to a DAC to be converted to analog form. The converted analog error signal, which is in staircase form, then passes through an analog low-pass filter. The filtered analog signal is then converted to digital form as described above.

Comparing the two schemes above, we note that both methods need two ADCs and one DAC. The major difference is the following: in the second scheme, the received signal is converted to digital form before the echo is removed. As a consequence, the ADC performing the conversion needs a wider dynamic range or higher precision than the ADC used in the first scheme. On the other hand, for the first scheme, the analog-to-digital and digital-to-analog conversions are included in the echo canceller's coefficient-updating loop, which may degrade performance if the conversions are not implemented properly.

DIGITAL SUBTRACTION/DIGITAL RATE CONVERSION

It can be seen from the above description that, in either of the analog rate conversion schemes, an ADC/DAC pair has to be used to convert a digital signal from one rate to another. It is well known from the Nyquist sampling theorem that the digital samples of an analog signal contain all the original information of the analog signal

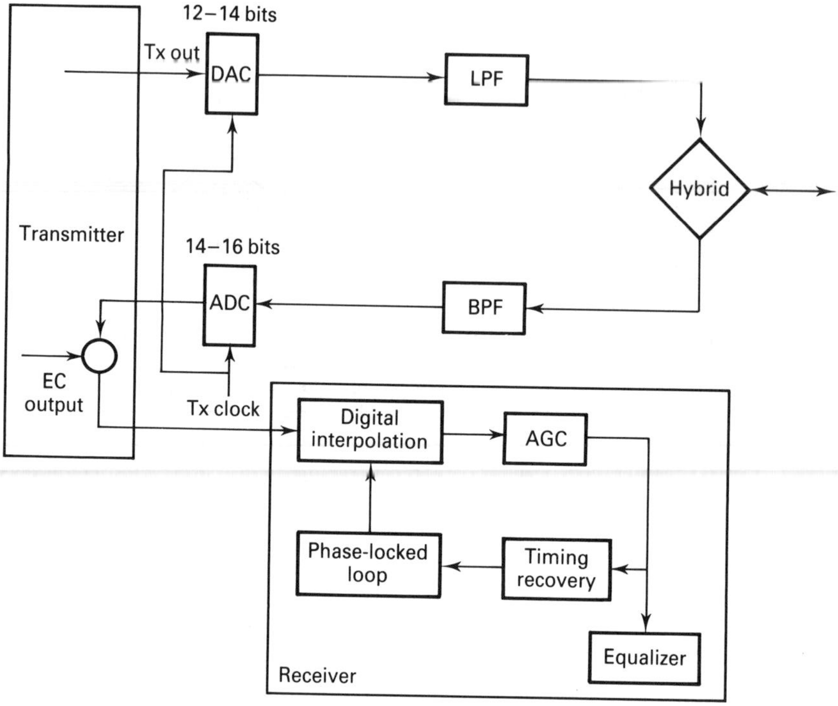

Figure 10.16 An echo cancellation modem with digital subtraction/digital rate conversion

as long as the sampling rate is above the Nyquist rate of the analog signal. Thus, it is possible to convert a digital sample sequence at a first sampling rate to a new digital sample sequence at another rate using only digital signal processing techniques. Such digital rate conversion techniques have been discussed in the literature. Interested readers may find [28], [29] and [30] useful for understanding the principle of digital rate conversion. Below we describe how digital rate conversion works and how such schemes may be used for the echo cancellation modem.

Roughly speaking, a digital rate converter is realized using a 1:N digital interpolator that consists of a bank of N FIR subfilters. Each of the subfilters has the same ideal low-pass magnitude frequency response but a different group delay, which differs by a small amount from one to another. By selecting the proper subfilter for each output sample, we can generate an output sequence that is sampled at an arbitrary time instant and at an arbitrary sampling rate, provided the number N of the subfilters is large enough. However, increasing the number of subfilters means increasing the required memory storage. Practically, a small error introduced in the conversion will always be allowed as long as the error is below the noise level allowed by the receiver. Thus, to generate a certain output sample, we can use the subfilter that has the delay closest to the desired delay of the sample. Normally, N can be chosen in the range from 16 to 128. To reduce the number of subfilters further, we can interpolate the outputs of two adjacent subfilters to obtain one output sample such that the desired delay is between the delays of these two interpolated samples. The desired output can be computed using linear interpolation of the samples. By using linear interpolation, $N = 8$ suffices for modem rate conversion. A detailed analysis can be found in [30].

Figure 10.16 shows a block diagram of an echo cancellation modem using digital echo subtraction/digital rate conversion. The echo subtraction operation is the same as the digital subtraction/analog rate conversion method described above. After the digital error samples are generated, they are fed into a digital rate converter, as described above. The digital rate converter is controlled by the time-recovery circuitry. Thus, at the output of the digital rate converter, we obtain the echo-removed received signal samples at the rate and sampling phase required by the receiver. An echo cancellation modem using the digital interpolation method is described in [31].

Since the digital rate conversion uses the input data sequence as a reference, it is only accurate when the input samples are generated by a stable clock. If the input sampling clock is not stable, jitters in the input clock will cause jitters at the output samples, which degrade the performance of the modem and even cause instability. Compensation techniques for such jitters have been described in [32].

10.4.3 Fast echo canceller training

Above, we have shown that, to achieve a high level of echo cancellation, the estimate of the echo path response must be accurate enough before useful data transmission can start. Conventionally, during initialization, some training data are sent and the LMS adaptation algorithm is used to obtain proper coefficients for the echo canceller. Unfortunately, as a result of the slow convergence speed of the LMS adaptation algorithm, a long training period of a few thousand data symbol intervals is usually required to obtain a satisfactory level of echo cancellation. In the past few years, research efforts have been devoted to reducing this initial training time and various techniques have been suggested. In this section, we provide a survey of these techniques and give the related references.

It is known that the LS algorithms (Chapter 5–8) can achieve a much faster convergence than the LMS algorithm. Actually, for echo cancellation, the LS algorithms provide the optimal initial convergence. Echo canceller training schemes using LS algorithms have been described in [33–35]. The main disadvantage of the training of LS echo cancellers is their computational complexity. Even though we use the most efficient fast RLS algorithm with prewindowed data, its computational complexity is still four times that of the LMS algorithm. It should be noted that the convergence of this algorithm is not optimal due to the prewindowing.

To reduce the training time as well as computational complexity, we can use some special training sequences that have special, i.e. impulse-like, correlation properties. In [36, 37] it was proposed to transmit pseudorandom (PN) or 'chirp' sequences, instead of random data, for echo canceller training. Since the schemes using special training sequences are equivalent to the LS method, they are also optimal in training time. On the other hand, the impulse-like correlation properties of these sequences can be used to reduce the computational complexity to the level of the LMS algorithm. However, the methods using the special training sequences described previously can only be used for those echo cancellers whose input and output are either both real or both complex valued, and they are not directly applicable to the widely used Nyquist real passband echo cancellers with a complex input and a real output.

All the fast training methods mentioned above are inefficient when the far echo exists. In [38], a more efficient estimation method was proposed for the fast simultaneous training of both a near and far echo canceller of a Nyquist echo canceller based on the discrete Fourier transform (DFT). Although the problem of simultaneous training of near and far echo cancellers is solved in principle in this paper, the algorithm proposed there is only correct in special cases. Furthermore, the DFT-based algorithm assumes that the received signal is the real part of an analytic signal. Similar to the analytic echo canceller, the transmitter filter of the modem must completely reject the negative frequencies to achieve satisfactory echo canceller training. It is also known that algorithms based on the DFT may potentially have poor numerical properties.

In [39] another fast training scheme is presented for the Nyquist real passband echo canceller. This method uses a special complex periodic training sequence that not only has an impulse-like autocorrelation function, but also the unique property that its real and imaginary parts are orthogonal. The echo canceller coefficients are estimated by directly correlating a segment of the training sequence with the real echo samples of the sequence. Owing to the orthogonality between the real and imaginary parts of the sequence, this method does not require the complete rejection of the negative frequencies for the transmitter filter and is well suited for training real passband Nyquist echo cancellers. Furthermore, an effective algorithm is proposed for the computation of the required period of the training sequence for simultaneously training the near and far echo cancellers with arbitrary lengths and for an arbitrary far echo delay. This paper also proposes an efficient scheme for the fast and accurate initialization of parameters for the phase locked loop for far echo phase-roll compensation.

10.4.4 Topics related to ISDN echo cancellation

Echo cancellation for ISDN subscriber loops is very similar to that for modem signals. Thus, the problems we have discussed for modem echo cancellers can also be applied to ISDN echo cancellers. However, they do differ in some aspects due to the differences in their cancellation requirements, symbol (baud) rates, signal constellation and synchronization methods. In this section, we provide a brief discussion on these aspects. References are also given for further study.

NON-LINEAR ECHO CANCELLATION

Although non-linear echo cancellation would also be beneficial to modem echo cancellers, it is used mostly in ISDN echo cancellers mainly for two reasons. First of all, since non-linear echo cancellers are usually more complex than linear echo cancellers and the complexity increases exponentially with the size of the possible data symbol alphabet, they are more feasible for ISDN data transmission, which has a simpler signal constellation. Secondly, ISDN has a much higher data rate and needs a higher cancellation performance (e.g. 70 dB) than that for modem echo cancellers (e.g. 55 dB). As a consequence, it becomes necessary to employ non-linear echo cancellers for ISDN data transmission.

The simplest non-linearity appearing in the signals is the non-linearity which only depends on the current signal magnitude. We call this *non-linearity without memory*. Such a non-linearity often arises from the non-linear behaviour of ADCs and can be corrected by using a device that has a non-linear characteristic that undoes the non-linear effect. Such a compensation method is conceptually simple and easy to implement. Unfortunately, non-linearities encountered in real life are more complicated and usually have memory. These non-linearities require more sophisticated techniques to compensate.

A frequently discussed non-linear compensation method involves a Volterra series expansion [40], which uses a number of non-linear filters to model a non-linear system. The number of filters depends on the memory and the order of the non-linearity to be modelled and easily becomes very large. On the other hand, we can always stop adding more filters if the cancellation requirement has been met when a certain number of filters is used. Note that the complexity of a Volterra series expansion does not depend on the signal constellation, unlike the RAM-based table-lookup compensation techniques described below.

The table-lookup non-linear echo canceller is conceptually very simple. Assuming that the non-linearity spans L bauds, the (non-linear) echo at each sampling time is completely determined by the sequence of L previously transmitted symbols. If the signal constellation consists of M distinct symbols, there exist M^L such possible sequences. Thus, it is possible, at least theoretically, to determine M^L coefficients, each of which is associated with such a sequence. Thus, each coefficient is equal to the echo value that is generated by the associated sequence. For each possible sequence, an echo estimate is generated using the corresponding coefficient. Obviously, the larger the number of M and/or L, the more coefficients are needed. The number of coefficients may become prohibitively large for modem signals that have a large symbol alphabet. On the other hand, the signals for ISDN only have a small symbol alphabet and it is adequate to use such a non-linear echo canceller. Since the coefficients are usually stored in a RAM and the address of each coefficient is determined by the sequence transmitted, such an echo canceller is called a RAM-based table-lookup non-linear echo canceller.

The RAM-based echo canceller was described in [41]. The analysis of its convergence when it is jointly adapted with a linear transversal echo canceller has been discussed in [42].

IIR ECHO CANCELLATION

Since ISDN transmission employs a much higher data rate than modem signalling, it requires more echo canceller coefficients to cover the same echo span than that for modem echo cancellers. On the other hand, because the long echo tail decays exponentially, it is possible to model it by an IIR transfer function. Thus, to reduce the number of echo canceller coefficients, in addition to FIR and non-linear echo cancellers, IIR echo cancellers are often used for ISDN echo cancellation to cover the long echo tail.

IIR echo cancellers can be implemented as adaptive filters or as compromise fixed filters. To implement a usable adaptive IIR filter, precautions have to be taken to ensure its stability and global convergence. These stability issues have been addressed in a number of papers including [43] and [44]. Fixed compromised IIR echo cancellers can always be made stable and have been shown to be adequate for ISDN echo canceller implementation. Of course, their performance will not be as good as that of a properly implemented adaptive IIR echo canceller.

JOINT EQUALIZATION AND ECHO CANCELLATION

In modems, echoes are cancelled first before the signal enters the receiver. As discussed in section 10.2, the residual echo mainly consists of the received signal, which behaves like a strong noise to modem echo cancellers. Such a strong noise interferes with the operation of the echo canceller and results in a slow tracking speed and a requirement for a longer computer wordlength during the data mode. As shown in [13], removing the received signal at the echo canceller output, by using combined echo cancellation and equalization, can greatly improve the tracking and numerical performance of the echo canceller.

Such a combined implementation of echo canceller and equalizer is especially suitable for ISDN echo cancellers whose transmitter and receiver clocks are synchronous. Nevertheless, removing the received signal from echo signals is still very complex when a conventional linear equalizer or decision feedback equalizer (DFE) is used. To simplify implementation, a joint echo canceller and equalizer for ISDN usually uses a fixed compromise feedforward filter and an adaptive pure DFE. Such a scheme was described in [45] and a simplified block diagram of the equalizer/echo canceller combination is given in Figure 10.17. Since the echo canceller has removed most of the echo and the DFE has removed most of the intersymbol interference, the summer output is simply data symbols plus noise. We then subtract the detected symbol from it, which is called *reference cancellation.* The error signal after reference cancellation contains only noise and can be used to update both the echo canceller and the adaptive DFE.

THE PROBLEM OF JITTER IN THE TRANSMITTER CLOCK

As shown above, the synchronous transmitter and receiver clocks of the ISDN terminal simplify the implementation of its echo canceller. However, while we can assume that, at the central office end, the transmitter clock is derived from a stable network clock, at the user end the transmitter clock is recovered from the received signal and may

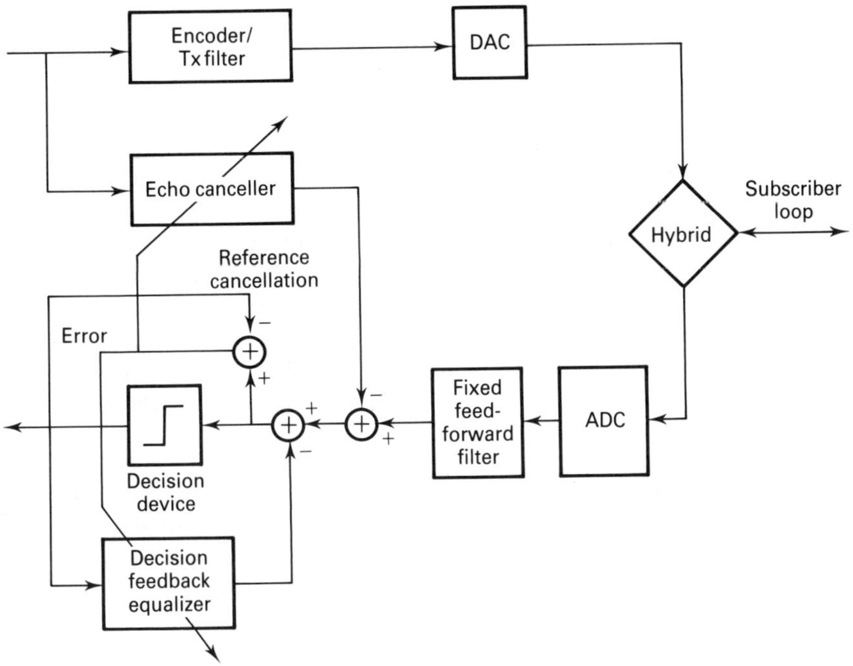

Figure 10.17 Combined echo cancellation and decision feedback equalization for ISDN transceiver

be jittery. It has been shown that such a jittery clock will cause serious problems in echo canceller implementation. In particular, when the ratio of the echo to the far end signal is large, the performance of the echo canceller is very sensitive to the jitter in the transmitter clock.

The effect of clock jitter has been thoroughly analyzed in [46]. It is possible to compensate for the jitter effect by using sophisticated echo canceller algorithms. On the other hand, another simpler but effective solution is to generate a more stable clock by improving the performance of the timing recovery and clock generation circuitry.

VLSI IMPLEMENTATION OF AN ISDN U-INTERFACE CHIP

Because of performance and cost requirements, the ISDN transceivers, called ISDN U-interfaces, are usually implemented using VLSI technology. The ISDN standard is specified in [10]. Its basic data rate is 144 kbits per second (kbps), which is divided as two 64 kbps and one 16 kbps data channels, or *2B + D channels*. In addition, there is a control data channel at 16 kbps. Thus, the ISDN data rate is equal to 160 kbps. The data transmission signalling is of a four-level PAM format. Each of the symbols carries two data bits. Such a signalling format is called the *2B1Q format*. The ISDN data symbols rate is equal to 80×10^3 symbols per second and, hence, its symbol period is 12.5 μs. There exist a number of U-interface VLSI chips on the market today. Below, we use the Motorola ISDN U-interface chip MC145472, jointly developed by Motorola and BNR, as an example.

As described in [47], the functions and architecture of MC145472 are as follows. The U-interface chip has an analog transmitter, a sigma/delta ADC, a symbol receiver unit which performs receiving and echo cancellation functions, a framer and deframer unit, and other supporting circuitry.

The analog transmitter converts the 2 bit 2B1Q digital symbols into filtered analog voltage levels to drive the digital subscriber loop through a hybrid. The sigma/delta ADC converts the analog signal from the hybrid into digital form. The resulting digital signal is processed by the symbol receiver to generate a 2B1Q symbol data sequence as well as a receiver sampling clock. The framer/deframer unit performs format conversion between the scrambled 2B1Q data symbols and the 2B + D binary data for the transmitter and receiver.

The symbol receiver performs many signal processing functions for recovering the 2B1Q data symbols from the received signal. The functional blocks are an adaptive transversal echo cancellation (TEC); an adaptive IIR echo cancellation (IIR-EC); an adaptive reference compensation (ARC); an adaptive decision feedback equalization (DFE); a fixed feedforward equalization (FE); a timing recovery and signal quantization for decision making.

Most of the functions performed by the symbol receiver have been discussed above in this section. These functions could be implemented using a single microcoded DSP unit with minimum complexity and maximum flexibility. However, all these functions are computationally intense and it is difficult to implement them using a single DSP. To reduce the computational burden of the main DSP unit, two special-purpose coprocessing units, the MEC and the LMS units are provided to assist the main DSP unit. Their functions are described below.

The main DSP unit controls the coprocessing units and performs the timing recovery, quantization, fixed feedforward equalization and adaptive IIR echo cancellation. The LMS unit executes updates and convolutions for the DFE and TEC, which are the LMS adaptive filters; it also implements the ARC function. The fast multiplier inside the LMS unit can be used by the main DSP unit whenever it becomes necessary. The MEC is a special-purpose coprocessor for implementing the non-linear memory-based echo canceller.

By using sophisticated echo cancellation functions, the ISDN chip can achieve an echo reduction ratio of about 70 dB. This ensures satisfactory performance for data communication over ISDN subscriber loops.

10.4.5 Topics related to voice telephone echo cancellation

A voice echo canceller is also similar to a data canceller in structure [48]. They are both usually implemented as FIR filters and the LMS algorithm is employed for coefficient updating. However, because the speech signal has different characteristics

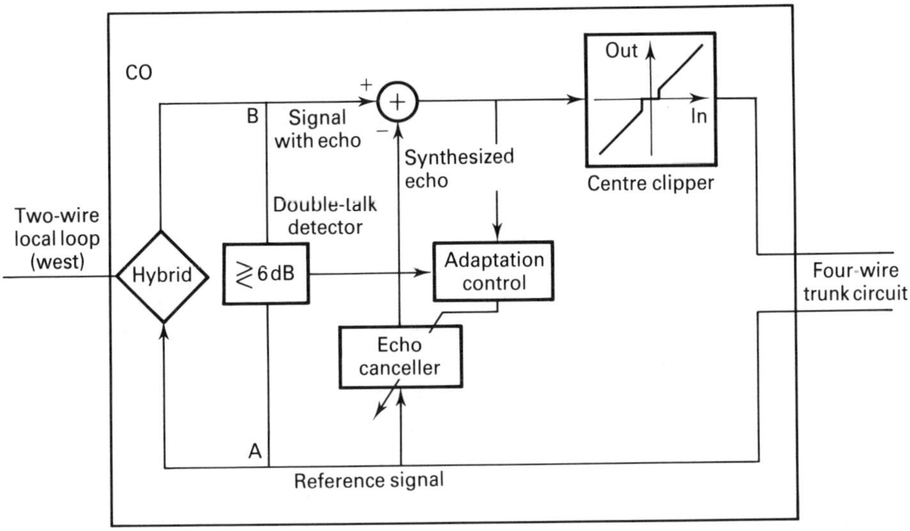

Figure 10.18 A voice echo canceller with double-talk detector and centre clipper

from the data signal, some special problems must be considered when implementing a voice telephone echo canceller.

DOUBLE-TALK DETECTION

Since there is no special training sequence for a voice echo canceller, it must use the normal speech signal to train its coefficients. As in the data echo canceller, during double-talk and while it still computes the synthesized echo and performs echo subtraction, a voice echo canceller must reduce its updating step size or completely stop its adaptation. On the other hand, to model the channel response accurately and to track the channel variations, it should keep updating its coefficients whenever double-talk stops. Thus, double-talk detection is an important part of effective voice echo cancellation.

Figure 10.18 shows a block diagram of a voice telephone echo canceller with a double-talk detector and a centre clipper that will be discussed below. Normally, the hybrid provides an attenuation of more than 6 dB to the remote talker speech signal. Thus, the signal level at point B is at least 6 dB lower than the signal level at point A if the near speaker is not talking. However, during double-talk, the signal at point B also contains the near speaker's voice and its level would become higher. The simplest double-talk detector monitors p_B, the signal level at point B, and compares it with p_A, the signal level at point A. If p_B is at least 6 dB lower than p_A, the detector assumes that there is no double-talk and lets the echo canceller update the coefficients as normal. The updating stops whenever the difference becomes less than 6 dB or p_B is higher than p_A.

The simple double-talk detector may not work properly if the hybrid fails to provide an attenuation of 6 dB or the near speaker does not speak loud enough. More sophisticated techniques are used to monitor the occurrence of double-talk more accurately. These techniques are discussed in a number of papers including [49].

CENTER CLIPPING

Since the echo canceller cannot be trained perfectly for various reasons and some residual echo often remains, centre clippers are often used to remove the weak residual echo.

A centre clipper is a kind of non-linear processor. It monitors the signal after echo cancellation. When only the far speaker is talking, the centre clipper sets the signal to be zero if it is below a certain threshold that may be either fixed or variable depending on the level of the remote speech signal. If there is no remote speech signal or the near end speech is present, the centre clipper is disabled [48].

Speech telephone echo cancellation requirements are specified in the CCITT standard G.165 [8]. Interested readers are referred to this document for further information.

10.5 Concluding remarks

In this chapter, we have discussed the generation mechanism of echoes in telephone networks and how echoes affect voice and data communications over telephone networks. We have shown that such echoes can be modelled as linear transformed transmitted signals. As a result, echo cancellation techniques can be used to eliminate such echoes to enhance telephone voice and data communications.

Various aspects of telephone echo cancellers, including echo canceller structures, adaptive echo cancellation algorithms, their learning and tracking characteristics and finite wordlength effects, have been discussed in this chapter. Many practical implementation issues were also considered. In deriving these conclusions, emphasis is placed on practical usefulness rather than theoretical rigour. Furthermore, we have used approximations that are justifiable in practical applications to simplify derivations. Nevertheless, the conclusions obtained in this chapter are theoretically sound. For instance, the results on the finite wordlength given in sections 10.3.1 and 10.3.2 are obtained from assumptions and approximations, but they agree well with the more rigorous analysis given in [22]. Some results and/or their derivations presented are new. Notably, the results on the effect of a non-ideal Tx filter on the performance of echo cancellers have not been seen elsewhere by the author. In this chapter, most of the discussion and analysis is based on data modem echo cancellers. However, the basic results can also be applied to voice echo cancellers, subscriber echo cancellers, as well as acoustic echo cancellers.

Although echo cancellers have been used in practical applications for only about 10 years, they have had an important impact on voice and data communications. Many of the new high speed communications standards, such as the V.32, V.32 bis modem standards, ISDN standards and other standards under discussion, are based on echo cancellation techniques. Voice echo cancellation also plays an important role in new packet-switched digital network designs for voice and data transmission, because of the long delay on such networks.

Because of the importance and advantages of echo cancellation technology, much progress has been made in this area during the past two decades. Today, there exists a rich body of technical literature on different aspects of echo cancellation. This chapter covers limited but fundamental material on this topic. It may be used by researchers and application engineers as a starting point for understanding and applying echo cancellation techniques, and to make further advances in this area.

Notes

1. This may not be true in today's digital network environment. In such a case, the local loop can be a portion of a packet-switched digital network. As a result, a large delay, e.g. as large as 100 ms, may be introduced by the network.
2. By definition, 1 dBm is a power measurement unit equal to the power of 1 milliwatt on a 600 ohm load.
3. Rigorously speaking, these two independent assumptions are not valid in this case. However, they are widely used in the analysis of LMS algorithms and the analytical results so obtained agree well with simulations and measurements. We adopt these two assumptions in the analysis without further justification.
4. It can be shown that if the signal magnitude is non-zero anywhere within the passband, the sign cannot change within the positive passband.
5. In some of the literature, the error due to truncation is also called roundoff error.
6. In some cases the data symbols may be represented exactly. We will not make such an assumption here.
7. Rigorously speaking, even when (10.63) is satisfied for the RMS values, occasional updating may still occur as the quantities involved reach their peak values. However, since such updating becomes infrequent and the adaptation becomes very slow, we can consider the adaptation has stopped.

References

[1] A.B. Clark, 'Telephone transmission over long cable circuits,' *Bell Syst. Tech. J.*, vol. 2, pp. 67–94, 1923.

[2] P.H. Wittke, S.R. Penstone and R.J. Keightley, 'Measurements of echo parameters pertinent to high-speed full-duplex data transmission on telephone circuits', *IEEE J. Selected Area Commun.*, vol. SAC-2, pp. 703–10, 1984.

[3] F.P. Duffy, G.K. McNees, I. Nasell and T.W. Thatcher Jr, 'Echo performance of toll telephone connections in the United States', *Bell Syst. Tech. J.*, vol. 54, pp. 209–43, 1975.

[4] A.L. Bonner, J.L. Garrison and W.J. Kopp, 'The E6 negative impedance repeaters', *Bell Syst. Tech. J.*, vol. 36, pp. 1455–504, 1960.

[5] T.H. Curtis, S.J. D'Ambra, R.H. Tegethoff and L.E. Ashkenazi, 'Use of a digital echo canceller in the AT&T Domsat Intertoll Network', *Proc. 5th Int. Conf. on Digital Satellite Communications, Genoa, Italy, Mar., 1981.*

[6] M.M. Sondhi and D.A. Berkley, 'Silencing echoes on the telephone network', *Proc. IEEE*, vol. 68, no. 8, pp. 948–63, 1980.

[7] S.B. Weinstein, 'Echo cancellation in the telephone network', *IEEE Commun. Mag.*, pp. 9–15, 1977.

[8] 'Recommendation G. 165, Echo cancellers', *CCITT Blue Book*, Fascicle VIII. 1, pp. 221–43, 1989.

[9] V.32 – A family of 2–wire, duplex modems operating at data signalling rates of up to 9600 bit/s for use on the general switched telephone network and on leased telephone-type circuits', *CCITT Blue Book*, Fascicle VIII. 1, pp. 221–38, 1989.

[10] 'Integrated services digital network – basic access interface for use on metallic loops for applications on the network side of the NT–Layer 1 specification', ANSI T1.601–1988.

[11] K.H. Mueller, 'A new digital echo canceller for two-wire full-duplex data transmission', *IEEE Trans. Commun.,* vol. COM-24, pp. 956–67, 1976.

[12] S.B. Weinstein, 'A passband data-driven echo canceller for full duplex transmission on two-wire circuits', *IEEE Trans. Commun.* vol. COM–25, pp. 654–66, 1977.

[13] D.D. Falconer, K.M. Mueller and S.B. Weinstein, 'Echo cancellation techniques for full-duplex data transmission on two-wire lines,' *Proc. NTC, Dallas, Dec., 1976.*

[14] J.J. Werner, 'An echo-cancellation-based 4800 bit/s full duplex DDD modem', *IEEE J. Selected Areas Commun*, vol. SAC-2, pp. 722–30, 1984.

[15] E.A. Lee and D.G. Messerschmitt, *Digital Communication*, Kluwer: Boston, 1988.

[16] J.E. Mazo, 'On the independence theory of equalizer convergence', *Bell Syst. Tech. J.*, vol. 58, pp., 963–93, 1979.

[17] K.H. Mueller and J.J. Werner, 'A hardware efficient passband equalizer structure for data transmission', *IEEE Trans. Commun.* vol. COM–30, pp. 538–41, 1982.

[18] F. Ling and S.U. Qureshi, 'Convergence and steady state behavior of a phase-splitting fractionally-spaced equalizer', *IEEE Trans. Commun.,* vol. COM-38, pp. 418–25, 1990.

[19] A.V. Oppenheim and R.W. Schafer, *Digital Signal Processing*, Prentice Hall: Englewood Cliffs, NJ, 1975.

[20] J.M. Cioffi and J.J. Werner, 'Effects of biases on digitally implemented data-driven echo cancellers', *AT&T Tech. J.*, vol. 64, pp. 115–39, 1985.

[21] R.D. Gitlin and S.B. Weinstein, 'The effects of large interference on the tracking capability of digitally implemented echo cancellers', *IEEE Trans. Comm.*, vol COM-26, pp. 833–9, 1978.

[22] N.J. Bershad, 'Nonlinear Quantization effects in the LMS and block LMS Adaptive algorithms – A comparison', IEEE Trans. Acoust., Speech, Signal Process. vol. ASSP-37, pp. 1504–12, 1989.

[23] D. Messerchmitt, D. Hedberg, C. Cole, A. Haoui and P. Winship, 'Digital voice echo canceller with a TMS32020', in *Digital Signal Processing Applications with the TMS320 Family*, Texas Instruments Inc., 1986.

[24] R.D. Gitlin and J.S. Thompson, 'A phase adaptive structure for echo cancellation', *IEEE Trans. Commun.*, vol. COM -26, pp. 1211–20, 1978.

[25] J.D. Wang and J.J. Werner, 'Performance analysis of an echo cancellation arrangement that compensates for frequency offset in the far echo', *IEEE Trans. Commun*, vol. COM-36, pp. 364–73, 1988.

[26] J.J. Werner, 'Effects of channel impairments on the performance of an in-band data-driven

echo canceller', *AT&T Tech. J.*, vol. 64, pp. 91–113, 1985.

[27] J.M. Cioffi, 'A fast echo canceller initialization method for the CCITT V.32 modem', *IEEE Trans. Commun.*, vol. COM–38, pp. 629–38, 1990.

[28] R.E. Crochiere and L.R. Rabiner, 'Optimum FIR digital filter implementations for decimation, interpolation, and narrowband filtering', *IEEE Trans. Acoust., Speech, Signal Process.* vol. ASSP-23, pp. 444–56, 1975.

[29] R.E. Crochiere and L.R. Rabiner, *Multirate Digital Signal Processing*, Prentice Hall: Englewood Cliffs, N.J., 1983.

[30] J.G. Proakis, C. Nikias, C. Rader and F. Ling, *Advanced Digital Signal Processing*, Macmillan: New York, 1992.

[31] A. Haoui, H.H. Lu and D. Hedberg, 'An all-digital timing recovery scheme for voiceband data modems', *Proc ICASSP–87 Conf.*, pp. 44.5.1–4, 1987.

[32] S. Qureshi and F. Ling, 'Digital sampler', US Patent no. 4,989, 221, Jan., 1991.

[33] J. Salz, 'On the start-up problem in digital echo cancellers', *Bell Syst. Tech. J.* vol. 62, pp. 1353–64, 1983.

[34] M.L. Honig, 'Echo cancellation of voiceband data signals using recursive least squares and stochastic gradient algorithms', *IEEE Trans. Commun.* vol. COM–33, pp. 65–73, 1985.

[35] J.M. Cioffi and T. Kailath, 'An efficient RLS data-driven echo canceller for fast initialization of full duplex data transmission', *IEEE Trans. Commun.* vol. COM–33, pp. 601–11, 1985.

[36] T. Kamitake, 'Fast start-up of an echo canceller in a 2–wire full-duplex modem', *IEEE Proc. ICC'84, Amsterdam Holland, May, 1984*, pp. 360–4

[37] V. Kanchan and E. Gibson, 'Measurement of echo path response', *IEEE Trans. Acoust. Speech, Signal Process.*, vol. ASSP–36, 1008–10, 1988.

[38] J.M. Cioffi, 'A fast echo canceller initialization method for the CCITT V. 32 modem', *IEEE Proc. GLOBECOM'87, Tokoyo, Japan, Nov., 1987*, pp. 1950–4.

[39] Ling, F. and Long, G., 'Correlation-based fast training of data-driven Nyquist inband echo cancellers', *IEEE Proc. ICC/SUPERCOM'90, Atlanta, GA, Apr., 1990.*

[40] M. Schetzen, 'Nonlinear system modeling based on the Wiener theory', *Proc. IEEE*, vol. 69, pp. 1557–73, 1981.

[41] N. Holte and S Stueflotten, 'A new digital echo canceller for two-wire subscriber lines', *IEEE Trans. Commun.*, vol. COM–29, 1981.

[42] K. Yamazaki, S. Aly and D. Falconer, 'Convergence behaviour of a jointly-adaptive transversal and memory based echo canceller', *IEE Proc.* pt F, 1991.

[43] 'Adaptive IIR filtering", *IEEE ASSP Mag.*, vol. 6, pp. 4–21, 1981.

[44] C.R. Johnson Jr, 'Adaptive IIR filtering: Current results and open issues', *IEEE Trans. Inf. Theory*, vol. IT–30, pp. 237–50, 1984.

[45] P.F. Adams, S.A. Cox, R.B.P. Carpenter and N.G. Cole, 'A long reach digital subscriber loop transceiver', *Proc. GLOBECOM'86, Dec., 1986*, pp. 2.1.1–5.

[46] D.D. Falconer, 'Timing jitter effects on digital subscriber loop echo cancellers: Part I and Part II', *IEEE Trans. Commun.*, vol. COM–33, pp. 826–38, 1985.

[47] L.S. Bonet, S.R. McCaslin, W.V. Knenust, T.A. Williams, S. Aly, B. Sayer, P. Hung, O. Bahgat and Deczky, 'Architecture of the 2B1Q symbol receiver in the MC145472 ISDN U transceiver', *Proc. ISCAS'89, May, 1989*, pp. 614–17,.

[48] G.S. Fang, 'Voice channel echo cancellation', *IEEE Commun. Mag.* vol. 71, pp. 11–14, 1983.

[49] H. Chang and B.P. Agrawal, 'A DSP-based echo canceller with two adaptive filters', *Proc. GLOBECOM'86, Dec. 1986*, pp. 46.8.1–5.

11

Interference Rejection and Channel Estimation for Spread-Spectrum Communications

*Ronald A. Iltis**

11.1 Definition of spread-spectrum communications

11.1.1 Introduction

The subject of spread-spectrum communications includes a wide range of digital modulation formats, which share the common property of using more bandwidth than would ordinarily be necessary to transmit the basic information. Among the advantages of spread-spectrum communications are low probability of intercept (LPI),

* This work was supported in part by the Office of Naval Research, under contract N00014-82-K-0376, the SDIO/IST program managed by the Office of Naval Research, Contract No. N00014-85-K-0551, and in part by the UC MICRO Program and Sonatech, Inc.

resistance to jamming and multipath, and multiple access capability. While these attributes have long made spread-spectrum communication desirable for military applications, the worsening congestion of the RF spectrum has motivated increased interest in commercial applications, in which multiple users must be accommodated simultaneously in the same frequency band. For example, spread-spectrum code-division multiple access (CDMA) has been proposed for the next generation of digital mobile cellular telephones, local-area wireless networks (LAWNs), and for personal communication networks (PCNs).

Although the use of excess bandwidth by itself provides anti-jam and anti-multipath capability, it has been shown that adaptive filtering techniques can provide significant performance improvements in spread-spectrum receivers [1–3]. In general, adaptive filters provide a way to approximate maximum likelihood spread-spectrum receivers, which implicitly reject interference and demodulate the signal using a filter matched to the channel characteristics [2,4]. Here, we will review the use of the LMS algorithm to reject interference in both direct-sequence and frequency-hopped systems. It will be shown that the problem of joint estimation of interferer and channel parameters is a specific case of ARX system identification. While the LMS algorithm can be used to solve the ARX identification problem, it is well known that RLS provides faster convergence. Furthermore it can be shown [5] that the RLS parameter estimates are exactly equivalent to the maximum likelihood interferer and channel estimates required in a generalized likelihood ratio test (GLRT) receiver. Thus, the performance of a direct-sequence (DS) receiver using the RLS algorithm and a RAKE matched filter will be reviewed for applications where both jamming and multipath are present.

The performance of a coherent spread-spectrum receiver is critically dependent on proper synchronization, or estimation of the signal time of arrival. For example, in direct-sequence spread-spectrum communications, the signal is described by a sequence of binary rectangular pulses, or 'chips', with duration $T_c \ll T_b$ s, where T_b is the bit duration. However, if the time-of-arrival estimation error is greater than T_c, it can be shown that the signal-to-noise ratio at the output of a matched filter or correlator tends to zero, since the DS signal correlation function is approximately a triangle of width $2T_c$ s. Despite the importance of proper code synchronization, the problem of time of arrival estimation for the DS waveform is usually viewed as separate from the tasks of interference rejection and multipath combining. Here, synchronization is accomplished by viewing the problems of time of arrival, interference, and channel estimation as a single, joint estimation problem. The extended Kalman filter will be introduced as an effective, recursive joint estimation technique for this application, which leads naturally to a purely digital DS spread-spectrum receiver.

11.1.2 Spread-spectrum signal models

The mathematical definitions of spread-spectrum signals are now introduced. The first modulation format, direct-sequence (DS) spread-spectrum, essentially uses a wideband binary waveform as a subcarrier for narrowband modulation. In the following, $s_T(t)$ corresponds to the actual transmitted RF waveform, and $s(t)$ to its low-pass equivalent:

$$s(t) = \sqrt{\frac{2E_b}{T_b}} \sum_{n=-\infty}^{\infty} d_n \sum_{k=0}^{L_{PN}-1} c_k P_{T_c}(t - kT_c - nT_b) \tag{11.1}$$

where d_n is a possibly complex-valued information sequence and c_k is a binary ± 1 pseudonoise (PN) sequence. The sequence c_k is chosen in a pseudorandom manner, such that the c_k are approximately uncorrelated. The function $P_{T_c}(t)$ is a rectangular pulse of unit height and duration T_c seconds. (These individual pulses, multiplied by the c_k, are sometimes referred to as 'chips'.) The factor E_b represents energy per bit, and T_b is the bit duration. Typically, $T_b \gg T_c$, and thus the bandwidth of the information sequence, proportional to $1/T_b$ is expanded to $1/T_c$. The bandwidth expansion factor T_b/T_c is the 'processing gain', which provides the anti-jam and anti-multipath capability of spread-spectrum signals. We will assume binary phase-shift keying (BPSK) throughout our discussion of DS systems, and thus $d_n = \pm 1$. The actual transmitted signal $s_T(t)$ has a carrier frequency f_c Hz, thus

$$s_T(t) = \text{Re}\{s(t) \exp(j2\pi f_c t)\}. \tag{11.2}$$

In frequency-hopped communications, information is represented by a sequence of sinusoidal waveforms with pseudorandomly chosen frequencies (or 'hops') f_l. The overall signal resembles an M-ary frequency-shift-keyed waveform, or just a sequence of tones with frequency changing every T_h seconds. The following discussion of interference rejection will be restricted to binary fast-frequency hopping (FFH), in which L different sinusoids, of duration T_h seconds each, represent each information symbol. For FFH, the low-pass equivalent transmitted signal is given by

$$s(t) = \sum_{n=-\infty}^{\infty} \sqrt{\frac{2E_b}{LT_h}} \sum_{l=0}^{L-1} \exp[j(2\pi f_l(n)t + \theta_l)] P_{T_h}(t - lT_h - nT_b) \tag{11.3}$$

In the above, L is the number of hopping frequencies per bit, $f_l(n)$ is the hopping frequency sequence on the nth bit, and θ_l is a random phase independent from hop to hop, and uniformly distributed in $[0, 2\pi]$. For binary modulation, one of two hopping sequences $f_l(n) = f_{l,M}$, or $f_l(n) = f_{l,S}$, is transmitted every $T_b = LT_h$ seconds. These sequences, commonly referred to as MARK and SPACE, are offset by multiples of $1/T_h$ Hz, and thus the binary signals are orthogonal in frequency.

The most important characteristic of both DS and FH signals, from the standpoint of interference rejection, is that they are approximately uncorrelated at certain lag values. It is simpler to illustrate this property in the case of DS waveforms, by first assuming that the PN sequence c_k is purely random, such that

$$P(c_k = +1) = P(c_k = -1) = \tfrac{1}{2}$$

and

$$E\{c_k c_l\} = \delta_{k,l}$$

where $\delta_{k,l}$ is the Kronecker delta function. Now consider the autocorrelation function of $s(t)$ at lag values of T_c, the duration of each binary chip, where an infinitely long PN sequence is assumed:

$$\begin{aligned}
E[s(t - nT_c)s(t - mT_c)] \\
&= \frac{2E_b}{T_b} \sum_{k=-\infty}^{\infty} \sum_{l=-\infty}^{\infty} E[c_k c_l] P_{T_c}[t - (n + k)T_c] P_{T_c}[t - (m + l)T_c] \\
&= \frac{2E_b}{T_b} \sum_{k=-\infty}^{\infty} P_{T_c}[t - (n + k)T_c] P_{T_c}[t - (m + kT_c] \qquad (11.4)
\end{aligned}$$

However, note that the pulse functions $P_{T_c}(t)$ are non-overlapping, and thus the above expectation is zero unless n equals m. Thus, we conclude that

$$\rho_m = E[s(t)s(t - mT_c)] = \begin{cases} 2E_b/T_b & \text{for } m = 0 \\ 0 & \text{otherwise} \end{cases} \qquad (11.5)$$

where ρ_m is the autocorrelation function of the transmitted signal.

It is more difficult to show that the FH signal is approximately uncorrelated. First, assume that the f_l hopping frequencies are selected in a statistically equally likely manner, such that [6]

$$Pr\left\{f_l = \frac{2n}{T_h} + \frac{1}{T_h}\right\} = \frac{1}{2N_h} \qquad (11.6)$$

for $n = 0, 1, \ldots, 2N_h - 1$. That is, there are $2N_h$ possible hopping frequencies, representing binary FSK modulation, with N_h possible MARK and N_h SPACE frequencies. Then, in [6], it is shown that the resulting FH signal in (11.3) is statistically uncorrelated at lag values equal to $T_h/4N_h$, as follows:

$$\rho_m = E[s(t)s(t - mT_h/4N_h)] = \begin{cases} 2E_b/LT_h & \text{for } m = 0 \\ 0 & \text{otherwise} \end{cases} \qquad (11.7)$$

It will be seen that since the spread-spectrum signals can be viewed as white uncorrelated sequences, they are mean square unpredictable, and hence narrowband interferers can be estimated and rejected even in the presence of the desired signal.

11.2 Interference rejection using the Wiener filter and the LMS algorithm

11.2.1 Interference rejection in a direct-sequence receiver

The application of the optimum Wiener filter and LMS adaptive algorithm to interference rejection is now considered in both the DS and FH spread-spectrum

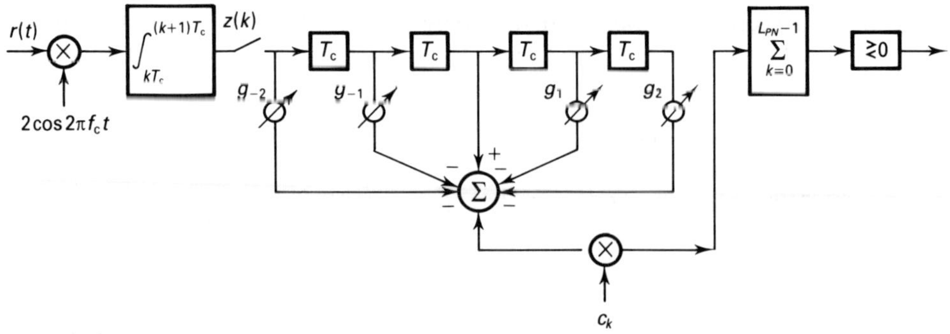

Figure 11.1 Direct sequence receiver with perfect synchronization and two-sided interface rejection filter

applications. First, a non-adaptive interference rejection filter which is optimum in the mean square sense will be derived for the DS signal case. Let $r(t)$ represent the received signal as follows:

$$r(t) = \text{Re}\{s(t + \tau) \exp[j(2\pi f_c t + \theta)]\} + j_{BP}(t) + \eta_w(t) \tag{11.8}$$

where $\eta_w(t)$ is additive white Gaussian noise with spectral density $N_0/2$. The bandpass interferer is denoted by $j_{BP}(t)$, and its low-pass equivalent will be denoted as $j(t)$. In the following, it is assumed that perfect phase and code synchronization is obtained, and thus τ, the signal arrival time, and θ, the carrier phase, are known exactly. (The perfect synchronization assumption will be removed in section 11.4.) Thus, the receiver of Figure 11.1 can be used to demodulate the waveform $r(t)$ and determine the information sequence d_n. The samples at the output of the chip duration integrate-and-dump form a discrete time data sequence $z(k)$ as follows:

$$z(k) = \int_{kT_c}^{(k+1)T_c} r(t) 2 \cos(2\pi f_c t) \, dt \tag{11.9}$$

$$z(k) = s(k) + j(k) + \eta(k) \tag{11.10}$$

where

$$s(k) = \sqrt{\frac{2E_b}{T_b}} T_c c_k$$

$j(k)$ represents the low-pass interferer samples, and $\eta(k)$ is a white Gaussian sequence, such that

$$E[\eta(k)\eta(l)] = \sigma_\eta^2 \delta_{k,l}$$

It will be assumed throughout that $j(k)$ is a wide-sense stationary, zero-mean Gaussian random process, with covariance function given by

$$E[j(n)j(m)] = \rho_{jj}(n - m)$$

The signal samples, $s(k)$, are the scaled binary-valued chips of the PN sequence, c_k, and are thus mutually uncorrelated, so that

$$E[s(n)s(m)] = \begin{cases} (2E_b/T_b)T_c^2 & \text{for } m = n \\ 0 & \text{otherwise} \end{cases}$$

The filter design problem is now to estimate and reject the interferer samples $j(k)$, while preserving the spread-spectrum signal samples $s(k)$ as much as possible. The resulting interference rejection problem is best presented in system identification form as follows. Following the notation of Chapter 2, the desired response is denoted by the sequence $y(k)$. In the interference rejection problem, the desired response is $y(k) = j(k)$, or just the interferer itself. The estimates of $y(k)$ are given in terms of the past and future observations $z(k)$ by

$$\hat{j}(k) = \hat{y}(k) = \sum_{n=-N, n\neq 0}^{N} g_n z(k-n) \tag{11.11}$$

where the g_n represent the rejection filter coefficients. (We have chosen a non-causal two-sided filter based on the analysis of [4], in which it was shown that this filter structure best approximates the maximum likelihood receiver. The non-causality is easily removed by inserting a delay of $N/2$ samples into the rejection filter. Issues related to the design of such linear phase filters are also discussed in [7].) Again, in the context of system identification, $\hat{y}(k)$ is parameterized by the rejection filter coefficients, g_n, and has the general form

$$\hat{y}(k \mid \boldsymbol{\theta}) = \boldsymbol{\theta}^{\mathrm{T}} \boldsymbol{\phi}(k) \tag{11.12}$$

The unknown parameter vector is the set of rejection filter coefficients,

$$\boldsymbol{\theta}^{\mathrm{T}} = [g_{-N}, g_{-N+1}, \ldots, g_{-1}, g_1, \ldots, g_N] \tag{11.13}$$

The regression vector is the set of past and future observations $z(k)$ used to estimate $j(k)$, thus

$$\boldsymbol{\phi}(k)^{\mathrm{T}} = [z(k-N), z(k-N+1), \ldots, z(k-1), z(k+1), \ldots, z(k+N)] \tag{11.14}$$

and the resulting error signal to be minimized is given by

$$\begin{aligned} e(k \mid \boldsymbol{\theta}) &= y(k) - \hat{y}(k \mid \boldsymbol{\theta}) \\ &= j(k) - \sum_{n=-N, n\neq 0}^{N} g_n z(k-n) \end{aligned} \tag{11.15}$$

If $\hat{y}(k \mid \boldsymbol{\theta})$ was exactly equal to $j(k)$, the interferer, then the interference rejection filter output would be

$$z(k) - \hat{y}(k \mid \boldsymbol{\theta}) = s(k) + \eta(k) \tag{11.16}$$

that is, the rejection filter output would consist solely of the desired signal plus thermal noise, and the interferer would be completely excised.

In forming the error signal $e(k \mid \boldsymbol{\theta})$ in (11.15), it was implicitly assumed that the interferer itself, $j(k)$, was available. However, if $j(k)$ were known, then there would be no need for an interference rejection filter. To circumvent the need for an exact reference, it will be seen shortly that the same estimate of the parameter vector $\boldsymbol{\theta}$ is obtained when $z(k) = y(k)$ is used as the desired response, instead of $j(k)$. That is, we claim that minimization of

$$V(\boldsymbol{\theta}) = E\left[\left| j(k) - \sum_{n=-N, n\neq 0}^{N} g_n z(k-n) \right|^2\right] \tag{11.17}$$

with respect to $\boldsymbol{\theta} = \mathbf{g}$ yields the same solution as minimization of

$$V'(\boldsymbol{\theta}) = E\left[\left| z(k) - \sum_{n=-N, n\neq 0}^{N} g_n z(k-n) \right|^2\right] \tag{11.18}$$

To verify this claim, compare the two Wiener solutions for $\boldsymbol{\theta}$ resulting from minimization of $V(\boldsymbol{\theta})$ and $V'(\boldsymbol{\theta})$, with respect to $\boldsymbol{\theta}$. From (2.37), we have

$$R\boldsymbol{\theta} = \mathbf{d} \tag{11.19}$$

where

$$R = E[\boldsymbol{\phi}(k)\boldsymbol{\phi}(k)^{\mathrm{T}}] \tag{11.20}$$

For minimization of $V(\boldsymbol{\theta})$, with the actual interferer $j(k)$ available as a reference, the cross-correlation vector $\mathbf{d}$ is given by

$$\mathbf{d} = E[j(k)\boldsymbol{\phi}(k)] \tag{11.21}$$

The elements of R are given by

$$\rho_{n,m} = \rho_{jj}(n-m) + \left(\frac{2E_{\mathrm{b}}}{T_{\mathrm{b}}} T_{\mathrm{c}}^2 + \sigma_\eta^2\right) \delta_{n,m} \tag{11.22}$$

for $n,m = -N, \ldots, -1, 1, \ldots, N$. The cross-correlation vector elements are

$$d_n = \rho_{jj}(n) \tag{11.23}$$

for $n = -N, \ldots, -1, 1, \ldots, N$. It is assumed throughout that the interferer samples $j(k)$ are mutually uncorrelated with the signal plus thermal noise, $s(k) + \eta(k)$.

Next, consider the solution $\boldsymbol{\theta}$ when $V'(\boldsymbol{\theta})$ is minimized. Clearly, the covariance matrix R remains unchanged. Let $\mathbf{d}'$ represent the cross-correlation vector obtained using $z(k)$ as the reference signal. Then the elements of $\mathbf{d}'$ are given by

$$\begin{aligned} d'_n &= E[z(k)z(k-n)] \\ &= \rho_{jj}(n) + E[s(k)s(k-n)] + E[\eta(k)\eta(k-n)] \\ &= \rho_{jj}(n) \end{aligned} \tag{11.24}$$

for $n = -N, -N+1, \ldots, -1, 1, \ldots, N$. The last line in (11.24) follows from the fact that both $s(k)$, the spread-spectrum signal samples, and $\eta(k)$, the thermal noise samples, are uncorrelated, and the index n excludes the term $n = 0$. That, is, since $s(k)$ and $\eta(k)$ are unpredictable, the cross-correlation vector $\mathbf{d}'$ equals $\mathbf{d}$, and the same solution for $\boldsymbol{\theta} = \mathbf{g}$ results if $z(k)$ is substituted for $j(k)$. Thus, in constructing an adaptive filter, the current received sample in Figure 11.1 can be used as a reference for updating the rejection filter coefficients.

11.2.2 Performance and analysis of the optimum interference rejection filter

The solution for $\boldsymbol{\theta} = \mathbf{g}$ given by (11.19) provides the optimum choice of rejection filter coefficients, in terms of minimizing total filter output power. From (11.24), it was seen that minimization of output power is exactly equivalent to subtracting the mean square estimate of the jammer from the current received sample $z(k)$. The performance of Wiener filters is typically evaluated by computing the residual mean square error, $V(\boldsymbol{\theta}_{\text{opt}})$, that results for the optimum choice of parameter vector. However, in digital communications, residual mean square error by itself does not sufficiently characterize receiver performance. Rather, the system designer is primarily interested in symbol error rate, or bit error rate (BER), when binary modulation is assumed.

The BER of the spread-spectrum receiver in Figure 11.1 is readily evaluated once the Wiener–Hopf equations (11.19) have been solved. First, it can be shown that the receiver output variable is Gaussian, conditioned on a specific choice of the rejection filter coefficients. For convenience, we define the prewhitening filter coefficients a_n in terms of the g_n as follows:

$$a_n = -g_n$$

for $n = -N, -N+1, \ldots, -1, 1, \ldots, N$, and with $a_0 = 1$. Again, the output of the prewhitening filter represents the received samples with the optimum estimate of the jammer $j(k)$ subtracted. The input to the rejection filter is given by $z(k)$, and thus the input to the correlator/accumulator during the zeroth bit interval, $[0, T_b]$ in Figure 11.1, is

$$\begin{aligned} e(k) &= \sum_{n=-N}^{N} a_n z(k-n) \\ &= d_0 \sqrt{\frac{2E_b}{T_b}}\, T_c \sum_{n=-N}^{N} a_n c_{k-n} + \sum_{n=-N}^{N} a_n [j(k-n) + \eta(k-n)] \end{aligned} \tag{11.25}$$

where $d_0 = \pm 1$ is the binary information symbol. Note that the rejection filter output $e(k)$ depends on the PN sequence c_k and the residual jammer and thermal noise. The correlator output accumulated over the zeroth bit interval is

$$U(T_b) = \sum_{k=0}^{L_{PN}-1} c_k e(k)$$

$$= d_0 \sqrt{\frac{2E_b}{T_b}}\, T_c \sum_{k=0}^{L_{PN}-1} c_k \sum_{n=-N}^{N} a_n c_{k-n}$$

$$+ \sum_{k=0}^{L_{PN}-1} c_k \sum_{n=-N}^{N} a_n[j(k-n) + \eta(k-n)] \tag{11.26}$$

The first term on the right-hand side of $U(T_b)$ is a deterministic quantity, since it depends on the filter coefficients, which are assumed known from the Wiener–Hopf equations, and the PN sequence c_k. The second term is random, since it depends on the jammer and thermal noise. If we assume that the $j(k)$ and $\eta(k)$ are Gaussian random processes, then $U(T_b)$, the receiver decision variable, is just a linear transformation of Gaussian random variables, and hence is itself conditionally Gaussian.

The probability of bit error for binary signalling is just

$$P_e = \tfrac{1}{2}P(U(T_b) > 0 \mid d_0 = -1) + \tfrac{1}{2}P(U(T_b) < 0 \mid d_0 = +1) \tag{11.27}$$

The conditional probabilities forming P_e are error function integrals, with parameters given by the mean and variance of $U(T_b)$. Specifically, we have

$$E[U(T_b) \mid d_0 = +1] = \sqrt{\frac{2E_b}{T_b}}\, T_c \sum_{k=0}^{L_{PN}-1} \sum_{n=-N}^{N} a_n c_k c_{k-n} \tag{11.28}$$

and

$$\mathrm{Var}\{U(T_b)\} = \sum_{k=0}^{L_{PN}-1} \sum_{k'=0}^{L_{PN}-1} c_k c_{k'} \sum_{n=-N}^{N} \sum_{n'=-N}^{N} a_n a_{n'}[\rho_{jj}(k-k'+n-n') + \sigma_\eta^2 \delta_{k-k'+n-n',0}] \tag{11.29}$$

Note that $\rho_{jj}(k)$ is the jammer covariance function at the output of the chip duration integrate-and-dump. Since the error rate is symmetric for $d_0 = \pm 1$, the final BER is given by

$$P_e = \frac{1}{2}\,\mathrm{erfc}\left(\frac{E[U(T_b)]}{\sqrt{2\mathrm{Var}\{U(T_b)\}}}\right) \tag{11.30}$$

Figure 11.2 is an example of BER performance obtained using the optimum rejection filter coefficients [4]. Two types of jammers were considered: the first, a narrowband Gaussian process with a Lorentz spectrum, and the second, a tone jammer with Gaussian distributed inphase and quadrature components. Specifically, the covariance function for the tone jammer is

$$\rho_{jj}(k) = P\cos(\delta\omega k T_c)$$

where $\delta\omega$ is the frequency offset from the carrier f_c, in radians per second, and P is the power of the jammer samples. For the Lorentz spectrum jammer,

$$\rho_{jj}(k) = P\exp(-\kappa\,|kT_c|)\cos(\delta\omega k T_c)$$

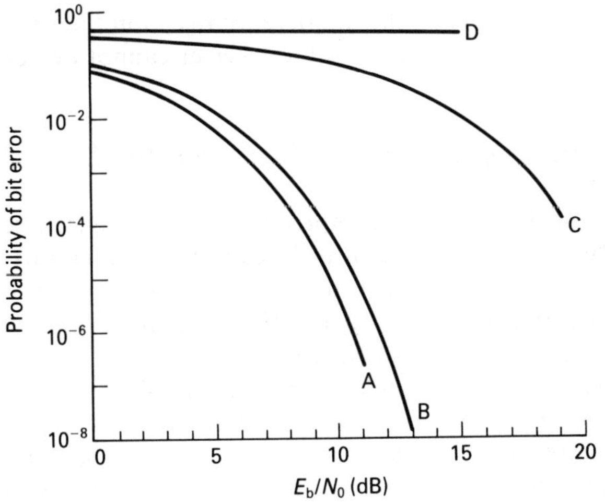

Figure 11.2 Analytic BER performance for the DS receiver, using Wiener filter. Jammer centre frequency $= 2\pi/7T_c$, Lorentz jammer bandwidth $= 2\pi/10T_c$, jammer power $= 32$. Curve A, ideal BPSK BER; curve B, nine-tap suppression filter, tone jammer; curve C, nine-tap suppression filter with Lorentz jammer; curve D, no suppression filter

The parameter κ represents the 3 dB bandwidth of the process $j(k)$ (that is, the point at which the power spectral density is half its peak value). The following signal and jammer parameters were used to obtain the results of Figure 11.2:

$L_{PN} = 7$, the number of chips per bit.

$N = 4$, the number of coefficients on each side of the rejection filter.

$N_0 = 0.125$, the spectral density of thermal noise.

$\delta\omega = 2\pi/7T_c$, the normalized jammer centre frequency.

$J/S = 32$, the jammer to signal power ratio.

$\kappa = 0.05/T_c$, the normalized bandwidth of the Lorentz jammer.

It can be seen that for the tone jammer, the receiver performance is only about 1 dB worse than ideal BPSK at a BER of 10^{-4}. However, for the Lorentz jammer with the same power, the performance is about 10 dB worse than BPSK. This difference in performance is readily explained as follows. For the tone jammer, the rejection filter forms a very narrow notch, and thus little signal energy is lost. However, since the Lorentz jammer occupies 10% of the DS bandwidth, the rejection filter removes an appreciable fraction of the signal energy in whitening the filter output, and thus

BER performance suffers. When the interference rejection filter is not used, it is seen that the BER approaches 1/2, and thus the receiver cannot demodulate the data.

11.2.3 An adaptive DS receiver using the LMS algorithm

The optimum parameter vector $\boldsymbol{\theta}$, comprising the interference rejection filter coefficients, is found by solving the system of Wiener–Hopf equations (11.19). However, the actual covariance matrix R and vector $\mathbf{d}$ are rarely known a priori, as they depend on the jammer covariance function $\rho_{jj}(n)$. However, $\rho_{jj}(n)$ is a function of jammer centre frequency and bandwidth, which represent parameters unknown to the receiver in general. An adaptive receiver could estimate R and $\mathbf{d}$ using the sample covariance matrix and vector of (2.22), and then solve the system $R\boldsymbol{\theta} = \mathbf{d}$. However, the computational complexity of such an approach may be unacceptable, especially at the wide bandwidths (tens of megahertz) often used in spread-spectrum systems. Thus, recursive algorithms are preferred for estimating the interference rejection filter coefficients, in order to minimize both storage and computational requirements. In direct-sequence applications, the LMS algorithm has been considered most frequently for interference rejection [1,8], and has actually been implemented in charge-coupled devices, as, for example, in [9]. The LMS algorithm for interference rejection is now discussed in more detail.

The LMS algorithm recursively minimizes the error

$$V(\boldsymbol{\theta}) = E[e^2(k \mid \boldsymbol{\theta})] \tag{11.31}$$

$$V(\boldsymbol{\theta}) = E[\,|\,y(k) - \boldsymbol{\theta}^{\mathrm{T}}\boldsymbol{\phi}(k)\,|^2] \tag{11.32}$$

where it is recalled that the parameter vector, $\boldsymbol{\theta}$, is given by

$$\boldsymbol{\theta}^{\mathrm{T}} = [g_{-N}, g_{-N+1}, \ldots, g_{-1}, g_1, \ldots, g_N]$$

and the regression vector is

$$\boldsymbol{\phi}(k)^{\mathrm{T}} = [z(k-N), z(k-N+1), \ldots, z(k-1), z(k+1), \ldots, z(k+N)] \tag{11.33}$$

The LMS algorithm is given by (2.72)

$$\boldsymbol{\theta}(k) = \boldsymbol{\theta}(k-1) + \mu e(k \mid \boldsymbol{\theta}(k-1))\boldsymbol{\phi}(k) \tag{11.34}$$

where $\boldsymbol{\theta}(k)$ represents the updated estimate of the parameter vector. Substitution of the parameter and regression vectors for the interference rejection problem yields, for each rejection coefficient g_n,

$$g_n(k) = g_n(k-1) + \mu\left(z(k) - \sum_{l=-N,l\neq 0}^{N} g_l(k-1)z(k-l)\right)z(k-n) \tag{11.35}$$

for $n = -N, -N+1, \ldots, -1, 1, \ldots, N$. Equation (11.35) provides the updated estimate of the rejection filter coefficients at the kth sample, given the data $z(k)$. In vector form, the algorithm can be written as

$$\mathbf{g}(k) = \mathbf{g}(k-1) + \mu e(k)\mathbf{z}(k) \tag{11.36}$$

where

$$e(k) = \left(z(k) - \sum_{l=-N, l \neq 0}^{N} g_l(k-1)z(k-l) \right)$$

and $\mathbf{z}(k) = \boldsymbol{\phi}(k)$ is the regression, or data, vector.

The LMS algorithm is particularly attractive for spread-spectrum applications due to its simplicity. At spread-spectrum bandwidths, it is often difficult to implement an adaptive filter, except in analog circuitry, and thus a fixed step size μ is preferred. The penalty for using a fixed step-size parameter is that the estimates g_n of the rejection filter coefficients are 'noisy', and cannot converge to the Wiener solution of (11.19). That is, let $\mathbf{g}_{\text{opt}}$ represent the vector of optimum coefficients found by solving (11.19), and let $\mathbf{g}(k)$ represent the time-varying estimate computed by the LMS algorithm. Then the error vector $\mathbf{v}$ is defined by

$$\mathbf{v}(k) = \mathbf{g}(k) - \mathbf{g}_{\text{opt}}$$

Then, clearly, $\mathbf{v}(k)$ is a random vector process. In Figure 11.3, the LMS adaptive filter has been decomposed into the Wiener filter, with coefficient vector $\mathbf{g}_{\text{opt}}$, and a 'misadjustment filter' (this decomposition was first considered in [10]). The misadjustment filter output, $\tilde{\mathbf{e}}(k)$, is given by

$$\tilde{e}(k) = \mathbf{v}(k)^{\mathrm{T}}\mathbf{z}(k) \tag{11.37}$$

The LMS filter output is seen to be the sum $e(k) + \tilde{e}(k)$, and thus differs from the optimum Wiener filter error process. Clearly, the additional noise represented by $\tilde{e}(k)$ will degrade the BER performance. In general, the statistics of the misadjustment error and error vector $\mathbf{v}(k)$ cannot be determined exactly, since $\mathbf{v}(k)$ depends on a non-linear transformation of Gaussian random vectors. However, in [8], it is shown that under the assumptions of slow adaptation and Gaussian interference, the misadjustment noise $\tilde{e}(k)$ can be approximated as Gaussian, and, thus, the BER analysis of section 11.2.2 can be applied. We now examine the statistics of the misadjustment filter in more detail to justify this Gaussian approximation.

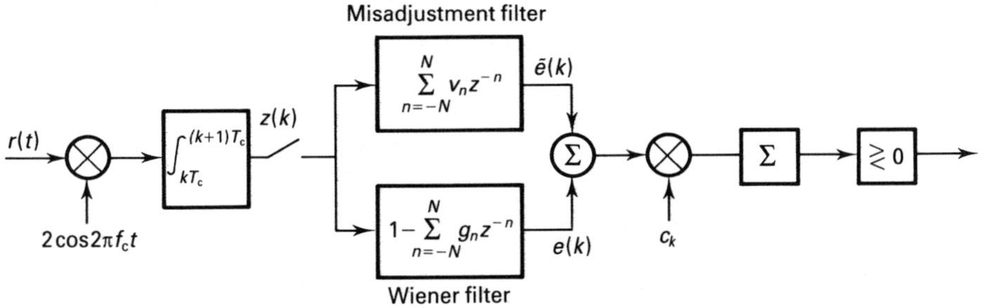

Figure 11.3 Adaptive DS receiver showing partitioning of Wiener and misadjustment filters

11.2.4 Statistics of the misadjustment filter

It is readily shown that the estimate $\mathbf{g}(k)$ of the Wiener filter coefficient vector $\mathbf{g}_{\text{opt}}$ is unbiased [11]. That is,

$$E[\mathbf{g}(k)] = \mathbf{g}_{\text{opt}}$$
$$\Rightarrow E[\mathbf{v}(k)] = 0 \tag{11.38}$$

and the misadjustment filter coefficients are thus zero-mean random processes. In order to evaluate the covariance matrix of $\mathbf{v}(k)$, and later justify the Gaussian approximation, it is convenient to perform an eigendecomposition of the data covariance matrix as follows [11]:

$$E[\mathbf{z}(k)\mathbf{z}(k)^{\mathrm{T}}] = \rho_{zz} = Q\Lambda Q^{\mathrm{T}} \tag{11.39}$$

It is implicitly assumed that $z(k)$ is a wide-sense stationary random process in (11.39). The columns of the matrix $Q \in \mathscr{R}^{2N,2N}$ are the eigenvectors of ρ_{zz}, and $\Lambda \in \mathscr{R}^{2N,27}$ is the diagonal matrix of eigenvalues. The matrices Q and Q^{T} are orthonormal, such that $Q^{\mathrm{T}}Q = I$. The following transformations [8] are now used:

$$\mathbf{z} = Q\mathbf{z}' \qquad \mathbf{g} = Q\mathbf{g}' \qquad \mathbf{v} = Q\mathbf{v}' \tag{11.40}$$

An equation for the propagation of $\mathbf{v}'(k)$ is obtained by premultiplying (11.36) by Q^{T} and subtracting $\mathbf{g}'_{\text{opt}}$ from both sides. Thus, in the primed coordinate system, the misadjustment vector propagates according to

$$\mathbf{v}'(k) = \mathbf{v}'(k-1) + \mu e(k)\mathbf{z}'(k) \tag{11.41}$$

In order to obtain the statistics of $\tilde{e}(k)$, it is first necessary to find the steady-state tap weight covariance matrix of $\mathbf{v}'(k)$, denoted K_v. Equation (11.41) is multiplied by its transpose and, after taking expectations, becomes in steady state [8]

$$E[\mathbf{v}'\mathbf{v}'^{\mathrm{T}}] = E[(I - \mu\mathbf{z}'\mathbf{z}'^{\mathrm{T}})\mathbf{v}'\mathbf{v}'^{\mathrm{T}}(I - \mu\mathbf{z}'\mathbf{z}'^{\mathrm{T}})^{\mathrm{T}}] + \mu^2 E[e_{\text{opt}}^2\mathbf{z}'\mathbf{z}'^{\mathrm{T}}] \tag{11.42}$$

where e_{opt} represents the error in the steady-state Wiener filter. The frequent assumption that the process $\mathbf{v}(k)$ is independent of $\mathbf{z}(k)$ is now used (valid for slow adaptation). Letting $\gamma_{n,m}$ represent the n,m element of K_v, we obtain the following expression for the elements of the steady-state covariance matrix of $\mathbf{v}'$ by direct evaluation of (11.42) [8]:

$$\begin{aligned}\gamma_{n,m} = \gamma_{n,m} - \mu(\lambda_n + \lambda_m)\gamma_{n,m} + \mu^2 \sum_l \sum_k \gamma_{l,k} E[z'_n z'_m z'_l z'_k] \\ + \mu^2 E[e_{\text{opt}}^2 z'_n z'_m]\end{aligned} \tag{11.43}$$

for $n, m = 1,2,\ldots, 2N$. The terms λ_n are the eigenvalues of Λ, the data covariance matrix, and z'_n is the nth element of the transformed data vector $Q^{\mathrm{T}}\mathbf{z}$.

When the data vector is zero-mean Gaussian, it is readily shown [12] that K_v is a diagonal matrix, with elements given by

$$\gamma_{i,i} = \frac{\mu(\psi_{\min} + \Sigma_n \gamma_{n,n}\lambda_n)}{2(1 - \mu\lambda_i)} \tag{11.44}$$

The minimum mean square error, resulting when $\mathbf{g} = \mathbf{g}_{\text{opt}}$, is denoted by $\psi_{\min}$ and is given by

$$\psi_{\min} = \rho_{jj}(0) + \frac{2E_{\text{b}}}{T_{\text{b}}} T_{\text{c}}^2 + \sigma_\eta^2 - \sum_{l=-N, l \neq 0}^{N} g_{\text{opt},l}\rho_{jj}(-l) \tag{11.45}$$

where $g_{\text{opt},l}$ is the lth component of $\mathbf{g}_{\text{opt}}$. For the Gaussian input case, it is possible to find a closed-form solution for the excess mean square error, ψ_{excess}, defined by

$$\psi_{\text{excess}} = E[\tilde{e}^2] = E\left[\left|\sum_{n=1}^{2N} v_n(k)z_n(k)\right|^2\right] \tag{11.46}$$

where $z_n(k)$ denotes the nth component of the data vector $\mathbf{z}(k)$. From [12] and [8], we obtain

$$\psi_{\text{excess}} = \mu/2 \frac{\psi_{\min}[\Sigma_n \lambda_n/(1 - \mu\lambda_n)]}{[1 - \mu/2\Sigma_n \lambda_n/(1 - \mu\lambda_n)]} \tag{11.47}$$

In the DS receiver, recall that $z(k)$ is the sum of a binary sequence (representing the PN chips c_k) and an additive, zero-mean Gaussian noise $\eta_k = j(k) + \eta(k)$. Thus, we can write

$$z(k) = \sqrt{P_{\text{s}}}c_k + \eta_k \tag{11.48}$$

where P_{s} is the signal power. It can be readily shown that the first and second moments of $z(k)$ are identical for both Gaussian c_k and binary-valued c_k. However, note that for a general binary sequence c_k,

$$E[c_k^2] = E[c_k^4] = 1$$

whereas, if c_k were Gaussian, then $E[c_k^4] = 3$. In general, the fourth moment of the binary component c_k can be expressed as the Gaussian fourth moment minus a correction term – this result was employed in [8] to obtain a system of linear equations for the steady-state covariance matrix elements $\gamma_{n,m}$. When the data vector $\mathbf{z}$ consists of a binary plus Gaussian vector, it is shown in [8] that K_v is no longer diagonal. The final set of equations for $\gamma_{n,m}$ is

$$\begin{aligned}\gamma_{n,m} &= \frac{\mu}{(\lambda_n + \lambda_m)}[2\gamma_{n,m}\lambda_n\lambda_m] \\ &\quad - \frac{2\mu P_{\text{s}}^2}{(\lambda_n + \lambda_m)} \sum_l \sum_k \gamma_{l,k} \sum_i Q_{i,l}Q_{i,k}Q_{i,n}Q_{i,m} \\ &\quad - \frac{2\mu P_{\text{s}}^2}{(\lambda_n + \lambda_m)} \sum_i g_i^2 Q_{i,n}Q_{i,m}\end{aligned} \tag{11.49}$$

for $n, m = 1, 2, \ldots, 2N$ and $n \neq m$, and where $Q_{n,m}$ denotes the n,mth element of the orthogonal matrix Q. The following equation holds for the diagonal elements of K_v:

$$\gamma_{n,n} = \frac{1}{(1 - \mu\lambda_n)} \left(\frac{\mu}{2}\psi_{\text{min}} + \frac{\mu}{2}\psi_{\text{excess}} - \frac{\mu P_s^2}{\lambda_n} \sum_l \sum_k \gamma_{l,k} \sum_i Q_{i,l}Q_{i,k}Q_{i,n}^2 - \frac{\mu P_s^2}{\lambda_n} \sum_i g_i^2 Q_{i,n}^2 \right) \tag{11.50}$$

where ψ_{excess} is given by (11.46).

Thus far, we have determined the steady-state covariance matrix of K_v in the DS adaptive receiver. It still must be shown that the misadjustment filter output, $\tilde{e}(k)$, is asymptotically Gaussian. Two central limit theorem arguments are used as follows. First, by iterating (11.41), the following expression for $\mathbf{v}'(k)$ is obtained in terms of the data vectors $\mathbf{z}(k)$ and initial conditions:

$$\begin{aligned}\mathbf{v}'(k) &= \prod_{n=0}^{k-1} [I - \mu\mathbf{z}'(n)\mathbf{z}'(n)^{\mathrm{T}}]\mathbf{v}'(0) \\ &+ \mu \sum_{n=1}^{k-1} \left[\left(\prod_{l=1}^{n} [I - \mu\mathbf{z}'(k-l)\mathbf{z}'(k-l)^{\mathrm{T}}] \right) e(k-l-1)\mathbf{z}'(k-l-1) \right] \\ &+ \mu e(k-1)\mathbf{z}(k-1)\end{aligned} \tag{11.51}$$

For sufficiently small μ, it is argued in [8], using the central limit theorem, that $\mathbf{v}'(k)$ is asymptotically a sequence of i.i.d. random vectors, and hence Gaussian in steady state. That is, $\mathbf{v}'_\infty$ is approximately

$$\mathbf{v}'_\infty \approx \lim_{k\to\infty,\,\mu\to 0} \sum_{n=0}^{k} e(n)\mathbf{z}'(n) \tag{11.52}$$

Given that $\mathbf{v}'$ is asymptotically a Gaussian random vector, it still remains to show that the misadjustment filter output in steady state, $\mathbf{v}^{\mathrm{T}}\mathbf{z}$, is itself Gaussian. Using the Lyapunov version of the central limit theorem, it is argued in [8] that for a sufficiently large number of filter coefficients N, and for sufficiently bounded eigenvalues λ_n, that

$$\tilde{e} = \sum_{n=1}^{2N} v'_n(k) z'_n(k)$$

is an asymptotically Gaussian random variable. Note that $z'_n(k)$ denotes the nth component of the transformed data vector $\mathbf{z}'(k)$. For Gaussian inputs this is readily shown, since K_v is diagonal, and hence the v'_n are uncorrelated. Furthermore, if asymptotically Gaussian, the v'_n are therefore independent. The covariance matrix of the transformed vector $\mathbf{z}'(k)$ is also diagonal, and equal to Λ; thus for Gaussian $\mathbf{z}'$, the components z'_n are likewise independent. Thus $\tilde{e}$ is approximately a sum of i.i.d. random variables in steady state.

It is now possible to develop an approximate expression for the BER of the adaptive DS receiver that includes the effects of misadjustment noise. From Figure

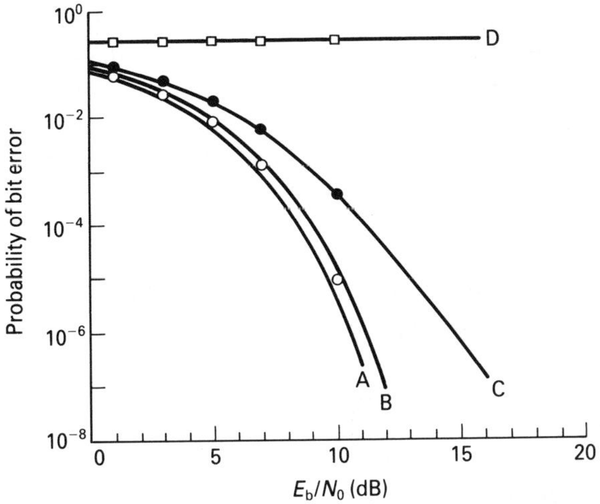

Figure 11.4 BER results for adaptive DS receiver, $L_{PN} = 7$, fading tone jammer, $\delta\omega = 2\pi/7T_c$, jammer power = 32. Curve A, ideal BPSK; curve B, nine-tap Wiener filter; curve C, Gaussian approximation for adaptive receiver BER, nine-tap filter; curve D, no interference rejection filter. Symbols represent simulated BER results

11.3, it is evident that the expected value of the decision variable, $E[U(T_b)]$, is identical to (11.28), since the misadjustment filter output is approximately a zero-mean Gaussian random process in steady state. The only change to the BER expression in (11.27) is an additional term to the variance, dependent on $\gamma_{n,m}$. The resulting expression for the variance used in [8] is

$$\begin{aligned}\mathrm{Var}\{U(T_b)\} = &\sum_{k=0}^{L_{PN}-1}\sum_{k'=0}^{L_{PN}-1} c_k c_{k'} \sum_{n=-N}^{N}\sum_{n'=-N}^{N} a_n a_{n'} \,[\rho_{jj}(k-k'+n-n')\\ &+ \sigma_\eta^2 \delta_{k-k'+n-n',0}] + P_s \sum_{k=0}^{L_{PN}-1}\sum_{l=0}^{L_{PN}-1} c_k c_l \sum_{n=1}^{2N}\sum_{m=1}^{2N} c_{k-n} c_{l-m} \gamma_{n,m}\\ &+ \sum_{k=0}^{L_{PN}-1}\sum_{l=0}^{L_{PN}-1} c_k c_l \sum_{n=1}^{2N}\sum_{m=1}^{2N} \gamma_{n,m}\,[\rho_{jj}(k-l+m-n) + \sigma_\eta^2 \delta_{k-l+m-n,0}]\end{aligned} \tag{11.53}$$

where the $a_n = -g_n$ correspond to the optimum prewhitening filter coefficients.

The performance of the adaptive receiver can now be evaluated in terms of BER. To test the validity of the Gaussian approximation for the misadjustment noise, the BER is obtained by both simulations of the receiver and evaluation of (11.27), (11.28) and (11.53). Figure 11.4 illustrates BER performance with the fading tone jammer of Figure 11.2. Again, nine total taps ($N = 4$) were used in the simulation and analysis. It can be seen that the Gaussian BER approximation closely matches the simulation results down to a BER of about 5×10^{-4}. Thus, by approximating the misadjustment noise by a Gaussian random process, we have shown that the degradation in BER due to the adaptive filter can be accurately predicted.

11.2.5 Interference rejection in frequency-hopped spread-spectrum systems

As discussed earlier, when sampled at an appropriate rate, a frequency-hopped waveform is approximately statistically uncorrelated. Thus, as in the DS system, the FFH waveform is mean square unpredictable, and hence narrowband interferers can be estimated and rejected independently of the desired signal. In the following discussion, fast frequency hopping is assumed, in which a sequence of L hops comprise each bit of information, and a total of $2N_{\mathrm{h}} > 2L$ total frequencies are available to the pseudorandom sequence generator. The two binary signals are commonly referred to as MARK and SPACE. Thus, to transmit a MARK, a pseudorandom sequence of L sinusoids with frequencies $f_{l,\mathrm{M}}$ is transmitted. To represent a SPACE, a sequence of L sinusoids with frequencies $f_{l,\mathrm{S}}$, is chosen which is orthogonal to the MARK sequence.

The mathematical representation of the low-pass equivalent complex-valued waveform is given, in the case of a MARK, by

$$s(t) = \sqrt{\frac{2E_{\mathrm{b}}}{LT_{\mathrm{h}}}} \sum_{l=0}^{L-1} \exp[j(2\pi f_{l,\mathrm{M}}t + \theta_l)]\; P_{T_{\mathrm{h}}}(t - lT_{\mathrm{h}}) \tag{11.54}$$

where $f_{l,\mathrm{M}}$ is the lth hopping frequency, relative to the start of the hopping band, when a MARK is transmitted, θ_l is the random phase on the lth hop, E_{b} is the energy per bit, T_{h} is the duration of each hop, and L is the number of hops per bit.

Again, it can be shown [6] that $s(t)$ is statistically uncorrelated at lag values of $T_{\mathrm{h}}/4N_{\mathrm{h}}$. The received low-pass equivalent signal is then given by

$$r(t) = s(t) + \sum_{k=1}^{N_{\mathrm{J}}} j_k(t) + \eta(t) \tag{11.55}$$

where $j_k(t)$ represents the kth narrowband interferer, and $\eta(t)$ is wideband thermal noise. The correlation function of the kth interferer is assumed to be

$$\rho_{J_k}(\tau) = 2J_k \frac{\sin \omega_{J_k}\tau}{\omega_{J_k}\tau} \exp(j2\pi\delta f_k\tau) \tag{11.56}$$

where $\omega_{J_k}/2\pi$ is the bandwidth and J_k represents the jammer power. This model for the jammer statistics corresponds to a wide-sense stationary, circular Gaussian process, with an ideal bandpass spectrum, of width $\omega_{J_k}/2\pi$ Hz and centred at

δf_K Hz. It is assumed that each δf_k coincides with one of the hopping frequencies, $f_{l,\mathrm{M}}$ or $f_{l,\mathrm{S}}$. Note that N_J is the total number of jammers, and thus $N_\mathrm{J} = 2N_\mathrm{h}$ corresponds to full-band jamming.

The optimum Wiener filter is designed in an almost identical manner to that for the DS receiver, with the exception that the coefficient vector **g** is now complex valued. Specifically, the regression vector is given by the signal samples $r(kT_\mathrm{s})$, where $T_\mathrm{s} = T_\mathrm{h}/4N_\mathrm{h}$. Thus

$$z(k) = r(kT_\mathrm{s}) \tag{11.57}$$

$$\boldsymbol{\phi}(k)^\mathrm{T} = [r((k-N)T_\mathrm{s}), \ldots, r((k-1)T_\mathrm{s}), r((k+1)T_\mathrm{s}), \ldots, r((k+N)T_\mathrm{s})]$$

As in the DS receiver development, the parameter vector consists of the interference rejection filter coefficients, g_n:

$$\theta^\mathrm{T} = \mathbf{g}^\mathrm{T} = [g_{-N}, g_{-N+1}, \ldots, g_{-1}, g_1, \ldots, g_N] \tag{11.58}$$

It should be emphasized that both $\boldsymbol{\phi}(k)$ and $\boldsymbol{\theta}$ are now complex vectors. Since the FFH signal is unpredictable, the current signal sample, corresponding to $z(k) = r(kT_\mathrm{s})$, can be used as the reference sample. Thus, the Wiener filter minimizes the following complex mean square error:

$$V(\boldsymbol{\theta}) = E\left[\left| r(kT_\mathrm{s}) - \sum_{n=-N, n\neq 0}^{N} g_n r((k-n)T_\mathrm{s}) \right|^2\right] \tag{11.59}$$

where $r(kT_\mathrm{s})$ has been substituted for $z(k)$ in (11.18). The Wiener solution for $\boldsymbol{\theta} = \mathbf{g}$ that minimizes $V(\boldsymbol{\theta})$ is again given by

$$R\boldsymbol{\theta} = \mathbf{d} \tag{11.60}$$

However, R is now a complex covariance matrix given by

$$R = E[\boldsymbol{\phi}(t)\boldsymbol{\phi}^\mathrm{H}(t)] \tag{11.61}$$

where H indicates complex conjugate (Hermitian) transpose. Similarly, the cross-covariance vector is given by

$$\mathbf{d} = E[r(kT_\mathrm{s})\boldsymbol{\phi}^*(k)] \tag{11.62}$$

The specific form of R and **d** for the FFH received signal $r(t)$ is given as follows [6]:

$$\rho_{n,m} = \rho_{jj}((n-m)T_\mathrm{h}/4N_\mathrm{h}) + 2N_0 B\delta_{n,m} + \frac{2E_\mathrm{b}}{LT_\mathrm{h}}\delta_{n,m} \tag{11.63}$$

where $\rho_{jj}(k)$ denotes the jammer covariance function and N_0 is the thermal noise spectral density. The total receiver bandwidth is denoted by B and equals $4N_\mathrm{h}/T_\mathrm{h}$. Recall that both the desired FFH signal and the thermal noise are statistically uncorrelated, and thus their correlation functions are given by the Kronecker delta terms. A typical frequency response for the optimum Wiener filter is shown in Figure 11.5. In this case, there are $2N_\mathrm{h} = 30$ total hopping frequencies available, and the interferer occupies the first two frequency slots. It is evident that since the FFH signal

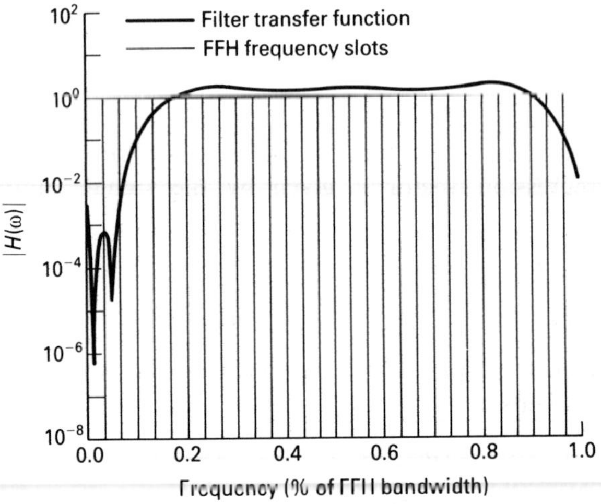

Figure 11.5 Interference rejection filter frequency response for the FFH receiver. $N_h = 15$ hops, two frequency slots jammed, $2N + 1 = 17$ filter coefficients

is wideband, the loss of signal energy due to the interference rejection filter notch is not very great.

Using the above definitions of the regression and parameter vectors, the complex LMS algorithm [13] can be readily developed for the FFH receiver, from (11.35). The error signal corresponding to $e(k \mid \boldsymbol{\theta})$ in (2.72) is given by

$$e(k \mid \boldsymbol{\theta}(k-1)) = \left(r(kT_s) - \sum_{n=-N, n \neq 0}^{N} g_n(k-1) r((k-n)T_s) \right) \tag{11.64}$$

Substitution of g_n for θ_n in (2.72), and taking into account that the regression and parameter vectors are complex valued, yields the following form of the complex LMS algorithm:

$$g_n(k) = g_n(k-1) + \mu \left(r(kT_s) - \sum_{l=-N, l \neq 0}^{N} g_l(k-1) r((k-l)T_s) \right) r^*((k-n)T_s) \tag{11.65}$$

for $n = -N, -N+1, \ldots, -1, 1, \ldots, N$. By employing (11.65), an adaptive FFH receiver is obtained which can reject interference with an a priori unknown covariance function.

11.2.6 FFH receiver BER analysis

The overall FFH receiver, together with the complex interference rejection filter, is shown in Figure 11.6. In each T_h second interval, the local oscillator multiplies the

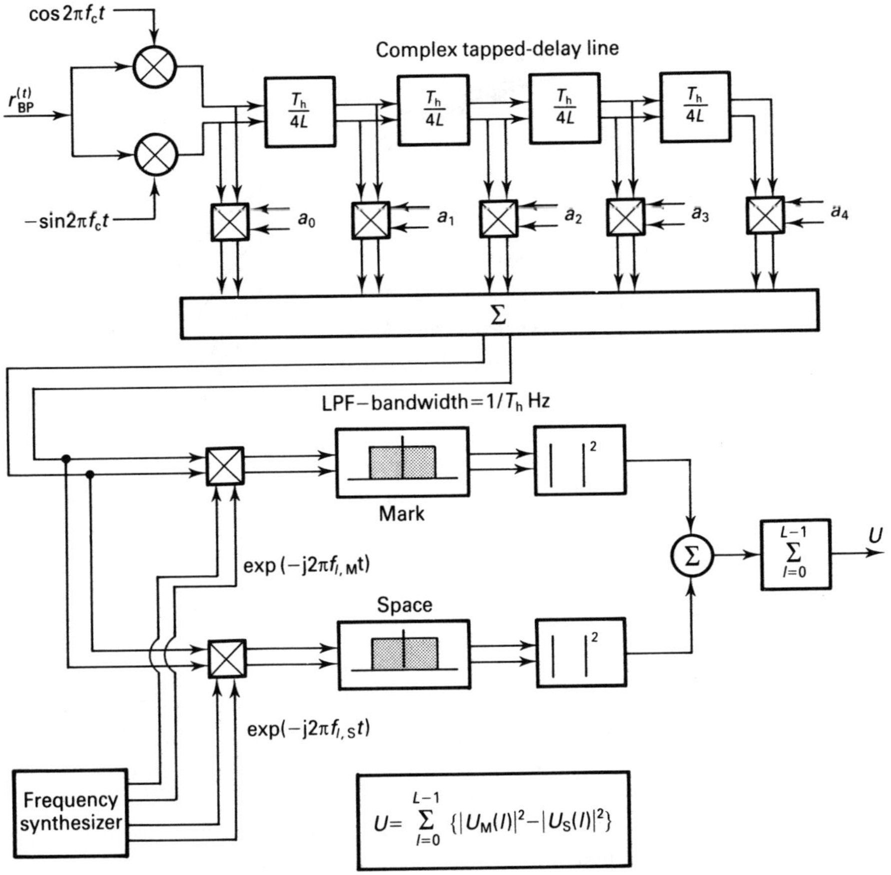

Figure 11.6 FFH receiver with interference rejection filter

incoming waveform $r(t)$ by two complex sinusoids, $\exp(-j2\pi f_{l,M}t)$ in the upper branch, and $\exp(-j2\pi f_{l,S}t)$ in the lower branch. Note that $f_{l,M}$ represents the frequency on the lth hop in the MARK slot. The combination of local oscillator and low-pass filter in Figure 11.6 approximates a matched filter for the FFH signal. Thus, if a MARK is transmitted, the upper filter output in the FFH receiver will be proportional to $\sqrt{E_b}$, and the lower filter output will be zero in the absence of noise. Let $U_M(l)$ represent the output of the upper (MARK) matched filter, sampled during the lth hop, and similarly let $U_S(l)$ represent the space filter output. The receiver output in Figure 11.6 is then

$$U = \frac{1}{2} \sum_{l=0}^{L-1} (|U_M(l)|^2 - |U_S(l)|^2) \tag{11.66}$$

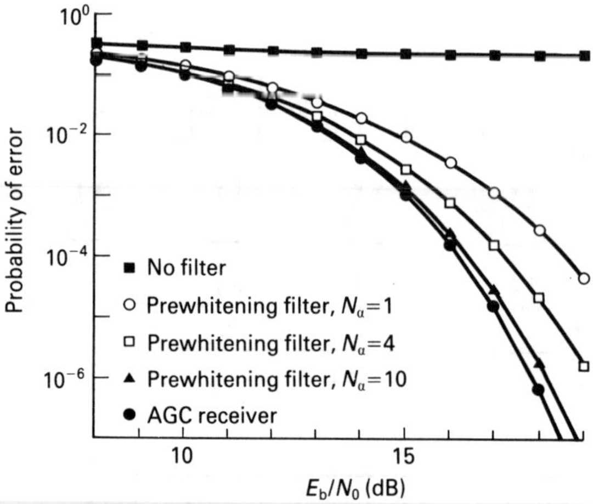

Figure 11.7 Comparison of the AGC receiver with Wiener-filter-based FFH receiver. $L = 7$ hops per bit, $N_h = 15$ hops, $E_b/_{\eta_J} = -10$dB

If a MARK is transmitted, and the noise is negligible, then, clearly, U should be positive; likewise, if a SPACE is transmitted, then U should be negative. In general, U is a sum of chi-squared random variables when the thermal noise and interferer are assumed to be Gaussian. However, the probability density function of U, which determines the BER, cannot be written in closed form. Thus, unlike the DS receiver case, the BER for the FFH system must be evaluated using numerical contour integration [6] of the characteristic function of U.

The sensitivity to interference of the FFH receiver is evident from the form of the decision variable U. If just one frequency slot, l, is jammed, then $U_M(l)$ ($U_S(l)$ can take on a large positive or negative value, and a decision error can easily occur, since U may undergo an undesired sign change. An obvious way to reject the interference is to eliminate, or at least de-emphasize, terms $U(l)$ in (11.66) which are subject to jamming. For example, assume that the actual noise plus interferer power in each hop was known a priori. The following so-called AGC receiver [14] weighs each $U(l)$ by the inverse of the noise power, and thus de-emphasizes jammed hops:

$$U = \frac{1}{2} \sum_{l=0}^{L-1} \left(\frac{|U_M(l)|^2}{\mathrm{Var}\{U_M(l)\}} - \frac{|U_S(l)|^2}{\mathrm{Var}\{U_S(l)\}} \right) \tag{11.67}$$

Note that $\mathrm{Var}\{U(l)\}$ denotes the variance, or noise power, of the matched filter sampled at the lth hop. In practice, the jammer centre frequency and power is seldom known, and thus $\mathrm{Var}\{U(l)\}$ is likewise unknown. Thus, in [15], an approximation to the AGC receiver, known as the self-normalizing receiver, was proposed which replaces the average noise power in each hop by its instantaneous value. In this receiver, U is given by

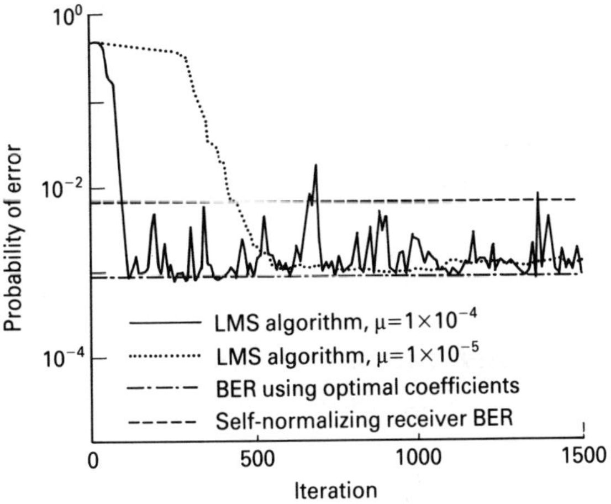

Figure 11.8 BER versus adaptations of the complex LMS algorithm-based receiver. $E_b/N_o = 16$ dB, $L = 7$ hops per bit, $E_b/_{\eta_I} = 0$ dB. Two frequency slots jammed

$$U = \frac{1}{2}\sum_{l=0}^{L-1}\left(\frac{|U_M(l)|^2}{|U_M(l)|^2 + |U_S(l)|^2} - \frac{|U_S(l)|^2}{|U_M(l)|^2 + |U_S(l)|^2}\right) \tag{11.68}$$

Thus, hops which are jammed are likely to have a large instantaneous power $|U(l)|^2$, and therefore these hops are de-emphasized. We will compare the performance of both the AGC and self-normalizing techniques with the Wiener-filter-based and adaptive LMS FFH receivers.

The results of the numerical BER analysis are shown in Figure 11.7, when the filter coefficients g_n are fixed at the Wiener solution given by (11.60). Here, $N_\alpha = N$ is the number of coefficients on each side of the centre tap, so there are a total of $2N_\alpha + 1$ coefficients g_n. The interferer has the ideal bandpass spectrum corresponding to the rejection filter response shown in Figure 11.5, with the first two frequency slots jammed. The interferer power spectral density equals η_j in the jammed hops. With a total of 21 taps, it can be seen that the Wiener-filter-based receiver performs nearly as well as the AGC receiver. (We can view the AGC receiver as providing near optimum performance, since it assumes exact knowledge of the interferer power in each hop.) Performance degrades as the number of filter coefficients, and hence sharpness of the rejection notch, is reduced. Without the interference rejection filter, the FFH receiver of Figure 11.6 is useless, with BER of one-half.

The LMS-based FFH receiver, like the self-normalizing receiver, does not require a priori knowledge of the jammer centre frequencies and bandwidths. In Figure 11.8, we have therefore compared the performance of these two interference rejection methods. At each iteration of the LMS algorithm, the rejection filter is essentially deterministic, and thus an analytic BER can be obtained by numerical integration. Thus, as the LMS algorithm adapts, the improvement in BER is plotted, forming a type of learning curve. As shown in Figure 11.8, a small value of μ provides relatively slow convergence. However, the misadjustment error is correspondingly small, and the BER performance approaches that of the receiver using a fixed Wiener filter. By increasing μ, convergence is speeded up, at the expense of increased misadjustment, and hence poorer BER performance. It is evident that the LMS FFH receiver provides superior performance to that of the self-normalizing method, as long as the gain μ is sufficiently small.

11.3 Joint channel estimation and interference rejection using the RLS algorithm

11.3.1 Multipath channel and interferer model

Multipath interference occurs in many spread-spectrum applications, including high-frequency (HF) troposcatter communications [16], mobile digital radio systems [17], and local-area wireless networks. In a multipath channel, the received signal can be described by a sum of delayed, amplitude-scaled, and phase-shifted versions of the transmitted waveform. These delayed replicas are generated by ionospheric scattering in the case of HF communications, and reflections from buildings and terrain in mobile radio, for example. It is well known that a wide variety of such multipath channels can be represented by a tapped delay line, or, equivalently, a finite impulse response filter, in which the coefficients are time varying. The low-pass equivalent multipath channel impulse response used here is given by ([18], chapter 7),

$$h(\tau, t) = \sum_{l=0}^{N_f-1} f_l(t)\delta(\tau - lT_s) \tag{11.69}$$

where the $f_l(t)$ are complex random processes representing the amplitude and phase shifts of the paths, and T_s is the Nyquist sampling interval for the transmitted signal. The function $\delta(t)$ is the Dirac delta function. For the direct-sequence spread-spectrum signal, $T_s = T_c/2$, or half the chip interval.

The following assumptions are made with regard to the multipath channel and spread-spectrum signal characteristics:

1. The multipath spread, equal to $T_m = N_f T_s$ in (11.69) is much shorter than the bit duration, T_b, but greater than the chip duration, T_c ($T_c < T_m \ll T_b$). Thus, intersymbol interference is negligible, even though the multipath spans multiple chips.
2. The channel Doppler spread, denoted f_D (or bandwidth of the individual processes $f_l(t)$) is much smaller than N/T_b, a multiple of the information rate ($f_D \ll N/T_b$). Thus, a sufficiently long observation interval, spanning N bits, is available in which to estimate the f_l, which are effectively constant over NT_b seconds.

In addition to multipath spread, spread-spectrum systems may simultaneously encounter narrowband interference, since in both commercial and military systems the spread-spectrum band must often be 'overlaid' on to existing narrowband users. The received low-pass equivalent DS signal in this case can be written as

$$r(t) = \sqrt{\frac{2E_b}{T_b}} \sum_{n=-\infty}^{\infty} d_n \sum_{l=0}^{N_f-1} f_l PN_{lp}(t - lT_s - nT_b) + j(t) + \eta(t) \tag{11.70}$$

Thus, the spread-spectrum waveform is distorted both by the channel with coefficients f_l and the additive narrowband interference $j(t)$.

Note that the received signal is band limited to $1/T_c$ Hz, the approximate DS signal bandwidth, and the low-pass filtered pseudonoise signal, $PN_{lp}(t)$, is therefore given by

$$PN_{lp}(t) = \left(\sum_{k=0}^{L_{PN}-1} c_k PT_c(t - kT_c) \right) * h_{lp}(t) \tag{11.71}$$

where

$$h_{lp}(t) = \frac{\sin\ 2\pi t/T_c}{\pi t}$$

Performing the above convolution yields the following representation for the low-pass filtered DS signal:

$$PN_{lp}(t) = \sum_{k=0}^{L_{PN}-1} c_k \frac{1}{\pi} \left[\mathrm{Si}\left(2\pi \frac{(t - kT_c)}{T_c}\right) - \mathrm{Si}\left(2\pi \frac{(t-(k+1)T_c)}{T_c}\right) \right] \tag{11.72}$$

where

$$\mathrm{Si}(x) \equiv \int_0^x \frac{\sin\ y}{y}\, \mathrm{d}y$$

In order to demodulate the information sequence d_n, an algorithm is required that can jointly estimate the channel parameters f_l and the narrowband interference. The problem of multipath estimation and optimum demodulation was first solved in [19], which introduced the RAKE receiver. The RAKE concept can be explained as follows. Let $S(f)$ represent the Fourier transform of the desired signal. Then the optimum demodulator for a white Gaussian noise channel is just a matched filter, with frequency response $S^*(f)\exp(-j2\pi fT)$, where T is the signal duration. When the multipath is present, the transmitted signal Fourier transform is multiplied by the channel frequency response, denoted $H_c(f)$. Thus, the optimum receiver should be matched to $S(f)H_c(f)$, instead of just $S(f)$, and has a transfer function given by $S^*(f)H_c^*(f)\exp(-j2\pi fT)$. This matched filter can be implemented using a tapped delay line with coefficients f_l^* when the channel impulse response is given by (11.69).

When narrowband interference is present, with a known power spectral density (PSD), the RAKE receiver is readily modified by inserting a prewhitening filter. Thus,

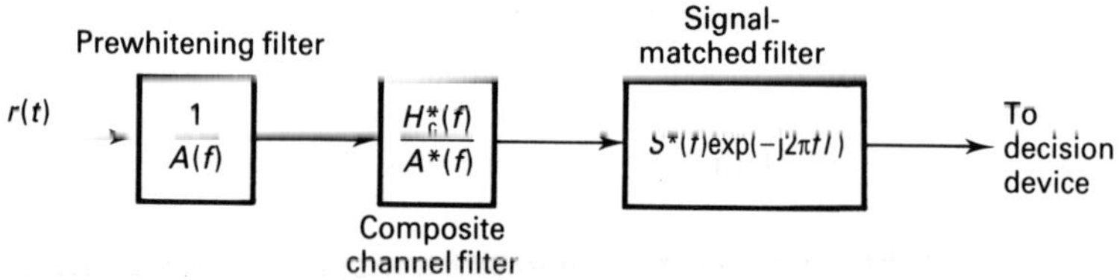

Figure 11.9 RAKE receiver architecture for channel with transfer function $H_c(f)$ and interferer PSD $A(f)A^*(f)$

if the interferer PSD has a spectral factorization

$$\Phi_{jj}(f) = A(f)A^*(f),$$

then the optimum modified RAKE response (for an infinite observation interval) $H_R(f)$ is given by

$$H_R(f) = \frac{1}{A(f)} S^*(f) \frac{1}{A^*(f)} H_c^*(f)\exp(-j2\pi fT)$$

Therefore, for a fixed channel with coefficients f_l, and for a wide-sense stationary interferer with known PSD, the optimum receiver is readily designed by synthesizing a filter with frequency response $H_R(f)$. A block diagram of this generic RAKE is shown in Figure 11.9. It will be shown that the filter $1/A(f)$ can be implemented using a tapped delay line, when the interferer can be modelled by a finite-order autoregressive process. The block $H_c^*(f)/A^*(f)$ can be thought of as the convolution of the channel with the prewhitening filter, and $S^*(f)$ is just the conventional signal matched filter.

The modified RAKE receiver shown in Figure 11.10 approximates the optimum receiver $H_R(f)$. The prewhitening filter, with coefficients α_n, rejects the narrowband interference, and thus approximately has a transfer function equal to $1/A(f)$. The tapped delay line with coefficients β_m^* attempts to combine coherently the multipath components and represents the composite channel $(1/A^*(f))H_c^*(f)$, consisting of the time reverse of the prewhitening filter, convolved with the multipath channel. Each delayed version of the signal is likewise multiplied by $PN(t)$, the direct-sequence signal, and integrated over the bit duration T_b, to implement the matched filter response $S^*(f)$. (Note that the incoming waveform is correlated with the rectangular pulse sequence, $PN(t)$, as opposed to the low-pass filtered version, $PN_{lp}(t)$, in order to simplify the receiver design and analysis.) However, in practice, the channel is time varying and unknown a priori, as is the interference. Thus, to implement this modified RAKE receiver, a method for jointly estimating the channel and interference is required.

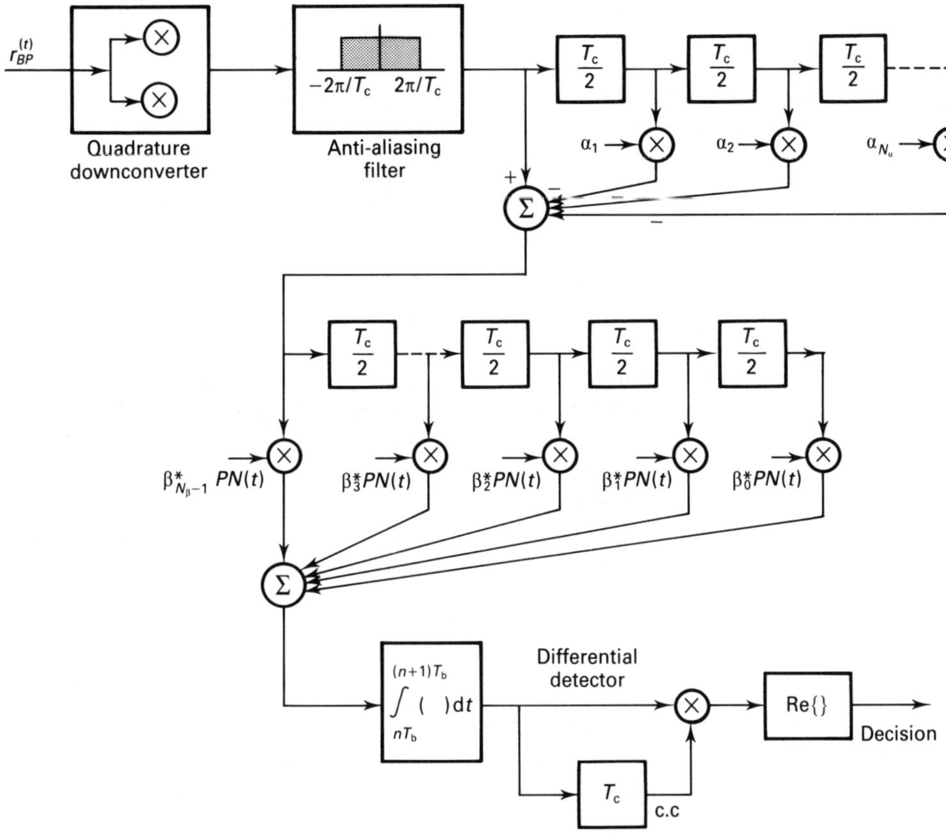

Figure 11.10 Modified RAKE receiver with prewhitening filter

11.3.2 Joint estimation of channel and interferer parameters

To solve the estimation problem, a suitable parameterization of the interferer is required, such that the received signal, which includes a coloured noise component, can be transformed into a new waveform consisting of a deterministic signal plus white additive noise. Such a parameterization is readily found by assuming that the interferer plus thermal noise is a finite-order autoregressive process [5]:

$$j(kT_s) + \eta(kT_s) = \sum_{n=1}^{N_\alpha} \alpha_n[j((k-n)T_s) + \eta((k-n)T_s)] + e(k) \tag{11.73}$$

The process $e(k)$ is assumed to be a white circular Gaussian sequence. Under this AR assumption, we claim that the following prewhitened version of $r(kT_s)$, denoted $x(k)$, is a *sufficient statistic* for determining the information sequence d_n:

$$x(k) = \sum_{n=0}^{N_\alpha} a_n r((k-n)T_s) \tag{11.74}$$

The prewhitening filter coefficients a_n are given by $a_n = -\alpha_n$, for $n = 1,2,\ldots, N_\alpha$, with $a_0 = 1$. Thus, $x(k)$ can be rewritten as

$$x(k) = \sum_{n=0}^{N_\alpha} a_n \sum_{l=0}^{N_f-1} f_l s(k-l-n) + \sum_{n=0}^{N_\alpha} a_n[j(k-n) + \eta(k-n)] \tag{11.75}$$

where

$$s(k) = \sqrt{\frac{2E_b}{T_b}} \sum_{n=-\infty}^{\infty} d_n PN_{1p}(kT_s - nT_b)$$

is the sequence of Nyquist samples of the DS signal. In practical applications, the information sequence d_n is unknown. In order to develop the joint estimator, it will be assumed that d_n is available either through a training sequence, or that *decision directed* adaptation is used. In the latter case, the sequence d_n is replaced by decisions $\hat{d}_n$ obtained by a RAKE receiver.

It is now shown that $x(k)$ consists of a deterministic signal plus a circular white-noise sequence. By using (11.73) in Equation (11.75), we see that

$$x(k) = \sum_{l=0}^{N_f-1} f_l s(k-l) - \sum_{n=1}^{N_\alpha} \alpha_n \sum_{l=0}^{N_f-1} f_l s(k-l-n) + e(k) \tag{11.76}$$

where $e(k)$ is the AR driving process, and thus by definition a white-noise sequence, in (11.73). The problem now is to estimate jointly the channel parameters f_l and AR coefficients α_n. If the sequence $x(k)$ was linear in these parameters, and under the assumption that $e(k)$ was a white Gaussian sequence, then the complex version of the RLS algorithm could be employed to obtain maximum likelihood estimates of the f_l and α_n. However, it is seen that $x(k)$ depends on the product of the α_n and f_l in (11.76). To obtain a linear measurement model, we first define a composite channel with coefficients β_m, as follows:

$$\beta_m = f_m - \sum_{n=1}^{N_\alpha} \alpha_n f_{m-n} \tag{11.77}$$

for $m = 0,1,\ldots, N_\alpha + N_f$. In terms of the β_m, the data sequence $x(k)$ can be rewritten as

$$x(k) = \sum_{m=0}^{N_\beta-1} \beta_m s(k-m) + e(k) \tag{11.78}$$

where $N_\beta = N_\alpha + N_f$. At this point, we have shown that $x(k)$, the prewhitened version of the received signal samples $r(kT_s)$, is the sum of the DS signal $s(k)$, filtered by the composite channel β_m, and a white circular Gaussian sequence $e(k)$. Thus, the optimum receiver in Figure 11.10 consists of a prewhitening filter with coefficients α_n, followed by a replica of the composite channel and a PN correlator. Since the coefficients α_n and β_m are unknown a priori, we seek maximum likelihood estimates of these quantities.

Define a data sequence $r(k)$ by $r(kT_s)$. Then using (11.74) and (11.76), the following measurement model is obtained:

$$r(k) = \sum_{n=1}^{N_\alpha} \alpha_n r(k-n) + \sum_{m=0}^{N_\beta - 1} \beta_m s(k-m) + e(k) \tag{11.79}$$

The above equation for $r(k)$ is seen to have the form of an ARX model (2.31), where the parameter vector is given by

$$\boldsymbol{\theta}^{\mathrm{T}} = [\alpha_1, \alpha_2, \ldots, \alpha_{N_\alpha}, \beta_0, \beta_1, \ldots, \beta_{N_\beta - 1}] \tag{11.80}$$

The regression vector includes both past values of the received signal and the DS signal samples $s(k)$, as follows:

$$\boldsymbol{\phi}(k)^{\mathrm{T}} = [r(k-1), r(k-2), \ldots, r(k-N_\alpha), s(k), s(k-1), \ldots, s(k-N_\beta-1)] \tag{11.81}$$

Again, the sequence $s(k)$, which depends on the information symbols d_n, is assumed known, using either a training sequence or receiver decisions d_n.

Under the assumption that $e(k)$ is a circular white Gaussian sequence, it is well known that the maximum likelihood estimates of α and β are found by minimizing the following least-square error cost function in the time interval $[0, N_s]$:

$$\begin{aligned} V_{N_s}(\boldsymbol{\theta}) &= \sum_{k=0}^{N_s} |r(k) - \boldsymbol{\phi}(k)^{\mathrm{T}}\boldsymbol{\theta}|^2 \\ &= \sum_{k=0}^{N_s} \left| r(k) - \sum_{n=1}^{N_\alpha} \alpha_n r(k-n) - \sum_{m=0}^{N_\beta-1} \beta_m s(k-m) \right|^2 \end{aligned} \tag{11.82}$$

When the channel and interferer parameters are time varying, a 'forgetting factor' λ is included in the least-square error, which de-emphasizes the influence of past data. Thus the RLS cost function is

$$\begin{aligned} V_{\mathrm{RLS}}(\boldsymbol{\theta}) &= \sum_{k=0}^{N_s} \lambda^{N_s-k} |r(k) - \boldsymbol{\phi}(k)^{\mathrm{T}}\boldsymbol{\theta}|^2 \\ &= \sum_{k=0}^{N_s} \lambda^{N_s-k} \left| r(k) - \sum_{n=1}^{N_\alpha} \alpha_n r(k-n) - \sum_{m=0}^{N_\beta-1} \beta_m s(k-m) \right|^2 \end{aligned} \tag{11.83}$$

The solution for the α_n and β_m is now obtained recursively using a complex version of the RLS algorithm given in (2.53)–(2.56). The complex RLS update is readily derived from the real-valued version of [18] (chapter 6), and is summarized as follows [5]:

$$\begin{aligned} \boldsymbol{\theta}(k) &= \boldsymbol{\theta}(k-1) - \mathbf{w}(k)e(k) \\ e(k) &= r(k) - \boldsymbol{\phi}(k)^{\mathrm{T}}\boldsymbol{\theta}(k-1) \\ \mathbf{w}(k) &= \frac{-1}{\lambda + \boldsymbol{\phi}(k)^{\mathrm{T}}P(k-1)\boldsymbol{\phi}^*(k)} P(k-1)\boldsymbol{\phi}(k)^* \\ P(k) &= \frac{1}{\lambda} P(k-1) - \frac{1}{\lambda} \mathbf{w}(k)\boldsymbol{\phi}(k)^{\mathrm{T}}P(k-1) \end{aligned} \tag{11.84}$$

where $\boldsymbol{\theta}$ and $\boldsymbol{\phi}(k)$ are given by (11.80) and (11.81) respectively. Note that, unlike (2.53)–(2.56), the parameter and data vectors are complex valued, and that the forgetting factor is explicitly included.

The version of the RLS algorithm presented here, and used in the simulations that follow, was chosen to illustrate the concept of joint interference and channel estimation. However, no attempt was made to minimize the computational complexity of the algorithm. In many spread-spectrum applications, however, computational efficiency is critical, because of the wide signal bandwidths used. In these cases, the stabilized fast RLS versions in their complex multichannel form, as suggested in Chapter 2, may be used instead.

11.3.3 Optimum prewhitening filter and composite channel estimates

Before proceeding to the BER analysis of the receiver, we consider the form of the optimum solution for the AR coefficients, α_n, and composite channel coefficients β_m. The optimum solution for the parameter vector $\boldsymbol{\theta}$ is again given by

$$R\boldsymbol{\theta} = \mathbf{d}$$

where $R = E[\boldsymbol{\phi}(k)\boldsymbol{\phi}(k)^{\mathrm{H}}]$ and $\mathbf{d} = E[r(k)\boldsymbol{\phi}(k)^{\mathrm{H}}]$. The solution for $\boldsymbol{\theta}$ is more readily visualized by partitioning R into submatrices, representing the correlation matrix of the received samples, $r(k)$, and cross-correlations of $r(k)$ with $s(k)$, the transmitted samples. Thus, for the regression vector in (11.81), the Wiener–Hopf equations can be written in the following form:

$$\begin{bmatrix} \rho_{rr} & \rho_{rs} \\ \rho_{sr} & \rho_{ss} \end{bmatrix} \begin{bmatrix} \alpha \\ \beta \end{bmatrix} = \begin{bmatrix} d_r \\ d_s \end{bmatrix} \tag{11.85}$$

Note that $\rho_{rr} \in C^{N_\alpha, N_\alpha}$, $\rho_{sr} \in C^{N_\beta, N_\alpha}$ and $\rho_{ss} \in C^{N_\beta, N_\beta}$.

The elements of the submatrices ρ_{rr} and ρ_{sr} are determined as follows:

$$\begin{aligned} \rho_{rr}(n,m) &= E[r(k-n)r^*(k-m)] \\ &= \frac{2E_{\mathrm{b}}}{T_{\mathrm{b}}} \sum_{l=0}^{N_{\mathrm{f}}-1} f_l \sum_{l'=0}^{N_{\mathrm{f}}-1} f_{l'}^* \rho_{PN}(l'-l+m-n) + \rho_{jj}(m-n) \\ &\quad + \rho_{\eta\eta}(m-n) \end{aligned} \tag{11.86}$$

for $n, m = 1, 2, \ldots, N_\alpha$. Since sampling is at the Nyquist rate of $2/T_{\mathrm{c}}$, the samples of the PN signal are no longer uncorrelated, and we have

$$\rho_{PN}(k) = \delta_{k,0} + \tfrac{1}{2}[\delta_{k+1,0} + \delta_{k-1,0}] \tag{11.87}$$

Again, $\rho_{jj}(k)$ represents the correlation function of the interferer, and $\rho_{\eta\eta}(k)$ is the correlation function of the Gaussian thermal noise. The cross-correlation submatrix is given by

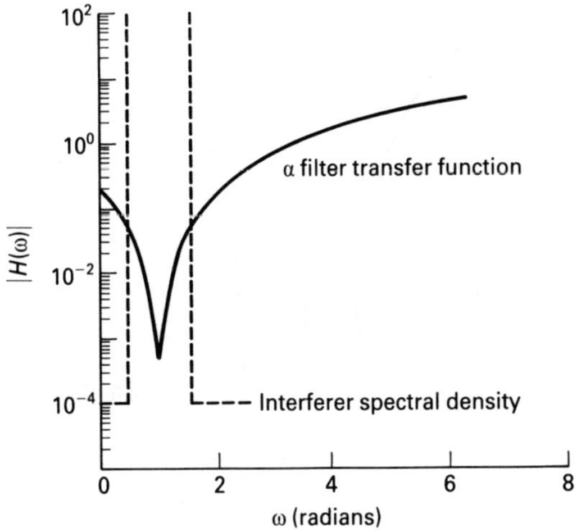

Figure 11.11 Interferer power spectral density and prewhitening filter transfer function from partitioned Wiener–Hopf equations

$$\rho_{sr}(n, m) = E[s(k-n+1)r^*(k-m)]$$
$$= \frac{2E_b}{T_b}\rho_{PN}(m-n+1) \tag{11.88}$$

for $n = 1, 2, \ldots, N_\beta$, and $m = 1, 2, \ldots, N_\alpha$. Finally, the cross-correlation vector, **d**, is given by

$$\mathbf{d}(n) = E[r(k)r^*(k-n)] \tag{11.89}$$

for $n = 1,2,\ldots, N_\alpha$, and

$$\mathbf{d}(n) = E[r(k)s^*(k-n+N_\alpha+1)] \tag{11.90}$$

for $n = N_\alpha + 1,\ldots, N_\alpha + N_\beta$. These correlations have the same form as (11.86) and (11.88).

The equations (11.85) were solved for a four-ray channel (three spectral nulls) and an interferer with a rectangular spectral density. The interferer covariance function is given by

$$\rho_{jj}(k) = 2J\frac{\sin\ \omega_j k T_s}{\omega_j k T_s}\exp(j\delta\omega k T_s) \tag{11.91}$$

where ω_j is the jammer bandwidth in radians per second, and $\delta\omega$ is the frequency offset from the carrier frequency. The power spectral density of the interferer, and transfer function of the prewhitening filter, with coefficients α_n, are shown in Figure 11.11. Note that 2π represents the one-sided DS signal bandwidth, with T_c normalized

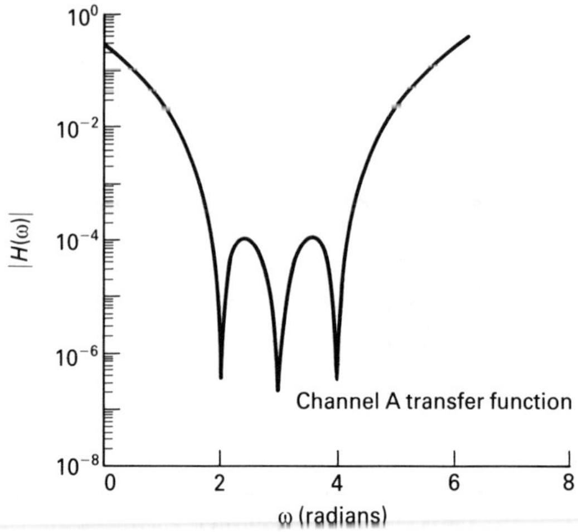

Figure 11.12 Four-ray channel transfer function

to unity. The channel transfer function is illustrated in Figure 11.12. The composite channel estimate, with coefficients β_m, is shown in Figure 11.13. Clearly, the composite channel transfer function is a product of the prewhitening filter function (represented by the additional null at 1 radian/s) and the channel transfer function, and thus represents the convolution of the prewhitening filter and original multipath channel.

11.3.4 BER analysis of the RLS-based receiver

We now consider the BER performance of the modified RAKE receiver in Figure 11.10 when the RLS algorithm is used to estimate the AR and composite channel coefficients jointly. The input to the prewhitening filter is given by

$$r(t) = \sqrt{\frac{2E_b}{T_b}} \sum_{n=-\infty}^{\infty} d_n \sum_{l=0}^{N_f-1} f_l PN_{lp}(t - lT_c/2 - nT_b)e^{j\theta} + j(t) + \eta(t) \qquad (11.92)$$

where $PN_{lp}(t)$ represents the pseudonoise waveform, bandlimited to $1/T_c$ Hz, θ is the random carrier phase, and $j(t)$ and $\eta(t)$ the narrowband interferer and thermal noise respectively. The random phase, $e^{j\theta}$, can be absorbed into the coefficients f_l without loss of generality.

Recall that the RAKE receiver approximates a correlator matched to the prewhitened version of the multipath signal. As discussed in [5], such a correlator implements the maximum likelihood receiver for this problem. From (11.92) and the form of the prewhitening filter, it is clear that the correlator reference waveform should approximate

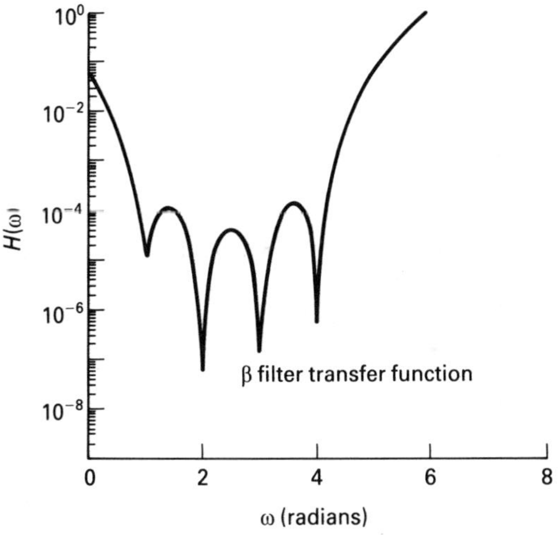

Figure 11.13 Composite channel estimate, β coefficients

$$s_c(t) \approx \left(\sum_{n=0}^{N_\alpha} a_n \sum_{l=0}^{N_f-1} f_l PN_{lp}(t - lT_c/2 - nT_c/2) \right)^* \tag{11.93}$$

in the interval $[0, T_b]$. However, recall that the coefficients β_m represent the convolution of the prewhitening filter with coefficients α_n and the channel, represented by the f_l. Thus, $s_c(t)$ can be rewritten as

$$s_c(t) = \sum_{l=0}^{N_\beta - 1} \beta_m^* PN(t - mT_c/2) \tag{11.94}$$

where $PN(t)$ is given by

$$PN(t) = \sum_{k=0}^{L_{PN}-1} c_k P_{T_c}(t - kT_c)$$

Again, while correlation with $PN_{lp}(t)$ would actually be optimum, we use $PN(t)$, the non-filtered waveform, in order to simplify the receiver design and analysis. The output of the RAKE combiner in the kth bit interval can then be expanded as

$$X(k) = \int_{kT_b}^{(k+1)T_b} s_c(t - kT_b) \sum_{n=0}^{N_\alpha} a_n r(t - nT_c/2) \, dt \tag{11.95}$$

Substitution of the explicit form of $s_c(t)$ in (11.94) into $X(K)$ gives the following representation for the RAKE combiner output [5]:

$$X(k) = \sum_{m=0}^{N_\beta - 1} \beta_m^* \sum_{l=0}^{L_{PN}-1} c_l \sum_{n=0}^{N_\alpha} a_n \int_{kT_b + lT_c + mT_c/2}^{kT_b + (l+1)T_c + mT_c/2} r(t - nT_c/2) \, dt \tag{11.96}$$

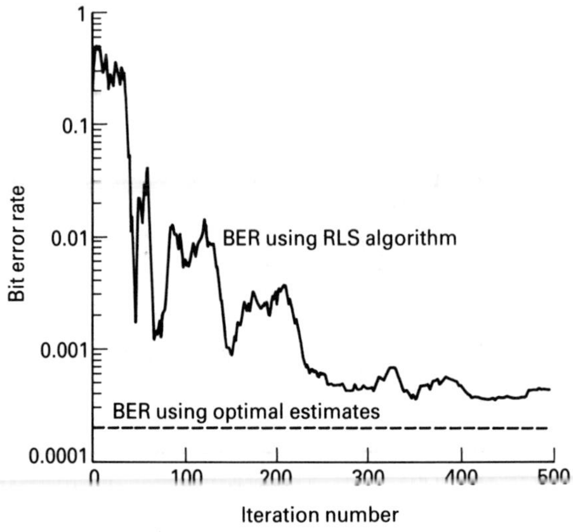

Figure 11.14 RLS and optimum BER for the modified RAKE receiver, $E_b/N_o = 10$ dB, $J/S = 20$ dB, $L_{PN} = 15$, $N_\alpha = 3$, $N_\beta = 6$

The information sequence is demodulated by comparing $X(k)$ with the combiner output at the previous bit, $X(k-1)$. Differential detection is employed to remove the random phase θ and minimize sensitivity to phase errors in the estimates of β_m. Thus, the decision variable is

$$U(k) = \text{Re}\{X(k)X^*(k-1)\} \tag{11.97}$$

In the noiseless case, $U(k)$ will be positive if the current and previous information symbol have the same sign, otherwise $U(k)$ will be negative. The probability of error is then given by

$$P_e = \tfrac{1}{2}P(U(k) > 0 \mid d_k \neq d_{k-1}) + \tfrac{1}{2}P(U(k) < 0 \mid d_k = d_{k-1}) \tag{11.98}$$

A closed-form representation for P_e can be found using the method of [18] (Appendix 4b). Specifically,

$$P_e = Q(a,b) - \frac{v_2/v_1}{1 + v_2/v_1} I_0(a,b) \exp\left(-\frac{a^2 + b^2}{2}\right) \tag{11.99}$$

where $Q(a,b)$ is the Marcum-Q function, and $I_0(x)$ is the modified Bessel function. The terms a,b,v_1 and v_2 have a fairly complicated form, but are completely defined in terms of the means and variances of $X(k)$. However, the forms of the mean and variance of the RAKE combiner output are rather complicated, and the interested reader is directed to [5] for a detailed derivation.

The RLS algorithm (11.84) was simulated using the same interferer and channel combination shown in Figures 11.11 and 11.12. The BER was evaluated analytically at the end of every bit interval, corresponding to once every 30 Nyquist samples. Figure 11.14 illustrates the analytic BER for the case of $E_b/N_0 = 10$ dB, $J/S = 20$ dB and $L_{PN} = 15$ chips per bit. The lower line illustrates the optimum BER resulting when the Wiener solution for the α_n and β_m coefficients (11.85) is employed. It is seen that the RLS bit error rate approaches the optimum after 300 iterations, corresponding to only 10 bits.

11.4 Joint estimation of PN code delay, multipath and interference using the extended Kalman filter

11.4.1 Parameterization of the interferer, channel and code delay

Thus far, we have developed an RLS algorithm that jointly estimates narrowband interference and multipath. However, the problem of code synchronization has not been considered in the receivers analyzed. We now present a complete digital spread-spectrum receiver that uses an extended Kalman filter (EKF) simultaneously to provide channel and interferer estimates, along with the code delay. In fact, we will show that the estimated signal computed by the EKF directly provides a reference signal for a digital RAKE receiver.

The code synchronization problem is seen to be highly non-linear by considering the form of the received signal samples, $r(k)$. Here, τ represents the unknown time of arrival of the waveform, written in low-pass equivalent form:

$$r(k) = \sum_{l=0}^{N_f-1} f_l s((k-l)T_s + \tau) + j(k) + \eta(k) \tag{11.100}$$

where

$$s(t) = \sqrt{\frac{2E_b}{T_b}} \sum_{n=-\infty}^{\infty} d_n PN_{lp}(t - nT_b) \tag{11.101}$$

The interferer samples are denoted by $j(k)$, the thermal noise by $\eta(k)$, and the low-pass-filtered PN waveform by $PN_{lp}(t)$. In order to implement the EKF, it will be seen that $PN_{lp}(t)$ must be first-order differentiable. From Equation (11.72) defining $PN_{lp}(t)$ and [20], the low-pass equivalent transmitted signal is given by

$$s(t) = \sqrt{\frac{2E_b}{T_b}} \sum_{n=-\infty}^{\infty} d_n \sum_{k=0}^{L_{PN}-1} c_k \frac{1}{\pi}$$
$$\times \left[\mathrm{Si}\left(2\pi \frac{(t - kT_c - nT_b)}{T_c}\right) - \mathrm{Si}\left(2\pi \frac{[t-(k+1)T_c - nT_b]}{T_c}\right)\right] \tag{11.102}$$

Clearly, from the definition of the Si(x) function, $s(t)$ is differentiable.

As in the previous section, a joint parameterization of the interferer and channel is required. Again, the interferer plus thermal noise is modelled as an N_α order AR process, with

$$j(k) + \eta(k) \approx \sum_{n=1}^{N_\alpha} \alpha_n [j(k-n) + \eta(k-n)] + e(k) \tag{11.103}$$

Replacing $j(k) + \eta(k)$ in (11.100) yields the following form for $r(k)$:

$$r(k) = \sum_{l=0}^{N_f - 1} f_l s((k-l)T_s + \tau) + \sum_{n=1}^{N_\alpha} \alpha_n \left(r(k-n) - \sum_{l=0}^{N_f-1} f_l s((k-l-n)T_s + \tau) \right) + e(k) \tag{11.104}$$

Unfortunately, the channel coefficients f_l and AR parameters α_n are coupled in the above representation. Thus, a composite channel β_m is again defined, such that

$$\beta_m = f_m - \sum_{n=1}^{N_\alpha} \alpha_n f_{m-n} \tag{11.105}$$

The composite channel β_m is seen to be the convolution of a prewhitening filter and the multipath channel. Rewriting $r(k)$ in terms of the β_m yields

$$r(k) = \sum_{m=0}^{N_\beta - 1} \beta_m s((k-m)T_s + \tau) + \sum_{n=1}^{N_\alpha} \alpha_n r(k-n) + e(k) \tag{11.106}$$

If the code delay, τ, were known a priori, the coefficients α_n and β_m, which are linear parameters, could be estimated using the RLS algorithm. However, the delay is a highly non-linear parameter, and can only be estimated recursively using a linearization procedure. The extended Kalman filter [21] provides a systematic linearization of $r(k)$ in terms of τ and, furthermore, generates quasi-optimal joint estimates of the channel and interferer parameters. In the next section, we review the extended Kalman filter and derive a specific form of the algorithm for the estimation of real parameters using complex measurements.

11.4.2 Extended Kalman filter for real parameters and complex measurements

The received samples $r(k)$ represent a non-linear measurement model of the following form:

$$r(k) = \phi(k,\boldsymbol{\theta}) + \eta(k) \tag{11.107}$$

where $\phi(\cdot)$ is a non-linear complex-valued function of the parameter vector $\boldsymbol{\theta}$. In contrast, the conventional linear, scalar measurement model is restricted to the form

$$r(k) = \boldsymbol{\phi}(k)^{\mathrm{T}}\boldsymbol{\theta} + \eta(k) \tag{11.108}$$

with $\boldsymbol{\phi} \in C^N$, and $\eta(k)$ a circular white Gaussian measurement noise, with variance σ_η^2.

In the problem considered here, $\boldsymbol{\theta}$ is a mixture of real and complex parameters, which are both linear and non-linear in nature. Specifically,

$$\boldsymbol{\theta} = [\tau, \beta_0, \dots, \beta_{N_\beta - 1}, \alpha_1, \dots, \alpha_{N_\tau}]^{\mathrm{T}} \tag{11.109}$$

In this case, τ is a real-valued non-linear parameter, whereas the α_n and β_m are complex-valued and linear.

We first review the general complex-valued Kalman filter for a linear parameter vector $\boldsymbol{\theta}$, given by the measurement model of (11.108). A dynamic model for the parameters is assumed, such that $\boldsymbol{\theta}$ is a first-order autoregressive vector Gaussian process. Specifically,

$$\boldsymbol{\theta}(k+1) = F\boldsymbol{\theta}(k) + \eta_{\mathrm{p}}(k) \tag{11.110}$$

where $\boldsymbol{\theta} \in C^N$, $F \in C^{N,N}$ and $\eta_{\mathrm{p}}(k) \in C^N$ represents a vector white Gaussian process noise. The covariance matrix of the process noise is given by

$$Q = E[\eta_{\mathrm{p}}(k)\eta_{\mathrm{p}}(k)^{\mathrm{H}}]$$

with $\boldsymbol{\phi} \in C^N$, and $\eta(k)$ a circular white Gaussian measurement noise, with variance σ_η^2. The Kalman filter computes the following minimum variance filtered and predicted estimates:

$$\begin{aligned} \hat{\boldsymbol{\theta}}(k \mid k) &= E[\boldsymbol{\theta}(k) \mid r(k), r(k-1), \dots, r(0)] \\ \hat{\boldsymbol{\theta}}(k+1 \mid k) &= E[\boldsymbol{\theta}(k+1) \mid r(k), r(k-1), \dots, r(0)] \end{aligned} \tag{11.111}$$

The error covariance matrices are defined by

$$\begin{aligned} P(k \mid k) &= E[\{\boldsymbol{\theta}(k) - \hat{\boldsymbol{\theta}}(k \mid k\}\{\boldsymbol{\theta}(k) - \hat{\boldsymbol{\theta}}(k \mid k)\}^{\mathrm{H}}] \\ P(k+1 \mid k) &= E[\{\boldsymbol{\theta}(k+1) - \hat{\boldsymbol{\theta}}(k+1 \mid k\}\{\boldsymbol{\theta}(k+1)\hat{\boldsymbol{\theta}}(k+1 \mid k)\}^{\mathrm{H}}] \end{aligned} \tag{11.112}$$

The complex Kalman filter equations are derived in [22] (chapter 5), and are summarized as follows:

Measurement update:

$$\begin{aligned} \hat{\boldsymbol{\theta}}(k \mid k) = \hat{\boldsymbol{\theta}}(k \mid k-1) &+ \frac{1}{\boldsymbol{\phi}(k)^{\mathrm{T}}P(k \mid k-1)\phi(k)^* + \sigma_\eta^2} P(k \mid k-1)\phi(k)^* \\ &\times [r(k) - \boldsymbol{\phi}(k)^{\mathrm{T}}\hat{\boldsymbol{\theta}}(k \mid k-1)] \end{aligned} \tag{11.113}$$

Covariance update:

$$\begin{aligned} P(k \mid k) = &\left(I - \frac{1}{\phi(k)^{\mathrm{T}}P(k \mid k-1)\boldsymbol{\phi}(k)^* + \sigma_\eta^2} P(k \mid k-1)\phi(k)^*\phi(k)^{\mathrm{T}}\right) \\ &\times P(k \mid k-1) \end{aligned} \tag{11.114}$$

One-step prediction:

$$\hat{\theta}(k+1 \mid k) = F\hat{\theta}(k \mid k) \tag{11.115}$$

$$P(k+1 \mid k) = FP(k \mid k)F^{H} + Q \tag{11.116}$$

Next, we consider the problem of estimating a mixture of real and complex parameters, which are in general non-linear in the observations $r(k)$. First, partition $\boldsymbol{\theta}$ into

$$\boldsymbol{\theta} = [\boldsymbol{\theta}_R, \boldsymbol{\theta}_C]^T$$

where $\boldsymbol{\theta}_R$ is a vector of real-valued parameters and $\boldsymbol{\theta}_C$ is a vector of complex parameters. However, when estimating $\boldsymbol{\theta}_R$ from complex-valued measurements, the filtered and predicted estimates of these parameters will likewise be complex. To develop a consistent version of the EKF, we assume that $\boldsymbol{\theta}_R$ and $\boldsymbol{\theta}_C$ evolve as real and complex Gaussian AR processes respectively:

$$\begin{aligned} \hat{\boldsymbol{\theta}}_R(k+1) &= F_R\hat{\boldsymbol{\theta}}(k) + \eta_{p,R}(k) \\ \hat{\boldsymbol{\theta}}_C(k+1) &= F_C\hat{\boldsymbol{\theta}}(k) + \eta_{p,C}(k) \end{aligned} \tag{11.117}$$

where $F_R \in \mathscr{R}^{N_R, N_R}$, $F_C \in C^{N_C, N_C}$, and $\eta_{p,R}(k) \in \mathscr{R}^{N_R}$, $\eta_{p,C}(k) \in C^{N_C}$. (It should be noted that the derivation of the complex Kalman filter in [22], and as summarized in the previous section, admits a parameter vector $\boldsymbol{\theta}$ which is a mixture of real and complex elements. The estimate $\hat{\boldsymbol{\theta}}(k \mid k)$, which is in general complex valued, can still be shown to be the best linear least-squares estimate.)

The measurement model, in terms of the real and complex parameters, is written as

$$r(k) = \boldsymbol{\phi}(k, \boldsymbol{\theta}_R(k), \boldsymbol{\theta}_C(k)) + \eta(k) \tag{11.118}$$

The EKF is derived [21] by expanding $\boldsymbol{\phi}(\cdot)$ about the one-step prediction $\boldsymbol{\theta}(k \mid k-1)$ at iteration k. However, while $\boldsymbol{\phi}(\cdot)$ may be analytic in $\hat{\boldsymbol{\theta}}_C$ (differentiable with respect to the complex vector), it is clearly not differentiable w.r.t. $\hat{\boldsymbol{\theta}}_R$. However, an expansion about $\text{Re}\{\hat{\boldsymbol{\theta}}_R(k \mid k-1)\}$ is valid, and corresponds to the ordinary Taylor series approximation. Thus, $r(k)$ is approximated as follows:

$$\begin{aligned} r(k) \approx\ & \boldsymbol{\phi}(k, \text{Re}\{\hat{\boldsymbol{\theta}}_R(k \mid k-1)\}, \boldsymbol{\theta}_C(k \mid k-1)) \\ &+ \frac{\partial \boldsymbol{\phi}(k)}{\partial \text{Re}\{\hat{\boldsymbol{\theta}}_R(k \mid k-1)\}} [\boldsymbol{\theta}_R(k) - \text{Re}\{\hat{\boldsymbol{\theta}}_R(k \mid k-1)\}] \\ &+ \frac{\partial \boldsymbol{\phi}(k)}{\partial \hat{\boldsymbol{\theta}}_C(k \mid k-1)} [\boldsymbol{\theta}_C(k) - \boldsymbol{\theta}_C(k \mid k-1)] \end{aligned} \tag{11.119}$$

To implement the EKF, $r(k)$ must be expressed as a linear transformation of both $\hat{\boldsymbol{\theta}}_R(k \mid k-1)$ and $\hat{\boldsymbol{\theta}}_C(k \mid k-1)$. However, $r(k)$ in the above equation is a linear function of $\text{Re}\{\hat{\boldsymbol{\theta}}_R(k \mid k-1)\}$ only. A linear representation in terms of the entire complex one-step prediction $\hat{\boldsymbol{\theta}}(k \mid k-1)$ can be obtained by adding and subtracting a correction term as follows:

$$
\begin{aligned}
r(k) \approx\ & \boldsymbol{\phi}(k, \operatorname{Re}\{\hat{\boldsymbol{\theta}}_{\mathrm{R}}(k \mid k-1)\}, \hat{\boldsymbol{\theta}}_{\mathrm{C}}(k \mid k-1)) \\
& + \frac{\partial \boldsymbol{\phi}(k)}{\partial \operatorname{Re}\{\hat{\boldsymbol{\theta}}_{\mathrm{R}}(k \mid k-1)\}} [\boldsymbol{\theta}_{\mathrm{R}}(k) - \hat{\boldsymbol{\theta}}_{\mathrm{R}}(k \mid k-1)] \\
& + \frac{\partial \boldsymbol{\phi}(k)}{\partial \hat{\boldsymbol{\theta}}_{\mathrm{C}}(k \mid k-1)} [\boldsymbol{\theta}_{\mathrm{C}}(k) - \hat{\boldsymbol{\theta}}_{\mathrm{C}}(k \mid k-1)] \\
& + \frac{\partial \boldsymbol{\phi}(k)}{\partial \operatorname{Re}\{\hat{\boldsymbol{\theta}}_{\mathrm{R}}(k \mid k-1)\}} \operatorname{Im}\{\hat{\boldsymbol{\theta}}_{\mathrm{R}}(k \mid k-1)\} \qquad (11.120)
\end{aligned}
$$

The EKF measurement model can now be written in a more compact form as follows:

$$
\begin{aligned}
r(k) =\ & \boldsymbol{\phi}'(k)[\boldsymbol{\theta}(k) - \hat{\boldsymbol{\theta}}(k \mid k-1)] \\
& + \boldsymbol{\phi}(k, \operatorname{Re}\{\hat{\boldsymbol{\theta}}_{\mathrm{R}}(k \mid k-1)\}, \hat{\boldsymbol{\theta}}_{\mathrm{C}}(k \mid k-1)) \\
& + \frac{\partial \boldsymbol{\phi}(k)}{\partial \operatorname{Re}\{\hat{\boldsymbol{\theta}}_{\mathrm{R}}(k \mid k-1)\}} \operatorname{Im}\{\hat{\boldsymbol{\theta}}_{\mathrm{R}}(k \mid k-1)\} \qquad (11.121)
\end{aligned}
$$

The row vector $\boldsymbol{\phi}'(k)$ is the gradient with respect to the one-step predictions, given by

$$
\boldsymbol{\phi}'(k) \equiv \left(\frac{\partial \boldsymbol{\phi}(k)}{\partial \operatorname{Re}\{\hat{\boldsymbol{\theta}}_{\mathrm{R}}(k \mid k-1)\}}, \frac{\partial \boldsymbol{\phi}(k)}{\partial \hat{\boldsymbol{\theta}}_{\mathrm{C}}(k \mid k-1)} \right) \qquad (11.122)
$$

The linearized measurement model (11.121) can be substituted directly into the ordinary complex Kalman filter equations (11.113). The last two terms in (11.121) correspond to prior means which are subtracted from the innovations, yielding the following form of the measurement update:

EKF measurement update:

$$
\begin{aligned}
\hat{\boldsymbol{\theta}}(k \mid k) =\ & \hat{\boldsymbol{\theta}}(k \mid k-1) + \frac{1}{\boldsymbol{\phi}'(k) P(k \mid k-1) \boldsymbol{\phi}'^{\mathrm{H}}(k) + \sigma_\eta^2} P(k \mid k-1) \boldsymbol{\phi}'(k)^{\mathrm{H}} \\
& \times [r(k) - \boldsymbol{\phi}(k, \operatorname{Re}\{\hat{\boldsymbol{\theta}}_{\mathrm{R}}(k \mid k-1)\}, \hat{\boldsymbol{\theta}}_{\mathrm{C}}(k \mid k-1)) - \gamma(k)] \qquad (11.123)
\end{aligned}
$$

EKF covariance update:

$$
\begin{aligned}
P(k \mid k) =\ & \left(I - \frac{1}{\boldsymbol{\phi}'(k) P(k \mid k-1) \boldsymbol{\phi}'^{\mathrm{H}}(k) + \sigma_\eta^2} P(k \mid k-1) \boldsymbol{\phi}'(k)^{\mathrm{H}} \boldsymbol{\phi}'(k) \right) \\
& \times P(k \mid k-1) \qquad (11.124)
\end{aligned}
$$

Note that the one-step predictions for the EKF remain unchanged, since the process model is still linear. The correction term $\gamma(k)$ is seen to equal

$$
\gamma(k) = \frac{\partial \boldsymbol{\phi}(k)}{\partial \operatorname{Re}\{\hat{\boldsymbol{\theta}}_{\mathrm{R}}(k \mid k-1)\}} \operatorname{Im}\{\hat{\boldsymbol{\theta}}_{\mathrm{R}}(k \mid k-1)\} \qquad (11.125)
$$

The effect of this correction term is to force the imaginary part of the measurement update $\hat{\boldsymbol{\theta}}_R(k \mid k-1)$ to equal zero, corresponding to the fact that the process model for $\boldsymbol{\theta}_R(k)$ is real valued.

11.4.3 EKF interference, multipath and delay estimator

The EKF equations (11.123) and (11.124) can now be directly applied to the code delay, interferer and multipath joint estimation problem. The measurement function is given by

$$\phi(k,\boldsymbol{\theta}_R,\boldsymbol{\theta}_C) = \sum_{m=0}^{N_\beta-1} \beta_m s((k-m)T_s + \tau) + \sum_{n=1}^{N_\alpha} \alpha_n r(k-n) + e(k) \qquad (11.126)$$

The real part of the parameter vector, $\boldsymbol{\theta}_R$, is a scalar equal to τ. The imaginary part contains the coefficients α_n and β_m. Thus, the gradient vector is given by

$$\boldsymbol{\phi}'(k) = \left[\frac{\partial \hat{s}_f(k)}{\partial \hat{\tau}(k)}, s(kT_s + \hat{\tau}), s((k-1)T_s + \hat{\tau}), \right.$$

$$\left. \ldots, s((k - N_\beta + 1)T_s + \hat{\tau}), r(k-1), \ldots, r(k - N_\alpha)\right] \qquad (11.127)$$

The estimated signal $\hat{s}_f(k)$ corresponds to $\phi(k, \hat{\boldsymbol{\theta}}(k \mid k-1))$, and is given by

$$\hat{s}_f(k) = \sum_{m=0}^{N_\beta-1} \hat{\beta}_m(k \mid k-1) s((k-m)T_s + \hat{\tau}(k)) \qquad (11.128)$$

The EKF equations can be implemented once the correction term is determined, which for the code delay estimation problem equals

$$\gamma(k) = \frac{\partial \hat{s}_f(k)}{\partial \hat{\tau}(k)} \operatorname{Im}\{\hat{\tau}(k \mid k-1)\} \qquad (11.129)$$

The operation of the joint estimator is most readily visualized by examining the measurement update equation:

$$\begin{bmatrix} \hat{\tau}(k \mid k) \\ \hat{\beta}_0(k \mid k) \\ \vdots \\ \hat{\beta}_{N_\beta-1}(k \mid k) \\ \hat{\alpha}_1(k \mid k) \\ \vdots \\ \hat{\alpha}_{N_\alpha}(k \mid k) \end{bmatrix} = \begin{bmatrix} \hat{\tau}(k \mid k-1) \\ \hat{\beta}_0(k \mid k-1) \\ \vdots \\ \hat{\beta}_{N_\beta-1}(k \mid k-1) \\ \hat{\alpha}_1(k \mid k-1) \\ \vdots \\ \hat{\alpha}_{N_\alpha}(k \mid k-1) \end{bmatrix} + \frac{1}{\sigma(k \mid k-1)} P(k \mid k-1) \times$$

$$
\times \begin{bmatrix} \partial\hat{s}_{\mathrm{f}}(k)/\partial\hat{\tau}(k) \\ s(kT_{\mathrm{s}} + \hat{\tau}) \\ s((k-1)T_{\mathrm{s}} + \hat{\tau}) \\ \vdots \\ s((k - N_{\beta} + 1)T_{\mathrm{s}} + \hat{\tau}) \\ r(k \quad 1) \\ \vdots \\ r(k - N_{\alpha}) \end{bmatrix}^{*} \left(r(k) - \sum_{n=1}^{N_{\alpha}} \hat{\alpha}_n(k \mid k-1) r(k-n) - \hat{s}_{\mathrm{f}}(k) - \gamma(k) \right) \tag{11.30}
$$

where the innovations covariance is given by

$$
\sigma(k \mid k-1) = \boldsymbol{\phi}'(k) P(k \mid k-1) \boldsymbol{\phi}'(k)^{\mathrm{H}} + \sigma_{\eta}^{2}
$$

Note that the error term is equivalent to the output of the prewhitening filter, with coefficients α_n, with the signal component and correction term subtracted. Thus, the effect of the interferer is minimized by subtracting its current estimated value from the innovations.

Consider the update for the prewhitening filter coefficients, for example. Ignoring the Kalman gain term, we see that the new estimate of α_n is given by the previous estimate, plus the nth previous sample $r(k)$ multiplied by an error term. This update for the prewhitening coefficients closely resembles the LMS algorithm update in (11.36), when the coefficient vector $\mathbf{g}(k)$ is replaced by α. The time delay estimate is updated by adding the product of the signal derivative and the error signal to the previous estimate. This same time delay estimator structure is found throughout the signal processing literature, although it is more often derived by minimization of instantaneous squared error, resulting in an LMS-type update [23]. The role of the Kalman gain (error covariance matrix divided by the innovations variance) is to provide a near optimal gain sequence, as opposed to the constant step-size parameter μ used in the LMS algorithm. Furthermore, the error covariance matrix $P(k \mid k-1)$ in the Kalman gain compensates for correlations between time delay, channel and interferer estimation errors.

11.4.4 Digital RAKE receiver and BER analysis

The performance of the EKF is evaluated, as in the previous RLS-based receiver, by computing analytic BER versus iterations of the estimator. The receiver to be analyzed is shown in Figure 11.15, and is equivalent to a digital RAKE combiner, in which the received samples are first prewhitened and then correlated with a replica of the transmitted samples, filtered by the composite channel estimate.

It is next shown that the predicted signal, $\hat{s}_{\mathrm{f}}(k)$, generated by the EKF, provides an appropriate reference for the RAKE correlator. First, compare the output of the prewhitening filter and (11.128) for the estimated signal, as follows. The transmitted low-pass waveform, $s(t)$, is received in the presence of multipath, and prewhitened using the current AR estimates, $\hat{\alpha}_n$. Assume that the interferer plus thermal noise is

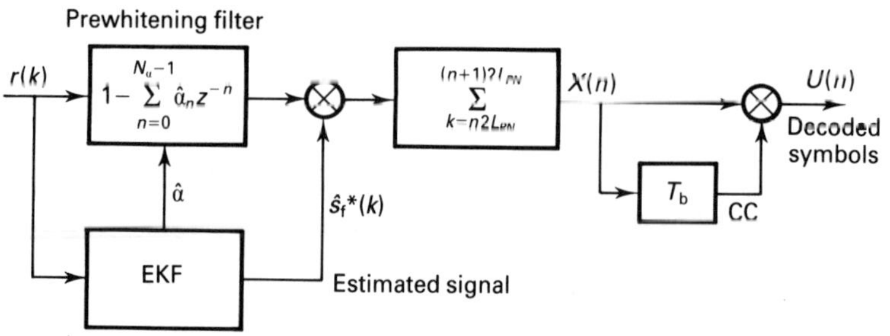

Figure 11.15 Digital RAKE receiver using interferer and channel estimates from the EKF

exactly an N_α-order AR process, that the AR estimates are exact, and thus the prewhitener output consists of a filtered version of $s(t)$ plus a circular white Gaussian process $e(k)$. Thus, the samples entering the correlator are given by

$$x(k) = \sum_{l=0}^{N_f-1} f_l \left(s((k-l)T_s + \tau) - \sum_{n=1}^{N_\alpha} \alpha_n s((k-l-n)T_s + \tau) \right) + e(k) \tag{11.131}$$

However, it can be seen that $x(k)$ is equivalent to the transmitted signal, filtered by the composite channel β, plus a white Gaussian noise process, and can be rewritten as

$$x(k) = \sum_{m=0}^{N_\beta-1} \beta_m s((k-m)T_s + \tau) + e(k) \tag{11.132}$$

Thus, the predicted waveform $\hat{s}_f(k)$ approximately matches the deterministic component of $x(k)$, where β_m and τ are replaced by their EKF predicted values.

The digital RAKE receiver can be derived in a manner similar to the continuous time version in section 11.3.4. Assume that the correlating waveform, $s_c(t)$, and received signal $r(t)$ in (11.95) are both bandlimited to $1/T_c$ Hz. Then from the sampling theorem, it is well known that the continuous time correlation can be replaced by a discrete time correlation as follows:

$$\begin{aligned} X(n) &= \int_{nT_b}^{(n+1)T_b} s_c(t-nT_b) \sum_{n=0}^{N_\alpha} a_n r(t-nT_c/2)\, dt \\ &\approx \sum_{k=nT_b/T_s}^{(n+1)T_b/T_s} s_c(kT_s - nT_b) \sum_{n=0}^{N_\alpha} a_n r((k-n)T_s) \end{aligned} \tag{11.133}$$

(The above approximation would be exact if the integration interval was $[-\infty, \infty]$.) We recognize that $\hat{s}_f(k)$ is a discrete time version of $s_c(t)$. Thus, replacing the prewhitening coefficients a_n with the rejection filter coefficients α_n, the discrete time correlation becomes

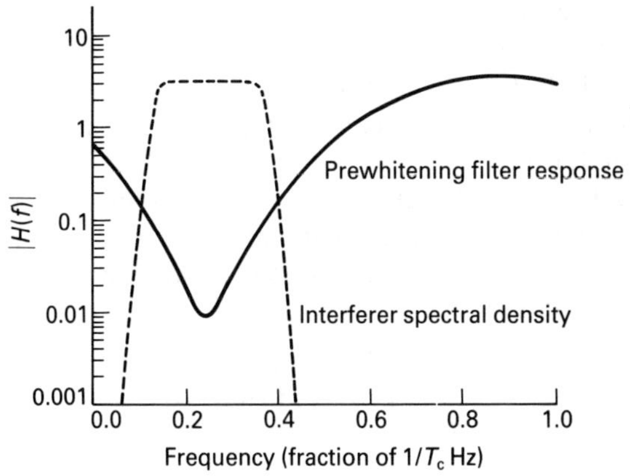

Figure 11.16 Prewhitening filter frequency response and interferer power spectral density, bandwidth $= 0.1/T_c$, $J/S = 10$ dB, $N_\alpha = 3$

$$X(n) = \sum_{k=n2L_{PN}}^{(n+1)2L_{PN}-1} \left(r(k) - \sum_{l=1}^{N_\alpha} \hat{\alpha}_l r(k-l) \right)^* \hat{s}_f(k) \qquad (11.134)$$

and the decision variable, as in the RLS-based receiver, is of the form

$$U(n) = \mathrm{Re}\{X(n)X(n-1)^*\} \qquad (11.135)$$

It is apparent from (11.133) that the RAKE receiver of section 11.3.4 could also have been implemented using a discrete time correlator. Such an implementation would

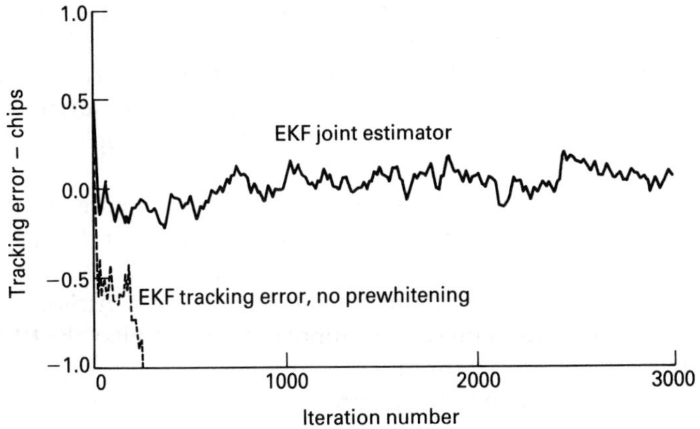

Figure 11.17 Tracking error, $E_b/N_o = 10$ dB, $J/S = 10$ dB, four-ray channel

provide performance nearly equivalent to that of the continuous time implementation, with a slight degradation due to the finite duration assumed for the correlation in (11.133). In Figure 11.10, the correlating waveforms $\beta_m PN(t)$ would be replaced by the filtered and Nyquist sampled versions $\beta_m PN_{\text{lp}}(kT_s)$, and the integrator with a discrete time accumulator.

The BER for differential detection is given by (11.99), where the parameters are again completely determined by the means and covariances of the RAKE output. Since the receiver operates in discrete time, the means and variances of $X(n)$ can be written compactly as follows:

$$E[X(n)) = \bar{X}(n) = \sum_{k=n2L_{PN}}^{(n+1)2L_{PN}-1} \left[\sum_{l=0}^{N_f-1} f_l \left(s((k-l)T_s + \tau) - \sum_{n=1}^{N_\alpha} \hat{\alpha}_n s((k-l-n)T_s + \tau) \right) \right]^* \hat{s}_f(k) \tag{11.136}$$

$$E[\,|X(n) - \bar{X}(n)|^2] = \sum_{k=n2L_{PN}}^{(n+1)2L_{PN}-1} \sum_{k'=n2L_{PN}}^{(n+1)2L_{PN}-1} \rho_{ee}(k-k')\hat{s}_f(k)\hat{s}_f^*(k') \tag{11.137}$$

where $\rho_{ee}(k)$ is the covariance function of the output of the prewhitening filter.

Some examples of the EKF-based receiver performance are considered next. In the following, the four-ray multipath channel of Figure 11.12 was used. The interferer was generated in the simulations by a fourth-order elliptic filter with a bandpass frequency response such that the spectral density was approximately rectangular. Both the signal-to-noise, E_b/N_0, and jammer to signal power ratios were fixed at $j/s = 10$ dB. The response of the estimated prewhitening filter, with coefficients $\hat{\alpha}_n(k\,|\,k-1)$, is illustrated in Figure 11.16, after 300 iterations of the EKF.

A timing error trajectory defined by $\tau - \hat{\tau}(k\,|\,k-1)$ is shown in Figure 11.17, with the four-ray channel ($N_f = 4$). It can be seen that the EKF estimate rapidly diverges when the prewhitening filter is not used. That is, the EKF is constructed with N_α assumed to equal zero, and thus the interferer is not excised in the innovations process. In contrast, with $N_\alpha = 3$, it can be seen that the timing error quickly converges to near zero, since the interferer is rejected prior to the measurement update. (In order to estimate the multipath, N_β was set to 7 in this example, corresponding to $N_\beta = N_\alpha + N_f$.)

The analytic BER is evaluated versus EKF iterations in Figure 11.18. The four-ray channel (Figure 11.12) is used, with an SNR of 10 dB. The lower dashed line indicates optimum BER, obtained when the values of α_n are fixed at the Wiener solution, and the β_m are computed using the true channel coefficient values, f_l. Although convergence is rapid (after 1000 iterations, representing approximately 33 bits), the BER can exceed the optimum by as much as a factor of ten. This phenomenon is explained as follows. The BER is extremely sensitive to the composite channel estimates, $\hat{\beta}_m$, as well as the prewhitening filter estimates. Even though the Kalman gain decreases with time, the Gaussian interferer can still generate large excursions in the measurement update equations. That is, unless zero process noise is assumed, with Q set to zero in the

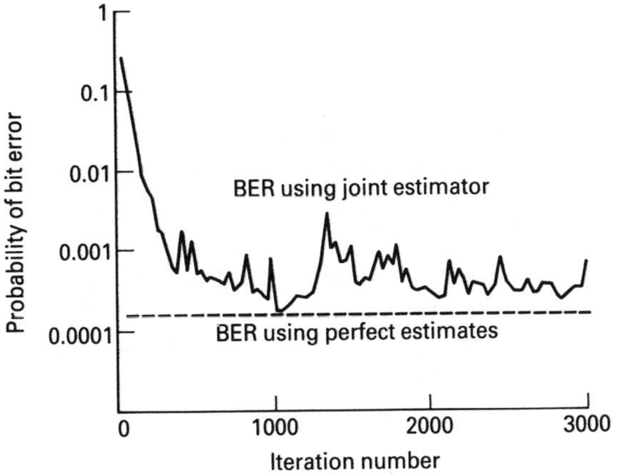

Figure 11.18 Bit error rate, $E_b/N_o = 10$ dB, $L_{PN} = 15$, $J/S = 10$ dB, four-ray channel; $N_\alpha = 3$, $N_\beta = 7$

prediction updates (and thus the Kalman gain eventually decays to zero), the gain sequence remains finite, and residual interference can still adversely affect the channel and interferer estimates in (11.130).

11.5 Summary

We have presented an overview of adaptive algorithms for interference rejection, channel estimation, and synchronization. Interference rejection, by itself, is easy to accomplish in spread-spectrum communications, since the transmitted signal is approximately statistically uncorrelated and is seen as white noise by the adaptive algorithm. Thus, minimization of total prewhitening filter output power was shown to be exactly equivalent to subtraction of the least-squares interferer estimate from the received signal. The LMS algorithm provides adequate interference rejection in both direct-sequence and frequency-hopped systems for stationary interferers. However, when estimates of the channel and signal code delay (time of arrival) are required, as in most practical spread-spectrum systems, more sophisticated adaptive filtering algorithms must be employed, based on the RLS and EKF algorithms.

The adaptive interference rejection filters discussed in this chapter have certain limitations. Primarily, it is assumed that the interferer is locally statistically stationary, so that the LMS algorithm has time to converge to the vicinity of the Wiener solution. However, when hostile jamming (for example, swept tones) is present, the statistics of the interference can become highly non-stationary. The LMS algorithm may be able to track some hostile interferers by increasing the step size μ, but the increased misadjustment noise and potential instabilities in the algorithm may prohibit this approach. Nevertheless, there are a wide variety of quasistationary interferers in both military and commercial systems that can be successfully excised using least-squares methods.

The RLS algorithm was seen to provide an effective technique for jointly estimating interferer and channel parameters. The underlying channel estimation problem was shown to be highly non-linear due to coupling between the AR and channel coefficients. The concept of the composite channel (convolution of the prewhitening filter and multipath channel) was introduced to alleviate this difficulty. It was then seen that the problem of joint estimation of interferer AR coefficients and the composite channel was linear, and of the ARX form. Thus, the RLS algorithm could provide the parameter estimates required to implement a modified RAKE receiver. The analytic bit error rate of such an adaptive receiver was evaluated versus iterations of the RLS algorithm, and it was found that the BER converged to the optimum (using the Wiener estimates) within 300 iterations. However, in the spread-spectrum system considered, there were 30 Nyquist samples per bit, and thus convergence occurred within only 10 bits, which is quite rapid in the context of communications applications.

An overall solution to the spread-spectrum demodulation problem was presented in the form of a digital RAKE receiver. The channel, interferer, and code delay estimates were provided simultaneously by an extended Kalman filter. To obtain a suitable parameterization, the same composite channel model was employed as in the RLS algorithm. The code delay was viewed as a non-linear parameter, which could be estimated recursively by the extended Kalman filter. The BER was again evaluated analytically, for a given estimation trajectory, providing a learning curve for EKF operation appropriate for communications system design. While the EKF-based digital RAKE is computationally complex, especially for wideband spread-spectrum applications, it is highly attractive, since it provides a complete solution to the problems of synchronization and channel/interferer estimation. As the capabilities of VLSI signal processing devices expand, the digital spread-spectrum receiver architectures should attract increased attention.

References

[1] F. M. Hsu and A. A. Giordano, 'Digital whitening techniques for improving spread-spectrum communications performance in the presence of narrowband jamming and interference', *IEEE Trans. Commun.*, vol. COM-26, pp. 209–16, 1978.

[2] J. W. Ketchum and J. G. Proakis, 'Adaptive algorithms for estimating and suppressing narrow-band interference in PN spread-spectrum systems', *IEEE Trans. Commun.*, vol. COM-30, pp. 913–23, 1982.

[3] L. Li and L. B. Milstein, 'Rejection of narrow-band interference in PN spread-spectrum systems using transversal filters', *IEEE Trans. Commu.*, vol. COM-30, pp. 925–36, 1982.

[4] R. A. Iltis and L. B. Milstein, 'Performance analysis of narrow-band interference rejection techniques in DS spread-spectrum systems', *IEEE Trans. Commun.*, vol. COM-32, pp. 1169–77, 1984.

[5] R. A. Iltis, 'A GLRT-based spread-spectrum receiver for joint channel estimation and interference suppression', *IEEE Trans. Commun.*, vol. COM-37, pp. 277–88, 1989.

[6] R. A. Iltis, J. A. Ritcey and L. B. Milstein, 'Interference rejection in FFH systems using least-squares estimation techniques', *IEEE Trans. Commun.*, vol. 38, pp. 2174–83, 1990.

[7] S. Theodoridis, N. Kalouptsidis, J. Proakis and G. Koyas, 'Interference rejection in PN spread-spectrum systems with LS filters with linear phase', *IEEE Trans. Commun.*, vol. COM-37, pp. 991–5, 1989.

[8] R. A. Iltis and L. B. Milstein, 'An approximate statistical analysis of the Widrow LMS algorithm with application to narrow-band interference rejection', *IEEE Trans. Commun.*, vol. COM-33, pp. 121–30, 1985.

[9] G. J. Saulnier, P. Das and L. B. Milstein, 'Suppression of narrow-band interference in a PN spread-spectrum receiver using a CTD-based adaptive filter', *IEEE Trans. Commun.*, vol. COM-32, pp. 1227–32, 1984.

[10] C. M. Anderson, E. H. Satorius and J. R. Zeidler, 'Adaptive enhancement of finite bandwidth signals in white Gaussian noise', *IEEE Trans. Acoust., Speech, Signal Process.*, vol. ASSP-31, pp. 17–27, 1983.

[11] B. Widrow, J. M. McCool, M. G. Larimore and C. R. Johnson, 'Stationary and nonstationary learning characteristics of the LMS adaptive filter', *Proc. IEEE*, vol. 64, pp. 1151–61, 1976.

[12] L. L. Horowitz and K. D. Senne, 'Performance advantage of complex LMS for controlling narrow-band adaptive arrays', *IEEE Trans. Circuits Syst.*, vol. CAS-28, June, 1981.

[13] B. Widrow, J. McCool and M. Ball, 'The complex LMS algorithm', *Proc. IEEE*, vol. 63, pp. 719–920, 1975.

[14] J. S. Lee, L. E. Miller and Y. K. Kim, 'Probability of error analyses of a BFSK frequency-hopping system with diversity – Part II', *IEEE Trans. Commun.*, vol. COM-32, pp. 1243–50, 1984.

[15] L. E. Miller, J. S. Lee and A. P. Kadrichu, 'Probability of error analyses of a BFSK frequency-hopping system with diversity under partial band jamming interference –Part III: Performance of a square-law self-normalizing soft decision receiver', *IEEE Trans. Commun.*, vol. COM-34, pp. 669–75, 1986.

[16] S. Stein, 'Fading channel issues in system engineering', *IEEE J. Selected Areas Commun.*, vol. SAC-5, pp. 68–89, 1987.

[17] G. L. Turin, 'Introduction to spread-spectrum antimultipath techniques and their application to urban digital radio', *Proc. IEEE*, vol. 68, pp. 328–52, 1980.

[18] John G. Proakis, *Digital Communications*, 2nd edn, McGraw-Hill: Singapore, 1989.

[19] R. Price and P. E. Green, 'A communications technique for multipath channels', *Proc. IRE*, vol. 46, pp. 555–70, 1958.

[20] R. A. Iltis, 'Joint estimation of PN code delay and multipath using the extended Kalman filter', *IEEE Trans. Commun.*, vol. COM-38, pp. 1677–85, 1990.

[21] B. D. O. Anderson and J. B. Moore, *Optimal Filtering*, Prentice Hall: Englewood Cliffs, NJ, 1979.

[22] A. A. Giordano and F. M. Hsu, *Least-Square Estimation with Applications to Digital Signal Processing*, Wiley: New York, 1985.

[23] H. Messer and Y. Bar-Ness, 'Closed-loop least-mean square time-delay estimator', *IEEE Trans. Acoust., Speech, Signal Process.*, vol. ASSP-35, pp. 413–24, 1987.

12

Neural Networks for Adaptive Signal Processing

Simon Haykin and Andrew Ukrainec

12.1 Introduction

A *neural network* consists of the interconnection of a *large* number of *simple* computing elements commonly referred to as *neurons* that operate in *parallel.* This description applies to natural as well as artificial forms of neural networks.

Interest in artificial neural networks may be traced back to the pioneering work of McCullough and Pitts [1]. In their classic paper published in 1943, McCullough and Pitts introduced a memoryless model for the neuron that consists of a linear combiner followed by a threshold unit. The *McCullough–Pitts model* formed the basis of much of the neural network literature that was published in the 1950s and 1960s. Notable among the papers published during this early stage of neural network development were those of Rosenblatt [2] on the *perceptron* and Widrow and Hoff [3] on the *Adaline* that is closely allied to the perceptron. However, serious interest in neural network research started to decline in the late 1960s largely because of some fundamental limitations in the computing capability of the perceptron. These limitations were highlighted by Minsky and Papert [4] in a book first published in 1969.

Lack of interest in artificial neural networks continued through the 1970s and the early 1980s. Nevertheless, a serious commitment to artificial neural networks was maintained by a handful of researchers, notably Anderson [5], Cooper [6] and Grossberg [7] in the United States of America, Kohonen [8] in Finland, and Amari [9] in Japan.

Interest in artificial neural networks was revived in 1982 through the publication of a paper by Hopfield [10] on an associative memory using a *recurrent network*. This new form of a neural network soon became known as the *Hopfield network*. Indeed, the Hopfield network has become the subject of intensive research for solving the travelling salesman problem [11–13] and as a content addressable memory [14,15].

The Hopfield network is based on a generalization of *Hebb's postulate of learning*, so named in honour of Hebb [15]. This important learning rule may be viewed as a correlational type of learning, in the sense that the weight of a synapse (connection) is changed by an amount proportional to the product of the presynaptic (input) signal and the postsynaptic (output) signal. Indeed, Hebb's postulate of learning is basic to the operation of other neural networks, namely *correlation matrix memories* [5,17], *self-organized feature map classifiers* [18] and *feedforward multilayer adaptive networks* [19].

An important by-product of the early work by Widrow and Hoff on the Adaline was the development of the ubiquitous *least mean square* (LMS) *algorithm*. The LMS algorithm is well suited for the design of linear adaptive filters consisting of a single neuron. In order to deal with a multitude of neurons arranged in the form of two or more layers, as in the *multilayer perceptron*, we may use the *back-error propagation* (BP) *algorithm* that represents a generalization of the LMS algorithm. The basic idea of the back-error propagation algorithm owes its origin to a PhD thesis written in 1974 by Werbos [20]. It was subsequently rediscovered by Rumelhart *et al.* [21], Parker [22] and leCun [23]. The use of the back-error propagation algorithm was popularized by the publication of the two-volume book written by members of the PDP Group and co-edited by Rumelhart and McClelland [24].

During the past 6 years or so, interest in the theory and design of artificial neural networks and their applications has intensified. Indeed, there are now a number of journals and many international conferences devoted exclusively to neural networks. Moreover, it is safe to say that neural networks are here to stay and will undoubtedly have an impact on future developments in non-linear adaptive signal processing.

In this chapter we will present a discussion of what neural networks are and why members of the signal processing community have to pay attention to their use. Specifically, in section 12.2 we describe the basic operation of a simple neuron, and in section 12.3 we argue the case for neural networks in adaptive signal processing. Then in section 12.4 we discuss different types of learning rules and classes of neural networks. Next, in section 12.5, we describe the essence of the back-error propagation algorithm. This is followed by a discussion of radial basis function (RBF) networks in section 12.6 that builds on the extensive knowledge developed for the design of linear adaptive filters. We conclude the chapter in section 12.8 by emphasizing the multidisciplinary nature of neural networks and identifying a number of areas in adaptive signal processing that may benefit from the use of neural networks.

Undoubtedly, the reader will have noted from these introductory remarks that neural networks constitute an interdisciplinary subject with many facets and applications. In the limited space available to us in this chapter, it is therefore impossible to cover the subject in detail. Our primary purpose here is to provide a source of motivation for the reader. For more details on the subject, the interested reader is referred to the papers listed in the References, and the special issue of the *Proceedings of the IEEE* [25]. A more complete list of references is found in [26]. The relevant journals on the subject include the *IEEE Transactions on Neural Networks* published by the Institute of Electrical and Electronic Engineers, *Neural Networks* published by the International Neural Network Society, *Neural Computation* published by MIT Press, and *Network-Computation in Neural Systems* published by the Institute of Physics, UK. For an introductory treatment of neural networks, the reader is referred to Lippmann's paper [27] and the book by Haykin [28].

12.2 Simplified model of a neuron

Figure 12.1 shows the *model* of an artificial neuron. It consists of a *linear combiner* followed by a *non-linear unit.* The linear combiner itself consists of a set of *synaptic weights* connected to respective input terminals, and whose weighted outputs are combined in a *summing* junction. An external bias plus the linear combiner output constitute the net input of the non-linear unit.

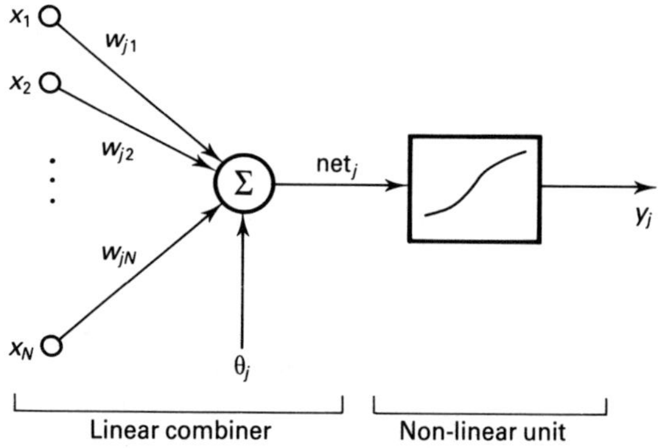

Figure 12.1 A simplified model of a neuron

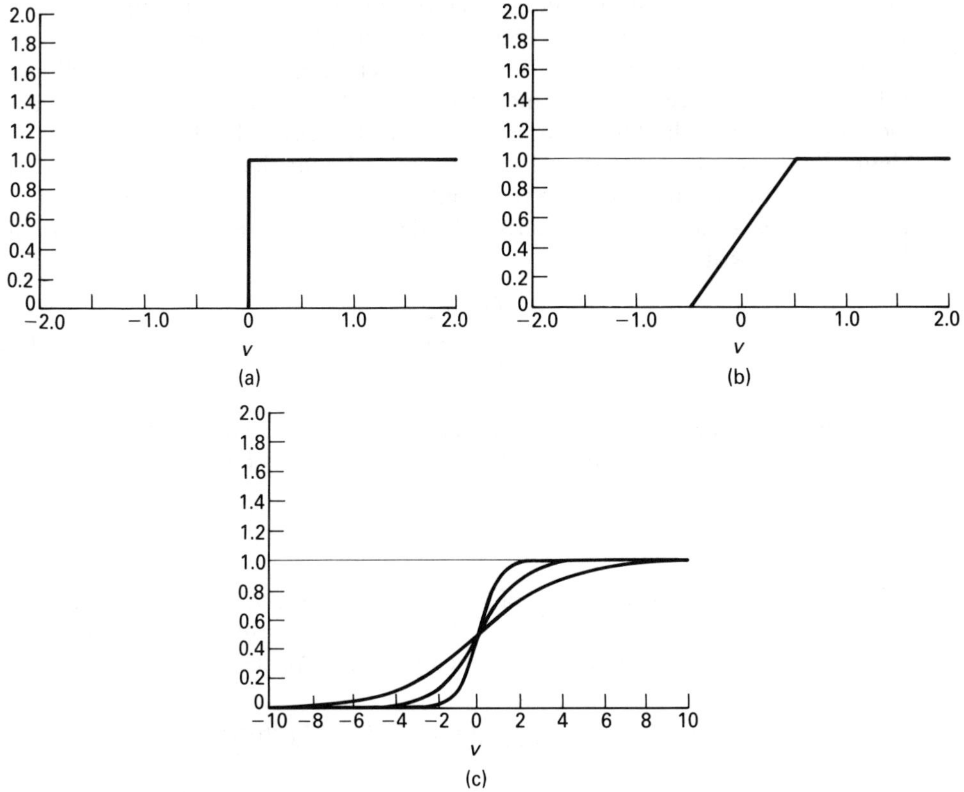

Figure 12.2 (a) Threshold function; (b) piecewise linear function; (c) sigmoidal function

We may distinguish four basic types of artificial neurons, depending on the exact description of the non-linear unit:

1. **Linear model:** In this model the non-linear unit is replaced by a *direct connection*, with the result that the output of the neuron is a weighted sum of its inputs. This special form of a neuron is basic to the operation of linear adaptive filters, on which much of our present knowledge of adaptive signal processing is based.
2. **McCullough–Pitts model:** In this second model of a neuron the non-linear unit is characterized by a *threshold function* as depicted in Figure 12.2(a). As mentioned previously, the McCullough–Pitts model is basic to the operation of Rosenblatt's perceptron. It is also basic to the operation of the discrete version of the Hopfield network.
3. **Piecewise linear model:** The input–output characteristic of the non-linear unit for this model of a neuron is described in Figure 12.2(b). The piecewise linear model includes the linear model and the McCullough–Pitts model as special cases. If the

linear region of the input–output characteristic in Figure 12.2(b) is made infinitely wide, we get the linear model. If, on the other hand, it is made infinitely narrow (i.e. the slope of the linear region is made infinitely large), we get the McCullough–Pitts model. The piecewise linear model is used in the construction of an associative memory known as the *brain-state-in-a-box* (BSB) *network* [29].

4. **Sigmoidal model:** This last model of a neuron is by far the most widely used one in practice. It is so called because the input–output characteristic of the non-linear unit is *S-shaped.* Let this characteristic be denoted by the non-linear function $f(\cdot)$. We may then write

$$f(\text{net}) = \begin{cases} 1 & \text{net} = \infty \\ \frac{1}{2} & \text{net} = 0 \\ 0 & \text{net} = -\infty \end{cases} \tag{12.1}$$

where net is the sum of the linear combiner output plus the bias. A highly popular form of sigmoidal non linearity is the *logistic function*, defined by

$$f(\text{net}) = \frac{1}{1 + \mathrm{e}^{-\text{net}}} \tag{12.2}$$

Figure 12.2(c) shows a depiction of the sigmoidal non-linearity. The maximum slope of the sigmoidal function of (12.2) equals one-quarter. We can change the slope of the logistic function by introducing a factor β, such that

$$f(\text{net}) = \frac{1}{1 + \mathrm{e}^{-\beta\text{net}}} \qquad f'(\text{net}) = \beta \mathrm{e}^{-\beta\text{net}} f^2(\text{net}) \tag{12.3}$$

When the value of β is increased beyond unity and made infinitely large, the slope is made infinitely large as well, and the sigmoidal model of a neuron reduces to the McCullough–Pitts neuron. The model of a neuron using the sigmoidal non-linearity is basic to the operation of the highly popular back-error propagation algorithm.

12.3 Why neural networks for adaptive signal processing?

Much of our present knowledge of adaptive signal processing is built around linear adaptive filters [30,31]. For the operation of a linear adaptive filter we may use the simple *least mean square* (LMS) *algorithm* or more complex *recursive least-squares* (RLS) *algorithm.* These two families of algorithms differ in their rates of convergence, sensitivity to variations in the eigenvalue spread of the input data, tracking characteristics, roundoff error characteristics, and computational complexities [30]. However, they are both *limited in their capacity to extract information from input data.* This limitation is exemplified by the fact that they respond to second-order statistics of the input data, but no higher!

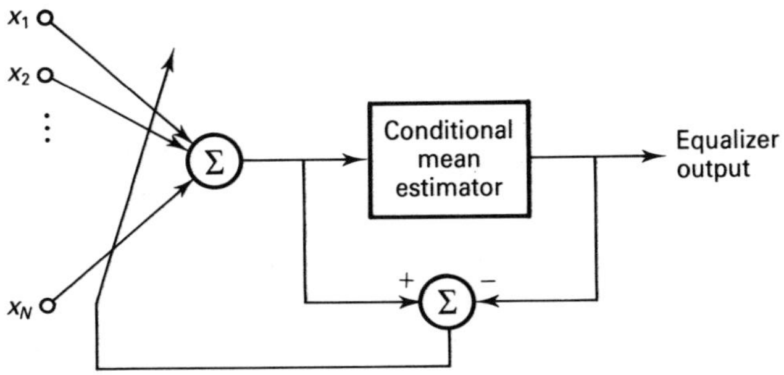

Figure 12.3 Block diagram of Bussgang algorithm for blind equalization

A good example to illustrate this important point is to consider the *blind deconvolution problem.* In the conventional form of deconvolution, we are given the input and output signals of a linear system (channel), and the requirement is to estimate the impulse response of the system. The blind deconvolution problem is more difficult to solve in the sense that we only have physical access to the output signal of the system; the only information we have available about the system input is in the form of a *statistical model.* The term 'blind deconvolution' is used to emphasize the fact there is no direct access to the input data. The only way in which this problem can be solved is for (a) the environment in which the system operates to be non-Gaussian, and (b) the adaptive filter for solving the problem to be non-linear [32].

Indeed, one way of solving the blind deconvolution problem is to use the *Bussgang algorithm* [33,34]. To be specific, Figure 12.3 shows the basic elements of a Bussgang type of *blind equalizer* that consists of a linear combiner followed by a *Bayes conditional mean estimator*. The output of the latter estimator provides the *desired response*, and the difference between its input and output signals provides the *error signal* that is used to adjust the weights in the linear combiner. The important point to note here is that the conditional mean estimator consists of a memoryless non-linear device, which itself may be very closely realized by a sigmoidal non-linearity [30]. In other words, to construct a Bussgang type of blind equalizer we essentially need a single neuron with a sigmoidal non-linearity!

For another example to illustrate the role of neural networks for adaptive signal processing, consider the problem of classifying primary radar returns in an air traffic control environment. The classes of radar returns of interest include aircraft echoes, weather clutter, clutter due to migrating flocks of birds in the vicinity, and ground clutter. The motivation is to identify areas of weather disturbances (e.g. clouds and storms) and migrating birds for the purpose of vectoring aircraft around them. Aircraft echoes and ground clutter tend to have a similar spectral spread, but the centre frequency of the spectrum of an aircraft echo can be larger than that of ground clutter due to the Doppler shift produced by radial motion of the aircraft with respect to the radar. Moreover, these two forms of radar returns usually have non-Gaussian statistics. On the other hand, weather clutter and bird clutter exhibit a wider spectral spread, and bird clutter tends to experience greater variability [35]. In so far as statistics are concerned, weather clutter is well approximated as a Gaussian process, whereas bird clutter may exhibit some form of non-Gaussian behaviour. Based on these observations, we may successfully classify these different classes of radar returns using power spectra or related parameters such as reflection coefficients. Indeed, this has been demonstrated using real-life data, with the feature classifier consisting of a reflection coefficient computer based on Burg's algorithm that relates to second-order statistics [35]. It has been further demonstrated that the classification accuracy can be improved by expanding the feature vector to include higher-order statistics such as skewness and kurtosis [36].

The main points to note from this discussion are the following:

1. Conventional adaptive filtering algorithms are usually *phase blind.* That is, they yield minimum phase solutions in the sense that they do not respond to phase information contained in the input signal in excess of a minimum-phased amplitude characteristic. This points to a major limitation of conventional adaptive filtering algorithms whose design is based on second-order statistics.
2. To exploit the full information content of the input signal, we need to use non-linear signal processing techniques. However, for such techniques to be of practical value, the input signal must have a non-Gaussian distribution.
3. Neural networks provide a practical means for efficient extraction of the information content of incoming data. They are able to outperform conventional adaptive filters by virtue of three factors:
   (a) the elaborate structure of neural networks;
   (b) the non-linearity built into the construction of artificial neurons; and
   (c) the ability of neural networks to learn from a non-linear environment through training.

12.4 Classification of neural networks

We may identify three basic classes of neural networks, depending on how the desired response (target signal) for the learning process is provided. Specifically, we have the following:

1. **Supervised learning:** In this form of learning, an *external teacher* is built into the design of the neural network for the purpose of providing the desired response. The LMS algorithm and the back-error propagation algorithm are examples of supervised learning algorithms.
2. **Reinforcement learning:** In this second form of learning, a critic is built into the design of the neural network [37]. The critic helps the neural network learn about its environment through a *punish and reward mechanism.* The network is rewarded for making a correct decision and punished for making a wrong decision.

Reinforcement learning is rooted in psychology. It finds application in adaptive control [38].

3. **Self-organized learning:** In this final form of learning, the neural network has no teacher. However, the network is designed in such a way that it can learn about its environment (i.e. discover new classes automatically) without external supervision. To do this, the network usually requires a greater volume of data than a supervised one. A blind equalizer represents a very simple example of a self-organized network. Other examples of self-organized networks include those designed using Hebb's postulate of learning [17,19].

It is of interest to note that we may hybridize supervised and self-organized networks for adaptive pattern classification. Specifically, the self-organized network is employed as the preprocessor (i.e. feature extractor) for the classifier-type, supervised neural network.

12.5 Back-propagation networks

The network structure for the *back-error propagation algorithm* is a *multilayer perceptron* (MLP) that consists of an *input layer* of source nodes, one or more *hidden layers* of computational nodes, and an *output layer* also of computational nodes. The source nodes provide physical access points for the application of input signals. The neurons in the hidden layers act as feature detectors. Finally, the neurons in the output layer present to a user the conclusions reached by the network in response to the input signals.

Figure 12.4 depicts a multilayer perceptron with a pair of input nodes, a single layer of hidden nodes, and a single output node. Two features of such a structure are immediately apparent from this figure:

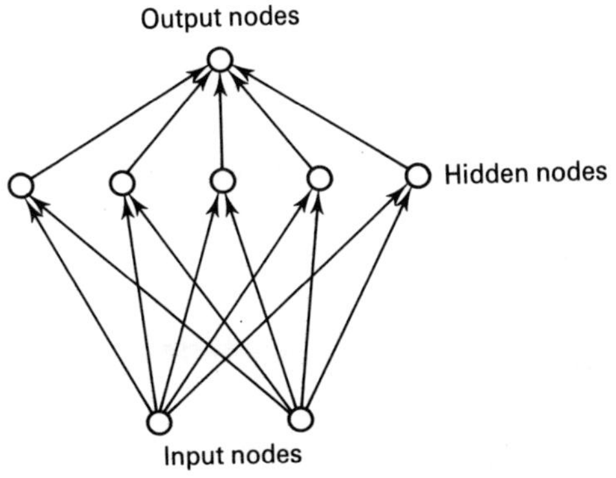

Figure 12.4 Signal flow graph of multilayer feedforward network; only a single hidden layer is shown

1. A multilayer perceptron is a *feedforward network* in the sense that the input signals produce a response at the outputs of the network by propagating in the forward direction only. There is *no* feedback in the network.
2. The network may be *fully connected*, as shown in Figure 12.4, in that each node in a layer of the network is connected to every other node in the layer above it. Alternatively, the network may be *partially connected* in that some of the synaptic connections may be missing. Locally connected networks represent an important type of partially connected networks; the term 'local' refers to the connectivity of a neuron in a hidden layer of the network only to a subset of possible inputs.

The number of source nodes in the input layer is determined by the dimensionality of the observation space that is responsible for the generation of the input signals. The number of computational nodes in the output layer is determined by the required dimensionality of the desired response. Thus the design of a multilayer perceptron requires that we address three issues:

1. The determination of the number of hidden layers.
2. The determination of the number of neurons in each of the hidden layers.
3. The specification of the synaptic weights that interconnect the neurons in the different layers of the network.

Issues 1 and 2 relate to *neural* (model) *complexity* [39]. As such, these two issues represent the weakest link in our present knowledge of how to design a multilayer perceptron. More will be said on network complexity later in the section. To resolve issue 3, we may use the back-error propagation algorithm, also referred to in the literature as the back-propagation algorithm or simply the backprop (BP) algorithm.

The following sections are concerned with the derivation and application of two types of back-propagation algorithm which are of interest to those who wish to use it for signal processing. First, the derivation of the complex back-error propagation algorithm is presented, followed by the real version of the BP algorithm. The real version of the BP algorithm is shown to be a special case of the complex algorithm.

12.5.1 Complex back-error propagation network

The back-error propagation algorithm may be viewed as a generalization of the LMS algorithm. Likewise, the complex back-error propagation (BP) algorithm is a generalization of the complex LMS algorithm, with an appropriately chosen (non-linear) activation function. The criterion used to determine an appropriate activation function will be discussed herein. In any event, there are two passes of signals in the implementation of the BP algorithm:

1. **Forward pass:** In the forward pass, also termed the function level adaptation, the synaptic weights are *fixed*, and the response of the network is computed by subjecting it to a prescribed set of input signals. The forward pass in the BP algorithm is analogous to the filtering process in the LMS algorithm [30].

2. **Backward pass:** In the backward pass, also termed the parameter level adaptation, the adjustments to the synaptic weights are computed for the purpose of minimizing the cost function defined as a sum of error squares. In particular, we start by computing the error signals in the output layer, and then work *backwards* through the network, layer by layer, until the complete network is covered. The BP algorithm derives its name from the backward nature of the error computations described herein. Note also that the backward pass in the BP algorithm is analogous to the adaptive process in the LMS algorithm [30].

The derivation of the BP algorithm is usually presented for real-valued data [24,27]. How do we handle the use of complex-valued data? This need often arises in processing coherent data signals, e.g. those found in radar, sonar and communications fields. We may accommodate the use of complex data in one of two ways:

1. The real and imaginary parts of each member of the input set of data are treated as two separate entities; similarly, the real and imaginary parts of each member of the network output are treated as two separate entities. The synaptic weights of the network are then computed in accordance with the *real* (conventional) form of the BP algorithm [36].
2. The synaptic weights are assigned complex values, and their computations are performed using the *complex* form of the BP algorithm [40–47].

We can show that the two approaches are equivalent. Let us say that the desired mapping is

$$\mathbf{z} \mapsto f(\mathbf{z}) = f(\mathbf{z}_I + j\mathbf{z}_Q) = u(\mathbf{z}_I, \mathbf{z}_Q) + jv(\mathbf{z}_I, \mathbf{z}_Q) \tag{12.4}$$

where

$$(\mathbf{z}_I, \mathbf{z}_Q) \mapsto u(\mathbf{z}_I, \mathbf{z}_Q) \quad \text{and} \quad (\mathbf{z}_I, \mathbf{z}_Q) \mapsto v(\mathbf{z}_I, \mathbf{z}_Q) \tag{12.5}$$

u and v being real functions of the complex input vector $\mathbf{z}$ [48][1]. Two real-valued feedforward networks (or one real-valued feedforward network with two outputs) can thus be used to compute the resultant mapping, one giving the real part of the mapping, the other the imaginary part of the mapping. There is, however, an advantage in using a network with complex weights. Referring to the linear combiner section of Figure 12.5, the complex linear function can be written in vector notation as

$$\begin{aligned} y = f(\mathbf{x}) = \mathbf{x}^H\mathbf{w} &= (\mathbf{x}_I + j\mathbf{x}_Q)^H(\mathbf{w}_I + j\mathbf{w}_Q) \\ &= (\mathbf{x}_I^T\mathbf{w}_I + \mathbf{x}_Q^T\mathbf{w}_Q) + j(-\mathbf{x}_Q^T\mathbf{w}_I + \mathbf{x}_I^T\mathbf{w}_Q) \end{aligned} \tag{12.6}$$

where $\mathbf{x} = \mathbf{x}_I + j\mathbf{x}_Q$ is an $n \times 1$ complex input vector, likewise $\mathbf{w} = \mathbf{w}_I + j\mathbf{w}_Q$ is an $n \times 1$ complex weight vector, and y is the complex scalar output. The equivalent real-valued combiner can also be constructed using only real-valued vectors, so that

$$\begin{aligned} f\left(\begin{bmatrix} \mathbf{x}_I \\ \mathbf{x}_Q \end{bmatrix}\right) &= [\mathbf{x}_I^T \; \mathbf{x}_Q^T]\begin{bmatrix} \mathbf{u}_I & \mathbf{v}_I \\ \mathbf{u}_Q & \mathbf{v}_Q \end{bmatrix} \\ &= [\mathbf{x}_I^T\mathbf{u}_I + \mathbf{x}_Q^T\mathbf{u}_Q \;\; \mathbf{x}_I^T\mathbf{v}_I + \mathbf{x}_Q^T\mathbf{v}_Q] = [y_I \; y_Q], \end{aligned} \tag{12.7}$$

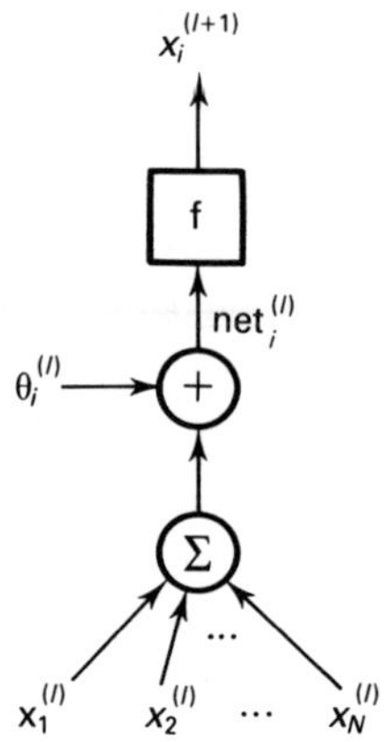

Figure 12.5 Adaptive non-linear combiner

where the weight matrix consists of real vectors $\mathbf{u}_I$, $\mathbf{u}_Q$, $\mathbf{v}_I$ and $\mathbf{v}_Q$. The resultant real vector $[y_I \, y_Q]$ contains the real and imaginary components of y. Comparing (12.6) and (12.7) we observe that

$$\begin{aligned} \mathbf{u}_I &= \mathbf{w}_I & \mathbf{v}_I &= -\mathbf{w}_I \\ \mathbf{u}_Q &= \mathbf{w}_Q & \mathbf{v}_Q &= \mathbf{w}_Q \end{aligned} \tag{12.8}$$

and therefore

$$\mathbf{u}_I = -\mathbf{v}_I \qquad \mathbf{u}_Q = \mathbf{v}_Q \tag{12.9}$$

It is apparent that a network with real-valued weights has more degrees of freedom than absolutely necessary to solve the complex mapping problem. In general, the real-valued learning algorithm treats all the weights as independent parameters, adjusting them to decrease the cost function. In the case of a complex-valued mapping, symmetries exists which are not taken advantage of by the learning algorithm[2]. In other words, the network with complex-valued weights gives a parsimonious solution as compared with the network with real-valued weights. Other considerations include those of convergence. It has been shown [49] that for the LMS algorithm, superior performance is achieved for the complex LMS algorithm, over that of the real version of the LMS algorithm. The algorithm is more stable and mean-squared convergence is almost twice that of the real LMS algorithm. Since the BP algorithm is a generalization of the LMS algorithm, we may hypothesize that this behaviour could carry over to the feedforward neural network as well.

12.5.2 The complex back-error propagation algorithm

We now present a detailed derivation of the complex form of the back-error propagation algorithm. The complex algorithm was developed independently by several

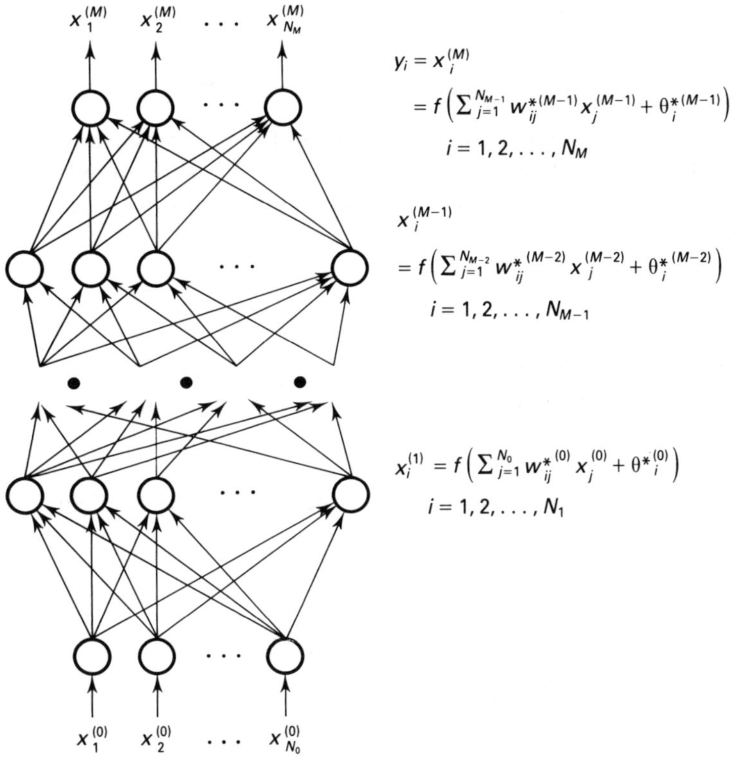

Figure 12.6 Multilayer perception

researchers [40–47]. All the approaches are fundamentally a generalization of the complex least mean squares (LMS) algorithm [3] to a network with multiple layers of multiple linear combiners with non-linearities. The introduction of non-linearity into the network raises the question 'What form does the complex activation function take?' The answer requires a consideration of the nature of differentiable functions of complex variables. After developing the complex algorithm, the conventional (real) form of the BP algorithm is presented as a special case of the complex version.

A multilayer perceptron, shown in Figure 12.6, consists of many adaptive linear combiners with a non-linearity at the output; such a combiner is shown in Figure 12.5. The input–output relationship of such a unit is characterized by the non-linear difference equation

$$x_i^{(l+1)} = f\left(\sum_{j=1}^{N} w_{ij}^{*(l)} x_j^{(l)} + \theta_i^{*(l)}\right) \tag{12.10}$$

with the output being the ith node in the $(l+1)$th layer. The parameter θ is a bias term, equivalent to a weight with a constant $+1$ input. Equation (12.10) is generalized to all units in the multilayer perceptron as shown in Figure 12.6.

The error signal is defined to be the difference between the desired response and the output of the network. Therefore

$$\varepsilon_i(n) = d_i(n) - y_i(n) \quad i = 1,2,\ldots, N_M \tag{12.11}$$

where $d_j(n)$ is the desired response at the jth node of the output layer at time n, $y_j(n)$ is the output at the jth node of the output layer, and N_M is the number of neurons in the Mth layer of the neural network. Hence, the sum of error squares produced by the network is

$$E(n) = \sum_{i=1}^{N_M} \varepsilon_i(n)\varepsilon_i^*(n) = \sum_{i=1}^{N_M} |\varepsilon_i(n)|^2 \tag{12.12}$$

The BP algorithm minimizes the cost function $E(n)$ by recursively altering the complex weights using the gradient descent technique. The weight update equation is

$$w_{ij}^{(l)}(n+1) = w_{ij}^{(l)}(n) + \Delta w_{ij}^{(l)}(n) \tag{12.13}$$

The weights are changed in proportion to the negative of the gradient. The update term is defined to be

$$\Delta w_{ij}^{(l)}(n) \stackrel{\text{def}}{=} -\tfrac{1}{2}\mu \nabla_{\mathrm{w}ij}^{(l)} E(n) \tag{12.14}$$

We must first find the partial derivative of $E(n)$ with respect to the coefficients of the $(M-1)$th layer, and then extend it to the coefficients of all the hidden layers. The gradient of the real function $E(n)$ with respect to the complex weights in the $(M-1)$th layer is thus defined as

$$\nabla_{wij}^{(M-1)} E(n) \stackrel{\text{def}}{=} \frac{\partial E(n)}{\partial \mathrm{w}_{\mathrm{I}ij}^{(M-1)}(n)} + \mathrm{j}\frac{\partial E(n)}{\partial \mathrm{w}_{\mathrm{Q}ij}^{(M-1)}(n)} \tag{12.15}$$

where the complex weight connecting the jth node to the ith node for layer $(M-1)$ at time n is given to be

$$\mathrm{w}_{ij}^{(M-1)}(n) = \mathrm{w}_{\mathrm{I}ij}^{(M-1)}(n) + \mathrm{j}\mathrm{w}_{\mathrm{Q}ij}^{(M-1)}(n) \tag{12.16}$$

The output $y_i(n)$ is therefore[3]

$$y_i = x_i^{(M)} = f(\mathrm{net}_i^{(M-1)}) \tag{12.17}$$

where

$$\begin{aligned} \mathrm{net}_i^{(M-1)} &= \mathrm{net}_{\mathrm{I}i}^{(M-1)} + \mathrm{j}\,\mathrm{net}_{\mathrm{Q}i}^{(M-1)} \\ &= \sum_{j=1}^{N_{M-1}} \mathrm{w}_{ij}^{*(M-1)} x_j^{(M-1)} + \theta_i^{*(M-1)} \end{aligned} \tag{12.18}$$

Assume that $f(\mathrm{net}_i^{(M-1)})$ is a suitable complex activation function; hence let

$$\begin{aligned} f(\mathrm{net}_i) &= f(\mathrm{net}_{\mathrm{I}i} + \mathrm{j}\,\mathrm{net}_{\mathrm{Q}i}) \\ &= u(\mathrm{net}_{\mathrm{I}i}, \mathrm{net}_{\mathrm{Q}i}) + \mathrm{j}v(\mathrm{net}_{\mathrm{I}i}, \mathrm{net}_{\mathrm{Q}i}) \end{aligned} \tag{12.19}$$

where u and v are real functions. The derivative of the activation function, if it exists, is

$$f'(\text{net}_i) = \frac{\mathrm{d}f(\text{net}_i)}{\mathrm{d}\ \text{net}_i} \tag{12.20}$$

The partial derivatives of u and v are defined as

$$u'_{\mathrm{I}i} = \frac{\partial u_i}{\partial \text{net}_{\mathrm{I}i}} \quad u'_{\mathrm{Q}i} = \frac{\partial u_i}{\partial \text{net}_{\mathrm{Q}i}} \quad v'_{\mathrm{I}i} = \frac{\partial v_i}{\partial \text{net}_{\mathrm{I}i}} \quad v'_{\mathrm{Q}i} = \frac{\partial v_i}{\partial \text{net}_{\mathrm{Q}i}} \tag{12.21}$$

Next, we need to find expressions for $\partial E(n)/\partial w_{\mathrm{I}ij}^{(M-1)}$ and $\partial E(n)/\partial w_{\mathrm{Q}ij}^{(M-1)}$. Using the chain rule,

$$\frac{\partial E(n)}{\partial w_{\mathrm{I}ij}^{(M-1)}} = \frac{\partial E}{\partial u_i}\left(\frac{\partial u_i}{\partial \text{net}_{\mathrm{I}i}}\frac{\partial \text{net}_{\mathrm{I}i}}{\partial w_{\mathrm{I}ij}} + \frac{\partial u_i}{\partial \text{net}_{\mathrm{Q}i}}\frac{\partial \text{net}_{\mathrm{Q}i}}{\partial w_{\mathrm{I}ij}}\right) + \frac{\partial E}{\partial v_i}\left(\frac{\partial v_i}{\partial \text{net}_{\mathrm{I}i}}\frac{\partial \text{net}_{\mathrm{I}i}}{\partial w_{\mathrm{I}ij}} + \frac{\partial v_i}{\partial \text{net}_{\mathrm{Q}i}}\frac{\partial \text{net}_{\mathrm{Q}i}}{\partial w_{\mathrm{I}ij}}\right)$$

$$\frac{\partial E(n)}{\partial w_{\mathrm{Q}ij}^{(M-1)}} = \frac{\partial E}{\partial u_i}\left(\frac{\partial u_i}{\partial \text{net}_{\mathrm{I}i}}\frac{\partial \text{net}_{\mathrm{I}i}}{\partial w_{\mathrm{Q}ij}} + \frac{\partial u_i}{\partial \text{net}_{\mathrm{Q}i}}\frac{\partial \text{net}_{\mathrm{Q}i}}{\partial w_{\mathrm{Q}ij}}\right) + \frac{\partial E}{\partial v_i}\left(\frac{\partial v_i}{\partial \text{net}_{\mathrm{I}i}}\frac{\partial \text{net}_{\mathrm{I}i}}{\partial w_{\mathrm{Q}ij}} + \frac{\partial v_i}{\partial \text{net}_{\mathrm{Q}i}}\frac{\partial \text{net}_{\mathrm{Q}i}}{\partial w_{\mathrm{Q}ij}}\right) \tag{12.22}$$

Solving for $\partial \text{net}_{\mathrm{I}i}/\partial w_{\mathrm{I}ij}$,

$$\frac{\partial \text{net}_{\mathrm{I}i}}{\partial w_{\mathrm{I}ij}} = \frac{\partial}{\partial w_{\mathrm{I}ij}}[w_{\mathrm{I}ij}x_{\mathrm{I}j} + w_{\mathrm{Q}ij}x_{\mathrm{Q}j} + \theta_{\mathrm{I}i}] = x_{\mathrm{I}j} \tag{12.23}$$

The other partial derivatives of (12.18) can be found in a similar manner. Summarizing,

$$\frac{\partial \text{net}_{\mathrm{I}i}}{\partial w_{\mathrm{I}ij}} = x_{\mathrm{I}j} \quad \frac{\partial \text{net}_{\mathrm{Q}i}}{\partial w_{\mathrm{I}ij}} = x_{\mathrm{Q}j}$$
$$\frac{\partial \text{net}_{\mathrm{I}i}}{\partial w_{\mathrm{Q}ij}} = x_{\mathrm{Q}j} \quad \frac{\partial \text{net}_{\mathrm{Q}i}}{\partial w_{\mathrm{Q}ij}} = -x_{\mathrm{I}J} \tag{12.24}$$

Substituting (12.21) and (12.24) into (12.22), the partial derivatives in the two lines of the latter equation can be expressed respectively as

$$\frac{\partial E(n)}{\partial w_{\mathrm{I}ij}^{(M-1)}} = \frac{\partial E(n)}{\partial u_i^{(M-1)}}(u'_{\mathrm{I}i}x_{\mathrm{I}j} + u'_{\mathrm{Q}i}x_{\mathrm{Q}j}) + \frac{\partial E(n)}{\partial v_i^{(M-1)}}(v'_{\mathrm{I}i}x_{\mathrm{I}j} + v'_{\mathrm{Q}i}x_{\mathrm{Q}j})$$

$$\frac{\partial E(n)}{\partial w_{\mathrm{Q}ij}^{(M-1)}} = \frac{\partial E(n)}{\partial u_i^{(M-1)}}(u'_{\mathrm{I}i}x_{\mathrm{Q}j} - u'_{\mathrm{Q}i}x_{\mathrm{I}j}) + \frac{\partial E(n)}{\partial v_i^{(M-1)}}(v'_{\mathrm{I}i}x_{\mathrm{Q}j} - v'_{\mathrm{Q}i}x_{\mathrm{I}j}) \tag{12.25}$$

For the weights belonging to layer $(M-1)$, the partial derivatives of the cost function can be readily found, as follows:

$$\frac{\partial E(n)}{\partial u_i^{(M-1)}} = -2[d_{\mathrm{I}i}(n) - y_{\mathrm{I}i}(n)] = -2\varepsilon_{\mathrm{I}i}(n)$$
$$\frac{\partial E(n)}{\partial v_i^{(M-1)}} = -2[d_{\mathrm{Q}i}(n) - y_{\mathrm{Q}i}(n)] = -2\varepsilon_{\mathrm{Q}i}(n) \tag{12.26}$$

Substituting (12.26) into (12.25) and simplifying, we find

$$\nabla_{w_{ij}}^{(M-1)} E(n) = x_j(n)\left(\frac{\partial E(n)}{\partial u_i^{(M-1)}}(u_{\mathrm{I}i} - \mathrm{j}u_{\mathrm{Q}i}) + \frac{\partial E(n)}{\partial v_i^{(M-1)}}(v_{\mathrm{I}i} - \mathrm{j}v_{\mathrm{Q}i})\right) \tag{12.27}$$

$$\nabla_{w_{ij}}^{(M-1)} E(n) = -2x_j(n)[\varepsilon_{\mathrm{I}i}(n)(u_{\mathrm{I}i} - \mathrm{j}u_{\mathrm{Q}i}) + \varepsilon_{\mathrm{Q}i}(n)(v_{\mathrm{I}i} - \mathrm{j}v_{\mathrm{Q}i})] \tag{12.28}$$

and the weight update rule therefore becomes

$$w_{ij}^{(M-1)}(n+1) = w_{ij}^{(M-1)}(n) + \mu x_j^{(M-1)}(n)[\varepsilon_{\mathrm{I}i}(n)(u'_{\mathrm{I}i} - \mathrm{j}u'_{\mathrm{Q}i}) + \varepsilon_{\mathrm{Q}i}(n)(v'_{\mathrm{I}i} - \mathrm{j}v'_{\mathrm{Q}i})] \tag{12.29}$$

The update rule for the bias term θ can be derived in a similar manner. We will just state it here to be

$$\theta_i^{(M-1)}(n+1) = \theta_i^{(M-1)}(n) + \mu[\varepsilon_{\mathrm{I}i}(n)(u'_{\mathrm{I}i} - \mathrm{j}u'_{\mathrm{Q}i}) + \varepsilon_{\mathrm{Q}i}(n)(v'_{\mathrm{I}i} - \mathrm{j}v'_{\mathrm{Q}i})] \tag{12.30}$$

Classification tasks often require a mapping from a multidimensional feature space to a class label. Feature data belonging to a class C_k are trained to map to a constant value at an output node k in the neural network. For this application, a bounded, non-linear activation function at the output is desirable. However, if a continuous mapping is required, e.g. for a one-step ahead non-linear prediction problem, it is necessary to remove the non-linearity before the output unit and have the final layer operate as a linear combiner. This allows the output to vary in an unbounded fashion [50]. In this case, we let

$$f(\mathrm{net}_i) = \mathrm{net}_i \tag{12.31}$$

The partial derivatives of this function reduce to

$$u'_{\mathrm{I}i} = 1 \qquad v'_{\mathrm{I}i} = 0$$
$$u'_{\mathrm{Q}i} = 0 \qquad v'_{\mathrm{Q}i} = 1 \tag{12.32}$$

Substituting these values into the weight update rule of (12.29), we find that

$$w_{ij}^{(M-1)}(n+1) = w_{ij}^{(M-1)}(n) + \mu x_j^{(M-1)}(n)e_i^*(n) \tag{12.33}$$

which corresponds to the familiar complex LMS weight update result.

We have now shown the process for updating the $(M-1)$th layer of weights. The next step is to derive the relations necessary to update the weights in the hidden layers. The main idea is to find expressions that relate the error in the (l) layer to the $(l-1)$ layer. In this way we may *back-propagate* the error, stepping from layer (M) back towards layer (0).

Restating (12.25) in terms of the hidden layer, the expression for the partial derivatives of the cost function in reference to the weights in layer $(M-2)$ is therefore

$$\begin{aligned}\frac{\partial E(n)}{\partial w_{\mathrm{I}ij}^{(M-2)}} &= \frac{\partial E(n)}{\partial u_i^{(M-2)}}(u'_{\mathrm{I}i}x_{\mathrm{I}i} + u'_{\mathrm{Q}i}x_{\mathrm{Q}i}) \\ &\quad + \frac{\partial E(n)}{\partial v_i^{(M-2)}}(v'_{\mathrm{I}i}x_{\mathrm{I}i} + v'_{\mathrm{Q}i}x_{\mathrm{Q}i}) \\ \frac{\partial E(n)}{\partial w_{\mathrm{Q}ij}^{(M-2)}} &= \frac{\partial E(n)}{\partial u_i^{(M-2)}}(u'_{\mathrm{I}i}x_{\mathrm{Q}ij} - u'_{\mathrm{Q}i}x_{\mathrm{I}i}) \\ &\quad + \frac{\partial E(n)}{\partial v_i^{(M-2)}}(v'_{\mathrm{I}i}x_{\mathrm{Q}ij} - v'_{\mathrm{Q}i}x_{\mathrm{I}i})\end{aligned} \tag{12.34}$$

Using the chain rule,

$$\begin{aligned}\frac{\partial E(n)}{\partial u_i^{(M-2)}} &= \sum_k \frac{\partial E(n)}{\partial U_k^{(M-1)}}\left(\frac{\partial u_k^{(M-1)}}{\partial \mathrm{net}_{\mathrm{i}k}}\frac{\partial \mathrm{net}_{\mathrm{I}k}}{\partial u_i^{(M-2)}} + \frac{\partial u_k^{(M-1)}}{\partial \mathrm{net}_{\mathrm{Q}k}}\frac{\partial \mathrm{net}_{\mathrm{Q}k}}{\partial u_i^{(M-2)}}\right) \\ &\quad + \sum_k \frac{\partial E(n)}{\partial v_k^{(M-1)}}\left(\frac{\partial v_k^{(M-1)}}{\partial \mathrm{net}_{\mathrm{I}k}}\frac{\partial \mathrm{net}_{\mathrm{I}k}}{\partial u_i^{(M-2)}} + \frac{\partial v_k^{(M-1)}}{\partial \mathrm{net}_{\mathrm{Q}k}}\frac{\partial \mathrm{net}_{\mathrm{Q}k}}{\partial u_i^{(M-2)}}\right)\end{aligned} \tag{12.35}$$

$$\begin{aligned}\frac{\partial E(n)}{\partial v_i^{(M-2)}} &= \sum_k \frac{\partial E(n)}{\partial u_k^{(M-1)}}\left(\frac{\partial u_k^{(M-1)}}{\partial \mathrm{net}_{\mathrm{I}k}}\frac{\partial \mathrm{net}_{\mathrm{I}k}}{\partial v_i^{(M-2)}} + \frac{\partial u_k^{(M-1)}}{\partial \mathrm{net}_{\mathrm{Q}k}}\frac{\partial \mathrm{net}_{\mathrm{Q}k}}{\partial v_i^{(M-2)}}\right) \\ &\quad + \sum_k \frac{\partial E(n)}{\partial v_k^{(M-1)}}\left(\frac{\partial v_k^{(M-1)}}{\partial \mathrm{net}_{\mathrm{I}k}}\frac{\partial \mathrm{net}_{\mathrm{I}k}}{\partial v_i^{(M-2)}} + \frac{\partial v_k^{(M-1)}}{\partial \mathrm{net}_{\mathrm{Q}k}}\frac{\partial \mathrm{net}_{\mathrm{Q}k}}{\partial v_i^{(M-2)}}\right)\end{aligned} \tag{12.36}$$

The partial derivatives of the cost function are now expressed in terms of the partial derivatives of the previous layer $(M-1)$. We now need only to determine the partial derivatives of

$$\begin{aligned}\mathrm{net}_k^{(M-1)} &= \sum_i w_{ki}^{*(M-1)} f(\mathrm{net}_i^{(M-2)}) + \theta_k^{*(M-1)} \\ &= \sum_i (u_i^{(M-2)}w_{\mathrm{I}ki}^{(M-1)} + v_i^{(M-2)}w_{\mathrm{Q}ki}^{(M-1)} + \theta_{\mathrm{I}k}^{(M-1)}) \\ &\quad + \mathrm{j}(v_i^{(M-2)}w_{\mathrm{I}ki}^{(M-1)} - u_i^{(M-2)}w_{\mathrm{Q}ki}^{(M-1)} - \theta_{\mathrm{Q}k}^{(M-1)})\end{aligned} \tag{12.37}$$

with respect to u and v of the previous layer $(M-2)$. Summarizing,

$$\frac{\partial \mathrm{net}_{\mathrm{I}k}^{(M-1)}}{\partial u_i^{(M-2)}} = w_{\mathrm{I}ki}^{(M-1)} \qquad \frac{\partial \mathrm{net}_{\mathrm{Q}k}^{(M-1)}}{\partial u_i^{(M-2)}} = -w_{\mathrm{Q}ki}^{(M-1)}$$

$$\frac{\partial \mathrm{net}_{\mathrm{I}k}^{(M-1)}}{\partial v_i^{(M-2)}} = w_{\mathrm{Q}ki}^{(M-1)} \qquad \frac{\partial \mathrm{net}_{\mathrm{Q}k}^{(M-1)}}{\partial v_i^{(M-2)}} = w_{\mathrm{I}ki}^{(M-1)} \tag{12.38}$$

Substituting (12.38) into (12.35) and (12.36), the partial derivatives can be expressed as

$$\frac{\partial E(n)}{\partial u_i^{(M-2)}} = \sum_k \frac{\partial E(n)}{\partial u_k^{(M-1)}} (u_{\mathrm{I}k}^{\prime(M-1)} w_{\mathrm{I}ki}^{(M-1)} - u_{\mathrm{Q}k}^{\prime(M-1)} w_{\mathrm{Q}ki}^{(M-1)})$$
$$+ \sum_k \frac{\partial E(n)}{\partial v_k^{(M-1)}} (v_{\mathrm{I}k}^{\prime(M-1)} w_{\mathrm{I}ki}^{(M-1)} - v_{\mathrm{Q}k}^{\prime(M-1)} w_{\mathrm{Q}ki}^{(M-1)}) \tag{12.39}$$

$$\frac{\partial E(n)}{\partial v_i^{(M-2)}} = \sum_k \frac{\partial E(n)}{\partial u_k^{(M-1)}} (u_{\mathrm{I}k}^{\prime(M-1)} w_{\mathrm{Q}ki}^{(M-1)} + u_{\mathrm{Q}k}^{\prime(M-1)} w_{\mathrm{I}ki}^{(M-1)})$$
$$+ \sum_k \frac{\partial E(n)}{\partial v_k^{(M-1)}} (v_{\mathrm{I}k}^{\prime(M-1)} w_{\mathrm{Q}ki}^{(M-1)} + v_{\mathrm{Q}i}^{\prime(M-1)} w_{\mathrm{I}ki}^{(M-1)})$$

Simplifying further,

$$\frac{\partial E(n)}{\partial u_i^{(M-2)}} + \mathrm{j}\,\frac{\partial E(n)}{\partial v_i^{(M-2)}}$$
$$= \sum_k w_{ki}^{(M-1)} \left(\frac{\partial E(n)}{\partial u_k^{(M-1)}} (u'_{\mathrm{I}k} + \mathrm{j}u'_{\mathrm{Q}k}) + \frac{\partial E(n)}{\partial v_k^{(M-1)}} (v'_{\mathrm{I}k} + \mathrm{j}v'_{\mathrm{Q}k}) \right) \tag{12.40}$$

Using induction, we can extend this relationship to the other hidden layers, $(M-3), \ldots, (0)$. Equation (12.40) gives us the means to back-propagate the error from the output layer, (M), to the input layer (0). After the values for $\partial E(n)/\partial u$ and $\partial E(n)/\partial v$ for the particular layer have been determined, Equation (12.27) gives the gradient, and hence the weight update values.

One of the difficulties encountered in extending the real BP algorithm to the complex domain involves the appropriate choice of activation function. The straightforward extension of the sigmoid from the real domain to the complex domain is inadequate, due to the fact that it has singularities, such that

$$\frac{1}{1+\mathrm{e}^{-z}} \to \infty \quad \text{for } z = \pm\mathrm{j}\,\frac{2k+1}{2\pi} \quad k \text{ any integer} \tag{12.41}$$

For practical implementations of the neural network, it is required that the activation function be bounded. Without such guarantees there is a risk of arithmetic overflow in software simulations. Hardware implementations would suffer in an analogous manner, with unbounded outputs resulting in possible clipping at node outputs. Singularities in an activation function should therefore be avoided.

Georgiou and Koutsougeras have developed a set of criteria which a complex activation function must meet in order to be useful in a back-error propagation network. Summarizing these properties [43] we have:

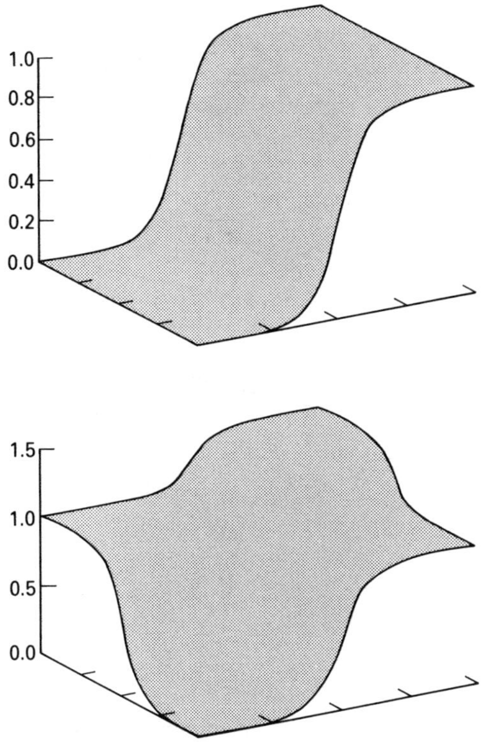

Figure 12.7 Real part (top) and magnitude (bottom) of the activation function $f(z) = c/(1 + e^{-kz_r}) + jc/(1 + e^{-kz_i})$

1. $f(z)$ should be non-linear in z_I and z_Q, otherwise there is no advantage in having a multilayer network. A multilayer linear network can always be collapsed to an equivalent single-layer network. The motivation here is to have a non-linear network that can compute a more general set of functions than a linear network.
2. The function $f(z)$ should be bounded. The computation of the forward pass of the network is required to be bounded, otherwise clipping or numerical overflow can occur.
3. The partial derivatives of $f(z)$ should exist and be bounded. The learning phase updates the weights proportional to the partial derivatives, so they also need to be bounded.
4. The function $f(z)$ should not be an entire function. Entire functions are defined as complex functions which are analytic everywhere in the complex domain. A function is defined to be *analytic* at some point z_0 if it is differentiable in some neighbourhood of z_0. By Liouville's theorem [48], we know that a bounded, entire function is constant. Clearly, a function that is entire is not a suitable choice, for the reasons stated in property 1.

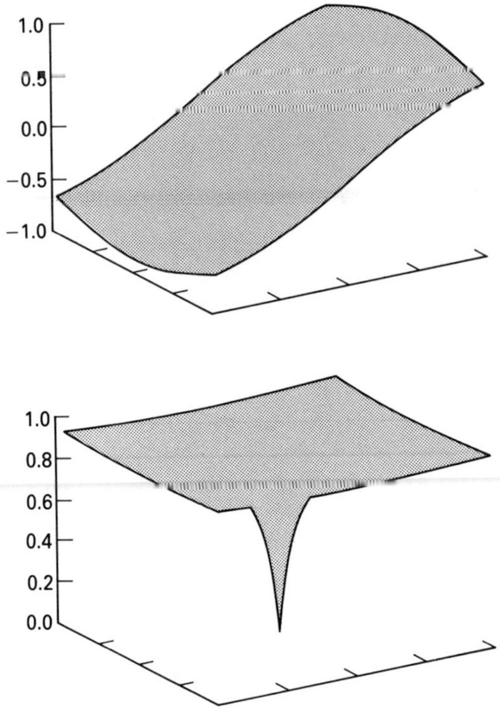

Figure 12.8 Real part (top) and magnitude (bottom) of the activation function $f(z) = z/[c + \frac{1}{r}|z|]$

5. $\nabla_w E \neq 0$ for $x_j(n) \neq 0$ and $\partial E/\partial u + j\partial E/\partial v \neq 0$. For this condition to hold, the partial derivatives of $f(z)$ should not satisfy the relation $u_I v_Q \equiv u_Q v_I$. This relationship can be found by simultaneously setting the real and imaginary parts of (12.27) equal to zero. Should the partial derivatives of the activation function satisfy the above relation, this would imply that in the presence of both non-zero input and error, it would be possible that $\nabla_w E = 0$, and therefore a stationary point would be reached. No further learning could take place, since the weight update is proportional to the gradient.

Figures 12.7 and 12.8 show two possible choices for the complex activation function. Figure 12.7 shows the complex activation function suggested by Benvenuto and Piazza [40]. The function is a superposition of real and imaginary sigmoids, such that

$$f(z) = \frac{c}{1 + e^{-kz_r}} + j\frac{c}{1 + e^{-kz_i}} \tag{12.42}$$

Another possible activation function is suggested by Georgiou and Koutsougeras [43]. It is also a sigmoid-like function, as shown in Figure 12.8, with

$$f(z) = \frac{z}{c + (1/r)\,|z|} \tag{12.43}$$

This function maps the z domain to an open disk $|z| < r$, the activation function effectively squashing the range of $|f(z)|$ to the interval $[0, r)$.

We end this development by summarizing the complex back-error propagation algorithm in Table 12.1.

Table 12.1 Summary of the complex back-error propagation algorithm

1. *Initialize weights and offsets*
   Set all weights and biases to small complex random values
2. *Present input and desired outputs*
   Present input vector $\mathbf{x}(0), \mathbf{x}(1), \cdots, \mathbf{x}(n)$ and desired response $\mathbf{d}(0), \mathbf{d}(1), \cdots, \mathbf{d}(n)$ where n is the number of training patterns
3. *Calculate actual outputs*
   Use the formulae in Figure 12.6 to calculate output $y_1, y_2, \cdots, y_{N_M}$
4. *Adapt weights and biases*

$$\Delta w_{ij}^{(l-1)}(n) = -\mu x_j^{(l-1)}(n)\,\frac{1}{2}\left(\frac{\partial E(n)}{u_i^{(l-1)}}(u'_{\mathrm{I}i} - \mathrm{j}u'_{\mathrm{Q}i}) + \frac{\partial E(n)}{\partial v_i^{(l-1)}}(v'_{\mathrm{I}i} - \mathrm{j}v'_{\mathrm{Q}i})\right)$$

$$\Delta \theta_i^{(l-1)}(n) = -\mu\,\frac{1}{2}\left(\frac{\partial E(n)}{\partial u_i^{(l-1)}}(u'_{\mathrm{I}i} - \mathrm{j}u'_{\mathrm{Q}i}) + \frac{\partial E(n)}{\partial v_i^{(l-1)}}(v'_{\mathrm{I}i} - \mathrm{j}v'_{\mathrm{Q}i})\right)$$

where

$$\frac{\partial E(n)}{\partial u_i^{(l-1)}} + \mathrm{j}\,\frac{\partial E(n)}{\partial v_i^{(l-1)}}$$

$$= \begin{cases} -2[d_i(n) - y_i(n)] & l = M \\ \Sigma_k w_{ki}^{(l)}\left(\dfrac{\partial E}{\partial u_k^{(l)}}(u'_{\mathrm{I}k} + \mathrm{j}u'_{\mathrm{Q}k}) + \dfrac{\partial E}{\partial v_k^{(l)}}(v'_{\mathrm{I}k} + \mathrm{j}v'_{\mathrm{Q}k})\right) & 1 \leqslant l < M \end{cases}$$

where $x_j(n)$ = output of node j or input to node i

Note that a momentum term can also be added to the above formula, as described in [24].

12.5.3 Special case of back-error propagation algorithm: real parameters

The common development of the back-error propagation algorithm, such as found in [24], is with real-valued data and parameters. We now show that this is a special case of the more general, complex network algorithm developed in the previous section.

We proceed by considering all the parameters to be real valued, including the input and desired output data. In terms of the complex-valued neural network, the quadrature components are all set to zero. Applying this principle to (12.22), we observe that only the first term survives, so that

$$\frac{\partial E(n)}{\partial w_{ij}} = \frac{\partial E}{\partial u_i} \frac{\partial u_i}{\partial \text{net}_i} \frac{\partial \text{net}_i}{\partial w_{ij}} \tag{12.44}$$

Note that the inphase designation, I, is dropped from the variables in this equation. Since there is no longer a quadrature signal component to consider, it is a redundant notation. We also replace all occurrences of u with

$$f(\text{net}_i) = u(\text{net}_i, 0) \tag{12.45}$$

and

$$f'(\text{net}_i) = u'_{\text{I}i} = \frac{\partial u_i}{\partial \text{net}_{\text{I}i}} \tag{12.46}$$

Equation (12.44) can now be rewritten as

$$\frac{\partial E(n)}{\partial w_{ij}} = \frac{\partial E}{\partial f(\text{net}_i)} \frac{\partial f(\text{net}_i)}{\partial \text{net}_i} \frac{\partial \text{net}_i}{\partial w_{ij}} \tag{12.47}$$

The activation function, $f(\text{net})$, can be any bounded, differentiable, monotonically increasing function. The sigmoid function is often the function of choice.

We now define a new variable

$$\delta_i^{(l-1)}(n) \stackrel{\text{def}}{=} -\frac{\partial E(n)}{\partial \text{net}_i^{(l-1)}} \tag{12.48}$$

For the case of the output layer $(l) = (M)$

$$\delta_i^{(M-1)}(n) = -\frac{\partial E}{\partial f(\text{net}_i)} \frac{\partial f(\text{net}_i)}{\partial \text{net}_i} = 2f'(\text{net}_i^{(M-1)}(n))\varepsilon_i(n) \tag{12.49}$$

When $1 \leqslant l < M$,

$$\delta_i^{(l-1)}(n) = f'(\text{net}_i^{(l-1)}(n)) \sum_k w_{ki}(n)\delta_k^l(n) \tag{12.50}$$

after substituting for the terms in (12.40). The variable δ is interpreted as a back-propagated error term which can be recursively computed for each layer of the neural network.

A summary of the algorithm is presented in Table 12.2.

Table 12.2 Summary of the real back-error propagation algorithm

1. *Initialize weights and offsets*
   Set all weights and biases to small, real random values
2. *Present input and desired outputs*
   Present input vector $\mathbf{x}(0), \mathbf{x}(1), \cdots, \mathbf{x}(n)$ and desired response $\mathbf{d}(0), \mathbf{d}(1), \cdots, \mathbf{d}(n)$ where n is the number of training patterns
3. *Calculate actual outputs*
   Use the formulae in Figure 12.6 to calculate output $y_1, y_2, \cdots, y_{N_M}$
4. *Adapt weights and biases*

$$\Delta w_{ij}^{(l-1)}(n) = \mu\tfrac{1}{2}x_j(n)\delta_i^{(l-1)}(n)$$
$$\Delta \theta_i^{(l-1)}(n) = \mu\tfrac{1}{2}\delta_i^{(l-1)}(n)$$

where

$$\delta_i^{(l-1)}(n) = \begin{cases} 2f'(\text{net}_i^{(l-1)}(n))[d_i(n) - y_i(n)] & l = M \\ f'(\text{net}_i^{(l-1)}(n))\Sigma_k w_{ki}\delta_k^l(n) & 1 \leqslant l < M \end{cases}$$

where $x_j(n)$ = output of node j or input to node i

Note that a momentum term can also be added to the above formula, as described in [24].

12.5.4 Non-linear prediction example

The back-propagation neural network architecture is able to encapsulate general non-linear mappings through supervised learning. Signal processing applications in areas such as prediction, equalization, cancellation, beamforming and system identification can benefit from a signal processor that can model non-linearities present in a time series. A difficult non-linear prediction problem [50] is presented herein in order to highlight the capability of the neural network to model higher-order statistical dependencies of a time series

The time series on which the network is to operate is created using a discrete Volterra time series model [51]. The Volterra series has the form

$$n_t = \sum_i g_i a_{t-i} + \sum_i \sum_j g_{ij} a_{t-i} a_{t-j} + \sum_i \sum_j \sum_k g_{ijk} a_{t-i} a_{t-j} a_{t-k} + \ldots \tag{12.51}$$

where $g_i \ldots$ are the Volterra series coefficients, a_t is a white, independently distributed Gaussian noise sequence, and n_t is the resultant Volterra series. The first term on the right-hand side of (12.51) is the familiar linear moving average (MA) time series model, and the remaining terms are non-linear components of ever increasing order. In general, the estimation of the coefficients is regarded as difficult, primarily due to the high number of coefficients and their non-linear relationship to the data. A simple model form is chosen for our illustrative example:

$$n_t = a_t + \beta a_{t-1} a_{t-2} \tag{12.52}$$

Analysis of the first two moments of the time series reveals a mean of zero and a variance equal to

$$\sigma_n^2 = E[n_t^2] = E[a_t^2] + \beta^2 E[a_{t-1}^2 a_{t-2}^2] = \sigma_a^2 + \beta^2 \sigma_a^4 \tag{12.53}$$

Solving for the autocorrelation components, we observe that

$$\begin{aligned} E[n_t n_{t+k}] &= E[a_t a_{t+k}] + \beta E[a_t a_{t+k-1} a_{t+k-2}] \\ &\quad + \beta E[a_{t+k} a_{t-1} a_{t-2}] + \beta^2 E[a_{t+k-1} a_{t+k-2} a_{t-1} a_{t-2}] \\ &= 0 \quad k \neq 0 \end{aligned} \tag{12.54}$$

The time series is uncorrelated, and therefore has a white spectrum. If we were to base our analysis only on these first two moments of the time series, we would conclude that the optimum predictor is given by the mean of the process or, in our case, the constant zero. However, since the time series samples are not independent of each other, a higher-order predictor can be constructed. Solving for this predictor, we find

$$\begin{aligned} \hat{n}_{t+1} &= E[n_{t+1} \mid n_t, n_{t-1}, \ldots] \\ &= E[a_{t+1}] + \beta E[a_t a_{t-1}] \\ &= \beta E[a_t] E[a_{t-1}] = \beta a_t a_{t-1} \end{aligned} \tag{12.55}$$

The above formula for the prediction of $\hat{n}_{t+1}$ is given in terms of a_t, rather than n_t. In order to re-express the predictor in terms of past samples of n_t, the model would have to be inverted. The fact that no obvious closed-form solution presents itself makes this a difficult problem.

The configuration of the neural network architecture used for the predictor is shown in Figure 12.9. A *tapped-delay line* structure is present at the input, followed by one or two hidden layers with sigmoid non-linearities, and a linear combiner at the output. The time series is real, and therefore a real back-error propagation learning algorithm is used to train the network. It is also possible to add tapped-delay lines in the hidden layers as well, creating a so-called *time delay neural network* architecture. For this problem, the architecture is kept relatively simple, and tapped-delay lines are only present in the input. The performance of the neural net predictor is compared with that of a linear predictor with the equivalent number of delayed inputs and training algorithm, but using only a single, linear, layer. The parameters of the neural network simulation are given in Table 12.3. Note that only a single hidden layer is used. The coefficients μ and α refer to the values of the back-propagation learning coefficient and the momentum term respectively.

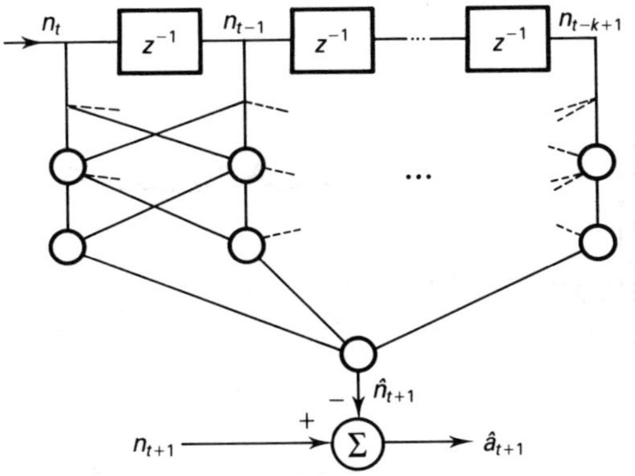

Figure 12.9 Neural network predictor architecture

Table 12.3 Back-propagation neural network configuration

Units in the input layer	6
Units in the hidden layer	16
μ	0.001
α	0.9
Total samples processed	100 000
Samples per epoch	1000

Before looking at the results of the training experiment, some possible predictor solutions are investigated. The white-noise input variance to the model is set to one ($\sigma_a^2 = 1$). From (12.53) we can calculate that the output variance of the predictor (when at a global minimum) should be equal to ($\sigma^2_{\hat{n}_{t+1}} = 0.25$). In addition, a local minimum condition exists in the cost surface which corresponds to the linear predictor solution. As discussed above, the linear solution is equal to a constant, with zero variance at the output. These possible variance conditions at the output of the predictor are summarized in Table 12.4.

Table 12.4 Local and global minimum solutions

Minima	$\sigma^2_{\hat{a}_{t+1}}$	$\sigma^2_{\hat{n}_{t+1}}$
Local	1.25	0.00
Global	1.00	0.25

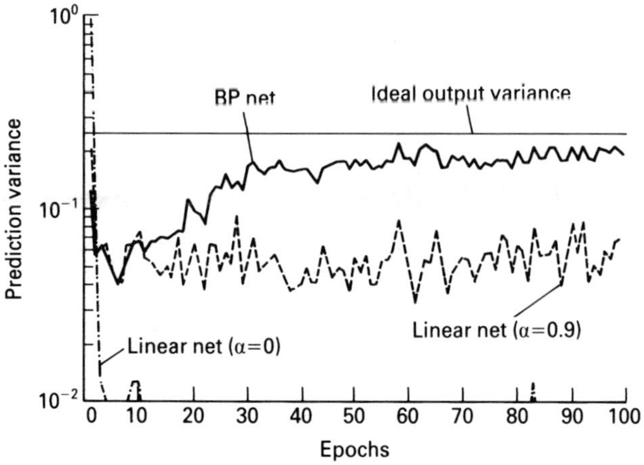

Figure 12.10 Learning curves of the network output $\hat{n}_{t+1}$

The learning curves for the experiment are shown in Figure 12.10. The ideal output variance of the predictor is marked with a constant dotted line. Observing the plot of the variance of the neural network output, the network seems first to converge to the aforementioned local minimum condition, closely following the linear predictor solution. After some time, however, the higher-order structure is captured by the network and the output increases and converges to the theoretical limit. By comparison, the linear predictor ($\alpha = 0.9$) output variance never rises significantly above the local minima condition. It is clear from the plot of the linear predictor variance (the momentum term turned down to zero ($\alpha = 0$)) that the linear network converges to the local minimum solution of a constant zero output. Table 12.5 summarizes the final results of the experiment.

Table 12.5 Results of experiment

	$\hat{\sigma}^2_{\hat{a}_{t+1}}$	$\hat{\sigma}^2_{\hat{n}_{t+1}}$
BP net *w*/linear output	1.088	0.173 29
Linear net ($\alpha = 0.9$)	1.282	0.027 20
Linear net ($\alpha = 0$)	1.260	0.005 328

This non-linear predictor experiment shows that the back-error propagation neural network can be used for non-linear signal processing, in situations where higher-order statistics are present. In situations of this kind, it can deliver a superior performance compared with linear adaptive filters.

12.5.5 Issues in learning

The computations involved in the application of the back-error propagation algorithm are terminated when the *training* set of data is completely accounted for. Then the synaptic weights of the network are *fixed* at their final values, and the network is frozen in that condition. In effect, the information contained in the training set is transplanted in the synaptic connections of the network. It is for this reason that neurocomputing is also referred to in the literature as *connectionism.* In any event, the training mode of operation is particularly important, because it determines the success or failure of a multilayer perceptron as a neurocomputer. We say that the network design is successful if the multilayer perceptron is capable of a good *generalization* by producing a response close enough to the true value when it is subjected to *test* data that have *not* been seen by the network before. Otherwise, we have a design failure.

For a preselected network structure, we may optimize the network design by following the so-called *optimal brain damage* (OBD) strategy [52]. The important steps of this strategy may be summarized as follows:

1. Train the network using regular batch or online update, long enough for it to converge to a satisfactory solution. It is assumed that the error surface at a stationary point (local or global minimum) is approximately estimated by a quadratic function.
2. Compute the *Hessian matrix* for all parameters of the network, assuming that it is a diagonal matrix.
3. Use this Hessian matrix to compute the *saliencies* for all parameters of the network. For the kth parameter at iteration n, the saliency is defined by

$$s_k(n) = \tfrac{1}{2} h_{kk} u_k^2(n) \tag{12.56}$$

   where h_{kk} is the kth element of the diagonal Hessian matrix, and $u_k(n) = w_{ji}(n)$, the synaptic weight from node i to node j, for all $j, i \in V_k$ where V_k is a set of index pairs.
4. Sort the network parameters by their saliencies; hence, delete those parameters that have a small saliency.
5. Iterate.

The essence of the optimal brain damage strategy is to start with a multilayer back-propagation network larger than an 'optimal' network size, and seek optimality by *pruning* the network in an orderly fashion.

We may arrive at an optimal design by proceeding in the converse fashion to the optimal brain damage strategy. Specifically, we go through a *growth* process in the sense that we start with a small multilayer perceptron, small for the problem at hand, and add a new hidden neuron or a new layer of hidden neurons only when we are unable to meet the design specification. The *cascade-correlation learning architecture* [53] is an example of this latter approach. The procedure begins with a minimal network that has some inputs and one or more output nodes as indicated by input–output considerations, but no hidden nodes. The LMS algorithm, for example, may be used to train this network. The hidden nodes are added to the network one by one, thereby creating a multilayer structure. Each new hidden node receives a synaptic connection from each of the input nodes and also from each pre-existing hidden node. When a new hidden node is added, the synaptic weights on the input side of that node are frozen; only the synaptic weights on the output side are trained repeatedly. This hidden node then becomes a permanent feature detector in the network. The cascade-correlation architecture offers the following advantages over the back-propagation architecture [53]:

1. It learns very quickly.
2. The architecture determines its own network size and topology.
3. When the training set changes, the algorithm retains the structure it has already built.
4. The algorithm requires no back-propagation of error signals through the synaptic connections of the neural network.

In yet another approach described in [54], a third level of computation termed the *structure level adaptations* is added to the forward pass (function level adaptation) and backward pass (parameter level adaptation). In this third level of computation the structure of the neural network is adapted by changing the number of neurons and the structural relationship among neurons in the network. The criterion used here is that when the estimation error (after stabilization) is larger than a desired value, then a new neuron is added to the network and it should be positioned where it is most needed. The desirable position for the new neuron may be found by monitoring the learning behaviour of the network. In particular, if, after a long period of parameter adaptation (training), the synaptic weight vector pertaining to the inputs of a neuron continues to fluctuate significantly, it may then be inferred that the neuron in question does not have enough representation power to learn its proper share of the problem. The structure level adaptation described in [54] also includes neuron annihilation. Specifically, a neuron is annihilated when it is not a functioning element of the network or it is a redundant element of the network.

Naturally, there is no guarantee that the network-pruning (e.g. optimal brain damage) and network-growing (e.g. cascade-correlation learning architecture) procedures will yield an identical solution. Moreover, at this point in time, we do not know which of the two procedures, network pruning and network growing, is the preferred one.

12.6 Radial basis function networks

The training process of a neural network may be viewed as one of *curve fitting*. In particular, we are given a set of points in the observation space defined by the specified values of input signals and desired response (target signals), and the requirement is to find an input–output mapping that passes through these points. Correspondingly, the generalization process may be viewed as one of *interpolation* in that the network is called upon to express its response to test data never seen before. This kind of viewpoint is exploited in the design of another important class of neural networks known as *radial basis function* (RBF) *networks*, as described next.

RBF networks have been successfully applied by researchers to difficult problems in signal processing [55–63]. The network consists of a hidden layer of non-linear RBF units, followed by a linear layer of weights. The linear output layer is trained in a supervised fashion using one of the many possible linear least-squares methods available. The choice of the parameters of the RBF units is more challenging. Various procedures have been experimented with to learn the centres and widths (or spread) of the hidden layer of RBF units.

RBF networks differ from back-error propagation networks in the following respects:

1. RBF networks have a single hidden layer, whereas back-error propagation networks may have one or more hidden layers.
2. In RBF networks, the links connecting the input layer to the hidden layer are non-linear, and those connecting the hidden layer to the output layer are linear. In back-error propagation networks, the hidden layer-connecting links are non-linear, and the output layer links can be non-linear or linear, depending on the application.
3. Each hidden neuron of an RBF network computes the Euclidean distance between the input vector and the centre of the radial basis (Gaussian) function associated with that neuron. On the other hand, each hidden neuron of a back-error propagation network computes the inner product (dot product) of the input vector applied to that neuron and the vector of associated synaptic weights.

RBF networks and back-error propagation networks do, however, share a common property: they are both universal approximators of the feedforward type [64]; naturally, they perform their approximation in different ways. The back-error propagation algorithm is a global approximation method, whereas the RBF algorithm is a local approximation method.

DESIGN STRATEGIES

Several non-adaptive strategies have been used to determine the RBF centres and spread. The most straightforward choice for the location of the centres is to place them on an evenly spaced grid, spanning the input space. A very large number of RBF units may be needed, since the number of units required grows exponentially with the dimensionality of the input space. However, as the dimensionality of the input grows, most of the input space becomes devoid of samples, and therefore a large percentage of the centres lie in an area where there are no data. Another more effective choice for the RBF centres is to set them equal to a random sampling of input data. This strategy ensures that centres are located only in areas where there are data. It has been shown that as long as a sufficiently large number of centres are used, good prediction performance on a chaotic time series is achieved [55]. In either case, the RBF spreads are chosen using some heuristic.

Supervised adaptation of the RBF centres, spread, and output weights using optimization techniques (such as quasi-Newton, or conjugate gradient methods) have been used [65,60]. This strategy can give a minimal RBF network configuration. Some disadvantages of using optimization techniques are considerable computational cost (offline processing), poor scaling of learning as network complexity grows, and the presence of suboptimal local minimum solutions. Lowe [65] points out that the same final error performance can be achieved with a network with a larger number of non-adaptive centres, with the same generalization performance.

Recent investigations of hybrid unsupervised/supervised training schemes have shown promise [60–62,66,63]. Some of the advantages are computational efficiency, good scaling of learning as network size grows, and faster convergence. The hybrid procedure consists of two stages of learning: an unsupervised clustering algorithm is used to determine the parameters of the hidden layer, followed by a supervised least-squares solution to the linear output weights. Moody and Darken [60,61] suggested the use of the k-means clustering algorithm to find more suitable positions for the centres. As a result, a smaller number of RBF units are required. After clustering, heuristics were used to choose the spread of the RBF units. The k-means algorithm is an approximate version of the maximum likelihood (ML) solution for determining the location of the means of a mixture density of component Gaussian densities. The expectation maximization (EM) algorithm can be used to find the exact ML solution for the means and covariances of the density. A comparison of these two learning strategies on a classification problem was done by Nowlan [66,67], with the EM algorithm shown to be superior. Saha and Keeler also studied the use of the k-means clustering for the adjustment of RBF centres, and suggested an approach which they termed *extended metric clustering* [62], where clustering is done in an augmented input–output space. Once learning is complete, the cluster locations are projected back on to the input space, and used as the RBF unit centres. Ukrainec and Haykin have applied this principle to signal processing applications, showing that extended metric clustering using the EM algorithm provides superior unsupervised training for signal processing problems.

12.6.1 Some preliminaries

As mentioned previously, the RBF network consists of a set of inputs connected to a non-linear hidden layer, followed by a linear output layer connected to a set of outputs. The hidden layer is made of radially symmetric basis functions. These functions can be chosen to be one of many different possible forms. Only the Gaussian form will be considered, where the nodes in the hidden layer are given by

$$\phi_j(\mathbf{x}) = \exp[-\tfrac{1}{2}(\mathbf{x} - \mathbf{c}_j)^{\mathrm{T}}\mathbf{S}_j(\mathbf{x} - \mathbf{c}_j)] = \exp[-\tfrac{1}{2} \| \mathbf{x} - \mathbf{c}_j \|^2_{\mathrm{M}}] \tag{12.57}$$

where $\mathbf{x} \stackrel{\text{def}}{=} (x_1, \ldots, x_{n_x})^{\mathrm{T}}$, $\phi_j(\mathbf{x})$ is the jth radial basis function evaluated at the input vector $\mathbf{x}$, $\mathbf{c}_j$ is the jth RBF centre, and $\mathbf{S}_j$ is the jth multidimensional width, or spread. The term in the exponential is known as the Mahalanobis metric. The functional form of the output of the network is therefore given by

$$y_i = \sum_{j=1}^{n_\mathrm{h}} w_j \phi_j(\mathbf{x}) + w_0 \tag{12.58}$$

where y_i is the ith output function evaluated for the input vector $\mathbf{x}$, and w_j is the linear output weight corresponding to the jth hidden unit. The hidden layer has a total of n_h radial basis functions. The weight w_0 is the bias term. Given a set of input data vectors and output data, $\{\mathbf{x}_i, y_i^{(j)} \mid i = 1, \ldots, n, j = 1, \ldots, n_\mathrm{o}\}$,

$$\mathbf{X} = \begin{pmatrix} \mathbf{x}_1 \\ \mathbf{x}_2 \\ \vdots \\ \mathbf{x}_n \end{pmatrix} \quad \mathbf{Y} = \begin{pmatrix} y_1^{(1)} & y_1^{(2)} & \cdots & y_1^{(n_\mathrm{o})} \\ y_2^{(1)} & \ddots & & \vdots \\ \vdots & & & \\ y_n^{(1)} & \cdots & & y_n^{(n_\mathrm{o})} \end{pmatrix}$$

$$\boldsymbol{\Phi} = \begin{pmatrix} \phi_1(\mathbf{x}_1) & \phi_2(\mathbf{x}_1) & \cdots & \phi_{n_\mathrm{h}}(\mathbf{x}_1) \\ \phi_1(\mathbf{x}_2) & \ddots & & \vdots \\ \vdots & & & \\ \phi_1(\mathbf{x}_n) & \cdots & & \phi_{n_\mathrm{h}}(\mathbf{x}_n) \end{pmatrix}$$

where n_o is the number of outputs.

Once the parameters of the RBF units have been determined, in some fashion, the layer of output weights can be found. The input data must first be fed forward to the output of the hidden layer. Then, the linear weights can be solved for by using the method of least squares. A robust solution is via the singular value decomposition (SVD), which is used to find the pseudoinverse $\boldsymbol{\Phi}^+$, such that

$$\boldsymbol{\Phi}\hat{\mathbf{W}} = \mathbf{Y} \tag{12.59}$$

$$\hat{\mathbf{W}} = \boldsymbol{\Phi}^+\mathbf{Y} \tag{12.60}$$

Alternatively, any one of the many available adaptive filter algorithms [30] can be used to compute an iterative, online least-squares solution for the linear weights.

The RBF network can be made to handle complex signals as well. This extension is accomplished in a straightforward manner. All the parameters of the network are now complex (centres, spreads and output weights), and the metric is defined to be a complex version of the Mahalanobis metric. Therefore,

$$\phi_j(\mathbf{x}) = \exp[-\tfrac{1}{2}(\mathbf{x} - \mathbf{c}_j)^{\mathrm{H}}\mathbf{S}_j(\mathbf{x} - \mathbf{c}_j)] = \exp[-\tfrac{1}{2} \| \mathbf{x} - \mathbf{c}_j \|_{\mathrm{Mc}}^2] \tag{12.61}$$

The output is calculated by a complex linear combiner. Restating (12.58) in complex terms,

$$y_i = \sum_{j=1}^{n_\mathrm{h}} w_j^* \phi_j(\mathbf{x}) + w_0^* \tag{12.62}$$

The weights are solved for in the same fashion as for the real network, the only difference being that we now use the complex version of the least squares algorithm.

12.6.2 Unsupervised learning of hidden layer parameters

The expectation maximization (EM) algorithm is a general approach to iteratively computing the maximum likelihood (ML) estimate of parameters of mixture density problems. This algorithm has had broad application in the areas of study of ML estimates from incomplete data [68], estimating mixture densities [69] and unsupervised clustering [70]. Here we will concentrate on the application of the EM algorithm for unsupervised clustering, to learn the RBF centres and spread.

A mixture distribution of component (Gaussian) densities is given by

$$p(\mathbf{x}_k \mid \boldsymbol{\theta}) = \sum_{j=1}^{n_h} P(j)p(\mathbf{x}_k \mid j, \boldsymbol{\theta}_j) \tag{12.63}$$

$$p(\mathbf{x}_k \mid j, \boldsymbol{\theta}_j) = \frac{1}{(2\pi)^{d/2}\,|\Sigma_j|^{1/2}} \exp[-\tfrac{1}{2}(\mathbf{x}_k - \boldsymbol{\mu}_j)^{\mathrm{T}}\Sigma_j^{-1}(\mathbf{x}_k - \boldsymbol{\mu}_j)] \tag{12.64}$$

where $\boldsymbol{\theta} = (\boldsymbol{\theta}_1, \ldots, \boldsymbol{\theta}_{n_h})$ is the vector of parameters (means and covariances) to be estimated, d is the dimensionality of the multivariate Gaussian density, $\boldsymbol{\mu}_j$ is the mean vector and Σ_j is the covariance matrix. The EM algorithm iteratively converges to a maximum of the likelihood function, yielding an estimate of the parameters of the component densities. Although the algorithm is guaranteed to converge, there is no guarantee that it will converge to a global maximum as it is possible for the algorithm to be trapped at a local minimum. The update equations are given as follows:

$$\hat{P}(i) = \frac{1}{n}\sum_{k=1}^{n} \hat{P}(i \mid \mathbf{x}_k, \hat{\boldsymbol{\theta}}) \tag{12.65}$$

$$\hat{\boldsymbol{\mu}}_i = \frac{\sum_{k=1}^{n}\hat{P}(i \mid \mathbf{x}_k, \hat{\boldsymbol{\theta}})\mathbf{x}_k}{\sum_{k=1}^{n}\hat{P}(i \mid \mathbf{x}_k, \hat{\boldsymbol{\theta}})} \tag{12.66}$$

$$\hat{\Sigma}_i = \frac{\sum_{k=1}^{n}\hat{P}(i \mid \mathbf{x}_k, \hat{\boldsymbol{\theta}})(\mathbf{x}_k - \hat{\boldsymbol{\mu}}_i)(\mathbf{x}_k - \hat{\boldsymbol{\mu}}_i)^{\mathrm{T}}}{\sum_{k=1}^{n}\hat{P}(i \mid \mathbf{x}_k, \hat{\boldsymbol{\theta}})} \tag{12.67}$$

$$\hat{P}(i \mid \mathbf{x}_k, \hat{\boldsymbol{\theta}}) = \frac{\hat{P}(i)p(\mathbf{x}_k \mid i, \hat{\boldsymbol{\theta}}_i)}{\sum_{j=1}^{n_h}\hat{P}(j)p(\mathbf{x}_k \mid j, \hat{\boldsymbol{\theta}}_j)} \tag{12.68}$$

These equations describe a batch processing algorithm, where all the data are used for each iteration. An online version of the EM algorithm was suggested by Nowlan [67], where the density parameters can be continuously updated as new data become available.

We observe that the function given in (12.57) for the RBF unit, and the Gaussian component density in (12.64), have almost the same form. The hypothesis is that the individually learned $\boldsymbol{\mu}_j$ of the component densities should give a good location for the centres of the RBF units. Likewise, the estimated covariance matrices can give the required spread of the RBF units. Experimentation with the algorithm shows that it tends to learn localized representations, underestimating (for the purposes of regression) the spread of the individual RBF units in the network. A heuristic factor α is introduced to increase the spread for all the units by a constant factor, improving the interpolation performance. The hybrid learning procedure is therefore given as follows:

1. Choose the number of RBF units (and hence the number of component densities).
2. Initialize the density parameters.
3. Iterate the EM algorithm until convergence.
4. Transplant the estimated parameters of the component densities into the RBF units so that $\mathbf{c}_j \leftarrow \hat{\boldsymbol{\mu}}_j$, $\mathbf{S}_j \leftarrow \alpha\hat{\Sigma}_j^{-1}$, where $0 < \alpha \leqslant 1$.
5. Forward-propagate the input data to the output of the RBF hidden layer.
6. Compute the least-squares solution to the linear weight layer.

There are two approaches that are explored in the application of unsupervised clustering. The first approach is to use the EM algorithm to cluster on the input space, adapting both $\boldsymbol{\mu}_j$ and Σ_j. The second is to use the concept of extended metric clustering combined with the EM algorithm. Previous studies used unsupervised learning on the input space only. Although input space clustering seems reasonable, closer investigation reveals some pitfalls. Indeed, for the purposes of classification, the strategy often works quite well. This is a result of the type of mapping that the network is being asked to learn. In classification problems, samples belonging to a class may often have a peculiar distribution, but are likely to be clustered in a particular area of input space, so that samples close to each other in input space map to the same area in output space. The required classifier output is binary, and therefore the desired mapping is from a volume of input space to a point in output space. For the continuous mappings required in signal processing, the problem is somewhat more complex. Samples which are close together in input space do not necessarily map to a similar location in output space. Therefore, clustering on information only in the input space can be misleading. Important details in the mapping may be lost, or blurred. This is especially a problem when a network has a relatively small number of RBF units.

12.6.3 Extended metric clustering

To gain an understanding of why the extended metric clustering strategy should be a more effective method, it is instructive to take a look at the regression function which is to be estimated, in terms of the expectation of the conditional probability density. The aim is to find an estimate of the regression function of y_n on the input space $[x_n, x_{n-1}, \ldots, x_{n-k}]$. The regression estimate $\hat{y}_n$ is defined to be equal to [71]

$$\begin{aligned}\hat{y}_n &= E[y_n \mid x_n, x_{n-1}, \ldots, x_{n-k}] \\ &= \int_{-\infty}^{\infty} y_n p(y_n \mid x_n, x_{n-1}, \ldots, x_{n-k})\, dy_n \end{aligned} \tag{12.69}$$

Using Bayes' rule, we can express the conditional probability density in terms of the joint probability density, such that

$$p(y_n \mid x_n, x_{n-1}, \ldots, x_{n-k}) = \frac{p(y_n, x_n, x_{n-1}, \ldots, x_{n-k})}{p(x_n, x_{n-1}, \ldots, x_{n-k})} \tag{12.70}$$

Restating (12.69),

$$\hat{y}_n = \frac{\int_{-\infty}^{\infty} y_n p(y_n, x_n, x_{n-1}, \ldots, x_{n-k})\, dy_n}{\int_{-\infty}^{\infty} p(y_n, x_n, x_{n-1}, \ldots, x_{n-k})\, dy_n} \tag{12.71}$$

The estimate of the regression function can therefore be viewed as a function of the joint probability distribution.

The principle of extended metric clustering is to learn, in an unsupervised manner, a mixture density representation of the joint probability density of input and output spaces. The clustering is performed in the augmented high-dimensional space consisting of a union of input and output spaces. The sample data vector that is input to the EM algorithm is now equal to $\mathbf{x}_k' = [\mathbf{x}_k \ y_k]$. After the unsupervised clustering stage is complete, the output y is removed from the input of the network, and the corresponding data field is truncated from the vectors and matrices of the learned parameters. Therefore, only the input field is used for the supervised learning stage. This is equivalent to 'collapsing' the component density estimates of the joint distribution function on to the space spanned by the input space. Although the output space is no longer present, the parameters of the component density (after training) are left with improved representations in areas of the input space where the joint density is greater. This approach is analogous to a weighted least-squares approach, where a priori knowledge is used to weight the input space. From an intuitive point of view, it is satisfying to create a better representation in areas of the input space where the joint distribution of input and output samples are more likely to be located.

12.6.4 Chaotic time series prediction problem

To illustrate the operation of the unsupervised/supervised learning algorithms, a chaotic time series prediction problem based on the quadratic map is used. The RBF network is trained to perform a one-step ahead prediction on the time series. The mapping that the network is required to learn is of low order, smooth, and highly non-linear. The network architecture is configured as shown in Figure 12.11. In the learning phase, the network is trained on a representative data set. An independent data set is used for testing, to show how well the network generalizes with respect to unseen data. The logarithm to base 10 of the normalized output error is used as an index of performance.

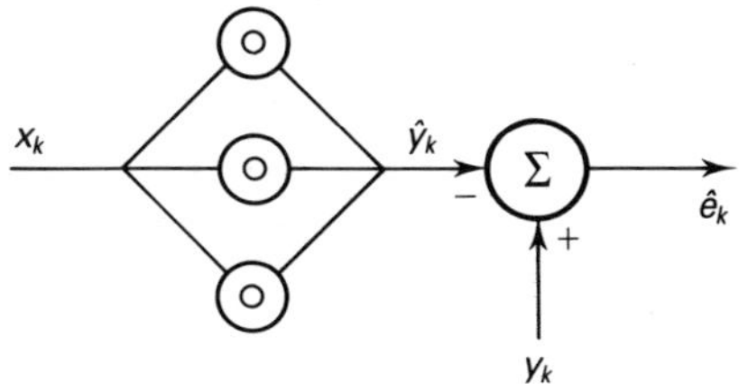

Figure 12.11 RBF network architecture used for training

The prediction of chaotic time series is recognized as a difficult signal processing problem, requiring non-linear mapping. The one-dimensional chaotic map,

$$x(n + 1) = 4x(n)[1 - x(n)] \tag{12.72}$$

commonly known as the quadratic, or logistic, map, is used to generate the time series for this experiment. The RBF network is assigned the task of providing an accurate prediction $\hat{x}(n + 1)$ given the sample $x(n)$. The training set and the testing set consist of 500 samples each. The clustering is stopped when no element of $\Delta\boldsymbol{\mu}_j$ exceeds 10^{-5}. The component density parameters were initialized by setting the means to span the input space, and by setting the covariances to a value determined by a heuristic. The component densities were adapted by the clustering algorithm. In the case of the EM algorithm with extended metric clustering (EMX), the covariances were constrained to have diagonal covariance matrices:

$$\Sigma_j = \sigma_j^2 \mathbf{I} \tag{12.73}$$

The prediction performance is calculated using the normalized error measure

$$\varepsilon = \frac{\sqrt{\Sigma_{k=1}^{n} \| \mathrm{y}_k - \hat{\mathrm{y}}_k \|^2}}{\sigma} \tag{12.74}$$

where σ is the standard deviation of the time series.

Figure 12.12 compares the error performance of the RBF network trained on the input space as well as on the augmented input–output space. The number of RBF units in the hidden layers varied from 3 to 15. The normalized error is generally lower, as expected, for the EMX-trained networks, and the generalization to the testing data is very good. After seven centres, the error performance flattens out at approximately $10^{-0.8}$. The network trained only on the input space fares less well. Good performance is realized only for the case of eight centres, which is a result of chance rather than design. The pattern in the EM error curve shows the suboptimal learning characteristic of the centres. Normally, one would expect that as the number of centres is increased, the error performance would improve, or flatten out (as in the case of EMX learning). Generalization is also not as good for the clustering of the input space only, showing a greater variation between trained and testing data.

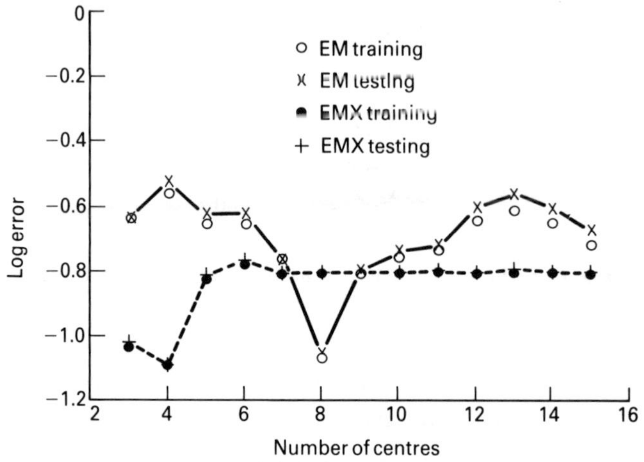

Figure 12.12 Prediction error performance for quadratic map problem ($\alpha = 1.0$)

The error performance is lowest for the EMX-trained network with less than five units. This can be explained by examining the internal representation in the hidden units. The clustering algorithm is more apt to learn localized representations than global ones. As the number of units increases in the network, the input space is more likely to have RBF units with narrow widths. Figure 12.13 shows the desired and learned mapping for the network with four units trained using EMX. The dotted

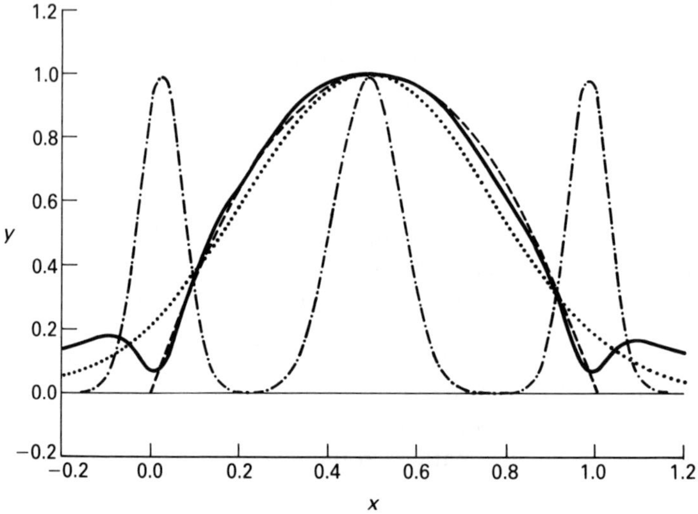

Figure 12.13 Learned mapping compared with desired quadratic map. Net with 4 units trained using EMX clustering

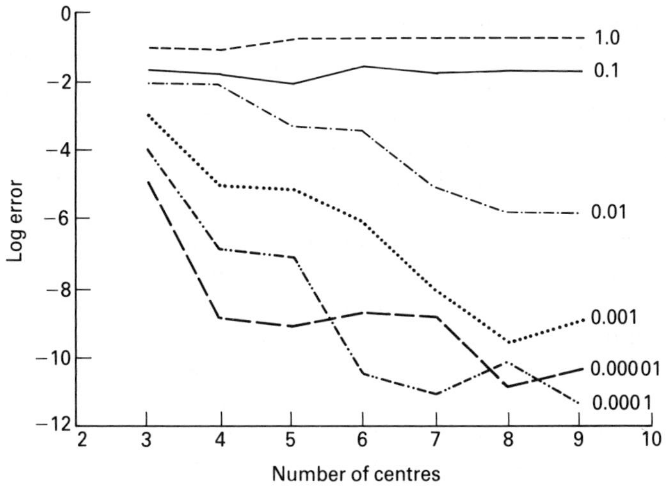

Figure 12.14 The effect of varying α on error performance for the quadratic map predictor network

lines show the RBF units, the solid line the learned mapping, and the dashed line the desired mapping for the predictor. The unit in the middle with the widest response is able to match the desired mapping quite well. When the number of units is increased, the individual unit widths converge to more local representations, which for this case does not give the best performance.

The performance can be improved for the RBF network if the learned spreads are expanded, making the interpolation smoother. Figure 12.14 shows the effect of the factor α on the prediction error performance of the EMX-trained network. Only the testing set results are shown. As α decreases, the RBF unit representations are made wider and a better interpolation is achieved. When α is made too small ($\alpha = 10^{-5}$), the detail in the mapping cannot be captured and the performance deteriorates.

12.7 Concluding remarks

Our present-day knowledge of the learning characteristics of linear adaptive filters is well developed [30,31]. We know how to evaluate the rate of convergence of this class of adaptive filters, their misadjustment, sensitivity to variations in the eigenvalue spread of the underlying covariance matrix, and their tracking behaviour when operating in a non-stationary environment. However, as of yet, the same cannot be said of back-propagation and radial basis function networks. This may not be surprising. A conventional adaptive filter consists of a single linear computing element.

On the other hand, a neural network consists of a large interconnection of computing units, most (if not all) of which are non-linear. Hence, the presence of one or more layers of hidden neurons, combined with the non-linear input–output characteristics of the neurons make a statistical learning analysis neural network a very difficult mathematical task indeed. Nevertheless, some useful results have been published on the learning behaviour of neural networks [72, 73]. There is, however, much to be done yet!

Notwithstanding their theoretical limitations, neural networks provide a powerful tool for solving adaptive signal processing problems that arise when operating in stationary or non-stationary environments with unknown statistics. The features that make such neural networks as back–error propagation and radial basis function networks a powerful tool are twofold. First, these networks contain one or more layers of hidden neurons (computing units), so called because there is no direct physical access to them from either the input end or the output end of the network; hidden neurons may be viewed as feature detectors. Second, most (if not all) of the neurons are designed to have non-linear transfer characteristics. Accordingly, a neural network is able to construct highly complex decision boundaries or input–output mapping functions that are beyond the reach of conventional linear adaptive filters. Moreover, it can extract information represented by second- as well as higher-order statistics of the input.

Another important point to note from our brief look at neural networks in this chapter is that they represent a highly interdisciplinary subject. There is much to be learned from neuroscience that deals with the functions of a human brain [74]. In the human brain we have a physical proof of the enormous computing power of a (natural) neural network. Moreover, the tissue of the central nervous system appears to be homogeneous; this suggests that statistical physics techniques may be useful in solving neural-network-related problems. The usefulness of well-established ideas in statistical mechanics has already been demonstrated in the development of the Hopfield network [75], simulated annealing [76], the Boltzmann machine [77] and statistical learning theory [72]. Simulated annealing is an optimization technique that is less dependent on the starting point than traditional methods. If the system to be optimized has a unique minimum, simulated annealing is usually able to find it; if the system is frustrated (i.e. with no global minimum), simulated annealing finds a good minimum point and stops there. The Boltzmann machine represents a generalization of the Hopfield network.

In the context of adaptive signal processing, we may mention the following areas of application that are natural for neural networks:

1. **Blind deconvolution**, which of course includes *blind equalization. Self-organized neural networks* may have a great deal to offer here by virtue of their ability to operate in an unsupervised fashion, which is the essence of blind deconvolution.
2. **Non-linear prediction**, where the need for the use of non-linearity is imposed by the nature of the problem at hand. The benefits of non-linear prediction using neural networks have already been demonstrated in [78,79].

3. **Non-linear modelling** of physical phenomena that are intrinsically non-linear. An example that we may mention here is that of sea non-linear modelling of clutter produced by radar backscatter from an ocean surface [80].
4. **Adaptive beamforming** for the purpose of interference cancellation, where again the physical nature of the environment of interest necessitates the use of non-linear processing.
5. **Adaptive pattern classification** that requires the construction of complex decision boundaries. For real-world examples demonstrating the application of the back-error propagation to adaptive pattern classification, see Haykin and Deng [36] and Casselman [81] who deal with surveillance radar and passive sonar problems respectively.

In conclusion, it would be wise for the adaptive signal processing community to add neural networks to their kit of design tools. The potential for their use is limited only by our physical understanding of the environment of interest and mathematical understanding of neural networks. Indeed, neural networks possess the computing power to solve large-scale processing problems that are currently intractable.

Notes

1. The subscripts I and Q follow the communication system convention of denoting the real and imaginary components of a signal to be the *in-phase* and *quadrature* components respectively.
2. It is possible to constrain the weights so that the aforementioned symmetry exists between the real-valued weights. However, the usual gradient descent learning algorithm does not make use of this information in the calculation of the updates, so information is still lost.
3. For the sake of readability, the time index n and layer designation $(M - 1)$ are sometimes omitted. It should also be clear from context what the omitted time and layer designations are.

References

[1] W. McCullough and W.H. Pitts, 'A logical calculus of the ideas imminent in nervous activity', *Bull. Math. Biophys.*, vol. 5, pp. 115–33, 1943.

[2] F. Rosenblatt, 'The perceptron: A probabilistic model for information storage and organization in the brain', *Psychol. Rev.*, vol. 65, pp. 386–408, 1958.

[3] B. Widrow and M.E. Hoff, 'Adaptive switching circuits', *WESCON Conv. Rec.*, pp. 96–104, 1960.

[4] M.L. Minsky and S.A. Papert, *Perceptrons: An introduction to computational geometry*, Cambridge, MA: MIT Press, 1969.

[5] J.A. Anderson, 'A simple neural network generating an interactive memory', *Math. Biosci.*, vol. 14, pp. 197–220, 1972.

[6] L.N. Cooper, 'A possible organization of animal memory and learning', in *Nobel Symp. on Collective Property of Physical Systems, 24th, Sweden, 12–16 June*, vol. 24, pp. 387–96, 1973.

[7] S. Grossberg, 'Classical and instrumental learning by neural networks', in *Progress in Theoretical Biology* eds. R. Rosen and F. Snell, vol. 3, New York: Academic Press, 1974.

[8] T. Kohonen, 'Correlation matrix memories', *IEEE Trans. Comput.*, vol. C-21, pp. 353–9, 1972.

[9] S.-I. Amari, 'Learning patterns and pattern sequences by self-organizing nets of threshold elements', *IEEE Trans. Comput.*, vol. C-21, pp. 1197–206, 1972.

[10] J.J. Hopfield, 'Neural networks and physical systems with emergent collective computational abilities', *Proc. Natl Acad. Sci.*, vol. 79, pp. 2554–8, 1982.

[11] J.J. Hopfield and D.W. Tank, 'Neural computation of decisions in optimization problems', *Biol. Cybern.*, vol. 52, pp. 1–25, 1985.

[12] V. Wilson and G.S. Lawley, 'On the stability of the TSP problem algorithm of Hopfield and Tank', *Biol. Cybern.*, vol. 58, pp. 63–70, 1988.

[13] S.V. Aiyer, N. Niranjan and F. Fallside, 'A theoretical investigation into the performance of the Hopfield model', *IEEE Trans. Neural Networks*, vol.1, pp. 204–15, 1990.

[14] Y.S. Abu-Mostafa and J.-M. S. Jacques, 'Information capacity of the Hopfield model', *IEEE Trans. Inf. Theory*, vol. IT-33, pp. 461–87, 1985.

[15] D.O. Hebb, *The Organization of Behavior*, New York: Springer, 1949.

[16] J.E. Mazo, 'On the independence theory of equalizer convergence', *Bull. Syst. Tech. J.*, vol. 58, pp. 963–93, 1979.

[17] T. Kohonen, *Self-organization of Behavior*, New York: Springer, 1987.

[18] T. Kohonen, 'Automatic formulation of topological maps of patterns in a self-organizing system', *in Second Scandinavian Conf. on Image Analysis, Helsinki, 15–17 June*, pp. 214–20, 1981.

[19] R. Linsker, 'Self-organization in a perceptional network', *Computer*, vol. 21, pp. 105–17, 1988.

[20] P.J. Werbos, 'Beyond regression: New tools for prediction and analysis in the behavioral sciences', *PhD Thesis*, Harvard University, Cambridge, MA, 1974.

[21] D.E. Rumelhart, G.E. Hinton and R.J. Williams, 'Learning internal representations by error propagation', Tech. Rep. 8506, University of California, San Diego, Institute for Cognitive Science, San Diego, CA, 1985.

[22] D.B. Parker, 'Learning–logic: Casting the cortex of the human brain in silicon', Tech. Rep. TR-87, Center for Computational Research in Economics and Management, Cambridge, MA, 1985.

[23] Y. leCun, 'Modeles connexionnists de l'appxentissage', *PhD Thesis*, University Pierre et Marie Curie, Paris, 1987.

[24] D.E. Rumelhart and J.L. McClelland (eds.), *Parallel Distributed Processing: Explorations in the Microstructure of Cognition*, vols 1 and 2, Cambridge, MA: MIT Press, 1986.

[25] G.G.Y. Lau and B. Widrow (eds.), 'Special issue on neural networks', *IEEE Proc.*, vol. 1, Part I: September, and Part II: October, 1990.

[26] C.C. Klimasauskas (ed.), *The Neuro-computing Bibliography*, 2nd edn, Cambridge, MA: MIT Press, 1989.

[27] R. Lippmann, 'An introduction to computing with neural nets', *IEEE ASSP Mag.*, vol. 4, pp. 4–22, 1987.

[28] S. Haykin, *Neural Networks*, New York: Macmillan, 1992.

[29] R.M. Golden, 'The brain-state-in-a-box neural model is a gradient descent algorithm', *J. Math. Psych.*, vol. 30, pp. 73–80, 1986.

[30] S. Haykin, *Adaptive Filter Theory*, 2nd edn, Prentice Hall: Englewood Cliffs, NJ Prentice Hall: 1991.

[31] B. Widrow and S. Stearns, *Adaptive Signal Processing*, Prentice Hall: Englewood Cliffs, NJ, 1985.

[32] A. Benveniste, 'Robust identification of a nonminimum phase system: Blind adjustment of a linear equalizer in data communications', *IEEE Trans. Autom. Control*, vol. AC-25, pp. 385–99, 1980.

[33] R. Godfrey and F. Rocca, 'Zero-memory non-linear deconvolution', *Geophys. Prospect.*, vol. 29, pp. 189–228, 1981.

[34] S. Bellini, 'Bussgang techniques for blind equalization', *Proc. GlobeCom '86, Houston, Texas*, pp. 1634–49, 1986.

[35] S. Haykin *et al.*, 'Classification of radar clutter in an air traffic control environment', *Proc. IEEE*, vol. 79, no. 6, 1991.

[36] S. Haykin and C. Deng, 'Classification of radar clutter using neural networks', *IEEE Trans. Neural Networks*, vol. 2, pp. 589–600, 1991.

[37] R.S. Sutton, 'Temporal credit assignment in reinforcement learning', *PhD Thesis*, University of Massachusetts, Amherst, MA, 1984.

[38] A. G. Barto, 'Connectionist learning for control', in *Neural Networks for Control* (eds. W.T. Miller, R.S. Sutton and P.J. Werbos), Cambridge, MA: MIT Press, 1990.

[39] E.-B. Baum and D. Jaussler, 'What size net gives valid generalization?', *Neural Comput.*, vol. 1, pp. 151–60, 1988.

[40] N. Benvenuto and F. Piazza, 'On the complex backpropagation algorithm', *IEEE Trans. Signal Process.*, vol. 40, pp. 967–9, 1992.

[41] D.L. Birx and S.J. Pipenberg, 'Chaotic oscillators and complex mapping feed forward networks (CMFFNS) for signal detection in noisy environments', in *IJCNN Int. Joint Conf. on Neural Networks, June, 1992*, pp. II-881–8, IEEE and INNS.

[42] T. L. Clarke, 'Generalization of neural networks to the complex plane', *Proc. Int. Joint Conf. on Neural Networks, San Diego, CA, June, 1990,* vol. II, pp. 435–40.

[43] G.M. Georgiou and C. Koutsougeras, 'Complex domain backpropagation', *IEEE Transactions on Circuits and Systems Part II: Analog and Digital Signal Processing, May, 1992*, vol. 39.

[44] J. Henseler and P.J. Braspenning, 'Training complex multi-layer neural networks', Tech. Rep. CS 90-02, University of Limburg, Department of Computer Science, PO Box 616, 6200 MD Maastricht, The Netherlands, 1990.

[45] M.S. Kim and C.C. Guest, 'Modification of back propagation networks for complex–vaued signal processing in frequency domain', *Proc. Int. Joint Conf. on Neural Networks, San Diego, CA, June, 1990,* vol. III, pp. 27–31.

[46] H. Leung and S. Haykin, 'The complex backpropagation algorithm', *IEEE Trans. Acoust., Speech, Signal Process.*, vol. ASSP-39, pp. 2101–4, 1991.

[47] G.R. Little, S.C. Gustafson and R.A. Senn, 'Generalization of the backpropagation neural network learning algorithms to permit complex weights', *Appl. Opt.*, vol. 29, pp. 1591–2, 1990.

[48] S. Lang, *Complex Analysis*, 2nd edn, New York: Springer, 1985.

[49] L.L. Horowitz and K.D. Senne, 'Performance advantage of complex lms for controlling narrow-band adaptive arrays', *IEEE Trans. Acoust. Speech, Signal Process.*, vol. ASSP-29, pp. 722–36, 1981.

[50] A. Ukrainec, S. Haykin and J. McGregor, 'A neural network nonlinear predictor', *Int. Joint Conf. on Neural Networks, Washington, DC, June 1989,* pp. II–622.

[51] M.B. Priestley, *Spectral Analysis and Time Series*, London: Academic Press, 1981.

[52] Y. leCun, 'Optimal brain damage', in *Neural Information Processing Systems*, ed. D.S.

Touretzky, vol. 2, pp. 598–605, Morgan Kaufman: 1973.

[53] S.E. Faulman and C. Lebiere, 'The cascade–correlation learning architecture', in *Neural Information Processing Systems*, ed. D.S. Touretzky, vol. 2, pp. 524–32, Morgan Kaufman. 1973.

[54] T.C. Lee, A.M. Peterson and J.C. Tsai, 'A multi-layer feed-forward neural network with dynamically adjustable structures', in *1990 IEEE Conf. on Systems, Man, and Cybernetics, Los Angeles, CA*, pp. 367–9, 1990.

[55] D.S. Broomhead and D. Lowe, 'Multivariable functional interpolation and adaptive networks', *Complex Syst.*, vol. 2, pp. 321–55, 1988.

[56] M. Casdagli, 'Nonlinear prediction of chaotic time series', *Physica D*, vol. 35, pp. 335–56, 1989.

[57] R.D. Jones, Y.C. Lee, C.W. Barnes, G.W. Flake, K. Lee, P.S. Lewis and S. Qian, 'Function approximation and time series prediction with neural networks', Tech. Rep., Los Alamos National Laboratory, Los Alamos, NM, Dec., 1989.

[58] D. Lowe and A. Webb, 'Adaptive networks, dynamical systems, and the predictive analysis of time series', *First IEE Int. Conf. Artificial Neural Networks*, pp. 95–9, 1989.

[59] D. Lowe and A. Webb, 'Time series prediction by adaptive networks: A dynamical systems perspective', Tech. Rep., Royal Signals and Radar Establishment, Great Malvern, Worcs, UK, Mar., 1990.

[60] J. Moody and C. J. Darken, 'Fast learning in networks of locally-tuned processing units', *Neural Comput.*, vol. 1, pp. 281–94, 1989.

[61] J. Moody and C. Darken, 'Learning with localized receptive fields', *Proc. of the 1988 Connectionist Models Summer School, San Mateo, CA,* eds. D. Touretzky, G. Hinton and T. Sejnowski, pp. 133–43, Morgan Kaufmann: 1989.

[62] A. Saha and J. Keeler, 'Algorithms for better representation and faster learning in radial basis function networks', Tech. Rep. ACT-NN-028-90, Microelectronics and Computer Technology Corporation, Austin, TX, Jan., 1990.

[63] A. Ukrainec and S. Haykin, 'Signal processing with radial basis function networks using expectation maximization algorithm clustering', *SPIE 36th Int. Symp. on Optical and Optoelectronic Applied Science and Engineering, July, 1991.*

[64] T. Poggio and F. Girosi, 'Networks for approximation and learning', *Proc. IEEE*, vol. 78, pp. 1481–96, 1990.

[65] D. Lowe, 'Adaptive radial basis function nonlinearities, and the problem of generalization', *First IEE Int. Conf. on Artificial Neural Networks*, 1989.

[66] S.J. Nowlan, 'Maximum likelihood competitive learning', in *Advances in Neural Information Processing Systems*, ed. D.S. Touretzky, vol. 2, pp. 574–82, Morgan Kaufman, 1990.

[67] S.J. Nowlan, 'Max likelihood competition in RBF networks', Tech. Rep. CRG-TR-90-2, Connectionist Research Group, University of Toronto, Toronto, Canada, Feb., 1990.

[68] A.P. Dempster, N.M. Laird and D.B. Rubin, 'Maximum likelihood from incomplete data via the EM algorithm', *Proc. R. Statist. Soc.*, vol. 39, pp. 1–38, 1977.

[69] R.A. Redner and H.F. Walker, 'Mixture densities, maximum likelihood and the EM algorithm', *SIAM Rev.*, vol. 26, pp. 195–239, 1984.

[70] R.O. Duda and P.E. Hart, *Pattern Classification and Scene Analysis*, New York: Wiley, 1973.

[71] A. Papoulis, *Probability, Random Variables, and Stochastic Processes*, 2nd edn, New York: McGraw-Hill, 1984.

[72] E. Levin, N. Tishby and S.A. Solla, 'A statistical approach to learning and generalization in layered neural networks', in *COLT'89*, pp. 245–60, 1989.

[73] G. Bilbro and D.E. Van den Bout, 'Learning theory and experiments with competitive networks', in *Neural Information Processing Systems*, eds. R.P. Lippmann, J.E. Moody and D.S. Touretzky, vol. 3, pp. 846–52, Morgan Kaufman: 1991.

[74] S.W. Kuffler, J.G. Nicholls and A.R. Martin, *From Neuron to Brain*, 2nd edn, Sunderland, MA: Sinauer, 1984.

[75] J.J. Hopfield, 'Neurons with graded response have collective computational properties like those of two-state neurons', *Proc. Natl. Acad. Sci.*, vol. 81, pp. 3088–92, 1984.

[76] S. Kirkpatrick *et al.*, 'Optimization by simulated annealing', *Science*, vol. 220, pp. 671–80, 1983.

[77] D.H. Ackley, G.E. Hinton and T.J. Sejnowksi, 'A learning algorithm for Boltzmann machines', *Cognitive Sci.*, vol. 9, pp. 147–69, 1985.

[78] A. Lepedes and R. Farber, How neural nets work', in *Neural Information Processing Systems*, ed. D.A. Anderson, pp. 442–56, New York: American Institute of Physics, 1988.

[79] S. Haykin and H. Leung, 'Neural network modeling of radar backscatter from an ocean surface', in *IEEE Conf. on Neural Networks for Oceanic Engineering, Houston, Texas*, pp. 1634–49, 1986.

[80] H. Leung and S. Haykin, 'Is there a radar clutter attractor?', *Appl. Phys. Lett.*, vol. 56, pp. 593–5, 1989.

[81] F.L. Casselman, 'A neural network-based passive sonar detection and classification design with a low false alarm rate', *IEEE Oceans '91, Hawaii*, pp. 1701–6, 1991.

Index